Advanced Time Series Data Analysis

Advanced Time Series Data Analysis

Forecasting Using EViews

I Gusti Ngurah Agung
The Ary Suta Center
Jakarta, Indonesia

This edition first published 2019

Registered Offices
John Wiley & Sons, Inc., 111 River Street, Hoboken, NJ 07030, USA
John Wiley & Sons Ltd, The Atrium, Southern Gate, Chichester, West Sussex, PO19 8SQ, UK

Editorial Office
9600 Garsington Road, Oxford, OX4 2DQ, UK

For details of our global editorial offices, customer services, and more information about Wiley products visit us at www.wiley.com.

Wiley also publishes its books in a variety of electronic formats and by print-on-demand. Some content that appears in standard print versions of this book may not be available in other formats.

Library of Congress Cataloging-in-Publication Data

Names: Agung, I Gusti Ngurah, author.
Title: Advanced time series data analysis : forecasting using EViews / I. Gusti Ngurah Agung.
Description: Hoboken, NJ : John Wiley & Sons, 2019. | Includes bibliographical references and index. |
Identifiers: LCCN 2018040053 (print) | LCCN 2018057198 (ebook) | ISBN 9781119504733 (Adobe PDF) | ISBN 9781119504740 (ePub) | ISBN 9781119504719 (hardcover)
Subjects: LCSH: Time-series analysis. | Econometric models.
Classification: LCC QA280 (ebook) | LCC QA280 .A38 2019 (print) | DDC 519.5/5–dc23
LC record available at https://lccn.loc.gov/2018040053

Cover Design: Wiley
Cover Images: © Butsaya/ Getty Images; © monsitj /Getty Images

Set in 10/12pt Warnock by SPi Global, Pondicherry, India

Printed in Singapore by C.O.S. Printers Pte Ltd

10 9 8 7 6 5 4 3 2 1

Dedicated to my wife, Anak Agung Alit Mas; our children, Martiningsih, Ratnaningsih, and Darma Putra; as well as all our Generation.

Contents

About the Author

I have a Ph.D. degree in Biostatistics (1981) and a Master's degree in Mathematical Statistics (1977) from the North Carolina University at Chapel Hill, NC, USA; a Master's degree in Mathematics from New Mexico State University, Las Cruces, NM, USA; a degree in Mathematical Education (1962) from Hasanuddin University, Makassar, Indonesia; and a certificate from "Kursus B-I/B-II Ilmu Pasti" (B-I/B-II Courses in Mathematics), Yogyakarta, which is a five-year non-degree program in advanced mathematics. So, I would say that I have a good background knowledge in mathematical statistics as well as applied statistics. In my dissertation on biostatistics, I presented new findings, namely the Generalized Kendall's tau, Generalized Pair Charts, and Generalized Simon's Statistics, based on the data censored to the right.

Supported by my knowledge in mathematics, mathematical functions in particular, I can evaluate the limitations, hidden assumptions, or the unrealistic assumption(s) of all regression functions, such as the fixed effects models based on panel data, which are in fact ANCOVA models. As a comparison, Agung (2011a) presents several alternative acceptable ANCOVA models, in the statistical sense, and the worst ANCOVA models, in both theoretical and statistical senses.

Furthermore, based on my exercises and experiments in doing data analyses of various fields of study; such as finance, marketing, education, and population studies since 1981 when I worked at the Population Research Center, Gadjah Mada University, 1985–1987; and while I have been at the University of Indonesia from 1987 up to 2018, I have found unexpected or unpredictable statistical results based on various time series, cross-section, and panel data models, which have been presented with special notes and comments in Agung (2014, 2009a, 2011a), compared to models that are commonly applied.

Similarly, based on my exercises and experiments in doing forecasting using EViews, I can present various alternative models using the same sets of variables, such as based on a single time series Y_t, bivariate time series (X_t,Y_t), multivariate time series $(X1_t,X2_t,Y1_t,Y2_t)$, and $(X1_t,X2_t,X3_t,Y1_t,Y2_t,Y3_t,Z1_t)$, without or with alternative time variables, based on monthly, quarterly, and annual time series. Aside from good forecast models, worse forecast models also are presented as illustrative examples. Many of those models have not been presented in other books, specifically in *Business Forecasting* by Hankle and Reitch (1992), and by Wilson and Keating (1994), or in the books by Gujarati (2003), Wooldridge (2002), and Tsay (2002).

Preface

It is well-known that forecasting is one of the best inputs for decision-making. However, we never know what type of model gives a perfect forecast values beyond the sample period, since there are a lot of possible models that can be developed to forecast any selected endogenous time series that are acceptable in the statistical sense. In addition, in-sample forecast values are highly dependent on the data that happens to be selected by or available to researchers.

This book presents many alternative multiple regression models of a monthly, quarterly, and annual endogenous time series with specific growth patterns, starting with the simplest up to the most advanced time series models so that those models can show their differential in sample forecast values of the endogenous variable. Hence, the main objectives of this book are to present (i) various general specific equation of forecast models, which in fact are multiple time series regression models; (ii) various illustrative statistical results based on selected specific equations, with special notes and comments; and (iii) comparative studies between a set of special type of models using the same set of variables, such as additive models, interaction models, and heterogeneous regression models, without trend and with various alternative trends. The best possible fit forecasting model of an endogenous time series, in the statistical sense, is presented based on alternative specific growth cures of the time series. Furthermore, as a comparison, several alternative models of the same endogenous time series are also presented with illustrative examples.

EViews provides the object/option "*Forecast*," which can directly be used to conduct the forecasting, while the estimate of a regression of a time series appears on the screen. I am very confident that all regressions of a time series presented in various books, such as Agung (2009a), Gujarati (2003), Wooldridge (2002), Tsay (2002), Hankle and Reitch (1992), and Wilson and Keating (1994), as well as presented in various journals, could be used to forecast their dependent variables. This book mainly presents forecasting data analysis based on various interaction models, such as the *lag-variable* models: LV(1) models, based on a single time series, say Y_t, bivariate (X_t, Y_t) or $(Y1_t, Y2_t)$, and triple time series $(X1_t, X2_t, Y_t)$ *or* $(Y1_t, Y2_t, Y2_t)$, since it is found that $Y_{t-1} = Y(-1)$ is the best predictor for all-time series, Y_t. In addition, those models are extended to *lag-variable-autoregressive-moving-average* models: LVARMA(p,q,r) for non-negative integers $p \geq 1$, $q \geq 0$, and $r \geq 0$, which should be selected using the *trial-and-error* method in order to obtain acceptable in-sample forecasting values. Over 350 general equation specifications of various models, with over 200 illustrative examples of the statistical results of specific models based on the same set of variables are presented, with special notes and comments so that the readers can be well informed on the limitations of a model compared to others in the set. Aside from the good fit forecast models, worse unexpected forecast models are also presented.

The models presented in this book in fact are the extension of my first book: *Time Series Data Analysis Using EViews* (Agung 2009a). The models also can be considered as modifications of the panel data models presented in the first part of Agung (2014). For this reason, it is recommended that readers also use the models in those books to conduct forecasting using their own data sets.

This book contains seven chapters. Chapter 1 presents various alternative models of a single monthly time series Y_t, with a specific growth curve, namely a systematic growth curves by *@Year,* such as basic and special LV(*p*), LVAR(*p,q*), ARMA(*q,r*), and TGARCH(*a,b,c*) models with illustrative examples of the statistical results based on selected models. This chapter also presents residual analysis with special notes and comments, such as the BPG Heteroskedasticity Test, the Harvey Test, and Glejser Test, the White Heteroskedasticity Tests, the ARCH Heteroskedasticity Tests, Custom Test Wizard, the Homogeneity test, and the Breusch–Godfrey Serial Correlation LM Test. In addition, this chapter discusses the application of the White and the HAC (Newey–West) Covariances.

Chapter 2 presents various models based on a monthly time series using three possible time predictors, such as *@Month, @Year,* and the time variable t = *@Trend* or t = *@Trend* + 1, as the extension of all models presented in Chapter 1. Special LV(12) interaction models are presented, as in the first part of this chapter, to demonstrate heterogeneous regressions models by *@Year* and alternative testing hypotheses, such as the Omitted Variables Test (OVT) and Redundant Variables Test (RVT). In addition, alternative heterogeneous classical growth models are also presented along with the reduced heterogeneous regression models, alternative ANCOVA models, and fixed-effects models, with special notes and comments.

Chapter 3 presents alternative continuous forecast models. As the simplest model presented is a two-way interaction LV(1) model with *"Y C Y(-1) t t*Y1(-1)"* as its equation specification. In practice, however, based on a data set, there are four alternative reduced models that can be obtained as a good fit. Then each of those models could be extended to the *lag-variable-autoregressive-moving-average*: LVARMA(*p,g,r*) models for various integers $p \geq 1$, $q \geq 0$, and $r \geq 0$, and models with alternative time variables. In addition, this chapter also presents translog-linear models with linear trend or logarithmic trend, translog interaction models, and alternative nonlinear models.

Chapter 4 presents various models based on bivariate time series (X_t, Y_t) and $(Y1_t, Y2_t)$ as the extension or modification of each model presented in previous chapters, depending on the growth patterns of the endogenous variables. Two of the simplest LV(1) two way interaction models with the equation specifications *"Y C Y(-1) X(-1) Y(-1)*X(-1)"* and *"Y C Y(-1) X Y(-1)*X"* are presented as the preliminary models. Then each of these can easily be modified to more advanced models, such as LVARMA(*p,q,r*) of *LNY* = *log(Y)* or *LNYul* = *log((U_L)/(U-Y)*, those with alternative trends that are presented in Table 4.1, and heterogeneous regression models by *@Year* or *@Month*, as presented in previous chapters. The application of the object VAR is also presented. In addition, based on $(Y1_t, Y2_t)$ six *two-way-interaction lag-variables models* are presented, namely TWI_LVM(*p1,p2*) where four of them are *reciprocal causal effect models* (CRE). Finally, this chapter also presents special notes and comments, referring to unbelievable and unexpected statistical results.

Chapter 5 presents various models based on triple time series $(X1_t, X2_t, Y_t)$ and $(Y1_t, Y2_t, Y3_t)$ as the extension or modification of each model presented in Chapter 4. Initially, a set of four translog-linear LV(1) models are presented, depending on the growth patterns of the endogenous variables. Then each of the translog-linear models can be extended to LVARMA(*p,g,r*), those with alternative trends, and heterogeneous regression models. As a comparative study, the same translog-linear models are presented based on $(X1_t, X2_t, Y1_t)$ and $(X1_t, X2_t, Y2_t)$ with *Y1* and *Y2* having different growth patterns. Then various two-way and three-way interaction models are also presented using the original time series. Furthermore, various special sets of *triangular and circular effects multivariate lag-variables models* are presented based on $(Y1_t, Y2_t, Y3_t)$, such as additive, two- and three-way interaction models, as the alternative applications of the objects *VAR* and *System*. Finally, referring to a lot of possible models based on the monthly time series, it is important to consider special notes by Tukey (1962) and Bezzecri (1973 in Gifi, 1991).

Chapter 6 presents various models based on a quarterly time series. In fact, all models presented in the previous five chapters can easily be applied for a quarterly time series, conditional on the growth curve of an endogenous time series. However, note that all models with the time variable *@Month* as an independent variable should be transformed to those with the time variable *@Quarter* as an independent variable.

Chapter 7 presents various models based on unstacked annually panel data, namely the annual time series by states. Agung (2014) has presented various models, such as the VAR and System Equation Model (or SCM = Seemingly Causal Models) based on unstacked panel data, POOLG7.wf1 in the EViews data file, with a small number of *N*. Referring to the multiple OLS regression analyses presented in previous chapters, all data analyses using the VAR and SCM models presented in Agung (2014) can be reanalyzed using sets of equation specifications in order to compute the in-sample forecast values of each of their dependent variables. However, this chapter presents other alternative models as their modifications or extended time series models, starting with the simple alternative two-way interaction lag-variable models based on *(Y1_1,Y1_2)*. Then these are extended to those models with exogenous variables *(X1_1,X1_2)* and an environmental variable *Z1*, and those based on triple endogenous variables *(Y1_1,Y1_2,Y1_3)* with exogenous variables *(X1_1,X1_2,X1_3)*, and an environmental variable *Z1*, Based on the last set of seven variables, in order to forecast *Y1_1*, a set of seven successive or interrelated interaction models of each variables are introduced, instead of the single model of *Y1_1* with a large number of independent variables.

I wish to express my gratitude to the Graduate School of Management, Faculty of Economics and Business, University of Indonesia, and The Ary Suta Center, Jakarta, for providing a rich, intellectual environment and facilities that were indispensable for writing this text. I also would like to thank Tridianto Subagio, the best computing staff at the Graduate School of Management, who has given me great help whenever I have problems with my PC.

Finally, I would like to thank the reviewers, editors, and all staff at John Wiley & Sons, Ltd for their hard work in getting this book to publication.

1

Forecasting a Monthly Time Series

1.1 Introduction

It is recognized that all possible models of a single time series, Y_t, can easily be applied to a forecast, Y_t. Refer to Agung (2009a), which presents a number of time series models that could easily be extended to many more possible models, as well as models based on panel data presented in Agung (2014), which can be used in forecasting. So, a researcher should never have to present the best possible forecasting, which should be highly dependent on his/her subjective expert judgment.

This chapter specifically presents forecasting based on a single monthly time series, namely Y_t, without taking into account the effects of exogenous variables, except for any lags or the time variable. More alternative and advanced models will be presented in the following chapters. For illustration, this chapter only presents selected illustrative forecasting based on the data in House.wf1, which contains only one single time series variable, namely HS_t, with 604 time-observations from 1946M01 to 1999M04. In addition, for comparison, illustrative examples are presented based on other selected data sets.

1.2 Forecasting Using LV(*p*) Models

1.2.1 Basic or Regular LV(*p*) Models

It is well known that the LV(p) model of a time series variable, Y_t, has the following general form.

$$Y_t = C(1) + C(2) * Y_{t-1} + \cdots + C(p+1) * Y_{t-p} + \mu_t \quad (1.1)$$

therefore, a forecast of any transformed variable $G(Y_t)$ can easily be done using the following equation specification, for any integer p, which should be highly dependent on the data used as well as the subjective interest of the researchers. However, this section only presents a few illustrative examples.

$$G(Y)\,C\,G(Y(-1))\;G(Y(-2))\cdots\;G(Y(-p)) \quad (1.2)$$

where $G(Y)$ can be any functions of Y_t without a parameter, such as the original time series Y_t, $log(Y_t)$, and $log((Y_t - L)/(U - Y))$, with L and U are the fixed lower and upper bounds of Y_t.

Example 1.1 A Dynamic Forecast Using the Simplest Model in (1.2)

As a preliminary forecast data analysis, this example presents the graphical presentation of the variable HS_t, as presented in Figure 1.1. Based on these figures, the following notes and comments are presented.

Advanced Time Series Data Analysis: Forecasting Using EViews, First Edition. I Gusti Ngurah Agung.

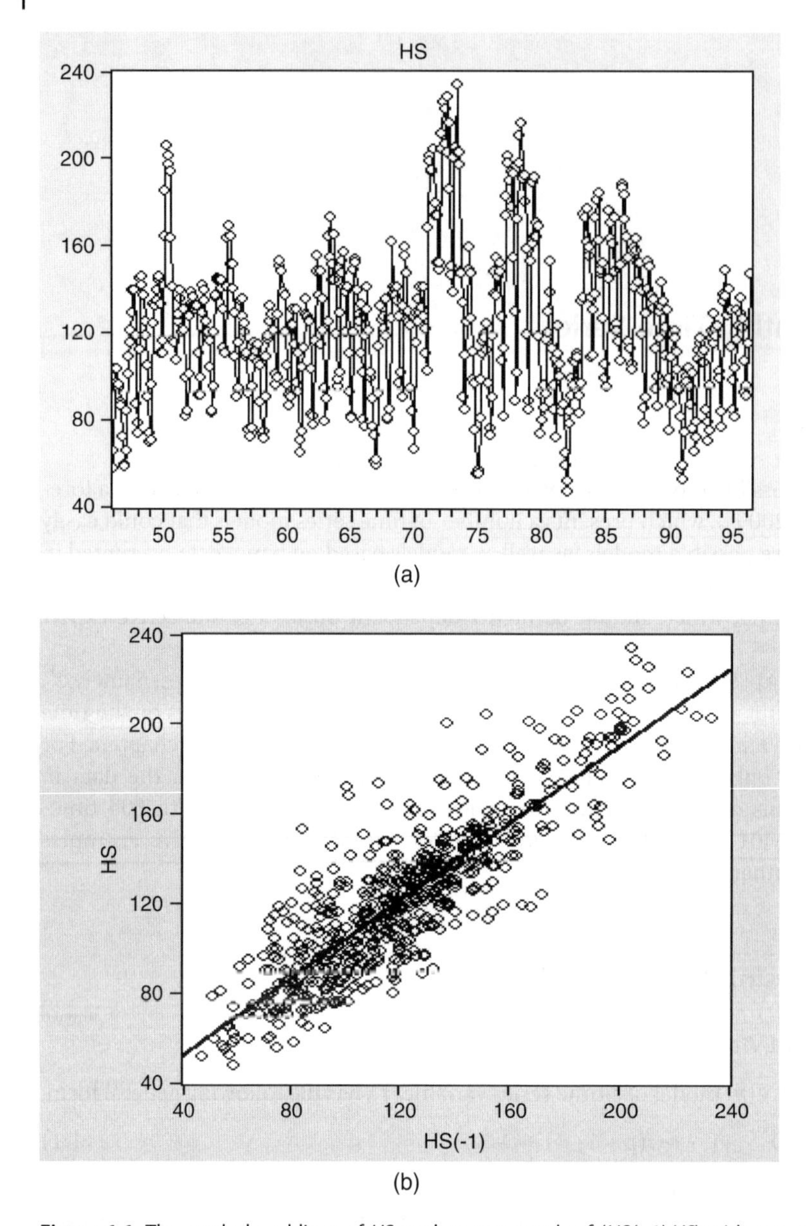

Figure 1.1 The symbol and lines of *HS*, and scatter graph of *(HS(−1),HS)* with regression lines.

1) Figure 1.1a presents a line and dot graph of the variable HS_t that clearly shows a seasonal pattern. This is similar for the graph of $HS_{t-1} = HS(-1)$. How would you forecast HS_t, in order to take into account its seasonal pattern?
2) Figure 1.2a presents statistical results using the simplest model in (1.2), namely the LV(1) model. Based on these results the following findings and notes are presented.
   2.1 The LS regression has the following equation, which can be used to forecast the scores of *HS* in the sample, as well as out of the sample or after the time of observation.

$$H\hat{S} = \hat{C}(1) + \hat{C}(2) * HS(-1) = 17.235685 + 0.861125HS(-1) \tag{1.3}$$

Dependent Variable: HS
Method: Least Squares
Date: 12/19/13 Time: 10:23
Sample (adjusted): 1946M02 1996M04
Included observations: 603 after adjustments

Variable	Coefficient	Std. Error	t-Statistic	Prob.
C	17.23569	2.624067	6.568310	0.0000
HS(-1)	0.861125	0.020527	41.95013	0.0000

R-squared	0.745427	Mean dependent var	123.1819
Adjusted R-squared	0.745003	S.D. dependent var	34.64025
S.E. of regression	17.49236	Akaike info criterion	8.564717
Sum squared resid	183895.6	Schwarz criterion	8.579317
Log likelihood	-2580.262	Hannan-Quinn criter.	8.570399
F-statistic	1759.814	Durbin-Watson stat	1.399409
Prob(F-statistic)	0.000000		

(a)

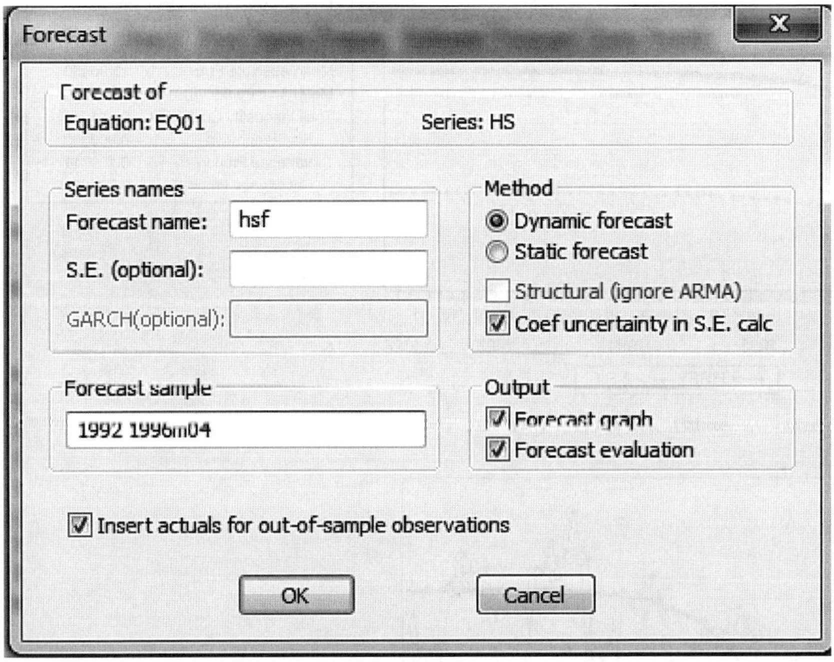

Forecast
Forecast of
Equation: EQ01 Series: HS
Series names
Forecast name: hsf
S.E. (optional):
GARCH(optional):
Method
Dynamic forecast
Static forecast
Structural (ignore ARMA)
Coef uncertainty in S.E. calc
Forecast sample
1992 1996m04
Output
Forecast graph
Forecast evaluation
Insert actuals for out-of-sample observations
OK Cancel

(b)

Figure 1.2 Statistical results of an LV(1) model of *HS* with the forecast dialog.

2.2 Since this regression has $R^2 = 0.745\,447$, and adjusted $R^2 = 0.745\,003$, then $HS(-1) = HS_{t-1}$ can be considered as a good predictor for HS_t. Therefore, the forecast should be an acceptable forecast. However, it might not the best forecast of *HS*. Note that referring to the relative small value of DW statistic, then a LV(p) model, for $p > 1$ should be explored. Do this as an exercise.

3) With Figure 1.2a on the screen, by clicking the option "*Forecast*," then Figure 1.2b is shown on the screen, then we can select a forecast sample. In this case, I select only a sample from

1992 to @last, namely 1996M04, as presented in the figure. Then by clicking "OK," the graph and statistics in Figure 1.3a are obtained. In addition, the forecast variable, *HSF*, is added to the work file. Based on these results, the following notes should be considered.

3.1 Figure 1.3a presents the graph of the forecast: HSF ± 2 S.E., from 1992 to @last, and its evaluation statistics, such as follows:

- *Root Mean Squared Error (RMSE)* and *Mean Absolute Error (MAE)* are relative measures to compare forecasts for the same series across different models; the smaller the error, the better the forecasting ability of that model according to that criterion. So the *RMSE* = 20.96267 cannot be used to evaluate goodness of fit of the forecasting, it should be compared to RMSEs of other forecasting results.

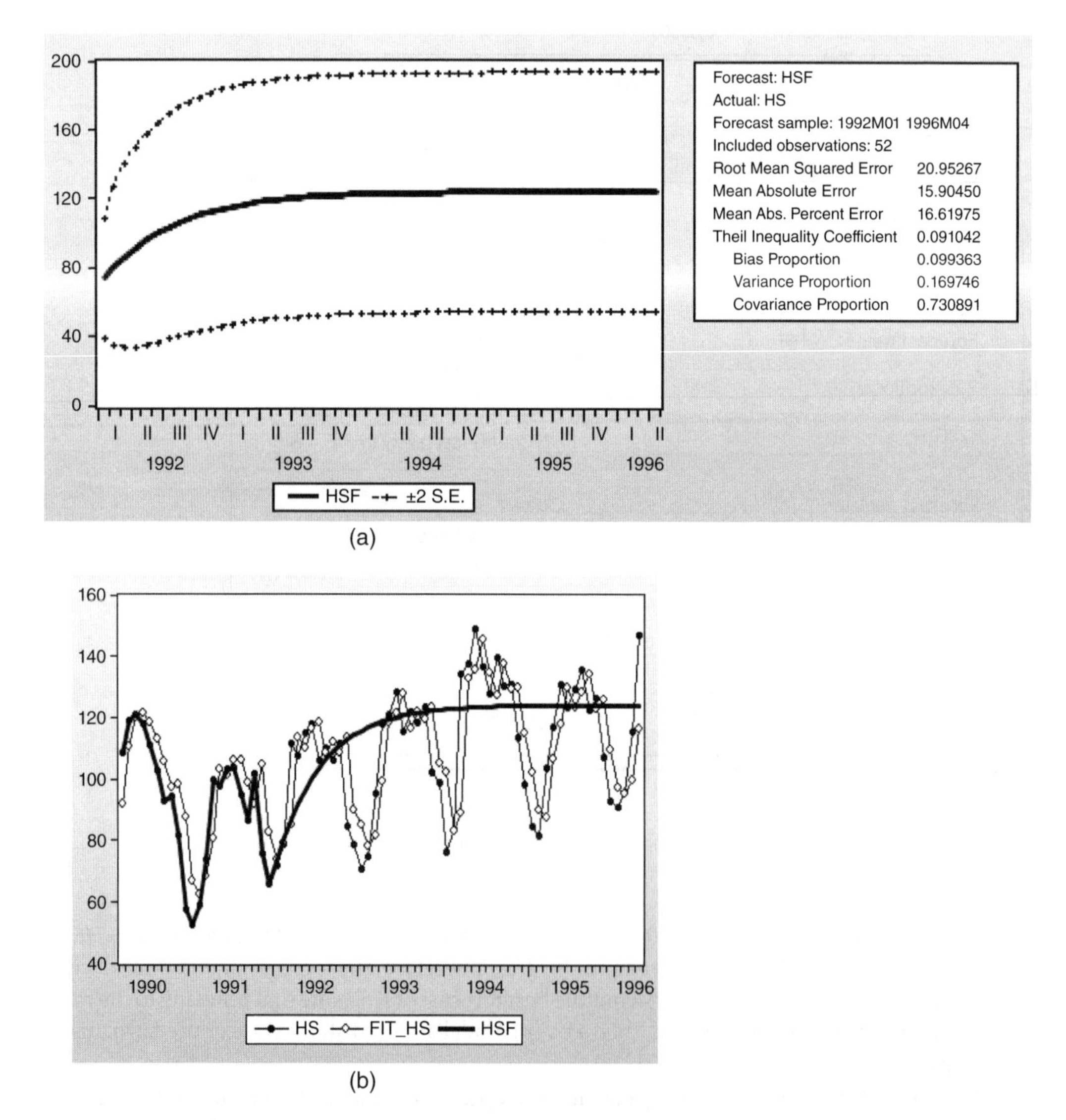

Figure 1.3 Forecast evaluation and graphs of a dynamic forecast of *HS* using the LV(1) model.

- *Mean Absolute Percentage Error (MAPE)* and *Theil Inequality Coefficient (TIC)* are scale invariant. The TIC always lies between zero and one, where zero indicates a perfect fit. Note that Figure 1.3a shows a TIC = 0.091042.
- In addition, the mean squared forecast error can be decomposed into *Bias Proportion (BP), Variance Proportion (VP),* and *Covariance Proportion (CP),* which have a total of one. The *BP* indicates how far the mean of the forecast is from the mean of the actual scores, the *VP* indicates how far the variance of the forecast is from the variance of the actual scores, and the *CP* measures the remaining unsystematic forecasting error.

3.2 The forecast *HSF* indicates that the LV(1) model is not an appropriate forecasting, since it presents a smooth curve without a seasonal pattern.

3.3 Note the limitation of the regression function (1.3), it represents the regression of a cross-section variable HS_i on $HS(-1)_i$ with $HS(-1)_i \leq HS(-1)_{i+1}$ for all $i = 2,\ldots,T$, with $T = 604$. For this reason, I would consider that the forecast using C in (1.3) is worse, even without looking its evaluation analysis. We can have many other possible models to forecast *HS*, such as LVAR(p,q) models for selected integers p and q (refer to the LVAR(p,0) in (1.1)), and other models, by taking into account the categorical variables *Month*, or *Year*, and the numerical time t. More so if there are additional relevant exogenous or environmental variables. However, a researcher never knows which one would give the best possible forecast, which is highly dependent on the data used, and he/she could never try all possible models.

4) Furthermore, Figure 1.3b presents the graphs of actual observed *HS*, fitted values of *HS*, namely *Fit_HS*, and the forecast variable *HSF*. Based on this graph, the following notes are presented.

4.1 These graphs are developed for the subsample from 1990 to 1996M04.

4.2 The bold curve is the curve of *HSF*, which is divided into two parts. The first part from 1990M01 to 1991M12 is representing the observed scores of *HS*, and the second part from 1992M01 to 1996M04 is representing the dynamic forecasts of *HS*, namely *HSF*, which are computed as follows:

$$
\begin{aligned}
H\hat{S}(1992m01) &= \hat{C}(1) + \hat{C}(2) * HS(1991m12) = 17.235685 + 0.861125HS(1991m12)) \\
H\hat{S}(1992m02) &= H\hat{S}(1992m01 + 1) = \hat{C}(1) + \hat{C}(2) * H\hat{S}(1992m01) \\
&= \hat{C}(1) + \hat{C}(2) * \{C(1) + \hat{C}(2) * HS(1991m12)\} \\
&= \hat{C}(1)\{1 + \hat{C}(2)\} + \hat{C}(2)^2 * HS(1991m12) \\
H\hat{S}(1992m01 + 2) &= \hat{C}(1)\{1 + \hat{C}(2) + \hat{C}(2)^2\} + \hat{C}(2)^3 * HS(1991m12) \\
&\textit{In general,} \\
H\hat{S}(1992m01 + k) &= \hat{C}(1) + \hat{C}(2) * H\hat{S}(1992m01 + (k-1)) \\
&= \hat{C}(1) * \sum_{i=0}^{k} \hat{C}(2)^i + \hat{C}(2)^{(k+1)} * HS(1991m12), \\
&\quad \textit{forall time of observations } k \geq 1, \textit{ after } 1992m01
\end{aligned}
\tag{1.4}
$$

4.3 Do the forecasts using LV(2), and LV(3) models as exercises.

Example 1.2 A Static Forecast Using the Simplest Model in (1.2)

As a comparison, Figure 1.4 presents the statistical results of a static forecast of *HS* using the LV(1) model. Based on these results, the following findings and notes are presented.

Forecast: HSF_STATIC	
Actual: HS	
Forecast sample: 1992M01 1996M04	
Included observations: 52	
Root Mean Squared Error	14.05771
Mean Absolute Error	10.86165
Mean Abs. Percent Error	10.17136
Theil Inequality Coefficient	0.062506
Bias Proportion	0.001445
Variance Proportion	0.034841
Covariance Proportion	0.963714

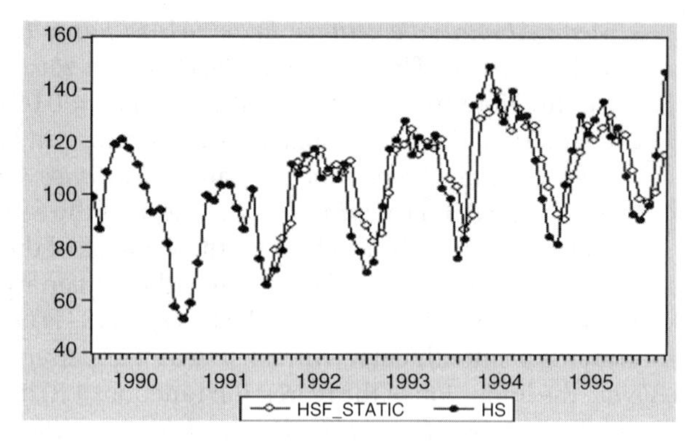

Figure 1.4 Statistical results of a static forecast of HS using the LV(1) model.

1) The forecast evaluation is obtained by selecting the static forecast method, and inserting the name *HSF_Static*, as the forecast series.
2) Compared to the dynamic forecast, this static forecast is better, since it has a much smaller RMSE specific to the in-sample forecast values.
3) This static forecast method cannot be used to compute the out of the sample period, namely after the last sample point 1996m04, if an observed score of *HS*(−1) is not available after 1996m04, because EViews computes the forecast as follows:

$$H\hat{S}(k) = \hat{C}(1) + \hat{C}(2) * HS(k-1) \tag{1.5}$$

which always uses the actual value of *HS*(*k*−1) for all observation points $k > 1$.

1.2.2 Special LV(*p*) Models

Corresponding to the basic LV(*p*) models in (1.2), Agung (2009a) introduced special or specific LV(*p*) models using the general equation specification as follows:

$$G(Y)\ C\ \ G(Y(-p)) \tag{1.6}$$

for the monthly, semi-annual, and quarterly time series Y_t, with p = 12, 6, and 4, respectively.

Example 1.3 Forecast of HS Based on a Special LV(12) Model

Since *HS* is a monthly time series, then a special LV(12) can be applied. The main objective of this example is to predict the monthly scores of a year using the monthly scores of previous year. The model considered has the following equation specification, with the dynamic forecast graph and its evaluation presented in Figure 1.5a, from 1992 to @last.

$$HS\ C\ HS(-12) \tag{1.7}$$

1) Compared to the dynamic forecast based on the LV(1) model in Figure 1.2, this dynamic forecast is a better forecast for *HS* because it has a smaller *RMSE,* as presented in Figure 1.5a.
2) Figure 1.5b presents the graphs of *HS*, *Fit_HS_Lag12*, and *HSF_Lag12*, which can be developed using the same method as presented in previous example. Note that this figure presents two graphs before 1992, those are the graphs of the observed scores of *HS* and its fitted values,

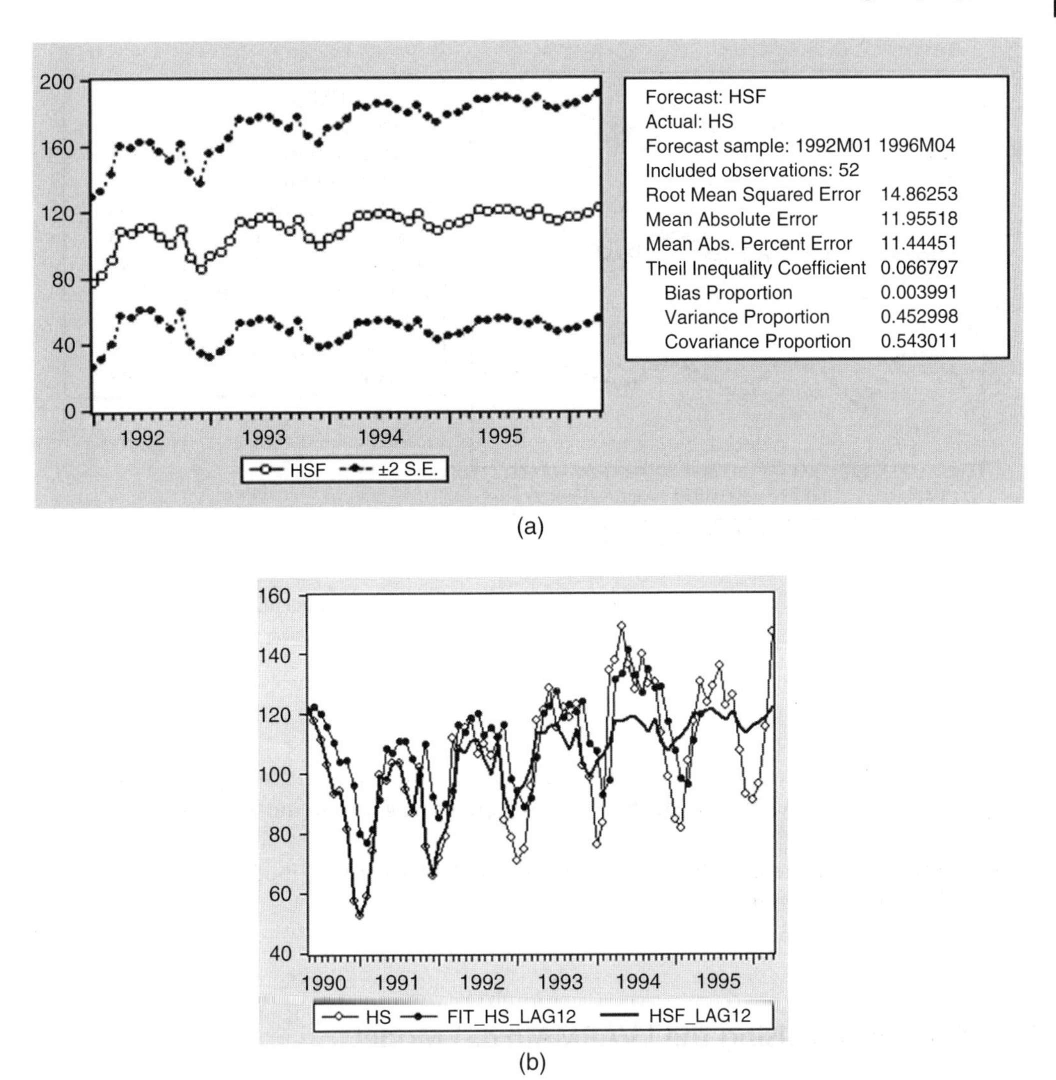

Figure 1.5 Statistical results of a dynamic forecast based on a special LV(12) models in (1.7).

namely *Fit_HS_Lag12* and three graphs starting from 1992M01, including the forecast of *HS*, namely *HSF_Lag12*.

3) A lot of possible LV(12) models can easily be applied, if additional relevant exogenous variables or predictors are available. Refer to all time series models presented in Agung (2009a).
4) As a comparison, Figure 1.6 presents the statistical results of a dynamic forecast using the regular or basic LV(12) model with the following equation specification.

$$\begin{aligned} &hs\, c\, hs(-1)\, hs(-2)\ hs(-3)\, hs(-4)\, hs(-5)\, hs(-6) \\ &hs(-7)\, hs(-8)\, hs(-9)\, hs(-10)\ hs(-11)\, hs(-12) \end{aligned} \tag{1.8}$$

Note that this forecasting has a greater RMSE than the forecasting in Figure 1.5. So it can be said that the special LV(12) model is better than the basic LV(12) model in (1.8) to conduct the forecasting. However, this model has a DW statistic of 2.100311, compared to the DW

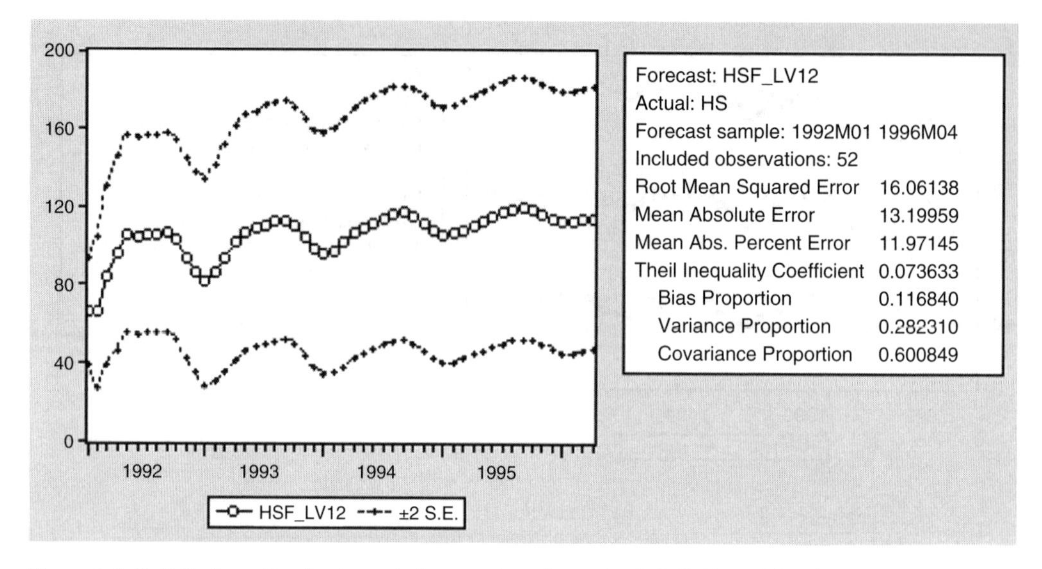

Figure 1.6 Statistical results of a dynamic forecast based on a basic LV(12) model in (1.4).

statistic of 0.268672 for the model in (1.7). However, it can be increased by using additional lag(s), or the terms AR(q), which should be selected using a trial-and-error method.

5) As additional comparisons, the statistical results of the basic LV(p) models, for all $p < 12$ can easily be obtained. Do it as an exercise.

1.3 Forecasting Using the LVARMA(p,q,r) Model

As the extension LV(p), I propose *lag-variable-autoregressive-moving-average* models, namely LVARMA(p,q,r) models for non-negative integers $p \geq 0$, $q \geq 0$, and $r \geq 0$, which should be selected using the trial-and-error method, in order to obtain acceptable or suitable model. Hence, we would have LVARMA(p,0,0) = LV(p), LVARMA(p,q,0) = LVAR(p,q), LVARMA (p,0,r) = LVMA(p,r), and LVARMA(0,q,r) =ARMA(q,r) = ARIMA(q,r), with the following general equation specification.

$$Y\ C\ Y(-1)\cdots Y(-p)\ AR(1)\cdots AR(q)\ MA(1)\cdots MA(r) \tag{1.9}$$

This model is extended to various models with alternative specific trends, such as linear, quadratic, cubic, and logarithmic trends, models with exogenous variables, and heterogeneous regression models by *Month*. However, for multivariate models, the term MA(r) cannot be applied. Hence, for multivariate models, we have lag-variable autoregressive models: MLVAR (p,q) models, with special forms, such as triangular and circular MLVAR($p.q$) models, which are presented in the following chapters. In addition, note that the Basic VAR models in fact are specific MLVAR(p,q) models, where all regressions have the same set of independent variables.

1.3.1 Special Notes on the ARMA Model

Autoregressive-Moving-Average (ARMA/ARIMA) models have been presented by Brooks (2008), Tsay (2002), Enders (1995), and Hankle and Reitch (1992). In fact, these are the *Lag-Variables-Moving-Average models.* For this reason, in this book, the ARMA/ARIMA(p,r) models are presented as LVARMA(p,0,r) or LVMA/LVAMA(p,r) models. As an illustration, Figure 1.7 presents the statistical results of an ARMA(1,1) model of the time series *HS* in this book, which is different from the ARMA/ARIMA model presented in other books. Based on these results, the following notes and comments are presented.

Dependent Variable: HS
Method: Least Squares
Date: 10/29/13 Time: 13:01
Sample (adjusted): 1946M02 1996M04
Included observations: 603 after adjustments
Convergence achieved after 10 iterations
MA Backcast: 1946M01

Variable	Coefficient	Std. Error	t-Statistic	Prob.
C	123.8187	4.355430	28.42858	0.0000
AR(1)	0.797759	0.027636	28.86627	0.0000
MA(1)	0.292670	0.044143	6.630037	0.0000

R-squared	0.767431	Mean dependent var	123.1819
Adjusted R-squared	0.766656	S.D. dependent var	34.64025
S.E. of regression	16.73321	Akaike info criterion	8.477630
Sum squared resid	168000.1	Schwarz criterion	8.499530
Log likelihood	-2553.005	Hannan-Quinn criter.	8.486153
F-statistic	989.9425	Durbin-Watson stat	1.904926
Prob(F-statistic)	0.000000		

Inverted AR Roots	.80
Inverted MA Roots	-.29

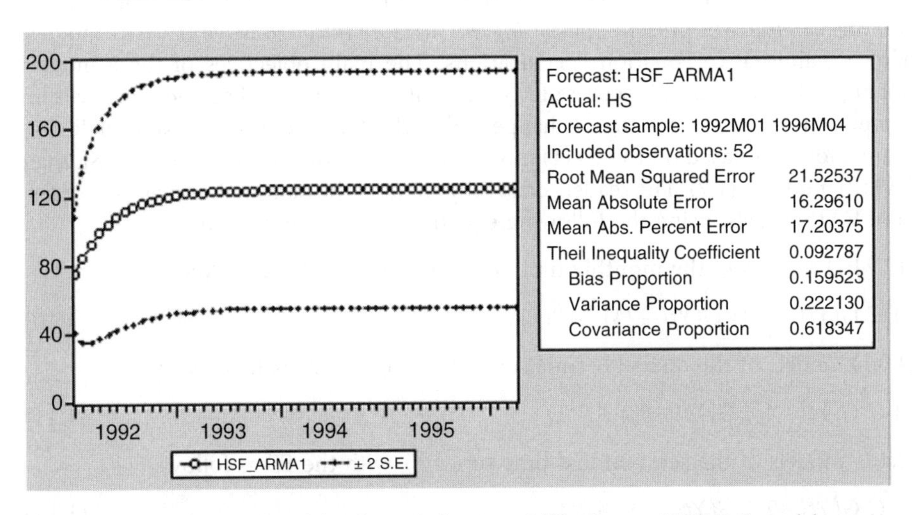

Figure 1.7 Statistical results of the dynamic forecast of *HS* using an ARMA(1,1) model.

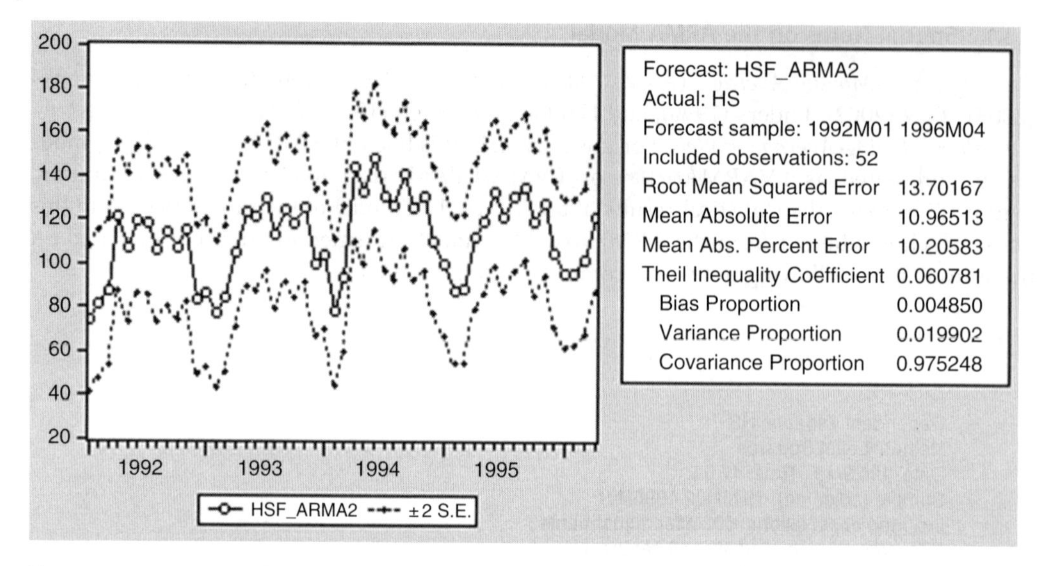

Figure 1.8 Statistical results of the static forecast of *HS* using an ARMA(1,1) model.

1) Note that the results show a regression of *HS* without the lag of *HS* as an independent variable.
2) Compared to the dynamic forecast previously presented, this dynamic forecast has the greatest value of RMSE = 21.525 37, and the dynamic forecast in Figure 1.5 has the smallest RMSE.
3) As a comparison, Figure 1.8 presents the static forecast using the same ARMA model, which gives a smaller value of RMSE = 13.70167, even it is smaller than the static forecast in Figure 1.4.

1.3.2 Application of Special LVAR Models

It has been found that the previous lagged variable models (1.2) and (1.3) have small DW statistics, so the term AR(p) should be added to the models, at least for $p = 1$. Do it as an exercise.

In addition, this section will present special *lagged-variable autoregressive* (LVAR) models, with the following alternative equation specifications. The main objectives of these models are to forecast or predict the monthly scores of *HS* in recent year using its observed scores within the last two previous years as predictors. It happens that the four models have smaller RMSE than the dynamic forecast based on the LV(1) model in (1.2), but they have greater RMSE than the special LV(12) model in (1.7). The readers can easily conduct the data analysis based on alternative special LVAR models, using the following equation specifications (ES).

i) LVAR(12,24;1) models of the monthly time series Y_t, with the ES as follows:

$$G(Y_t)\ C\ G(Yt(-12))\ G(Yt(-24))\ AR(1) \tag{1.10}$$

ii) LVAR(4,8;1) models of the quarterly time series Y_t, with the ES as follows

$$G(Y_t)\ C\ G(Yt(-4))\ G(Yt(-8))\ AR(1) \tag{1.11}$$

iii) LVAR(2,4;1) models of the semi-annual time series Y_t, with the ES as follows

$$G(Y_t)\ C\ G(Yt(-2))\ G(Yt(-4))\ AR(1) \tag{1.12}$$

iv) LVAR(1,2;1) models of the annually time series Y_t, with the ES as follows

$$G(Y_t)\ C\ G(Yt(-1))\ G(Yt(-2))\ AR(1) \tag{1.13}$$

Example 1.4 Forecast Using the Special LVAR in (1.10)

As an illustration, Figure 1.9 presents the statistical results of the special LVAR(12,24;1) model of HS in (1.10). Based on these results, the following findings and notes are presented.

1) The equation of the model is as follows:

$$\begin{aligned} HS_t &= C(1) + C(2) * HS_{t-12} + C(3) * HS_{t-24} + \mu_t \\ \mu t &= \rho_1 * \mu_{t-1} + \varepsilon_t = C(4) * \mu_{t-1} + \varepsilon_t \end{aligned} \tag{1.14}$$

where the error terms ε_t, for t = 1,...,T are assumed to have identical independent normal distribution. Note that this is a theoretical and abstract mathematical statistics, which is valid for the sample space and it should not be tested using the only one sampled data – refer to Agung (2009a, 2011a, b).

Dependent Variable: HS
Method: Least Squares
Date: 11/22/13 Time: 12:59
Sample (adjusted): 1948M02 1996M04
Included observations: 579 after adjustments
Convergence achieved after 7 iterations

Variable	Coefficient	Std. Error	t-Statistic	Prob.
C	25.44032	6.515362	3.904667	0.0001
HS(-12)	0.443410	0.039470	11.23421	0.0000
HS(-24)	0.357261	0.039430	9.060639	0.0000
AR(1)	0.904809	0.018102	49.98325	0.0000

R-squared	0.879357	Mean dependent var	124.3138
Adjusted R-squared	0.878728	S.D. dependent var	34.57983
S.E. of regression	12.04212	Akaike info criterion	7.821583
Sum squared resid	83382.28	Schwarz criterion	7.851713
Log likelihood	-2260.348	Hannan-Quinn criter.	7.833330
F-statistic	1397.049	Durbin-Watson stat	2.201170
Prob(F-statistic)	0.000000		

Inverted AR Roots	.90

(a)

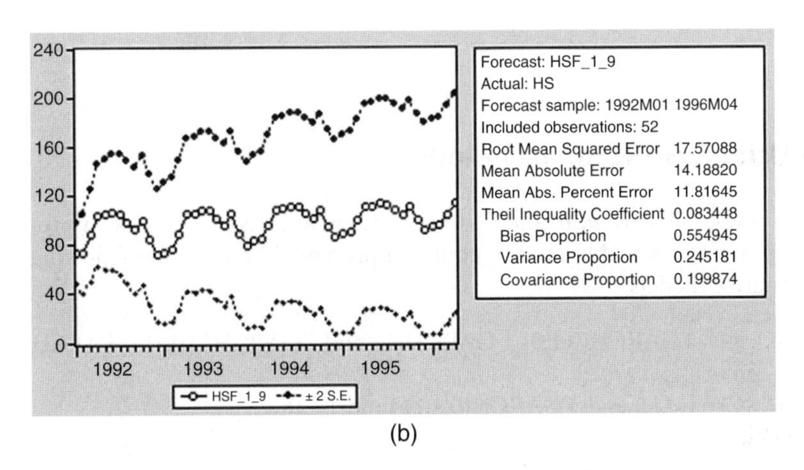

(b)

Figure 1.9 Statistical results of the LVAR(12,24;1) model in (1.10).

2) Then the forecast values can easily be computed for all t starting from t = 1992 M01 = 553 to the last sample point (t = 1996 M04 = 640), as well as for the time points after the last sample point. For instance, as follows:

$$\begin{aligned} H\hat{S}_t &= \hat{C}(1) + \hat{C}(2) * HS(-12) + \hat{C}(3) * HS(-24) \\ &= 25.4403162266 + 0.443410013085 * \mathrm{HS}(-12) \\ &\quad + 0.357261323168 * \mathrm{HS}(-24) \\ \hat{\mu}_t &= \hat{C}(4) * \hat{\mu}_t = 0.904808948651 * \hat{\mu}_{t-1} \end{aligned} \tag{1.15a}$$

for t = 553 up to t – 12 = 552 or t = 564,

$$\begin{aligned} H\hat{S}_t &= \hat{C}(1) + \hat{C}(2) * H\hat{S}(-12) + \hat{C}(3) * HS(-24) \\ \hat{\mu}_t &= \hat{C}(4) * \hat{\mu}_t = 0.904808948651 * \hat{\mu}_{t-1} \end{aligned} \tag{1.15b}$$

for t – 12 = 552 (t = 564) up to t – 24 = 552 (t = 576), and for $t > 576$

$$\begin{aligned} H\hat{S}_t &= \hat{C}(1) + \hat{C}(2) * H\hat{S}(-12) + \hat{C}(3) * H\hat{S}(-24) \\ \hat{\mu}_t &= \hat{C}(4) * \hat{\mu}_t = 0.904808948651 * \hat{\mu}_{t-1} \end{aligned} \tag{1.15c}$$

3) Finally, to forecast beyond the last sample period, say $t = T + j =$ *1966 M04* $+ j$, for j = 1,2,…; the forecast values are

$$\begin{aligned} H\hat{S}_t &= \hat{C}(1) + \hat{C}(2) * H\hat{S}(-12) + \hat{C}(3) * H\hat{S}(-24) \\ \hat{\mu}_t &= \hat{C}(4) * \hat{\mu}_t = 0.904808948651 * \hat{\mu}_{t-1} \end{aligned} \tag{1.16}$$

4) In a statistical sense, the LS regression is an acceptable model, since each of the independent variables as well as the AR(1) have significant effects. Therefore, the LV(12,24;1) model of *HS* also can be considered an acceptable forecast model. However, it might not be the best forecast, because we would never conduct data analysis based on all possible models.

   As a comparison, this forecast has a RMSE of 17.57088, which is greater than the special LV(12) model in (1.7), with RMSE = 14.86253. Therefore, it can be said that this model is worse than the model in (1.7) to present the forecasting of *HS*.

1.4 Forecasting Using TGARCH(*a,b,c*) Models

Various TGARCH(*a,b,c*) models, where *a*, *b*, and *c* are non-negative integers, have been presented in Agung (2009a). A lot of possible models could be presented here, so the following shows the simplest alternative models.

i) TGARCH(1,0,0) = ARCH(1), TGARCH(0,1,0) = GARCH(1), and TGARCH(0,0,1) = TARCH (1) models,
ii) TGARCH(1,1,0), TGARCH(1,0,1), and TGARCH(0,1,1) models, and
iii) TGARCH(1,1,1) model.

For data analysis, the following notes and comments are presented.

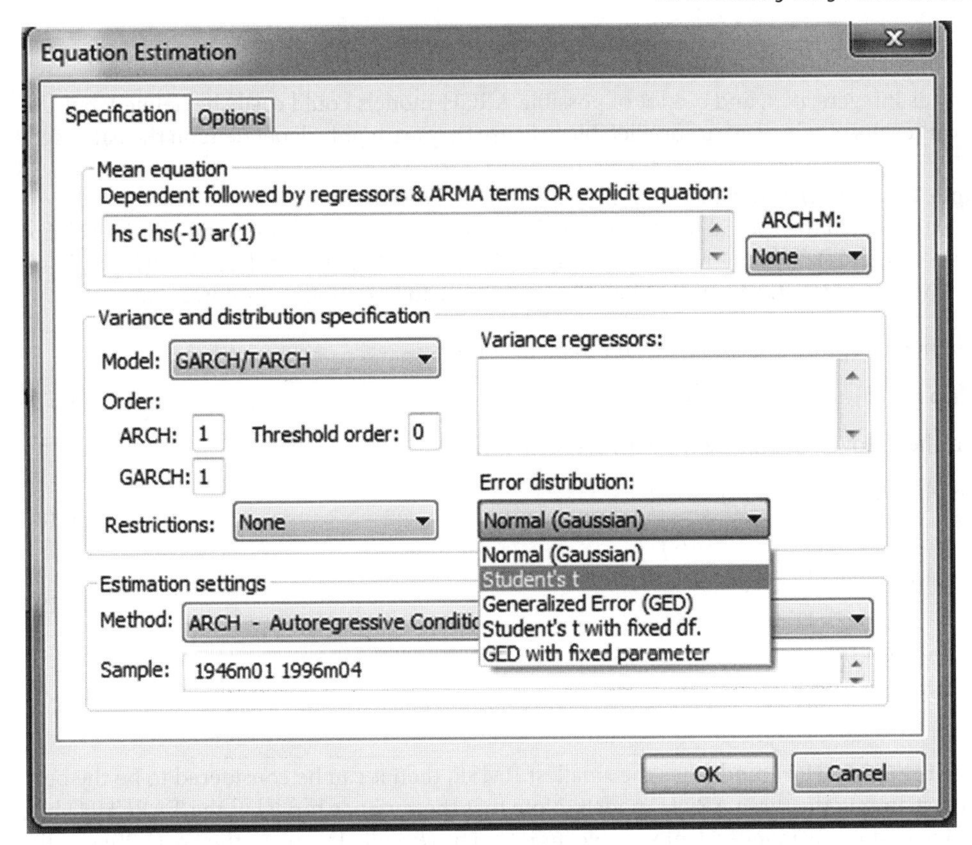

Figure 1.10 Dialog for conducting the ARCH estimation setting.

1) After inserting the mean equation, by selecting the ARCH estimation setting, then the dialog in Figure 1.10 is shown on the screen. Then note a lot of the following alternative options.
   - The ARCH, GARCH, and Threshold orders can easily be changed
   - The ARCH-M
   - Restriction
   - Four alternative error distributions and
   - Sample
2) Furthermore, we can insert additional variance regressors, which should be subjectively selected by researchers. I would say that the variance regressors should be selected from the exogenous (external, source, cause, upstream, or independent) variables of the mean model, since the exogenous variables should have effects on both the mean and variance of the endogenous variable.
3) Referring to a lot of possible options indicated previously, we can never say that a forecast model is the best possible model, because we will not try to apply all possible combinations of options. On the other hand, the data has only limited number of observed or measured exogenous variable, such as the House.wf1 that only has a single time series variable HS_t. Agung (2011a) presented special notes and comments on the validity and reliability forecasting and predicted probabilities of binary choice models.

1.4.1 Application of ARCH(*a*), GARCH(*b*), and TARCH(*c*) Models

For various integers, *a*, *b*, and *c*, a lot of possible ARCH models could easily be defined, but the statistical results obtained are unpredictable because they are highly dependent on the data used.

Example 1.5 Application of the Simplest Models, Namely for $a = b = c = 1$
As the extension of the special model in (1.3), Figures 1.11–1.13 present the statistical results of the three simplest TGARCH models, namely ARCH(1), GARCH(1), and TARCH(1) models.
Based on these statistical results, the following notes and comments are presented.

1) The models applied have the mean model as follows:

$$HS = C(1) + C(2) * HS(-12) + [AR(1) = C(3)] \tag{1.17}$$

with the following variance models, respectively.

$$\begin{aligned} GARCH &= C(4) + C(5) * RESID(-1)\hat{}2 \\ \sigma_t^2 &= C(4) + C(5) * \varepsilon_{t-1}^2 \end{aligned} \tag{1.18}$$

$$\begin{aligned} GARCH &= C(4) + C(5) * RESID(-1)\hat{}2 \\ \sigma_t^2 &= C(4) + C(5) * \sigma_{t-1}^2 \end{aligned} \tag{1.19}$$

$$\begin{aligned} GARCH &= C(4) + C(5) * RESID(-1)\hat{}2 \\ \sigma_t^2 &= C(4) + C(5) * \varepsilon_{t-1}^2 * \left(\varepsilon_{t-1}^2 < 0\right) \end{aligned} \tag{1.20}$$

2) Since the GARCH(1) model has the smallest RMSE, then it can be considered to be the best forecast among the three ARCH models. Note that the variance model of the TARCH(1) has a group or dummy independent variable indicated by ($Resid(-1) < 0$), so the Eq. (1.20) should also present $TARCH = C(4)$ for $Resid(-1) \geq 0$. In other words, the score of *TARCH* is assumed to be invariant or constant over time for the $Resid(-1) \geq 0$.
3) In order to study the characteristics of more advanced models, do the data analysis based on the ARCH(*a*), GARCH(*b*), and TARCH(*c*) models, for $a > 1$, $b > 1$, and $c > 1$.

1.4.2 Application of TGARCH(*a,b,0*) Models

The following example only presents a forecast based on the simplest model, namely TGARCH (1,1,0), the others, as well as various TGARCH(a,0,c) and TGARCH(0,b,c), are recommended as exercises.

Example 1.6 Application of the Simplest Model, for $a = b = 1$
As an extension of the TGARCH(1,0,0) previously, Figure 1.14 presents the statistical results of the TGARCH(1,1,0) model with the mean model in (1.17), and the variance model is as follows:

$$GARCH = C(4) + C(5) * RESID(-1)\hat{}2 + C(6) * GARCH(-1) \tag{1.21a}$$

$$\sigma_t^2 = C(4) + C(5) * \varepsilon_{t-1}^2 + C(6) * \sigma_{t-1}^2 \tag{1.21b}$$

Based on these results the following findings and notes are presented.

1) Since each of the exogenous variables of the mean and variance models has a significant effect, then the model is an acceptable model in a statistical sense.

Dependent Variable: HS
Method: ML - ARCH (Marquardt) - Normal distribution
Date: 10/18/13 Time: 14:05
Sample (adjusted): 1947M02 1996M04
Included observations: 591 after adjustments
Convergence achieved after 13 iterations
Presample variance: backcast (parameter = 0.7)
GARCH = C(4) + C(5)*RESID(-1)^2

Variable	Coefficient	Std. Error	z-Statistic	Prob.
C	37.59669	5.877540	6.396671	0.0000
HS(-12)	0.698464	0.025226	27.68812	0.0000
AR(1)	0.876018	0.018466	47.43957	0.0000
		Variance Equation		
C	138.8242	9.989689	13.89675	0.0000
RESID(-1)^2	0.154785	0.059853	2.586088	0.0097

R-squared	0.862617	Mean dependent var	123.9640
Adjusted R-squared	0.862150	S.D. dependent var	34.47252
S.E. of regression	12.79901	Akaike info criterion	7.932969
Sum squared resid	96322.99	Schwarz criterion	7.970041
Log likelihood	-2339.192	Hannan-Quinn criter.	7.947410
F-statistic	923.0038	Durbin-Watson stat	2.180799
Prob(F-statistic)	0.000000		
Inverted AR Roots	.88		

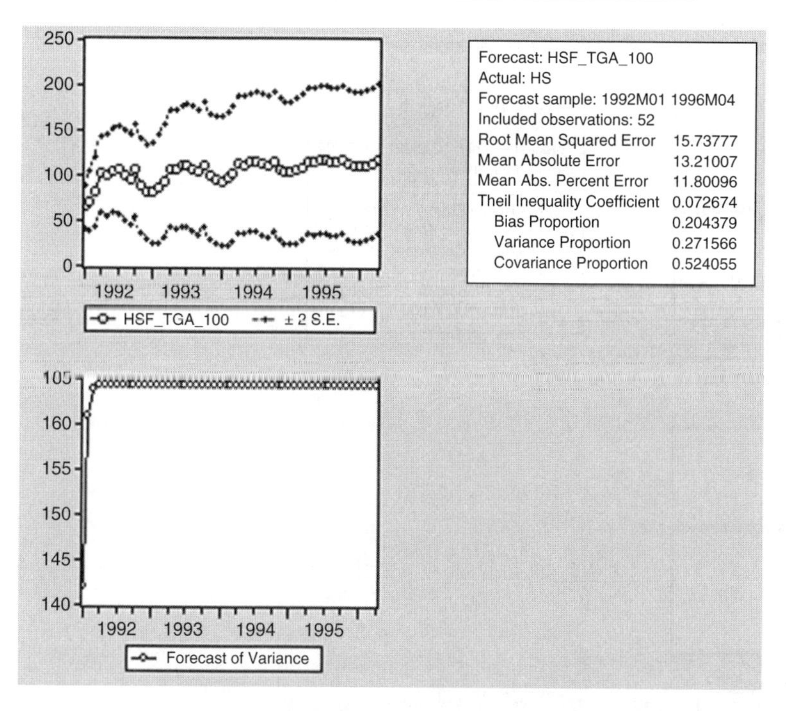

Figure 1.11 Statistical results of an ARCH(1) model with estimation Eqs. (1.17) and (1.18).

Dependent Variable: HS
Method: ML - ARCH (Marquardt) - Normal distribution
Date: 10/18/13 Time: 14:10
Sample (adjusted): 1947M02 1996M04
Included observations: 591 after adjustments
Convergence achieved after 12 iterations
Presample variance: backcast (parameter = 0.7)
GARCH = C(4) + C(5)*GARCH(-1)

Variable	Coefficient	Std. Error	z-Statistic	Prob.
C	39.77074	5.864759	6.781309	0.0000
HS(-12)	0.687236	0.025550	26.89724	0.0000
AR(1)	0.866420	0.018795	46.09722	0.0000
	Variance Equation			
C	296.0894	73.53546	4.026485	0.0001
GARCH(-1)	-0.816907	0.438105	-1.864637	0.0622

R-squared	0.862706	Mean dependent var	123.9640
Adjusted R-squared	0.862239	S.D. dependent var	34.47252
S.E. of regression	12.79487	Akaike info criterion	7.947922
Sum squared resid	96260.65	Schwarz criterion	7.984993
Log likelihood	-2343.611	Hannan-Quinn criter.	7.962363
F-statistic	923.6968	Durbin-Watson stat	2.148834
Prob(F-statistic)	0.000000		

Inverted AR Roots	.87

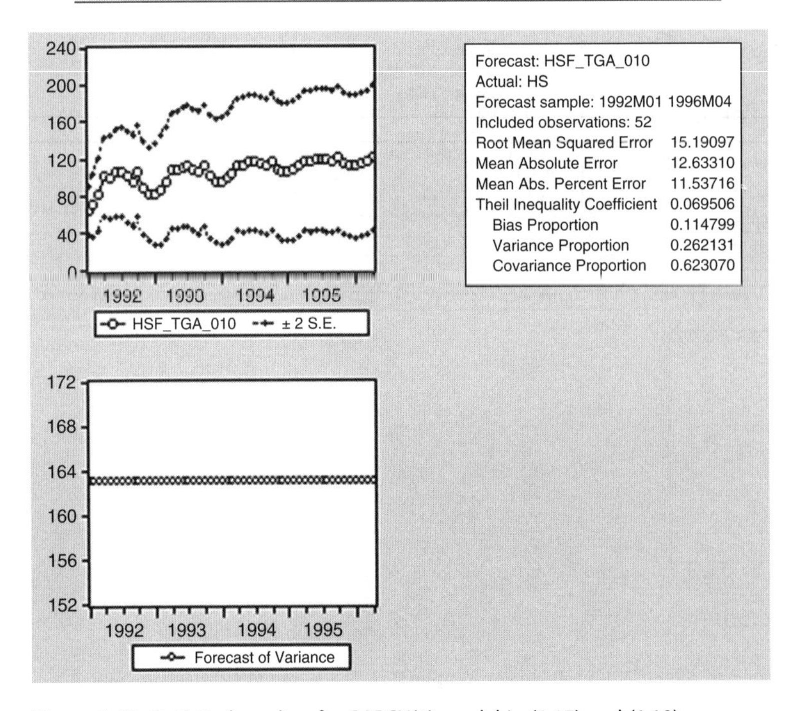

Figure 1.12 Statistical results of a GARCH(1) model in (1.17) and (1.19).

Dependent Variable: HS
Method: ML - ARCH (Marquardt) - Normal distribution
Date: 10/18/13 Time: 14:13
Sample (adjusted): 1947M02 1996M04
Included observations: 591 after adjustments
Convergence achieved after 13 iterations
Presample variance: backcast (parameter = 0.7)
GARCH = C(4) + C(5)*RESID(-1)^2*(RESID(-1)<0)

Variable	Coefficient	Std. Error	z-Statistic	Prob.
C	38.27065	6.031711	6.344908	0.0000
HS(-12)	0.694183	0.025300	27.43823	0.0000
AR(1)	0.871585	0.018887	46.14747	0.0000
	Variance Equation			
C	156.7698	9.191473	17.05600	0.0000
RESID(-1)^2*(RESID(-1)<0)	0.078885	0.069856	1.129247	0.2588

R-squared	0.862675	Mean dependent var	123.9640
Adjusted R-squared	0.862208	S.D. dependent var	34.47252
S.E. of regression	12.79634	Akaike info criterion	7.945332
Sum squared resid	96282.76	Schwarz criterion	7.982403
Log likelihood	-2342.846	Hannan-Quinn criter.	7.959773
F-statistic	923.4509	Durbin-Watson stat	2.167160
Prob(F-statistic)	0.000000		
Inverted AR Roots	.87		

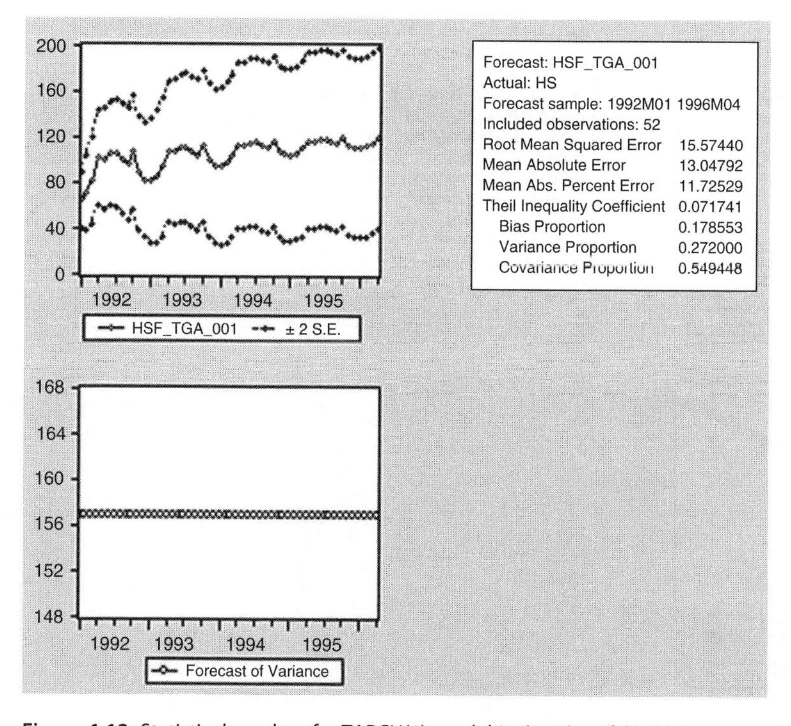

Figure 1.13 Statistical results of a TARCH(1) model in (1.17) and (1.20).

Dependent Variable: HS
Method: ML - ARCH (Marquardt) - Normal distribution
Date: 10/16/13 Time: 06:50
Sample (adjusted): 1947M02 1996M04
Included observations: 591 after adjustments
Convergence achieved after 17 iterations
Presample variance: backcast (parameter = 0.7)
GARCH = C(4) + C(5)*RESID(-1)^2 + C(6)*GARCH(-1)

Variable	Coefficient	Std. Error	z-Statistic	Prob.
C	35.43918	5.253119	6.746313	0.0000
HS(-12)	0.714933	0.022029	32.45481	0.0000
AR(1)	0.876102	0.021038	41.64461	0.0000
	Variance Equation			
C	3.408938	1.998712	1.705567	0.0881
RESID(-1)^2	0.077787	0.022676	3.430426	0.0006
GARCH(-1)	0.904729	0.026295	34.40748	0.0000

R-squared	0.862457	Mean dependent var	123.9640
Adjusted R-squared	0.861989	S.D. dependent var	34.47252
S.E. of regression	12.80649	Akaike info criterion	7.887796
Sum squared resid	96435.59	Schwarz criterion	7.932281
Log likelihood	-2324.844	Hannan-Quinn criter.	7.905125
F-statistic	737.4036	Durbin-Watson stat	2.197931
Prob(F-statistic)	0.000000		
Inverted AR Roots	.88		

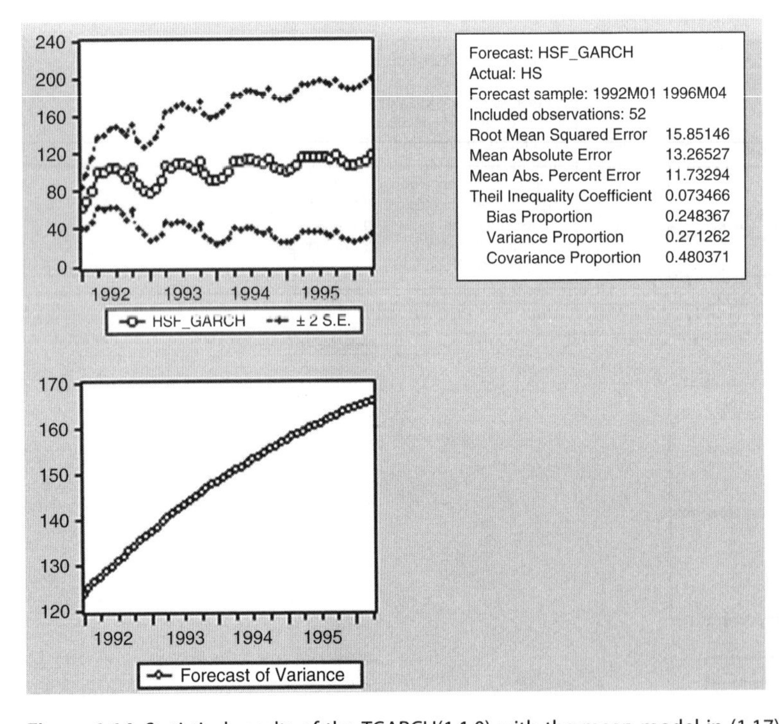

Figure 1.14 Statistical results of the TGARCH(1,1,0) with the mean model in (1.17).

Dependent Variable: HS
Method: ML - ARCH (Marquardt) - Normal distribution
Date: 10/16/13 Time: 06:19
Sample (adjusted): 1947M02 1996M04
Included observations: 591 after adjustments
Convergence achieved after 17 iterations
Presample variance: backcast (parameter = 0.7)
GARCH = C(4) + C(5)*RESID(-1)^2 + C(6)*RESID(-1)^2*(RESID(-1)<0) +
C(7)*GARCH(-1)

Variable	Coefficient	Std. Error	z-Statistic	Prob.
C	36.31424	5.511856	6.588387	0.0000
HS(-12)	0.712911	0.023432	30.42523	0.0000
AR(1)	0.874298	0.021251	41.14094	0.0000
	Variance Equation			
C	3.019998	1.759030	1.716854	0.0860
RESID(-1)^2	0.089591	0.028984	3.091040	0.0020
RESID(-1)^2*(RESID(-1)<0)	-0.031808	0.037587	-0.846236	0.3974
GARCH(-1)	0.911070	0.022851	39.86925	0.0000

R-squared	0.862511	Mean dependent var	123.9640
Adjusted R-squared	0.862043	S.D. dependent var	34.47252
S.E. of regression	12.80396	Akaike info criterion	7.889791
Sum squared resid	96397.58	Schwarz criterion	7.941691
Log likelihood	-2324.433	Hannan-Quinn criter.	7.910008
F-statistic	614.7839	Durbin-Watson stat	2.192327
Prob(F-statistic)	0.000000		
Inverted AR Roots	.87		

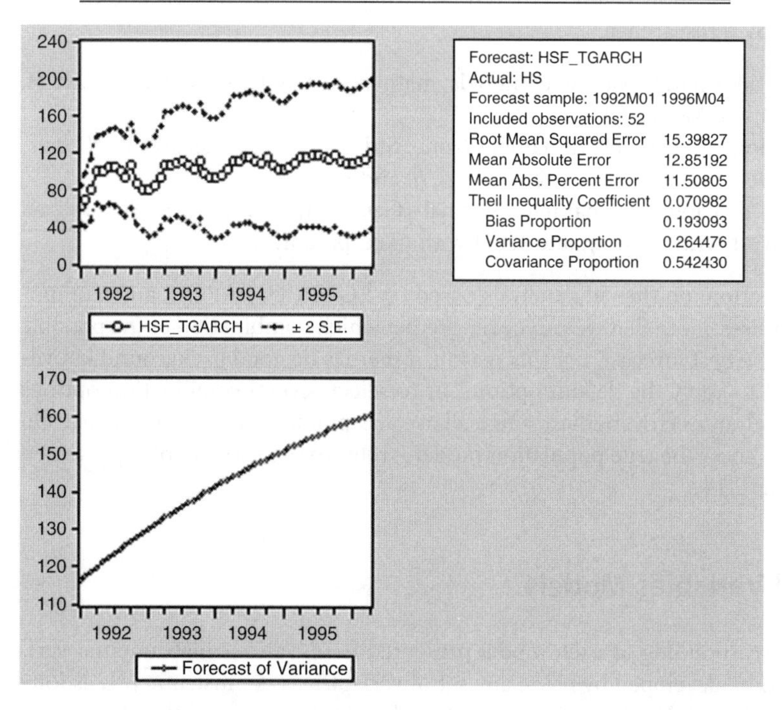

Figure 1.15 Statistical results of the TGARCH(1,1,1) with the mean model in (1.17).

2) The forecast evaluation using the *Normal* error distribution presents a RMSE = 15.85146. As a comparison, using the other error distributions, namely the *Student's t, GED, Student's with fixed df,* and *GED with fixed parameters*, respectively, are obtained the forecast evaluation with the RMSEs of 14.91643, 15.63520, 16.16500, and 15.440899. So we have three better forecasts, with the best being the *Student's t* error distribution, which is really unpredictable. And we never know the true error distribution for the corresponding population – refer to the special notes presented in Agung (2009a, Section 2.14).

1.4.3 Application of TGARCH(*a,b,c*) Models

Example 1.7 Application of the Simplest Model

As an extension of the TGARCH(1,1,0) before, Figure 1.15 presents the statistical results of the TGARCH(1,1,1) model with the mean model in (1.17) and the variance model as follows:

$$GARCH = C(4) + C(5) * Resid(-1)\hat{}2 + C(6) * \mathrm{Resid}(-1)\hat{}2* \\ (Resid(-1) < 0) + C(7) * GARCH(-1) \tag{1.22a}$$

$$\sigma_t^2 = C(4) + C(5) * \varepsilon_{t-1}^2 + C(6) * \varepsilon_{t-1}^2 * \left(\varepsilon_{t-1}^2 < 1\right) + C(7) * \sigma_{t-1}^2 \tag{1.22b}$$

1.4.4 Other Alternative Models

Referring to the dialog presented in Figure 1.10, the following alternative models and options could easily be selected by a researcher.

1) General equation of alternative variance models, namely GARCH/TARCH, EGARCH, PARCH, and Component ARCH(1,1).
2) Three alternative options for ARCH-M, namely None, Std. Dev., Variance, and Log(Var).
3) The order of the models, namely various integers *a*, *b*, and *c*.
4) Five alternative error distributions; namely, Normal (Gaussian), Student's *t*, Generalized Error (GED), Student's t with fixed *df*, and GED with fixed parameters.

As an additional illustration, do the data analysis based on EGARCH, PARCH, and Component ARCH(1,1), as exercises. Note that we can never predict which combination of the possible options that would give the best forecast. For this reason, if there is no good background knowledge, it is recommended to apply the default options, or to select several combination options subjectively using the trial-and-error method, since all are acceptable in a statistical sense. On the other hand, we never know the true population model – refer to the special notes presented in Agung (2009a, Section 2.14).

1.5 Instrumental Variables Models

It is recognized that, corresponding to each model presented here, various instrumental variables models (IVMs) can be developed by using any set of variables as the instruments, as long as the number of the instrumental variables is greater than the number of parameters in the corresponding mean model. For this reason, Agung (2014, 2009a) has recommended researchers not to apply an IVM, because a better model without instruments can easily be applied. On the other hand, it is impossible to select the best set of instrumental variables because a lot of possible sets of variables can be used as instrumental variables, and we can never apply all of those possible sets. See the unpredictable impact of an instrument list presented in Agung (2014, 2009a), as well as in the following examples. On the other hand, I have found that the statistical results can be obtained based on all IVMs, by using the intercept C as a single instrument variable, with the option for ***the lagged dependent and regressors added to the instrument list.***

1.5.1 Application of the GMM Estimation Method

Agung (2014) presents various statistical results using the GMM estimation method. For all models presented previously, the GMM estimation method can easily be applied. See the following illustrative examples, with special notes.

1.5.1.1 Statistical Results of Special LVAR(12,*q*) Using Simple Instrument

Example 1.8 Forecast of HS Using a Special LVAR(12,1) IVMs

Figure 1.16 presents two dialogs of the equation estimation for the GMM in EViews 8 SV, based on a special LVAR(12,1), using the following ES. Based on the options in Figure 1.16, the following findings and notes are presented.

$$HS\ C\ HS(-12)\ AR(1) \tag{1.23}$$

Note that both dialogs look the same, but unexpectedly they give different outcomes, such as follows:

1) By using the options *"Include a constant,"* and *"Include lagged regressors for linear eqs with ARMA"* as presented in Figure 1.16a, an error message is obtained *"Equation or VAR specification is incomplete."*

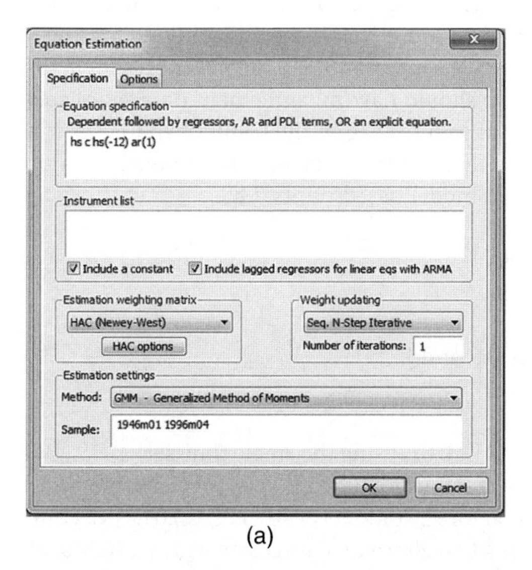

(a)

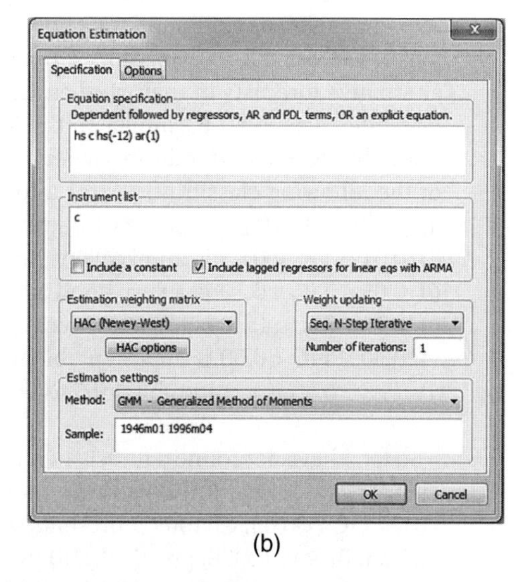

(b)

Figure 1.16 Two alternative dialogs for the model in (1.23), using the GMM.

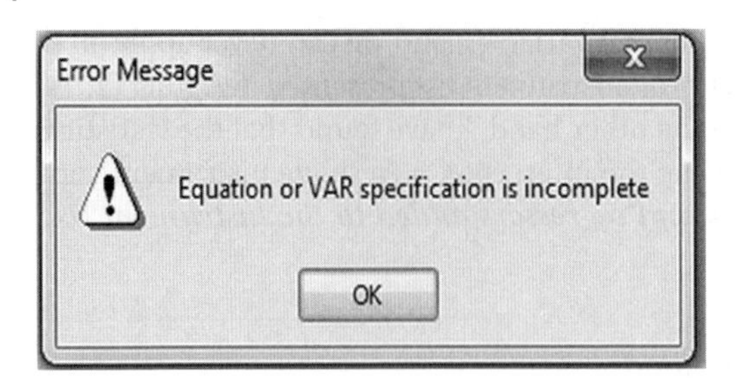

2) By using a single instrument list "*C*" instead of the option "*Include a constant,*" as presented in Figure 1.16b, then the statistical results are obtained in Figure 1.17a. Based on these results, the following findings and notes are presented.
3) Using EViews 8 SV, the results in Figure 1.17a present the instrument specification *C*, with the "*Lagged dependent variable & regressors added to the instrument list,*" where each of the regressors has such a large probability > 0.99. Why? And what should be done to modify the model?
4) Even though each regressor has very large probabilities (*p values*), I would also like to study how it is forecast. For this reason, Figure 1.18 presents its forecast evaluation. Based on this forecast evaluation, the following findings and notes are presented.

4.1 This forecast has a TIC = 0.071212, which is sufficiently small compared to the TIC = 0, which could be considered as a perfect forecast. However, based on my experiments and findings, I would say there is no model with a perfect forecast.

4.2 This TIC is smaller than the forecast in Figure 1.3 based on an LV(1) LS regression, with a TIC = 0.091042, and the forecast based on a basic LV(12) LS regression in Figure 1.6, with a TIC = 0.073 633, but it is greater than the forecast in Figure 1.5 based on a special LV(12) LS regression, with a TIC = 0.066797. In addition, GARCH(*a,b,c*) models presented in previous examples also have TICs < 0.10, where some of them have smaller TICs than this forecast.

4.3 Since all forecasts mentioned here have TICs < 0.10, would you say that all forecasts are good forecasts, in a statistical sense? If they are not good forecasts, what should be the maximum TIC upper bound for a good forecast? Moreover, is the upper bound for excellent or very good forecasts? Since students, as well as researchers, can never present many alternative forecasts in their studies, then we should have an agreement on a set of specific intervals for a TIC similar to the level of significance in the testing hypothesis; it is not only TIC = 0 that indicates a perfect forecast. For this reason, in this book, I use a set of the following classifications for a forecast model based on its TIC.

i) TIC < 0.01(1%) is a perfect forecast,
ii) 0.01 ≤ TIC < 0.05 is an excellent forecast,
iii) 0.05 ≤ TIC < 0.10 is a very good forecast,
iv) 0.10 ≤ TIC < 0.30 is a good forecast,
v) 0.30 ≤ TIC < 0.50 is an acceptable forecast, and
vi) TIC ≥ 0.50 is a miserable forecast.

So far, I have not found a forecast with a TIC < 0.01, using the single time series HS_t. But Chapter 2 presents some illustrative models of HS_t using time independent variables, with 0.01 ≤ TICs < 0.05. Chapter 7 presents several forecast models with TIC < 0.01. As a comparison, Brooks (2008, pp. 257–258) presents two alternative forecasts with a TIC > 0.50, namely TIC = 0.535810 and TIC = 0.509415, respectively.

Dependent Variable: HS
Method: Generalized Method of Moments
Date: 12/19/13 Time: 11:58
Sample (adjusted): 1947M02 1996M04
Included observations: 591 after adjustments
Sequential 1-step weighting matrix & coefficient iteration
Estimation weighting matrix: HAC (Bartlett kernel, Newey-West fixed
bandwidth = 6.0000)
Standard errors & covariance computed using estimation weighting matrix
Convergence achieved after 39 iterations
Instrument specification: C
Lagged dependent variable & regressors added to instrument list

Variable	Coefficient	Std. Error	t-Statistic	Prob.
C	50.83140	782194.8	6.50E-05	0.9999
HS(-12)	0.597357	6304.569	9.47E-05	0.9999
AR(1)	0.859175	240.3873	0.003574	0.9971

R-squared	0.860564	Mean dependent var	123.9640
Adjusted R-squared	0.860090	S.D. dependent var	34.47252
S.E. of regression	12.89429	Sum squared resid	97762.42
Durbin-Watson stat	2.027639	J-statistic	3.270614
Instrument rank	3		

Inverted AR Roots	.86

(a). Using EViews 8 SV

Dependent Variable: HS
Method: Generalized Method of Moments
Date: 12/19/13 Time: 13:33
Sample (adjusted): 1947M02 1996M04
Included observations: 591 after adjustments
Kernel: Bartlett, Bandwidth: Fixed (5), No prewhitening
Simultaneous weighting matrix & coefficient iteration
Convergence not achieved after: 499 weight matrices, 500 total coef
iterations
Instrument list: C
Lagged dependent variable & regressors added to instrument list

Variable	Coefficient	Std. Error	t-Statistic	Prob.
C	145.8495	126.5980	1.152068	0.2498
HS(-12)	-0.169547	0.966871	-0.175356	0.8609
AR(1)	0.897667	0.033613	26.70586	0.0000

R-squared	0.670276	Mean dependent var	123.9640
Adjusted R-squared	0.669155	S.D. dependent var	34.47252
S.E. of regression	19.82830	Sum squared resid	231179.0
Durbin-Watson stat	1.366615	J-statistic	-1.87E-16

Inverted AR Roots	.90

(b). Using EViews 6

Figure 1.17 Statistical results of the model in (1.23), using the GMM in Figure 1.16b.

5) Since each of the regressors in Figure 1.17a have very large p values > 0.99, then a reduced model should be explored. Figure 1.19 presents the dynamic forecast of an acceptable reduced model, with a TIC = 0.116367, which is greater than all previous forecasts. Therefore, this forecast could be considered the worst forecast compared to the others.
6) Since my previous draft had been written using EViews 6, then as a comparison, I present the statistical results in Figure 1.17b with a note "*Convergence not achieved after 499 weight*

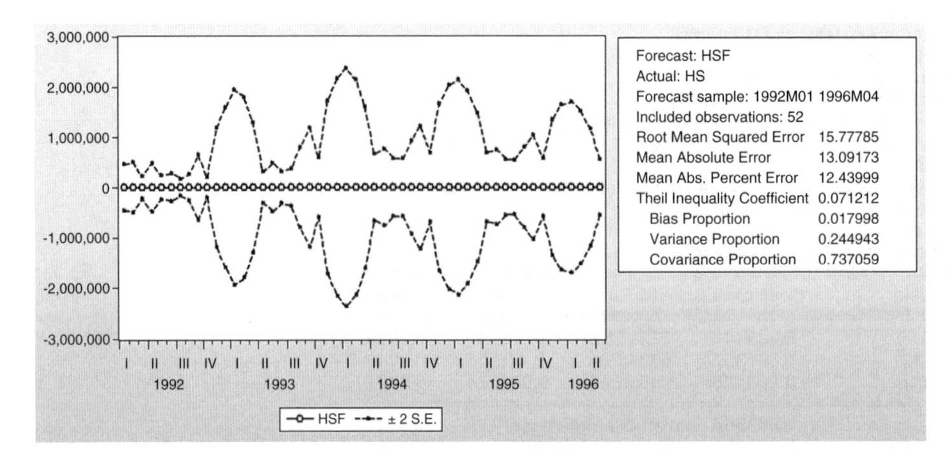

Figure 1.18 Forecast evaluation of the LV(12,1) model in Figure 1.17a.

matrices," using the default options, only by inserting the instrument list *C*. In fact, the convergence also cannot be achieved by using up to 2000 iterations. So, I have found different problems using EViews 6.

Dependent Variable: HS
Method: Generalized Method of Moments
Date: 12/19/13 Time: 16:20
Sample (adjusted): 1947M02 1996M04
Included observations: 591 after adjustments
Sequential 1-step weighting matrix & coefficient iteration
Estimation weighting matrix: HAC (Bartlett kernel, Newey-West fixed bandwidth = 6.0000)
Standard errors & covariance computed using estimation weighting matrix
Convergence achieved after 8 iterations
Instrument specification: C
Lagged dependent variable & regressors added to instrument list

Variable	Coefficient	Std. Error	t-Statistic	Prob.
HS(-12)	1.008857	0.032987	30.58351	0.0000
AR(1)	0.863655	0.019812	43.59203	0.0000

R-squared	0.836018	Mean dependent var	123.9640
Adjusted R-squared	0.835739	S.D. dependent var	34.47252
S.E. of regression	13.07140	Sum squared resid	114972.8
[illegible]	[illegible]	[illegible]	[illegible]
Instrument rank	3	Prob(J-statistic)	0.067776
Inverted AR Roots	.86		

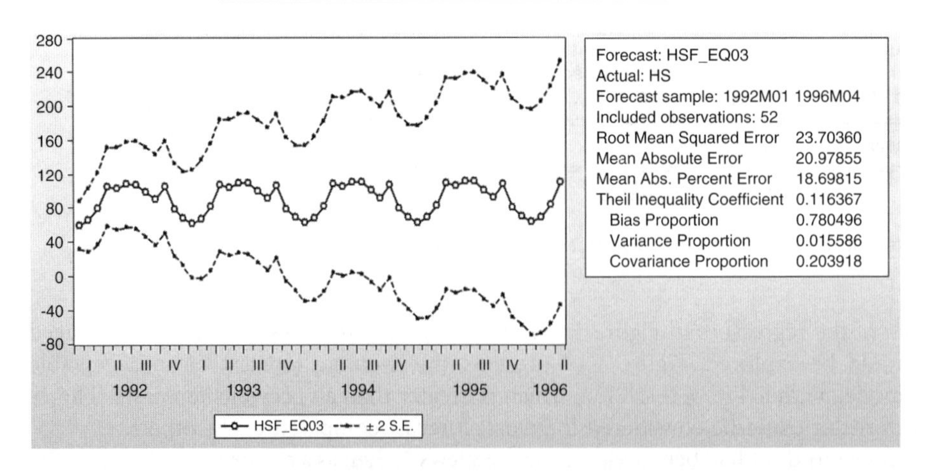

Figure 1.19 Statistical results of the dynamic forecast of an acceptable reduced model.

Example 1.9 Forecast of HS Using a Special LVAR(12,2) IVMs

Corresponding to the statistical results in Figure 1.17a, where each of the regressors has such a large *p* value > 0.99, Figure 1.20 presents two statistical results of a special LVAR(12,2) IVM, using EViews 6, and EVIEWS 8 SV. Based on these results the following notes and comments are presented.

Dependent Variable: HS
Method: Generalized Method of Moments
Date: 10/29/13 Time: 16:24
Sample (adjusted): 1947M03 1996M04
Included observations: 590 after adjustments
Kernel: Bartlett, Bandwidth: Fixed (5), No prewhitening
Simultaneous weighting matrix & coefficient iteration
Convergence achieved after: 8 weight matrices, 9 total coef iterations
Instrument list: C
Lagged dependent variable & regressors added to instrument list

Variable	Coefficient	Std. Error	t-Statistic	Prob.
C	-0.873560	8.670625	-0.100749	0.9198
HS(-12)	1.021399	0.064140	15.92453	0.0000
AR(1)	0.718012	0.052614	13.64687	0.0000
AR(2)	0.166183	0.055500	2.994287	0.0029

R-squared	0.838942	Mean dependent var	124.0639
Adjusted R-squared	0.838117	S.D. dependent var	34.41597
S.E. of regression	13.84716	Sum squared resid	112361.8
Durbin-Watson stat	2.012562	J-statistic	0.006808

Inverted AR Roots	.90	-.18

(a). Using EViews 6

Dependent Variable: HS
Method: Generalized Method of Moments
Date: 12/19/13 Time: 14:04
Sample (adjusted): 1947M03 1996M04
Included observations: 590 after adjustments
Sequential 1-step weighting matrix & coefficient iteration
Estimation weighting matrix: HAC (Bartlett kernel, Newey-West fixed bandwidth = 6.0000)
Standard errors & covariance computed using estimation weighting matrix
Convergence achieved after 15 iterations
Instrument specification: C
Lagged dependent variable & regressors added to instrument list

Variable	Coefficient	Std. Error	t-Statistic	Prob.
C	-1.003355	8.908127	-0.112634	0.9104
HS(-12)	1.022589	0.065619	15.58366	0.0000
AR(1)	0.718584	0.052966	13.56696	0.0000
AR(2)	0.166617	0.055995	2.975557	0.0030

R-squared	0.838752	Mean dependent var	124.0639
Adjusted R-squared	0.837926	S.D. dependent var	34.41597
S.E. of regression	13.85531	Sum squared resid	112494.2
Durbin-Watson stat	2.013751	J-statistic	3.982370
Instrument rank	5	Prob(J-statistic)	0.045979

Inverted AR Roots	.90	-.18

(b). Using EViews 8 SV

Figure 1.20 Statistical results of a special LVAR(12,2) model, using GMM.

1) In my previous experiment using EViews 6, I have also found acceptable statistical results, as presented in Figure 1.20a, in comparison with the statistical results using EViews 8 SV presented in Figure 1.20b. These two outputs again show that EViews 6 and EViews 8 do have different results, even though they are not as bad as the two outputs presented in Figure 1.17. I am very confident that EViews 8 should be an improvement on EViews 6. Therefore, I decide to present this book using EViews 8 SV, which is not so costly for students.

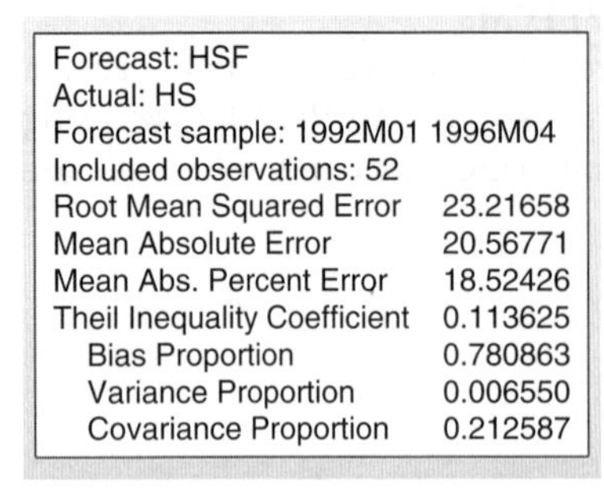

Forecast: HSF	
Actual: HS	
Forecast sample: 1992M01 1996M04	
Included observations: 52	
Root Mean Squared Error	23.21658
Mean Absolute Error	20.56771
Mean Abs. Percent Error	18.52426
Theil Inequality Coefficient	0.113625
Bias Proportion	0.780863
Variance Proportion	0.006550
Covariance Proportion	0.212587

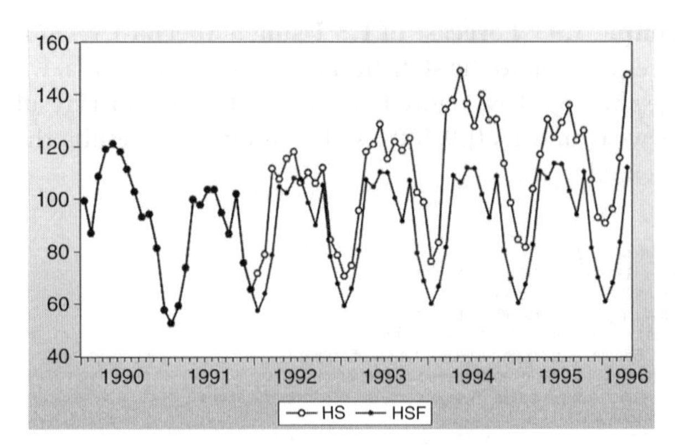

Figure 1.21 Forecast evaluation and a graph of *HS* and *HSF* from the model in Figure 1.20b.

2) Figure 1.21 presents a dynamic forecast of the model in Figure 1.20b using EViews 8 SV. Note this forecast has a TIC = 0.113625, which is greater than the forecast in Figure 1.18 with a TIC = 0.071212 based on a LVAR(12,1) model, but it is a little bit smaller than the forecast in Figure 1.19 with a TIC = 0.116367, based on a reduced model of the LVAR(12,1) model.
3) Since the intercept has such a large p value, then the model could be reduced by deleting the intercept C. Do it as an exercise.

1.5.1.2 Statistical Results of Special LVAR(12,q) using Any Instrument List

Referring to the statistical results of the model using the simplest instrument, namely the intercept C with "*the lagged dependent variable & regressors added to the instrument list*," it has been found that statistical results of the models also can easily be obtained by using any instrument list, or an unintentionally or random instrument list, as long as the instrument variables are in the data set, including their lags, with the option: "*the lagged dependent variable & regressors added to the instrument list.*" See the following illustrative statistical results.

It is recognized that, for a single dependent variable, all of its possible lags can be used an instrument list. However, we will never know the best possible instrument list for any defined model. The following illustrative examples, as well as the examples in Agung (2014, 2009a), present statistical results using an unintentional or unexpected instruments list. So, it can be said that a researcher can use any instrument list to obtain the statistical results of an IVM. Therefore, all researchers should not have any difficulty in doing data analysis using this option.

Example 1.10 The Model in (1.23), with *C*, *HS*(−1), and *HS*(−2) as Instruments

As a modification of the model in Figure 1.17a, use the simplest instrument list *C*, where each of the regressors has such a large p value > 0.99, Figure 1.22 presents the statistical results of a model, using *C*, *HS*(−1), and *HS*(−2) as instruments, with the option "*Lagged dependent variable & regressors added to instrument list.*" Based on these results, the following findings and notes are presented.

1) In a statistical sense, the LS regression is an acceptable model, since each of the variables *HS* (−12) and *AR*(1) has a significant adjusted effect. Compare these parameter estimates with

Dependent Variable: HS
Method: Generalized Method of Moments
Date: 12/20/13 Time: 13:57
Sample (adjusted): 1947M02 1996M04
Included observations: 591 after adjustments
Sequential 1-step weighting matrix & coefficient iteration
Estimation weighting matrix: HAC (Bartlett kernel, Newey-West fixed bandwidth = 6.0000)
Standard errors & covariance computed using estimation weighting matrix
Convergence achieved after 14 iterations
Instrument specification: C HS(-1) HS(-2)
Lagged dependent variable & regressors added to instrument list

Variable	Coefficient	Std. Error	t-Statistic	Prob.
C	14.87943	11.10036	1.340446	0.1806
HS(-12)	0.889183	0.089457	9.939767	0.0000
AR(1)	0.861916	0.020007	43.08073	0.0000

R-squared	0.852194	Mean dependent var	123.9640
Adjusted R-squared	0.851692	S.D. dependent var	34.47252
S.E. of regression	13.27564	Sum squared resid	103630.7
Durbin-Watson stat	2.286961	J-statistic	3.410128
Instrument rank	4	Prob(J-statistic)	0.064797

Inverted AR Roots	.86

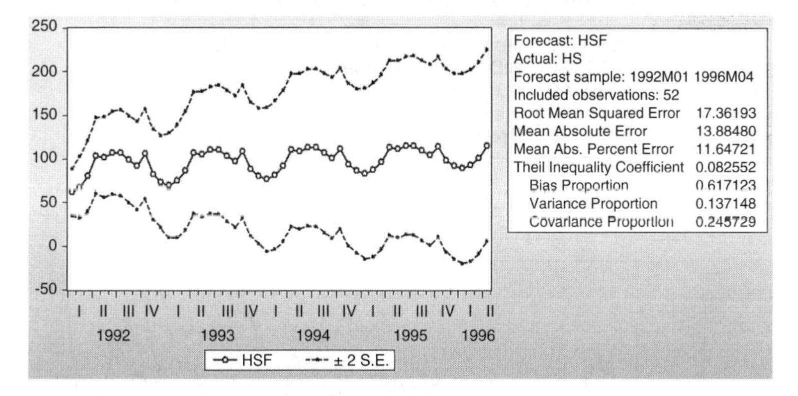

Figure 1.22 Statistical results of a special LVAR(12,1) model, with *C*, *HS(−1)*, and *HS(−2)* as instruments.

the estimates of the same regression in Figure 1.17a, where each has a large *p* value. These results have demonstrated the unpredictable impact of an instrument list. The readers are recommended to experiment with any instrument list.

2) Compared to the forecast of *HS* using the simplest instrument list *C* in Figure (1.19) with the TIC = 0.071212, this forecast has a greater TIC = 0.082552. So the forecast in Figure 1.8, based on the regression in (1.18a), is better than the forecast of this LV(12,1) IVM.

Example 1.11 Application of LVAR(12,2) Models, Using Any Instrument List

Corresponding to the models with an intercept in Figure 1.20b, Figure 1.23 presents two illustrative statistical results of LVAR(12,2) IVMs without the intercept or through the origin. Based on these results the following notes are presented.

1) If the option "*the lagged dependent variable & regressors added to the instrument list*" was not applied then the error message "*Order condition violated – Insufficient instruments*" would be shown on the screen.
2) It is surprising and unexpected; Figure 1.24 presents exactly the same statistical results as the statistical results in Figure 1.23. These statistical results again show the unpredictable impact of an instrument list.

Dependent Variable: HS
Method: Generalized Method of Moments
Date: 12/20/13 Time: 14:52
Sample (adjusted): 1947M03 1996M04
Included observations: 590 after adjustments
Sequential 1-step weighting matrix & coefficient iteration
Estimation weighting matrix: HAC (Bartlett kernel, Newey-West fixed
bandwidth = 6.0000)
Standard errors & covariance computed using estimation weighting matrix
Convergence achieved after 12 iterations
Instrument specification: C HS(-1)
Lagged dependent variable & regressors added to instrument list

Variable	Coefficient	Std. Error	t-Statistic	Prob.
HS(-12)	1.017163	0.037896	26.84125	0.0000
AR(1)	0.717211	0.052320	13.70816	0.0000
AR(2)	0.168075	0.055221	3.043663	0.0024

R-squared	0.839669	Mean dependent var	124.0639
Adjusted R-squared	0.839122	S.D. dependent var	34.41597
S.E. of regression	13.80409	Sum squared resid	111854.5
Durbin-Watson stat	2.011166	J-statistic	4.045034
Instrument rank	5	Prob(J-statistic)	0.132322

Inverted AR Roots	.90	-.19

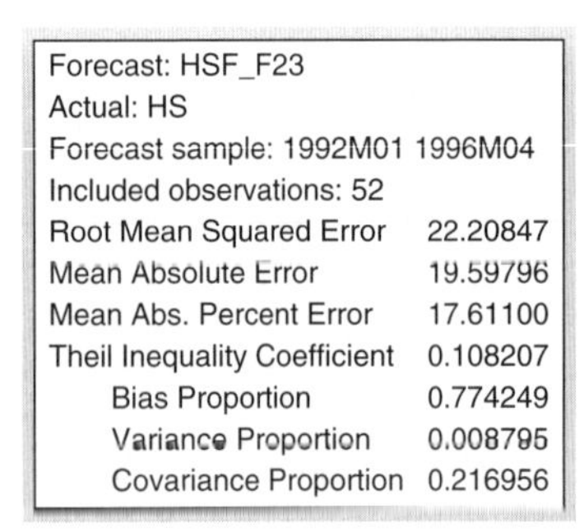

Forecast: HSF_F23
Actual: HS
Forecast sample: 1992M01 1996M04
Included observations: 52

Root Mean Squared Error	22.20847
Mean Absolute Error	19.59796
Mean Abs. Percent Error	17.61100
Theil Inequality Coefficient	0.108207
Bias Proportion	0.774249
Variance Proportion	0.008795
Covariance Proportion	0.216956

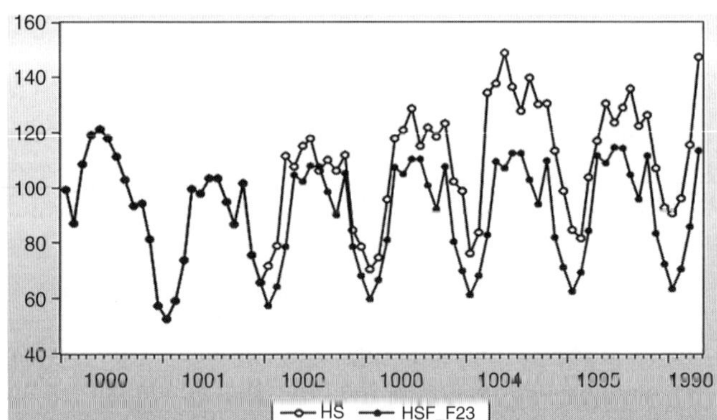

Figure 1.23 Statistical results of a LVAR(12,2) IVMs using an unintentionally instrument list.

1.5.1.3 Application of IVMs without the Option "Lagged..."

In order to study the idea or meaning of the option "*the lagged dependent and regressors added to the instrument list*," this section presents alternative lists of instruments. Agung (2014, 2009a, 2011a) had recommended to not apply an IVM, because it is not an easy task to select appropriate lists or sets of instruments. So, we have to use the trial-and-error method. Corresponding to the option before, I have been doing experiments using selected lists of instruments, such as the lag(s) of dependent variable *HS*, namely *HS*(−*i*), for selected $i = 1,2,\ldots$; and the lag(s) of regressor *HS*(−12), namely *HS*(−12−j), for selected $j = 1,2,\ldots$; with the intercept *C*, as alternative instrument lists.

During the experiment, I found the error messages of "*Near singular matrix*" by using alternative lists of instruments, such as "*C HS*(−1) *HS*(−13)," which use the first lag of the dependent variable *HS* and the first lag of the regressor *HS*(−12), "*C HS*(−1)," "*C HS*(−13)," "*C HS*(−1) *HS*(−2)," and "*C HS*(−13) *HS*(−14)." The following example presents statistical results of two special LVAR(12,2) IVMs, which can be considered unexpected instrumental models.

Figure 1.24 The same statistical results of two LVAR (12,2) IVMs as in Figure 1.23.

Dependent Variable: HS
Method: Generalized Method of Moments
Date: 12/20/13 Time: 15:18
Sample (adjusted): 1947M03 1996M04
Included observations: 590 after adjustments
Sequential 1-step weighting matrix & coefficient iteration
Estimation weighting matrix: HAC (Bartlett kernel, Newey-West fixed bandwidth = 6.0000)
Standard errors & covariance computed using estimation weighting matrix
Convergence achieved after 9 iterations
Instrument specification: C HS(-1) HS(-13)
Lagged dependent variable & regressors added to instrument list

Variable	Coefficient	Std. Error	t-Statistic	Prob.
HS(-12)	1.017163	0.037895	26.84127	0.0000
AR(1)	0.717211	0.052320	13.70816	0.0000
AR(2)	0.168075	0.055221	3.043664	0.0024

R-squared	0.839669	Mean dependent var	124.0639
Adjusted R-squared	0.839123	S.D. dependent var	34.41597
S.E. of regression	13.80409	Sum squared resid	111854.5
Durbin-Watson stat	2.011166	J-statistic	4.045040
Instrument rank	5	Prob(J-statistic)	0.132322
Inverted AR Roots	.90	-.19	

Dependent Variable: HS
Method: Generalized Method of Moments
Date: 12/20/13 Time: 15:21
Sample (adjusted): 1947M03 1996M04
Included observations: 590 after adjustments
Sequential 1-step weighting matrix & coefficient iteration
Estimation weighting matrix: HAC (Bartlett kernel, Newey-West fixed bandwidth = 6.0000)
Standard errors & covariance computed using estimation weighting matrix
Convergence achieved after 9 iterations
Instrument specification: C HS(-1) HS(-2) HS(-13)
Lagged dependent variable & regressors added to instrument list

Variable	Coefficient	Std. Error	t-Statistic	Prob.
HS(-12)	1.017163	0.037895	26.84127	0.0000
AR(1)	0.717211	0.052320	13.70816	0.0000
AR(2)	0.168075	0.055221	3.043664	0.0024

R-squared	0.839669	Mean dependent var	124.0639
Adjusted R-squared	0.839123	S.D. dependent var	34.41597
S.E. of regression	13.80409	Sum squared resid	111854.5
Durbin-Watson stat	2.011166	J-statistic	4.045040
Instrument rank	5	Prob(J-statistic)	0.132322
Inverted AR Roots	.90	-.19	

Example 1.12 Unexpected Statistical Results

Figure 1.25 presents unexpected statistical results of a special LV(12,2) instrumental model of *HS*, using the following ES. Based on these results, the following findings and notes are presented.

$$HS\ C\ HS(-12)\ AR(1)AR(2)\ @\ C\ HS(-1)HS(-2)HS(-13) \qquad (1.24)$$

1) This figure shows that all regressors, *C, HS(−12), AR(1),* and *AR(2),* have such large *p* values so that the regression is not an acceptable model in a statistical sense. What should we do? In general, one would delete one of the regressors. Do it as an exercise and compare its TIC with the TIC in Figure 1.25.
2) Adding or modifying the instrument list is unexpected, I obtain some regressions with each of the dependent variables, namely *HS*(−12), *AR*(−1), and *AR*(2), which has a small *p* value < 0.01, such as that presented in Figure 1.26. The other instrument lists I have applied are (i) *C, HS*(−1), *HS*(−2), and *HS*(−3); (ii) *C, HS*(−1), *HS*(−2), *HS*(−3), and *HS*(−13); (iii) *C, HS*(−1), *HS* (−2), *HS*(−3), *HS*(−13), and *HS*(−14); and (iv) *C, HS*(−1), *HS*(−2), *HS*(−3), *HS*(−13), *HS*(−14),

Dependent Variable: HS
Method: Generalized Method of Moments
Date: 12/21/13 Time: 16:57
Sample (adjusted): 1947M03 1996M04
Included observations: 590 after adjustments
Sequential 1-step weighting matrix & coefficient iteration
Estimation weighting matrix: HAC (Bartlett kernel, Newey-West fixed bandwidth = 6.0000)
Standard errors & covariance computed using estimation weighting matrix
Convergence achieved after 58 iterations
Instrument specification: C HS(-1) HS(-2) HS(-13)
Lagged dependent variable & regressors not added to instrument list

Variable	Coefficient	Std. Error	t-Statistic	Prob.
C	144.2762	3607007.	4.00E-05	1.0000
HS(-12)	-0.163221	29198.51	-5.59E-06	1.0000
AR(1)	1.163277	3407.226	0.000341	0.9997
AR(2)	-0.359432	3416.959	-0.000105	0.9999

R-squared	0.722721	Mean dependent var	124.0639
Adjusted R-squared	0.721302	S.D. dependent var	34.41597
S.E. of regression	18.16884	Sum squared resid	193442.6
Durbin-Watson stat	2.015076	J-statistic	1.957694
Instrument rank	4		

Inverted AR Roots	.58-.15i	.58+.15i

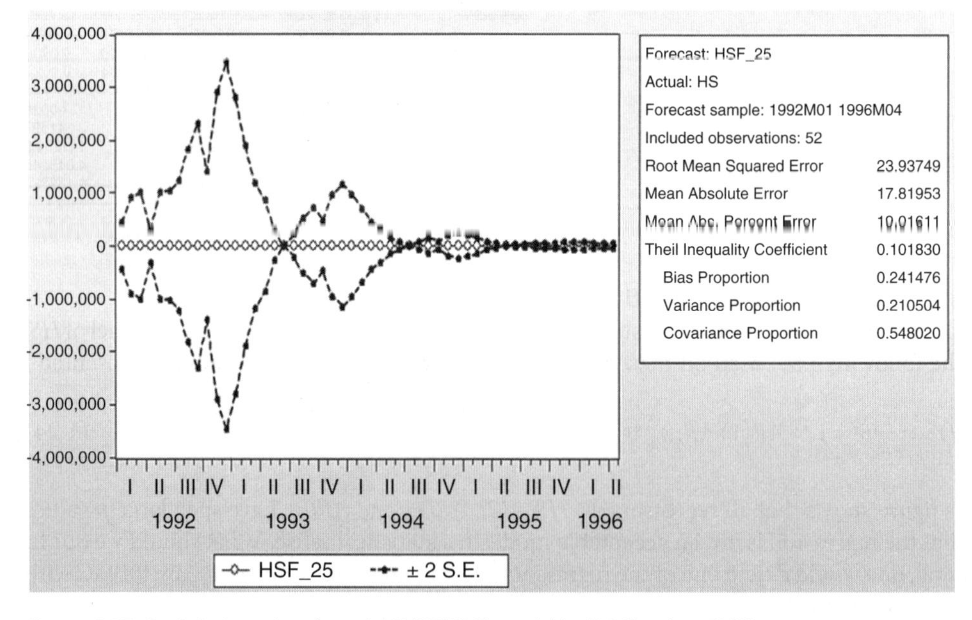

Figure 1.25 Statistical results of special LVAR(12,2) model in (1.24), using GMM.

Dependent Variable: HS
Method: Generalized Method of Moments
Date: 12/21/13 Time: 17:47
Sample (adjusted): 1947M03 1996M04
Included observations: 590 after adjustments
Sequential 1-step weighting matrix & coefficient iteration
Estimation weighting matrix: HAC (Bartlett kernel, Newey-West fixed bandwidth = 6.0000)
Standard errors & covariance computed using estimation weighting matrix
Convergence achieved after 8 iterations
Instrument specification: C HS(-1) HS(-2) HS(-13) HS(-14)
Lagged dependent variable & regressors not added to instrument list

Variable	Coefficient	Std. Error	t-Statistic	Prob.
C	278.1850	45.51632	6.111764	0.0000
HS(-12)	-1.245547	0.360053	-3.459347	0.0006
AR(1)	1.240618	0.028357	43.75074	0.0000
AR(2)	-0.444039	0.034294	-12.94816	0.0000

R-squared	0.106378	Mean dependent var	124.0639
Adjusted R-squared	0.101803	S.D. dependent var	34.41597
S.E. of regression	32.61714	Sum squared resid	623432.4
Durbin-Watson stat	2.075504	J-statistic	0.624141
Instrument rank	5	Prob(J-statistic)	0.429512

Inverted AR Roots	.62-.24i	.62+.24i

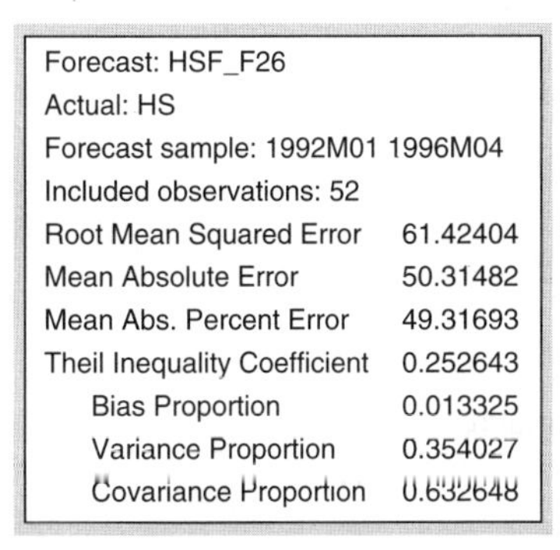

Forecast: HSF_F26
Actual: HS
Forecast sample: 1992M01 1996M04
Included observations: 52

Root Mean Squared Error	61.42404
Mean Absolute Error	50.31482
Mean Abs. Percent Error	49.31693
Theil Inequality Coefficient	0.252643
Bias Proportion	0.013325
Variance Proportion	0.354027
Covariance Proportion	0.632648

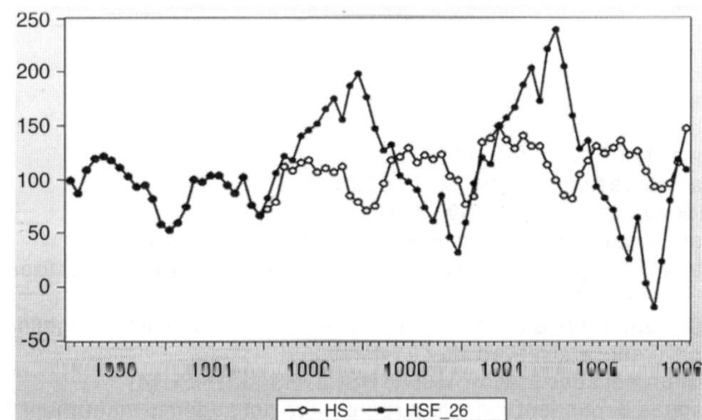

Figure 1.26 Statistical results of the model in Figure 1.25 using an additional instrument variable.

and $HS(-15)$. The readers can easily do the exercise with many other instrument lists using various alternative regression models.

3) Furthermore, note that the forecast in Figure 1.26 has a TIC = 0.252643, which is much greater than the forecast in Figure 1.25 with a TIC = 0.101830. So, in a statistical sense, the forecast in Figure 1.25 is better, even though it uses an unacceptable regression function. If you have to choose one of the two statistical results as a whole, which one is your preference?

Example 1.13 Another Unexpected Statistical Results

This is really unexpected, Figure 1.27a presents exactly the same statistical results as Figure 1.20 by using the following ES with the "*Lagged dependent variable & regressors not added to the instrument list.*" Based on these results, the following findings and notes are presented.

$$HS\ C\ HS(-12)\ AR(1)AR(2)\ @\ C\ HS(-1)HS(-2)HS(-13)HS(-14) \tag{1.25}$$

Dependent Variable: HS
Method: Generalized Method of Moments
Date: 12/22/13 Time: 09:01
Sample (adjusted): 1947M03 1996M04
Included observations: 590 after adjustments
Sequential 1-step weighting matrix & coefficient iteration
Estimation weighting matrix: HAC (Bartlett kernel, Newey-West fixed bandwidth = 6.0000)
Standard errors & covariance computed using estimation weighting matrix
Convergence achieved after 12 iterations
Instrument specification: C HS(-1) HS(-2) HS(-13) HS(-14)
Lagged dependent variable & regressors not added to instrument list

Variable	Coefficient	Std. Error	t-Statistic	Prob.
C	-1.003355	8.908127	-0.112634	0.9104
HS(-12)	1.022589	0.065619	15.58366	0.0000
AR(1)	0.718584	0.052966	13.56695	0.0000
AR(2)	0.166617	0.055995	2.975557	0.0030

R-squared	0.838752	Mean dependent var	124.0639
Adjusted R-squared	0.837926	S.D. dependent var	34.41597
S.E. of regression	13.85531	Sum squared resid	112494.2
Durbin-Watson stat	2.013751	J-statistic	3.982369
Instrument rank	5	Prob(J-statistic)	0.045979

Inverted AR Roots	.90	-.18

(a)

Dependent Variable: HS
Method: Generalized Method of Moments
Date: 12/22/13 Time: 09:06
Sample (adjusted): 1947M03 1996M04
Included observations: 590 after adjustments
Sequential 1-step weighting matrix & coefficient iteration
Estimation weighting matrix: HAC (Bartlett kernel, Newey-West fixed bandwidth = 6.0000)
Standard errors & covariance computed using estimation weighting matrix
Convergence achieved after 11 iterations
Instrument specification: HS(-1) HS(-2) HS(-13) HS(-14)
Lagged dependent variable & regressors not added to instrument list

Variable	Coefficient	Std. Error	t-Statistic	Prob.
HS(-12)	-0.823466	0.247607	-3.325698	0.0009
AR(1)	1.339134	0.029551	45.31644	0.0000
AR(2)	-0.350210	0.028291	-12.37865	0.0000

R-squared	0.308604	Mean dependent var	124.0639
Adjusted R-squared	0.306248	S.D. dependent var	34.41597
S.E. of regression	28.66567	Sum squared resid	482349.9
Durbin-Watson stat	1.949399	J-statistic	1.753444
Instrument rank	4	Prob(J-statistic)	0.185444

Inverted AR Roots	.98	.36

(b)

Figure 1.27 Statistical results of a special LV(12,2) IVM, and its reduced model using GMM.

1) The statistical results in Figure 1.20 has shown the option the "*Lagged dependent & regressors added to the instrument list*," but without the option, Figure 1.27, presents exactly the same statistical results. Therefore, the "*Lagged dependent & regressors added to the instrument list*" can be replaced by the variables *HS*(−1), *HS*(−2), *HS*(−13), and *HS*(−14). To study whether these findings are valid for other monthly data sets, see the following examples.
2) Since the intercept *C* has a very large *p* value = 0.9104, then we may present a reduced model through the origin with the statistical results presented in Figure 1.27b, but the forecast based on the model has a large TIC = 0.584782 compared to the TIC = 0.113625 for the forecast in Figure 1.21.

Dependent Variable: HS
Method: Generalized Method of Moments
Date: 12/13/13 Time: 12:12
Sample (adjusted): 1946M07 1996M04
Included observations: 598 after adjustments
Kernel: Bartlett, Bandwidth: Fixed (5), No prewhitening
Simultaneous weighting matrix & coefficient iteration
Convergence achieved after: 1 weight matrix, 2 total coef iterations
Instrument list: C HS(-1) HS(-2) HS(-3) HS(-4) HS(-5) HS(-6)

Variable	Coefficient	Std. Error	t-Statistic	Prob.
C	21.19082	3.190366	6.642128	0.0000
HS(-1)	1.136662	0.037031	30.69526	0.0000
HS(-2)	-0.155662	0.065143	-2.389557	0.0172
HS(-3)	-0.224144	0.066593	-3.365869	0.0008
HS(-4)	-0.128558	0.054388	-2.363692	0.0184
HS(-5)	0.327971	0.056065	5.849794	0.0000
HS(-6)	-0.127431	0.037459	-3.401909	0.0007

R-squared	0.791309	Mean dependent var	123.4376
Adjusted R-squared	0.789190	S.D. dependent var	34.64705
S.E. of regression	15.90785	Sum squared resid	149558.2
Durbin-Watson stat	1.982004	J-statistic	4.67E-43

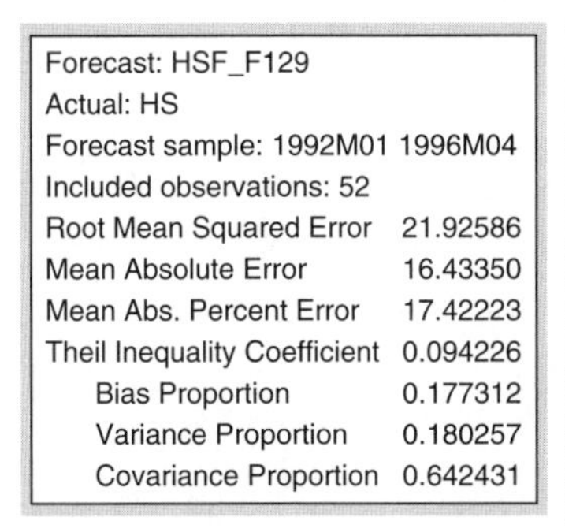

Forecast: HSF_F129	
Actual: HS	
Forecast sample: 1992M01 1996M04	
Included observations: 52	
Root Mean Squared Error	21.92586
Mean Absolute Error	16.43350
Mean Abs. Percent Error	17.42223
Theil Inequality Coefficient	0.094226
Bias Proportion	0.177312
Variance Proportion	0.180257
Covariance Proportion	0.642431

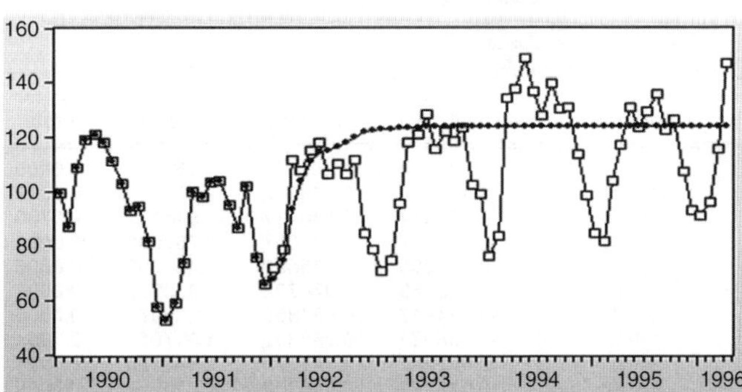

Figure 1.28 Statistical results of the basic LV(6) IVM in (1.26) using GMM.

Example 1.14 Application of Basic or Common LV(*p*) GMM Models

As an illustration, Figure 1.28 presents the statistical results of a common LV(6) IVM (IVM), using the following ES. Corresponding to these results, the following findings and notes are presented.

$$HS\ C\ HS(-1)HS(-2)...\ HS(-6)@\ C\ HS(-1)...HS(-6) \tag{1.26}$$

Dependent Variable: HS
Method: Generalized Method of Moments
Date: 12/28/13 Time: 15:30
Sample (adjusted): 1946M11 1996M04
Included observations: 594 after adjustments
Sequential 1-step weighting matrix & coefficient iteration
Estimation weighting matrix: HAC (Bartlett kernel, Newey-West fixed bandwidth = 6.0000)
Standard errors & covariance computed using estimation weighting matrix
Convergence achieved after 9 iterations
Instrument specification: C
Lagged dependent variable & regressors added to instrument list

Variable	Coefficient	Std. Error	t-Statistic	Prob.
C	96.21234	10.49500	9.167449	0.0000
HS(-1)	1.118249	0.078382	14.26669	0.0000
HS(-2)	-0.254917	0.140437	-1.815172	0.0700
HS(-3)	-0.714657	0.124027	-5.762091	0.0000
HS(-4)	0.228599	0.065550	3.487390	0.0005
HS(-5)	0.445363	0.074722	5.960262	0.0000
HS(-6)	-0.584913	0.051850	-11.28078	0.0000
AR(1)	-0.138921	0.085909	-1.617065	0.1064
AR(2)	0.077262	0.047097	1.640496	0.1014
AR(3)	0.648890	0.045782	14.17353	0.0000
AR(4)	0.324557	0.074840	4.336671	0.0000

R-squared	0.841528	Mean dependent var	123.6561
Adjusted R-squared	0.838810	S.D. dependent var	34.65913
S.E. of regression	13.91511	Sum squared resid	112886.5
Durbin-Watson stat	2.038808	J-statistic	7.60E-23
Instrument rank	11		

Inverted AR Roots	.97	-.32-.78i	-.32+.78i	-.47

Dependent Variable: HS
Method: Generalized Method of Moments
Date: 12/28/13 Time: 16:01
Sample (adjusted): 1946M11 1996M04
Included observations: 594 after adjustments
Sequential 1-step weighting matrix & coefficient iteration
Estimation weighting matrix: HAC (Bartlett kernel, Newey-West fixed bandwidth = 6.0000)
Standard errors & covariance computed using estimation weighting matrix
Convergence achieved after 5 iterations
Instrument specification: C HS(-1) HS(-2) HS(-3) HS(-4) HS(-5) HS(-6) HS(-7) HS(-8) HS(-9) HS(-10)
Lagged dependent variable & regressors not added to instrument list

Variable	Coefficient	Std. Error	t-Statistic	Prob.
C	96.21234	10.49500	9.167449	0.0000
HS(-1)	1.118249	0.078382	14.26669	0.0000
HS(-2)	-0.254917	0.140437	-1.815172	0.0700
HS(-3)	-0.714657	0.124027	-5.762091	0.0000
HS(-4)	0.228599	0.065550	3.487390	0.0005
HS(-5)	0.445363	0.074722	5.960262	0.0000
HS(-6)	-0.584913	0.051850	-11.28078	0.0000
AR(1)	-0.138921	0.085909	-1.617065	0.1064
AR(2)	0.077262	0.047097	1.640496	0.1014
AR(3)	0.648890	0.045782	14.17353	0.0000
AR(4)	0.324557	0.074840	4.336671	0.0000

R-squared	0.841528	Mean dependent var	123.6561
Adjusted R-squared	0.838810	S.D. dependent var	34.65913
S.E. of regression	13.91511	Sum squared resid	112886.5
Durbin-Watson stat	2.038808	J-statistic	6.42E-24
Instrument rank	11		

Inverted AR Roots	.97	-.32-.78i	-.32+.78i	-.47

Figure 1.29 A pair of the same statistical results of the model in (1.27) using different instrument lists.

1) In a statistical sense, the model is acceptable because each of the independent variables has a significant adjusted effect on *HS*. However, by observing the graph of the observed scores and the in-sample forecast values, I would say the model is a worse forecast model. On the other hand, this forecast has smaller values of RMSE and TIC than the model in (1.26).
2) In fact it has been found that the basic or common LV(p) IVMs, for various $p = 1, 2, \ldots, 12$, give acceptable statistical results with some of the lags $HS(-p)$ having insignificant adjusted effects with p values > 0.10. Do this as an exercise.

Example 1.15 Unexpected Statistical Results Based on a Pair of LVAR(6,4) IVMs

As an extension of the LV(6) IVM in Figure 1.28, Figure 1.29 presents a pair of the same statistical results of two basic LVAR(6,4) models, using the simplest instrument specification *C* with the option *"Lagged dependent variable & regressors added to instrument list,"* and a specific instrument specification with the option: *"Lagged dependent variable & regressors not added to instrument list"* (Figure 1.30). The ES applied is as follows:

$$HS\ C\ HS(-1)\,HS(2)\ldots\ HS(6)\,AR(1)\,AR(2)\,AR(3)\,AR(4) \tag{1.27}$$

Based on these results the following findings and notes are presented.

1) It is unexpected; by doing the trial-and-error method, I obtain a pair of the same statistical results of the basic or common LVAR(6,4) in (1.27) using two different instrument lists. Even though AR(1) and AR(2) have p values > 0.10, at the $\alpha = 0.10$ level of significance AR(1) has a significant negative effect with a p value $= 0.1064/2 = 0.0532 < 0.10$, and AR(2) has a significant negative effect with a p value $= 0.1014/2 = 0.0507 < 0.10$. So the model in (1.27) can be considered a good fit model for the data of *HS*.
2) As an additional analysis, Figure 1.31 presents their forecast evaluation. Based on this figure, the following findings and comments are presented.
   2.1 Compared to the forecast in Figure 1.28, this forecast has smaller values of RSME and TIC, so this forecast should be considered a better forecasting for *HS* within the period 1992M01 to 1996M04. In fact, this forecast can be considered to be the best forecast for *HS* compared to all forecasts previously presented.
   2.2 Note that this figure also presents the same graphs of *HS* and *HSF_LVAR64* for the period 1990M01 to 1992M01.

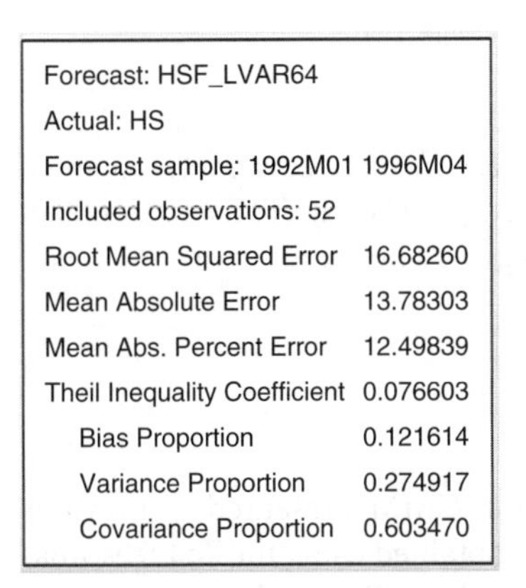

Forecast: HSF_LVAR64	
Actual: HS	
Forecast sample: 1992M01 1996M04	
Included observations: 52	
Root Mean Squared Error	16.68260
Mean Absolute Error	13.78303
Mean Abs. Percent Error	12.49839
Theil Inequality Coefficient	0.076603
Bias Proportion	0.121614
Variance Proportion	0.274917
Covariance Proportion	0.603470

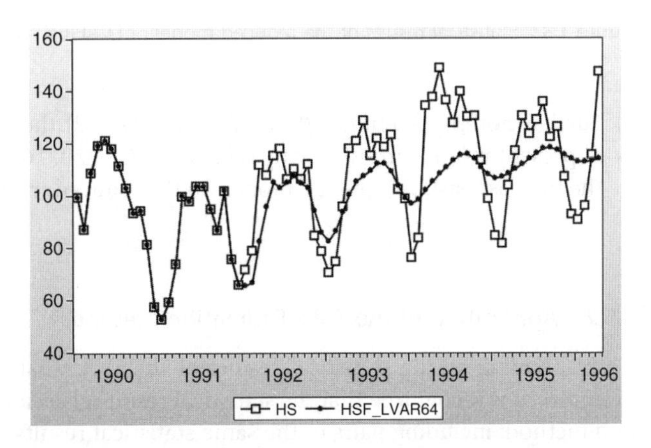

Figure 1.30 Forecast evaluation based on the regressions in Figure 1.29.

Dependent Variable: HS
Method: Generalized Method of Moments
Date: 12/28/13 Time: 17:18
Sample (adjusted): 1946M11 1996M04
Included observations: 594 after adjustments
Sequential 1-step weighting matrix & coefficient iteration
Estimation weighting matrix: HAC (Bartlett kernel, Newey-West fixed bandwidth = 6.0000)
Standard errors & covariance computed using estimation weighting matrix
Convergence achieved after 5 iterations
Instrument specification: C
Lagged dependent variable & regressors added to instrument list

Variable	Coefficient	Std. Error	t-Statistic	Prob.
C	99.01003	10.05464	9.847201	0.0000
HS(-1)	1.122682	0.083011	13.52449	0.0000
HS(-2)	-0.312522	0.178164	-1.754124	0.0799
HS(-3)	-0.666231	0.156662	-4.252653	0.0000
HS(-4)	0.225597	0.067830	3.325911	0.0009
HS(-5)	0.422095	0.086292	4.891468	0.0000
HS(-6)	-0.578739	0.055801	-10.37157	0.0000
AR(3)	0.647276	0.051293	12.61927	0.0000
AR(4)	0.258061	0.048276	5.345565	0.0000

R-squared	0.834301	Mean dependent var	123.6561
Adjusted R-squared	0.832035	S.D. dependent var	34.65913
S.E. of regression	14.20455	Sum squared resid	118034.9
Durbin-Watson stat	2.346067	J-statistic	1.00E-20
Instrument rank	9		

Inverted AR Roots	.97	-.30+.79i	-.30-.79i	-.37

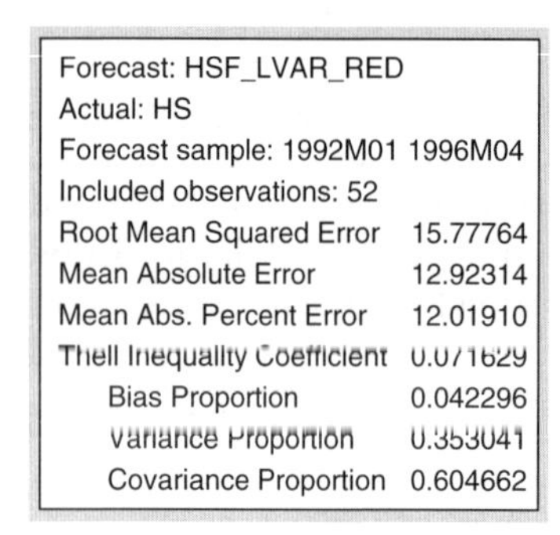

Forecast: HSF_LVAR_RED
Actual: HS
Forecast sample: 1992M01 1996M04
Included observations: 52

Root Mean Squared Error	15.77764
Mean Absolute Error	12.92314
Mean Abs. Percent Error	12.01910
Theil Inequality Coefficient	0.071629
Bias Proportion	0.042296
Variance Proportion	0.353041
Covariance Proportion	0.604662

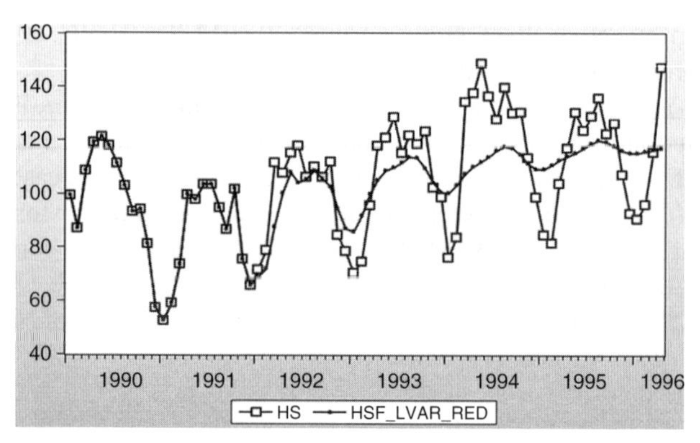

Figure 1.31 Statistical results of the reduced model of LVAF(6,4) IVM in Figure 1.29.

3) Furthermore, by deleting the terms AR(1) and AR(2), the forecast of a reduced model is obtained as presented in Figure 1.31, with a smaller TIC = 0.071629 compared to the TIC = 0.076603 of the forecast based on the full model. So the reduced model gives a better forecast.

1.5.2 Application of the TSLS Estimation Method

Corresponding to the statistical results of the IVMs using the GMM presented in previous examples, it is recognized that the statistical results also can be obtained using the TSLS estimation method, including pairs of the same statistical results using two different instrument lists. Do it as an exercise. However, for illustrative comparisons, selected statistical results are presented in the following two examples.

Dependent Variable: HS
Method: Two-Stage Least Squares
Date: 12/28/13 Time: 17:53
Sample (adjusted): 1946M11 1996M04
Included observations: 594 after adjustments
Convergence achieved after 10 iterations
Instrument specification: C
Lagged dependent variable & regressors added to instrument list

Variable	Coefficient	Std. Error	t-Statistic	Prob.
C	-5.658915	3.435152	-1.647355	0.1000
HS(-1)	0.922423	0.123102	7.493130	0.0000
HS(-2)	1.146324	0.221622	5.172416	0.0000
HS(-3)	-0.816441	0.170873	-4.778057	0.0000
HS(-4)	-0.629284	0.093988	-6.695337	0.0000
HS(-5)	-0.159114	0.121412	-1.310528	0.1905
HS(-6)	0.582653	0.086678	6.722073	0.0000
AR(3)	-0.631749	0.045712	-13.82027	0.0000
AR(4)	-0.060178	0.040496	-1.486011	0.1378

R-squared	0.591808	Mean dependent var	123.6561
Adjusted R-squared	0.586225	S.D. dependent var	34.65913
S.E. of regression	22.29459	Sum squared resid	290773.6
Durbin-Watson stat	2.092527	J-statistic	2.12E-22
Instrument rank	9		

Inverted AR Roots	.46+.74i	.46-.74i	-.10	-.82

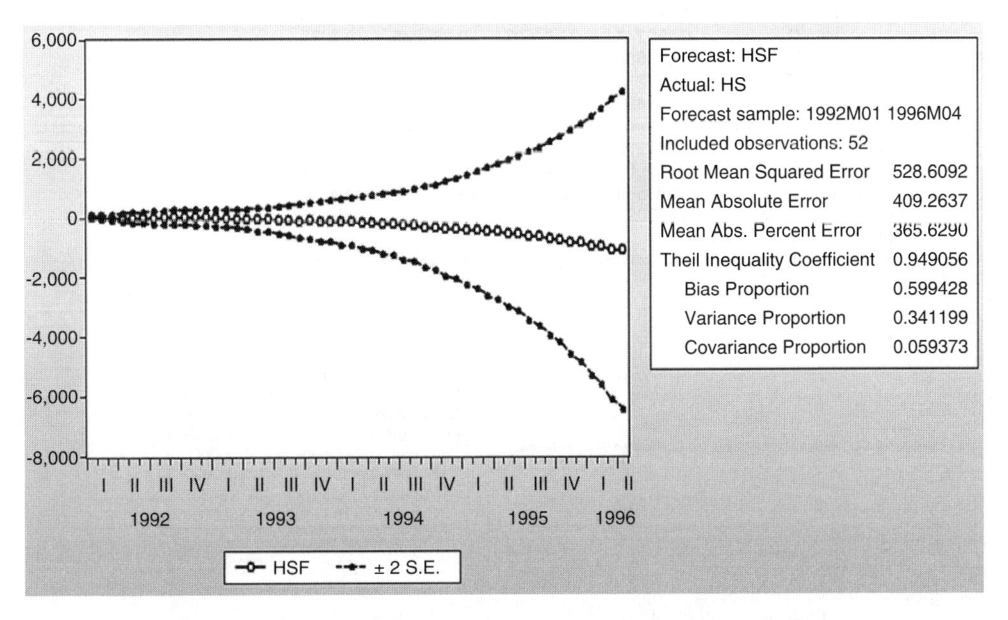

Figure 1.32 Statistical results of the reduced model in Figure 1.31 using TSLS estimates.

Example 1.16 Application of the Reduced Model in Figure 1.31, Using TSLS
Corresponding to the reduced model, in Figure 1.31, Figure 1.32 present the statistical results of the same model but using the TSLS estimation setting. Based on these results, the following notes and comments are made.

1) Compared to the forecast in Figure 1.31 using GMM, this forecast has such a large TIC = 0.949056 where the maximum value of TIC = 1 indicates the worst forecast. So, the trial-and-error method should be applied to obtain a better forecast.
2) Figure 1.33 presents the worst forecast with TIC = 1.000000, which is obtained by deleting *AR*(4) from the model in Figure 1.32. On the other hand, by deleting *HS*(−5) instead of *AR*(4), since it has greater p value, the forecast with a TIC = 0.962867 is obtained and, by deleting both, the forecast with a much smaller TIC = 0.136945 is obtained, as presented in Figure 1.34. Then, if you have to choose one out of the four forecasts, would you consider

Dependent Variable: HS
Method: Two-Stage Least Squares
Date: 12/28/13 Time: 18:23
Sample (adjusted): 1946M10 1996M04
Included observations: 595 after adjustments
Convergence achieved after 7 iterations
Instrument specification: C
Lagged dependent variable & regressors added to instrument list

Variable	Coefficient	Std. Error	t-Statistic	Prob.
C	-62.43562	27.88940	-2.238687	0.0256
HS(-1)	4.417738	1.201760	3.676057	0.0003
HS(-2)	-2.322031	1.122354	-2.068893	0.0390
HS(-3)	-1.271426	0.652519	-1.948489	0.0518
HS(-4)	-0.150241	0.394597	-0.380745	0.7035
HS(-5)	1.638686	0.591072	2.772399	0.0057
HS(-6)	-0.807217	0.345872	-2.333859	0.0199
AR(3)	-0.242490	0.089925	-2.696583	0.0072

R-squared	-1.954332	Mean dependent var	123.5928
Adjusted R-squared	-1.990662	S.D. dependent var	34.66434
S.E. of regression	59.93586	Sum squared resid	2108601.
Durbin-Watson stat	1.777296	J-statistic	1.05E-16
Instrument rank	8		

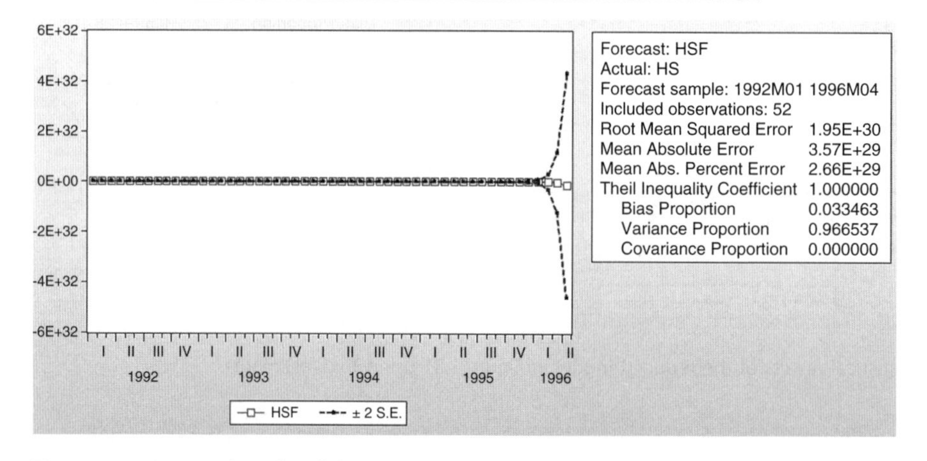

Figure 1.33 Statistical results of the reduced model in Figure 1.32.

Dependent Variable: HS
Method: Two-Stage Least Squares
Date: 12/29/13 Time: 05:51
Sample (adjusted): 1946M10 1996M04
Included observations: 595 after adjustments
Convergence not achieved after 19 iterations
WARNING: Singular covariance - coefficients are not unique
Instrument specification: C
Lagged dependent variable & regressors added to instrument list

Variable	Coefficient	Std. Error	t-Statistic	Prob.
C	16.01963	110160.3	0.000145	0.9999
HS(-1)	1.992229	23336.05	8.54E-05	0.9999
HS(-2)	-1.271026	52584.20	-2.42E-05	1.0000
HS(-3)	-0.155313	46182.82	-3.36E-06	1.0000
HS(-4)	0.511986	18244.27	2.81E-05	1.0000
HS(-6)	-0.206846	2178.459	-9.50E-05	0.9999
AR(3)	0.257880	13313.06	1.94E-05	1.0000

R-squared	0.628244	Mean dependent var	123.5928
Adjusted R-squared	0.624451	S.D. dependent var	34.66434
S.E. of regression	21.24303	Sum squared resid	265344.5
Durbin-Watson stat	3.105118	J-statistic	37.61583
Instrument rank	7		

Inverted AR Roots	.64	-.32+.55i	-.32-.55i

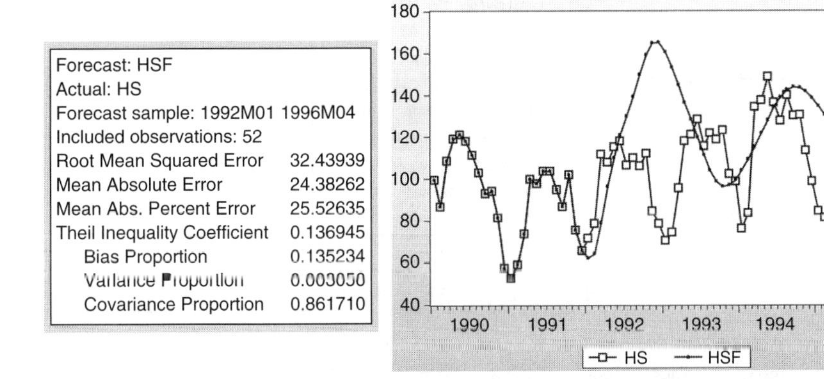

Figure 1.34 Statistical results based on a reduced model of the model in Figure 1.33.

the last one to be the best? Corresponding to these findings, I have to present alternative LVAR(p,q) IVMs, as shown in the following examples.

Example 1.17 Alternative LVAR(*p*,*q*) IVMs

As an additional illustration, I am considering alternative LVAR(*p*,*q*) IVMs. Finally, I obtain an acceptable forecast based on a LVAR(1,4) IVM, using the simplest instrument specification *C* with the specific option, namely, "*Lagged dependent variable & regressors added to instrument list,*" as presented in Figure 1.35. This forecast with a TIC = 0.090757 is the best forecast compared to the four in the previous example.

Furthermore, the same statistical results as in Figure 1.36 have also been found by using the instrument specification: *C HS*(−1) *HS*(−2) *HS*(−3) *HS*(−4) *HS*(−5), without the specific option. On the other hand, I also found several forecasts with a greater TICs, but still less than 0.10, based on LVAR(1,*q*) for $q \neq 4$. For instance, TIC = 0.094648 for $q = 3$ and TIC = 0.094119 for $q = 5$.

Dependent Variable: HS
Method: Two-Stage Least Squares
Date: 12/28/13 Time: 18:41
Sample (adjusted): 1946M06 1996M04
Included observations: 599 after adjustments
Convergence achieved after 4 iterations
Instrument specification: C
Lagged dependent variable & regressors added to instrument list

Variable	Coefficient	Std. Error	t-Statistic	Prob.
C	18.77337	3.921548	4.787234	0.0000
HS(-1)	0.848435	0.031331	27.07956	0.0000
AR(1)	0.264867	0.048212	5.493803	0.0000
AR(2)	0.082767	0.042864	1.930922	0.0540
AR(3)	-0.129598	0.042145	-3.075058	0.0022
AR(4)	-0.219424	0.043687	-5.022644	0.0000

R-squared	0.788003	Mean dependent var	123.3935
Adjusted R-squared	0.786215	S.D. dependent var	34.63492
S.E. of regression	16.01410	Sum squared resid	152075.7
Durbin-Watson stat	1.949780	J-statistic	3.45E-20
Instrument rank	6		

Inverted AR Roots	.58+.52i	.58-.52i	-.45+.40i	-.45-.40i

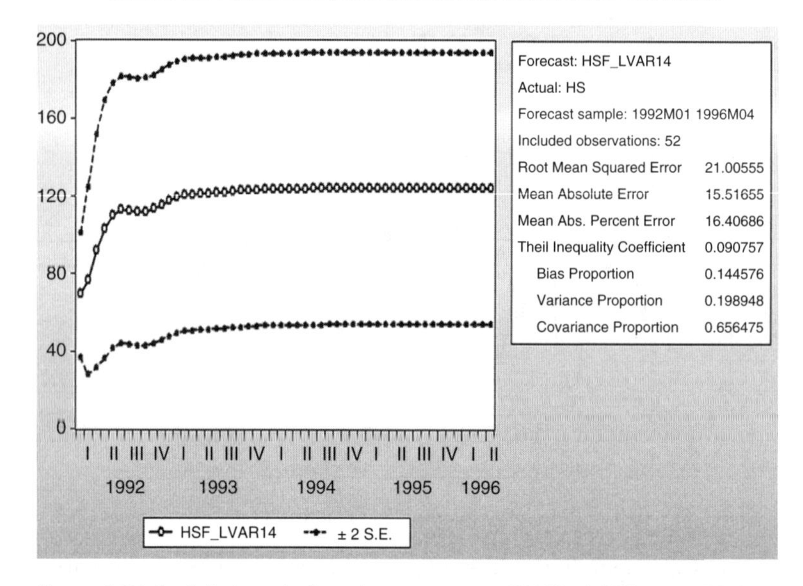

Figure 1.35 Statistical results based on a common LVAR(1,4) IVM, using TSLS.

Example 1.18 Application of the Model in (1.28), Using TSLS

Figure 1.37 presents the statistical results of the LV(6) IVM in (1.28), using the TSLS estimation setting. Based on these results, the following findings and notes are presented.

1) The result is unexpected: the forecast using this TSLS regression presents exactly the same values of statistics as the forecast using the GMM in Figure 1.28, even though they have different regression functions. This forecast has a TIC = 0.094228, which is greater than the forecast in Figure 1.36, with a TIC = 0.090757, but it has a greater R-squared value. So which one is a better model?

Dependent Variable: HS
Method: Two-Stage Least Squares
Date: 12/18/13 Time: 12:36
Sample (adjusted): 1946M07 1996M04
Included observations: 598 after adjustments
Instrument list: C HS(-1) HS(-2) HS(-3) HS(-4) HS(-5) HS(-6)

Variable	Coefficient	Std. Error	t-Statistic	Prob.
C	21.19082	3.061038	6.922757	0.0000
HS(-1)	1.136662	0.040857	27.82039	0.0000
HS(-2)	-0.155662	0.060623	-2.567694	0.0105
HS(-3)	-0.224144	0.060745	-3.689890	0.0002
HS(-4)	-0.128558	0.060708	-2.117652	0.0346
HS(-5)	0.327971	0.060551	5.416475	0.0000
HS(-6)	-0.127431	0.040734	-3.128337	0.0018

R-squared	0.791309	Mean dependent var	123.4376
Adjusted R-squared	0.789190	S.D. dependent var	34.64705
S.E. of regression	15.90785	Sum squared resid	149558.2
F-statistic	373.4900	Durbin-Watson stat	1.982004
Prob(F-statistic)	0.000000	Second-Stage SSR	149558.2

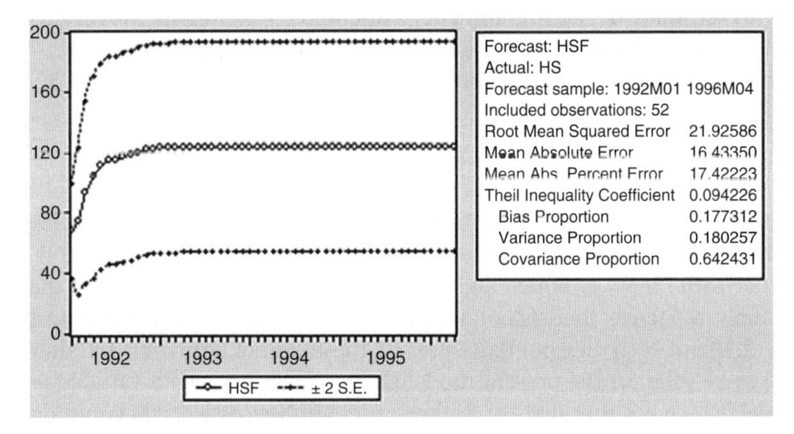

Figure 1.36 The dynamic forecast based on the model in (1.28) using the TSLS estimation method.

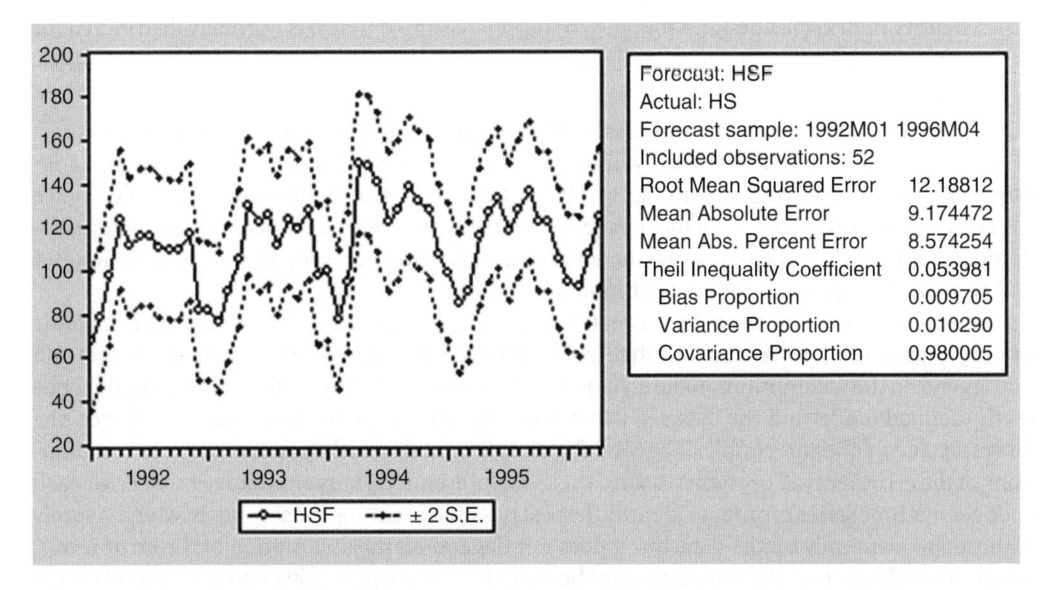

Figure 1.37 A static forecast of *HS* using the TSLS regression in Figure 1.35.

2) Note that each of the independent variables has a significant effect on *HS* adjusted for the other independent variables. In other words, each of the independent variables has a significant adjusted effect on *HS*.
3) As an additional illustration, corresponding to the dynamic forecast in Figure 1.36 with RMSE = 21.95826 and TIC = 0.094226, Figure 1.37 presents the static forecast with smaller RMSE = 12.18812 and TIC = 0.053981. Do data analysis based on other LV(p) IVMs, for alternative integers, p, as exercises. You should obtain unexpected forecast(s).

1.6 Special Notes and Comments on Residual Analysis

The basic assumptions of the error terms of any time series model, namely ε_t, have independent identical normal distribution, with $E(\varepsilon_t) = 0$, and $Var(\varepsilon_t) = \sigma^2$, for all $t = 1, 2, \ldots, T$. Agung (2009a, Section 2.14.3) presented special notes and comments with the title "*To Test or Not the assumptions of the error terms,*" with the following basic comments. For complete comment, the readers are advised to read the section. In addition, refer to Agung (2011a, Chapter 1), presenting "*Misinterpretation of Selected Theoretical Concepts of Statistics.*"

> ... the true values of $E(\varepsilon_t)$, $E\left(\varepsilon_t^2\right)$, $E(\varepsilon_t\varepsilon_{t-s}) = \mathrm{Cov}(\varepsilon_t\varepsilon_{t-s})$ are never known by the researchers. Hence, they could be considered as theoretical or abstract indicators. In practice, since only a single observation exists within each time period or at one time point t, then only one set of the estimated error terms, say $\{e_t, t = 1, 2, \ldots T\}$, is observed, where e_t is a constant or fixed number with $E(e_t) = e_t$, which highly depends on the sampled data and the model used in the analysis. Hence, there is not a sufficient number of observations to test the assumptions in (2.90) for each time point. As a result, these cannot be proven, but they should be assumed to be valid for the present model(s). By applying a lagged-variable or autoregressive model, which is either first- or higher-order autoregressive, it is common to assume that the error terms $\{\varepsilon_t\}$ are white noise processes.
>
> On the other hand, in order for the error term ε_t to have an expected value of zero for each time point t, $E(\varepsilon_t) = 0$ in particular, an assumption should be used that ε_t has a certain density or distribution function. In general, it is assumed that ε_t is normally distributed for each time point t. This normal density function also cannot be proved but is assumed.

Furthermore, note that the residual assumptions of the time series model should be accepted, in order to prove or derive the validity of the distributions of *Student's t*, *Snedecor's F*, and *Chi-square* statistics, for testing hypotheses on the partial effects of its independent variables. Since EViews provides several residual diagnostics tests, mainly the normality distribution, heteroskedasticity, and serial correlation tests, the following subsections present additional special notes and comments, supported by selected models.

Furthermore, note that the OLS estimates of all models are obtained without using the basic assumptions of the error terms. In other words, the OLS estimates of the model parameters do not depend on the assumptions of the error terms. So, a regression function obtained only depends on the defined model and the data set, which happens to be available to researchers. Hence, the forecast values, either in-sample or beyond the sample period, also do not depend on the assumptions of the error terms. For these reasons, this book presents various alternative models for each endogenous time series in order to identify the best possible fit among those models, which is surely highly dependent only on the data, but it does not depend on the assumption of the error terms. Finally, I would say that the forecast values beyond the sample period also do not depend on the assumptions of the error terms of the model – see the findings in Section 1.5.

Aside from the OLS estimates of the model parameters, as additional information and comparison the maximum likelihood estimators of the model parameters are obtained using the assumption that a random sample, X_i, for all $i = 1, 2, \ldots, n$, has an independent and identical normal distribution, which should be taken for granted, and it should not be proven based on a sample (Agung 2011a, p. 17–18; Graybill 1976; Neter and Wasserman 1974; Wilks 1962).

1.6.1 Specific Residual Analysis

As an illustration, this subsection presents residual analysis based on a simple LV(2) model of the monthly time series, *HS*, such as follows:

Dependent Variable: HS
Method: Least Squares
Date: 07/30/14 Time: 17:28
Sample (adjusted): 1946M03 1996M04
Included observations: 602 after adjustments

Variable	Coefficient	Std. Error	t-Statistic	Prob.
C	22.89711	2.553305	8.967637	0.0000
HS(-1)	1.160510	0.038437	30.19273	0.0000
HS(-2)	-0.345614	0.038323	-9.018560	0.0000

R-squared	0.774806	Mean dependent var	123.2786
Adjusted R-squared	0.774054	S.D. dependent var	34.58760
S.E. of regression	16.44078	Akaike info criterion	8.442378
Sum squared resid	161909.3	Schwarz criterion	8.464306
Log likelihood	-2538.156	Hannan-Quinn criter.	8.450913
F-statistic	1030.466	Durbin-Watson stat	2.076257
Prob(F-statistic)	0.000000		

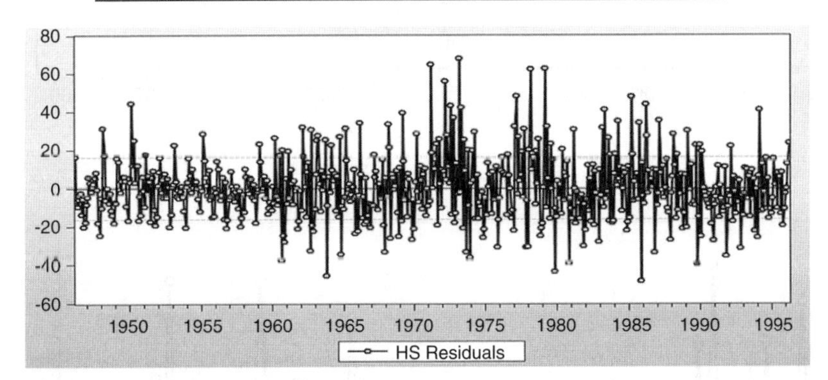

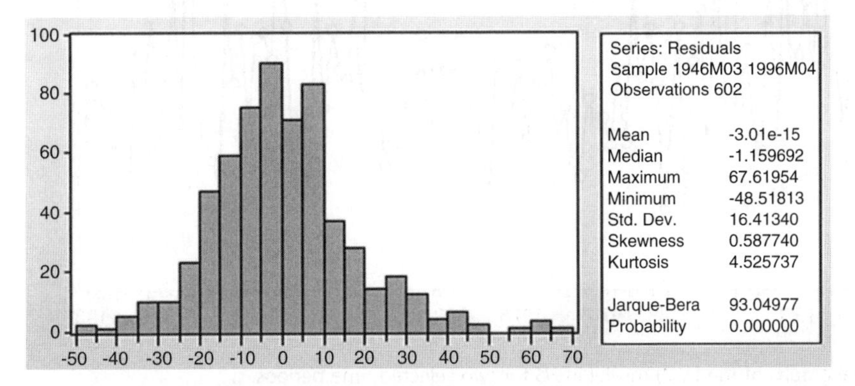

Figure 1.38 Statistical results of the LV(2) model in *HS* (1.28), and its residual graphs and statistics.

1.6.1.1 Residual Graphs and Descriptive Statistics

As an illustrative example, a simple LV(2) forecast model considered has the following ES, with a statistical results presented in Figure 1.38.

$$HS\ C\ HS(-1)\ HS(-2) \tag{1.28}$$

Based on these results, the following findings and special notes are presented.

1) In a statistical sense, the regression function is an acceptable regression, since each of the independent variables has a significant effect and its DW statistic = 2.076257. However, it

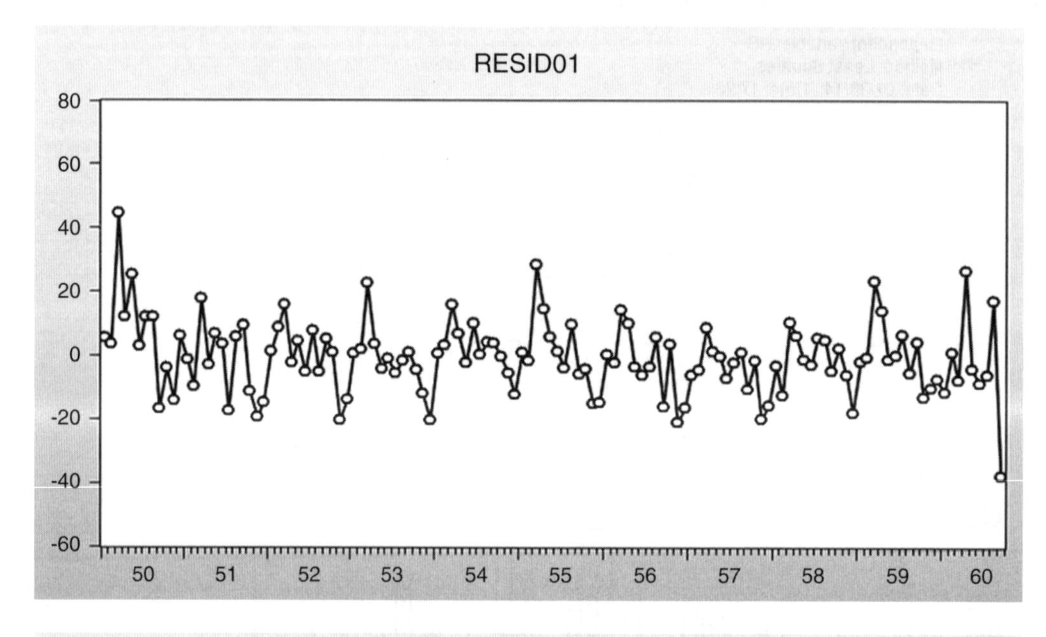

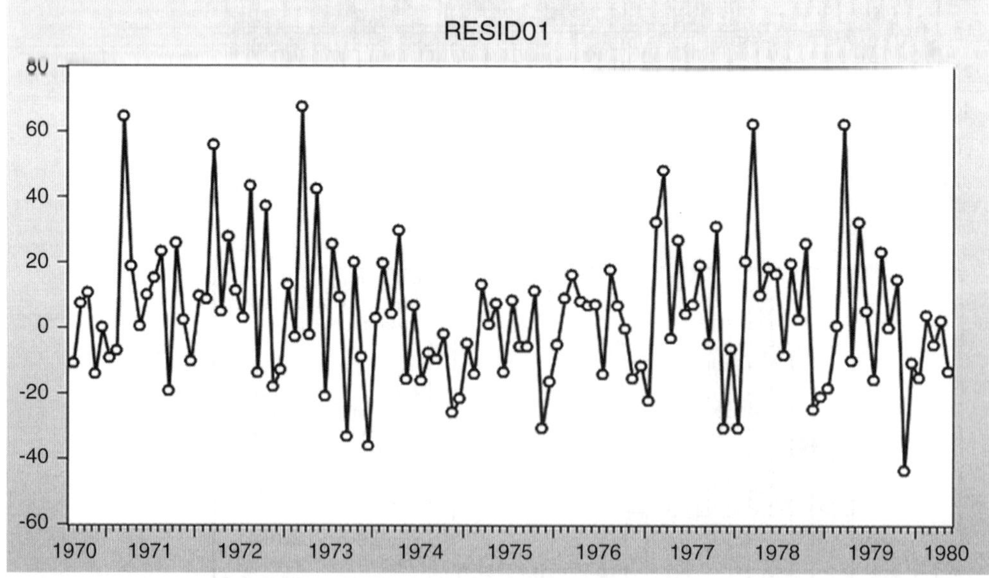

Figure 1.39 Residual graphs of the LV(2) model in *HS* for two selected time periods.

might not be the best forecast model for *HS*. Compare to previous alternative models and the models presented in Chapter 2.

2) The graph of residuals visually shows they might have heterogeneous variances by time periods, such as presented in Figure 1.39 for selected two time periods.
3) Based on the residual statistics, the following special notes and comments can be presented.

   3.1 The mean of the observed error terms, say e_t, for $t = @Trend + 1 = 3, 4, \ldots, T = 604$, is very closed to zero, that is −3.01e-15, and the null hypothesis $H_0: E(\varepsilon_t) = 0$ is accepted based on the t-statistic of $t_0 = -4.38\text{e-}15$, with $df = 601$ and a p value = 1.0000.

   3.2 The error terms e_t, for $t = 1, 2, \ldots, T$, has a single statistic of Std. Dev., as well as a single variance = $(\text{Std. Dev.})^2 = (16.41340)^2$. It does not indicate that each of the error terms has the same variance.

   3.3 Furthermore, I would say that a researcher can never show that all the error terms have independent identical normal distributions. Note that the Jarque–Bera test only shows whether or not the set of the observed values e_t, for $t = 3,4, \ldots, T$ has an approximate normal distribution. Refer to additional special notes presented in Agung (2009a).

1.6.1.2 Breucsh–Pagan–Godfrey Heteroskedasticity Test

Example 1.19 The BPG Heteroskedasticity Test

Figure 1.40 presents the statistical results of the BPG heteroskedasticity test, using the default option. Based on these results, the following findings and special notes are presented.

1) The BPG test is a Lagrange multiplier test of the null hypotheses of no heteroskedasticity against heteroskedasticity of the form $\sigma_t^2 = \sigma(z_t\alpha)$, where z_t is an independent vector. Usually, this vector contains the regressors from the original least squares regression, but it is not necessary, as presented in the EViews User Guide. This statement in EViews also provides "*Add equation regressors*" as an option, which would lead to uncertainty or contradictory conclusions of the BPG heteroskedasticity tests. See the following notes and illustrative example.
2) The test equation presents the regression of the LV(2) model as follows:

$$RESID\hat{\ }2 = C(1) + C(2) * HS(-1) + C(3) * HS(-2) + \nu_t \tag{1.29}$$

   Corresponding to this model, the null hypothesis H_0: $C(2) = C(3) = 0$ is accepted based on the F-statistic of $F_0 = 0.377577$, with $df = (2{,}599)$, and a p value = 0.685684, as presented in Figure 1.40a.

   So the conclusion of the testing is that *HS(−1)* and *HS(−2)* have insignificant joint effects on *RESID^2*. These results can easily be reconfirmed by doing LS regression analysis, using the following ES, where *RESID01* is the residual series of the model (1.28), with the statistical results in Figure 1.40b.

$$RESID01\hat{\ }2\ C\ HS(-1)\ HS(-2) \tag{1.30}$$

3) Note that the F-statistic presented in the BPG heteroskedasticity test is exactly the same as the F-statistic of the LS regression. Therefore, I would say that the three BPG statistical tests, namely the F-statistic, Obs*R-squared, and the Scaled explained SS, give the same conclusion; that is, *HS*(−1) and *HS*(−2) have insignificant joint effects on *RESID*^2. Therefore, it is not very clear, at least for myself, how these statistical tests could be leading to the heterogeneity or homogeneity of the set of $(T-2) = 602$ error terms (parameters) ε_t, $t = 3, 4, \ldots, T$. This is similar for the other heteroskedasticity tests. See the following additional illustrative examples.

Heteroskedasticity Test: Breusch-Pagan-Godfrey

F-statistic	0.377577	Prob. F(2,599)	0.6857
Obs*R-squared	0.757980	Prob. Chi-Square(2)	0.6846
Scaled explained SS	1.322934	Prob. Chi-Square(2)	0.5161

Test Equation:
Dependent Variable: RESID^2
Method: Least Squares
Date: 07/30/14 Time: 17:26
Sample: 1946M03 1996M04
Included observations: 602

Variable	Coefficient	Std. Error	t-Statistic	Prob.
C	204.3782	78.57631	2.601015	0.0095
HS(-1)	0.466349	1.182866	0.394254	0.6935
HS(-2)	0.058083	1.179350	0.049250	0.9607

R-squared	0.001259	Mean dependent var	268.9522
Adjusted R-squared	-0.002076	S.D. dependent var	505.4301
S.E. of regression	505.9544	Akaike info criterion	15.29574
Sum squared resid	1.53E+08	Schwarz criterion	15.31767
Log likelihood	-4601.018	Hannan-Quinn criter.	15.30428
F-statistic	0.377577	Durbin-Watson stat	2.021719
Prob(F-statistic)	0.685684		

(a). BPG Hetroskedasisty Test

Dependent Variable: RESID01^2
Method: Least Squares
Date: 08/01/14 Time: 19:25
Sample (adjusted): 1946M03 1996M04
Included observations: 602 after adjustments

Variable	Coefficient	Std. Error	t-Statistic	Prob.
C	204.3782	78.57631	2.601015	0.0095
HS(-1)	0.466349	1.182866	0.394254	0.6935
HS(-2)	0.058083	1.179350	0.049250	0.9607

R-squared	0.001259	Mean dependent var	268.9522
Adjusted R-squared	-0.002076	S.D. dependent var	505.4301
S.E. of regression	505.9544	Akaike info criterion	15.29574
Sum squared resid	1.53E+08	Schwarz criterion	15.31767
Log likelihood	-4601.018	Hannan-Quinn criter.	15.30428
F-statistic	0.377577	Durbin-Watson stat	2.021719
Prob(F-statistic)	0.685684		

(b). Output of a LS Regression in (1.30)

Figure 1.40 BPG heteroskedasticity test based on the model (1.28) and its corresponding LS regression.

4) In addition, note that the numerator of the F-statistic only has two degrees of freedom, which is not representative of the test of the basic assumption on the homogeneity of a set of $(T-2) = 602$ error term parameters ε_t, $t = 3, 4, \ldots, T$, with the statistical hypothesis:

$$H_0 : Var(\varepsilon_t) = \sigma^2, \forall t = 3,4,\ldots,602; \; versus \;\; H_1 : otherwise \tag{1.31}$$

In addition, I would say that this hypothesis cannot be tested because we have only a single observed values for each parameter, ε_t. Refer to the special notes and comments presented in Agung (2009a, Section 14.3). We can easily test the homogeneity of the error term parameters by time periods, such as by *@Year* or *@Month*, which can be tested easily by selecting *View/Descriptive Statistics & Tests/Equality Test by Classification* .... Do it as an exercise.

Example 1.20 BPG Tests with Additive Additional Equation Regressors

The BPG heteroskedasticity test in fact is one out of six alternative tests, as presented in Figure 1.41. It also presents an option "*Add equation regressors.*" Therefore, various additional equation regressors could be subjectively inserted by a researcher. By using the trial-and-error method, I have found unexpected or contradictory conclusions of the BPG heteroskedasticity tests, compared to the test in previous example, where two of them are presented in Figure 1.42. Based on these results, the following findings and notes are presented.

1) Figure 1.42a in fact presents the statistical results of a LV(4) model of RESID^2, with the following general equation. Based on these results, the following findings and notes are presented.

$$\begin{aligned} RESID\hat{}2 = {} & C(1) + C(2) * HS(-1) + C(3) * HS(-2) \\ & + C(4) * HS(-3) + C(5) * HS(-4) + \nu_t \end{aligned} \tag{1.32}$$

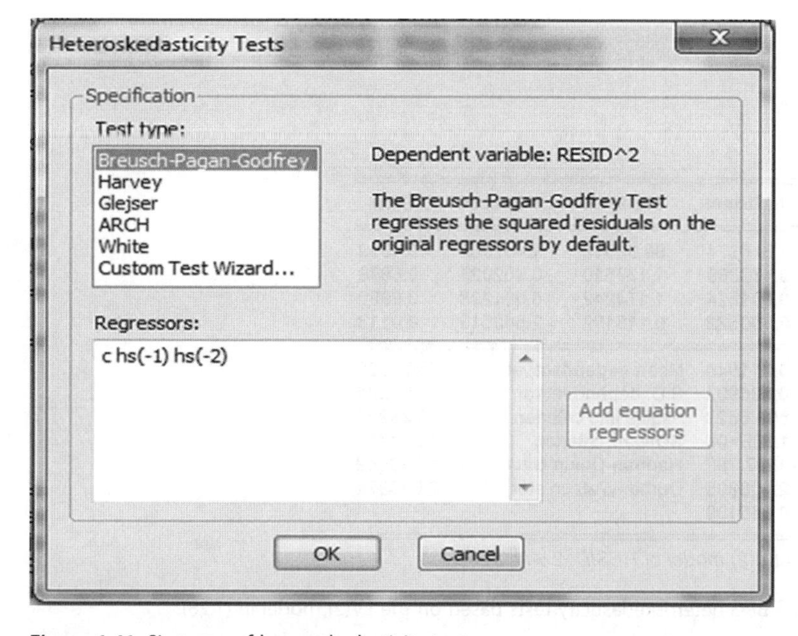

Figure 1.41 Six types of heteroskedasticity tests.

Heteroskedasticity Test: Breusch-Pagan-Godfrey

F-statistic	11.29952	Prob. F(4,595)	0.0000
Obs*R-squared	42.36010	Prob. Chi-Square(4)	0.0000
Scaled explained SS	74.15797	Prob. Chi-Square(4)	0.0000

Test Equation:
Dependent Variable: RESID^2
Method: Least Squares
Date: 08/04/14 Time: 14:14
Sample (adjusted): 1946M05 1996M04
Included observations: 600 after adjustments

Variable	Coefficient	Std. Error	t-Statistic	Prob.
C	-47.75673	86.74053	-0.550570	0.5821
HS(-1)	2.344381	1.228546	1.908257	0.0568
HS(-2)	-1.590952	1.836985	-0.866067	0.3868
HS(-3)	-5.855624	1.836189	-3.189009	0.0015
HS(-4)	7.678269	1.225290	6.266493	0.0000

R-squared	0.070600	Mean dependent var	269.3038
Adjusted R-squared	0.064352	S.D. dependent var	506.2004
S.E. of regression	489.6420	Akaike info criterion	15.23352
Sum squared resid	1.43E+08	Schwarz criterion	15.27017
Log likelihood	-4565.057	Hannan-Quinn criter.	15.24779
F-statistic	11.29952	Durbin-Watson stat	2.256461
Prob(F-statistic)	0.000000		

(a). Results of a LV(4) model of RESID^2

Heteroskedasticity Test: Breusch-Pagan-Godfrey

F-statistic	2.408805	Prob. F(3,598)	0.0661
Obs*R-squared	7.187893	Prob. Chi-Square(3)	0.0661
Scaled explained SS	12.54533	Prob. Chi-Square(3)	0.0057

Test Equation:
Dependent Variable: RESID^2
Method: Least Squares
Date: 08/04/14 Time: 14:44
Sample: 1946M03 1996M04
Included observations: 602

Variable	Coefficient	Std. Error	t-Statistic	Prob.
C	119.6374	85.02518	1.407082	0.1599
HS(-1)	0.473389	1.177510	0.402025	0.6878
HS(-2)	-0.001524	1.174242	-0.001298	0.9990
T	0.300522	0.118199	2.542513	0.0113

R-squared	0.011940	Mean dependent var	268.9522
Adjusted R-squared	0.006983	S.D. dependent var	505.4301
S.E. of regression	503.6623	Akaike info criterion	15.28831
Sum squared resid	1.52E+08	Schwarz criterion	15.31755
Log likelihood	-4597.782	Hannan-Quinn criter.	15.29969
F-statistic	2.408805	Durbin-Watson stat	2.043374
Prob(F-statistic)	0.066109		

(a). Results of a LV(2) model of RESID^2 with trend

Figure 1.42 Two alternative BPG heteroskedasticity tests based on the LV(2) model in (1.28).

1.1 The null hypothesis H_0: C(2) = C(3) = C(4) = C(5) = 0 is rejected based on the *F*-statistic $F_0 = 11.29952$, with *df* = (4,595), and a *p* value = 0.0000. So we reach the conclusion that *HS*(−1), *HS*(−2), *HS*(−3), and *HS*(−4) have a significant joint effects on RESID^2. If this would have been accepted as a conclusion of the significant heteroskedasticity of the error terms, then it is contradictory to the conclusion presented in previous example.

1.2 In addition, note each of the additional regressors *HS*(−3) and *HS*(−4) has a significant effect on *RESID^2*, and they should have a significant joint effects too, which have a great impact on the *F*-statistic.

1.3 Alternative conclusions would also be obtained based on the BPG test by adding equation regressors *HS*(−*p*) for a single integer *p* or several $p > 2$. Do it as an exercise.

2) As an additional illustration, Figure 1.42b presents the statistical results of a LV(2) model of RESID^2 with trend, with the following equation. Based on these results, the following findings and notes are presented.

$$RESID\hat{}2 = C(1) + C(2) * HS(-1) + C(3) * HS(-2) + C(4) * t + \vartheta_t \quad (1.33)$$

2.1 Similar to the additional regressors of the model (1.32), the time *t* also has a significant adjusted effect on RESID^2.

2.2 At the 5% level of significance, the null hypothesis H_0: *C*(2) = *C*(3) = *C*(4) = 0 is accepted based on the *F*-statistic $F_0 = 2.08805$, with *df* = (3,598), and a *p* value = 0.0661, but based on the Scaled explained SS the null hypothesis is rejected based on the Chi-square statistic of $\chi_0^2 = 12.54533$, with *df* = 3, and a *p* value = 0.0057. Which one would be acceptable as a final conclusion?

2.3 On the other hand, at the 10% level of significance, the three statistics would reject the null hypothesis, so there is no problem. See the additional illustration in the following example.

2.4 As more illustrative examples, the time *t* = *@trend + 1* could be replaced by *@Month* and *@Year*. Various forecast models with the time *t*, *@Month*, or *@Year* as predictors will be presented in Chapter 2. Furthermore, exogenous and environmental variables can also be inserted as additional equation regressors. Do this as an exercise.

Example 1.21 BPG Tests with Interactions as Additional Equations Regressors

As a more advanced of the test equations, Figure 1.43 presents statistical results of the BPG tests with interactions as additional regressors.

1) It is very important to note the differences between the sets of additional regressors of these models and the models in Figure 1.41. Each of the additional regressors of both tests in Figure 1.43 has an insignificant adjusted effect on RESID^2, but in Figure 1.41, each of the additional regressors has a significant adjusted effect. Referring to these findings, I have a question: Should I use only significant or insignificant additional equation regressors? Since there is no criterion yet for selecting relevant additional regressors.
2) At the 5% level of significance, the BPG test in Figure 1.43a shows the null hypothesis H_0: *C* (2) = *C*(3) = *C*(4) = *C*(5) = *C*(6) = 0 is accepted based on the *F*-statistic $F_0 = 1.542565$, with *df* = (5,596), and a *p* value = 0.1746, but based on the Scaled explained SS the null hypothesis is rejected based on the Chi-square statistic of $\chi_0^2 = 13.42332$, with *df* = 5, and a *p* value = 0.0197. So we have a problem in selecting which one would be accepted to present the final conclusion. If the *F*-statistic should be used, then we can conclude *HS*(−1), *HS*(−2), *t*, *t*∗*HS*(−1), and *t*∗*HS*(−2) have insignificant joint effects on *RESID^2*.

Heteroskedasticity Test: Breusch-Pagan-Godfrey

F-statistic	1.542565	Prob. F(5,596)	0.1746
Obs*R-squared	7.690942	Prob. Chi-Square(5)	0.1741
Scaled explained SS	13.42332	Prob. Chi-Square(5)	0.0197

Test Equation:
Dependent Variable: RESID^2
Method: Least Squares
Date: 08/05/14 Time: 13:14
Sample: 1946M03 1996M04
Included observations: 602

Variable	Coefficient	Std. Error	t-Statistic	Prob.
C	181.3034	177.5844	1.020942	0.3077
HS(-1)	-1.277545	2.832456	-0.451038	0.6521
HS(-2)	1.221204	2.800402	0.436082	0.6629
T	0.108979	0.501496	0.217308	0.8280
T*HS(-1)	0.005389	0.007920	0.680394	0.4965
T*HS(-2)	-0.003747	0.007832	-0.478451	0.6325

R-squared	0.012776	Mean dependent var	268.9522
Adjusted R-squared	0.004494	S.D. dependent var	505.4301
S.E. of regression	504.2933	Akaike info criterion	15.29411
Sum squared resid	1.52E+08	Schwarz criterion	15.33797
Log likelihood	-4597.527	Hannan-Quinn criter.	15.31118
F-statistic	1.542565	Durbin-Watson stat	2.044244
Prob(F-statistic)	0.174618		

(a)

Heteroskedasticity Test: Breusch-Pagan-Godfrey

F-statistic	2.450098	Prob. F(5,595)	0.0327
Obs*R-squared	12.12439	Prob. Chi-Square(5)	0.0331
Scaled explained SS	21.19642	Prob. Chi-Square(5)	0.0007

Test Equation:
Dependent Variable: RESID^2
Method: Least Squares
Date: 08/05/14 Time: 13:03
Sample (adjusted): 1946M04 1996M04
Included observations: 601 after adjustments

Variable	Coefficient	Std. Error	t-Statistic	Prob.
C	141.5375	175.2607	0.807583	0.4197
HS(-1)	1.367706	1.253330	1.091258	0.2756
HS(-2)	-3.082055	1.868905	-1.649123	0.0997
HS(-3)	1.973782	1.817109	1.086221	0.2778
T	0.041594	0.481054	0.086464	0.9311
T*HS(-3)	0.002200	0.004028	0.546205	0.5851

R-squared	0.020174	Mean dependent var	268.9539
Adjusted R-squared	0.011940	S.D. dependent var	505.8511
S.E. of regression	502.8222	Akaike info criterion	15.28828
Sum squared resid	1.50E+08	Schwarz criterion	15.33220
Log likelihood	-4588.129	Hannan-Quinn criter.	15.30538
F-statistic	2.450098	Durbin-Watson stat	2.105563
Prob(F-statistic)	0.032704		

(b)

Figure 1.43 Two BPG heteroskedasticity tests with specific interaction additional regressors.

3) At the 5% level of significance, the BPG test in Figure 1.43b shows the null hypothesis H_0: $C(2) = C(3) = C(4) = C(5) = C(6) = 0$ is rejected based on the *F*-statistic $F_0 = 2.450098$, with $df = (5{,}596)$, and a *p* value = 0.02704, as well as based on the other two statistics. So it can be concluded that *HS*(−1), *HS*(−2), *HS*(−3), *t*, and *t*HS*(−3) have significant joint effects on *RESID^2*.
4) Note that each of the additional equation regressors in both models has an insignificant adjusted effect on *RESID^2*.

1.6.1.3 The Harvey Test and Glejser Test

There are two other heteroskedasticity tests, the Harvey heteroskedasticity test with dependent variable *log*(*RESID^2*) and the Glejser heteroskedasticity test with a dependent variable *abs(RESID)*, where we have the option to insert additional equation regressors. So we would face the same problems as the BPG heteroskedasticity test. Do similar data analyses as exercises.

1.6.1.4 The White Heteroskedasticity Tests

Example 1.22 The Only Two Alternative White Heteroskedasticity Tests
As a comparison, Figure 1.44 presents the statistical results of both alternative White heteroskedasticity tests. The first test uses the default option; that is, using the option "*Include White cross terms*," and the second uses otherwise.

Note that in the first test equation, the Scaled explained SS gives a contradictory conclusion compared to the first two statistics at the 10% level of significance. On the other hand, in the second test equation, the three statistics give the same conclusions.

However, based on the *F*-statistic, both test equations show that all independent variables of each models have insignificant joint effects, which do not represent the homogeneity of all error parameters ε_t, $t = 3, 4, \ldots, T$ as presented in the hypothesis (1.31).

For the ARCH heteroskedasticity test, we have the option to modify the number of lags where 1 (one) is the default option. So we could use any number of lags, but we would never know the true population test equation (see Agung 2009a). However, the following example only presents two selected tests.

Example 1.23 Two Specific ARCH Heteroskedasticity Tests
Figure 1.45 presents two of the ARCH heteroskedasticity tests with RESID^2 as a dependent variable. These two statistical results are intentionally selected out of several results of alternative LV(*p*) test equations for the best possible illustration. Do the analyses based on other alternative test equations as exercises. Based on these results, the following notes and comments are presented.

1) At the 5% level of significance, the results in Figure 1.45a show that *RESID*^2(−1) and *RESID^2(−2)* have insignificant joint effects on *RESID*^2. On the other hand, the results in Figure 1.45b show that *RESID*^2(−1) up to *RESID*^2(−5) have significant joint effects on *RESID*^2.
2) The contradictory conclusions based on these two test equations show the uncertainty conclusion of a heteroskedasticity test. See again the special notes in point 4 of Example 1.19.

Heteroskedasticity Test: White

F-statistic	1.152442	Prob. F(5,596)	0.3314
Obs*R-squared	5.764488	Prob. Chi-Square(5)	0.3298
Scaled explained SS	10.06100	Prob. Chi-Square(5)	0.0735

Test Equation:
Dependent Variable: RESID^2
Method: Least Squares
Date: 08/05/14 Time: 16:44
Sample: 1946M03 1996M04
Included observations: 602

Variable	Coefficient	Std. Error	t-Statistic	Prob.
C	528.7033	229.6745	2.301968	0.0217
HS(-1)^2	-0.058938	0.042118	-1.399375	0.1622
HS(-1)*HS(-2)	0.095293	0.074197	1.284325	0.1995
HS(-1)	3.673015	5.099516	0.720267	0.4716
HS(-2)^2	-0.016839	0.041284	-0.407866	0.6835
HS(-2)	-8.251419	5.070877	-1.627217	0.1042

R-squared	0.009576	Mean dependent var	268.9522
Adjusted R-squared	0.001267	S.D. dependent var	505.4301
S.E. of regression	505.1099	Akaike info criterion	15.29735
Sum squared resid	1.52E+08	Schwarz criterion	15.34120
Log likelihood	-4598.501	Hannan-Quinn criter.	15.31442
F-statistic	1.152442	Durbin-Watson stat	1.903811
Prob(F-statistic)	0.331419		

Heteroskedasticity Test: White

F-statistic	0.595038	Prob. F(2,599)	0.5519
Obs*R-squared	1.193665	Prob. Chi-Square(2)	0.5506
Scaled explained SS	2.083354	Prob. Chi-Square(2)	0.3529

Test Equation:
Dependent Variable: RESID^2
Method: Least Squares
Date: 08/05/14 Time: 16:49
Sample: 1946M03 1996M04
Included observations: 602

Variable	Coefficient	Std. Error	t-Statistic	Prob.
C	228.1704	42.84849	5.325051	0.0000
HS(-1)^2	0.000805	0.004511	0.178524	0.8584
HS(-2)^2	0.001689	0.004504	0.374919	0.7079

R-squared	0.001983	Mean dependent var	268.9522
Adjusted R-squared	-0.001349	S.D. dependent var	505.4301
S.E. of regression	505.7710	Akaike info criterion	15.29502
Sum squared resid	1.53E+08	Schwarz criterion	15.31694
Log likelihood	-4600.800	Hannan-Quinn criter.	15.30355
F-statistic	0.595038	Durbin-Watson stat	2.020026
Prob(F-statistic)	0.551867		

Figure 1.44 Statistical results of two alternative White heteroskedasticity tests.

Heteroskedasticity Test: ARCH

F-statistic	1.835022	Prob. F(3,595)	0.1396
Obs*R-squared	5.491268	Prob. Chi-Square(3)	0.1392

Test Equation:
Dependent Variable: RESID^2
Method: Least Squares
Date: 08/06/14 Time: 14:07
Sample (adjusted): 1946M06 1996M04
Included observations: 599 after adjustments

Variable	Coefficient	Std. Error	t-Statistic	Prob.
C	249.0130	27.73213	8.979222	0.0000
RESID^2(-1)	-0.003645	0.040998	-0.088901	0.9292
RESID^2(-2)	0.094519	0.040808	2.316164	0.0209
RESID^2(-3)	-0.013921	0.041003	-0.339503	0.7344

R-squared	0.009167	Mean dependent var	269.6808
Adjusted R-squared	0.004172	S.D. dependent var	506.5392
S.E. of regression	505.4815	Akaike info criterion	15.29556
Sum squared resid	1.52E+08	Schwarz criterion	15.32491
Log likelihood	-4577.019	Hannan-Quinn criter.	15.30698
F-statistic	1.835022	Durbin-Watson stat	1.998196
Prob(F-statistic)	0.139578		

(a). LV(3) test equation of RESID^2

Heteroskedasticity Test: ARCH

F-statistic	2.953245	Prob. F(5,591)	0.0121
Obs*R-squared	14.55254	Prob. Chi-Square(5)	0.0125

Test Equation:
Dependent Variable: RESID^2
Method: Least Squares
Date: 08/05/14 Time: 17:50
Sample (adjusted): 1946M08 1996M04
Included observations: 597 after adjustments

Variable	Coefficient	Std. Error	t-Statistic	Prob.
C	210.1625	31.00134	6.779142	0.0000
RESID^2(-1)	-0.009436	0.040882	-0.230819	0.8175
RESID^2(-2)	0.090368	0.040820	2.213826	0.0272
RESID^2(-3)	-0.024019	0.040983	-0.586069	0.5581
RESID^2(-4)	0.054177	0.040826	1.327007	0.1850
RESID^2(-5)	0.112191	0.040888	2.743888	0.0063

R-squared	0.024376	Mean dependent var	270.3646
Adjusted R-squared	0.016122	S.D. dependent var	507.2469
S.E. of regression	503.1413	Akaike info criterion	15.28962
Sum squared resid	1.50E+08	Schwarz criterion	15.33376
Log likelihood	-4557.951	Hannan-Quinn criter.	15.30681
F-statistic	2.953245	Durbin-Watson stat	1.988991
Prob(F-statistic)	0.012107		

(b). LV(5) test equation of RESID^2

Figure 1.45 Statistical results of two specially selected ARCH heteroskedasticity tests.

1.6.1.5 Custom Test Wizard...

By selecting the Custom test Wizard ..., we have the option to conduct each of the three alternative tests, those are the BPG, Hervey, and Glejser heteroskedasticity tests, respectively, with dependent variables $RESID^\wedge2$, $LRESID^\wedge2 = log(RESID^\wedge2)$, and $ARESID = abs(RESID)$, as presented in Figure 1.46(a).

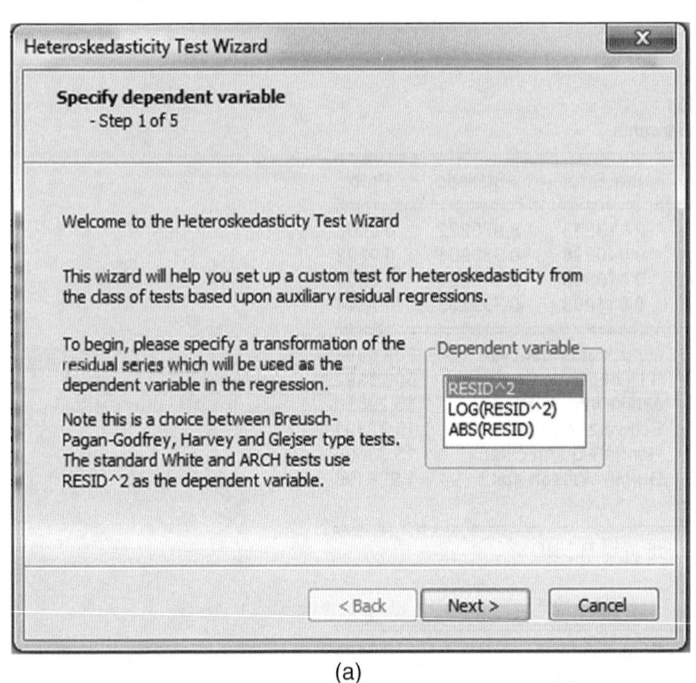

(a)

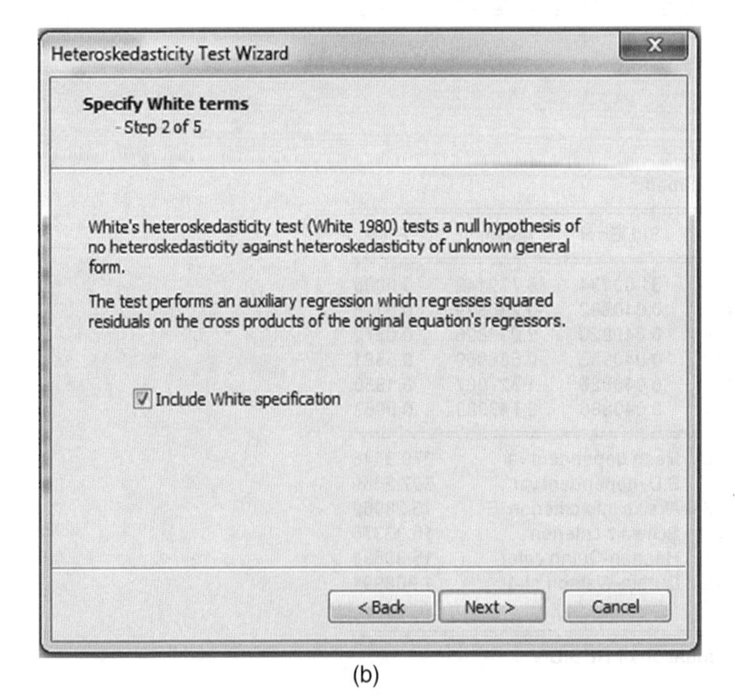

(b)

Figure 1.46 The first two steps of the Custom Test Wizard....

This figure presents the note "*Specify dependent variable*" that indicates Step 1 of 5 where *RESID^2* has been blocked as the first selected dependent variable. In addition to various test equations of *RESID^2* previously presented, this Custom Test Wizard provides the option to select a set of alternative independent variables of the test equation, at the last four steps, such as follows:

1) *Step-1:* With the statistical results of the LV(2) model on the screen, select *View/Residual Diagnostic/Heteroskedasticity Tests ... Custom Test Wizard,* Figure 1.46a shown on the screen, where *RESID^2* has been blocked, then click ... *Next* >.
2) *Step-2*: We have the option to specify the White terms, as presented in Figure 1.46b. By clicking the "*Include White specification*" ... *Next* >, EViews provides five possible types of the White tests as presented in Figure 1.47. Furthermore, if the *Custom Test* is selected, then we have the option to add equation regressors, to use cross terms, and to insert the key *@REGS* to specify every variable in the original regression. So, we can have a lot of possible test equations that could present contradictory conclusions. See previous examples.

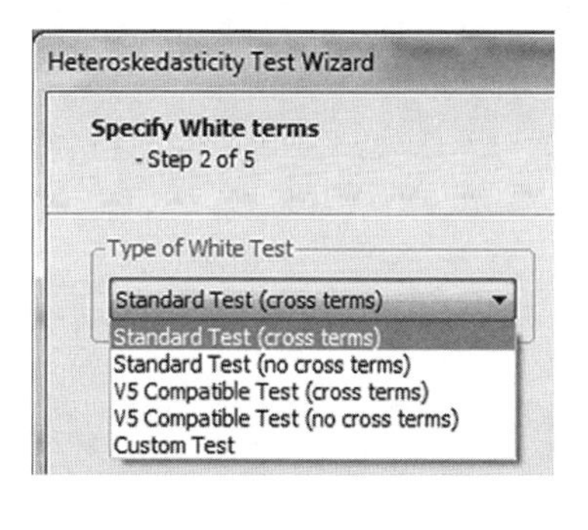

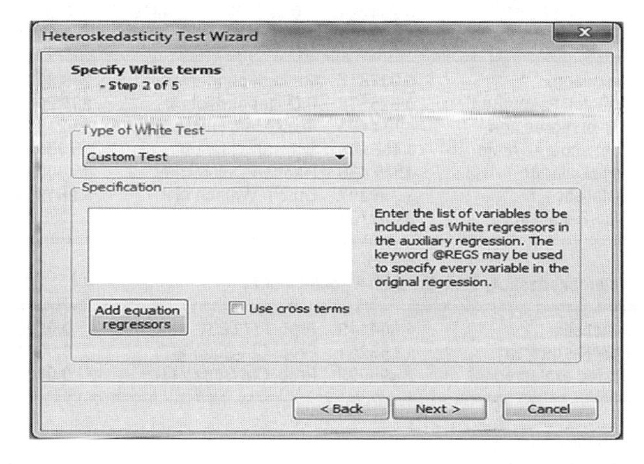

Figure 1.47 Additional options for the White Terms.

Figure 1.48 Final specification confirmation.

Heteroskedasticity Test: Breusch-Pagan-Godfrey

F-statistic	3.329397	Prob. F(7,592)	0.0017
Obs*R-squared	22.72605	Prob. Chi-Square(7)	0.0019
Scaled explained SS	39.78549	Prob. Chi-Square(7)	0.0000

Test Equation:
Dependent Variable: RESID^2
Method: Least Squares
Date: 08/06/14 Time: 15:02
Sample (adjusted): 1946M05 1996M04
Included observations: 600 after adjustments

Variable	Coefficient	Std. Error	t-Statistic	Prob.
C	251.8750	239.6425	1.051045	0.2937
RESID^2(-1)	0.402845	0.114527	3.517472	0.0005
RESID^2(-2)	0.154384	0.045812	3.369928	0.0008
HS(-1)^2	-0.439286	0.112254	-3.913328	0.0001
HS(-1)*HS(-2)	0.732642	0.192770	3.800592	0.0002
HS(-1)	18.94339	6.340152	2.987845	0.0029
HS(-2)^2	-0.294294	0.088362	-3.330565	0.0009
HS(-2)	-18.92163	5.775512	-3.276182	0.0011

R-squared	0.037877	Mean dependent var	269.3038
Adjusted R-squared	0.026500	S.D. dependent var	506.2004
S.E. of regression	499.4481	Akaike info criterion	15.27813
Sum squared resid	1.48E+08	Schwarz criterion	15.33675
Log likelihood	-4575.438	Hannan-Quinn criter.	15.30095
F-statistic	3.329397	Durbin-Watson stat	1.981904
Prob(F-statistic)	0.001736		

Heteroskedasticity Test: Breusch-Pagan-Godfrey

F-statistic	1.804149	Prob. F(7,593)	0.0839
Obs*R-squared	12.53251	Prob. Chi-Square(7)	0.0844
Scaled explained SS	21.90992	Prob. Chi-Square(7)	0.0026

Test Equation:
Dependent Variable: RESID^2
Method: Least Squares
Date: 08/11/14 Time: 05:59
Sample (adjusted): 1946M04 1996M04
Included observations: 601 after adjustments

Variable	Coefficient	Std. Error	t-Statistic	Prob.
C	373.1649	245.7676	1.518365	0.1295
RESID^2(-1)	0.298738	0.114522	2.608557	0.0093
HS(-1)	13.50960	6.407430	2.108428	0.0354
HS(-1)^2	-0.332072	0.112985	-2.939088	0.0034
HS(-1)*HS(-2)	0.562503	0.193779	2.902805	0.0038
HS(-2)	-14.89066	5.678081	-2.622482	0.0090
HS(-2)^2	-0.222846	0.089137	-2.500039	0.0127
@MONTH	-7.193358	7.560297	-0.951465	0.3418

R-squared	0.020853	Mean dependent var	268.9539
Adjusted R-squared	0.009295	S.D. dependent var	505.8511
S.E. of regression	503.4948	Akaike info criterion	15.29425
Sum squared resid	1.50E+08	Schwarz criterion	15.35280
Log likelihood	-4587.921	Hannan-Quinn criter.	15.31704
F-statistic	1.804149	Durbin-Watson stat	1.995069
Prob(F-statistic)	0.083891		

Figure 1.49 Two specially selected BPG heteroskedasticity tests.

Therefore, a researcher could have a different conclusion to others, and we will never know which one is the truth for the corresponding population.

3) *Step-3:* To select the option *"Include Breusch–Pagan Specification,"* and by clicking *Next* >, we have again the option to add equation regressors using cross terms or to insert *@REGS* to specify every variable in the original regression.
4) *Step-4:* To select the option *"Include ARCH Specification,"* and by clicking *Next* >, we have two options; to enter the *Number of lags* or *Custom of lags*, then ... *Next* >. Note the option *Custom of lags* gives us to insert any set of lags, such as 2, 4, 5 as illustrated in Figure 1.48. However, we cannot predict what set of lags would be appropriate, moreover the best possible set.
5) *Step-5:* The screen shows the final specification confirmation, as presented in Figure 1.48. In this final stage we still can modify the auxiliary regressors using the box, as well as to insert additional regressors. Refer to the illustrations presented earlier. Then click *Finish.*

Example 1.24 Custom Test Wizard ... of *RESID^2*

By selecting the RESID^2 as a dependent variable of the test equation, in fact we are doing the BPG heteroskedasticity test with many or a lot of possible additional equation regressors.

As an illustration, Figure 1.49 presents two out of many possible BPG heteroskedasticity tests using the Custom Test Wizard of *RESID^2*. Based on these intentionally selected tests, the following findings and notes are presented.

1) The test in Figure 1.49a is a BPG heteroskedasticity test, using the five steps from before without adding any new equation regressors. At the 1% level of significance the error terms are significantly heterogeneous, based on the *F*-statistic of $F_0 = 3.329937$, with $df = (7{,}592)$, and a *p* value = 0.0017, as well as based on the other two statistics. However, in fact the corresponding LS regression shows that the seven independent variables of the test equation have significant joint effects on *RESID^2*. Refer to the notes in the Example 1.19, especially the hypothesis (1.31) in point 4.
2) Note the regressors of the test equation are a combination of the regressors of the ARCH test equation with 2 (two) as the number of lags, indicated by *RESID^2(−1)* and *RESID^2(−2)*, and the regressors of the White test equation with cross terms.
3) On the other hand, the results in Figure 1.49b are obtained by using the trial-and-error method in order to present a contradictory conclusion at the 1% level of significance. Even at the 5% level of significance, based on the *F*-statistic of $F_0 = 1.804149$, with $df = (7{,}592)$, and a *p value* = 0.0839. So, it can be concluded that the error terms are insignificantly heterogeneous. In order words, the data supports the homogeneity of the error terms. However, in fact the corresponding LS regression shows that the seven independent variables of the test equation have insignificant joint effects on *RESID^2*.

   Refer again to the notes in the Example 1.19, especially the hypothesis (1.31) in point 4. See the following example.
4) Many alternative sets of additional equation regressors can be inserted. However, in this case I use *@Month* as an additional equation regressor in order to present a contradictory conclusion of the test. The time series models with *@Month* as an independent variable will be presented in more detail later.

Example 1.25 Custom Test Wizard ... of *log(RESID^2)*

By selecting the *log(RESID^2)* as a dependent variable of the test equation, in fact we are doing the Harvey heteroskedasticity test, with many of, or as many as possible, additional equation regressors. As an illustration, Figure 1.50 presents two out of many possible Harvey

Heteroskedasticity Test: Harvey

F-statistic	1.670045	Prob. F(8,590)	0.1026
Obs*R-squared	13.26381	Prob. Chi-Square(8)	0.1031
Scaled explained SS	15.02910	Prob. Chi-Square(8)	0.0586

Test Equation:
Dependent Variable: LRESID2
Method: Least Squares
Date: 08/11/14 Time: 05:31
Sample (adjusted): 1946M06 1996M04
Included observations: 599 after adjustments

Variable	Coefficient	Std. Error	t-Statistic	Prob.
C	6.107176	1.269783	4.809620	0.0000
LRESID2(-1)	0.058244	0.049318	1.180995	0.2381
LRESID2(-2)	-0.056372	0.041895	-1.345561	0.1790
LRESID2(-3)	0.047649	0.042546	1.119928	0.2632
HS(-1)^2	-5.98E-05	0.000239	-0.250281	0.8025
HS(-1)*HS(-2)	3.97E-06	0.000411	0.009653	0.9923
HS(-1)	0.008218	0.025126	0.327069	0.7437
HS(-2)^2	0.000199	0.000218	0.915670	0.3602
HS(-2)	-0.045502	0.024192	-1.880901	0.0605

R-squared	0.022143	Mean dependent var	4.066212
Adjusted R-squared	0.008884	S.D. dependent var	2.366628
S.E. of regression	2.356092	Akaike info criterion	4.566797
Sum squared resid	3275.190	Schwarz criterion	4.632836
Log likelihood	-1358.756	Hannan-Quinn criter.	4.592506
F-statistic	1.670045	Durbin-Watson stat	1.995931
Prob(F-statistic)	0.102626		

(a)

Heteroskedasticity Test: Harvey

F-statistic	2.343940	Prob. F(4,596)	0.0536
Obs*R-squared	9.307992	Prob. Chi-Square(4)	0.0538
Scaled explained SS	10.51197	Prob. Chi-Square(4)	0.0326

Test Equation:
Dependent Variable: LRESID2
Method: Least Squares
Date: 08/11/14 Time: 05:41
Sample (adjusted): 1946M04 1996M04
Included observations: 601 after adjustments

Variable	Coefficient	Std. Error	t-Statistic	Prob.
C	3.416993	0.286381	11.93161	0.0000
LRESID2(-1)	0.057947	0.040812	1.419850	0.1562
HS(-1)^2	-1.77E-05	2.12E-05	-0.834014	0.4046
HS(-2)^2	1.49E-05	2.20E-05	0.678112	0.4980
@MONTH	0.070591	0.030467	2.316925	0.0208

R-squared	0.015488	Mean dependent var	4.065742
Adjusted R-squared	0.008880	S.D. dependent var	2.362711
S.E. of regression	2.352197	Akaike info criterion	4.556861
Sum squared resid	3297.567	Schwarz criterion	4.593455
Log likelihood	-1364.337	Hannan-Quinn criter.	4.571106
F-statistic	2.343940	Durbin-Watson stat	1.994359
Prob(F-statistic)	0.053616		

(b)

Figure 1.50 Two specially selected Harvey heteroskedasticity tests.

heteroskedasticity tests using the Custom Test Wizard of *LRESID^2*, which are intentionally selected using the trial-and-error method to present contradictory conclusions, especially based on the *F*-statistic. Based on these intentionally selected test equations, the following findings and notes are presented.

1) The test in Figure 1.50a is a Harvey heteroskedasticity test, using the five steps from before, without adding any new equation regressors. At the 10% level of significance, the error terms are insignificantly heterogeneous based on the *F*-statistic of F_0 = 1.670045, with df = (8,590), and a p *value* = 0.102626. On the other hand, at the 10% level of significance, Figure 1.50b shows the error terms are significantly heterogeneous based on the *F*-statistic of F_0 = 2.343290, with df = (4,596), and a p value = 0.0536 < 0.10.
2) *As additional illustrations.* Note that at the 5% level of significance both test equations show that each set of their independent variables have insignificant joint effects on *RESID^2*. On the other hand, at the 15% level of significance (see Lapin 1973) both test equations show each set of their independent variables have significant joint effects on *RESID^2*.
3) Corresponding to each test equation, we can present a LS regression with the same set of independent variables to show that the *Harvey* test in fact presents the joint effects of the independent variables on *LRESID^2* = *log(RESID^2)*. Refer to the notes in Example 1.19, especially the hypothesis (1.31) in point 4.

Example 1.26 Custom Test Wizard ... of *ABS(RESID)*

By selecting the *abs(RESID)* as a dependent variable of the test equation, in fact we are doing the Glejser heteroskedasticity test, with many or a lot of possible additional equation regressors. As an illustration, Figure 1.51 presents two out of many possible Glejser heteroskedasticity tests using the Custom Test Wizard of *ARESID* = *abs*(*RESID*), which are intentionally selected using the trial-and-error method to present contradictory conclusions, especially based on the *F*-statistic. Based on these intentionally selected test equations, the following findings and notes are presented.

1) The test in Figure 1.51a is a Glejser heteroskedasticity test of *ARESID* = *abs*(*Resid*) with a set of independent variables, which are a combination of the independent variables of an ARCH test equation, namely *ARESID*(−1) and *ARESID*(−2), and the dependent variables of a standard White test equation with *cross terms*. At the 1% level of significance, the seven variables have significant joint effects on *ARESID* based on the *F*-statistic of F_0 = 3.254061, with df = (7,592), and a p value = 0.0021 < 0.01. So, based on the Glejser heteroskedasticity test, the data supports the homogeneity of the set of 600 error terms
2) The independent variables of the test equation in Figure 1.51b also are a combination of the independent variables of the same ARCH test equation and the dependent variables of a standard White test equation but with *no cross terms*, which is a reduced model of the test equation in Figure 1.51a. The test equation presents a contradictory conclusion, since at the 10% level of significance its four independent variables have insignificant joint effects on *ARESID* based on the *F*-statistic of F_0 = 1.550961, with df = (4,595), and a p value = 0.1860. So, based on the Glejser heteroskedasticity test, the data supports the homogeneity of the set of 600 error terms. However, the numerator of the *F*-statistic only has four degrees of freedom. See the following illustrative example.

Heteroskedasticity Test: Glejser

F-statistic	3.254061	Prob. F(7,592)	0.0021
Obs*R-squared	22.23087	Prob. Chi-Square(7)	0.0023
Scaled explained SS	26.62698	Prob. Chi-Square(7)	0.0004

Test Equation:
Dependent Variable: ARESID
Method: Least Squares
Date: 08/06/14 Time: 15:08
Sample (adjusted): 1946M05 1996M04
Included observations: 600 after adjustments

Variable	Coefficient	Std. Error	t-Statistic	Prob.
C	14.27362	5.416651	2.635138	0.0086
ARESID(-1)	0.252643	0.081014	3.118519	0.0019
ARESID(-2)	0.087573	0.042998	2.036682	0.0421
HS(-1)^2	-0.005759	0.001756	-3.278793	0.0011
HS(-1)*HS(-2)	0.009231	0.003023	3.053571	0.0024
HS(-1)	0.305860	0.124946	2.447944	0.0147
HS(-2)^2	-0.003122	0.001449	-2.155232	0.0315
HS(-2)	-0.389969	0.115755	-3.368903	0.0008

R-squared	0.037051	Mean dependent var	12.31500
Adjusted R-squared	0.025665	S.D. dependent var	10.85546
S.E. of regression	10.71525	Akaike info criterion	7.594456
Sum squared resid	67971.37	Schwarz criterion	7.653082
Log likelihood	-2270.337	Hannan-Quinn criter.	7.617278
F-statistic	3.254061	Durbin-Watson stat	1.971803
Prob(F-statistic)	0.002128		

(a). Full test equation

Heteroskedasticity Test: Glejser

F-statistic	1.550961	Prob. F(4,595)	0.1860
Obs*R-squared	6.191422	Prob. Chi-Square(4)	0.1853
Scaled explained SS	7.415764	Prob. Chi-Square(4)	0.1155

Test Equation:
Dependent Variable: ARESID
Method: Least Squares
Date: 08/12/14 Time: 14:39
Sample (adjusted): 1946M05 1996M04
Included observations: 600 after adjustments

Variable	Coefficient	Std. Error	t-Statistic	Prob.
C	10.12194	1.078373	9.386310	0.0000
ARESID(-1)	0.037064	0.042033	0.881787	0.3782
ARESID(-2)	0.067518	0.041383	1.631544	0.1033
HS(-1)^2	-5.34E-05	9.95E-05	-0.536578	0.5918
HS(-2)^2	0.000109	9.83E-05	1.106089	0.2691

R-squared	0.010319	Mean dependent var	12.31500
Adjusted R-squared	0.003666	S.D. dependent var	10.85546
S.E. of regression	10.83554	Akaike info criterion	7.611839
Sum squared resid	69858.32	Schwarz criterion	7.648480
Log likelihood	-2278.552	Hannan-Quinn criter.	7.626102
F-statistic	1.550961	Durbin-Watson stat	2.004501
Prob(F-statistic)	0.185957		

(b). Reduced test equation of (a)

Figure 1.51 Two specially selected Glejser heteroskedasticity tests.

1.6.2 Additional Special Notes and Comments

Contradictory conclusions to the alternative heteroskedasticity tests of the error terms of the LV(2) model of *HS* in (1.28) have been presented in previous examples. In other words, the conclusions are not consistent and are unpredictable. Based on these findings, I must present some special notes and comments as follows:

1) The illustrated test equations presented in previous examples show the conclusion of a heteroskedasticity test is highly dependent on the test equations with a selected dependent variable, such as *RESIDS^2*, *log*(*RESID^2*), or *abs*(*RESID*), and a lot of possible sets of independent variables, aside from the data set that happens to be selected by or available to the researchers. For these reasons, I would say that the heteroskedasticity tests are not reliable, since they do not give consistent conclusions for the error terms.
2) Note that the *F*-statistic presented in the output of a test equation in fact presents a test of the hypothesis on the joint effects of the independent variables on the corresponding dependent variables, which is considered or defined the heteroskedasticity test or homogeneity test of a set of (T–2) error terms, as presented by the hypothesis in (1.3).
3) Agung (2009a, Section 2.14.3) has presented a statement that the basic assumptions of the error terms of any model, namely ε_t, have independent identical normal distribution with $E(\varepsilon_t) = 0$, and $Var(\varepsilon_t) = \sigma^2$, for all $t = 3, 4, \ldots, T$; are theoretical concepts of statistics and they should not and cannot be tested. In addition, note that each of the observed error terms, namely e_t, for each time point $t = 3, 4, \ldots, T$, does not have a variance, moreover a distribution, since it is supported by a single observed score only. So I would say that the *F*- test statistic presented as the heteroskedasticity test of the error terms, as presented in previous examples, is not a reliable test for the homogeneity of the (T–2) error terms of the LV(2) model (1.28). Likewise for the residuals of all other models.
4) Furthermore, note that if the (T–2) error terms should have equal variances, then the set of error terms would be homogeneous by *@Year*, *@Month*, or any defined *time period*. On the other hand, if the set of error terms are heterogeneous by *@Year* or by *@Month* then the (T–2) error terms should be heterogeneous. On the other hand, it is impossible to conduct the homogeneity test by *@Year* and *@Month*, since each cell/class/group has a single observation only. See Example 1.27.
5) So I would recommend applying the test for equality of variances of *RESID01* by *@Year* or by *@Month* as an alternative heteroskedasticity test; namely, which one can give a fixed conclusion, based on three alternative statistics, such as Bartlett, Leven, and Brown–Forsythe test statistics? See the following illustration.

Example 1.27 Homogeneity Test of *RESID01* by *@Year or @Month*

As an illustration, Figure 1.52 presents two tests for equality of variances of *RESID01* of the LV(2) model (1.28). Based on these results or outputs, the following findings and notes are presented.

1) Based on the results in Figure 1.52a, we have the following findings:
   1.1 The set of 51 error terms have significant heterogeneous variances, based on the Bartlett Chi-square statistic of $\chi_0^2 = 115.4357$, with $df = 50$, and a p value = 0.0000, as well as based on the Levene *F*-statistic of $F_0 = 1.98441$, with $df = (50{,}551)$, and a p value = 0.0001, and the Brown–Forsythe *F*-statistic of $F_0 = 1.721710$ with $df = (50{,}551)$, and a p value = 0.0021.
   1.2 The three test statistics show consistent conclusions. So far, I am very sure there is no other homogeneity test of variances of *RESID01* by @Year, which would give contradictory conclusions.

Test for Equality of Variances of RESID01
Categorized by values of @YEAR
Date: 08/14/14 Time: 16:04
Sample (adjusted): 1946M03 1996M04
Included observations: 602 after adjustments

Method	df	Value	Probability
Bartlett	50	115.4357	0.0000
Levene	(50, 551)	1.984415	0.0001
Brown-Forsythe	(50, 551)	1.721710	0.0021

Category Statistics

@YEAR	Count	Std. Dev.	Mean Abs. Mean Diff.	Mean Abs. Median Diff.
1946	10	10.08187	6.494991	6.494991
1947	12	9.977209	7.505516	7.019823
1948	12	13.87806	9.581782	9.069899
1949	12	9.725094	7.870300	7.870300
1950	12	16.46329	11.43846	11.24708
....				
....				
1992	12	13.38042	9.348568	9.340928
1993	12	10.94062	8.901537	8.901537
1994	12	18.06019	14.05001	13.84874
1995	12	11.37903	10.04779	10.04779
1996	4	12.19303	9.776650	9.776650
All	602	16.41340	11.77543	11.61756

Bartlett weighted standard deviation: 16.47664

(a). Homogeneity test by @Year

Test for Equality of Variances of RESID01
Categorized by values of @MONTH
Date: 08/14/14 Time: 16:02
Sample (adjusted): 1946M03 1996M04
Included observations: 602 after adjustments

Method	df	Value	Probability
Bartlett	11	38.89053	0.0001
Levene	(11, 590)	3.067897	0.0005
Brown-Forsythe	(11, 590)	2.679197	0.0023

Category Statistics

@MONTH	Count	Std. Dev.	Mean Abs. Mean Diff.	Mean Abs. Median Diff.
1	50	11.06799	8.677640	8.669639
2	50	10.62252	7.706859	7.571133
3	51	17.77306	14.04260	13.85192
4	51	11.69905	9.317945	9.315530
5	50	14.21974	10.22608	9.846560
6	50	9.421078	7.597060	7.597060
7	50	9.700892	7.985695	7.981604
8	50	11.53323	8.415312	8.413839
9	50	11.45267	8.881339	8.869502
10	50	13.17299	10.73225	10.62927
11	50	10.95908	8.342798	8.184640
12	50	10.03737	7.553613	7.523128
All	602	16.41340	9.131761	9.046277

Bartlett weighted standard deviation: 12.02445

(b). Homogeneity test by @Month

Figure 1.52 Two homogeneity tests of *RESID01* by @Year and by @Month.
(a) Homogeneity test by @*Year*.
(b) Homogeneity test by @*Month*.

Test for Equality of Variances of RESID01
Categorized by values of @MONTH and @YEAR
Date: 08/15/14 Time: 15:34
Sample (adjusted): 1946M03 1996M04
Included observations: 602 after adjustments

Method	df	Value	Probability
Bartlett	601	NA	NA
Levene	(601, 0)	NA	NA
Brown-Forsythe	(601, 0)	NA	NA

Category Statistics

@YEAR	@MONTH	Count	Std. Dev.	Mean Abs. Mean Diff.	Mean Abs. Median Diff.
1946	1	0	NA	NA	NA
1946	2	0	NA	NA	NA
1946	3	1	NA	0.000000	0.000000
1946	4	1	NA	0.000000	0.000000
1946	5	1	NA	0.000000	0.000000

Figure 1.53 A homogeneity test of *RESID01*.

2) Figure 1.52b presents the homogeneity test of variances of *RESID01* by *@Month*, which also shows that the set of 12 error terms are significantly heterogeneous based on the three test statistics.
3) As an additional illustration, Figure 1.53 presents a part of the output of the test of equality of variances of *RESID01* by *@Month* and *@Year*, which shows NA (not applicable) for the Std. Dev., because each cell/class/ group has a single observation only. So we can never test the homogeneity of all error terms of any regression models. Note that the denominator of the *F*-statistic has a $df = 0$.

Example 1.28 Homogeneity Test of *RESID01* by *@Trend*

A final alternative homogeneity test of *RESID01* is the test of *RESID01* by @Trend, with various possible numbers of time periods or time intervals, which can easily be generated using the program. Figure 1.54 presents a specific option and the statistical results of a test equality of variances of *RESID01* by *@Trend*. Based on these results or outputs, the following findings and notes are presented.

1) Note that the variable *@Trend* is inserted as the *Series/Group to classify*, with the *Max. # of bins*: 100. However, the statistical results only present 61 time intervals, which are directly generated by the program. So the exact number of time intervals (TI) obtained are unpredictable. I have been using the trial-and-error method in order to obtain the number of the TI or time periods (TP) is greater than 51, as presented in Figure 1.52a.
2) As it is expected, the three test statistics also show that the error terms have significant heterogeneous variances. Even by using several other *Max. # of bins* $\leq$ 120, I have obtained the same conclusions. Do it as an exercise.
3) By doing many data analyses, I have found unexpected test statistics, as presented in Figure 1.55, by using 120 < *Max. # of bins* $\leq$ 200, which shows the Brown–Forsythe statistics present exactly the *F*-statistic of $F_0 = 0.891523$, with $df = (120{,}481)$, and a p value $= 0.7754$,

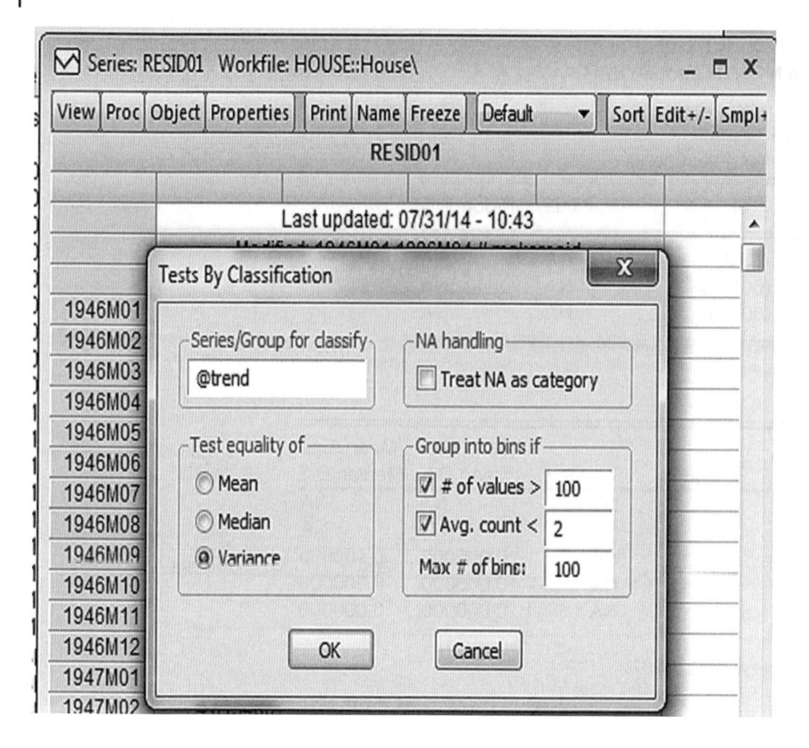

Test for Equality of Variances of RESID01
Categorized by values of @TREND
Date: 08/16/14 Time: 11:12
Sample (adjusted): 1946M03 1996M04
Included observations: 602 after adjustments

Method	df	Value	Probability
Bartlett	60	136.1479	0.0000
Levene	(60, 541)	2.023262	0.0000
Brown-Forsythe	(60, 541)	1.612788	0.0035

Category Statistics

@TREND	Count	Std. Dev.	Mean Abs. Mean Diff.	Mean Abs. Median Diff.
[0, 10)	8	9.048591	5.710364	5.015570
[10, 20)	10	9.253499	7.356416	7.045107
[20, 30)	10	16.54446	12.09286	12.09286
....				
....				
[570, 580)	10	19.35071	14.27988	14.27988
[580, 590)	10	11.26319	9.235025	8.908424
[590, 600)	10	11.86176	10.32951	9.825344
[600, 610)	4	12.19303	9.776650	9.776650
All	602	16.41340	11.72272	11.47379

Bartlett weighted standard deviation: 16.45427

Figure 1.54 Statistical results of a test of variance of *RESID01* by *@Trend*.

Test for Equality of Variances of RESID01
Categorized by values of @TREND
Date: 08/16/14 Time: 12:29
Sample (adjusted): 1946M03 1996M04
Included observations: 602 after adjustments

Method	df	Value	Probability
Bartlett	120	215.7960	0.0000
Levene	(120, 481)	2.058543	0.0000
Brown-Forsythe	(120, 481)	0.891523	0.7754

Category Statistics

@TREND	Count	Std. Dev.	Mean Abs. Mean Diff.	Mean Abs. Median Diff.
[0, 5)	3	13.58330	10.44809	8.016794
[5, 10)	5	4.362934	3.408439	3.069242
[10, 15)	5	10.46270	7.911014	7.783570
[15, 20)	5	2.816982	2.107522	2.014896
[20, 25)	5	13.33595	10.71368	10.11144

Figure 1.55 A homogeneous test of *RESID01* by *@Trend*, using Max. # of bins: 200.

compared to the other two test statistics with 0.0000 probabilities. I select the *Max. #of bins*: 200 in order to have at least three observations, that is 602/200 > 3, within each time interval. Based on these contradictory conclusions, and the heteroskedasticity tests of *RESID01* by *@Year* or *@Month* as presented in Example 1.27, I would consider the Brown–Forsythe test statistic is not reliable for a large number of classes.

4) If the residuals have heterogeneous variances, then it is recommended to apply the adjusted estimation method by using the *White heteroskedasticity-consistent standard errors & covariance*, as presented in Section 1.5.

1.6.3 Serial Correlation Tests

I have found that the residuals of a time series model do have serial correlations in most cases. So, the basic assumptions of its residuals with independent identical normal distributions cannot be accepted. Therefore, the common *t*, *F*, and *Chi-square* statistics for testing the hypotheses on the model parameters are incorrect. See the following section.

Example 1.29 Breusch–Godfrey Serial Correlation LM Test

As an illustration, Figure 1.56 presents two statistical results of the BG serial correlation LM tests, by using LV(1) and LV(2) test equations of *RESID*. Based on these results, the following findings and notes are presented.

1) Both tests show that *RESID* has ` significant serial correlation. Therefore, the *t*, *F*, and *Chi-square* statistics for testing the hypotheses on the parameters of the model (1.28) with the output presented in Figure 1.38 are inappropriate, since the output assumes the *RESID* are independent.

Breusch-Godfrey Serial Correlation LM Test:

F-statistic	8.392170	Prob. F(1,598)	0.0039
Obs*R-squared	8.331385	Prob. Chi-Square(1)	0.0039

Test Equation:
Dependent Variable: RESID
Method: Least Squares
Date: 08/18/14 Time: 15:33
Sample: 1946M03 1996M04
Included observations: 602
Presample missing value lagged residuals set to zero.

Variable	Coefficient	Std. Error	t-Statistic	Prob.
C	-5.164412	3.101289	-1.665247	0.0964
HS(-1)	0.299835	0.110326	2.717714	0.0068
HS(-2)	-0.258208	0.096929	-2.663894	0.0079
RESID(-1)	-0.340330	0.117480	-2.896924	0.0039

R-squared	0.013840	Mean dependent var	-3.01E-15
Adjusted R-squared	0.008892	S.D. dependent var	16.41340
S.E. of regression	16.34026	Akaike info criterion	8.431764
Sum squared resid	159668.5	Schwarz criterion	8.461001
Log likelihood	-2533.961	Hannan-Quinn criter.	8.443144
F-statistic	2.797390	Durbin-Watson stat	1.970903
Prob(F-statistic)	0.039470		

(a). LV(1) test equation of *RESID*

Breusch-Godfrey Serial Correlation LM Test:

F-statistic	7.660179	Prob. F(2,597)	0.0005
Obs*R-squared	15.06214	Prob. Chi-Square(2)	0.0005

Test Equation:
Dependent Variable: RESID
Method: Least Squares
Date: 07/30/14 Time: 17:35
Sample: 1946M03 1996M04
Included observations: 602
Presample missing value lagged residuals set to zero.

Variable	Coefficient	Std. Error	t-Statistic	Prob.
C	14.75499	8.214739	1.796160	0.0730
HS(-1)	-0.370600	0.278763	-1.329445	0.1842
HS(-2)	0.251235	0.217286	1.156239	0.2480
RESID(-1)	0.342519	0.285966	1.197760	0.2315
RESID(-2)	0.305544	0.116775	2.616514	0.0091

R-squared	0.025020	Mean dependent var	-3.01E-15
Adjusted R-squared	0.018488	S.D. dependent var	16.41340
S.E. of regression	16.26097	Akaike info criterion	8.423684
Sum squared resid	157858.3	Schwarz criterion	8.460231
Log likelihood	-2530.529	Hannan-Quinn criter.	8.437909
F-statistic	3.830090	Durbin-Watson stat	2.035307
Prob(F-statistic)	0.004395		

(b). LV(2) test equation of *RESID*

Figure 1.56 Two Breusch–Godfrey serial correlation LM tests of the residuals of the model in (1.28).

2) To conduct the testing hypotheses on the model parameters in (1.28) with the output presented in Figure 1.38, it is recommend to apply the *HAC standard errors and covariance (Bartlett kernel, Newey–West fixed bandwidth = 6.0000),* as presented in Section 1.7.

1.7 Statistical Results Using Alternative Options

Figure 1.57 presents alternative options for the equation estimation. It has been found that the basic or regular OLS estimates are the same as the estimates obtained by using the combination of options: *Covariance coefficient matrix: Estimation default* and *Weight Type: Non.*

In addition, by using all combinations of the other options, the same estimates of the model parameters as the OLS regression are obtained, but with different standard errors. So each of the parameters have different values of *t*-statistics. See the following examples.

1.7.1 Application of an Alternative Coefficient Covariance Matrix

Referring to the significant heteroskedasticity or serial correlation of the residuals of any time series model, the basic or common *t*, *F*, and *Chi-square* statistics obtained based on the OLS regression are inappropriate for testing the hypotheses on the model parameters. Then it is recommended to apply the White or the HAC (Newey–West) coefficient covariance matrix in order to adjust the three basic test statistics. See Example 1.30.

Example 1.30 Application of the White and the HAC Coefficient Covariance Matrices
As a comparative study, Figure 1.58 presents a summary of three statistical results of the model (1.28), by using the three options of the *coefficient covariance matrix*; namely, Default, White, and HAC (Newey–West) with *df* Adjustment.

Based on this summary, the following findings and notes are presented.

1) The three outputs present exactly the same estimates of the model parameters, namely *C(1), C(2),* and *C(3),* as well as R-squared, and some other statistics as presented in the output using the default estimation, so they will have the same set of residuals or error terms. Hence, all tests of the residual presented based on the estimates using the default coefficient covariance matrix are acceptable statistical results.
2) Note that only their standard errors and *t*-statistics have different values. In fact, they also should have different *Prob* (*t*-statistics), but the outputs present the same scores of 0.000 since they only use three decimal points.

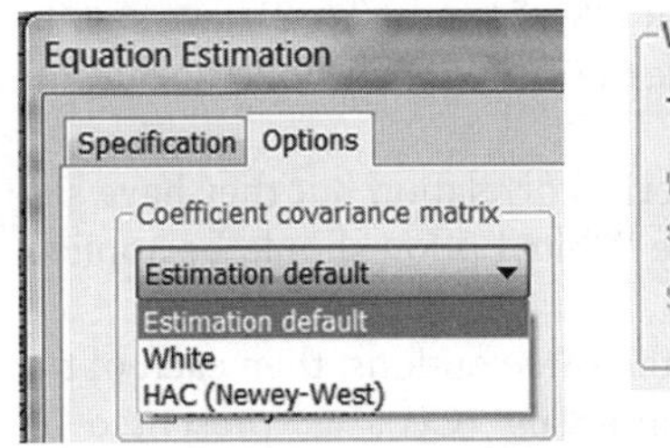

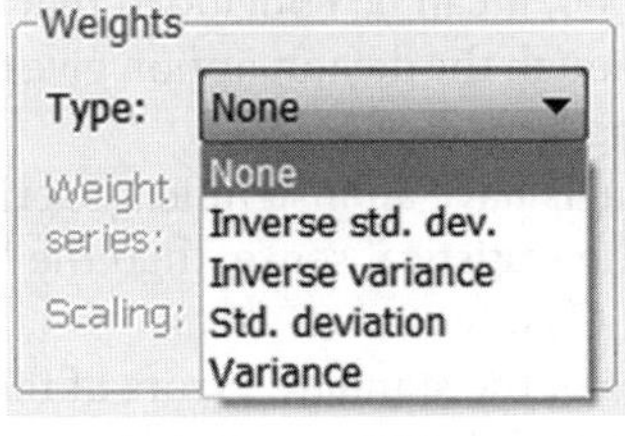

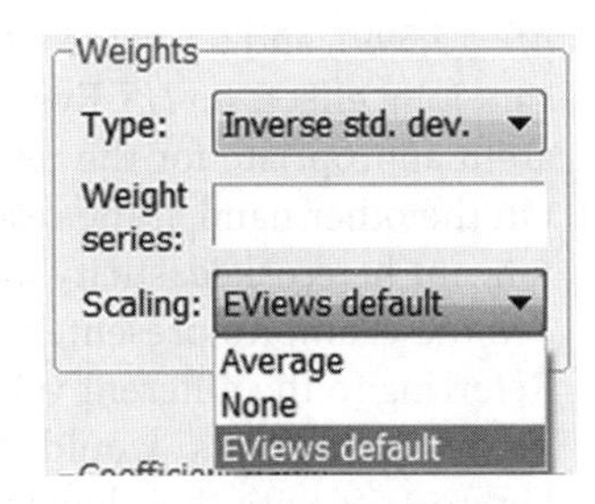

Figure 1.57 Alternative options for equation estimation.

Dependent Variable: HS									
Method: Least Squares									
Date: 08/19/14 Time: 16:17									
Sample (adjusted): 1946M03 1996M04									
Included observations: 602 after adjustments									
	Default			White hec s.e. & cov			HAC s.e & cov.		
Variable	Coef.	t-Stat.	Prob.	Coef.	t-Stat.	Prob.	Coef.	t-Stat.	Prob.
C	22.897	8.968	0.000	22.897	8.363	0.000	22.897	8.477	0.000
HS(-1)	1.161	30.193	0.000	1.161	32.991	0.000	1.161	33.619	0.000
HS(-2)	-0.346	-9.019	0.000	-0.346	-9.414	0.000	-0.346	-9.135	0.000
R-squared	0.775			0.775			0.775		
Adjusted R-squared	0.774			0.774			0.774		
F-statistic	1030.5			1030.5			1030.5		
Prob(F-statistic)	0.000			0.000			0.000		
DW Stat	2.076			2.076			2.076		
Wald F-statistic				1000.05			1096.69		
Prob(Wald F-stat)				0.000			0.000		

Figure 1.58 Statistical results summary of the LV(2) model in (1.28) using three options.

		Default				White				HAC			
		90% CI		95% CI		90% CI		95% CI		90% CI		95% CI	
Variable	Coef.	Low	High	Low	High	Low	High	Low	High	Low	High	Low	High
C	22.897	18.691	27.103	17.883	27.912	18.386	27.408	17.520	28.274	18.447	27.347	17.592	28.202
HS(-1)	1.161	1.097	1.224	1.085	1.236	1.103	1.218	1.091	1.230	1.104	1.217	1.093	1.228
HS(-2)	-0.346	-0.409	-0.282	-0.421	-0.270	-0.406	-0.285	-0.418	-0.274	-0.408	-0.283	-0.420	-0.271

	Default, CI Length		White, CI Length		HAC, CI Length	
Parameter	90%	95%	90%	95%	90%	95%
C(1)	8.413	10.029	9.021	10.755	8.900	10.610
C(2)	0.127	0.151	0.116	0.138	0.114	0.136
C(3)	0.126	0.151	0.121	0.144	0.125	0.149

Figure 1.59 Confidence intervals of the parameters using three options in Figure 1.58.

3) Since residuals (*RESID*) have significant serial correlations, as presented in Example 1.29, then the third option should be considered as the best one to conduct the testing hypotheses on the model parameters. For instance, based on the *Wald F*-statistic of $WF_0 = 1096.69$, with $df = (2,599)$, and a p value = 0.000, it can be concluded that *HS(−1)* and *HS(−2)* have significant joint effects on *HS*. Even though the default option give the same conclusion, the outputs are inappropriate for the testing hypothesis.
4) On the other hand, if the residuals have an insignificant serial correlation but they have significant heteroskedasticity in the statistical sense, then the White option should be applied. See the examples presented in Chapter 2.
5) Referring to the different values of the standard errors of the three options, then each of the model parameters would have different confidence intervals (CIs), as presented in Figure 1.59, with their length of each confidence interval. Based on this table, the following findings and notes are presented.

- I would consider the shortest *CI* of a parameter to be the best interval estimate of the corresponding parameter.
- Note that for the parameter *C*(1) the default option presents the best interval estimates. However, for *C*(2) the White option presents the best interval estimates and the HAC option presents the best interval estimate for *C*(3) only.
- So, I would say that the HAC option does not present the best interval estimates for all model parameters, even though this option is considered to be the best for estimating model parameters by taking into account the unknown heteroskedasticity and serial correlations of the residuals.

6) As additional statistical results, Figure 1.60 presents exactly the same values of the forecast evaluation statistics of the variables *HSF_Default*, *HSF_White*, and *HSF_HAC*.

Based on these findings, it can be concluded that the characteristics of the error terms of a model do not have any effect on the forecast evaluations. In addition, note that the OLS estimates of the model parameters do not in fact depend on the characteristics and assumptions of the error terms of the model. However, their forecast values are not exactly the same, as shown by their correlations in Figure 1.61. So, to forecast a time series, we do not need to conduct any test for the residual of the model. Therefore, it can be concluded that all forecasts presented in previous examples, as well as all forecasts presented in the following chapters, are acceptable in the statistical sense by using only the default coefficient covariance matrices.

Forecast: HSF_DEFAULT	
Actual: HS	
Forecast sample: 1946M01 1996M04 I...	
Included observations: 76	
Root Mean Squared Error	28.58000
Mean Absolute Error	22.49889
Mean Abs. Percent Error	26.63604
Theil Inequality Coefficient	0.124959
Bias Proportion	0.348954
Variance Proportion	0.239868
Covariance Proportion	0.411178

Forecast: HSF_WHITE	
Actual: HS	
Forecast sample: 1992M01 1996M04 I...	
Included observations: 76	
Root Mean Squared Error	28.58000
Mean Absolute Error	22.49889
Mean Abs. Percent Error	26.63604
Theil Inequality Coefficient	0.124959
Bias Proportion	0.348954
Variance Proportion	0.239868
Covariance Proportion	0.411178

Forecast: HSF_ HAC	
Actual: HS	
Forecast sample: 1946M01 1996M04 I...	
Included observations: 76	
Root Mean Squared Error	28.58000
Mean Absolute Error	22.49889
Mean Abs. Percent Error	26.63604
Theil Inequality Coefficient	0.124959
Bias Proportion	0.348954
Variance Proportion	0.239868
Covariance Proportion	0.411178

Figure 1.60 Forecast evaluations based on the three sets of statistical results in Figure 1.57.

Covariance Analysis: Ordinary
Date: 03/06/17 Time: 10:02
Sample: 1990M01 1996M04
Included observations: 76

Correlation t-Statistic Probability	HSF_DEFA...	HSF_WHITE
HSF_DEFAULT	1.000000 ----- -----	
HSF_WHITE	0.980780 43.24039 0.0000	1.000000 ----- -----
HSF_HAC	0.980780 43.24039 0.0000	1.000000 4.08E+08 0.0000

Figure 1.61 Correlation tests between the three forecast values.

	Combination_1			Combination_2			Combination_3		
Variable	**C**	**HS(-1)**	**HS(-2)**	**C**	**HS(-1)**	**HS(-2)**	**C**	**HS(-1)**	**HS(-2)**
Coef.	22.8971	1.1605	-0.3456	22.8971	1.1605	-0.3456	22.8971	1.1605	-0.3456
Std.Err.	2.5533	0.0384	0.0383	2.7380	0.0352	0.0367	2.7011	0.0345	0.0378
t-Stat.	8.9676	30.1927	-9.0186	8.3626	32.9914	-9.4136	8.4770	33.6188	-9.1349
Prob.	0.0000	0.0000	0.0000	0.0000	0.0000	0.0000	0.0000	0.0000	0.0000

Figure 1.62 The estimates of the LV(2) model (1.28) using three combinations of options.

1.7.2 Application of Selected Combinations of Options

Referring to the alternative options for equation estimation in Figure 1.57, we could have $3 \times 5 \times 3 = 45$ combinations of options. However, Example 1.31 presents statistical results using three selected combinations of the options.

Example 1.31 Application of Three Selected Options
Figure 1.62 presents the statistical result estimates of the LV(2) model (1.28) using three selected combinations of the Covariance coefficient matrix, Weight Type, and Scaling options, as follows:

Combination 1: Estimation default, Inverse Std. Dev., and Average
Combination 2: White, Inverse variance, and Average
Combination 3: HAC, Std. Dev., and Average

Note that the three combinations also present the same estimates of the model parameters as the basic OLS regression, as well as the three options presented in Figure 1.59. And it is also found that they present the same values of forecast evaluation statistics as presented in Figure 1.60. So, the results are not presented.

Example 1.32 Computing the Forecast Values
Based on the results in previous examples, we have the following regression function, which is used to compute the in-sample forecast values, However, it can be obtained directly by clicking the object/option "*FORECAST*."

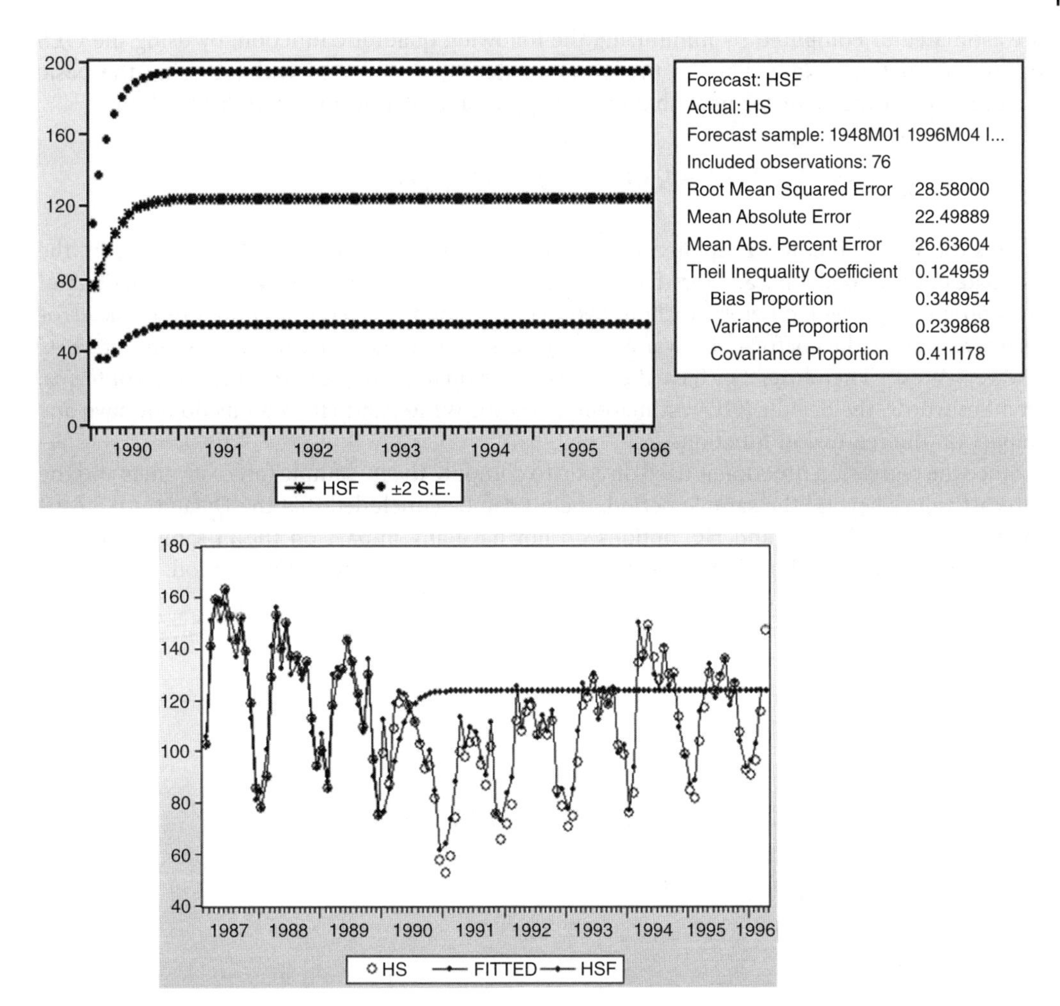

Figure 1.63 Forecast evaluation and the graphs showing *HS*, *Fitted*, and *HSF*.

$$H\hat{S}(t) = 22.8971115345 + 1.16050972982 * HS(t-1) - 0.345613835037 * HS(t-2)$$

Figure 1.63a presents its forecast evaluation for {@*Year*>1989} and Figure 1.63b presents the graphs of observed scores of *HS*, *Fitted* (Values), and *HSF* (in-sample forecast values).

1.7.3 Final Notes and Conclusions

Based on all findings presented in this chapter, I have the following final notes and conclusions. In fact, the estimate of the parameters of a LV(p) model for $p \geq 1$, or a dynamic model of a time series Y_t, are obtained by minimizing a function of the sum of its squared error terms. As an illustration, the parameters of a LV(1) model, namely:

$$Yt = \beta 0 + \beta 1 * Y_t(t-1) + \beta 2 * X1t + \beta 3 * X2t + \varepsilon t,$$

are estimated or computed by minimizing the following quadratic function, by using the OLS estimation method. Note that the model parameters are estimated without using the basic assumptions of the error terms with an independent and identical normal distribution.

$$Q = \sum_{t=2}^{T} \varepsilon_t^2 = \sum_{t=2}^{T} (Y - \beta_0 - \beta_1 * Y(-1) - \beta_2 * X1 - \beta_3 * X2)^2$$

This estimation method is applied for all combinations of the options. For this reason, the estimates of the model parameters have exactly the same values by using all combinations of the options, as presented before, without taking into account the basic assumptions of the error terms, as well as the coefficient covariance matrices, presented or assumed by White or HAC (Newey–West). Therefore, they have the same regression functions and residuals or error terms. In other words, the default (OLS estimation method), White, and HC options do not have any impact on the regression functions.

Since the regression function is used directly to compute the in-sample forecast values and the forecast values beyond the sample period, then it can be concluded that the Default (OLS estimation method), White, and HC options do not have any impact on the in-sample forecast values, as presented in Figure 1.60, and forecast values beyond the sample period.

2

Forecasting with Time Predictors

2.1 Introduction

Various models with the numerical time, t, or the time dummies as independent variables have been presented in Agung (2014, 2009a), Brooks (2008), Gujarati (2003), Wooldridge (2002), Hankle and Reitch (1992), and Wilson and Keating (1994), as well as the EViews User Guide. Now, referring to the forecast of a monthly time series, *HS*, presented in previous chapter, we have two additional time variables, namely *YEAR* and *MONTH*, which can be treated as additional numerical or ordinal variables in forecasting, as well as categorical variables. In general, for all monthly time series data, two time variables can easily be generated as the variables *YEAR* = *@Year* and *MONTH* = *@Month*, in addition to time t = *@Trend*, or t = (*@Trend* + *1*). However, the functions *@Year*, *@Month*, and *@Trend* can in fact be used directly in presenting the equation specification (ES) of the corresponding models. See the following examples.

As the extension of all LVAR(p,q) models of *HS* previously presented, we can easily conduct the forecasting based on a lot of LVAR(p,q) models of *HS* on the numerical or categorical variables *YEAR* and *MONTH*, either a basic or common LVAR(p,q) model, or a special LVAR(p,q) model. Note that for $q = 0$, and $p > 0$, we would have various LV(p) models, and various AR(q) models for $p = 0$, and $q > 0$. However, this section presents only a few illustrative examples of LVAR(p,q) models.

2.2 Application of LV(p) Models of *HS* on *MONTH* by *YEAR*

Several or many possible basic or common LV(p) models of a monthly time series *HS* on the numerical variable *MONTH=@Month* by *YEAR=@Year* can easily be applied, however, we cannot predict which one gives the best possible forecast, so we should be using the trial-and-error method. See the following examples for selected models only.

2.2.1 Special LV(12) Models of *HS* on *MONTH* by *YEAR*

Example 2.1 A Special LV(12) Model of *HS* on *MONTH* by *YEAR*
Referring to the curve of *HS* on the time t = *@Trend* + *1* = 1, 2, … , 604, which shows a parabolic curve within each year, then I am considering a special LV(12) model of HS_t, using the following ES, for the time $t > 12$.

$$\begin{aligned} &hs\ hs(-12)\ (@month)\hat{}2*@expand(@year) \\ &@month*@expand(@year)\ @expand(@year) \end{aligned} \tag{2.1}$$

Advanced Time Series Data Analysis: Forecasting Using EViews, First Edition. I Gusti Ngurah Agung.

Table 2.1 Summary of a part of the statistical results of the LV(12) model in (1.27).

	HS(−12)		MONTH^2		MONTH		Intercept	
Year	Coef.	Prob.	Coef.	Prob.	Coef.	Prob.	Coef.	Prob.
1947	0.489	0.0000	−0.917	0.0009	17.078	0.0000	2.702	0.7922
1948	0.489	0.0000	−1.582	0.0000	17.566	0.0000	31.926	0.0018
1949	0.489	0.0000	−0.486	0.0888	11.782	0.0020	14.105	0.1696
1950	0.489	0.0000	−2.433	0.0000	28.186	0.0000	49.268	0.0000
…								
…								
1990	0.489	0.0000	−0.451	0.1033	3.456	0.3474	45.240	0.0000
1991	0.489	0.0000	−0.864	0.0016	14.048	0.0001	−8.640	0.4159
1992	0.489	0.0000	−0.630	0.0222	7.807	0.0340	41.989	0.0000
1993	0.489	0.0000	−0.745	0.0066	12.136	0.0009	19.874	0.0558
1994	0.489	0.0000	−1.152	0.0000	15.148	0.0000	32.844	0.0015
1995	0.489	0.0000	−0.580	0.0377	8.668	0.0203	28.498	0.0062
1996	0.489	0.0000	4.693	0.3372	−10.407	0.6750	55.810	0.0424

R-squared	0.9404	Mean dependent variable	123.853
Adjusted R-squared	0.9201	S.D. dependent variable	34.550
S.E. of regression	9.7643	Akaike info criterion	7.611
Sum squared residual	42,046.010	Schwarz criterion	8.729
Log likelihood	−2101.864	Hannan–Quinn criterion	8.047
Durbin–Watson stat	1.9990		

where the function *@Month* represents a numerical variable with the scores 1 to 12, and the function *@Year* in the function *@Expand(@Year)* represents an ordinal variable, but it is treated as a nominal categorical variable and as a set of year-dummies: (*@Year*=1947) = D1947 to (*@Year*=1964) = D1964.

The summary of part of the statistical results is presented in Table 2.1, with its forecast evaluation presented in Figure 2.1. Based on these results, the following findings and notes are presented.

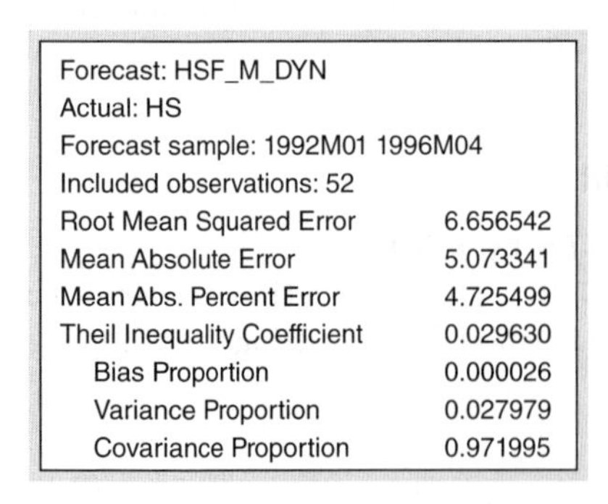

Forecast: HSF_M_DYN
Actual: HS
Forecast sample: 1992M01 1996M04
Included observations: 52

Root Mean Squared Error	6.656542
Mean Absolute Error	5.073341
Mean Abs. Percent Error	4.725499
Theil Inequality Coefficient	0.029630
Bias Proportion	0.000026
Variance Proportion	0.027979
Covariance Proportion	0.971995

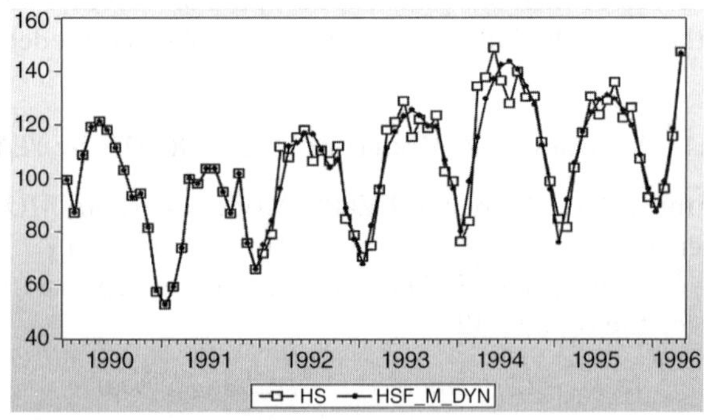

Figure 2.1 Dynamic forecast evaluation based on the special LV(12) model in (1.27).

1) Compared to the other forecasts of *HS* presented in previous chapter, this forecast has the smallest TIC = 0.029630. Therefore, this forecast can be considered to be the best forecast of *HS*.
2) The numerical variables *MONTH^2*, and *MONTH* have different slopes over the years, so that the model in (2.1) is a heterogeneous regressions model of *HS* on *MONTH*, by *YEAR*.
3) On the other hand *HS(−12)* has the same or homogeneous slopes, so the model in (2.1) is a homogeneous regressions model of *HS* on *LH(−12)*, by *YEAR*.
4) Since there are only four months in 1996, then it is better to conduct data analysis based on a subsample starting from 1947M01 up to 1995M12, or a selected *M01 up to 1995M12. So, we can have a complete set of parabolic regressions by *YEAR*. For the analysis, the ES in (2.1) also can be applied directly. Do this as an exercise.
5) However, if the joint effects of all independent variables and slopes or effects differences of *MONTH^2* or *MONTH* between selected years are to be tested, it is recommended to apply the following ES. Based on the subsample *@Year*<1996, the result that the joint effects of all independent variables are significant would be obtained, based on the *F*-statistic of $F_0 = 47.06336$ with a p value = 0.000000.

$$\begin{aligned} &hs\ C\ hs(-12)\ (@month)\hat{\ }2*@expand(@year) \\ &@month*@expand(@year)\ @expand(@year,@dropfirst) \end{aligned} \tag{2.2}$$

Then, to test the slopes differences of MONTH^2 between selected years from 1947 up to 1995, the parameters *C(3)* for the year 1947 up to *C(51)* for the year 1995, should be considered – refer to the output obtained by selecting *View/Representations*. For instance, to test of the slopes differences within the last 5 years, namely 1991 up to 1995, the null hypothesis is H_{01}: *C(47) = C(48) = C(49) = C(50) = C(51)* is accepted based on the *F*-statistic of $F_0 = 0.726545$ with *df* = (4,440), and *p value* = 0.5742, but the null hypothesis on the slope differences with the first 5 years, namely H_{02}: *C(3) = C(4) = C(5) = C(6) = C(7)*, is rejected based on the *F*-statistic of $F_0 = 15.77652$ with *df* = (4,440), and *p value* = 0.0000.

2.2.2 Application of the Omitted Variables Test – Likelihood Ratio

As a more advanced data analysis based on the model in (2.1), the following examples present the application of the omitted variables tests (OVT) corresponding to the numerical variables *@month^2*, *@month*, and *HS(−12)*.

Example 2.2 Testing the Hypothesis on the Effect of *@month^2*@expand(@year)*
To test this hypothesis, we have to consider the reduced model (Red_M) of the full model in (2.1), with the following ES.

$$\begin{aligned} &Red\_M1: hs\ C\ hs(-12)\ @month*@expand(@year) \\ &\qquad @expand(@year,@dropfirst) \end{aligned} \tag{2.3}$$

Then the statistical hypothesis can be written as follows:

$$H_0: The\ Red\_M1, \text{ versus } H_1: The\ full\ model\ in\ (2.2) \tag{2.4}$$

Omitted Variables Test
Equation: UNTITLED
Specification: HS C HS(-12) @MONTH*@EXPAND(@YEAR)
@EXPAND(@YEAR,@DROPFIRST)
Omitted Variables: @MONTH^2*@EXPAND(@YEAR)

	Value	df	Probability
F-statistic	8.221550	(49, 440)	0.0000
Likelihood ratio	382.2125	49	0.0000

F-test summary:

	Sum of Sq.	df	Mean Squares
Test SSR	38486.61	49	785.4411
Restricted SSR	80521.76	489	164.6662
Unrestricted SSR	42035.15	440	95.53443
Unrestricted SSR	42035.15	440	95.53443

LR test summary:

	Value	df
Restricted LogL	-2280.685	489
Unrestricted LogL	-2089.579	440

Figure 2.2 Output for testing the hypothesis (2.3).

The steps of the data analysis are as follows:

1) Conduct the data analysis based on the reduced model in (2.3).
2) With the statistical results of the reduced model on screen, select *View/Coefficient Diagnostic/Omitted Variables Test-Likelihood Ratio ...,* then by inserting *@month^2*@expand (@year) ... OK,* the output in Figure 2.2 is obtained. Based on this output the following findings and notes are presented.

   - Note this output presents the ES of the reduced model, as well as the omitted variables.
   - The null hypothesis in (2.4) is rejected based on the *F*-statistic of $F_0 = 8.221550$ with $df = (49{,}440)$, and $p\ value = 0.0000$. So it can be concluded that the full model in (2.2) is an acceptable model.

Example 2.3 Testing the Hypothesis that *@month^2* has Homogeneous Slopes
To test the hypothesis that *@month^2* has the same or homogeneous slopes over the years, we have to consider the reduced model (Red_M2) of the full model in (2.2), with the following ES.

$$\begin{aligned} Red\_M2 : hs\ C\ hs(-12)\ @month\hat{\ }2\ @month*@expand(@year) \\ @expand(@year, @dropfirst) \end{aligned} \tag{2.5}$$

Then the statistical hypothesis can be written as follows:

$$H_0 : The\ Red\_M2,\ \text{versus}\ H_1 : The\ full\ model\ in\ (2.2) \tag{2.6}$$

The steps of the data analysis are as follows:

1) Conduct the data analysis based on the reduced model in (2.5). Compared to the model in (2.1) with the output in Table 2.1, this model only has a single parameter for the numerical variable *@month^2*.

Omitted Variables Test
Equation: EQ07
Specification: HS C HS(-12) @MONTH^2 @MONTH
*@EXPAND(@YEAR) @EXPAND(@YEAR,@DROPFIRST)
Omitted Variables: @MONTH^2*@EXPAND(@YEAR,@DROPFIRST)

	Value	df	Probability
F-statistic	4.405218	(48, 440)	0.0000
Likelihood ratio	230.7469	48	0.0000

F-test summary:

	Sum of Sq.	df	Mean Squares
Test SSR	20200.80	48	420.8500
Restricted SSR	62235.95	488	127.5327
Unrestricted SSR	42035.15	440	95.53443
Unrestricted SSR	42035.15	440	95.53443

LR test summary:

	Value	df
Restricted LogL	-2204.952	488
Unrestricted LogL	-2089.579	440

Figure 2.3 Output for testing the hypothesis (2.6).

2) With the statistical results of the reduced model on the screen, select *View/Coefficient Diagnostic/OVT-Likelihood Ratio ...*, then by inserting
3) *@month^2∗@expand(@year,@dropfirst)... OK*, the output in Figure 2.3 is obtained. Based on this output, the following findings and notes are presented.

- Note this output presents the ES of the *Red_M2*, as well as the omitted variables.
- The null hypothesis in (2.6) is rejected based on the *F*-statistic of $F_0 = 4.405218$ with $df = (48,440)$, and p *value* $= 0.0000$. So it can be concluded the quadratic variable *@Month^2* has significantly different effects on *HS* over the years.

Example 2.4 Testing the Hypothesis that *@month* has Homogeneous Slopes
Similar to the testing presented in previous example, to test this hypothesis, the following reduced model (Red_M3) should be considered.

$$Red\_M3: hs\ C\ hs(-12)\ @month\ @month\hat{\ }2*\ @expand(@year)\ @expand(@year, @dropfirst) \tag{2.7}$$

Then the statistical hypothesis can be written as follows:

$$H_0: The\ Red\_M3, versus\ H_1:\ The\ full\ model\ in\ (2.2) \tag{2.8}$$

And the result is obtained that the null hypothesis is rejected based on the *F*-statistic of $F_0 = 3.991075$, with $df = (48,440)$, and p value $= 0.0000$.

Therefore, based on the conclusion of the testing hypotheses (2.6), and (2.8), we can reach the conclusion that the model in (2.1) using the subsample (*@Year*<1996) is a significant heterogeneous regression of *HS* on the both numerical variables *@Month^2*, and *@Month*.

Example 2.5 Testing the Hypothesis that *HS(−12)* Has Heterogeneous Slopes
In this case, the full model in (2.2) is considered a reduced model of an upper model (Up_M), with the ES.

$$\begin{gathered} Up\_M : hs\ C\ hs(-12)\ hs(-12)*@expand(@year,@dropfirst) \\ (@month)\hat{}2*@expand(@year) \\ @month*@expand(@year)@expand(@year,@dropfirst) \end{gathered} \tag{2.9}$$

which is obtained by inserting *hs(−12)*@expand(year,@dropfirst)* in the ES (2.1). Then the statistical hypothesis can be presented as follows:

$$H_0 : The\ full\ model\ in\ (2.2),\ \text{versus}\ H_1 : The\ Up\_M\ in\ (2.9) \tag{2.10}$$

The steps of the data analysis are as follows:

1) Conduct the data analysis based on the full model in (2.2).
2) With the statistical results of the full model on the screen, select *View/Coefficient Diagnostic/OVT-Likelihood Ratio ...*, then by inserting *HS(−12)*@expand(@year,@dropfirst) ... OK*, the output in Figure 2.4 is obtained. Based on this output the following findings and notes are presented.
   - Note this output presents the ES of the full model in (2.2), as well as the omitted variables.
   - The null hypothesis in (2.10) is rejected based on the *F*-statistic of $F_0 = 1.619663$, with $df = (48,392)$, and *p value* = 0.0076. So it can be concluded that *HS(−12)* has a significant effect on *HS* over the years.

Omitted Variables Test
Equation: UNTITLED
Specification: HS C HS(-12) (@MONTH)^2*@EXPAND(@YEAR)
@MONTH*@EXPAND(@YEAR) @EXPAND(@YEAR,@DROPFIRST)
Omitted Variables: HS(-12)*@EXPAND(@YEAR,@DROPFIRST)

	Value	df	Probability
F-statistic	1.619663	(48, 392)	0.0076
Likelihood ratio	106.3843	48	0.0000

F-test summary:

	Sum of Sq.	df	Mean Squares
Test SSR	6956.928	48	144.9360
Restricted SSR	42035.15	440	95.53443
Unrestricted SSR	35078.22	392	89.48526
Unrestricted SSR	35078.22	392	89.48526

LR test summary:

	Value	df
Restricted LogL	-2089.579	440
Unrestricted LogL	-2036.387	392

Figure 2.4 Output for testing the hypothesis (2.6).

2.2.3 Heterogeneous Model of *HS* on *HS(−12)* and *Month* by *YEAR*

Based on the findings presented previously, we can have the final model in (2.9), which is a heterogeneous regressions model of *HS* on the two numerical variables *HS(−12)* and *MONTH*, by *YEAR*.

Example 2.6 Forecast Evaluation Based on the Model in (2.9)
As a comparison with the forecast evaluation based on the model in (2.1) using the whole sample, and an extension of the forecast in Figure 1.2, Figure 2.5 presents the forecast evaluation based on the model in (2.9) also based on the whole sample. Based on these results, the following findings and notes are presented.

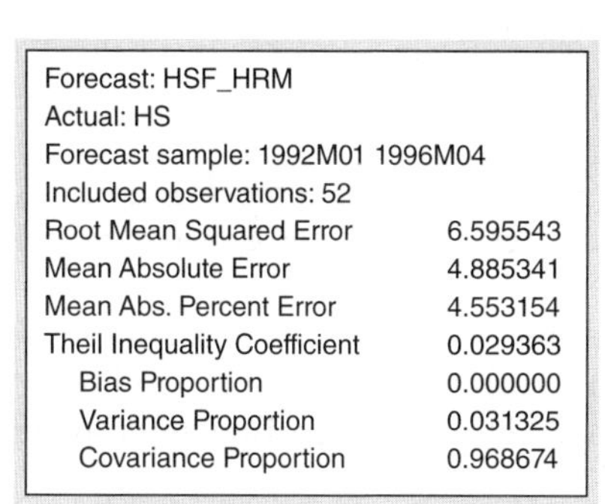

Forecast: HSF_HRM
Actual: HS
Forecast sample: 1992M01 1996M04
Included observations: 52

Root Mean Squared Error	6.595543
Mean Absolute Error	4.885341
Mean Abs. Percent Error	4.553154
Theil Inequality Coefficient	0.029363
Bias Proportion	0.000000
Variance Proportion	0.031325
Covariance Proportion	0.968674

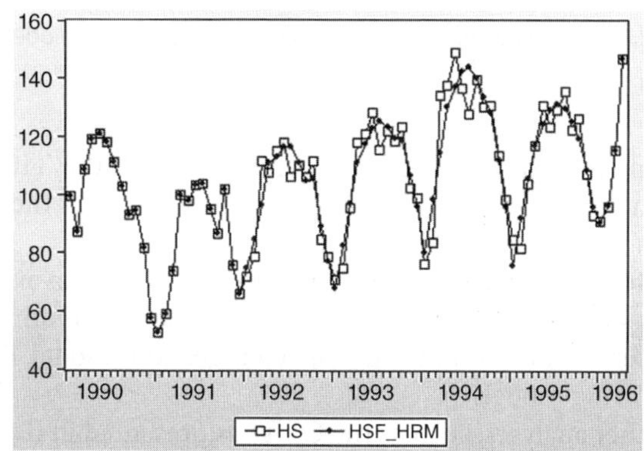

Figure 2.5 Forecast evaluation based on the model in (2.9) using the whole sample.

1) This forecast has a TIC = 0.029 383, which is a slightly smaller than the forecast in Figure 2.1, with the TIC = 0.029 630. So this forecast is a better forecast for *HS*, in a statistical sense.
2) So, a researcher would have a choice whether he/she want to apply a simpler LV(12) model in (2.1) with a little more TIC, or a more complex model in (2.9). Compare with the simplest LV(1) model presented in the following example, which has a slightly greater TIC = 0.031001 than the model (2.1).

2.3 Forecast Models of *HS* on *MONTH* by *YEAR*

2.3.1 Application of LV(1) Models of *HS* on *MONTH* by *YEAR*

Example 2.7 LV(1) Heterogeneous Model of *HS* on *MONTH* by *YEAR*
As additional comparison, Figure 2.6 presents the forecast evaluation of the simplest LV(p) heterogeneous model of *HS* on *@Month* by *@Year*, say for $p = 1$, using the following ES. Based on these statistical results, the following findings and notes are presented.

$$hs\ C\ hs(-12)\ (@month)\hat{\ }2*@expand(@year) \\ @month*@expand(@year)@expand(@year,@dropfirst) \tag{2.11}$$

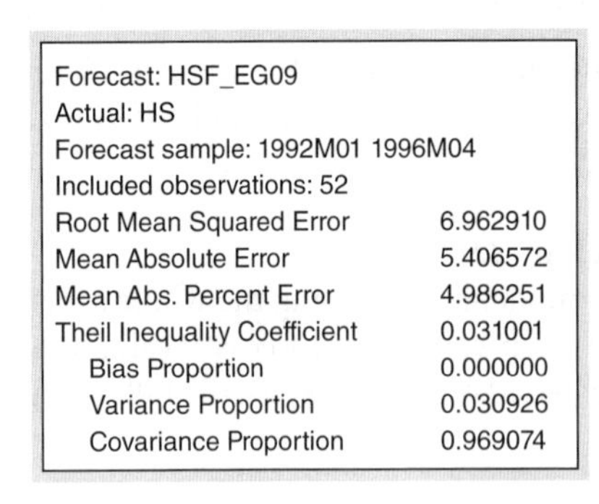

Forecast: HSF_EG09	
Actual: HS	
Forecast sample: 1992M01 1996M04	
Included observations: 52	
Root Mean Squared Error	6.962910
Mean Absolute Error	5.406572
Mean Abs. Percent Error	4.986251
Theil Inequality Coefficient	0.031001
Bias Proportion	0.000000
Variance Proportion	0.030926
Covariance Proportion	0.969074

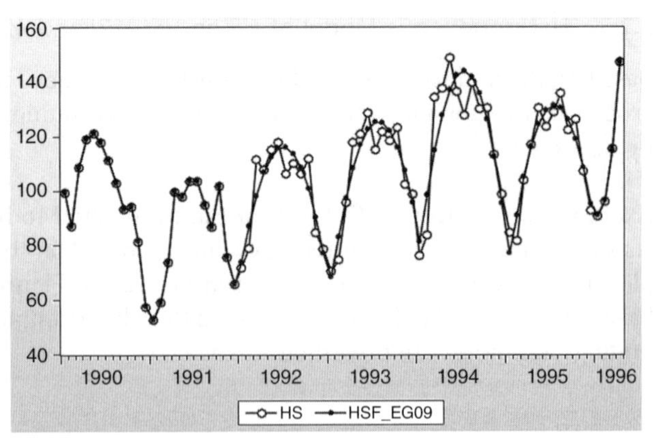

Figure 2.6 Forecast evaluation of the LV(1) heterogeneous model in (2.11).

1) Compared to the forecast in Figure 2.1, with the TIC = 0.029630, this forecast has only a slightly greater TIC = 0.031001, but it is based on the simplest LV(p) heterogeneous model of *HS* on *@Month* by @Year.
2) If a forecast with a TIC < 0.05 is considered to be an excellent or very good forecast, this forecast should be accepted as one of the best. Therefore, a researcher does not have to find or present a more complex LV(p) heterogeneous model. For additional comparison, see also the following models
3) Further analyses, such as the OVTs as presented in the previous example, are recommended to be done as exercises. However, the following examples present alternative reduced models without considering the omitted tests based on the model in (2.11), since the models are commonly applied in practice.

Example 2.8 A LV(1) ANCOVA Model of *HS* by *YEAR*, with Numerical Covariates *HS (−1)*, *MONTH^2*, and *MONTH*

As an additional illustration, Figure 2.7 presents the statistical results and forecast evaluation based on a reduced model of the LV(1) heterogeneous model in (2.11), namely an ANCOVA model of *HS* with *YEAR* = *@Year* as a categorical factor, and numerical covariates: *HS(−1)*, *@Month^2*, and *@Month*, using the following ES. Based on this model, and the results in Figure 2.7, the following findings and notes are presented.

$$hs\ C\ hs(-1)\ (@month)^{\wedge}2\ @month\ @expand(@year,@dropfirst) \qquad (2.12)$$

1) Referring the model based on panel data (Agung 2014), this ANCOVA model can be considered to be a year-fixed-effects model, or time-fixed-effects model (TFEM) in general. Note this ANCOVA model is a quadratic model of *HS* on the numerical variable *@Month*. However, it has been found ANCOVA additive models with linear covariate(s) have been presented in various books and journals, as well as theses and dissertations. See the following example.
2) The forecast has a TIC = 0.043214 < 0.05 – refer to the notes presented in the previous example and the set of intervals of TIC recommended in Chapter 1. The statistical results of the model are not presented. It can be obtained easily by the readers, if needed.

Dependent Variable: HS
Method: Least Squares
Date: 01/07/14 Time: 16:18
Sample (adjusted): 1946M02 1996M04
Included observations: 603 after adjustments

Variable	Coefficient	Std. Error	t-Statistic	Prob.
C	11.64289	4.354507	2.673757	0.0077
HS(-1)	0.464377	0.033962	13.67355	0.0000
(@MONTH)^2	-1.254737	0.070579	-17.77764	0.0000
@MONTH	15.58841	1.017861	15.31487	0.0000
Year Dummies	Yes			

R-squared	0.874459	Mean dependent var	123.1819
Adjusted R-squared	0.862340	S.D. dependent var	34.64025
S.E. of regression	12.85244	Akaike info criterion	8.030230
Sum squared resid	90686.65	Schwarz criterion	8.424431
Log likelihood	-2367.114	Hannan-Quinn criter.	8.183650
F-statistic	72.15250	Durbin-Watson stat	1.794374
Prob(F-statistic)	0.000000		

Forecast: HSF_EQ10
Actual: HS
Forecast sample: 1992M01 1996M04
Included observations: 52

Root Mean Squared Error	9.734810
Mean Absolute Error	7.923121
Mean Abs. Percent Error	7.289973
Theil Inequality Coefficient	0.043214
Bias Proportion	0.000419
Variance Proportion	0.018883
Covariance Proportion	0.980698

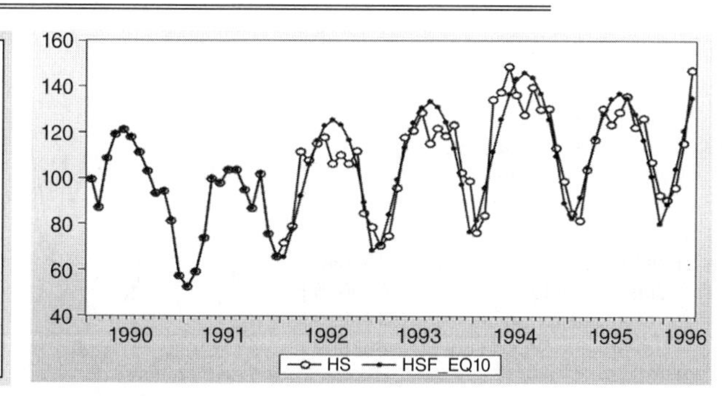

Figure 2.7 Statistical results and forecast evaluation of the LV(1) heterogeneous model in (2.12).

Example 2.9 A LV(1) ANCOVA Additive Model of *HS*

As an illustration, Figure 2.8 presents the statistical results and forecast evaluation of a LV(1) ANCOVA additive linear model of *HS* using the following ES. Based on these results, the following notes and comments are presented.

$$hs\ C\ hs(-1)\ @month\ @expand(@year, @dropfirst) \tag{2.13}$$

1) The parameter estimates show that the numerical independent variables have a significant adjusted effect on *HS*. Based on my observations, I would say most students and less experienced researchers would be satisfied with the estimates, moreover, to see such a large $R^2 = 0.8028129$.
2) It is recognized that the lagged variable models or the autoregressive models should have a large R-squared, it could be greater than 0.95.
3) Since the data shows the parabolic graph of *HS* on *@month* within each year, therefore, I would say that small *p* values of all independent variables or large R^2 do not directly mean the model is a good fit.
4) Furthermore, I would consider the model in (2.13) is not recommended but also not the worst model, compared to all previous models, even though its forecast has a TIC = 0.063640 < 0.10, because it uses the unrealistic assumption that *@Month* has a homogeneous of the

Dependent Variable: HS
Method: Least Squares
Date: 01/07/14 Time: 16:45
Sample (adjusted): 1946M02 1996M04
Included observations: 603 after adjustments

Variable	Coefficient	Std. Error	t-Statistic	Prob.
C	25.20865	5.376553	4.688627	0.0000
HS(-1)	0.892482	0.030033	29.71634	0.0000
@MONTH	-2.240364	0.218231	-10.26602	0.0000
Year Dummies	Yes			

R-squared	0.802189	Mean dependent var	123.1819
Adjusted R-squared	0.783487	S.D. dependent var	34.64025
S.E. of regression	16.11845	Akaike info criterion	8.481595
Sum squared resid	142892.5	Schwarz criterion	8.868497
Log likelihood	-2504.201	Hannan-Quinn criter.	8.632175
F-statistic	42.89288	Durbin-Watson stat	1.863621
Prob(F-statistic)	0.000000		

Forecast: HSF_EQ11
Actual: HS
Forecast sample: 1992M01 1996M04
Included observations: 52

Root Mean Squared Error	14.37300
Mean Absolute Error	11.24815
Mean Abs. Percent Error	11.41024
Theil Inequality Coefficient	0.063640
Bias Proportion	0.019025
Variance Proportion	0.175286
Covariance Proportion	0.805689

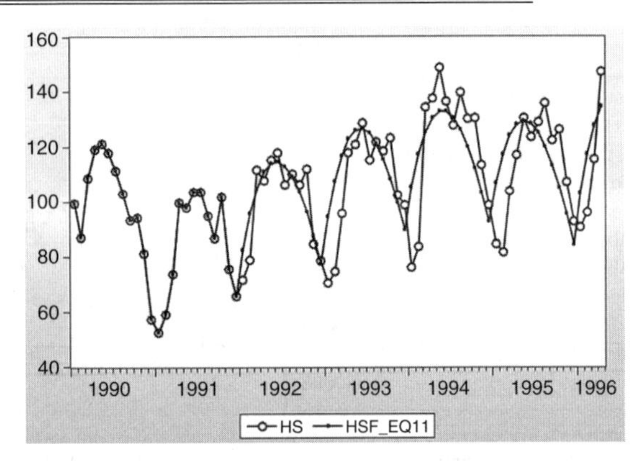

Figure 2.8 Statistical results and forecast evaluation of the ANCOVA additive model in (1.13).

same slope over the years. However, I have found many researchers present additive linear models. An extreme example is an additive model with eight distinct dichotomous (which are represented by eight dummies) and eight numerical independent variables presented by Li et al. (2010), and Brooks (2008) presents a model with two dummies and seven numerical variables. Moreover, for the LV(1) additive linear model with all numerical independent variables, including the model of *HS* on three numerical independent variables *HS(−1)*, *@Year*, and *@Month*, represented by the following ES. Refer to various alternative interaction models, such as heterogeneous, ANCOVA, or fixed-effects models, and random effects models with special notes and comments presented in Agung (2014, 2009a, 2011a).

$$Hs\ C\ hs(-1)\ @year\ @month \tag{2.14}$$

2.3.2 Application of Basic LV(*p*) Models of *HS* on *MONTH* by *YEAR*

As an extension of the general model in (1.1), this section presents the models using the numerical variable *@Month*, and categorical variable *@Year* as additional predictors. Therefore, similar

Dependent Variable: HS
Method: Least Squares
Date: 01/08/14 Time: 15:54
Sample (adjusted): 1946M07 1996M04
Included observations: 598 after adjustments

Variable	Coefficient	Std. Error	t-Statistic	Prob.
C	76.03754	137.1729	0.554319	0.5796
HS(-1)	0.116180	0.042063	2.762045	0.0060
HS(-2)	-0.023194	0.043062	-0.538624	**0.5904**
HS(-3)	-0.031986	0.038421	-0.832499	**0.4056**
HS(-4)	-0.265427	0.038095	-6.967556	0.0000
HS(-5)	0.162944	0.037166	4.384230	0.0000
HS(-6)	0.060485	0.039857	1.517568	0.1298
@Month^2*@Expand(@YEAR)	Yes			
@Month*@Expand(@YEAR)	Yes			
@Expand(@YEAR)	Yes			

R-squared	0.945322	Mean dependent var	123.4376
Adjusted R-squared	0.925643	S.D. dependent var	34.64705
S.E. of regression	9.447723	Akaike info criterion	7.552106
Sum squared resid	39184.91	Schwarz criterion	8.720301
Log likelihood	-2099.080	Hannan-Quinn criter.	8.006929
F-statistic	48.03694	Durbin-Watson stat	2.565305
Prob(F-statistic)	0.000000		

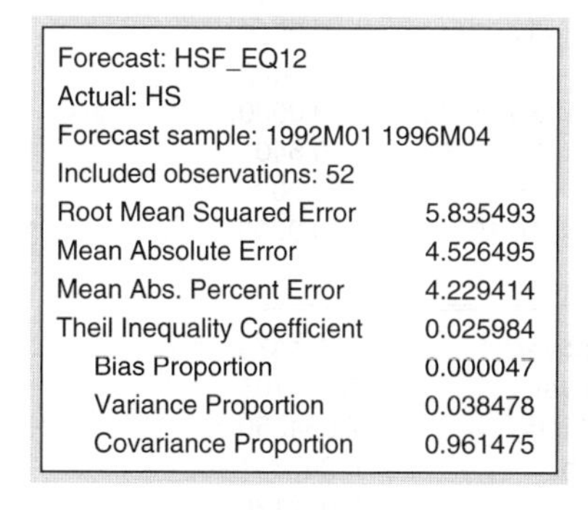

Forecast: HSF_EQ12
Actual: HS
Forecast sample: 1992M01 1996M04
Included observations: 52

Root Mean Squared Error	5.835493
Mean Absolute Error	4.526495
Mean Abs. Percent Error	4.229414
Theil Inequality Coefficient	0.025984
Bias Proportion	0.000047
Variance Proportion	0.038478
Covariance Proportion	0.961475

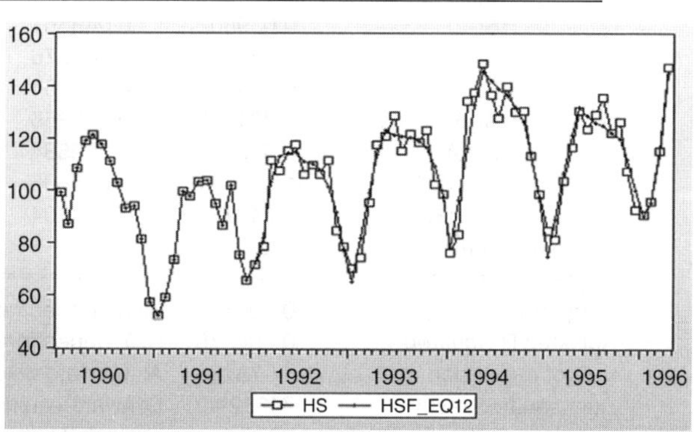

Figure 2.9 Statistical results with forecast evaluation based on the model in (2.15).

to all previous models, a lot of possible models can be applied. However, only a few selected models will be presented.

Example 2.10 Application of a Basic LV(6) Heterogeneous Model of *HS*
As an extension of the LV(1) model in (1.1), and a modification of the instrumental variables model in (1.25), Figure 2.9 presents the forecast evaluation of a basic LV(6) heterogeneous quadratic model of HS_t on *@Month* by *@Year* with the ES (2.15), for the time $t > 6$. Based on these results, the following findings and notes are presented.

$$\begin{aligned} &hs\ C\ hs(-1)...hs(-6)\ (@month)\hat{\ }2*@expand(@year) \\ &@month*@expand(@year)\ @expand(@year, @dropfirst) \end{aligned} \tag{2.15}$$

1) This forecast has the smallest TIC = 0.025 984 compared to previous forecast, so it can be said this forecast is the best among all forecasts presented in this chapter and Chapter 1.
2) However, the estimates show each one of *HS(−2)* and *HS(−3)* has a large p value > 0.4, even though the model can be considered to be a good fit. On the other hand, if these two lags are deleted from the model, then the remaining lags *HS(−1), HS(−4), HS(−5),* and *HS(−6)* have p values of 0.0149, 0.0000, 0.0000, and 0.0082, respectively, with a TIC = 0.026267 for its forecast.
3) On the other hand, almost all variables in *@Month^2∗@Expand(@Year)* and *@Month∗ @Expand(@Year)* have very small p values < 0.001, except for *@Year*=1946, and *@Year*=1996.

Dependent Variable: HS
Method: Least Squares
Date: 01/08/14 Time: 20:42
Sample (adjusted): 1946M07 1996M04
Included observations: 598 after adjustments

Variable	Coefficient	Std. Error	t-Statistic	Prob.
C	0.731189	6.876210	0.106336	0.9154
HS(-1)	0.551568	0.042978	12.83370	0.0000
HS(-2)	0.039802	0.048561	0.819631	**0.4128**
HS(-3)	-0.140073	0.046576	-3.007414	0.0028
HS(-4)	-0.174138	0.045764	-3.805154	0.0002
HS(-5)	0.261385	0.045255	5.775812	0.0000
HS(-6)	0.018517	0.039584	0.467797	**0.6401**
@MONTH^2	-1.312441	0.085606	-15.33115	0.0000
@MONTH	16.42785	1.157204	14.19616	0.0000
Year Dummies	Yes			

R-squared	0.895608	Mean dependent var	123.4376
Adjusted R-squared	0.884375	S.D. dependent var	34.64705
S.E. of regression	11.78128	Akaike info criterion	7.864350
Sum squared resid	74812.40	Schwarz criterion	8.297831
Log likelihood	-2292.441	Hannan-Quinn criter.	8.033120
F-statistic	79.72823	Durbin-Watson stat	2.053224
Prob(F-statistic)	0.000000		

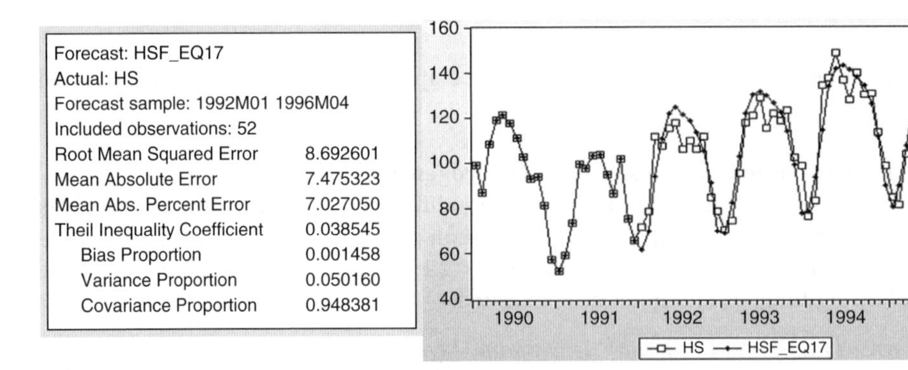

Figure 2.10 Statistical results with forecast evaluation based on the model in (2.16).

Example 2.11 The Basic LV(6) Year-Fixed-Effects Model of *HS*

As a reduced model of the model in (2.15), and an extension of the LV(1) year-fixed-effects model of *HS* in (2.12), Figure 2.10 presents the statistical results of a LV(6) year-fixed-effects model of *HS*, using the ES as follows:

$$hs\ C\ hs(-1)...hs(-6)\ @month\hat{\ }2\ @month\ @expand(@year,@dropfirst) \tag{2.16}$$

Based on these statistical results, the following findings and notes are presented.

1) Even though *HS(−2)* and *HS(−6)* have large *p* values, the model is an acceptable model as a basic LV(6) ANCOVA or TFEM, in a theoretical sense, under the assumption that all *HS(−p)*, *@Month^2*, and *@Month* have the same slopes over the years.
2) However, one might want to reduce the model to a basic LV(5) in the first stage. Do this as an exercise.

Dependent Variable: HS
Method: Least Squares
Date: 01/08/14 Time: 21:03
Sample (adjusted): 1946M06 1996M04
Included observations: 599 after adjustments

Variable	Coefficient	Std. Error	t-Statistic	Prob.
C	1.526673	5.791868	0.263589	0.7922
HS(-1)	0.572902	0.035401	16.18335	0.0000
HS(-3)	-0.125696	0.036994	-3.397697	0.0007
HS(-4)	-0.174882	0.045531	-3.840933	0.0001
HS(-5)	0.267958	0.035370	7.575944	0.0000
@MONTH^2	-1.286274	0.072445	-17.75513	0.0000
@MONTH	16.14163	1.003407	16.08682	0.0000
Year Dummies	Yes			

R-squared	0.895501	Mean dependent var	123.3935
Adjusted R-squared	0.884704	S.D. dependent var	34.63492
S.E. of regression	11.76036	Akaike info criterion	7.857667
Sum squared resid	74961.84	Schwarz criterion	8.275914
Log likelihood	-2296.371	Hannan-Quinn criter.	8.020494
F-statistic	82.94053	Durbin-Watson stat	2.098045
Prob(F-statistic)	0.000000		

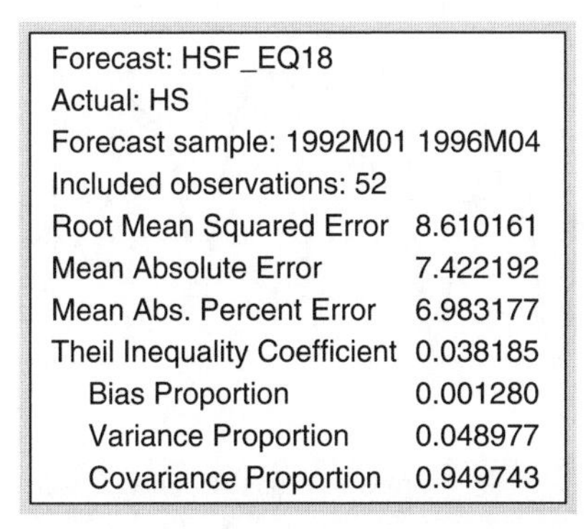

Forecast: HSF_EQ18	
Actual: HS	
Forecast sample: 1992M01 1996M04	
Included observations: 52	
Root Mean Squared Error	8.610161
Mean Absolute Error	7.422192
Mean Abs. Percent Error	6.983177
Theil Inequality Coefficient	0.038185
Bias Proportion	0.001280
Variance Proportion	0.048977
Covariance Proportion	0.949743

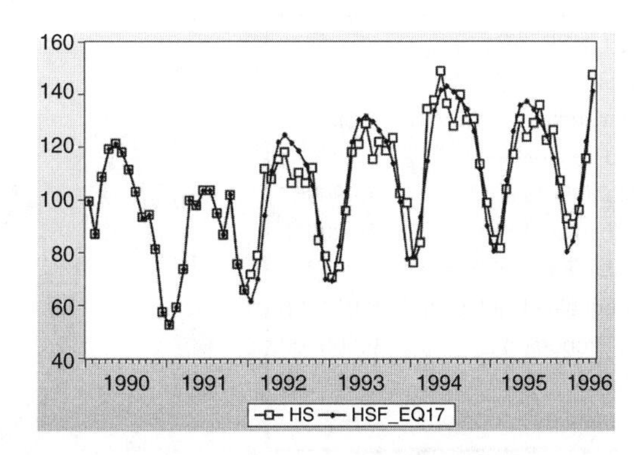

Figure 2.11 Statistical results with forecast evaluation based on a reduced model in (2.16).

3) Unexpected results are obtained by deleting both lags *HS(−2)* and *HS(−6)*, as presented in Figure 2.11, which shows its forecast has a slightly smaller RMSE = 8.610161 and TIC = 0.038158 compared to the RSME = 8.692601 and TIC = 0.038545 based on the full LV(6) model. Therefore, which one would you prefer?

2.3.3 Application of AR(q) Models of *HS* on *MONTH* by *YEAR*

As a modification of the LV(p) models, in this case, the AR(q) models considered would have the following general equation, for any integer $q > 0$, with all its alternative reduced models as presented for the LV(1) model in (2.11). However, only a few examples are presented in this section.

$$\begin{aligned} &hs\ C\ @month\hat{\ }2*@expand(@year)\ @month*@expand(@year) \\ &\qquad @expand(@year,@dropfirst)\ ar(1)...ar(q) \end{aligned} \tag{2.17}$$

Dependent Variable: HS
Method: Least Squares
Date: 01/07/14 Time: 17:17
Sample (adjusted): 1946M02 1996M04
Included observations: 603 after adjustments
Convergence achieved after 8 iterations

Variable	Coefficient	Std. Error	t-Statistic	Prob.
C	47.84389	17.19692	2.782120	0.0056
@MONTH^2*@EXPAND(@YEAR)	Yes			
@MONTH*@EXPAND(@YEAR)	Yes			
@EXPAND(@YEAR,@DROPFIRST)	Yes			
AR(1)	0.011939	0.049106	0.243128	**0.8080**

R-squared	0.922929	Mean dependent var	123.1819
Adjusted R-squared	0.896667	S.D. dependent var	34.64025
S.E. of regression	11.13527	Akaike info criterion	7.873997
Sum squared resid	55673.40	Schwarz criterion	8.998201
Log likelihood	-2220.010	Hannan-Quinn criter.	8.311529
F-statistic	35.14263	Durbin-Watson stat	1.994850
Prob(F-statistic)	0.000000		

Inverted AR Roots	.01

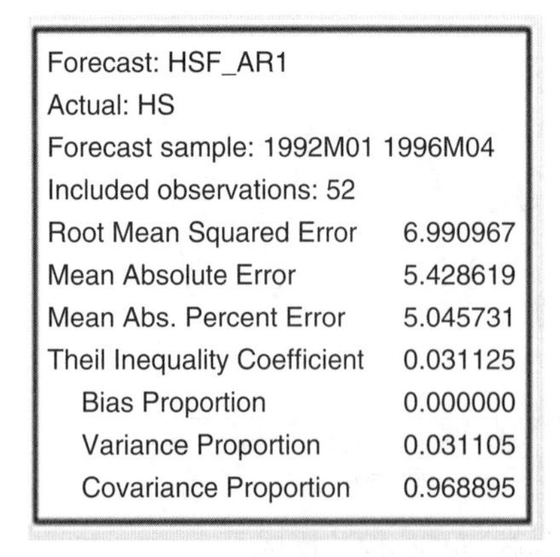

Forecast: HSF_AR1
Actual: HS
Forecast sample: 1992M01 1996M04
Included observations: 52

Root Mean Squared Error	6.990967
Mean Absolute Error	5.428619
Mean Abs. Percent Error	5.045731
Theil Inequality Coefficient	0.031125
Bias Proportion	0.000000
Variance Proportion	0.031105
Covariance Proportion	0.968895

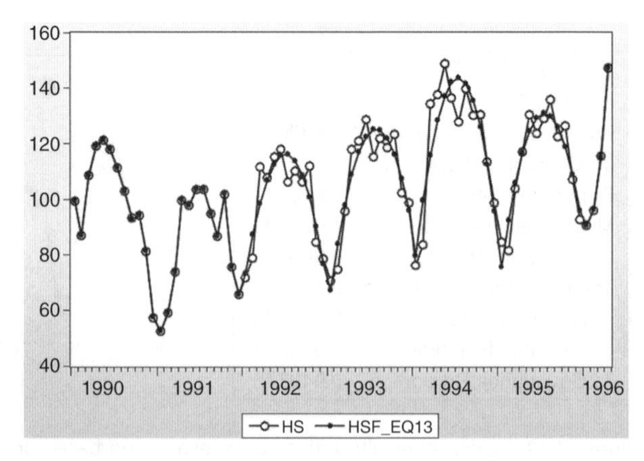

Figure 2.12 Statistical results with forecast evaluation based on model (2.17), for $q = 1$.

Example 2.12 The Simplest Model in (2.17)

Figure 2.12 presents the statistical results of the simplest AR(q) model in (2.17), namely an AR(1) model. Based on these results, the following notes and comments are presented.

1) The complete estimate of the parameters are not presented, because it is too large and is not important to the discussion.
2) The forecast has a relatively small TIC = 0.031126, compared to the previous models. Since the TIC < 0.05, then it can be considered an excellent forecast – refer to the recommended classification in Chapter 1. But the problem is the AR(1) has such a large p value = 0.8080. So, a modified model should be explored. See an alternative model presented in the following example.

Dependent Variable: HS
Method: Least Squares
Date: 01/09/14 Time: 14:51
Sample (adjusted): 1946M05 1996M04
Included observations: 600 after adjustments
Convergence achieved after 14 iterations

Variable	Coefficient	Std. Error	t-Statistic	Prob.
C	56.11879	9.131234	6.145805	0.0000
@MONTH^2*@EXPAND(@YEAR)	Yes			
@MONTH*@EXPAND(@YEAR)	Yes			
@EXPAND(@YEAR,@DROPFIRST)	Yes			
AR(1)	-0.318410	0.040477	-7.866528	0.0000
AR(2)	-0.468255	0.037313	-12.54951	0.0000
AR(3)	-0.458530	0.037128	-12.34983	0.0000
AR(4)	-0.548295	0.040701	-13.47126	0.0000

R-squared	0.956747	Mean dependent var	123.3595
Adjusted R-squared	0.941516	S.D. dependent var	34.61601
S.E. of regression	8.371361	Akaike info criterion	7.307483
Sum squared resid	31045.30	Schwarz criterion	8.458013
Log likelihood	-2035.245	Hannan-Quinn criter.	7.755362
F-statistic	62.81470	Durbin-Watson stat	2.378495
Prob(F-statistic)	0.000000		

Inverted AR Roots	.43+.82i	.43-.82i	-.59-.54i	-.59+.54i

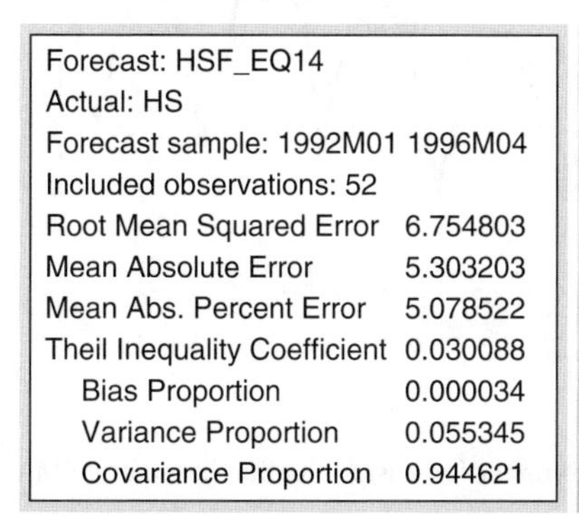

Forecast: HSF_EQ14
Actual: HS
Forecast sample: 1992M01 1996M04
Included observations: 52

Root Mean Squared Error	6.754803
Mean Absolute Error	5.303203
Mean Abs. Percent Error	5.078522
Theil Inequality Coefficient	0.030088
Bias Proportion	0.000034
Variance Proportion	0.055345
Covariance Proportion	0.944621

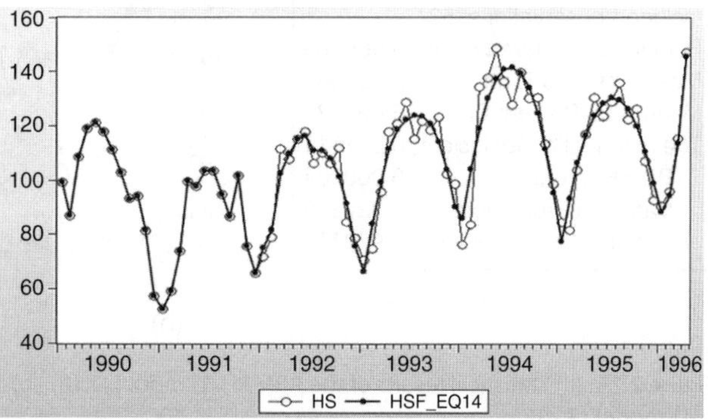

Figure 2.13 Statistical results with forecast evaluation based on model (2.17), for q = 4.

Example 2.13 An Unexpected AR(4) Model in (2.17)

By using the trial-and-error method, I obtained the unexpected statistical results presented in Figure 2.13 based on an AR(4) model in (2.17), where each of AR(1) to AR(4) has a p value = 0.0000. Based on these results, the following notes and comments are presented.

1) The forecast has a slightly smaller TIC compared to the AR(1) model. So, it can be considered to be a better model.

Dependent Variable: HS
Method: Least Squares
Date: 01/09/14 Time: 16:19
Sample (adjusted): 1946M02 1996M04
Included observations: 603 after adjustments
Convergence achieved after 19 iterations
MA Backcast: 1946M01

Variable	Coefficient	Std. Error	t-Statistic	Prob.
C	48.42677	17.11206	2.829978	0.0049
@MONTH^2*@EXPAND(@YEAR)	Yes			
@MONTH*@EXPAND(@YEAR)	Yes			
@EXPAND(@YEAR,@DROPFIRST)	Yes			
AR(1)	-0.793903	0.058253	-13.62848	0.0000
MA(1)	0.923695	0.036706	25.16466	0.0000

R-squared	0.925963	Mean dependent var	123.1819
Adjusted R-squared	0.900513	S.D. dependent var	34.64025
S.E. of regression	10.92607	Akaike info criterion	7.837153
Sum squared resid	53481.81	Schwarz criterion	8.968657
Log likelihood	-2207.902	Hannan-Quinn criter.	8.277526
F-statistic	36.38342	Durbin-Watson stat	2.118435
Prob(F-statistic)	0.000000		

Inverted AR Roots	-.79
Inverted MA Roots	-.92

(a)

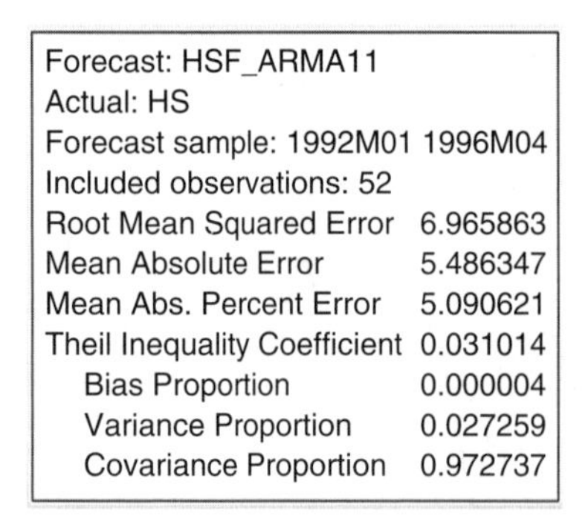

Forecast: HSF_ARMA11
Actual: HS
Forecast sample: 1992M01 1996M04
Included observations: 52

Root Mean Squared Error	6.965863
Mean Absolute Error	5.486347
Mean Abs. Percent Error	5.090621
Theil Inequality Coefficient	0.031014
Bias Proportion	0.000004
Variance Proportion	0.027259
Covariance Proportion	0.972737

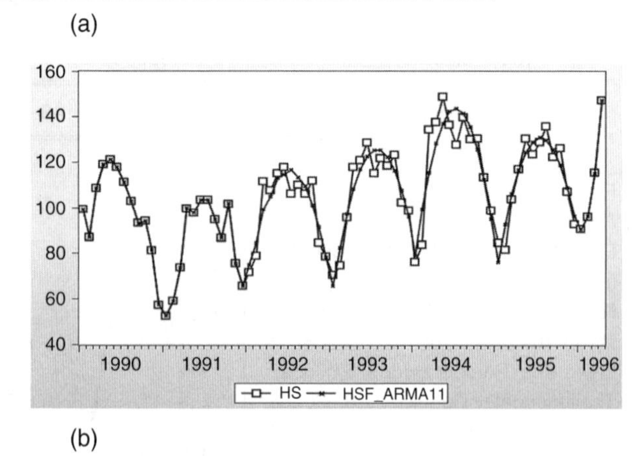

(b)

Figure 2.14 (a) Statistical results of the ARMA(1,1) model (2.18). (b) Dynamic forecast evaluation based on ARMA (1,1) model (2.18).

2) In fact, based on the AR(3), it was found AR(2) and AR(3) have p values = 0.0000, but AR(1) has a significant effect at a 10% level of significance, with a p value = 0.0903. However, I prefer to apply the AR(4) model to take into account the *cyclical deviation* (EViews 8, *User Guide II*, p. 637).

2.3.4 Application of ARMA(*q*,*r*) Models of *HS* on *MONTH* by *YEAR*

As an extension of AR(q) models in (2.17), in this section, the models considered are ARMA(q,r) with the following general equation, for any integer q and r, with all its alternative reduced models as presented for the LV(1) model in (2.11). However, an example will be presented in this section.

$$\begin{aligned} &hs\ C\ @month\hat{\ }2*@expand(@year)\ @month*@expand(@year) \\ &@expand(@year,@dropfirst)\ ar(1)...ar(q)\ ma(1)...ma(r) \end{aligned} \tag{2.18}$$

Example 2.14 The Simplest ARMA Model in (2.18)
Figure 2.14 presents the statistical results of the simplest ARMA(q,r) model in (2.18), namely an ARMA(1,1) model. Based on these results, the following notes and comments are presented.

1) The complete estimates of the parameters are not presented because they are too large and not important to the discussion.
2) The forecast has a relatively small TIC = 0.031014 compared to the previous models. Since the TIC < 0.05, then it can be considered an excellent forecast – refer to the recommended classification in Chapter 1.
3) Do analyses based on another ARMA(q,r) as an exercise.

2.3.5 Application of LVAR(*p*,*q*) Models of *HS* on *MONTH* by *YEAR*

As a combination of the LV(p) and of AR(q) models presented previously, this section presents LVAR(p,q) models with the following general equation, for any integers p and q, with all its alternative reduced models as presented for the LV(1) model (2.11).

$$\begin{aligned} &hs\ C\ hs(-1)...hs(-p)\ @month\hat{\ }2*@expand(@year) \\ &@month*@expand(@year)\ @expand(@year,@dropfirst)\ ar(1)...ar(q) \end{aligned} \tag{2.19}$$

Example 2.15 Application of a LVAR(2,4) Model in (2.19)
As an illustration, Figure 2.15 presents incomplete statistical results of one out of many possible models in (2.19), namely the LVAR(2,4) model, since the parameter's estimates of three sets of the independent variables are not presented because they are too long. Based on these results, the following findings and notes are presented.

1) Each of *HS(−1)*, *HS(−2)*, *AR(1)*, *AR(2)*, *AR(3)*, and *AR(4)* has a significant adjusted effect on *HS* with a p value = 0.0000, and the joint effects of all independent variables are significance based on the F-statistic of $F_0 = 77.02325$, with a p value = 0.000000.
2) The terms *AR(1)* to *AR(4)* are applied to take into account the *cyclical deviation* of the data.
3) The readers are recommended to experiment with other alternative LVAR(p,q) as an exercise, especially for other patterns/types of monthly time series variables.

Dependent Variable: HS
Method: Least Squares
Date: 01/10/14 Time: 14:11
Sample (adjusted): 1946M07 1996M04
Included observations: 598 after adjustments
Convergence achieved after 58 iterations

Variable	Coefficient	Std. Error	t-Statistic	Prob.
C	61.15692	17.85424	3.425345	0.0007
HS(-1)	0.710596	0.029892	23.77224	0.0000
HS(-2)	-0.546549	0.017859	-30.60306	0.0000
@MONTH^2*@EXPAND(@YEAR)	Yes			
@MONTH*@EXPAND(@YEAR)	Yes			
@EXPAND(@YEAR)	Yes			
AR(1)	-1.154369	0.052087	-22.16239	0.0000
AR(2)	-0.904833	0.080428	-11.25022	0.0000
AR(3)	-0.526347	0.079924	-6.585566	0.0000
AR(4)	-0.315537	0.050588	-6.237353	0.0000

R-squared	0.965183	Mean dependent var	123.4376
Adjusted R-squared	0.952652	S.D. dependent var	34.64705
S.E. of regression	7.539087	Akaike info criterion	7.100760
Sum squared resid	24951.81	Schwarz criterion	8.268956
Log likelihood	-1964.127	Hannan-Quinn criter.	7.555583
F-statistic	77.02325	Durbin-Watson stat	2.167609
Prob(F-statistic)	0.000000		

Inverted AR Roots	.09-.69i	.09+.69i	-.66+.46i	-.66-.46i

(a)

Forecast: HSF_LVAR24	
Actual: HS	
Forecast sample: 1992M01 1996M04	
Included observations: 52	
Root Mean Squared Error	5.753102
Mean Absolute Error	4.689345
Mean Abs. Percent Error	4.404847
Theil Inequality Coefficient	0.025604
Bias Proportion	0.000027
Variance Proportion	0.009665
Covariance Proportion	0.990308

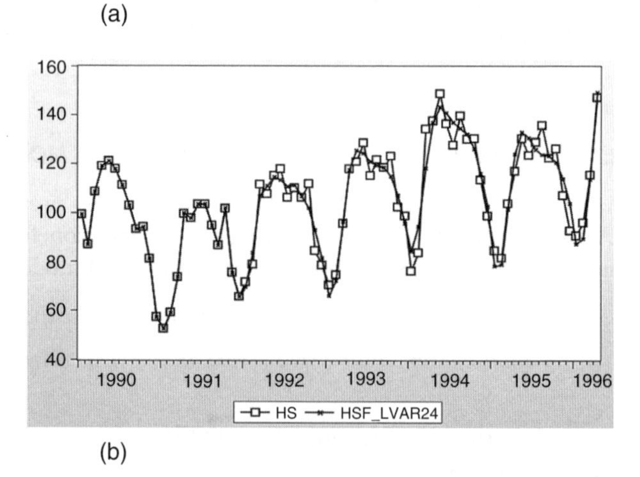

(b)

Figure 2.15 (a) Statistical results of a LVAR(2,4) model in (2.19). (b) Dynamic forecast evaluation based on the model in Figure 2.15.

4) The joint effects of *@Month^2∗@Expand(@Year)* can be tested using the Omitted Variables – Likelihood Ratio Test, as presented in Section 2.2.2. Do this as an exercise.
5) Similarly, for the joint effects of *@Month∗@Expand(@Year)* and the joint effects of both sets of variables.
6) The dynamic forecast has a TIC = 0.025604, which is the smallest TIC compared to all presented forecasts of *HS*. Therefore, this forecast can be considered to be the best, in a statistical sense.

Example 2.16 Application of a Special LVAR(12,4) Model in (2.19)
As an extension of the special LV(12) model in (1.3), and the special LVAR(12,2) model in Figure 1.20, and a modification of the special LV(12) heterogeneous model (2.1), Figure 2.16 presents the statistical results of a special LV(12,4) heterogeneous model of *HS* on the numerical variables *@Month^2* and *@Month,* by the year dummy variables *@Expand(@Year)* using the ES as follows:

Dependent Variable: HS
Method: Least Squares
Date: 01/10/14 Time: 13:54
Sample (adjusted): 1947M05 1996M04
Included observations: 588 after adjustments
Convergence achieved after 17 iterations

Variable	Coefficient	Std. Error	t-Statistic	Prob.
C	-7.826841	10.71394	-0.730528	0.4655
HS(-12)	0.249205	0.048495	5.138752	0.0000
@MONTH^2*@EXPAND(@YEAR)	Yes			
@MONTH*@EXPAND(@YEAR)	Yes			
@EXPAND(@YEAR)	Yes			
AR(1)	-0.269684	0.044265	-6.092514	0.0000
AR(2)	-0.402172	0.042303	-9.506923	0.0000
AR(3)	-0.423302	0.041280	-10.25439	0.0000
AR(4)	-0.466065	0.045641	-10.21153	0.0000

R-squared	0.958044	Mean dependent var	124.1713
Adjusted R-squared	0.943122	S.D. dependent var	34.42154
S.E. of regression	8.209246	Akaike info criterion	7.269621
Sum squared resid	29180.61	Schwarz criterion	8.423350
Log likelihood	-1982.269	Hannan-Quinn criter.	7.719150
F-statistic	64.20323	Durbin-Watson stat	2.298512
Prob(F-statistic)	0.000000		

Inverted AR Roots	.43+.79i	.43-.79i	-.57-.50i	-.57+.50i

(a)

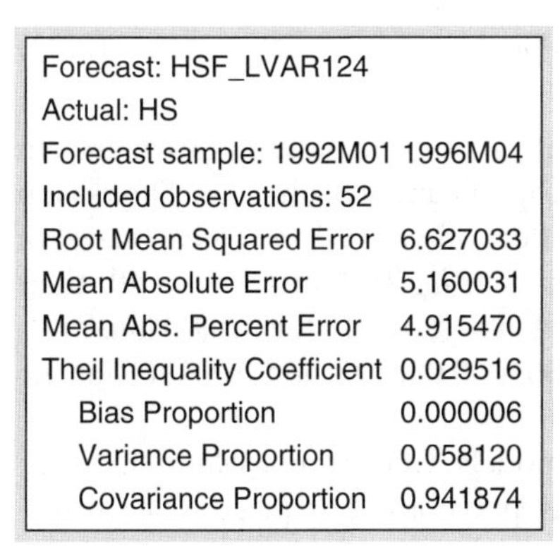

Forecast: HSF_LVAR124	
Actual: HS	
Forecast sample: 1992M01 1996M04	
Included observations: 52	
Root Mean Squared Error	6.627033
Mean Absolute Error	5.160031
Mean Abs. Percent Error	4.915470
Theil Inequality Coefficient	0.029516
Bias Proportion	0.000006
Variance Proportion	0.058120
Covariance Proportion	0.941874

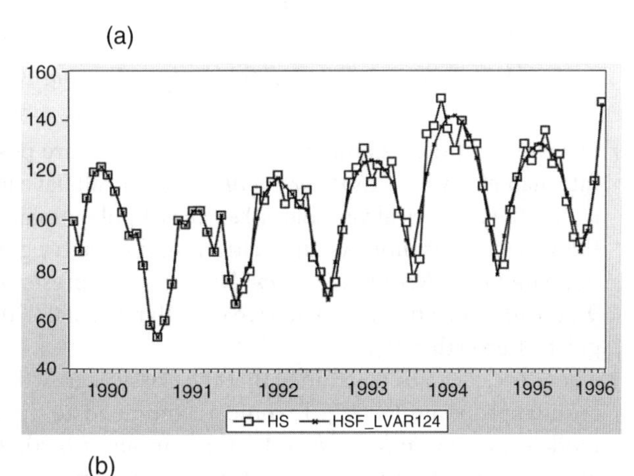

(b)

Figure 2.16 (a) Statistical results of a special LVAR(12,4) model in (2.20). (b) Dynamic forecast evaluation of the LVAR(12,4) model in (2.20).

$$hs\ C\ hs(-12)\ (@month)\hat{}2*@expand(@year)\ @month*@expand(@year)\ @expand(@year,@dropfirst)\ ar(1)\ ar(2)\ ar(3)\ ar(4) \quad (2.20)$$

1) The complete estimate of the parameters are not presented, because it is too large and is not important for the discussion.
2) The forecast has a relatively small TIC = 0.029516, compared to the previous models. Since the TIC < 0.05, then it can be considered to be an excellent forecast – refer to the recommended classification in Chapter 1.

2.3.6 Application of LVAR(*p,q*) Models of *HS* on *YEAR* by *MONTH*

It is recognized that the variable *HS* has different scores in January up to December within or over the years. So there is a reason to consider a set of 12 annual time series variables, for any monthly time series variables, in general. Referring to all possible growth models based on the time series by states presented in Agung (2014), such as the multivariate classical growth model, polynomial growth models, and piece-wise growth models. Then the variable HS_t can be considered an annual time series by *MONTH* and it can be represented as $HS(month,year) = HS_{my}$. Therefore, the set of 12 (Het_CGMs) would have the following basic equation.

$$Log(HS_{my}) = \beta_{0m} + \beta_{1m} * @Year + \mu_{my}, for\ m = 1,\ldots,12 \quad (2.21)$$

where *@Year* is a numerical variable, and β_{1m} is the parameter of the exponential annual growth rate of *HS* for each *m(month)* = 1, …, 12.

Example 2.17 Application of the Model (2.21)

Note that the model in (2.21) represents a set of 12 heterogeneous simple regressions of *log(HS)* on the numerical variable *@Year* by the categorical variable *@Expand(@Month)*, and the data analysis can be done using the following ES, with the statistical results or estimates and the forecast evaluation presented in Figure 2.17. Note this forecast is neither a dynamic nor a static forecast. Compared to the following examples. Based on these results the following findings and notes are presented.

$$Log(HS)\ @Year*@Expand(@Month)\ @Expand(@Month) \quad (2.22)$$

1) Eleven out of the 12 annual growth rates of *HS* are positive, but one has a negative growth rate, namely in September (*@Month* = 9), and all have insignificant growth rates. The reason is that the numerical variable *@Year* has a polynomial effect on *log(HS)* within each level of the variable *@Month*. As an illustration, the scatter graphs of *log(HS)* on *@Year*, with their nonparametric *Nearest Neighbor Fit Curves* are presented in Figure 2.18 for the *@Month* = 1, 9, and 10, respectively, which have growth rates of 0.000559, −0,000128, and 0.001908 (as the greatest growth rate).
2) Therefore, it can be concluded that the classical growth models (CGMs) by *@Month* in (2.22) are inappropriate for *HS*. I would recommend to apply polynomial or piece-wise growth models by *@Month* – refer to the various growth models presented in Agung (2014, 2009a). For instance, see a possible quadratic model for the *@Month* = 10.

Dependent Variable: LOG(HS)
Method: Least Squares
Date: 01/15/14 Time: 10:13
Sample: 1946M01 1996M04 IF @YEAR<1996
Included observations: 600

Variable	Coefficient	Std. Error	t-Statistic	Prob.
@YEAR*(MONTH=1)	0.000559	0.002159	0.258969	0.7958
@YEAR*(MONTH=2)	4.94E-05	0.002159	0.022861	0.9818
@YEAR*(MONTH=3)	0.001117	0.002159	0.517626	0.6049
@YEAR*(MONTH=4)	0.000274	0.002159	0.126957	0.8990
@YEAR*(MONTH=5)	0.000275	0.002159	0.127585	0.8985
@YEAR*(MONTH=6)	0.001061	0.002159	0.491680	0.6231
@YEAR*(MONTH=7)	0.000450	0.002159	0.208371	0.8350
@YEAR*(MONTH=8)	4.29E-05	0.002159	0.019864	0.9842
@YEAR*(MONTH=9)	-0.000128	0.002159	-0.059174	0.9528
@YEAR*(MONTH=10)	0.001908	0.002159	0.883765	0.3772
@YEAR*(MONTH=11)	0.000186	0.002159	0.085936	0.9315
@YEAR*(MONTH=12)	0.000792	0.002159	0.366897	0.7138
@MONTH=1	3.327513	4.254181	0.782175	0.4344
@MONTH=2	4.369223	4.254181	1.027042	0.3048
@MONTH=3	2.581249	4.254181	0.606756	0.5443
@MONTH=4	4.389329	4.254181	1.031768	0.3026
@MONTH=5	4.424676	4.254181	1.040077	0.2987
@MONTH=6	2.865634	4.254181	0.673604	0.5008
@MONTH=7	4.026878	4.254181	0.946569	0.3443
@MONTH=8	4.823607	4.254181	1.133851	0.2573
@MONTH=9	5.102717	4.254181	1.199459	0.2308
@MONTH=10	1.115582	4.254181	0.262232	0.7932
@MONTH=11	4.327676	4.254181	1.017276	0.3094
@MONTH=12	2.937437	4.254181	0.690482	0.4902

R-squared	0.443940	Mean dependent var	4.772676
Adjusted R-squared	0.421737	S.D. dependent var	0.289696
S.E. of regression	0.220295	Akaike info criterion	-0.148517
Sum squared resid	27.95331	Schwarz criterion	0.027360
Log likelihood	68.55508	Hannan-Quinn criter.	-0.080051
Durbin-Watson stat	0.160613		

(a)

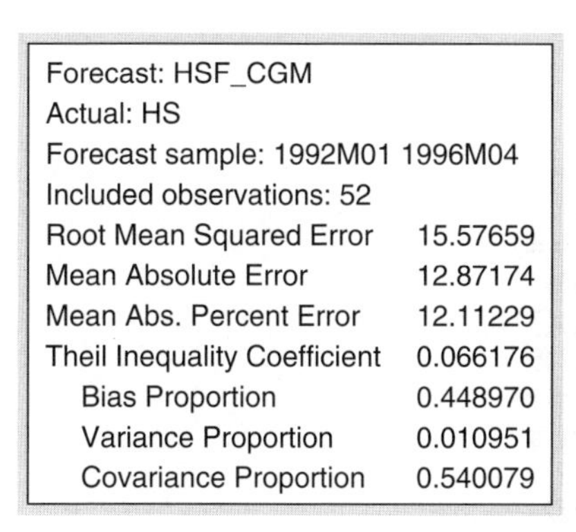

Forecast: HSF_CGM
Actual: HS
Forecast sample: 1992M01 1996M04
Included observations: 52

Root Mean Squared Error	15.57659
Mean Absolute Error	12.87174
Mean Abs. Percent Error	12.11229
Theil Inequality Coefficient	0.066176
Bias Proportion	0.448970
Variance Proportion	0.010951
Covariance Proportion	0.540079

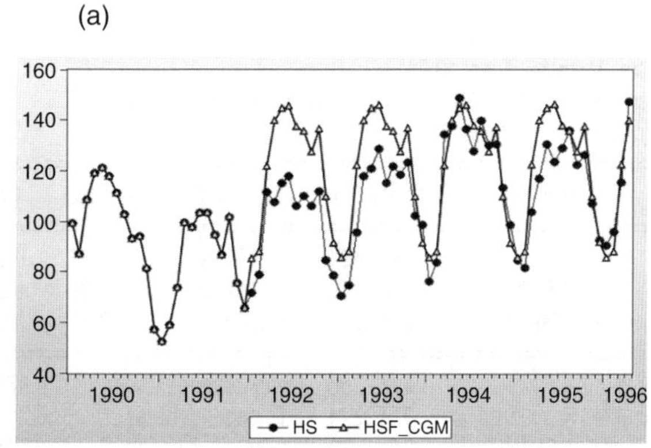

(b)

Figure 2.17 (a) Statistical results of the heterogeneous classical growth model in (2.22). (b) Forecast evaluation based on the heterogeneous classical growth model in (2.22).

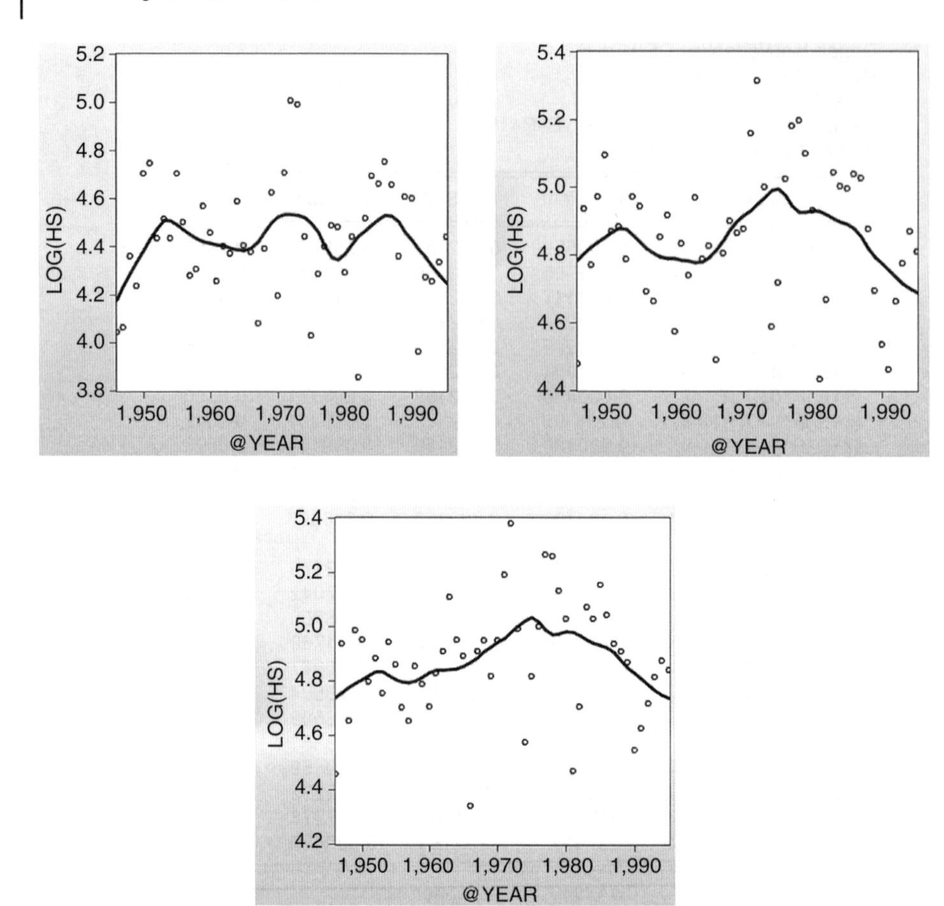

Figure 2.18 Scatter graphs of (@Year, log(HS)) with their nearest neighbor fit curves for the levels *@Month* = 1, 9, and 10.

Dependent Variable: LOG(HS)
Method: Least Squares
Date: 01/15/14 Time: 13:28
Sample: 1946M01 1996M04 IF @YEAR<1996 AND @MONTH=10
Included observations: 50

Variable	Coefficient	Std. Error	t-Statistic	Prob.
C	-1307.434	601.0702	-2.175177	0.0347
@YEAR	1.330119	0.610092	2.180196	0.0343
@YEAR^2	-0.000337	0.000155	-2.177080	0.0345

R-squared	0.107355	Mean dependent var	4.875179
Adjusted R-squared	0.069371	S.D. dependent var	0.211228
S.E. of regression	0.203770	Akaike info criterion	-0.285524
Sum squared resid	1.951547	Schwarz criterion	-0.170802
Log likelihood	10.13809	Hannan-Quinn criter.	-0.241837
F-statistic	2.826268	Prob(F-statistic)	0.069334

Figure 2.19 A quadratic annual growth model of HS in October (*@Month* = 10).

3) Even though *@Year* has insignificant linear effect on *log(HS)* by the *@Month*, it is recognized that its modified models, namely LV(*p*), AR(*q*), and LVAR(*p,q*) heterogeneous classical growth models (Het_CGMs), give acceptable forecasts. See the following sections.
4) Specific for *@Month* = 10, Figure 2.19 presents the estimates of a quadratic growth model, where each of the numerical variables *@Year* and *@Year^2* has a significant adjusted effect on *lg(HS)*.
5) The graphs in Figure 2.18 show a specific polynomial or piece-wise model would be applicable for each month. So it is recommended to conduct a time series data analysis for each month. See the forecasting presented in the following examples.

2.4 Heterogeneous Classical Growth Models

2.4.1 Forecasting Based on LV(*p*) Het_CGMs of *HS*

The LV(*p*) Het_CGMs considered would have the following general ES.

$$Log(HS)\ log(HS(-1))\dots log(HS(-p))$$
$$@Year*@Expand(@Month)\ @Expand(@Month) \tag{2.23}$$

Example 2.18 The Simplest LV(1) Het_CGM
As a modification of the Het_CGM in (2.1) and (2.2), Figure 2.20 presents the statistical results of a LV(1) Het_CGM, using the ES as follows:

$$log(HS)\ log(HS(-1))\ @Year*@Expand(@Month)\ @Expand(@Month) \tag{2.24}$$

Based on these results, the following findings and notes are presented.

1) This is unexpected when using the model of *log(HS)* in (2.24), the forecast evaluation presents the forecast of *HS*, namely *HSF_LV1_CGM*, as presented in Figure 2.20b, instead of the forecost of *log(HS)*. EViews provides the option for selecting either the series *HS* or *log(HS)* to forecast, as presented in Figure 2.21. Do this as an exercise.
2) As an alternative method, I have tried to apply a model using a transformed variable *LNHS* = *log(HS)*, using the following ES, with the forecast evaluation presented in Figure 2.22. Based on these results, the following findings and notes are presented.

$$LNHS\ LNHS(-1)\ @Year*@Expand(@Month)\ @Expand(@Month) \tag{2.25}$$

2.1 This forecast is presenting the forecast of $LNHS_{my} = log(HS_{my})$ with a TIC = 0.008607, which is much smaller than the TIC = 0.038423 for the forecast of the original variable HS_{my}, by using ES (2.23). However, note that these two TICs are not comparable, because they are the forecasts of two different variables.

2.2 In addition, note that the results in Figure 2.20 show the observed scores of HS_{my} have two different trends, negative and positive. So it can be said a linear effect of *@Trend* on *HS* is inappropriate. They should have a polynomial or piece-wise relationship. The models with *@Trend* will be considered later.

Dependent Variable: LOG(HS)
Method: Least Squares
Date: 01/14/14 Time: 21:32
Sample: 1946M01 1996M04 IF @YEAR<1996
Included observations: 599

Variable	Coefficient	Std. Error	t-Statistic	Prob.
LOG(HS(-1))	0.917208	0.016345	56.11425	0.0000
@YEAR*(@MONTH=1)	-0.001088	0.000873	-1.245714	0.2134
@YEAR*(@MONTH=2)	-0.000463	0.000847	-0.547188	0.5845
@YEAR*(@MONTH=3)	0.001072	0.000847	1.266053	0.2060
@YEAR*(@MONTH=4)	-0.000751	0.000847	-0.886421	0.3758
@YEAR*(@MONTH=5)	2.40E-05	0.000847	0.028394	0.9774
@YEAR*(@MONTH=6)	0.000809	0.000847	0.955045	0.3400
@YEAR*(@MONTH=7)	-0.000524	0.000847	-0.618297	0.5366
@YEAR*(@MONTH=8)	-0.000370	0.000847	-0.436542	0.6626
@YEAR*(@MONTH=9)	-0.000167	0.000847	-0.197286	0.8437
@YEAR*(@MONTH=10)	0.002025	0.000847	2.391198	0.0171
@YEAR*(@MONTH=11)	-0.001564	0.000847	-1.846019	0.0654
@YEAR*(@MONTH=12)	0.000622	0.000847	0.734345	0.4630
@MONTH=1	2.455456	1.721357	1.426465	0.1543
@MONTH=2	1.317202	1.669748	0.788863	0.4305
@MONTH=3	-1.426236	1.670389	-0.853835	0.3936
@MONTH=4	2.021786	1.669395	1.211089	0.2264
@MONTH=5	0.398750	1.670404	0.238714	0.8114
@MONTH=6	-1.192714	1.670428	-0.714017	0.4755
@MONTH=7	1.398495	1.669519	0.837664	0.4026
@MONTH=8	1.130123	1.670160	0.676656	0.4989
@MONTH=9	0.678466	1.670723	0.406091	0.6848
@MONTH=10	-3.564670	1.670945	-2.133326	0.0333
@MONTH=11	3.304456	1.668962	1.979947	0.0482
@MONTH=12	-1.031942	1.670361	-0.617796	0.5370

R-squared	0.913811	Mean dependent var	4.773894
Adjusted R-squared	0.910207	S.D. dependent var	0.288396
S.E. of regression	0.086419	Akaike info criterion	-2.018375
Sum squared resid	4.286784	Schwarz criterion	-1.834934
Log likelihood	629.5034	Hannan-Quinn criter.	-1.946960
Durbin-Watson stat	2.347535		

(a)

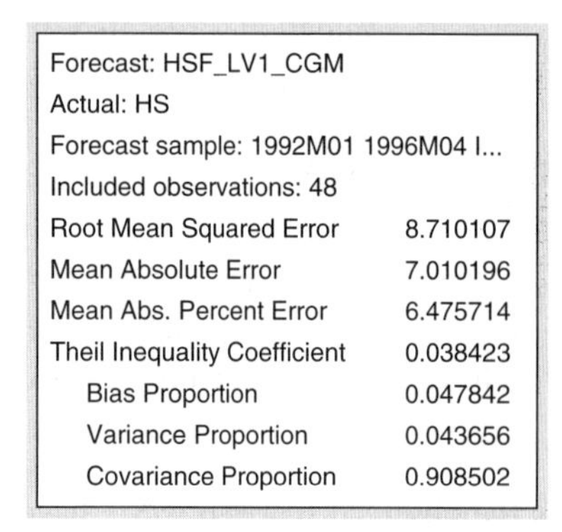

Forecast: HSF_LV1_CGM
Actual: HS
Forecast sample: 1992M01 1996M04 I...
Included observations: 48

Root Mean Squared Error	8.710107
Mean Absolute Error	7.010196
Mean Abs. Percent Error	6.475714
Theil Inequality Coefficient	0.038423
Bias Proportion	0.047842
Variance Proportion	0.043656
Covariance Proportion	0.908502

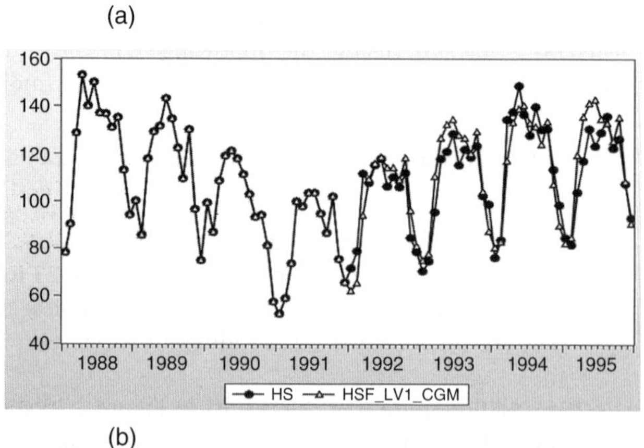

(b)

Figure 2.20 (a) Statistical results of the LV(1) Het_CGM (2.24). (b) Dynamic forecast evaluation based on the LV(1) Het_CGM (2.24).

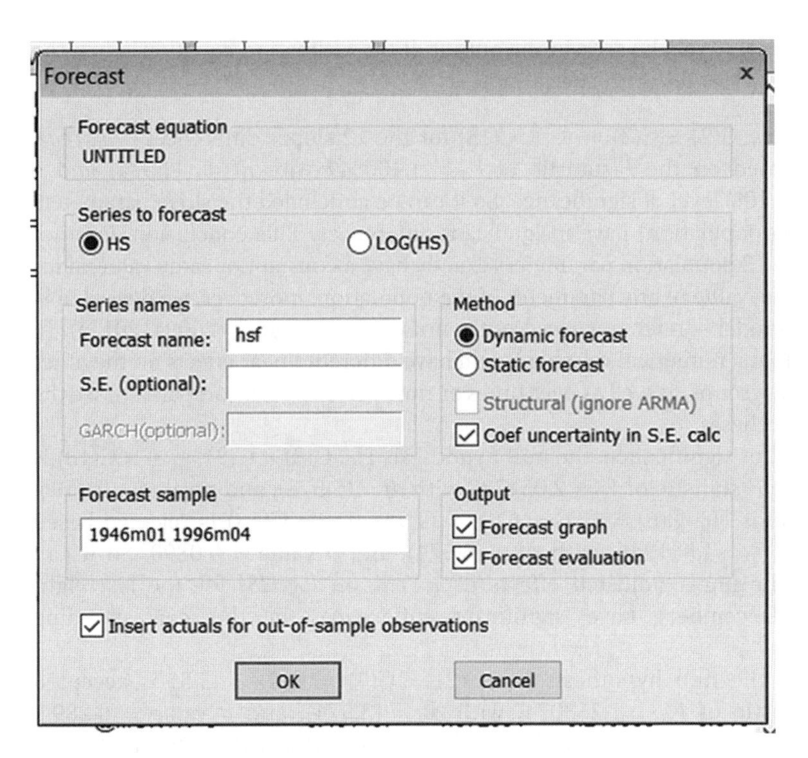

Figure 2.21 Option for a series to forecast.

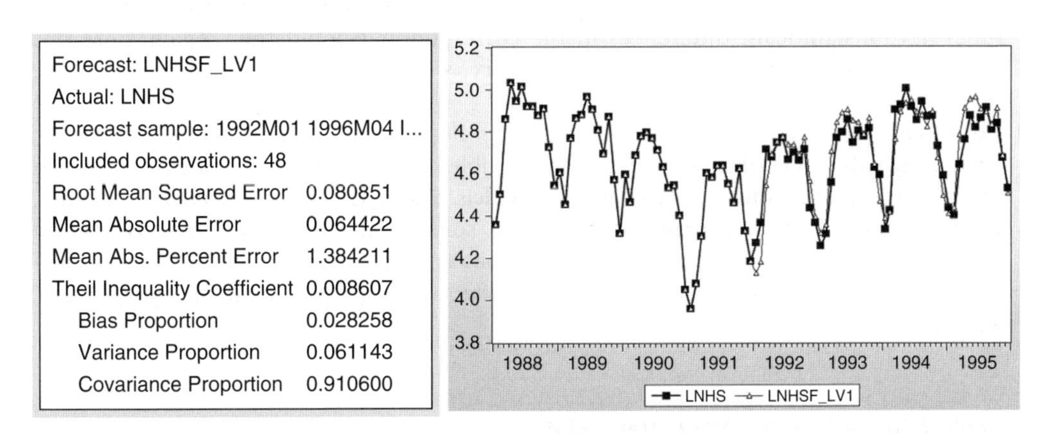

Figure 2.22 Dynamic forecast evaluation based on the LV(1) Het_CGM (2.25).

Example 2.19 Special Testing Hypotheses of the LV(1) Het_CGM in Figure 2.20
Corresponding to the various *p* values of the slopes of the numerical variable *@Year* over months, this example presents alternative conclusions for some selected hypotheses using the Wald test. In order to write and test the hypotheses, the parameters of the model in Figure 2.20 should be identified, such as follows:

- C(1) is the slope parameter of *log(HS(−1)*,
- C(2) to C(13) are the slopes parameters of *@Year* for January to December, or the linear effects parameters of *@Year* on log*(HS)*, and
- C(14) to C(22) are the intercept parameters of the model.

Note these parameters can easily be seen in the output of the estimation equation. Then the following hypotheses are presented with special notes.

1) The null hypothesis H_0: $C(2) = C(3) = \cdots = C(13)$ for the 12 slopes differences of *@Year*, which is accepted based on the *F*-statistic of $F_0 = 1.400842$ with $df = (11{,}574)$ and *p value* = 0.1681, at the 10% level of significance. So it can be concluded the slope parameters (in the corresponding population) have insignificant differences. This conclusion does not mean directly that the 12 population parameters exactly have a constant or same value, since we never know the true value of any parameter of the population, moreover, we never know the true population model – refer to special notes presented in Agung (2009a, 2011a). In general, I would say that a numerical variable would have different linear effects on the other numerical variables by groups, as well as over times or time periods. As a comparison, see the following testing hypotheses.
2) At the level 10% level of significance, the null hypothesis H_0: $C(8) = C(9) = \cdots = C(13)$ is rejected based on the *F*-statistic of $F_0 = 2.054723$ with $df = (5{,}574)$, and *p* value = 0.0695, and the null hypothesis H_0: $C(6) = C(7) = C(8) = C(9) = \cdots = C(13)$ is rejected based on the *F*-statistic of $F_0 = 1.841246$, with $df = (6{,}574)$, and *p* value = 0.0890. So it can be concluded that the linear adjusted effects of *@Year* on *log(HS)* for the last eight months (May to December) have significant differences in the corresponding population.
3) On the other hand, the null hypothesis H_0: $C(2) = C(3) = C(4) = C(5)$ is accepted based on the *F*-statistic of $F_0 = 1.253074$, with $df = (3{,}574)$, and *p* value = 0.2897, and H_0: $C(2) = C(3) = C(4) = C(5) = C(6) = C(7)$ is accepted based on the *F*-statistic of $F_0 = 1.025292$, with $df = (5{,}574)$, and *p* value = 0.4018. So, it can be concluded that the linear adjusted effects of *@Year* on *log(HS)* for the first six months (January to June) have insignificant differences in the corresponding population. Again, note this conclusion does not directly mean the six slope parameters have exactly the same value.
4) Corresponding to different conclusions of the slope parameters presented previously, I would say the LV(1) Het_CGM is an acceptable or recommended model, rather than the LV(1) homogeneous classical growth models (Hom_CGMs) or ANCOVA models, which in general are not recommended. Even some of the ANCOVA models, namely the N-way additive ANCOVA, are considered the worst statistical models: refer to the special notes on alternative ANCOVA or fixed-effects models presented in Agung (2014, 2009a, 2011a).

Example 2.20 A Special LV(12) Het_CGM

As an extension or modification of the LV(1) Het_CGM of *LNHS* in (2.14), Figure 2.23 presents the forecast evaluation of a special or uncommon LV(12) Het_CGM, using the following ES.

$$LNHS\ LNHS(-12)\ @Year*@Expand(@Month)\ @Expand(@Month) \tag{2.26}$$

Compared to forecast based on the LV(1) Het_CGM in (2.24) with TIC = 0.008607, this special LV(12) Het_CGM presents the forecast with a slightly smaller TIC = 0.008392. So this forecast can be considered to be a better forecast, in a statistical sense, and note that a TIC = 0 is defined as a perfect forecast.

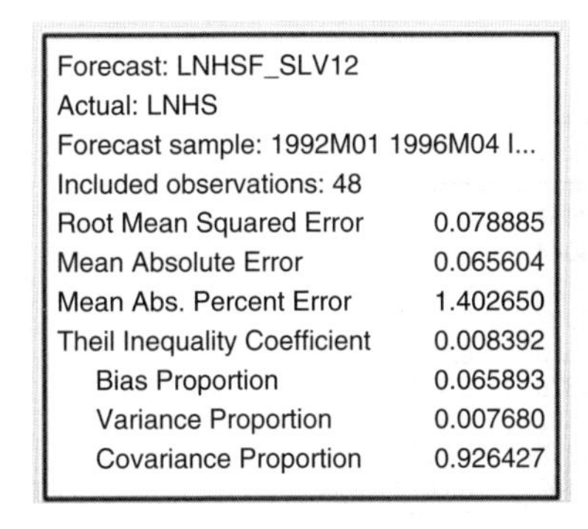

Forecast: LNHSF_SLV12	
Actual: LNHS	
Forecast sample: 1992M01 1996M04 I...	
Included observations: 48	
Root Mean Squared Error	0.078885
Mean Absolute Error	0.065604
Mean Abs. Percent Error	1.402650
Theil Inequality Coefficient	0.008392
Bias Proportion	0.065893
Variance Proportion	0.007680
Covariance Proportion	0.926427

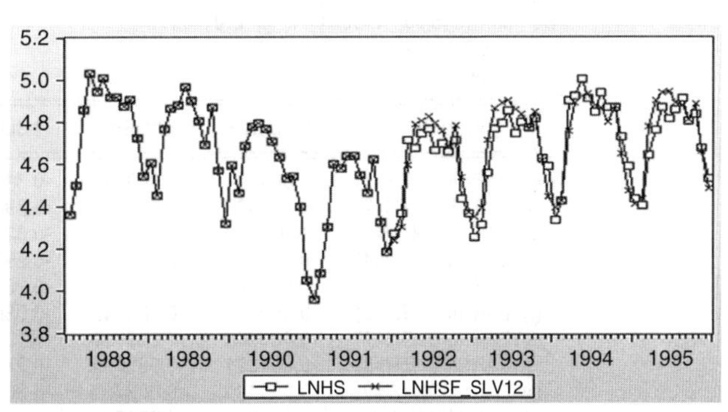

Figure 2.23 Dynamic forecast evaluation of the special LV(12) Het_CGM (2.26).

2.4.2 Forecasting Based on AR(q) Het_CGMs

The AR(q) Het_CGMs considered would have the following general ES.

$$Log(HS)\ @Year*@Expand(@Month)\ @Expand(@Month)\ AR(1)...AR(q) \tag{2.27}$$

Example 2.21 The Simplest AR(1) Het_CGM

As a modification of the Het_CGM in (2.1) and (2.2), Figure 2.24 only presents the forecast evaluation based on the AR(1) Het_CGM, in (2.27).

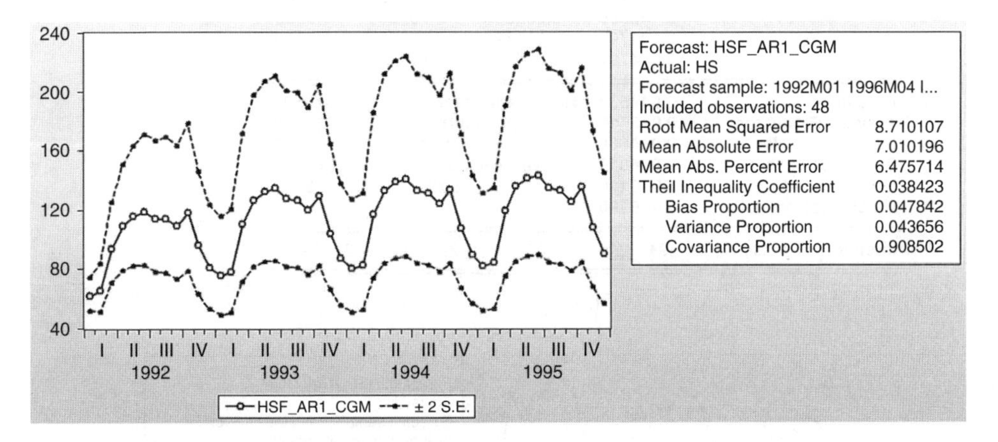

Forecast: HSF_AR1_CGM	
Actual: HS	
Forecast sample: 1992M01 1996M04 I...	
Included observations: 48	
Root Mean Squared Error	8.710107
Mean Absolute Error	7.010196
Mean Abs. Percent Error	6.475714
Theil Inequality Coefficient	0.038423
Bias Proportion	0.047842
Variance Proportion	0.043656
Covariance Proportion	0.908502

Figure 2.24 Dynamic forecast evaluation of the AR(1) Het_CGM (2.27).

Unexpectedly, the forecast evaluation presents exactly the same set of statistics as the forecast evaluation based on the LV(1) Het_CGM (2.24) in Figure 2.20.

Example 2.22 An AR(4) Het_CGM

As an extension of the AR(1) in (2.27), Figure 2.25 present the statistical results of an AR(4) Het_CGM, in (2.27). Based on these results, the following findings and notes are presented.

1) The terms AR(1) to AR(4) are applied in order to take into account the *cyclical deviation* of the time series. It happens each of the terms has a significant effect at the 10% level of significance.
2) Compared to the forecast based on the AR(1) in Figure 2.24, this forecast has a slightly smaller TIC, so this is a better forecast.

Dependent Variable: LOG(HS)
Method: Least Squares
Date: 01/16/14 Time: 16:38
Sample: 1946M01 1996M04 IF @YEAR<1996
Included observations: 596
Convergence achieved after 9 iterations

Variable	Coefficient	Std. Error	t-Statistic	Prob.
@YEAR*(MONTH=1)	-0.001194	0.003734	-0.319857	0.7492
@YEAR*(MONTH=2)	-0.001437	0.003742	-0.384044	0.7011
@YEAR*(MONTH=3)	-9.46E-05	0.003748	-0.025226	0.9799
@YEAR*(MONTH=4)	-0.001123	0.003755	-0.299097	0.7650
@YEAR*(MONTH=5)	-0.000988	0.003747	-0.263659	0.7921
@YEAR*(MONTH=6)	-0.000114	0.003743	-0.030370	0.9758
@YEAR*(MONTH=7)	-0.000686	0.003739	-0.183587	0.8544
@YEAR*(MONTH=8)	-0.001021	0.003734	-0.273340	0.7847
@YEAR*(MONTH=9)	-0.001131	0.003731	-0.303142	0.7619
@YEAR*(MONTH=10)	0.000958	0.003727	0.257032	0.7972
@YEAR*(MONTH=11)	-0.000710	0.003724	-0.190651	0.8489
@YEAR*(MONTH=12)	-5.34E-05	0.003721	-0.014348	0.9886
@MONTH=1	6.799468	7.366229	0.923060	0.3564
@MONTH=2	7.312715	7.380977	0.990752	0.3222
@MONTH=3	4.981733	7.393161	0.673830	0.5007
@MONTH=4	7.156217	7.406724	0.966178	0.3344
@MONTH=5	6.926382	7.389499	0.937328	0.3490
@MONTH=6	5.192754	7.380750	0.703554	0.4820
@MONTH=7	6.276947	7.373141	0.851326	0.3949
@MONTH=8	6.929891	7.364029	0.941046	0.3471
@MONTH=9	7.089275	7.356457	0.963681	0.3356
@MONTH=10	2.996659	7.349519	0.407735	0.6836
@MONTH=11	6.101067	7.343082	0.830859	0.4064
@MONTH=12	4.611713	7.337237	0.628535	0.5299
AR(1)	0.734255	0.041822	17.55684	0.0000
AR(2)	0.142390	0.051555	2.761896	0.0059
AR(3)	0.131797	0.051592	2.554613	0.0109
AR(4)	-0.074784	0.041681	-1.794212	0.0733

R-squared	0.917539	Mean dependent var	4.775503
Adjusted R-squared	0.913619	S.D. dependent var	0.287874
S.E. of regression	0.084608	Akaike info criterion	-2.055742
Sum squared resid	4.066010	Schwarz criterion	-1.849489
Log likelihood	640.6110	Hannan-Quinn criter.	-1.975427
Durbin-Watson stat	2.006290		

Inverted AR Roots	.94	.33	-.27+.40i	-.27-.40i

(a)

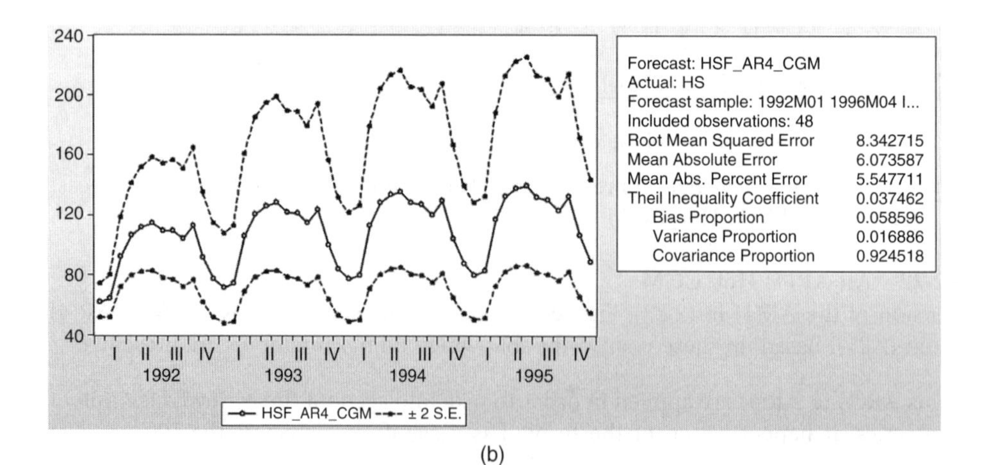

(b)

Figure 2.25 (a) Statistical results of an AR(4) Het_CGM in (2.27). (b) Forecast evaluation based on the AR(4) Het_CGM in Figure 2.25.

3) At the $\alpha = 0.05$ level of significance, each of AR(1), AR(2), and R(3) has a positive significant effect, but AR(4) has a negative significant effect, based on the t-statistic of $t_0 = -1.794212$ with $df = 568$, and p value $= 0.0733/2 = 0.03665 < 0.05$. So there is no reason to delete the AR(4) from the model.
4) Even though the *@Year* is insignificant over the months, the dynamic forecast can be considered to be very good with TIC = 0.037462 < 0.05 – refer to the recommended classification presented in Chapter 1.
5) As a comparison, based on two simpler models, namely the AR(2) and AR(3), it has been found that the dynamic forecasts have greater TICs of 0.037694 and 0.040061, respectively. So the AR(4) gives the best forecast compared to the AR(1), AR(2), and AR(3) models in (2.27).

2.4.3 Forecasting Based on LVAR(*p,q*) Het_CGMs

The LVAR(*p,q*) Het_CGMs considered would have the following general ES.

$$\begin{gathered} Log(HS)\ log(HS(-1))...log(HS(-p))\ \ @Year*@Expand(@Month) \\ @Expand(@Month)\ AR(1)...AR(q) \end{gathered} \tag{2.28}$$

Note that these models should be applicable for all monthly time series, Y. Therefore, we would have the following general ES for any monthly time series variables and, for each numerical variable Y, we could have various possible models. I would say readers can easily conduct the data analysis. For this reason, only special model(s) will be presented in the following examples

$$\begin{gathered} log(Y)\ log(Y(-1))...log(Y(-p))\ \ @Year*@Expand(@Month) \\ @Expand(@Month)\ AR(1)...AR(q) \end{gathered} \tag{2.29}$$

Example 2.23 A Special LVAR(12,4) Het_CGM

As a modification of the heterogeneous classical growth model (Het_CGM) in (2.1) and (2.2), Figure 2.26 only presents the forecast evaluation based on a special LVAR(12,4) Het_CGM, using the ES as follows:

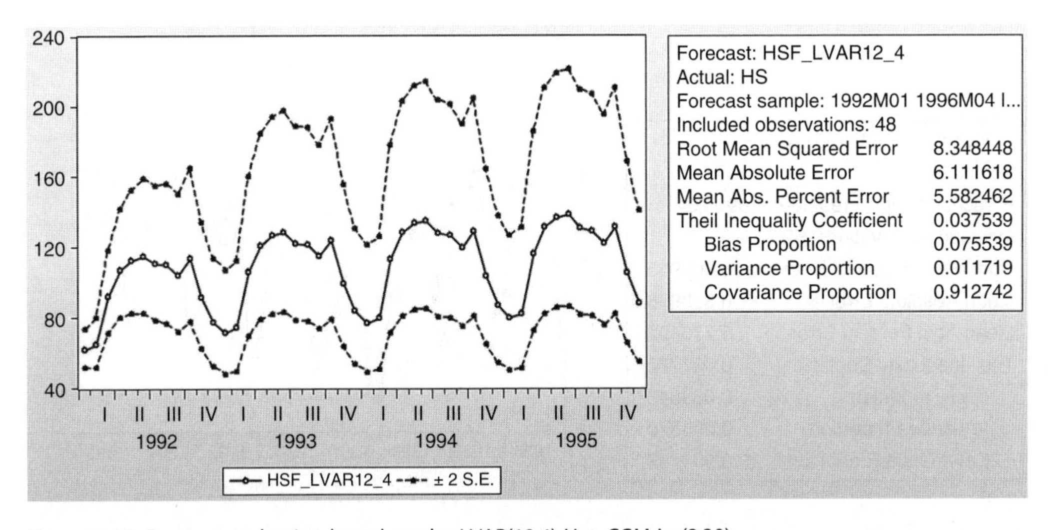

Figure 2.26 Forecast evaluation based on the LVAR(12,4) Het_CGM in (2.30).

$$log(HS)\ log(HS(-12))\ @Year*@Expand(@Month)\ @Expand(@Month)\ AR(1)\ AR(2)\ AR(3)\ AR(4) \tag{2.30}$$

Based on the complete statistical results of this model, the following notes and comments are presented.

1) This forecast has a TIC = 0.037539, which is smaller than the forecast based on the AR(1) Het_CGM with a TIC = 0.038324, and greater than the forecast based on the AR(4) Het-CGM with a TIC = 0.037462.
2) By using the Wald test, it is found that the slopes of *@Year* over the months have insignificant differences, or the null hypothesis H_0: C(2) = C(3) = ... = C(13) is accepted, based on the *F*-statistic of $F_0 = 0.970885$, with $df = (11,555)$, and *p value* = 0.4718. So, in a statistical sense, the heterogeneous model could be reduced to a homogeneous model of *log(HS)* on *log(HS(−1))* and *@Year* by the categorical variable *@Expand(@Month),* or *MONTH,* with the following ES. However, please refer to special notes on the conclusions of hypotheses testing, presented in Example 2.18

$$log(HS)\ log(HS(-12))\ @Year\ @Expand(@Month)\ AR(1)\ AR(2)\ AR(3)\ AR(4) \tag{2.31}$$

Note that this homogeneous model in fact is a special LVAR(12,4) One-Way ANCOVA model of *log(HS)* with a single factor *@Expand(@Month)* and covariates *log(HS(-12))* and *@Year.* Referring to various fixed-effects models based on panel data presented by various researchers, and Agung (2014), this model can be considered to be a fixed-effects model, namely a special LVAR(12,4) *month-fixed-effects model* (Month_FEM). As a comparison with the other forecasts, Figure 2.27 presents its forecast evaluation. This model has a TIC = 0.037772, which is slightly greater than the LV(12,4) Het_CGM in (2.30) with the TIC = 0.037539, and the AR(4) Het-CGM with a TIC = 0.037462. Despite this, in the statistical sense, this model is the worst model among the three. However, it is better than the AR(1) Het_CGM with a TIC = 0.038324.

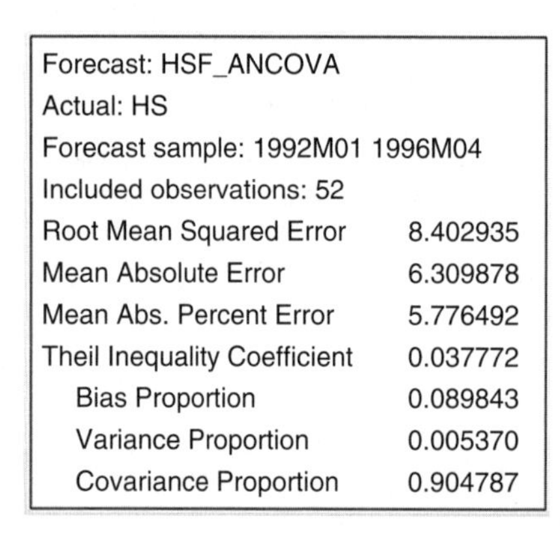

Forecast: HSF_ANCOVA
Actual: HS
Forecast sample: 1992M01 1996M04
Included observations: 52

Root Mean Squared Error	8.402935
Mean Absolute Error	6.309878
Mean Abs. Percent Error	5.776492
Theil Inequality Coefficient	0.037772
Bias Proportion	0.089843
Variance Proportion	0.005370
Covariance Proportion	0.904787

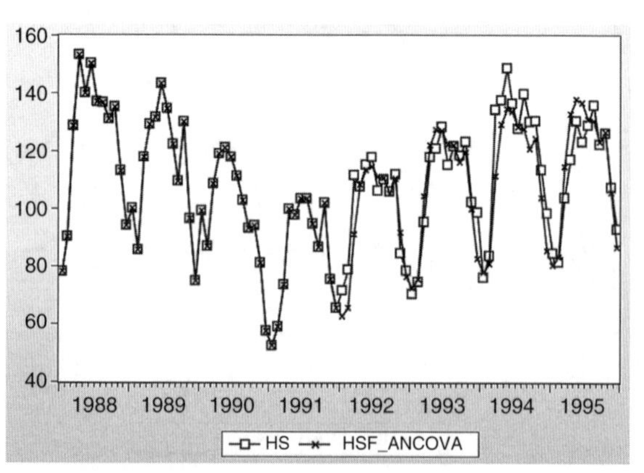

Figure 2.27 Forecast evaluation based on the special LVAR(12,4) Month_FEM in (2.31).

Omitted Variables Test
Equation: UNTITLED
Specification: LOG(HS) LOG(HS(-12)) @YEAR @EXPAND(@MONTH)
AR(1) AR(2) AR(3) AR(4)
Omitted Variables: LOG(HS(-12))*@EXPAND(@MONTH,@DROPFIRST)

	Value	df	Probability
F-statistic	0.593289	(11, 555)	0.8350
Likelihood ratio	6.827129	11	0.8129

Figure 2.28 An omitted variables test based on the model (2.31).

3) On the other hand, to study whether *log(HS(−12))* should have heterogeneous slopes over the months in the model (2.31), the OVT – Likelihood Ratio can be done, as presented in the following notes.
   - With the statistical results of the model (2.31) on the screen, select *View/Coefficient Diagnostics/OVT,* then insert a set of 11 variables in *log(HS(−12))∗ @Expand(@Month,@-Dropfirst) ... OK,* the out in Figure 2.28 is obtained
   - Note the term *log(HS(−12)∗@Expand(@Month,@Dropfirst)* indicates that January (@Month = 1(first) is used as the reference month. So the coefficient of *log(HS(−12)* represents the linear effect of *log(HS(−12))* in January.
   - Based on the *F*-statistic of $F_0 = 0.593289$ with $df = (11{,}544)$, and p value = 0.8350, it can be concluded that the set of 12 variables have insignificant effects, so they do not have to be inserted as additional variables.
4) Finally, based on these findings, then the additive ANCOVA or month-fixed-effects model in (2.31) is a good fit model, in a statistical sense. However, refer to the illustrated polynomial effects of *@Year* on *log(HS),* as presented in Figure 2.18, and a quadratic relationship in Figure 2.9 specific for *@Month* = 10. Figure 2.28 presents an omitted variables test based on the model in (2.31).

2.5 Forecast Models of *G* in Currency.wf1

It is recognized that a good or the best fit model obtained based on a data set or a time series might not become the best fit model based on other data set or time series. For this reason, I select the time series *G* in Currency.wf1 for additional illustration. Agung (2014, 2009a) presented a lot of models with the time t = *@Trend* + 1 = {1, 2, ..., T} as one of the independent variables, such as models with trend and models with *time-related-effects* (TRE). It is recognized that all of those models should be valid for forecasting. However, this section will present a few of their modifications for illustration.

In addition, referring to the model presented by Startz (2007, p. 213), illustrated in EViews, with the following ES where $G_t = 1200*dlog(curr)$ and $dlog(curr_t) = (log(curr_t) - log(curr_{t-1}))$, which in fact is the exponential growth rates of the time series $CURR_t$ in Currency.wf1.

$$G\ G(-1)\ @Trend\ @Expand(@Month) \tag{2.32}$$

This section presents modified or alternative models with *@Trend* as one of the numerical independent variables, as an extension of the Startz's model, and the modification of various models *HS* on *@Year,* by *@Month,* presented earlier.

2.5.1 LVAR(*p,q*) Additive Models of *G* by *@Month* with *@Trend*

As an extension of the Startz's model in (2.32), we could have more general LVAR(*p,q*) additive models of any monthly time series variable G_t, either the observed or measured variable or the transformed variable, which can be represented by the following ES.

$$G\ G(-1)...G(-p)\ @Trend\ @Expand(@Month)\ AR(1)...AR(q) \quad (2.33a)$$

It is recognized that exactly the same parameter estimates could be obtained using the time variable $t = @Trend + 1 = \{1, 2, ..., T\}$ similar to all models presented in Agung (2014, 2009a). Therefore, in order to be consistent with the models with trend presented in Agung (2014, 2009a) I prefer to use this time variable $t = @Trend + 1$ for all models in this book. So the ES in (2.32a) will be presented as follows.

$$Y\ Y(-1)...Y(-p)\ t\ @Expand(@Month)\ AR(1)...AR(q) \quad (2.33b)$$

For $p > 0$ and $q = 0$ we will have LV(p) models, and AR(q) models for $p = 0$ and $q > 0$ by *@Month* with trend. Therefore, we could present various alternative models, but we could never know which one is the true population model – refer to special notes presented in Agung (2009a, Section 2). However, this section presents very few illustrative examples, which can easily be extended to the models in (2.33a,b) with various integers of p and q, which should be obtained using the trial-and-error method to find a good fit model based on any monthly time series.

Example 2.24 Application of Startz's Model
As one of the simplest models in (2.33a,b), Figure 2.29 presents the statistical results of the Startz's model in (2.32), using the time variable $t = @Trend + 1$. Based on these results, the following findings and notes are presented.

1) The parameters' estimates obtained are exactly the same as the output presented in EViews Illustrated.
2) In this case, the forecast sample is 2001M01 2005M04, with a large TIC = 0.371234. And the graphs of the observed scores of *G* and its *GF_LV1* for the *YEAR* > 1998 show unexpected patterns or break points. It is found the data has four missing values, namely in 1933M03, 1933M04, 1999M12, and 2000M01. What should be done with the missing values? One of the methods is to estimate a missing value by using the CGMs based on a subsample around the missing value. Do it as an exercise.
3) Compared to the results presented by Startz, he did not observe missing values or the break points, where one of them is clearly presented in Figure 2.29.
4) For this reason, I would present an alternative analysis by using the Least Squares with Breakpoints. After finding the breakpoints, a modified model by a time-period (*TP*) can be applied. See the following section.
5) Note that this model in fact is a month-fixed-effects model, or an ANCOVA with a single factor *@Month*, and covariates *G(−1)*, and the time *t*, with the assumption that *G(−1)*, and the time *t* have constant linear effects on *G*. Agung (2006, 2009a, 2011a, 2014) considered ANCOVA models are not recommended, since a set of heterogeneous regressions are valid and more reliable in general. On the other hand, additive multi-way ANCOVA models, or multi-factorial-fixed-effects models are considered the worst within the group of ANCOVA models.
6) As a comparative study, I am considering a modified Startz's model, using the independent variable *@Year*, instead of *@Trend*, with the statistical results presented in Figure 2.30. Based on these results, the following findings and notes are presented.

Dependent Variable: G
Method: Least Squares
Date: 01/27/14 Time: 13:05
Sample (adjusted): 1917M10 2005M04
Included observations: 1045 after adjustments

Variable	Coefficient	Std. Error	t-Statistic	Prob.
G(-1)	0.469423	0.027485	17.07935	0.0000
T	0.001784	0.001011	1.764780	0.0779
@MONTH=1	-32.41308	1.334413	-24.29014	0.0000
@MONTH=2	0.475212	1.323574	0.359037	0.7196
@MONTH=3	9.413020	1.222929	7.697113	0.0000
@MONTH=4	3.315609	1.190559	2.784918	0.0055
@MONTH=5	2.160471	1.197076	1.804790	0.0714
@MONTH=6	5.739418	1.186273	4.838193	0.0000
@MONTH=7	6.100579	1.199755	5.084856	0.0000
@MONTH=8	-2.868662	1.211199	-2.368449	0.0180
@MONTH=9	7.018904	1.183580	5.930234	0.0000
@MONTH=10	3.721678	1.194653	3.115279	0.0019
@MONTH=11	8.319947	1.192758	6.975385	0.0000
@MONTH=12	17.40884	1.220072	14.26870	0.0000

R-squared	0.594056	Mean dependent var	6.093774
Adjusted R-squared	0.588937	S.D. dependent var	15.35868
S.E. of regression	9.847094	Akaike info criterion	7.425536
Sum squared resid	99971.19	Schwarz criterion	7.491876
Log likelihood	-3865.843	Hannan-Quinn criter.	7.450696
Durbin-Watson stat	2.128568		

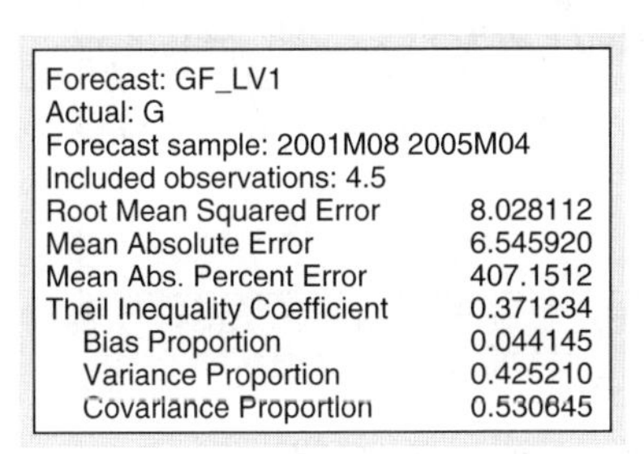
Forecast: GF_LV1
Actual: G
Forecast sample: 2001M08 2005M04
Included observations: 4.5

Root Mean Squared Error	8.028112
Mean Absolute Error	6.545920
Mean Abs. Percent Error	407.1512
Theil Inequality Coefficient	0.371234
Bias Proportion	0.044145
Variance Proportion	0.425210
Covariance Proportion	0.530645

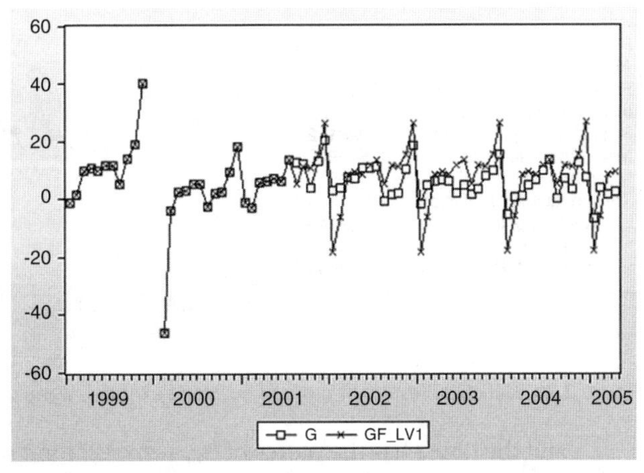

Figure 2.29 Statistical results based on Startz's model in (2.32).

6.1 The forecast evaluation of this model presents exactly the same statistical values, such the RMSE = 9.028112 and TIC = 0.37124, with the forecast evaluation of the original Startz's model in Figure 2.29. So this model can be used to replace the Starz's model to do the forecasting.

6.2 In addition note that the parameters of the numerical variables *G(−1)* and *@Year* have the probabilities of 0.0000 and 0.0779, respectively, which are exactly the same as the variables *G(−1)* and the time t = *@Trend* + 1, in Figure 2.29.

6.3 Note that the variables *@Year* and the time t (or *@Trend*) do have different scores, but within each of the 12 months, each of the variables has the same spacing, namely *@Year* has exactly the same set of scores of 1917, 1918, 1919, ... and so on; within each month,

Dependent Variable: G
Method: Least Squares
Date: 03/31/14 Time: 09:19
Sample (adjusted): 1917M10 2005M04
Included observations: 1045 after adjustments

Variable	Coefficient	Std. Error	t-Statistic	Prob.
G(-1)	0.469423	0.027485	17.07935	0.0000
@YEAR	0.021413	0.012134	1.764780	0.0779
@MONTH=1	-73.47322	23.76233	-3.092004	0.0020
@MONTH=2	-40.58314	23.87777	-1.699620	0.0895
@MONTH=3	-31.64354	23.85338	-1.326585	0.1849
@MONTH=4	-37.73917	23.81207	-1.584876	0.1133
@MONTH=5	-38.89252	23.80354	-1.633897	0.1026
@MONTH=6	-35.31179	23.80278	-1.483515	0.1382
@MONTH=7	-34.94885	23.79400	-1.468809	0.1422
@MONTH=8	-43.91630	23.78918	-1.846062	0.0652
@MONTH=9	-34.02695	23.80864	-1.429185	0.1533
@MONTH=10	-37.32240	23.78695	-1.569028	0.1169
@MONTH=11	-32.72234	23.78810	-1.375576	0.1693
@MONTH=12	-23.63166	23.77360	-0.994030	0.3204

R-squared	0.594056	Mean dependent var	6.093774
Adjusted R-squared	0.588937	S.D. dependent var	15.35868
S.E. of regression	9.847094	Akaike info criterion	7.425536
Sum squared resid	99971.19	Schwarz criterion	7.491876
Log likelihood	-3865.843	Hannan-Quinn criter.	7.450696
Durbin-Watson stat	2.128568		

Forecast: GF_YEAR
Actual: G
Forecast sample: 2001M08 2005M04
Included observations: 45

Root Mean Squared Error	8.028112
Mean Absolute Error	6.545920
Mean Abs. Percent Error	407.1512
Theil Inequality Coefficient	0.371234
Bias Proportion	0.044145
Variance Proportion	0.425210
Covariance Proportion	0.530645

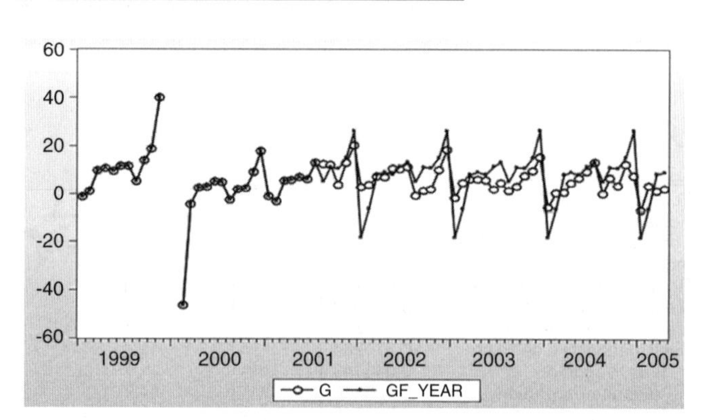

Figure 2.30 Statistical results of a modified Startz's model, using *@Year*, instead of *@Trend*.

and the time t has the scores of $\{t_m + d \times 12\}$ for $d = 0, 1, 2, \ldots$, and so on, where t_m is the starting score of the time t in the *mth* month. And the scores of both variables can be linearly transformed to two variables with the scores of 1, 2, 3, ..., and so on. For this reason, we can apply either one of the two models to present the forecasting.

Example 2.25 Application of the AR(1) Model in (2.33b)

As a comparison to Startz's model, Figure 2.31 presents the statistical results of an AR(1) model in (2.33b). Based on these results, the following findings and notes are presented.

1) It is found that the statistical results of both the regressions in the LV(1) model and AR(1) model present the same values of R-squared, Adjusted R-squared, and some other statistics. However, their parameters' estimates have different values.
2) On the other hand, the RMSE and TIC of the AR(1) model presents greater values than the LV(1) model. So it can be said that the LV(1) model is better model, in a statistical sense.

Dependent Variable: G
Method: Least Squares
Date: 01/29/14 Time: 13:21
Sample (adjusted): 1917M10 2005M04
Included observations: 1045 after adjustments
Convergence achieved after 3 iterations

Variable	Coefficient	Std. Error	t-Statistic	Prob.
T	0.003363	0.001897	1.773133	0.0765
@MONTH=1	-21.64476	1.554906	-13.92030	0.0000
@MONTH=2	-9.686926	1.555343	-6.228162	0.0000
@MONTH=3	4.864171	1.562659	3.112752	0.0019
@MONTH=4	5.597386	1.566407	3.573391	0.0004
@MONTH=5	4.786437	1.568823	3.050972	0.0023
@MONTH=6	7.984705	1.563242	5.107784	0.0000
@MONTH=7	9.847208	1.561830	6.304917	0.0000
@MONTH=8	1.752270	1.561916	1.121872	0.2622
@MONTH=9	7.839882	1.562631	5.017104	0.0000
@MONTH=10	7.400324	1.555652	4.757057	0.0000
@MONTH=11	11.79225	1.554053	7.588067	0.0000
@MONTH=12	22.94283	1.554865	14.75551	0.0000
AR(1)	0.469423	0.027485	17.07935	0.0000

R-squared	0.594056	Mean dependent var	6.093774
Adjusted R-squared	0.588937	S.D. dependent var	15.35868
S.E. of regression	9.847094	Akaike info criterion	7.425536
Sum squared resid	99971.19	Schwarz criterion	7.491876
Log likelihood	-3865.843	Hannan-Quinn criter.	7.450696
Durbin-Watson stat	2.128568		

Inverted AR Roots	.47

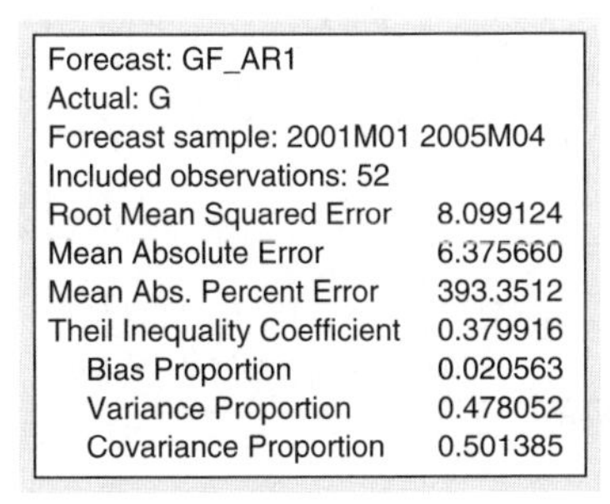

Forecast: GF_AR1
Actual: G
Forecast sample: 2001M01 2005M04
Included observations: 52

Root Mean Squared Error	8.099124
Mean Absolute Error	6.375660
Mean Abs. Percent Error	393.3512
Theil Inequality Coefficient	0.379916
Bias Proportion	0.020563
Variance Proportion	0.478052
Covariance Proportion	0.501385

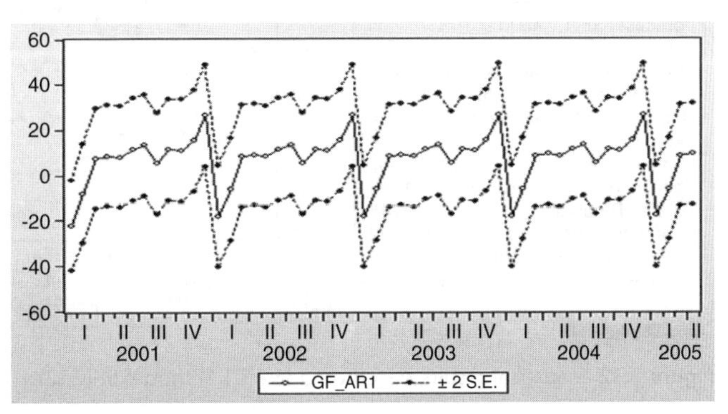

Figure 2.31 Statistical results based on an AR(1) model in (2.33b).

Example 2.26 Application of the LVAR(1,1) Model in (2.33b)

As an illustration, Figure 2.32 presents the statistical results of a LVAR(1,1) model in (2.33b), which is a combination of the LV(1), and AR(1) models. Even though this model has a greater R-squared, it has greater values of RMSE and TIC. So the forecast based on this model is worse than the forecasts based on both LV(1), and AR(1) models.

As a comparison, the following example presents a better forecast model, which is an extension of the special LV(12,4) model without trend presented in the previous chapter.

Dependent Variable: G
Method: Least Squares
Date: 01/29/14 Time: 13:12
Sample (adjusted): 1917M11 2005M04
Included observations: 1042 after adjustments
Convergence achieved after 5 iterations

Variable	Coefficient	Std. Error	t-Statistic	Prob.
G(-1)	0.651726	0.029968	21.74708	0.0000
T	0.001176	0.000788	1.491687	0.1361
@MONTH=1	-36.55446	1.338491	-27.31019	0.0000
@MONTH=2	4.424733	1.326027	3.336834	0.0009
@MONTH=3	11.05339	1.200421	9.207927	0.0000
@MONTH=4	2.630044	1.163729	2.260015	0.0240
@MONTH=5	1.156366	1.173026	0.985797	0.3245
@MONTH=6	4.997365	1.170487	4.269475	0.0000
@MONTH=7	4.713255	1.176690	4.005520	0.0001
@MONTH=8	-4.634808	1.191679	-3.889309	0.0001
@MONTH=9	6.712787	1.157328	5.800244	0.0000
@MONTH=10	2.115622	1.179515	1.793637	0.0732
@MONTH=11	7.004498	1.169897	5.987276	0.0000
@MONTH=12	15.23266	1.205030	12.64090	0.0000
AR(1)	-0.272317	0.038528	-7.068026	0.0000

R-squared	0.605362	Mean dependent var	6.115998
Adjusted R-squared	0.599982	S.D. dependent var	15.32274
S.E. of regression	9.691169	Akaike info criterion	7.394598
Sum squared resid	96454.56	Schwarz criterion	7.465839
Log likelihood	-3837.585	Hannan-Quinn criter.	7.421620
Durbin-Watson stat	1.990807		

Inverted AR Roots	-.27

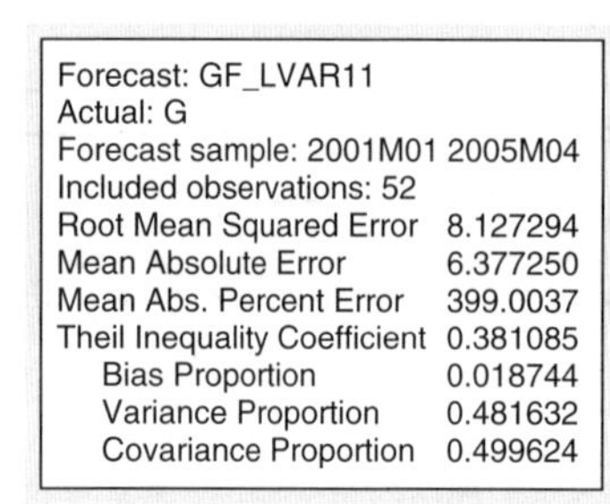

Forecast: GF_LVAR11	
Actual: G	
Forecast sample: 2001M01 2005M04	
Included observations: 52	
Root Mean Squared Error	8.127294
Mean Absolute Error	6.377250
Mean Abs. Percent Error	399.0037
Theil Inequality Coefficient	0.381085
Bias Proportion	0.018744
Variance Proportion	0.481632
Covariance Proportion	0.499624

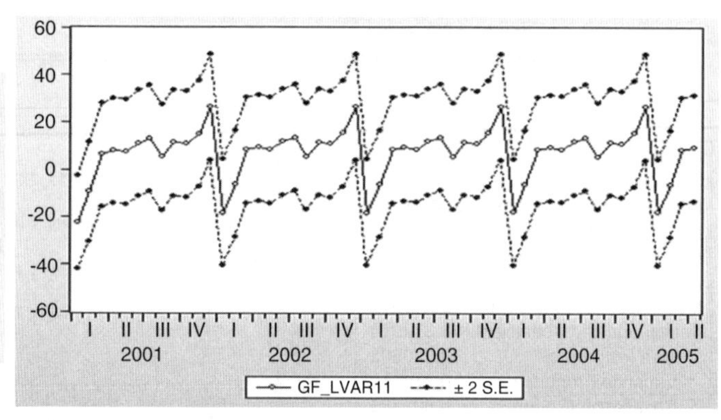

Figure 2.32 Statistical results based on a LVAR(1,1) model in (2.33b).

Example 2.27 Application of a Special LVAR(12,4) Additive Model

As an extension of the special LV(12) model without trend presented in Chapter 1, namely model (1.7), Figure 2.33 presents the statistical results of the LVAR(12,4) additive model of *G* with trend by *@Month*. Based on these results, the following findings and notes are presented.

1) This model in fact is obtained by using the trial-and-error method until I found a forecast with smaller values of RMSE and TIC, compared to the models presented in the last three examples. Refer to the special LVAR models presented in previous chapter and do the analysis based on other alternative models as exercises.
2) Note that the estimates of the regression present greater values of R-squared. So this model should be considered the best compared to the models in the last three examples.

Dependent Variable: G
Method: Least Squares
Date: 01/29/14 Time: 13:47
Sample (adjusted): 1919M01 2005M04
Included observations: 1012 after adjustments
Convergence achieved after 5 iterations

Variable	Coefficient	Std. Error	t-Statistic	Prob.
G(-12)	0.481253	0.028152	17.09456	0.0000
T	0.002517	0.002178	1.155725	0.2481
@MONTH=1	-11.75543	1.682965	-6.984951	0.0000
@MONTH=2	-5.347054	1.586577	-3.370182	0.0008
@MONTH=3	1.644388	1.579398	1.041149	0.2981
@MONTH=4	2.329535	1.584231	1.470451	0.1418
@MONTH=5	2.365081	1.582192	1.494813	0.1353
@MONTH=6	4.010094	1.593977	2.515779	0.0120
@MONTH=7	5.049493	1.603239	3.149557	0.0017
@MONTH=8	0.087463	1.583241	0.055243	0.9560
@MONTH=9	3.103247	1.590171	1.951518	0.0513
@MONTH=10	2.869854	1.586401	1.809034	0.0707
@MONTH=11	5.786432	1.608498	3.597414	0.0003
@MONTH=12	11.46182	1.702066	6.734062	0.0000
AR(1)	0.400030	0.033343	11.99740	0.0000
AR(2)	0.045036	0.035546	1.266963	0.2055
AR(3)	0.109229	0.035137	3.108642	0.0019
AR(4)	0.041249	0.031234	1.320615	0.1869

R-squared	0.706931	Mean dependent var	5.989905
Adjusted R-squared	0.701919	S.D. dependent var	15.17745
S.E. of regression	8.286401	Akaike info criterion	7.084735
Sum squared resid	68252.45	Schwarz criterion	7.172239
Log likelihood	-3566.876	Hannan-Quinn criter.	7.117973
Durbin-Watson stat	1.930608		

Inverted AR Roots	.75	-.04+.44i	-.04-.44i	-.28

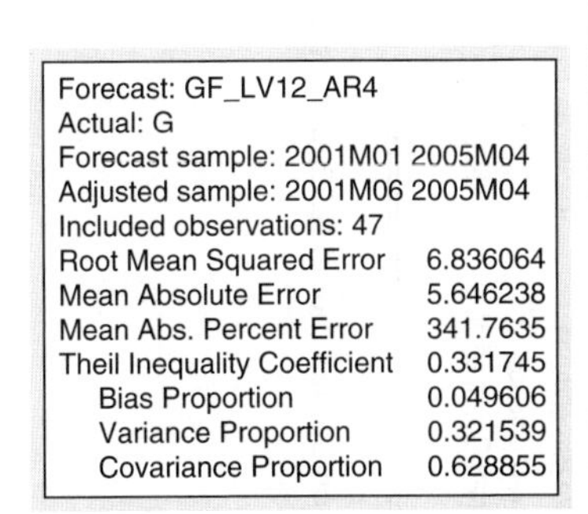

Forecast: GF_LV12_AR4	
Actual: G	
Forecast sample: 2001M01 2005M04	
Adjusted sample: 2001M06 2005M04	
Included observations: 47	
Root Mean Squared Error	6.836064
Mean Absolute Error	5.646238
Mean Abs. Percent Error	341.7635
Theil Inequality Coefficient	0.331745
Bias Proportion	0.049606
Variance Proportion	0.321539
Covariance Proportion	0.628855

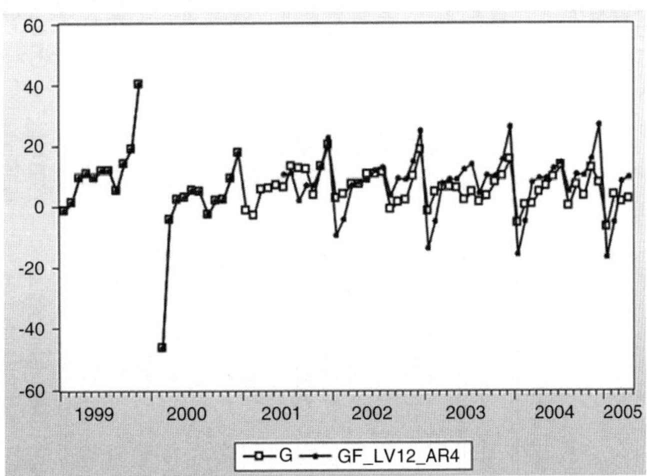

Figure 2.33 Statistical results of a special LVAR(12,4) model with trend.

Dependent Variable: G
Method: Least Squares
Date: 01/23/14 Time: 15:00
Sample (adjusted): 1917M10 2005M04
Included observations: 1045 after adjustments

Variable	Coefficient	Std. Error	t-Statistic	Prob.
G(-1)	0.469423	0.027485	17.07935	0.0000
@TREND	0.001784	0.001011	1.764780	0.0779
@MONTH=1	-32.41130	1.334063	-24.29519	0.0000
@MONTH=2	0.476996	1.323129	0.360506	0.7185
@MONTH=3	9.414805	1.222462	7.701508	0.0000
@MONTH=4	3.317393	1.190114	2.787459	0.0054
@MONTH=5	2.162256	1.196640	1.806939	0.0711
@MONTH=6	5.741203	1.185832	4.841496	0.0000
@MONTH=7	6.102364	1.199326	5.088160	0.0000
@MONTH=8	-2.866877	1.210778	-2.367798	0.0181
@MONTH=9	7.020689	1.183130	5.933996	0.0000
@MONTH=10	3.723463	1.194226	3.117887	0.0019
@MONTH=11	8.321732	1.192328	6.979395	0.0000
@MONTH=12	17.41063	1.219665	14.27493	0.0000

R-squared	0.594056	Mean dependent var	6.093774
Adjusted R-squared	0.588937	S.D. dependent var	15.35868
S.E. of regression	9.847094	Akaike info criterion	7.425536
Sum squared resid	99971.19	Schwarz criterion	7.491876
Log likelihood	-3865.843	Hannan-Quinn criter.	7.450696
Durbin-Watson stat	2.128568		

Omitted Variables Test
Equation: EQ03
Specification: G G(-1) @TREND @EXPAND(@MONTH)
Omitted Variables: @TREND*@EXPAND(@MONTH,@DROPLAST)

	Value	df	Probability
F-statistic	36.72264	(11, 1020)	0.0000
Likelihood ratio	348.6448	11	0.0000

Omitted Variables Test
Equation: EQ03
Specification: G G(-1) @TREND @EXPAND(@MONTH)
Omitted Variables: G(-1)*@EXPAND(@MONTH,@DROPLAST)

	Value	df	Probability
F-statistic	9.542375	(11, 1020)	0.0000
Likelihood ratio	102.3581	11	0.0000

Omitted Variables Test
Equation: EQ03
Specification: G G(-1) @TREND @EXPAND(@MONTH)
Omitted Variables: G(-1)*@EXPAND(@MONTH,@DROPLAST) @TREND *@EXPAND(@MONTH,@DROPLAST)

	Value	df	Probability
F-statistic	21.96851	(22, 1009)	0.0000
Likelihood ratio	408.9750	22	0.0000

Figure 2.34 Statistical results based on Startz's model (2.32) and omitted variables tests.

2.5.2 LV(1) Heterogeneous Models of *G* by *@Month*

Note that Starz's model is in fact an ANCOVA model with the assumption that *G(−1)* and *@Trend* (or the time *t*) have the same linear effects on *G* within the 12 months. I would say that *G(−1)* or the time *t* might have different linear effects over the 12 months. For this reason, I am considering possible heterogeneous models, as the extension of the Startz's model, by inserting additional independent variables *G(−1)*@Expand(@Month)*, *t*@Expand(@Month)*, or both. See the following example.

Example 2.28 Application of the Omitted Variables Tests
As a more advanced analysis based on Startz's model, Figure 2.34 presents the statistical results of the Startz's model in (2.32) with three alternative OVTs. Based on these results the following findings and notes are presented.

1) The OVTs can easily be done, with conclusions as follows:
   1.1 Having the statistical results of the model on the screen, by selecting *View/Coefficient Diagnostics/OVT – Likelihood Ratio ...*, then the dialog is shown on the screen.
   1.2 Insert the function or a set of variables *@Trend*@Expand(@Month,@Dropfirst) ... OK*, then the first test in Figure 1.36 as a part of the statistical results is shown on the screen. Based on the *F*-statistic of $F_0 = 36.72264$, with $df = (11{,}1020)$, and *p* value = 0.0000, it can be concluded that the model with heterogeneous trends is an acceptable model.
   1.3 The second test also shows that *G(−1)*@Expand(@Month,@Dropfirst)* has a significant effect on *G*, based on the *F*-statistic of $F_0 = 9.542375$, with $df = (11{,}1009)$, and *p* value = 0.0000.
   1.4 Finally, both sets of variables also have significant joint effects on *G*, based on the *F*-statistic of $F_0 = 21.96821$ with $df = (22{,}1009)$, and *p* value = 0.0000. So it can be concluded that the heterogeneous regression of *G* on both *G(−1)* and the time *t* is an acceptable model, in both theoretical and statistical senses.

2) Then it can be concluded that heterogeneous LV(1) models should be better than the additive Startz's model, in the theoretical and statistical sense, and these are represented in the following subsections.

2.5.2.1 Heterogeneous Model of *G* on both *G(−1)* and *@Trend*

The model can be presented using the ES as follows:

$$G\ G(-1)*@Expand(@Month)\ @Trend*@Expand(@Month)\ @Expand(@Month) \tag{2.34a}$$

$$G\ C\ G(-1)*@Expand(@Month)\ @Trend*@Expand(@Month)\ @Expand(@Month,@Droplast) \tag{2.34b}$$

Note that the ES (2.34b) has an intercept parameter *C*, which indicates that the output will present the *F*-statistic for testing the joint effects of all independent variables on *G*. In a statistical sense, an unexpected reduced model might be obtained as a good fit model, which is highly dependent on the data set, specifically the data within each of the 12 months.

However, there is a question as to whether or not we should reduce the model to forecast *G*, even though the model has been accepted in a theoretical sense. If the reduced model could be applied to forecasting, then *G* would be predicted by a single variable, either *G(−1)* or *@Trend*, for at least one specific month. Refer to the following Eq. (2.34c) and three of all possible reduced models are presented in the following subsections.

Furthermore, note that the model (2.34a) represents a set of 12 additive regression models G_{mt} on $G_{m,t-1}$ and t = *@Trend*, for each *@Month* = *m*, with the following general equation.

$$G_{mt} = \beta_{mo} + \beta_{m1} \times G_{m,t-1} + \beta_{m2} \times t + \varepsilon_{mt},\ m = 1,2...,12 \tag{2.34c}$$

2.5.2.2 Heterogeneous Model of G on *G(−1)* by *@Month* with *@Trend*

This model in fact is one of many possible reduced models of the models in (2.34a,b,c), which can be presented using the ES as follows:

$$G\ G(-1)\ )*@Expand(@Month)\ @Trend\ @Expand(@Month) \tag{2.35a}$$

$$G\ C\ G(-1)*@Expand(@Month)\ @Trend\ @Expand(@Month,@Droplast) \tag{2.35b}$$

2.5.2.3 LV(1) Model of G with Heterogeneous *@Trend* by *@Month*

This model in fact is another reduced model of the models in (2.34a,b,c), which can be presented using the ES as follows:

$$G\ G(-1)\ @Trend*@Expand(@Month)\ @Expand(@Month) \tag{2.36a}$$

$$G\ C\ G(-1)\ @Trend*@Expand(@Month)\ @Expand(@Month,@Droplast) \tag{2.36b}$$

2.5.2.4 The Startz's Model

Note that Startz's model (2.32) is in fact an additive reduced model of the models in (2.34a), which can be presented using an alternative ES with an intercept parameter as follows:

$$G\ C\ G(-1)\ @Trend\ @Expand(@Month,@Droplast) \tag{2.37}$$

Example 2.29 Application of the Model (2.34b)

Figure 2.35 presents the statistical results of the model in (2.34b). Based on these results, the following findings and notes are presented.

1) By using the Wald test, it was found that *G(−1)* has significant different effects on *G* between the 12 months, based on the *F*-statistic of F_0 = 5.451466, with *df* = (11,1009), and *p* value = 0.0000. And *@Trend* also has significantly different effects on *G* between the 12 months, based on the *F*-statistic of F_0 = 31.27872, with *df* = (11,1009), and *p* value = 0.0000. So there is no reason to reduce the model to a model with homogeneous slopes of *G(−1)* or homogeneous *@Trend* by *@Month*.
2) As expected, *G(−1)* has a significant effect on *G* within each of the 12 months. Based on my observations, *Y(−1)* always has a linear significant effect on *Y* for any times series Y_t.
3) However, the numerical variable *@Trend* or the time *t* might have an insignificant linear effect within some months. In this case, *@Trend* does have very large *p* values, such as 0.8974 and 0.8000 within *@Month* = 3,and *@Month* = 6, respectively, and it has a *p* value of 0.3735 in *@Month* = 10. So, in a statistical sense, the model could be reduced to a model without the *@Trend* for the third and sixth months. The data analysis can be easily done

Dependent Variable: G
Method: Least Squares
Date: 03/30/14 Time: 10:26
Sample (adjusted): 1917M10 2005M04
Included observations: 1045 after adjustments

Variable	Coefficient	Std. Error	t-Statistic	Prob.
C	23.55314	2.017962	11.67175	0.0000
G(-1)*(@MONTH=1)	0.336044	0.084589	3.972668	0.0001
G(-1)*(@MONTH=2)	0.562905	0.061214	9.195687	0.0000
G(-1)*(@MONTH=3)	0.396792	0.078496	5.054943	0.0000
G(-1)*(@MONTH=4)	0.750492	0.117686	6.377070	0.0000
G(-1)*(@MONTH=5)	0.520687	0.112081	4.645639	0.0000
G(-1)*(@MONTH=6)	0.821384	0.096900	8.476585	0.0000
G(-1)*(@MONTH=7)	0.780651	0.092782	8.413776	0.0000
G(-1)*(@MONTH=8)	0.592752	0.099257	5.971875	0.0000
G(-1)*(@MONTH=9)	0.951587	0.096409	9.870347	0.0000
G(-1)*(@MONTH=10)	0.655079	0.080752	8.112242	0.0000
G(-1)*(@MONTH=11)	0.300328	0.069919	4.295378	0.0000
G(-1)*(@MONTH=12)	0.697531	0.109514	6.369329	0.0000
@TREND*(@MONTH=1)	0.045008	0.003031	14.84741	0.0000
@TREND*(@MONTH=2)	-0.022850	0.003857	-5.925100	0.0000
@TREND*(@MONTH=3)	0.000438	0.002885	0.151706	0.8794
@TREND*(@MONTH=4)	0.010530	0.002885	3.649903	0.0003
@TREND*(@MONTH=5)	0.003573	0.003223	1.108507	0.2679
@TREND*(@MONTH=6)	0.000797	0.003144	0.253427	0.8000
@TREND*(@MONTH=7)	-0.004023	0.003081	-1.305851	0.1919
@TREND*(@MONTH=8)	-0.006748	0.002945	-2.291164	0.0222
@TREND*(@MONTH=9)	-0.013672	0.002939	-4.652415	0.0000
@TREND*(@MONTH=10)	-0.002845	0.003196	-0.890271	0.3735
@TREND*(@MONTH=11)	0.013410	0.003034	4.420097	0.0000
@TREND*(@MONTH=12)	-0.015829	0.003023	-5.235306	0.0000
@MONTH=1	-75.23634	3.683951	-20.42273	0.0000
@MONTH=2	-8.335264	3.693191	-2.256927	0.0242
@MONTH=3	-14.08481	2.790696	-5.047061	0.0000
@MONTH=4	-26.61239	2.765290	-9.623724	0.0000
@MONTH=5	-22.70464	2.695821	-8.422163	0.0000
@MONTH=6	-19.32391	2.681835	-7.205481	0.0000
@MONTH=7	-17.31230	2.704255	-6.401873	0.0000
@MONTH=8	-23.32717	2.832340	-8.236006	0.0000
@MONTH=9	-10.01961	2.740260	-3.656446	0.0003
@MONTH=10	-19.20341	3.068775	-6.257680	0.0000
@MONTH=11	-19.77742	2.909396	-6.797774	0.0000

R-squared	0.725527	Mean dependent var	6.093774
Adjusted R-squared	0.716006	S.D. dependent var	15.35868
S.E. of regression	8.184805	Akaike info criterion	7.076278
Sum squared resid	67593.95	Schwarz criterion	7.246865
Log likelihood	-3661.355	Hannan-Quinn criter.	7.140975
F-statistic	76.20396	Durbin-Watson stat	2.118470
Prob(F-statistic)	0.000000		

(a)

Figure 2.35 (a) Statistical results of the model (2.34b). (b) Forecast evaluation based on the LV(1) model in (a).

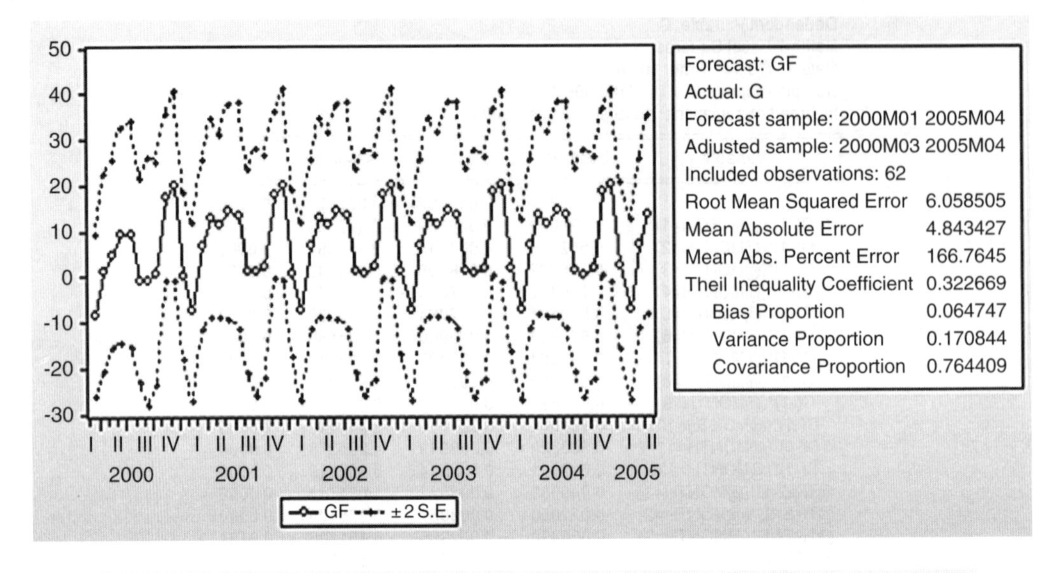

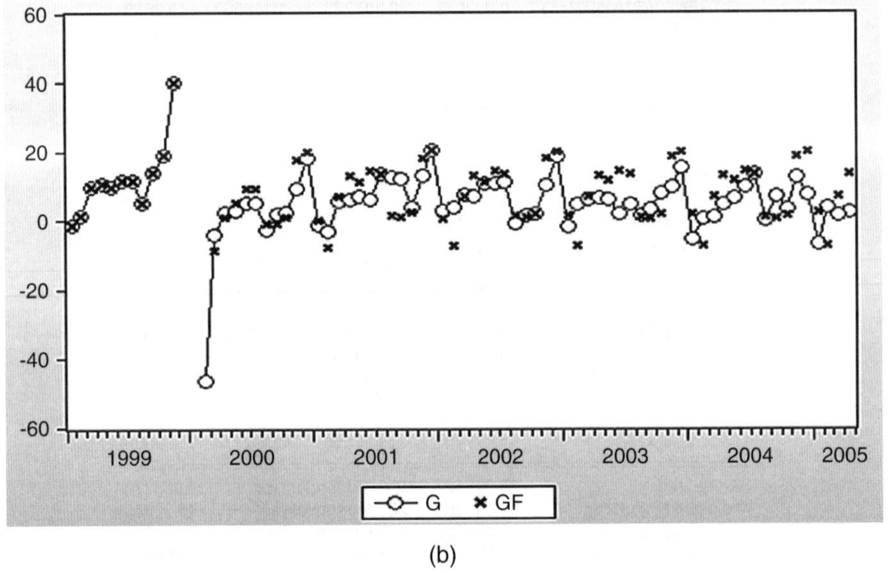

(b)

Figure 2.35 (Continued)

using the ES as follows. Do it as an exercise and note the use of the functions *@Drop(3)* and *@Drop(6)*.

$$G\ C\ G(-1)*@Expand(@Month)$$
$$@Trend*@Expand(@Month,@Drop(\mathbf{3}),@Drop(\mathbf{6}))@Expand(@Month,@Droplast) \tag{2.38}$$

On the other hand, by doing data analysis based on one of the models in (2.34a), and (2.34b), the slopes differences of *G(−1)* and *@Trend* can easily be tested using the Wald test. Do it as an exercise and see Example 2.41.

4) In addition, note that the coefficients of the two numerical variables *G(−1)* and *@Trend* within each month could be unpredicted, because of the impact of their significant correlation, based on the *t*-statistic of t_0 = 2.188026 with a *p* value = 0.0289. For instance, for *@Month* = 2, *@Trend* has a negative significant effect on *G*, but *@Trend* and *G* have a significant positive bivariate correlation based on the *t*-statistic of t_0 = 8.076291 with a *p* value = 0.0000.
5) On the other hand, the linear effect of *@Trend* on *G* within *@Month* = 3 has an insignificant linear effect with a very large *p* value = 0.8794. Why? Because it might have a polynomial or nonlinear trend.
6) Note that any model with *@trend* is an acceptable model, in a statistical sense, even though *@Trend* has a nonlinear effect on *G*.
7) As an illustration, Figure 2.36 presents three scatter graphs of *(@Trend,G)* with their nearest neighbor fit curves for the *@Month* = 1, 2, and 3, which show that *@Trend* has nonlinear effects on *G*. But the regression line of *G* on *@Trend* is almost horizontal for *@Month* = 3.

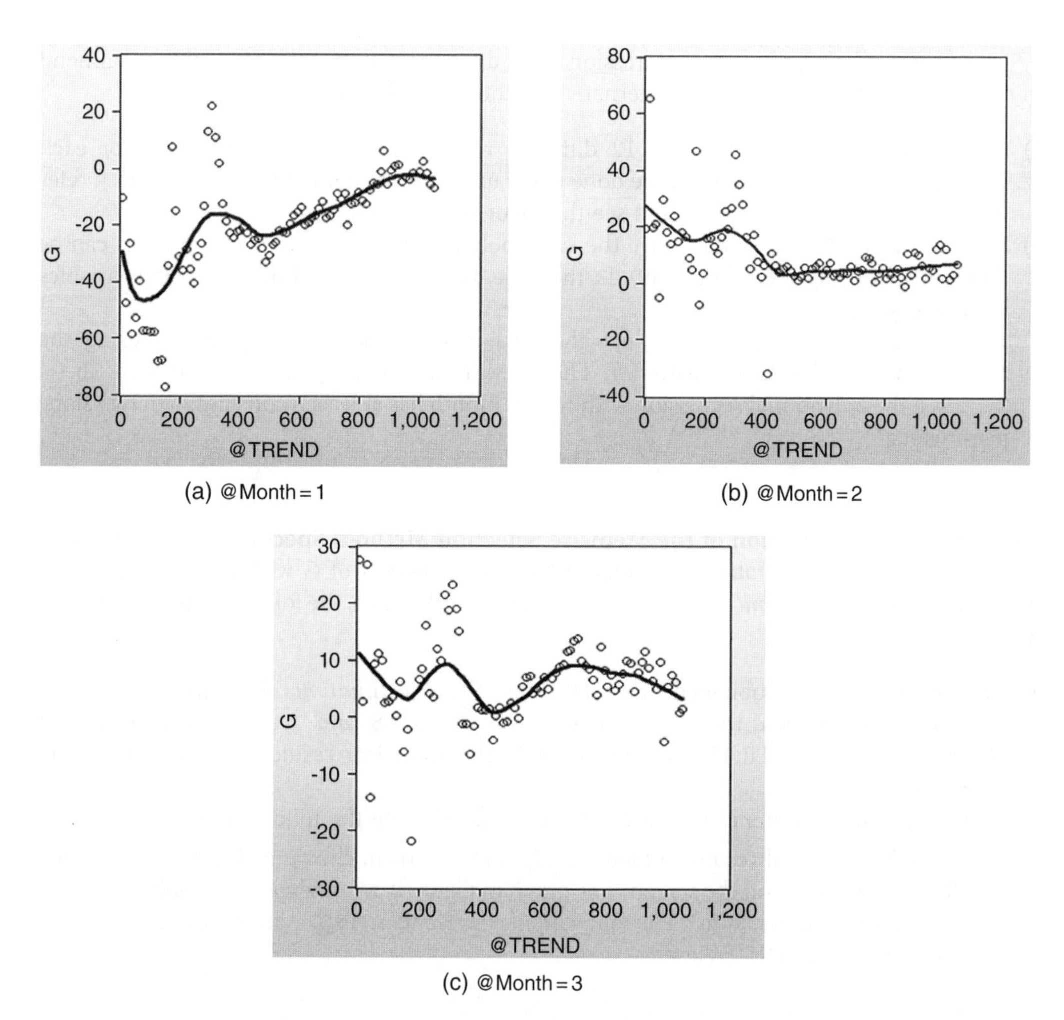

Figure 2.36 Nonparametric (polynomial) regressions of *G* on *@Trend*, within *@Month* = 1, 2, and 3.

2.6 Forecast Models of G on G(−1) and Polynomial Time Variables

Referring to the growth pattern of G, I consider that forecast models with polynomial time variables would be an appropriate forecast model of G as an extension of the previous models with trends. However, only two alternative models are presented in the following subsections.

2.6.1 Heterogeneous Model of G on G(−1) and Polynomial T by @Month

Note the models in (2.34a,b,c) are LV(1) heterogeneous regressions of G with linear trends by *@Month*, and this section presents heterogeneous regressions of G on *G(−1)* and the *k*-degree polynomial trends by *@Month*, with the general ES as follows:

$$\begin{aligned} &G\ G(-1)*@Expand(@Month)\ t*@Expand(@Month)...\\ &\quad t\hat{\ }k*@Expand(@Month)\ @Expand(@Month) \end{aligned} \tag{2.39}$$

Since, in general, the polynomial regressions have different patterns, then I would recommend to conduct data analysis using two alternative methods as follows:

1) *Recommended Method.* Conduct 12 different regression analyses, conditional for each *@Month* = 1, 2, …, 12, which can be done easily using the stepwise selection method – refer to Agung (2014, 2009a, 2011a) and see the following example.
2) However, if the 12 regressions have the same polynomial pattern, then the analysis can be done directly using the ES in (2.39). In this case, the results would present many variables with large *p* values.
3) *Not-Recommended Method.* Conduct the data analysis using an explicit ES using the 12 dummy variables of the *@Month*. The ES will have many parameters, at least 36 (= 3 × 12), if the polynomial regression within each month has at least a constant with two slope parameters.

Example 2.30 Application of the Stepwise Selection Method, Specific for *@Month* = 1
Figure 2.37 presents the statistical results of two LV(1) models of G with polynomial trends, using the variable *t* = *@Trend* + 1. Based on these statistical results, the following notes and comments are presented.

1) The estimates are obtained by selecting *Quick/Estimation Equation …/STEPLS* estimation method, and then insert "*G C G(−1)*" as the ES, and "t t^2 t^3 t^4" as the list of search regressors. Finally, by clicking *OK*, the statistical results in Figure 2.37a are obtained.
2) The forecast evaluations in Figure 2.37b are obtained using the following steps.
   2.1 With the statistical results in Figure 2.37a on the screen, click the object *Forecast*, then Figure 2.38 shown on the screen. The graph in Figure 2.37b unexpectedly only presents a single forecast point. Why? The following box presents an explanation from Gareth Thomas, Director of HIS EViews.

Dependent Variable: G
Method: Stepwise Regression
Date: 04/17/14 Time: 12:38
Sample: 1917M08 2005M04 IF @MONTH=1
Included observations: 87
Number of always included regressors: 2
Number of search regressors: 5
Selection method: Stepwise forwards
Stopping criterion: p-value forwards/backwards = 0.5/0.5

Variable	Coefficient	Std. Error	t-Statistic	Prob.*
C	-43.43206	8.195935	-5.299219	0.0000
G(-1)	0.479865	0.132328	3.626315	0.0005
T	-0.519602	0.154718	-3.358389	0.0012
T^2	0.004170	0.000909	4.585346	0.0000
T^4	1.15E-08	2.26E-09	5.098349	0.0000
T^5	-4.32E-12	8.48E-13	-5.099444	0.0000
T^3	-1.08E-05	2.17E-06	-4.971017	0.0000

R-squared	0.622461	Mean dependent var	-19.94558
Adjusted R-squared	0.594145	S.D. dependent var	19.16777
S.E. of regression	12.21116	Akaike info criterion	7.919615
Sum squared resid	11928.99	Schwarz criterion	8.118022
Log likelihood	-337.5033	Hannan-Quinn criter.	7.999508
F-statistic	21.98308	Prob(F-statistic)	0.000000

Selection Summary

Added T
Added T^2
Added T^4
Added T^5
Added T^3

(a)

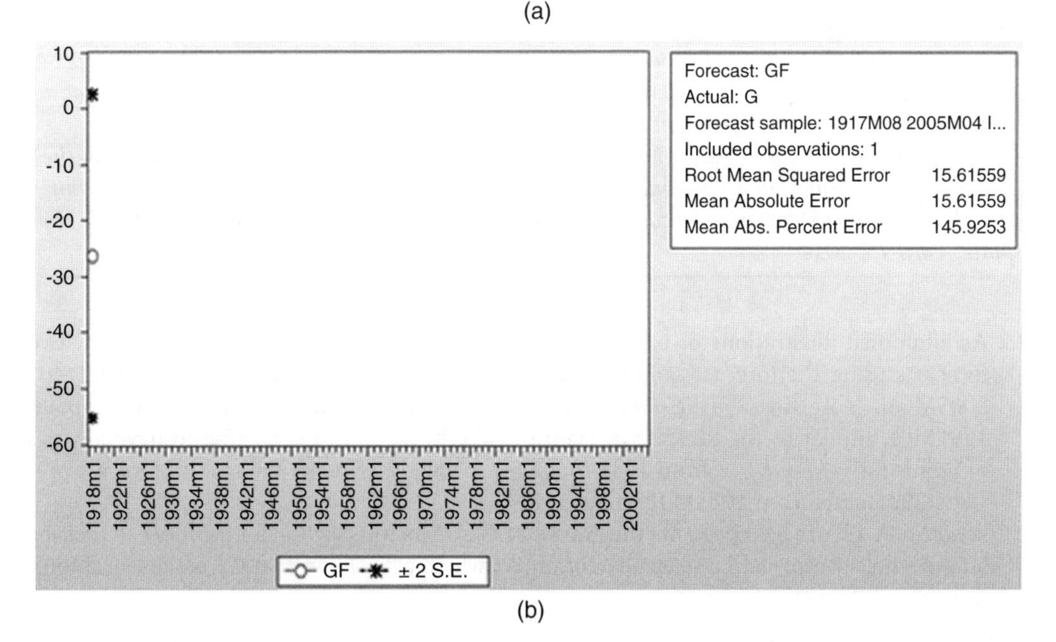

(b)

Figure 2.37 Statistical results of LV(1) regressions of *G* with fifth degree polynomial trend for *@Month* = 1 using stepwise regression with the default options.

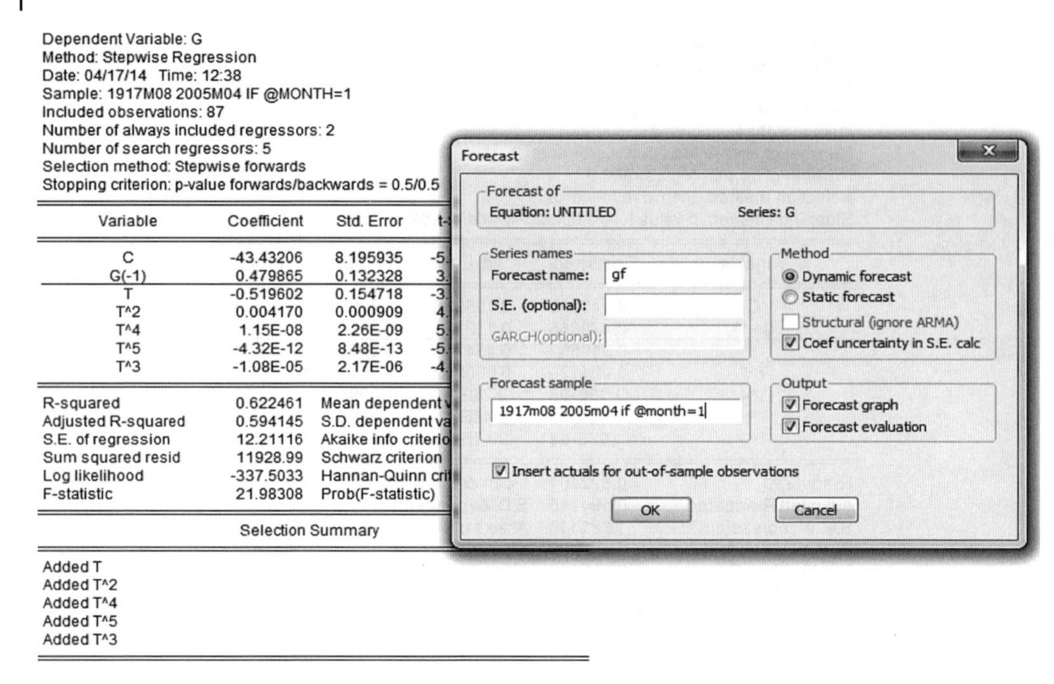
Dependent Variable: G
Method: Stepwise Regression
Date: 04/17/14 Time: 12:38
Sample: 1917M08 2005M04 IF @MONTH=1
Included observations: 87
Number of always included regressors: 2
Number of search regressors: 5
Selection method: Stepwise forwards
Stopping criterion: p-value forwards/backwards = 0.5/0.5

Variable	Coefficient	Std. Error	t-
C	-43.43206	8.195935	-5.
G(-1)	0.479865	0.132328	3.
T	-0.519602	0.154718	-3.
T^2	0.004170	0.000909	4.
T^4	1.15E-08	2.26E-09	5.
T^5	-4.32E-12	8.48E-13	-5.
T^3	-1.08E-05	2.17E-06	-4.

R-squared	0.622461	Mean dependent v
Adjusted R-squared	0.594145	S.D. dependent va
S.E. of regression	12.21116	Akaike info criterio
Sum squared resid	11928.99	Schwarz criterion
Log likelihood	-337.5033	Hannan-Quinn cri
F-statistic	21.98308	Prob(F-statistic)

Selection Summary

Added T
Added T^2
Added T^4
Added T^5
Added T^3

Figure 2.38 The options to forecast based on the regression in Figure 2.37(a).

> *Here's what going on...*
> *Your equation has a lagged dependent variable. Thus each forecast depends upon the previous month's value of the dependent variable. In a dynamic forecast it depends on the forecast of the previous month's dependent variable.*
> *In your case you are forecasting for January only. The forecast for January depends upon the forecast for December (in a dynamic forecast). But you told EViews not to forecast for December, thus the forecast for January cannot be computed.Gareth ThomasDirector, HIS EViewsE-mail: April 16, 2014*

2.2 As additional illustrations or experiments, Figures 2.39a and b, respectively, present two forecasts using the forecast samples {1981m01 2005m04 if *@Month* = 1} and {1981M01 2005M04}. Note that the first forecast only presents a single point for the time point 1981M01, similar to the forecast in Figure 2.37b. In fact, I am expecting to have a series of points for *@Month* = 1 from 1981 to 2005. On the other hand, the second forecast presents 290 points from 1981M01 to 2005M04.

2.3 In addition, Figure 2.40 presents the graphs of *G* and its forecast *GF* for *@Month* = 1, based on the whole sample and a selected subsample, which also shows that *GF* is the worst forecast.

For these reasons, I have done experiments by doing the forecast based on the full models in (2.39) for some selected integers of *k*, as presented in the following example.

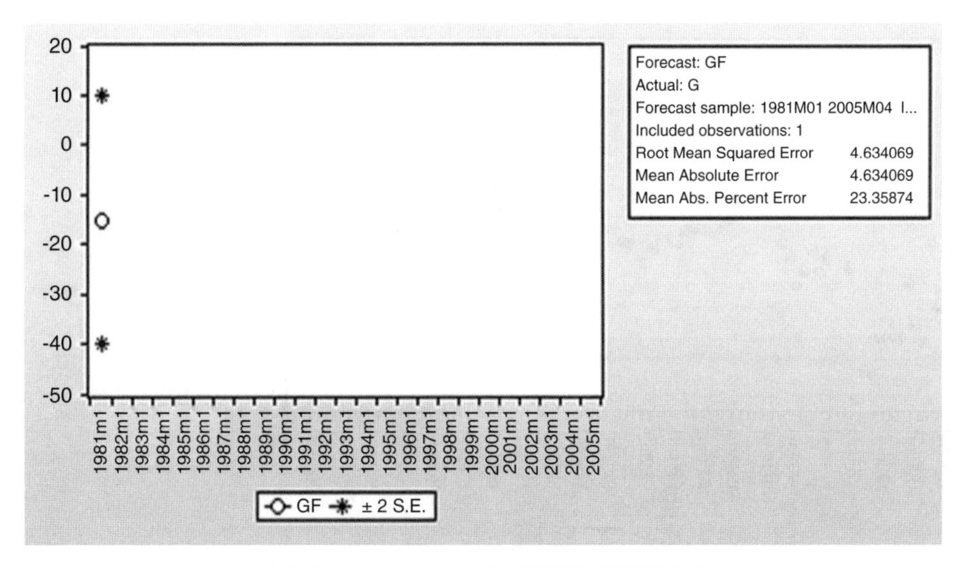

(a). Forecast sample {1981M01 2005M04}

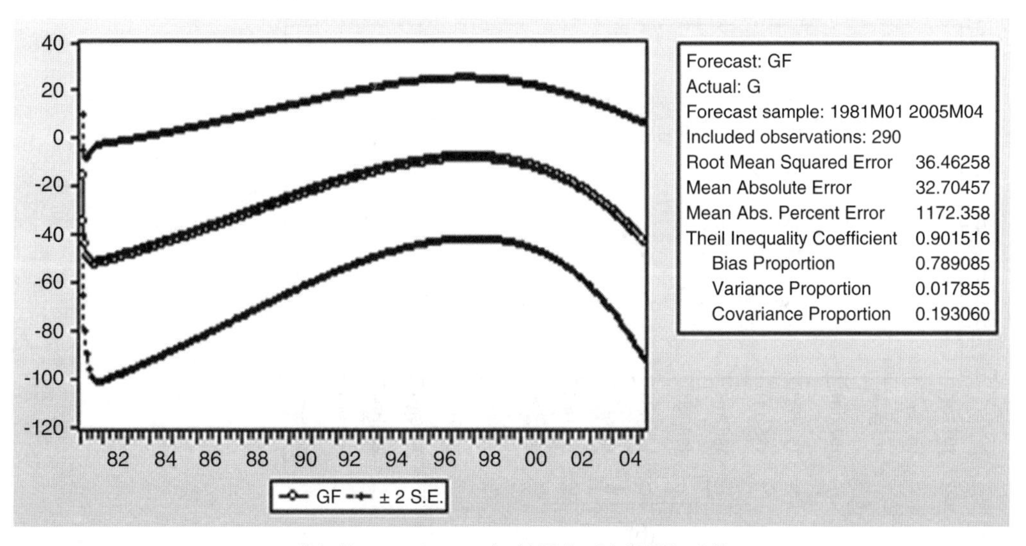

(b). Forecast sample {1981m01 2005m04}

Figure 2.39 Two alternative forecasts based on the regression in Figure 2.37a.

Example 2.31 Application of the Full Model for the Selected Degree *k*

Table 2.2 presents the summary of the statistical results of the LV(1) model in (2.39) for $k = 5$, with its forecast evaluation presented in Figure 2.41. Based on these results the following notes and comments are made.

1) Corresponding to the LV(1) model with the fifth degree polynomial trend for *@Month* = 1, in Figure 2.38, the summary of the statistical results in Table 2.2 is obtained by using the following ES, with its forecast evaluation presented in Figure 2.41.

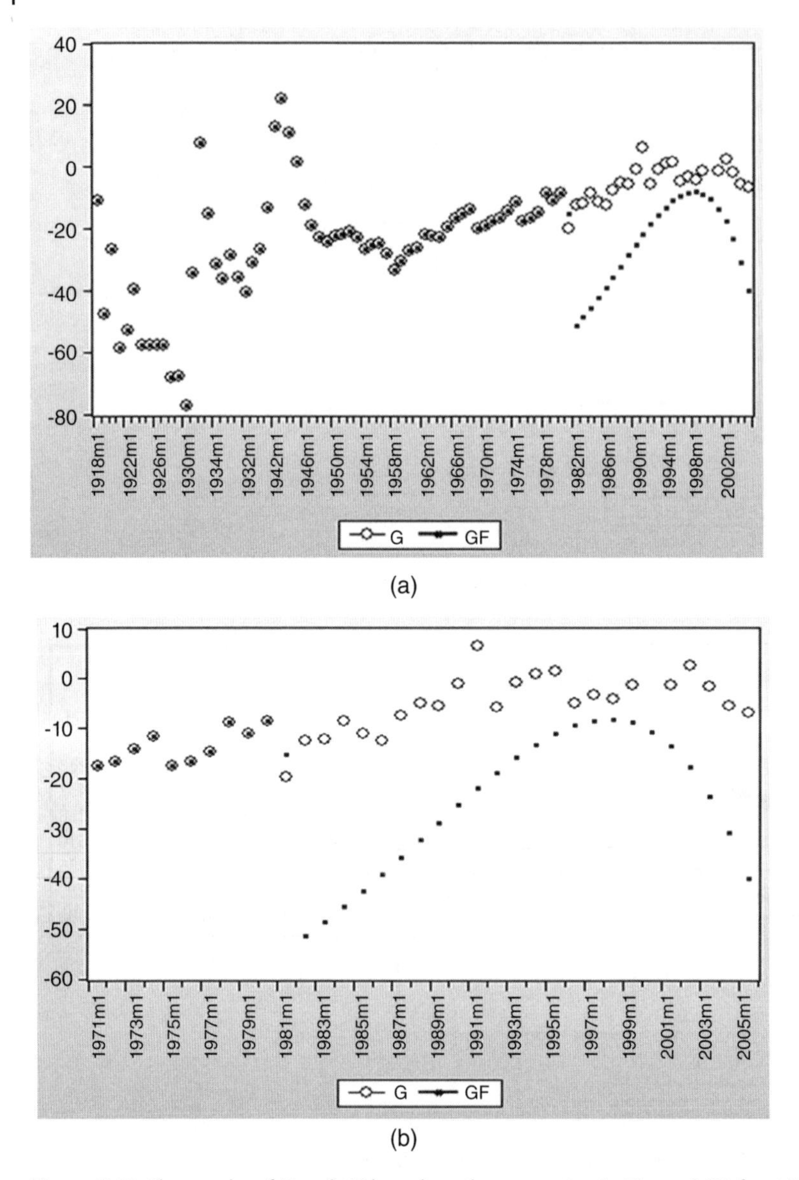

Figure 2.40 The graphs of *G* and *GF* based on the regression in Figure 2.37, for *@Month* = 1.

$$G\ @Expand(@Month)\ G(-1)*@Expand(@Month)\ t*@Expand(@Month)$$
$$t\hat{\ }2*@Expand(@Month)\ t\hat{\ }3*@Expand(@Month) \quad (2.40)$$
$$t\hat{\ }4*@Expand(@Month)\ t\hat{\ }5*@Expand(@Month)$$

2) The summary of the results shows all slopes of the time variables in five out of the 12 regressions have large p values, which are the unpredictable impacts of the multicollinearity within the five months, as presented using the gray background. In addition, the regression within *@Month* = 3 also has a similar problem. So, in a statistical sense, each of the six models should be reduced by deleting at least one of the time variables. Then the problem is which greatest

Table 2.2 Summary of the statistical results of the LV(1) mcdel in (2.39), for $k = 5$.

Dependent Variable: *G*

Method: Least Squares

Date: 04/15/14 Time: 16:32

Sample (adjusted): 1917M10 2005MM

Included observations: 1045 after adjustments

	Intercept		G(−l)		t		t^2		t^3		t^4		t^5	
Variable	**Coef.**	**Prob.**	**Coef.**	**Prob.**	**Coef.**	**Prob.**	**Coef.**	**Prob.**	**Coef.**	**Prob.**	**Coef.**	**Prob.**	**Coef.**	**Prob.**
@MONTH = 1	−43.432	0.000	0.480	0.000	−0.520	0.000	0.004	0.000	−1.08E-05	0.000	1.15E-08	0.000	−4.32E-12	0.000
@MONTH = 2	0.523	0.922	0.625	0.000	0.398	0.000	−0.002	0.000	5.01E-06	0.001	−4.79E-09	0.001	1.69E-12	0.003
@MONTH = 3	23.196	0.000	0.512	0.000	−0.172	0.071	0.001	0.274	−9.28E-07	0.483	7.33E-10	0.594	−2.55E-13	**0.621**
@MONTH = 4	−15.672	0.003	0.762	0.000	0.233	0.017	−0.001	0.038	2.54E-06	0.058	−2.38E-09	0.085	7.98E-13	0.123
@MONTH = 5	−6.846	0.190	0.511	0.000	0.067	0.505	0.000	0.995	−4.93E-07	0.722	8.19E-10	0.571	−3.78E-13	**0.488**
@MONTH = 6	−3.476	0.510	0.766	0.000	0.087	0.381	0.000	0.530	7.92E-07	0.566	-8.10E-10	0.575	2.97E-13	**0.586**
@MONTH = 7	3.671	0.496	0.838	0.000	0.092	0.358	−0.001	0.299	1.35E-06	0.329	−1.29E-09	0.370	4.51E-13	**0.407**
@MONTH = 8	15.072	0.006	0.656	0.000	−0.327	0.001	0.002	0.001	−4.62E-06	0.001	4.78E-09	0.001	−1.77E-12	0.001
@MONTH = 9	21.439	0.000	0.944	0.000	−0.051	0.626	0.000	0.938	2.86E-07	0.842	−3.31E-10	0.825	1.25E-13	**0.824**
@MONTH = 10	12.682	0.013	0.635	0.000	−0.180	0.045	0.001	0.060	−2.27E-06	0.079	2.24E-09	0.100	−8.02E-13	0.121
@MONTH = 11	−1.556	0.760	0.341	0.000	0.053	0.564	0.000	0.768	3.86E-07	0.768	-4.04E-10	0.768	1.3SE-13	**0.790**
@MONTH = 12	11.742	0.015	0.749	0.000	0.300	0.001	−0.002	0.001	4.18E-06	0.001	−3.98E-09	0.003	1.36E-12	0.008
R-squared			0.7818				Mean dependent variable				6.093774			
Adjusted R-squared			0.7629				S.D. dependent variable				15.35868			
S.E. of regression			7.4779				Akaike info criterion				6.938752			
Sum squared residual			53 738				Schwarz criterion				7.336789			
Log likelihood			−3541.5				Hannan–Quinn criterion				7.08971			
Durbin–Watson statistic			2.1448											

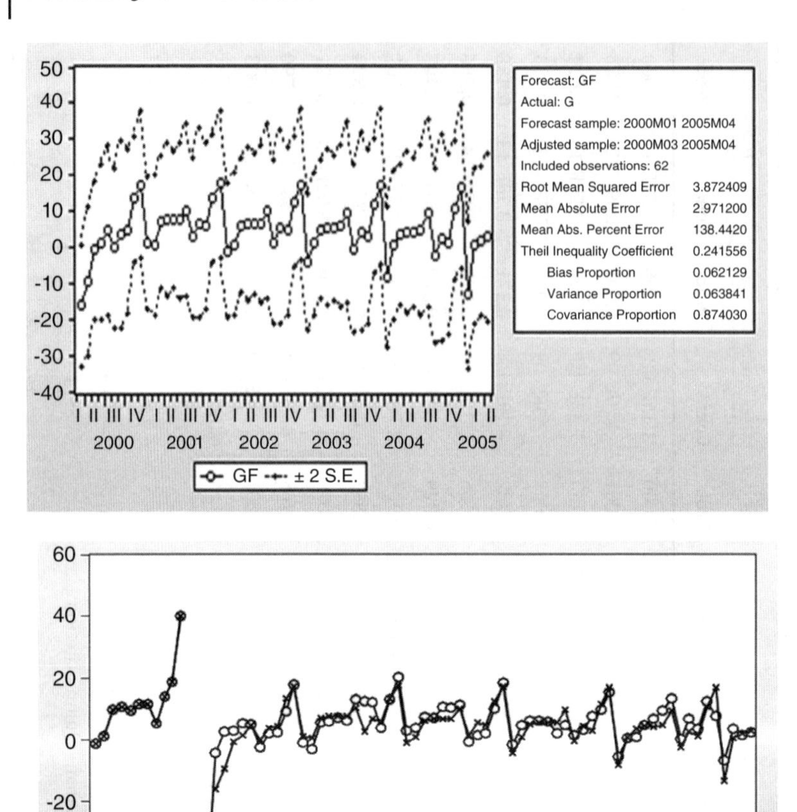

Figure 2.41 The graphs of the statistical results of the LV(1) model in (2.39), with $k = 5$.

degree of the time t should be kept within each of the six months. Note that the acceptable reduced model within each month, in general, would have different sets of the time independent variables, which are highly dependent on the data and they are unpredictable. See the following example.

Example 2.32 Developing Possible Reduced Models with Unexpected Statistical Results
Referring to the full regression in Table 2.2, with the best forecast of G, do we have to reduce the model to forecast G? In order to find out the impact of the reduced model for forecasting, as an illustration, this example presents four types of statistical results specific to the subsample *@Month* = 3, namely the statistical results of a correlation matrix, two stepwise selection methods, and a manual stepwise selection method or the trial-and-error method.

1) ***A Correlation Matrix***
Figure 2.42 presents a correlation matrix of the variables G, T, T^2, T^3, T^4, and T^5. Based on this matrix, the following notes and comments are presented.

Covariance Analysis: Ordinary
Date: 04/19/14 Time: 10:54
Sample (adjusted): 1918M03 2005M03
Included observations: 87 after adjustments
Balanced sample (listwise missing value deletion)

Correlation t-Statistic Probability	G	T	T^2	T^3	T^4	T^5
G	1.000000 ----- -----					
T	0.035950 0.331659 0.7410	1.000000 ----- -----				
T^2	0.047952 0.442607 0.6592	0.968138 35.64342 0.0000	1.000000 ----- -----			
T^3	0.038230 0.352722 0.7252	0.916356 21.10179 0.0000	0.986034 54.58541 0.0000	1.000000 ----- -----		
T^4	0.022615 0.208549 0.8353	0.865955 15.96350 0.0000	0.958426 30.96749 0.0000	0.992188 73.32698 0.0000	1.000000 ----- -----	
T^5	0.005987 0.055197 0.9561	0.820698 13.24286 0.0000	0.927247 22.83032 0.0000	0.975095 40.53405 0.0000	0.995007 91.91487 0.0000	1.000000 ----- -----

Figure 2.42 A correlation matrix of *G, T, T^2, T^3, T^4*, and *T^5*.

1.1 Unexpectedly, each of time variablse t^k, $k = 0, 1, \ldots, 5$ has an insignificant positive correlation with the dependent variable *G* with a very large *p* value. In other words, each t^k has an insignificant positive linear effect on *G* based on the regression model $G = \beta_0 + \beta_1 \times t^k + \varepsilon$. On the other hand, for the level *@Month* = 3 in Table 2.2, *t* has a significant negative adjusted effect on *G*, based on the *t*-statistic with a *p* value = 0.071/2 = 0.0355 < α = 0.05.

1.2 The two regression functions in Figure 2.43, which are the reduced models of the model in Table 2.2, conditional for *@Month* = 3, also show that t^k, for each selected *k*, has a significant adjusted effect on *G*.

1.3 These findings indicate the unpredictable impact of the multicollinearity between the time independent variables of the models, and *G(−1)*.

2) ***Application of the Basic Stepwise Selection Method***

Figure 2.43 presents statistical results of two alternative reduced regressions, which are obtained using the basic stepwise selection method, which can be considered to be the simplest selection method to develop a reduced model. However, Agung (2014, 2009a, 2011a) recommends not to apply this method specific for a set of several distinct independent variables, or for many distinct independent variables, but to apply the manual stepwise selection method or the manual multistage selection method.

Based on the results in Figure 2.43, the following notes and comments are presented.

2.1 The outputs are obtained using *G C G(−1)* as the ES and *t t^2 t^3 t^4 t^5* as the list of search regressors.

2.2 Even though the variable *t^5* has a *p* value = 0.2105, I would say that the regression in Figure 2.43a is an acceptable model, since the *p* value < 0.30, and at the α = 0.15 level of significance, *t^5* has a positive significant adjusted effect on *G* based on the *t*-statistic of $t_0 = 1.262216$ with a *p* value 0.2105/2 = 0.10 525 < 0.15. As a comparison, Figure 2.42b presents its reduced regression with a smaller R-squared.

2.3 Both regressions are the fifth degree polynomial of the time variable *t*. Compare these regressions with the fourth degree polynomial of the time variable *t* in Figure 2.44.

Dependent Variable: G
Method: Stepwise Regression
Date: 04/19/14 Time: 10:39
Sample: 1917M08 2005M04 IF @MONTH=3
Included observations: 87
Number of always included regressors: 2
Number of search regressors: 5
Selection method: Stepwise forwards
Stopping criterion: p-value forwards/backwards = 0.5/0.5
Stopping criterion: Number of search regressors = 4

Variable	Coefficient	Std. Error	t-Statistic	Prob.*
C	20.91267	2.720074	7.688272	0.0000
G(-1)	0.516994	0.052334	9.878742	0.0000
T^5	9.83E-14	7.79E-14	1.262216	0.2105
T^4	-2.24E-10	1.17E-10	-1.904760	0.0604
T	-0.111285	0.026469	-4.204307	0.0001
T^2	0.000223	6.41E-05	3.472081	0.0008

R-squared	0.574722	Mean dependent var	6.072747
Adjusted R-squared	0.548471	S.D. dependent var	7.504385
S.E. of regression	5.042641	Akaike info criterion	6.140209
Sum squared resid	2059.687	Schwarz criterion	6.310272
Log likelihood	-261.0991	Hannan-Quinn criter.	6.208688
F-statistic	21.89277	Prob(F-statistic)	0.000000

Selection Summary

Added T^5
Added T^4
Added T
Added T^2

(a)

Dependent Variable: G
Method: Stepwise Regression
Date: 04/19/14 Time: 11:25
Sample: 1917M08 2005M04 IF @MONTH=3
Included observations: 87
Number of always included regressors: 2
Number of search regressors: 5
Selection method: Stepwise forwards
Stopping criterion: p-value forwards/backwards = 0.5/0.5
Stopping criterion: Number of search regressors = 3

Variable	Coefficient	Std. Error	t-Statistic	Prob.*
C	13.63046	1.845083	7.387453	0.0000
G(-1)	0.464202	0.053346	8.701793	0.0000
T^5	-1.53E-13	3.09E-14	-4.940328	0.0000
T^4	1.68E-10	3.51E-11	4.779837	0.0000
T	-0.021978	0.006655	-3.302613	0.0014

R-squared	0.511428	Mean dependent var	6.072747
Adjusted R-squared	0.487595	S.D. dependent var	7.504385
S.E. of regression	5.371824	Akaike info criterion	6.255966
Sum squared resid	2366.233	Schwarz criterion	6.397685
Log likelihood	-267.1345	Hannan-Quinn criter.	6.313032
F-statistic	21.45900	Prob(F-statistic)	0.000000

Selection Summary

Added T^5
Added T^4
Added T

*Note: p-values and subsequent tests do not account for stepwise selection.

(b)

Figure 2.43 Statistical results of two reduced models, using the basic stepwise method, with four and three numbers of search regressors, respectively, and *p* values 0.5/0.5 as the stopping criterion.

3) ***Application of the Combinatorial Selection Method***
There are three other selection methods available, but in this case I would consider the application of the combinatorial selection method as the best for a comparative study. Do the others as an exercise. Figure 2.44 presents statistical results of two alternative reduced regressions, which are obtained using the combinatorial selection method, which can be considered to be a special selection method. Based on these results the following notes and comments are made.
   3.1 Compared to the reduced model in Figure 2.43a this model is a model with fourth degree polynomial of the time *t*, but *t^4* has an insignificant adjusted effect with a very large *p* value > 0.50. So it is an unacceptable or inappropriate model, in a statistical sense, and it is the worst reduced model compared to those in Figure 2.43.
   3.2 Figure 2.44b presents a model with the third degree polynomial of the time *t*, with the greatest $R^2 = 0.575535$, compared to the two models in Figure 2.43 with $R^2 = 0.574722$ and 0.511428. Then, which one would you consider to present the population model? However, referring to the polynomial nonparametric graph in Figure 2.36c, I would say that each of the models with the fifth degree polynomials in Figure 2.43 is a better fit model.
   3.3 So, in Figure 2.45, I present alternative statistical results of two reduced models with the fifth degree polynomial of the time *t*, using an ES *G C G(−1) t^5* and *t t^2 t^3 t^4* as a list of search regressors for the combinatorial selection method. Based on these results, the following findings and notes are presented.
   - Figure 2.45a presents a good fit model with the time independent variables *t t^2*, and *t^5*, with *t ^5* having a significant negative adjusted effect on *G*.
   - On the other hand, Figure 2.45b presents a worse model with the time independent variables *t t^2*, *t^3*, and *t^5* because *t^5* has an insignificant adjusted effect on *G* with a large *p* value = 0.5822.

4) ***Application of the Trial-and-Error Method***
Table 2.3 presents a summary of the statistical results of three alternative reduced regressions, which are obtained using the trial-and-error method. Based on these results the following notes and comments are presented.
   4.1 The main objective of this analysis is to find reduced models with the fifth degree polynomial of the time *t* using the trial-and-error method. So the first step is to delete one out of the four independent variables *t*, *t^2*, *t^3*, and *t^4*, then I obtain two good fit models, namely the Reduced Model 1 and Reduced Model 2, as presented in Table 2.3.
   4.2 Note that the Reduced Model 1 is obtained by deleting the time *t*, which has a smallest *p* value = 0.071, and it has a significant adjusted effect on *G*. The Reduced Model 2 is obtained by deleting *t^2*, which has the second smallest *p* value = 0.274. These results have shown the unpredictable impact of the multicollinearity. They are unexpected reduced models. And the models are not the same as the models obtained by using the software selection method.
   4.3 By deleting *t^3*, the reduced model obtained will be the same as the reduced model in Figure 2.43a and by deleting *t^4* the reduced model obtained will be a model with *t^5* and a *p* value = 0.5822. So, as an alternative reduced model, the Reduced Model 3 is presented by deleting both *t^3* and *t^4*. Then we have four alternative reduced models with *t^5* as one of the independent variables.

Dependent Variable: G
Method: Stepwise Regression
Date: 04/19/14 Time: 13:37
Sample: 1917M08 2005M04 IF @MONTH=3
Included observations: 87
Number of always included regressors: 2
Number of search regressors: 5
Selection method: Combinatorial
Number of search regressors: 4

Variable	Coefficient	Std. Error	t-Statistic	Prob.*
C	21.88882	2.993216	7.312812	0.0000
G(-1)	0.517127	0.051923	9.959499	0.0000
T^4	5.81E-11	9.23E-11	0.629349	0.5309
T	-0.134084	0.037425	-3.582762	0.0006
T^2	0.000353	0.000142	2.488053	0.0149
T^3	-2.92E-07	1.99E-07	-1.468342	0.1459

R-squared	0.577601	Mean dependent var	6.072747
Adjusted R-squared	0.551527	S.D. dependent var	7.504385
S.E. of regression	5.025547	Akaike info criterion	6.133418
Sum squared resid	2045.746	Schwarz criterion	6.303480
Log likelihood	-260.8037	Hannan-Quinn criter.	6.201897
F-statistic	22.15236	Prob(F-statistic)	0.000000

Selection Summary

Number of combinations compared: 5

*Note: p-values and subsequent tests do not account for stepwise selection.

(a)

Dependent Variable: G
Method: Stepwise Regression
Date: 04/19/14 Time: 13:38
Sample: 1917M08 2005M04 IF @MONTH=3
Included observations: 87
Number of always included regressors: 2
Number of search regressors: 5
Selection method: Combinatorial
Number of search regressors: 3

Variable	Coefficient	Std. Error	t-Statistic	Prob.*
C	20.74760	2.372633	8.744548	0.0000
G(-1)	0.513047	0.051327	9.995723	0.0000
T	-0.113669	0.018594	-6.113078	0.0000
T^2	0.000267	4.12E-05	6.487414	0.0000
T^3	-1.68E-07	2.57E-08	-6.525574	0.0000

R-squared	0.575535	Mean dependent var	6.072747
Adjusted R-squared	0.554830	S.D. dependent var	7.504385
S.E. of regression	5.007006	Akaike info criterion	6.115307
Sum squared resid	2055.749	Schwarz criterion	6.257026
Log likelihood	-261.0159	Hannan-Quinn criter.	6.172373
F-statistic	27.79614	Prob(F-statistic)	0.000000

Selection Summary

Number of combinations compared: 10

*Note: p-values and subsequent tests do not account for stepwise selection.

(b)

Figure 2.44 Statistical results of two reduced model, using the combinatorial selection method, with equation specification *"G C G(−1)"* and *"t t^2 t^3 t^4 t^5"* as a list of search regressors.

Dependent Variable: G
Method: Stepwise Regression
Date: 04/20/14 Time: 07:05
Sample: 1917M08 2005M04 IF @MONTH=3
Included observations: 87
Number of always included regressors: 3
Number of search regressors: 4
Selection method: Combinatorial
Number of search regressors: 2

Variable	Coefficient	Std. Error	t-Statistic	Prob.*
C	17.65144	2.147217	8.220614	0.0000
G(-1)	0.494550	0.051801	9.547127	0.0000
T^5	-4.93E-14	8.10E-15	-6.083925	0.0000
T	-0.067145	0.012994	-5.167310	0.0000
T^2	0.000105	1.83E-05	5.769528	0.0000

R-squared	0.555674	Mean dependent var	6.072747
Adjusted R-squared	0.533999	S.D. dependent var	7.504385
S.E. of regression	5.122813	Akaike info criterion	6.161038
Sum squared resid	2151.943	Schwarz criterion	6.302757
Log likelihood	-263.0052	Hannan-Quinn criter.	6.218104
F-statistic	25.63726	Prob(F-statistic)	0.000000

Selection Summary

Number of combinations compared: 6

*Note: p-values and subsequent tests do not account for stepwise selection.

(a)

Dependent Variable: G
Method: Stepwise Regression
Date: 04/20/14 Time: 07:07
Sample: 1917M08 2005M04 IF @MONTH=3
Included observations: 87
Number of always included regressors: 3
Number of search regressors: 4
Selection method: Combinatorial
Number of search regressors: 3

Variable	Coefficient	Std. Error	t-Statistic	Prob.*
C	21.64661	2.885475	7.501922	0.0000
G(-1)	0.516986	0.052036	9.935095	0.0000
T^5	1.92E-14	3.47E-14	0.552417	0.5822
T	-0.128635	0.032905	-3.909312	0.0002
T^2	0.000323	0.000109	2.968052	0.0039
T^3	-2.28E-07	1.13E-07	-2.027226	0.0459

R-squared	0.577129	Mean dependent var	6.072747
Adjusted R-squared	0.551025	S.D. dependent var	7.504385
S.E. of regression	5.028356	Akaike info criterion	6.134535
Sum squared resid	2048.033	Schwarz criterion	6.304598
Log likelihood	-260.8523	Hannan-Quinn criter.	6.203014
F-statistic	22.10952	Prob(F-statistic)	0.000000

Selection Summary

Number of combinations compared: 4

*Note: p-values and subsequent tests do not account for stepwise selection.

(b)

Figure 2.45 Statistical results of two reduced models using the combinatorial selection method with equation specification "*G C G(−1) t^5*" and "*t t^2 t^3 t^4*" as a list of search regressors.

Table 2.3 Statistical results summary of three reduced models, using the trial-and-error method.

Dependent Variable: *G*

Method: Least Squares

Date: 04/19/14 Time: 09:56

Sample: 1917M08 2005M04 IF @MONTH = 3

Included observations: 87

	Reduced Model-1			Reduced Model-2			Reduced Model-3		
Variable	**Coef.**	***t*-Stat**	**Prob.**	**Coef.**	***t*-Stat**	**Prob.**	**Coef.**	***t*-Stat**	**Prob.**
C	15.2095	8.0455	0.0000	19.1055	7.8225	0.0000	17.6514	8.2206	0.0000
G(−l)	0.5057	9.2922	0.0000	0.5153	9.7229	0.0000	0.4946	9.5471	0.0000
T				−0.0714	−4.2915	0.0000	−0.0671	−5.1673	0.0000
T^2	−0.0004	−3.6368	0.0005				0.0001	5.770	0.000
T^3	1.25E-06	3.2815	0.0015	4.96E-07	3.2107	0.0019			
T^4	−1.40E-09	−2.8348	0.0058	−7.04E-10	−2.5739	0.0119			
T^5	5.02E-13	2.3826	0.0195	2.67E-13	1.9937	0.0496	−4.93E-14	−6.084	0.000
R-squared	0.543			0.567			0.556		
Adjusted R-squared	0.514			0.540			0.534		
S.E. of regression	5.229			5.091			5.123		
Sum squared residuals	2214.726			2099.088			2151.943		
Log likelihood	−264.256			−261.923			−263.005		
F-statistic	19.226			21.178			25.637		
Prob. (*F*-Stat)	0.000			0.000			0.000		

5) ***Special Notes and Comments***

These examples have presented an unpredictable or unexpected reduced model can be obtained by using a specific selected method. Corresponding to the statistical results of the full model presented in Table 2.2, we have to develop an acceptable reduced model, namely each of the independent variables should have a p value < 0.30 since, at the 0.15 level of significance, the independent variable has a either positive or negative significant adjusted effect on G (Agung 2014, 2009a, 2011a; Lapin 1973).

Specifically, in order to obtain the reduced models of the six months, namely *@Month* = 3, 5, 6, 7, 9, and 11, of the full model in Table 2.2, it is recommended to apply the same selection method. The simplest method should be using the software stepwise regression for each of the month, then the results can be combined to a single reduced model for the 12 months by inserting the function *@Drop(∗)* in the ES (2.40). See the following example.

Example 2.33 A Reduced Model of the Full Model in (2.40), Using a Single-Stage STEPLS
Table 2.3 presents a summary of the basic stepwise selection, as presented in Figure 2.43a for the six months, namely *@Month* = 3, 5, 6, 7, 9, and 11, of the full model in Table 2.2. Based on this summary the following notes and comments are presented.

1) The results are obtained by using exactly the same stepwise selection method as presented in Figure 2.43a, specifically the "*Stopping criterion Number of search regressors = 4*," even though the final reduced model may have an independent variable with a p value > 0.30. In this case, I have found T^3 has a p value = 0.3974 > 0.30 within @Month = 9, as presented using the symbol (x) in Table 2.4. If we want to reduce again then we can insert the number of search regressors = 3. Do this as an exercise.
2) This table presents the symbol (∗) by *@Month* and the time variable *t^k*, which indicates that *t^k* should be deleted for the corresponding month, by inserting the function *@Drop(∗)*, as presented in Table 2.5. For instance, the IV *t∗@Expand(@Month)* of the full model should be replaced by *t∗@Expand(@Month,@Drop(3))*, and the IV *t^∗@Expand(@Month)* of the full model should be replaced by *t^2∗@Expand(@Month,@Drop(7),@Drop(11))*.
3) Then the reduced model of the full model in (2.40), can be presented using the following ES, with the summary of the statistical results presented in Table 2.6, but by using the scaled or transformed time variable *Sc_t* = *t*/100, in order to have greater slopes coefficients, compared to the slopes in Table 2.3, with its forecast evaluation presented in Figure 2.46. Based on these results, the following notes and comments area presented.

Table 2.4 Summary of the statistical results of the six reduced models, using the basic stepwise regressions for each month.

@Month	T	T^2	T^3	T^4	T^5
3	(∗)				
5				(∗)	
6			(∗)	(∗)	(∗)
7		(∗)		(∗)	(∗)
9			(x)	(∗)	(∗)
11		(∗)	(∗)	(∗)	

(∗) = *@Expand(@Month,@Drop(∗),...)*
(x) *p value* = 0.3974.

Table 2.5 Independent variables (IVs) of the full model in Table 2.2, and its reduced model, specific for the six months indicated in Table 2.4.

T^k	IV of the Full Model	IV of the Reduced Model
T	*t*@Expand(@Month)*	*t*@Expand(@Month, @Drop(3))*
T^2	*t^2*@Expand(@Month)*	*t^2*@Expand(@Month, @Drop(7), @Drop(11))*
T^3	*t^3*@Expand(@Month)*	*t^3*@Expand(@Month, @Drop(6), @Drop(11))*
T^4	*t^4*@Expand(@Month)*	*t^4*@Expand(@Month, @Drop(5),@Drop(6),@Drop(7), @Drop(9),@Drop(11))*
T^5	*t^5*@Expand(@Month)*	*t^5*@Expand(@Month, @Drop(6),@Drop(7),@Drop(9))*

$$\begin{gathered} G\ @Expand(@Month)\ G(-1)*@Expand(@Month)\ t*@Expand(@Month,@drop(3)) \\ t\hat{}2*@Expand(@Month,@Drop(7),@Drop(11)) \\ t\hat{}3*@Expand(@Month,@Drop(6),@Drop(11)) \\ t\hat{}4*@Expand(@Month,@Drop(5),@Drop(6),@Drop(7),@Drop(9),@Drop(11)) \\ t\hat{}5*@Expand(@Month,@Drop(6),@Drop(7),@Drop(9)) \end{gathered} \tag{2.41}$$

3.1 Note that there is only one independent variable with a p value = 0.4939 > 0.30. So that I would say that this reduced model is a good fit reduced model, out of several or many possible reduced options, which can be obtained by using other stepwise selection methods. Refer to the alternative stepwise selection methods presented in previous example.

3.2 However, unexpectedly, *SC_T^3* for the *@Month* = *9* has a large p value = 0.4939, which is greater than the p value = 0.3974 of the LS polynomial regression for the *@Month* = 9, as presented in Table 2.4. In a statistical sense, a researcher might want to reduce the model, which can be done easily. Note the following example also presents unexpected parameters' estimates.

3.3 In addition, Figure 2.47 presents the forecast evaluation of the model (2.41), with RMSE = 3.938116 and TIC = 0.245653, which are slightly greater than the values based on the full model in (2.39) as presented in Figure 2.41, RMSE = 3.872409, and TIC = 0.241556.

Example 2.34 A Reduced Model of the Full Model in (2.40), Using a Two-Stage STEPLS
In order to apply a two-stage STEPLS estimation method for each subsample, namely *@Month* = 3, 5, 6, 7, 9, and 11, I observe the four scatter graphs of G on the time t with their NNF Curves in Figure 2.47.

Since each of the NNF Curves is a polynomial function of the time t, then I divide the time independent variables *sc_t, sc_t^2, sc_t^3, sc_t^4,* and *sc_t^5* into two sets of variables, the first set is *sc_t^3, sc_t^4,* and *sc_t^5* and the second is *sc_t* and *sc_t^2,* to be the lists of search regressors for the two-stage STEPLS regressions.

Table 2.6 Summary of statistical results of the reduced model in (2.41).

Dependent Variable: *G*

Method: Least Squares

Date: 04/26/14 Time: 16:45

Sample (adjusted): 1917M10 2005M04

Included observations: 1045 after adjustments

	Intercept		*G(−1)*		*SC_T*		*SC_T^2*		*SC_T^3*		*SC_T^4*		*SC_T^5*	
Variable	**Coef.**	**Prob..**	**Coef.**	**Prob...**	**Coef.**	**Prob.**	**Coef.**	**Prob..**	**Coef.**	**Prob..**	**Coef.**	**Prob..**	**Coef.**	**Prob..**
@MONTH = 1	−43.4321	0.0000	0.4799	0.0000	−51.9602	0.0000	41.6979	0.0000	−10.8098	0.0000	1.1519	0.0000	−0.0432	0.0000
@MONTH = 2	0.5233	0.9218	0.6246	0.0000	39.8147	0.0000	−23.2057	0.0001	5.0122	0.0004	−0.4785	0.0013	0.0169	0.0026
@MONTH = 3	15.2095	0.0000	0.5057	0.0000			−3.6080	0.0108	1.2475	0.0214	−0.1397	0.0469	0.0050	0.0947
@MONTH = 4	−15.6721	0.0031	0.7620	0.0000	23.3183	0.0168	−11.6710	0.0377	2.5354	0.0568	−0.2382	0.0838	0.0080	0.1217
@MONTH = 5	−8.7212	0.0303	0.5061	0.0000	11.6177	0.0173	−3.1932	0.0511	0.2855	0.0980			−0.0007	**0.2036**
@MONTH = 6	−1.2390	0.5190	0.7624	0.0000	3.2124	0.0044	−0.2904	0.0041						
@MONTH = 7	8.6718	0.0001	0.8438	0.0000	−1.5928	0.0375			0.0112	0.0945				
@MONTH = 8	15.0724	0.0055	0.6559	0.0000	−32.7281	0.0014	19.1246	0.0011	−4.6192	0.0009	0.4784	0.0009	−0.0177	0.0011
@MONTH = 9	22.2099	0.0000	0.9390	0.0000	−7.0109	0.0108	0.7741	0.1977	−0.0255	**0.4939**				
@MONTH = 10	12.6823	0.0130	0.6347	0.0000	−18.0306	0.0443	10.0281	0.0591	−2.2747	0.0776	0.2243	0.0982	−0.0080	0.1192
@MONTH = 11	−0.3025	0.8944	0.3360	0.0000	2.6315	0.0000							−0.0001	0.0011
@MONTH = 12	11.7423	0.0145	0.7489	0.0000	30.0071	0.0010	−18.7030	0.0005	4.1782	0.0012	−0.3977	0.0032	0.0136	0.0075
R-squared			0.7806					Mean dependent variable			6.0938			
Adjusted R-squared			0.7648					S.D. dependent variable			15.3587			
S.E. of regression			7.4483					Akaike info criterion			6.9194			
Sum squared residuals			54,034.21					Schwarz criterion			7.2558			
Log likelihood			−3544.37					Hannan–Quinn criterion			7.0470			
Durbin–Watson statistic			2.1361											

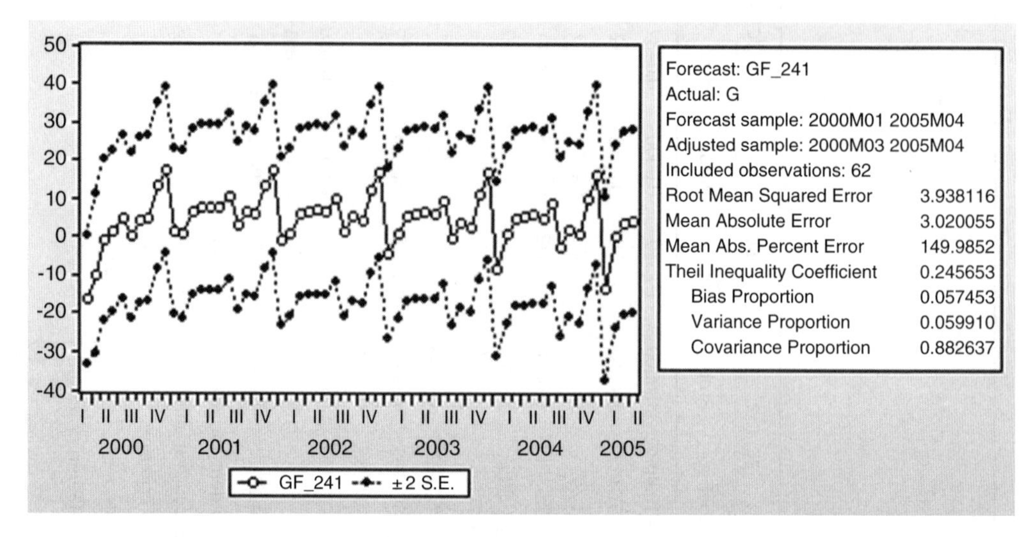

Forecast: GF_241
Actual: G
Forecast sample: 2000M01 2005M04
Adjusted sample: 2000M03 2005M04
Included observations: 62
Root Mean Squared Error 3.938116
Mean Absolute Error 3.020055
Mean Abs. Percent Error 149.9852
Theil Inequality Coefficient 0.245653
Bias Proportion 0.057453
Variance Proportion 0.059910
Covariance Proportion 0.882637
GF_241
±2 S.E.
2000
2001
2002
2003
2004
2005

Figure 2.46 Forecast evaluation based on the reduced model in (2.41).

As additional illustration, Figure 2.48 presents three out of the six regressions for *@Month* = 5, 7, and 11, using a two-stage stepwise selection method. Based on these results, the following notes and comments are presented.

1) The first stage of the six analyses are using the ES: *"G C G(−1)"* with *"SC_T^3 SC_T^4 SC_T^5"* as the list of search regressors in order to obtain a first stage regression, namely REG_1. Then at the second stage, REG_1 is apply as the ES with *"SC_T SC_T^2"* as the list of search regressors.
2) For each month, the REG_1 and the final regressions should be considered unexpected or unpredictable regressions, which are highly dependent on the data. Figure 2.48a–c presents specific STEPLS regressions that are very important and rare findings.
   - Note that REG_1 for the *@Month* = 5 has only one insignificant time variable, namely *T^5*, with a *p* value = 0.3500, but the final regression has three time-independent variables and each has a significant adjusted effect. So, the final regression can be considered to be an acceptable one, or a good fit model, but it is an unexpected result.
   - REG_1 for the *@Month* = 7 has three time-independent variables, and each has a significant adjusted effect, but the final regression has five time independent variables and each has a specific probability, namely $0.10 < Prob. < 0.20$. So that at the 10% level of significance, each time *T^k* has either a positive or negative significant adjusted effect on the dependent variable G with a $p\ value = Prob./2 < 0.10$. For this reason, I would consider the final model to be a good fit.
   - On the other hand, REG_1 for the *@Month* = 11, has three time-independent variables and each has a significant adjusted effect, but the final regression has four time independent variables and each has an insignificant adjusted effect. Then the REG_1 should be considered an acceptable regression or a good fit model.
   - Similar to the results in Table 2.4, Table 2.7 presents the summary of the statistical results of the six reduced models, using the two-stage stepwise selection method for each month. Note the symbol (∗) indicates that the time independent variable should be deleted for the corresponding month.

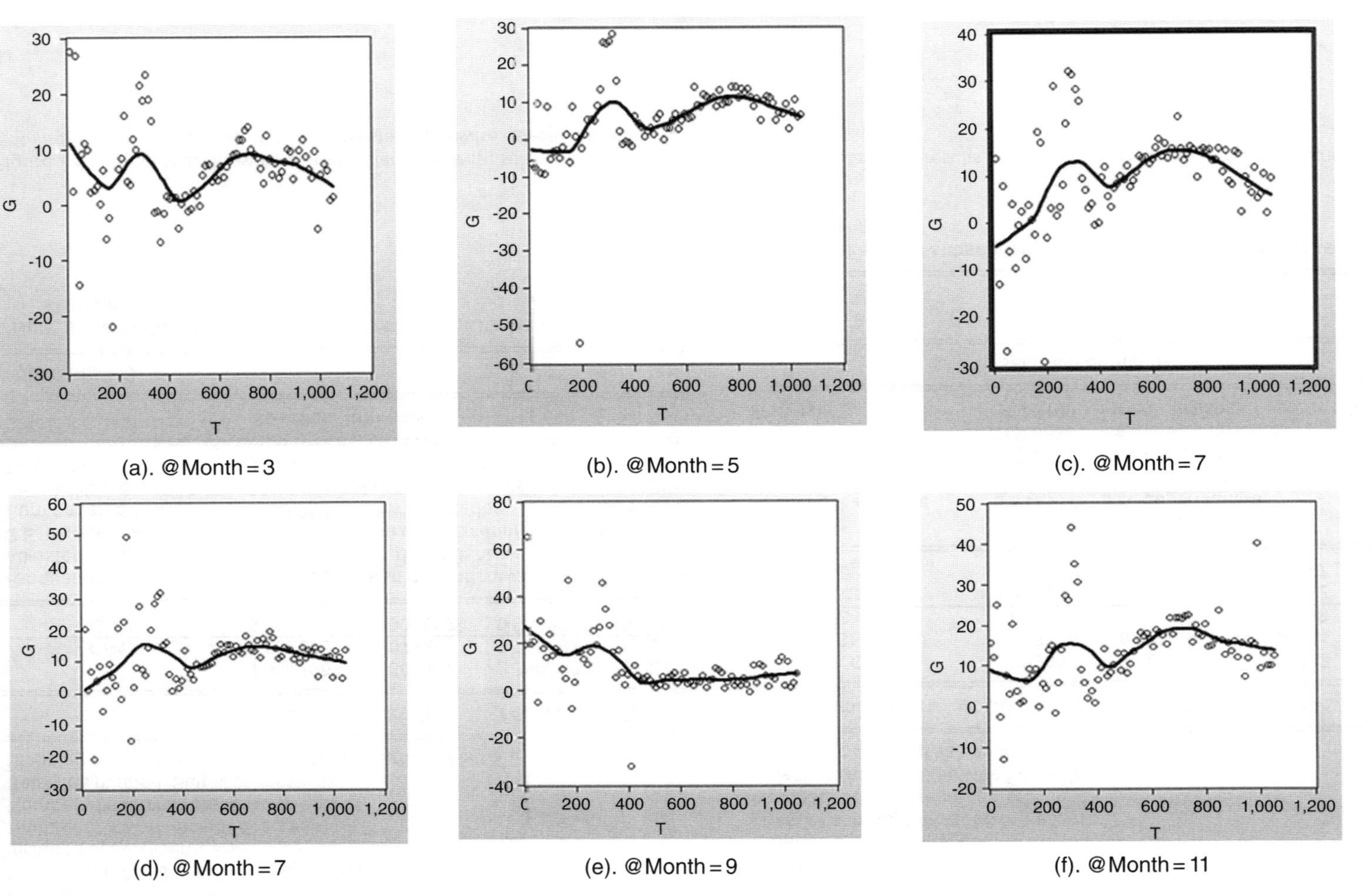

Figure 2.47 Scatter graphs of G on the time t with their *NNF* curves for the selected six months.

Dependent Variable: G
Method: Stepwise Regression
Date: 04/29/14 Time: 11:17
Sample: 1917M08 2005M04 IF @MONTH=5
Included observations: 86
Number of always included regressors: 2
Number of search regressors: 3
Selection method: Stepwise forwards
Stopping criterion: p-value forwards/backwards = 0.5/0.5

Variable	Coefficient	Std. Error	t-Statistic	Prob.*
C	1.992796	0.863407	2.308062	0.0235
G(-1)	0.552353	0.071264	7.750787	0.0000
SC_T^3	0.001773	0.001887	0.939828	0.3500

R-squared	0.462904	Mean dependent var	6.475848
Adjusted R-squared	0.449962	S.D. dependent var	7.364824
S.E. of regression	5.462086	Akaike info criterion	6.267800
Sum squared resid	2476.254	Schwarz criterion	6.353416
Log likelihood	-266.5154	Hannan-Quinn criter.	6.302256
F-statistic	35.76744	Prob(F-statistic)	0.000000

Selection Summary

Added SC_T^3

*Note: p-values and subsequent tests do not account for stepwise selection.

Dependent Variable: G
Method: Stepwise Regression
Date: 04/29/14 Time: 11:19
Sample: 1917M08 2005M04 IF @MONTH=5
Included observations: 86
Number of always included regressors: 3
Number of search regressors: 2
Selection method: Stepwise forwards
Stopping criterion: p-value forwards/backwards = 0.5/0.5

Variable	Coefficient	Std. Error	t-Statistic	Prob.*
C	-5.799966	2.267311	-2.558082	0.0124
G(-1)	0.538734	0.073434	7.336334	0.0000
SC_T^3	0.071871	0.026336	2.728998	0.0078
SC_T	6.470248	1.864172	3.470842	0.0008
SC_T^2	-1.262240	0.417936	-3.020177	0.0034

R-squared	0.539889	Mean dependent var	6.475848
Adjusted R-squared	0.517167	S.D. dependent var	7.364824
S.E. of regression	5.117535	Akaike info criterion	6.159604
Sum squared resid	2121.322	Schwarz criterion	6.302298
Log likelihood	-259.8630	Hannan-Quinn criter.	6.217032
F-statistic	23.76109	Prob(F-statistic)	0.000000

Selection Summary

Added SC_T
Added SC_T^2

*Note: p-values and subsequent tests do not account for stepwise selection.

(a)

Figure 2.48 (a) Statistical results of a two-stage stepwise selection method for *@Month* = 5. (b) Statistical results of a two-stage stepwise selection method for *@Month* = 7. (c) Statistical results of a two-stage stepwise selection method for *@Month* = 11.

Dependent Variable: G
Method: Stepwise Regression
Date: 04/30/14 Time: 11:40
Sample: 1917M08 2005M04 IF @MONTH=7
Included observations: 87
Number of always included regressors: 2
Number of search regressors: 3
Selection method: Stepwise forwards
Stopping criterion: p-value forwards/backwards = 0.5/0.5

Variable	Coefficient	Std. Error	t-Statistic	Prob.*
C	6.939181	1.049528	6.611714	0.0000
G(-1)	0.830068	0.057951	14.32372	0.0000
SC_T^3	-0.141350	0.059051	-2.393703	0.0190
SC_T^4	0.027034	0.013103	2.063178	0.0423
SC_T^5	-0.001320	0.000738	-1.787887	0.0775

R-squared	0.726646	Mean dependent var	11.45031
Adjusted R-squared	0.713311	S.D. dependent var	8.993554
S.E. of regression	4.815448	Akaike info criterion	6.037289
Sum squared resid	1901.460	Schwarz criterion	6.179008
Log likelihood	-257.6221	Hannan-Quinn criter.	6.094355
F-statistic	54.49423	Prob(F-statistic)	0.000000

Selection Summary

Added SC_T^3
Added SC_T^4
Added SC_T^5

*Note: p-values and subsequent tests do not account for stepwise selection.

Dependent Variable: G
Method: Stepwise Regression
Date: 04/30/14 Time: 11:45
Sample: 1917M08 2005M04 IF @MONTH=7
Included observations: 87
Number of always included regressors: 5
Number of search regressors: 2
Selection method: Stepwise forwards
Stopping criterion: p-value forwards/backwards = 0.5/0.5

Variable	Coefficient	Std. Error	t-Statistic	Prob.*
C	3.671151	3.447554	1.064857	0.2901
G(-1)	0.838050	0.060045	13.95709	0.0000
SC_T^3	1.348963	0.884377	1.525325	0.1311
SC_T^4	-0.128969	0.092137	-1.399757	0.1655
SC_T^5	0.004506	0.003474	1.296983	0.1984
SC_T^2	-6.021526	3.707380	-1.624200	0.1083
SC_T	9.235045	6.430094	1.436222	0.1548

R-squared	0.736347	Mean dependent var	11.45031
Adjusted R-squared	0.716573	S.D. dependent var	8.993554
S.E. of regression	4.787975	Akaike info criterion	6.047130
Sum squared resid	1833.976	Schwarz criterion	6.245536
Log likelihood	-256.0502	Hannan-Quinn criter.	6.127022
F-statistic	37.23821	Prob(F-statistic)	0.000000

Selection Summary

Added SC_T^2
Added SC_T

*Note: p-values and subsequent tests do not account for stepwise selection.

(b)

Figure 2.48 (Continued)

Dependent Variable: G
Method: Stepwise Regression
Date: 04/29/14 Time: 11:32
Sample: 1917M08 2005M04 IF @MONTH=11
Included observations: 88
Number of always included regressors: 2
Number of search regressors: 3
Selection method: Stepwise forwards
Stopping criterion: p-value forwards/backwards = 0.5/0.5

Variable	Coefficient	Std. Error	t-Statistic	Prob.*
C	2.881120	1.882046	1.530845	0.1296
G(-1)	0.329523	0.062681	5.257172	0.0000
SC_T^3	0.236295	0.085813	2.753613	0.0072
SC_T^4	-0.040713	0.019113	-2.130141	0.0361
SC_T^5	0.001799	0.001078	1.668139	0.0991

R-squared	0.358320	Mean dependent var	13.58153
Adjusted R-squared	0.327396	S.D. dependent var	8.826751
S.E. of regression	7.239036	Akaike info criterion	6.851993
Sum squared resid	4349.502	Schwarz criterion	6.992751
Log likelihood	-296.4877	Hannan-Quinn criter.	6.908701
F-statistic	11.58699	Prob(F-statistic)	0.000000

Selection Summary

Added SC_T^3
Added SC_T^4
Added SC_T^5

*Note: p-values and subsequent tests do not account for stepwise selection.

Dependent Variable: G
Method: Stepwise Regression
Date: 04/29/14 Time: 11:33
Sample: 1917M08 2005M04 IF @MONTH=11
Included observations: 88
Number of always included regressors: 5
Number of search regressors: 2
Selection method: Stepwise forwards
Stopping criterion: p-value forwards/backwards = 0.5/0.5

Variable	Coefficient	Std. Error	t-Statistic	Prob.*
C	-0.447531	3.315385	-0.134986	0.8930
G(-1)	0.336673	0.062774	5.363245	0.0000
SC_T^5	-3.56E-05	0.001851	-0.019215	0.9847
SC_T^4	-0.001657	0.037304	-0.044423	0.9647
SC_T^3	0.005736	0.207751	0.027609	0.9780
SC_T	2.681465	2.201757	1.217875	0.2268

R-squared	0.369721	Mean dependent var	13.58153
Adjusted R-squared	0.331289	S.D. dependent var	8.826751
S.E. of regression	7.218055	Akaike info criterion	6.856794
Sum squared resid	4272.226	Schwarz criterion	7.025703
Log likelihood	-295.6989	Hannan-Quinn criter.	6.924843
F-statistic	9.620203	Prob(F-statistic)	0.000000

Selection Summary

Added SC_T

*Note: p-values and subsequent tests do not account for stepwise selection.

(c)

Figure 2.48 (Continued)

Table 2.7 Summary of the statistical results of the six reduced models, using the two-stage stepwise selection method for each month.

@Month	T	T^2	T^3	T^4	T^5
3	(*)				
5				(*)	(*)
6	(*)	(*)			
7	(x)	(x)	(x)	(x)	(x)
9	(*)	(*)			
11	(*)	(*)			

(*) = *@Expand(@Month, @Drop(*),...)*
(x) 0.10 < *Prob.* < 0.20.

Table 2.8 Independent variables (IV) of the full model in Table 2.2, and its reduced model, specific for the six months indicated in Table 2.4.

T^k	IV of the Full Model	IV of the Reduced Model
T	*t*@Expand(@Month)*	*t*@Expand(@Month, @Drop(3),@Drop(6), @Drop(9),@Drop(11))*
T^2	*t^2*@Expand(@Month)*	*t^2*@Expand(@Month, @Drop(6), @Drop(9), @Drop(11))*
T^3	*t^3*@Expand(@Month)*	*t^3*@Expand(@Month)*
T^4	*t^4*@Expand(@Month)*	*t^4*@Expand(@Month, @Drop(5))*
T^5	*t^5*@Expand(@Month)*	*t^5*@Expand(@Month, @Drop(5))*

- Corresponding to the symbol (*) in Table 2.7, column 2 in Table 2.8 presents the five sets of IVs of the full model in Table 2.2, and the IVs of its reduced model presented in column 3. For instance, column 2 presents the symbols (*) for the *@Month* = 3, 6, 9, and 11, which indicate that the reduced model does not have the time t for the four months. So, the four *@Month* of the time t should be deleted/dropped from the full model by using the function *t*@Expand(@Month,@Drop(3),@Drop(6),@Drop(9),@Drop(11))* as presented in column 3 in order to get the reduced model.

3) Then, to obtain a reduced model of the full model in (2.40), the following ES should be applied, with the summary of the statistical results presented in Figure 2.49. Based on these results, the following notes and comments are presented.

$$\begin{gathered} G\ @Expand(@Month)\ G(-1)*@Expand(@Month) \\ Sc\_t*@Expand(@Month,@drop(3),@drop(6),@drop(9),@drop(11)) \\ Sc\_t\hat{\ }2*@Expand(@Month,@drop(6),@Drop(9),@Drop(11)) \\ Sc\_t\hat{\ }3*@Expand(@Month)\ Sc_t\hat{\ }4*@Expand(@Month,@Drop(5)) \\ Sc\_t\hat{\ }5*@Expand(@Month,@Drop(5)) \end{gathered} \tag{2.42}$$

Dependent Variable: G
Method: Least Squares
Date: 05/08/14 Time: 11:02
Sample (adjusted): 1917M10 2005M04
Included observations: 1045 after adjustments

	C		G(-1)		SC_T		SC_T^2		SC_T^3		SC_T^4		SC_T^5	
Variable	Coef.	t-Stat.	Coef.	t-Stat.	Coef.	t-Stat.	Coef.	t-Stat.	Coef.	t-Stat.	Coef.	t-Stat.	Coef.	t-Stat.
@MONTH=1	-43.432	-8.631	0.480	5.906	-51.960	-5.470	41.698	7.468	-10.810	-8.096	1.152	8.304	-0.043	-8.305
@MONTH=2	0.523	0.098	0.625	9.818	39.815	4.122	-23.207	-3.957	5.012	3.511	-0.479	-3.195	0.017	3.001
@MONTH=3	15.210	5.611	0.506	6.481			-3.608	-2.536	1.248	2.289	-0.140	-1.977	0.005	1.662
@MONTH=4	-15.672	-2.943	0.762	6.779	23.318	2.381	-11.671	-2.068	2.535	1.895	-0.238	-1.720	0.008	1.539
@MONTH=5	-5.800	-1.746	0.539	5.008	6.470	2.369	-1.262	-2.061	0.072	1.863				
@MONTH=6	2.297	1.418	0.786	8.775					0.164	1.826	-0.033	-1.621	0.002	1.414
@MONTH=7	3.671	0.680	0.838	8.913	9.235	**0.917**	-6.022	***-1.037***	1.349	**0.974**	-0.129	**-0.894**	0.005	**0.828**
@MONTH=8	15.072	2.767	0.656	6.944	-32.728	-3.186	19.125	3.252	-4.619	-3.311	0.478	3.304	-0.018	-3.249
@MONTH=9	14.665	8.525	0.918	10.321					-0.330	-3.721	0.064	3.209	-0.003	-2.825
@MONTH=10	12.682	2.472	0.635	8.037	-18.031	-2.001	10.028	1.877	-2.275	-1.755	0.224	1.644	-0.008	-1.549
@MONTH=11	2.881	1.478	0.330	5.076					0.236	2.659	-0.041	-2.057	0.002	1.611
@MONTH=12	11.742	2.434	0.749	7.095	30.007	3.288	-18.703	-3.473	4.178	3.222	-0.398	-2.931	0.014	2.664
R-squared	0.7786	Mean dependent var		6.0938										
Adjusted R-squared	0.7617	S.D. dependent var		15.3587										
S.E. of regression	7.4975	Akaike info criterion		6.9361										
Sum squared resid	54526.5	Schwarz criterion		7.2915										
Log likelihood	-3549.11	Hannan-Quinn criter.		7.0709										
Durbin-Watson stat	2.1318													

Figure 2.49 Summary of the statistical results of the model in (2.42).

3.1 Unexpectedly, the five independent time variables *SC_T^k*, $k = 1, \ldots, 5$; of the regression, for @Month = 7, have small t-statistics with $0.2999 \leq p \leq 0.4077$ (or $0.30 < p < 0.41$), compared to the multiple regression in Figure 2.48b with $0.10 < p$ values < 0.20.

3.2 In a statistical sense the model in (2.42) should be reduced. One of the possible reduced models, for the *@Month* = 7, is the model in Figure 2.48a, with *SC_T^3*, *SC_T^4*, and *SC_T^5* as the time independent variables. Do it as an exercise.

2.6.2 Forecast Model of *G* on *G(−1)* with Heterogeneous Polynomial Trend

Similar to the time series *HS*, Figure 2.47 shows that the time series *G* also has a cyclical growth pattern on *@Month* by *@Year*. Based on the growth pattern presented in Figure 2.47, I try to apply a forecast model of *G* on *G(−1)* with heterogeneous trends by *@Year*, using the following ES, which is a model with third degree polynomial of *@Month* by *@Year*. The summary of a part of the statistical results presented in Table 2.9, with its forecast evaluation presented in Figure 2.50. Based on these results, the following findings and notes are presented.

$$\begin{aligned} &G\ G(-1)@Month*@Expand(@Year)@Month\hat{\ }2*@Expand(@Year) \\ &\qquad @Month\hat{\ }3*@Expand(@Year)\ @Expand(@Year) \end{aligned} \tag{2.43}$$

1) By using the model (2.43) for the whole sample, the error message "*Near singular matrix. Regressors may be perfectly collinear*" was obtained. In fact, by selecting several other sub-samples, I have obtained the statistical results. In this case, however, I decided to use the

Table 2.9 Summary of a part of the statistical results of the model (2.43), for *@Year*>1980.

Dependent Variable: *G*

Method: Least Squares

Date: 05/15/14 Time: 11:37

Sample: 1917M08 2005M04 IF @YEAR>1980

Included observations: 289

	G(−1)			*MONTH*			*MONTH^2*			*MONTH^3*			*Intercept*		
Year	**Coef.**	**t-Stat.**	**Prob.**	**Coef.**	**t-Stat.**	**Prob.**	**Coef.**	**t-Stat.**	**Prob.**	**Coef.**	**t-Stat.**	**Prob.**	**Coef.**	**t-Stat.**	**Prob.**
@*YEAR*=1981	0.217	6.495	0.000	36.229	7.935	0.000	−5.938	−7.425	0.000	0.291	7.184	0.000	−56.842	−7.958	0.000
@*YEAR*=1982	0.217	6.495	0.000	30.009	6.568	0.000	−4.804	−6.001	0.000	0.232	5.714	0.000	−46.716	−6.530	0.000
@*YEAR*=1983	0.217	6.495	0.000	27.953	6.129	0.000	−4.524	−5.663	0.000	0.219	5.409	0.000	−40.699	−5.702	0.000
@*YEAR*=1984	0.217	6.495	0.000	27.022	5.922	0.000	−4.422	−5.533	0.000	0.215	5.307	0.000	−40.071	−5.612	0.000
@*YEAR*=1985	0.217	6.495	0.000	27.834	6.098	0.000	−4.509	−5.638	0.000	0.220	5.422	0.000	−42.420	−5.941	0.000
…															
…															
@*YEAR*=2001	0.217	6.495	0.000	10.119	2.211	0.028	−1.502	−1.873	0.063	0.073	1.790	0.075	−15.056	−2.100	0.037
@*YEAR*=2002	0.217	6.495	0.000	13.509	2.961	0.004	−2.480	−3.104	0.002	0.130	3.220	0.002	−14.162	−1.980	0.049
@*YEAR*=2003	0.217	6.495	0.000	9.083	1.990	0.048	−1.638	−2.050	0.042	0.088	2.172	0.031	−10.709	−1.497	0.136
@*YEAR*=2004	0.217	6.495	0.000	10.025	2.196	0.029	−1.407	−1.760	0.080	0.062	1.518	0.131	−16.338	−2.286	0.023
@*YEAR*=2005	0.217	6.495	0.000	82.073	1.478	0.141	−31.867	−1.300	0.195	3.843	1.180	0.239	−62.491	−1.721	0.087
R-squared				0.8301	Mean dependent variable					7.1368					
Adjusted R-squared				0.7398	S.D. dependent variable					8.5454					
S.E. of regression				4.3591	Akaike info criterion					6.0514					
Log likelihood				−773.4	Schwarz criterion					7.3328					
Durbin–Watson statistic				3.0073	Hannan–Quinn criterion					6.5648					

Forecast: GF_EQ08
Actual: G
Forecast sample: 2000M01 2005M04
Adjusted sample: 2000M03 2005M04
Included observations: 62

Root Mean Squared Error	2.521137
Mean Absolute Error	1.907164
Mean Abs. Percent Error	219.3609
Theil Inequality Coefficient	0.162779
Bias Proportion	0.000000
Variance Proportion	0.051981
Covariance Proportion	0.948019

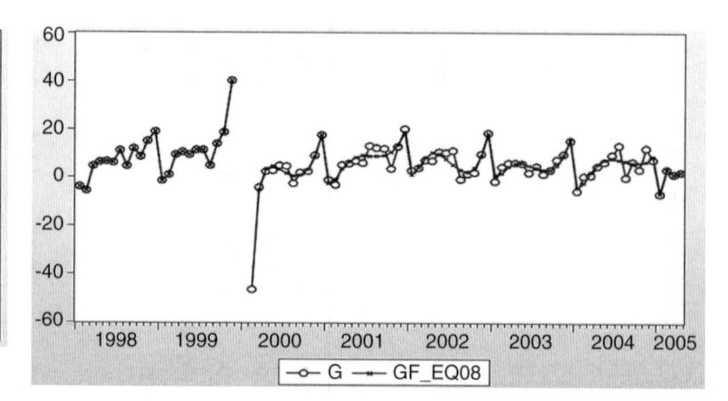

Figure 2.50 Forecast evaluation based on the LV(1) model in (2.43).

subsample @*Year*>1980. I would say that the model can be considered a good fit, since for only five out of 36 slopes do the time independent variables have insignificant effects with $p > 0.20$. So, all of the other slope parameters have either positive or negative significant effects on *G*.

2) By using the Wald test, the slopes differences of each of the time independent variable can be tested easily. For instance, the time *t* has significant different adjusted effects on *G* since H_0: C(2) = … = C(13) is rejected based on the *F*-statistic of $F_0 = 2.477772$, with $df = (11,188)$, and a *p* value = 0.0064. So the model cannot be reduced to the model with homogeneous (the same) slope for the time *t*. Do this as an exercise for the other slopes' parameters.
3) Compared to the forecasts of *G* in Figures 2.41 and 2.47, this forecast is the best since it has the smallest RMSE and TIC. And the curves of *G* and its forecast *GF_Eq08* are very close.
4) In addition, I have tried to insert heterogeneous slopes for *G(−1)* by using the following ES, based on several selected subsamples, but all gave the error message.

$$G\ G(-1)*@Expand(@Year)\ @Month*@Expand(@Year)@ \\ Month\char94 2*@Expand(@Year)\ @Month\char94 3*@Expand(@Year)\ @Expand(@Year) \tag{2.44}$$

5) However, the statistical results are obtained by deleting either one of the three sets of the time variables, with most of the variables *G(−1)*(@Year=k)* having large *p* values > 0.30 for all *k*. And based on the subsample @*Year*>1980, their forecasts have greater RMSE than the forecast based on the model (2.43). Do it as an exercise.

2.7 Forecast Models of CURR in Currency.wf1

As additional illustrative examples, I select the time series CURR in Currency.wf1, since it has different growth curve, as presented in Figure 2.51, which presents three scatter graphs of the time series *CURR* (currency in circulation) on time *t* based on the whole sample, subsample {*t*<700}, and subsample {*t*>600}, respectively, with their regression lines and Nearest Neighbor Fit (NNF) curves. Refer to the time series *G = 1200*dlog(Curr)* before.

So the forecast models of *CURR* could be very important to present as illustrative examples. Based on the graphs in Figure 2.51, the following notes and comments are made.

1) The simple linear regression (SLR) of $CURR_t$ on the numerical time *t* is an inappropriate model. By looking at the graph in Figure 2.51b, a fourth degree polynomial model or other

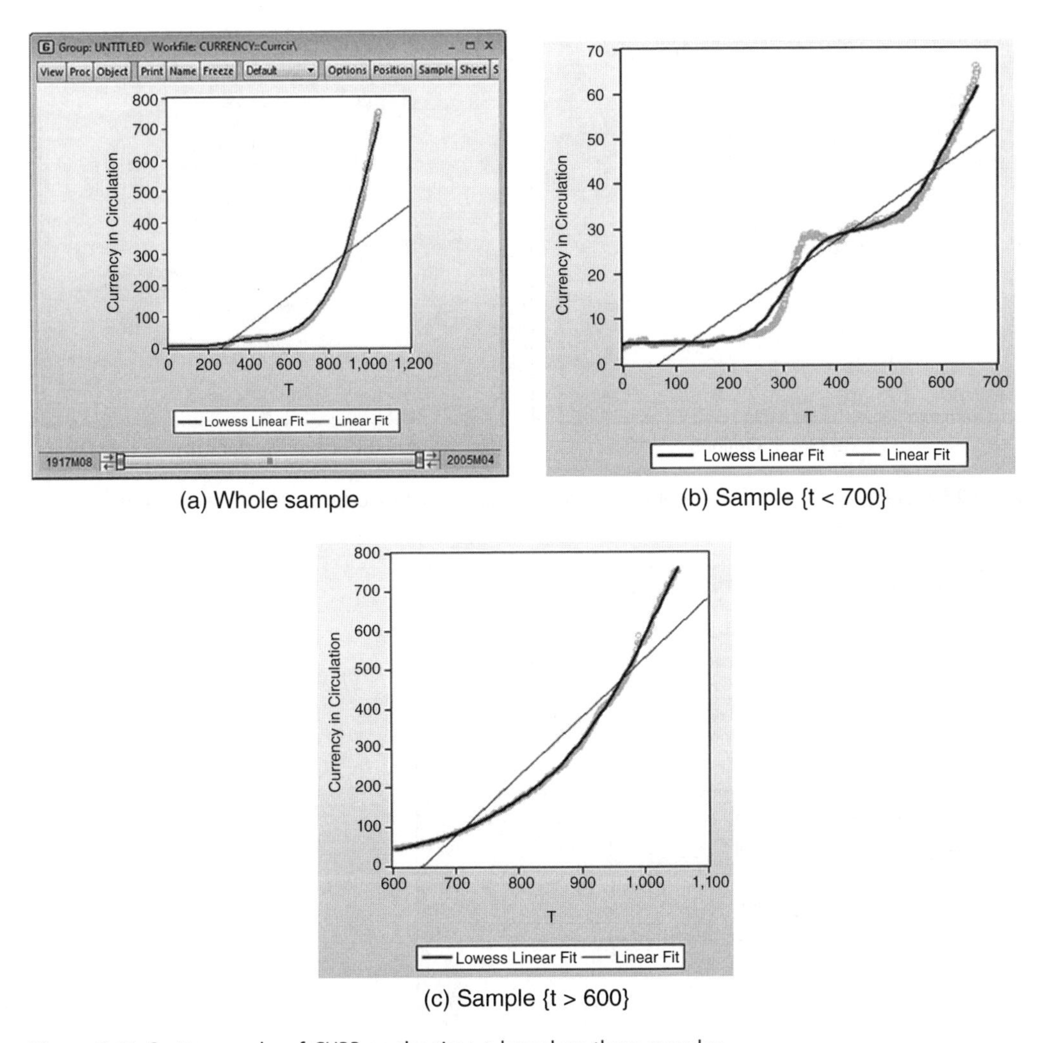

(a) Whole sample

(b) Sample {t < 700}

(c) Sample {t > 600}

Figure 2.51 Scatter graphs of *CURR* on the time *t*, based on three samples.

nonlinear model should be explored. Note the regression line has four intercepts with the NNF curve. See Example 2.35.

2) However, based on the graph in Figure 2.51a, a hyperbolic regression model might be a good fit model with a horizontal asymptote $Curr = \alpha$, or a vertical asymptote $t = t_0$ with the values α and t_0 should be subjectively selected, using the trial-and-error method. However, *CURR* would not increase without an upper bound. Hence, a model with an upper bound should be explored. See Section 3.3.1.

2.7.1 Developing Scatter Graphs with Regressions

The processes in developing a scatter graph with regressions, such as in Figure 2.51, based on the bivariate *(T,CURR),* are as follows:

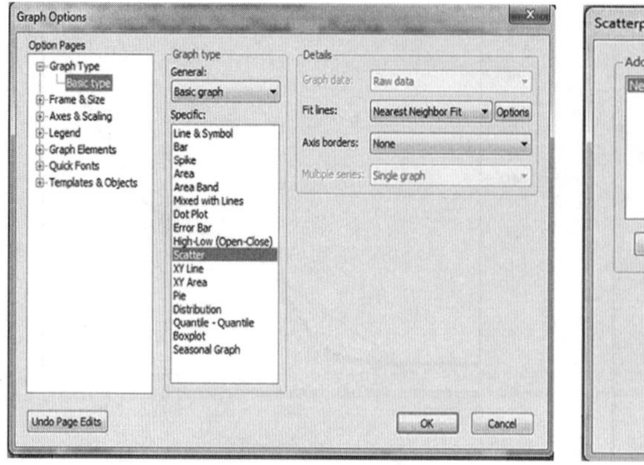

(a) Graph Options

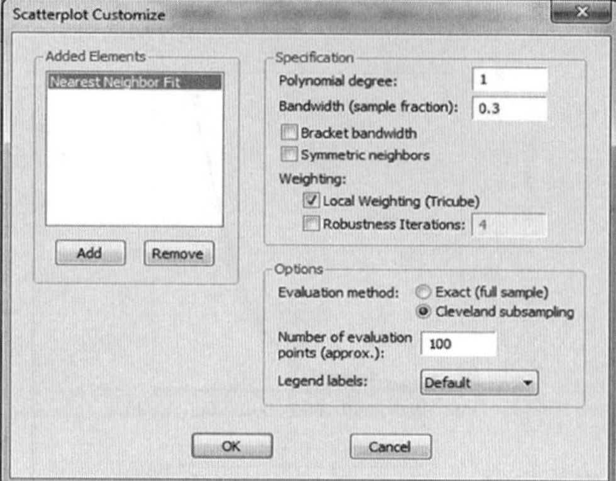

(b) Scatter Customize

Figure 2.52 Graph options and scatterplot customized in developing scatter plots with regressions.

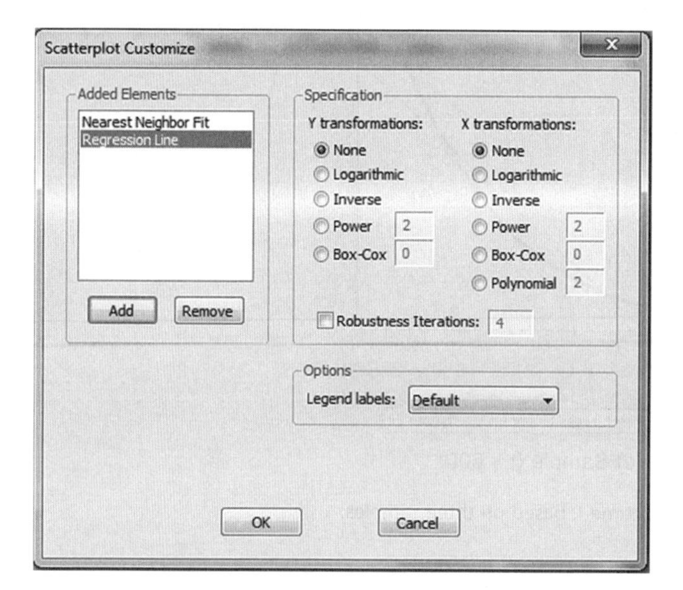

Figure 2.53 Options for the equation specification.

1) With the two variables t and *CURR* on the screen, select *View/Graph.../Scatter*, and then select *Nearest Neighbor Fit* for the fit lines, as presented in Figure 2.52a. Note that there are other possible fit lines that can be selected.
2) By clicking the *Options*, the *Scatter Customize* in Figure 2.52b, shown on the screen, then by clicking the option *Add...OK*, options the regression of $Y = CURR$ on $X = t$, as presented in Figure 2.53, show on the screen. Note that there are five options of Y and six options for X. However, the graphs in Figure 2.51a, based on the whole sample, are obtained by clicking *OK*.
3) Note the options at the bottom of the graph, which can be used to select a subsample based on the X-variable. Figure 2.51b is obtained for a lower subsample *{t<700}*, by moving the upper bound to the left and Figure 2.51c for an upper subsample *{t>600}* by moving the lower bound to the right.

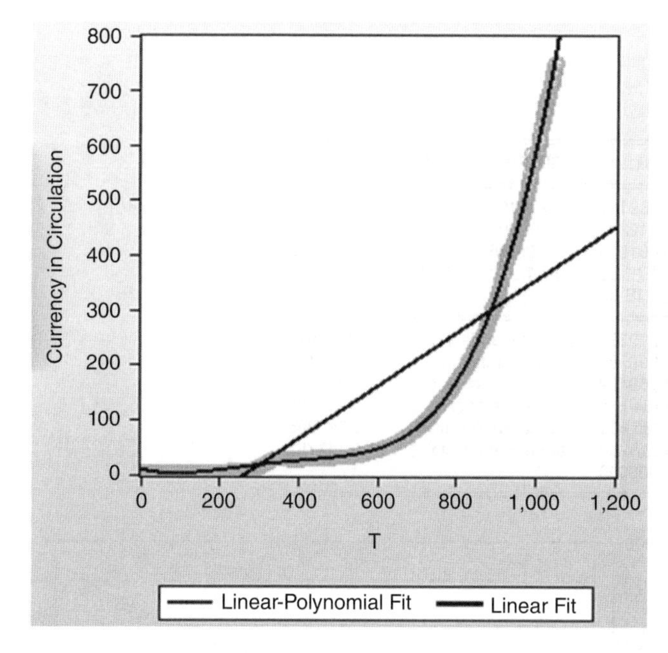

Figure 2.54 Scatter graph of *CURR* on *T*, with a fourth degree polynomial fit and a linear fit.

Example 2.35 Scatter Graph of $CURR_t$ on the time *t*, with a Fourth Degree Polynomial Fit
As a graphical representation, Figure 2.54 presents the scatter graph of $CURR_t$ on the time *t*, with a fourth degree polynomial fit and a linear fit. Based on this graph, the following findings and notes are presented.

1) The graph shows that the fourth degree polynomial regression of $CURR_t$ on the time *t* can be considered a good fit model. Note the SLR of *CURR* on *t* should be the worst forecast model. Note the scatter graph is presented in gray.
2) On the other hand the *SLR* of *CURR* on *t* should be the worst forecast model.
3) So the fourth degree polynomial of $CURR_t$ on the time *t* can be applied as a forecast model of *CURR*.
4) It has been found that several alternative forecast models could be subjectively defined or developed by researchers. However, we will never know the best possible forecast model. See the following example.

2.7.2 Additive Forecast Models of *CURR* with a Time Predictor

2.7.2.1 Forecast Model of *CURR* with Polynomial Trend

Example 2.36 A Fourth Degree Polynomial Forecast Model of $CURR_t$ on the Time *t*
Referring to the scatter graph of $CURR_t$ on the time *t* with its polynomial fit, Figure 2.55 presents the statistical results of the fourth degree polynomial regression of $CURR_t$ on the time *t* using the following ES. Based on these results, the following findings and notes are presented.

$$CURR\ C\ t\ t\hat{\ }2\ t\hat{\ }3\ t\hat{\ }4 \tag{2.45}$$

Dependent Variable: CURR
Method: Least Squares
Date: 06/08/14 Time: 14:42
Sample: 1917M08 2005M04
Included observations: 1051

Variable	Coefficient	Std. Error	t-Statistic	Prob.
C	9.989112	0.708193	14.10508	0.0000
T	-0.212298	0.009308	-22.80792	0.0000
T^2	0.001610	3.59E-05	44.86429	0.0000
T^3	-3.57E-06	5.11E-08	-69.73138	0.0000
T^4	2.74E-09	2.41E-11	113.7536	0.0000

R-squared	0.999383	Mean dependent var	125.6826
Adjusted R-squared	0.999381	S.D. dependent var	183.6484
S.E. of regression	4.569482	Akaike info criterion	5.881423
Sum squared resid	21840.65	Schwarz criterion	5.905008
Log likelihood	-3085.688	Hannan-Quinn criter.	5.890365
F-statistic	423742.0	Durbin-Watson stat	0.117782
Prob(F-statistic)	0.000000		

Forecast: CURRF
Actual: CURR
Forecast sample: 2001M01 2005M04
Included observations: 52

Root Mean Squared Error	8.007768
Mean Absolute Error	6.818115
Mean Abs. Percent Error	1.002189
Theil Inequality Coefficient	0.005900
Bias Proportion	0.005410
Variance Proportion	0.143873
Covariance Proportion	0.849717

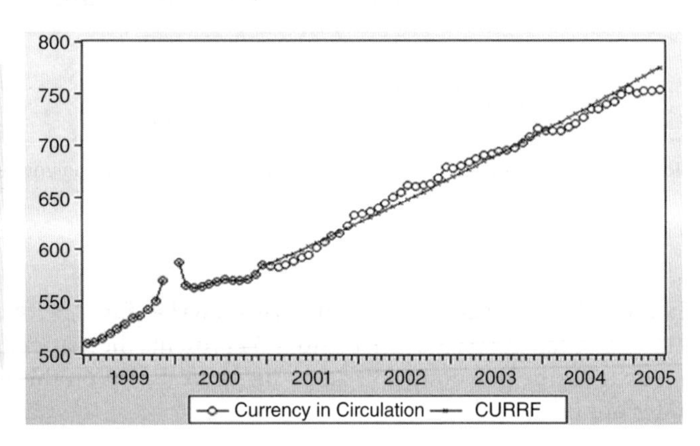

Figure 2.55 Forecast statistical results based on the model in (2.45).

1) Each of the time independent variables has a significant adjusted effect on *CURR*. So, in a statistical sense, the model is a good fit model. Note its forecast has a very small TIC of 0.005900. However, it might not the best forecast model of *CURR*.
2) On the other hand, the regression has a very small DW statistic of 0.117782. In addition, the *Breausch–Godfrey Serial Correlation LM Test* shows that its residuals have a significant autocorrelation, based on the *F*-statistic of $F_0 = 3078.810$, with $df = (2,1044)$, and a *p value* = 0.0000. In order to increase the DW statistic or to reduce the serial correlation problem, then alternative LVAR(*p,q*) models should be explored, which will be presented in the following subsection.
3) Referring to such very small coefficients of the independent variables T^3, and T^4, it is recommended to use the scaled time variable SC_T = T/100. Do it as an exercise.

Example 2.37 A Fourth Degree Polynomial Forecast Model of Log($CURR_t$) on the Time *t*
As a comparison with the forecast model (2.45), Figure 2.56 presents the statistical results of the fourth degree polynomial regression of log($CURR_t$) on the time *t* using the following ES.

$$log(Curr)\ c\ t\ t\hat{}2\ t\hat{}3\ t\hat{}4 \tag{2.46}$$

Dependent Variable: LOG(CURR)
Method: Least Squares
Date: 06/09/14 Time: 14:53
Sample: 1917M08 2005M04
Included observations: 1051

Variable	Coefficient	Std. Error	t-Statistic	Prob.
C	1.334428	0.034838	38.30405	0.0000
T	0.000728	0.000458	1.590235	0.1121
T^2	1.58E-05	1.77E-06	8.972700	0.0000
T^3	-2.25E-08	2.52E-09	-8.934416	0.0000
T^4	1.10E-11	1.18E-12	9.263396	0.0000

R-squared	0.980524	Mean dependent var	3.710540
Adjusted R-squared	0.980449	S.D. dependent var	1.607615
S.E. of regression	0.224784	Akaike info criterion	-0.142605
Sum squared resid	52.85225	Schwarz criterion	-0.119021
Log likelihood	79.93911	Hannan-Quinn criter.	-0.133663
F-statistic	13164.97	Durbin-Watson stat	0.003337
Prob(F-statistic)	0.000000		

Forecast: CURRF
Actual: CURR
Forecast sample: 2001M01 2005M04
Included observations: 52

Root Mean Squared Error	124.1628
Mean Absolute Error	106.6857
Mean Abs. Percent Error	15.18803
Theil Inequality Coefficient	0.084510
Bias Proportion	0.738294
Variance Proportion	0.248000
Covariance Proportion	0.013706

Figure 2.56 Statistical results based on the model (2.46).

Compared to the model (2.45) with a RMSE = 8.007768, this model is worse, since it has a much greater RMSE = 124.1628. So it is not a recommended model to apply.

2.7.2.2 LVAR(*p*,*q*) Forecast Model of *CURR* with Polynomial Trend

Referring to the regression in Figure 2.55 with a very small DW statistic, the following examples present selected LV(*p*,*q*) models with special notes and comments.

Example 2.38 LV(1), and AR(1) Models of *CURR*

Figure 2.57 presents the statistical results of a LV(1) and AR(1) models of *CURR* on a fourth degree polynomial of the time *t*, respectively, using the ES as follows:

$$Curr\ C\ Curr(-1)\ t\ t\hat{}2\ t\hat{}3\ t\hat{}4 \tag{2.47}$$

and

$$Curr\ C\ t\ t\hat{}2\ t\hat{}3\ t\hat{}4\ AR(1) \tag{2.48}$$

1) Based on the statistical results of the LV(1) model in Figure 2.57a, the following findings and notes are presented.
   1.1 Compared to the model (2.45) with a RMSE = 8.007768, this forecast has a slightly greater RMSE = 8.518987, but with a much greater DW statistic of 1.191359.
   1.2 Since each of the independent variables has a significant adjusted effect, at the 1% level of significance, then it can be said that the model is a good fit model. However, it might not be the best forecast of *CURR*. See the following example.

2) Based on the statistical results of the AR(1) model in Figure 2.57b, the following findings and notes are presented.

Dependent Variable: CURR
Method: Least Squares
Date: 06/09/14 Time: 15:27
Sample (adjusted): 1917M09 2005M04
Included observations: 1048 after adjustments

Variable	Coefficient	Std. Error	t-Statistic	Prob.
C	0.582790	0.264089	2.206791	0.0275
CURR(-1)	0.927183	0.010661	86.96612	0.0000
T	-0.011994	0.003913	-3.065462	0.0022
T^2	0.000102	2.12E-05	4.812184	0.0000
T^3	-2.38E-07	4.20E-08	-5.660456	0.0000
T^4	1.92E-10	3.04E-11	6.324036	0.0000

R-squared	0.999931	Mean dependent var	125.4722
Adjusted R-squared	0.999930	S.D. dependent var	183.2792
S.E. of regression	1.530909	Akaike info criterion	3.695309
Sum squared resid	2442.116	Schwarz criterion	3.723675
Log likelihood	-1930.342	Hannan-Quinn criter.	3.706065
F-statistic	3001058.	Durbin-Watson stat	1.191359
Prob(F-statistic)	0.000000		

Forecast: CURRF_LV1
Actual: CURR
Forecast sample: 2001M01 2005M04
Included observations: 52

Root Mean Squared Error	8.518987
Mean Absolute Error	7.247821
Mean Abs. Percent Error	1.085142
Theil Inequality Coefficient	0.006296
Bias Proportion	0.156711
Variance Proportion	0.011522
Covariance Proportion	0.831767

Dependent Variable: CURR
Method: Least Squares
Date: 06/09/14 Time: 16:22
Sample (adjusted): 1917M09 2005M04
Included observations: 1048 after adjustments
Convergence achieved after 1 iteration

Variable	Coefficient	Std. Error	t-Statistic	Prob.
C	10.14162	0.690175	14.69426	0.0000
T	-0.213931	0.009051	-23.63583	0.0000
T^2	0.001615	3.48E-05	46.37566	0.0000
T^3	-3.57E-06	4.96E-08	-72.06545	0.0000
T^4	2.74E-09	2.33E-11	117.5706	0.0000
AR(1)	0.002500	0.030565	0.081792	0.9348

R-squared	0.999429	Mean dependent var	125.4722
Adjusted R-squared	0.999427	S.D. dependent var	183.2792
S.E. of regression	4.388977	Akaike info criterion	5.801778
Sum squared resid	20072.17	Schwarz criterion	5.830144
Log likelihood	-3034.132	Hannan-Quinn criter.	5.812535
F-statistic	364946.0	Durbin-Watson stat	0.095634
Prob(F-statistic)	0.000000		
Inverted AR Roots	.00		

Forecast: CURRF_AR1
Actual: CURR
Forecast sample: 2001M01 2005M04
Included observations: 52

Root Mean Squared Error	7.975680
Mean Absolute Error	6.793266
Mean Abs. Percent Error	0.999257
Theil Inequality Coefficient	0.005878
Bias Proportion	0.001561
Variance Proportion	0.142255
Covariance Proportion	0.856185

Figure 2.57 Statistical results of the LV(1) and AR(1) models, in (2.47) and (2.48).

2.1 Compared to the model (2.45) with a RMSE = 8.007768, this forecast has a slightly smaller RMSE = 7.975680 with a greater DW statistic.
2.2 Unexpectedly, the term AR(1) has a very large p value = 0.9347, which indicates the residuals have insignificant first order autocorrelation. So, if this term should be deleted from the model, then we have the original model (2.45).
2.3 As an additional illustration, note the model in (2.48) has the following equation.

$$Curr_t = C(1) + C(2) * Curr_{t-1} + C(3) * t + C(4) * t\hat{\ }2 + C(5) * t\hat{\ }3 + C(6) * t\hat{\ }4 + \mu_t$$
$$\mu_t = \rho\mu_{t-1} + \varepsilon_t \tag{2.49}$$

Corresponding to the term AR(1) in (2.48), the null hypothesis H_0: $\rho = 0$ is accepted based on the t-statistic of $t_0 = 0.087192$, with $df = 1042$, and a p value = 0.9348.

Example 2.39 LV(2) Models of *CURR*
By using the trial-and-error method, I found acceptable statistical results based on the LV(2), model of *CURR* with the fourth degree polynomial trend, as presented in Figure 2.58, using the following ES, with its reduced model.

Curr c Curr(−1) Curr(−2) t t^2 t^3 t^4 (2.50)

Based on these results, the following findings and notes are presented.

1) Only the independent variable T has an insignificant adjusted effect on *CURR*, at the 10% level of significance, with a p value = 0.2430 < 0.30. Note that at the 15% level of significance (Lapin 1973), T has a significant negative effect with a p value = 0.2413/2 = 0.1215 < 0.15. Then, I would consider the model is an acceptable or a good fit model, to forecast *CURR*.
2) Compared to the forecasts based on the models (2.45) up to (2.49), this forecast has the smallest RMSE = 7.605676, with a greater DW statistic of 1.714462. So, it can be concluded this model is the best forecast model compared to the five previous models.
3) Unexpectedly, its reduced model presents the forecast with a smaller RMSE = 7.409729. Even though they have only a very small difference, its reduced model should be considered a better forecast model than the full version.

Example 2.40 LV(3) Models of *CURR*
As an extension of the LV(2) model in (2.50), Figure 2.59 presents the statistical results based on a LV(3) model of *CURR* with the fourth degree polynomial trend, using the following ES, with its reduced model.

Curr C Curr(−1) Curr(−2) Curr(−3) t t^2 t^3 t^4 (2.51)

Based on these results, the following findings and notes are presented.

1) Compared to the LV(2) model in (2.50) and its reduced model where the reduced model gives a better forecast, the results of the LV(3) model and its reduced model in Figure 2.59 present contradictory results, whereas the full model presents a better forecast.
2) So the full model would be considered a better forecast model, even though the time t in the full model has a large p value > 0.30.

Dependent Variable: CURR
Method: Least Squares
Date: 06/09/14 Time: 15:32
Sample (adjusted): 1917M10 2005M04
Included observations: 1045 after adjustments

Variable	Coefficient	Std. Error	t-Statistic	Prob.
C	0.198447	0.230180	0.862138	0.3888
CURR(-1)	1.207731	0.028210	42.81233	0.0000
CURR(-2)	-0.240078	0.027368	-8.772317	0.0000
T	-0.004007	0.003430	-1.168243	0.2430
T^2	4.00E-05	1.88E-05	2.131174	0.0333
T^3	-9.90E-08	3.74E-08	-2.649404	0.0082
T^4	8.39E-11	2.71E-11	3.096949	0.0020

R-squared	0.999949	Mean dependent var	125.2822
Adjusted R-squared	0.999949	S.D. dependent var	182.9596
S.E. of regression	1.305746	Akaike info criterion	3.378103
Sum squared resid	1769.763	Schwarz criterion	3.411272
Log likelihood	-1758.059	Hannan-Quinn criter.	3.390683
F-statistic	3416018.	Durbin-Watson stat	1.714462
Prob(F-statistic)	0.000000		

Forecast: CURRF_LV2
Actual: CURR
Forecast sample: 2001M01 2005M04
Included observations: 52

Root Mean Squared Error	7.605674
Mean Absolute Error	6.537197
Mean Abs. Percent Error	0.977670
Theil Inequality Coefficient	0.005607
Bias Proportion	0.000042
Variance Proportion	0.023282
Covariance Proportion	0.976676

Dependent Variable: CURR
Method: Least Squares
Date: 06/11/14 Time: 14:36
Sample (adjusted): 1917M10 2005M04
Included observations: 1045 after adjustments

Variable	Coefficient	Std. Error	t-Statistic	Prob.
C	-0.040023	0.106391	-0.376188	0.7069
CURR(-1)	1.211974	0.027980	43.31558	0.0000
CURR(-2)	-0.237521	0.027285	-8.705239	0.0000
T^2	1.95E-05	6.69E-06	2.916411	0.0036
T^3	-6.19E-08	1.97E-08	-3.139545	0.0017
T^4	5.96E-11	1.74E-11	3.431541	0.0006

R-squared	0.999949	Mean dependent var	125.2822
Adjusted R-squared	0.999949	S.D. dependent var	182.9596
S.E. of regression	1.305976	Akaike info criterion	3.377503
Sum squared resid	1772.089	Schwarz criterion	3.405934
Log likelihood	-1758.745	Hannan-Quinn criter.	3.388285
F-statistic	4097782.	Durbin-Watson stat	1.720461
Prob(F-statistic)	0.000000		

Forecast: CURRF
Actual: CURR
Forecast sample: 2001M01 2005M04
Included observations: 52

Root Mean Squared Error	7.409729
Mean Absolute Error	6.346927
Mean Abs. Percent Error	0.953438
Theil Inequality Coefficient	0.005465
Bias Proportion	0.004091
Variance Proportion	0.007480
Covariance Proportion	0.988429

Figure 2.58 Statistical results based on the LV(2) model in (2.50) and its reduced model.

Dependent Variable: CURR
Method: Least Squares
Date: 06/09/14 Time: 15:38
Sample (adjusted): 1917M11 2005M04
Included observations: 1042 after adjustments

Variable	Coefficient	Std. Error	t-Statistic	Prob.
C	0.125083	0.228095	0.548380	0.5835
CURR(-1)	1.288614	0.032374	39.80442	0.0000
CURR(-2)	-0.531783	0.046661	-11.39680	0.0000
CURR(-3)	0.215871	0.027605	7.819875	0.0000
T	-0.002422	0.003407	-0.710709	0.4774
T^2	2.97E-05	1.87E-05	1.585483	0.1132
T^3	-7.78E-08	3.72E-08	-2.088459	0.0370
T^4	6.90E-11	2.70E-11	2.556475	0.0107

R-squared	0.999952	Mean dependent var	125.0930
Adjusted R-squared	0.999952	S.D. dependent var	182.6421
S.E. of regression	1.270104	Akaike info criterion	3.323723
Sum squared resid	1668.012	Schwarz criterion	3.361718
Log likelihood	-1723.660	Hannan-Quinn criter.	3.338135
F-statistic	3075071.	Durbin-Watson stat	1.822500
Prob(F-statistic)	0.000000		

Forecast: CURRF_LV3
Actual: CURR
Forecast sample: 2001M01 2005M04
Included observations: 52

Root Mean Squared Error	7.744828
Mean Absolute Error	6.648196
Mean Abs. Percent Error	0.997570
Theil Inequality Coefficient	0.005719
Bias Proportion	0.072243
Variance Proportion	0.005399
Covariance Proportion	0.922358

Dependent Variable: CURR
Method: Least Squares
Date: 06/11/14 Time: 11:24
Sample (adjusted): 1917M11 2005M04
Included observations: 1042 after adjustments

Variable	Coefficient	Std. Error	t-Statistic	Prob.
C	-0.019154	0.104087	-0.184024	0.8540
CURR(-1)	1.290115	0.032297	39.94536	0.0000
CURR(-2)	-0.530694	0.046624	-11.38231	0.0000
CURR(-3)	0.217470	0.027507	7.906013	0.0000
T^2	1.72E-05	6.59E-06	2.613468	0.0091
T^3	-5.52E-08	1.94E-08	-2.843590	0.0045
T^4	5.41E-11	1.71E-11	3.168467	0.0016

R-squared	0.999952	Mean dependent var	125.0930
Adjusted R-squared	0.999952	S.D. dependent var	182.6421
S.E. of regression	1.269800	Akaike info criterion	3.322292
Sum squared resid	1668.827	Schwarz criterion	3.355538
Log likelihood	-1723.914	Hannan-Quinn criter.	3.334902
F-statistic	3589300.	Durbin-Watson stat	1.824198
Prob(F-statistic)	0.000000		

Forecast: CURRF_EQ251
Actual: CURR
Forecast sample: 2001M01 2005M04
Included observations: 52

Root Mean Squared Error	7.768116
Mean Absolute Error	6.665418
Mean Abs. Percent Error	1.001380
Theil Inequality Coefficient	0.005738
Bias Proportion	0.105031
Variance Proportion	0.000649
Covariance Proportion	0.894320

Figure 2.59 Statistical results based on the LV(3) model in (2.51) and its reduced model.

2.7.2.3 LVAR(*p,q*) Month-Fixed-Effects Model of *CURR* and Alternatives

Referring to the forecast models of *HS* and *G*, the variable *@Month* can be applied as a numerical or a categorical variable. This section only presents a few illustrative examples.

Example 2.41 LV(2) Month-Fixed-Effects Model of *CURR* with a Polynomial Trend
Referring to the forecast polynomial models of *CURR* presented before, I have been using the trial-and-error method to explore a better forecast model. Finally, I found a LV(2) month-fixed-effects or ANCOVA model with the statistical results presented in Figure 2.60, by using the following ES. Based on these results, the following findings and notes are presented.

Dependent Variable: CURR
Method: Least Squares
Date: 06/12/14 Time: 16:07
Sample (adjusted): 1917M10 2005M04
Included observations: 1045 after adjustments

Variable	Coefficient	Std. Error	t-Statistic	Prob.
C	-1.675169	0.146865	-11.40621	0.0000
CURR(-1)	1.256088	0.027230	46.12953	0.0000
CURR(-2)	-0.271812	0.026564	-10.23250	0.0000
T^2	1.16E-05	5.64E-06	2.052678	0.0404
T^3	-3.76E-08	1.66E-08	-2.261642	0.0239
T^4	3.74E-11	1.46E-11	2.555202	0.0108
@MONTH=2	0.819437	0.180503	4.539749	0.0000
@MONTH=3	1.950788	0.182705	10.67726	0.0000
@MONTH=4	1.870000	0.170961	10.93818	0.0000
@MONTH=5	1.740437	0.169530	10.26622	0.0000
@MONTH=6	1.933129	0.169411	11.41087	0.0000
@MONTH=7	2.048781	0.168482	12.16021	0.0000
@MONTH=8	1.088816	0.167705	6.492440	0.0000
@MONTH=9	1.693099	0.172878	9.793597	0.0000
@MONTH=10	1.574363	0.170059	9.257723	0.0000
@MONTH=11	2.550220	0.170232	14.98087	0.0000
@MONTH=12	2.807898	0.167048	16.80893	0.0000

R-squared	0.999965	Mean dependent var	125.2822
Adjusted R-squared	0.999964	S.D. dependent var	182.9596
S.E. of regression	1.094756	Akaike info criterion	3.035075
Sum squared resid	1232.049	Schwarz criterion	3.115630
Log likelihood	-1568.826	Hannan-Quinn criter.	3.065626
F-statistic	1822388.	Durbin-Watson stat	1.695834
Prob(F-statistic)	0.000000		

Forecast: CURRF_EQ252
Actual: CURR
Forecast sample: 2001M01 2005M04
Included observations: 52

Root Mean Squared Error	7.217026
Mean Absolute Error	6.192138
Mean Abs. Percent Error	0.924533
Theil Inequality Coefficient	0.005324
Bias Proportion	0.016861
Variance Proportion	0.034358
Covariance Proportion	0.948782

Redundant Variables Test
Equation: EQ12
Specification: CURR C CURR(-1) CURR(-2) T^2 T^3 T^4 @EXPAND(@MONTH,@DROPFIRST)
Redundant Variables: @EXPAND(@MONTH,@DROPFIRST)

	Value	df	Probability
F-statistic	40.96367	(11, 1028)	0.0000
Likelihood ratio	379.8374	11	0.0000

Figure 2.60 Statistical results based on the LV(2) model in (2.52) with a redundant variable test.

Omitted Variables Test
Equation: UNTITLED
Specification: CURR C CURR(-1) CURR(-2) T^2 T^3 T^4
@EXPAND(@MONTH,@DROPFIRST)
Omitted Variables: CURR(-1)*@EXPAND(@MONTH,@DROPFIRST)

	Value	df	Probability
F-statistic	109.9305	(11, 1017)	0.0000
Likelihood ratio	818.7106	11	0.0000

Omitted Variables Test
Equation: UNTITLED
Specification: CURR C CURR(-1) CURR(-2) T^2 T^3 T^4
@EXPAND(@MONTH,@DROPFIRST)
Omitted Variables: T*@EXPAND(@MONTH)

	Value	df	Probability
F-statistic	73.13396	(12, 1016)	0.0000
Likelihood ratio	650.6279	12	0.0000

Figure 2.61 Two omitted variables tests based on the model (2.52).

$$Curr\ C\ Curr(-1)\ Curr(-2)\ t\hat{}2\ t\hat{}3\ t\hat{}4\ @Expand(@Month, @dropfirst) \tag{2.52}$$

1) This model is the best forecast model, compared to all forecast models of *CURR* previously presented, since it has the smallest RMSE = 7.217026 and TIC = 0.005324.
2) In addition, Figure 2.60 presents a redundant variables test for the intercept parameters, namely *@EXPAND(@MONTH,@DROPFIRST)*, which shows that the 12 regressions have significantly different intercept parameters, based on the *F*-statistic of $F_0 = 40.96367$, with $df = (11{,}1028)$, and a p value = 0.0000.
3) This model can be considered to be a One-Way ANCOVA model or month-fixed-effects model, which might be extended to alternative heterogeneous regressions models by conducting OVTs, for instance, *Curr(−1)∗ @Expand(@Month,@dropfirst)* and *t∗@Expand (@Month)*, with the statistical results presented in Figure 2.61. Note the difference between the two functions, the first is using *@Dropfirst* because the variable *Curr(−1)* is in the model (2.52) already, but the main time variable *t* is not in the model.

These results show that each of the sets has a significant effect on *CURR*. So, in a statistical sense, each set can be inserted as additional independent variables of the model (2.52). However, I have found that worse forecasting would be obtained by inserting additional interaction variables. In other words, forecast models with greater RMSE or TIC would be obtained. Do it as an exercise.

Example 2.42 LVAR(1,2) Models of *CURR* with Quadratic Trends

Figure 2.62 presents the statistical results of two LVAR(1,2) models of *CURR* with quadratic trends, using the following ES, respectively. Based on these results, the following findings and notes are presented.

$$Curr\ C\ Curr(-1)\ t\ t\hat{}2\ AR(1)\ AR(2) \tag{2.53}$$

and

$$Curr\ C\ Curr(-1)\ t\ t\hat{}2\ @Expand(@Month, @dropfirst)\ AR(1)\ AR(2) \tag{2.54}$$

1) The LVAR(1,2) model (2.53) presents a single regression with the statistical results presented in Figure 2.62a, and Figure 2.62b presents the estimates of a set of 12 homogeneous regressions of the LVAR(1,2) model (2.54), which is a month-fixed-effects or One-Way ANCOVA model. Note the specific assumption of ANCOVA models, that is, all covariates have constant slope parameters.
2) The two models are nested models, so in a statistical sense one of them can be selected using either the OVT or the redundant variables test (RVT), with the results presented in Figure 2.63. Based on these, the following findings and notes are presented.
   2.1 Note OVT and RVT present exactly the same values of *F*-statistic, and *LR* statistic, which show that the 11 dummy variables in *@Expand(@Month,@Dropfirst)* have joint significant effects on *CURR* in both models, but they have different theoretical concepts.
   2.2 Based on the OVT, it can be concluded that *@Expand(@Month,@Dropfirst)* should be inserted as additional independent dummy variables of the model (2.53) to obtain a better fit model, which is the model (2.54).
   2.3 On the other hand, the RVT shows that the set of 11 dummy variables indicated by *@Expand(@Month,@Dropfirst)* have significant joint effects on *CURR*. In other words, the 12 intercept parameters of the homogenous regression have significant differences.

3) Compared to the forecast model of *CURR* in Figure 2.60, these models give a worse forecast since they have greater values of RMSE.
4) On the other hand, we have to note very large R-squared values and Adjusted R-squared values of the model, say, greater than 0.999. Moreover, this is the case for models with R-squared = 1, because the model should be an over fitted model. We have to use our subjective judgment whether to use the model or not.

Example 2.43 LVAR(2,3) Month-Fixed-Effects Model of *CURR* with Quadratic Trends
As an extension of previous models, Figure 2.64 presents the statistical results of a LVAR(2,3) month-fixed-effects models of *CURR* with quadratic trends, using the following ES. Based on these results, the following findings and notes are presented.

$$\begin{gathered} Curr\ C\ Curr(-1)\ Curr(-2)\ t\ t\hat{}2 \\ @Expand(@Month, @dropfirst)\ AR(1)\ AR(2)\ AR(3) \end{gathered} \tag{2.55}$$

1) Compared to previous models, this model is the best forecast model of *CURR*, since it has the smallest RMSE = 2.754519 and TIC = 0.002301. It might be the best compare it to a lot of models. However, we never know whether or not there is a better forecast model, since we would never apply all those possible models.
2) This model in fact is an unexpected model, which is obtained by using the trial-and-error method. I have found worse regression functions by using LVAR(*p*,*q*) models for several integers *p* and *q*. Do it as an exercise,
3) Note that, even though the terms AR(2), and AR(3) have $0.20 < p$ values < 0.25, but at the 15% level of significance (Lapin 1973) each of the terms has a significant negative effect with a *p* value = *Prob.*/2 < 0.15. Referring to Hosmer and Lemesshow (2000), they recommended that variables with $p < 0.25$ should be kept in the model. So, Model (2.55) can be considered to be a good fit.

Dependent Variable: CURR
Method: Least Squares
Date: 06/13/14 Time: 14:09
Sample (adjusted): 1917M11 2005M04
Included observations: 1042 after adjustments
Convergence achieved after 6 iterations

Variable	Coefficient	Std. Error	t-Statistic	Prob.
C	0.175142	0.152750	1.146593	0.2518
CURR(-1)	1.003017	0.000847	1183.944	0.0000
T	-0.001717	0.000866	-1.983401	0.0476
T^2	2.91E-06	1.13E-06	2.575831	0.0101
AR(1)	0.306134	0.032117	9.531737	0.0000
AR(2)	-0.220983	0.027550	-8.021115	0.0000

R-squared	0.999951	Mean dependent var	125.0930
Adjusted R-squared	0.999951	S.D. dependent var	182.6421
S.E. of regression	1.278128	Akaike info criterion	3.334412
Sum squared resid	1692.422	Schwarz criterion	3.362909
Log likelihood	-1731.229	Hannan-Quinn criter.	3.345221
F-statistic	4251210.	Durbin-Watson stat	1.831822
Prob(F-statistic)	0.000000		

Inverted AR Roots	.15-.44i	.15+.44i

Forecast: CURRF_LVAR12	
Actual: CURR	
Forecast sample: 2001M01 2005M04	
Included observations: 52	
Root Mean Squared Error	7.412516
Mean Absolute Error	6.375406
Mean Abs. Percent Error	0.957549
Theil Inequality Coefficient	0.005472
Bias Proportion	0.052721
Variance Proportion	0.002944
Covariance Proportion	0.944335

Dependent Variable: CURR
Method: Least Squares
Date: 06/13/14 Time: 14:24
Sample (adjusted): 1917M11 2005M04
Included observations: 1042 after adjustments
Convergence achieved after 6 iterations

Variable	Coefficient	Std. Error	t-Statistic	Prob.
C	-1.176869	0.190659	-6.172625	0.0000
CURR(-1)	1.003166	0.000836	1199.319	0.0000
T	-0.001656	0.000856	-1.934170	0.0534
T^2	2.80E-06	1.12E-06	2.510619	0.0122
@MONTH Dummies	Yes			
AR(1)	0.348629	0.033465	10.41779	0.0000
AR(2)	-0.136454	0.027626	-4.939420	0.0000

R-squared	0.999965	Mean dependent var	125.0930
Adjusted R-squared	0.999965	S.D. dependent var	182.6421
S.E. of regression	1.087411	Akaike info criterion	3.021656
Sum squared resid	1212.024	Schwarz criterion	3.102396
Log likelihood	-1557.283	Hannan-Quinn criter.	3.052282
F-statistic	1835398.	Durbin-Watson stat	1.813319
Prob(F-statistic)	0.000000		

Inverted AR Roots	.17+.33i	.17-.33i

Forecast: CURRF_LVAR12_M	
Actual: CURR	
Forecast sample: 2001M01 2005M04	
Included observations: 52	
Root Mean Squared Error	7.344293
Mean Absolute Error	6.269044
Mean Abs. Percent Error	0.938076
Theil Inequality Coefficient	0.005422
Bias Proportion	0.058143
Variance Proportion	0.022988
Covariance Proportion	0.918869

Figure 2.62 Statistical results based on the LVAR(1,2) models in (2.53) and (2.54).

Omitted Variables Test
Equation: EQ16_LVAR12
Specification: CURR C CURR(-1) T T^2 AR(1) AR(2)
Omitted Variables: @EXPAND(@MONTH,@DROPFIRST)

	Value	df	Probability
F-statistic	36.93364	(11, 1025)	0.0000
Likelihood ratio	347.8921	11	0.0000

Redundant Variables Test
Equation: EQ16_LVAR12_MONTH
Specification: CURR C CURR(-1) T T^2 AR(1) AR(2) @EXPAND(@MONTH,@DROPFIRST)
Redundant Variables: @EXPAND(@MONTH,@DROPFIRST)

	Value	df	Probability
F-statistic	36.93364	(11, 1025)	0.0000
Likelihood ratio	347.8921	11	0.0000

Figure 2.63 Omitted variables test, and redundant variables test, based on the models (2.53) and (2.54), respectively.

4) As a more advanced analysis, I do this as follows:

4.1 The LVAR(2,3) model with the following general equation.

$$Curr_t = C(1) + C(2) * Curr_{t-1} + C(3) * Curr_{t-2} + C(4) * t + C(5) * t\hat{}2 + \sum_{m=2}^{12} C(m+4) * D(@Month = m) + [AR(1) = C(17), AR(2) = C(18), AR(3) = C(19)] + \varepsilon_t \quad (2.56a)$$

or

$$Curr_t = C(1) + C(2) * Curr_{t-1} + C(3) * Curr_{t-2} + C(4) * t + C(5) * t\hat{}2 + \sum_{m=2}^{12} C(m+4) * D(@Month = m) + \mu_t \quad (2.56b)$$

$$\mu_t = \rho_1 \mu_{t-1} + \rho_2 \mu_{t-2} + \rho_3 \mu_{t-3} + \varepsilon_t = C(17)\mu_{t-1} + C(18)\mu_{t-2} + C(19)\mu_{t-3} + \varepsilon_t$$

4.2 Based on the Eq. (2.56a), by using the RVT of the joint effects of AR(1), AR(2), and AR(3), it is obtained the statistical results, which shows that the three terms AR(1), AR(2), and AR(3) have joint insignificant effects, based both *F* and *LR* statistics. However, by using the Wald test, the null hypothesis H_0: C(17) = C(18) = C(19) = 0 is accepted based on the *F*-statistic of $F_0 = 1.392982$, with $df = (3,33)$, and p value = 0.2622.

4.3 Based on the Eq. (2.56b), we can conclude that the residuals μ_{t-1}, μ_{t-2} & μ_{t-3} have insignificant joint effects on μ_t.

4.4 So, in a statistical sense, a reduced model should be explored by deleting one or two of the three terms. Note the impacts of the multicollinearity on the estimates of the parameters C(17), C(18), and C(19) are unpredictable. So an acceptable reduced model should be obtained using the trial-and-error method

Method: Least Squares
Date: 06/19/14 Time: 16:59
Sample: 2001M01 2005M04
Included observations: 52
Convergence achieved after 30 iterations

Variable	Coefficient	Std. Error	t-Statistic	Prob.
C	-3406.746	1502.302	-2.267683	0.0300
CURR(-1)	1.565815	0.174480	8.974191	0.0000
CURR(-2)	-0.695905	0.157796	-4.410148	0.0001
T	6.351234	2.832352	2.242388	0.0318
T^2	-0.002877	0.001306	-2.203719	0.0346
@MONTH=2	8.916186	2.172217	4.104647	0.0002
@MONTH=3	7.990855	1.806372	4.423704	0.0001
@MONTH=4	7.568603	1.756748	4.308304	0.0001
@MONTH=5	7.903430	1.601478	4.935087	0.0000
@MONTH=6	7.381353	1.711855	4.311904	0.0001
@MONTH=7	9.464068	1.616295	5.855409	0.0000
@MONTH=8	4.137452	1.501892	2.754828	0.0095
@MONTH=9	8.424431	1.698388	4.960252	0.0000
@MONTH=10	6.562817	1.788100	3.670273	0.0008
@MONTH=11	10.99296	1.785368	6.157251	0.0000
@MONTH=12	10.91681	1.663910	6.560939	0.0000
AR(1)	-0.429130	0.219967	-1.950886	0.0596
AR(2)	-0.274587	0.232300	-1.182035	0.2456
AR(3)	-0.270353	0.206840	-1.307066	0.2002

R-squared	0.999253	Mean dependent var	676.1145
Adjusted R-squared	0.998845	S.D. dependent var	52.80414
S.E. of regression	1.794232	Akaike info criterion	4.283065
Sum squared resid	106.2359	Schwarz criterion	4.996019
Log likelihood	-92.35968	Hannan-Quinn criter.	4.556394
F-statistic	2452.177	Durbin-Watson stat	2.064965
Prob(F-statistic)	0.000000		

Inverted AR Roots	.11+.64i	.11-.64i	-.65

Forecast: CURRF_LVAR23
Actual: CURR
Forecast sample: 2001M01 2005M04
Included observations: 52

Root Mean Squared Error	2.754519
Mean Absolute Error	2.244001
Mean Abs. Percent Error	0.330154
Theil Inequality Coefficient	0.002031
Bias Proportion	0.000238
Variance Proportion	0.003111
Covariance Proportion	0.996651

Figure 2.64 Statistical results based on the LV(2,3) month-fixed-effects model in (2.55).

4.5 For an important discussion, Table 2.10 presents the statistical results summary of three reduced models, namely RD1, RD2, and RD3, respectively, by deleting *AR(2)*, *AR(3)*, or *AR(1)* from the model (2.55). Based on these reduced models, the following notes are made.

- In general, students and less experienced researchers would delete a variable with the greatest p value, that is *AR(2)*. So I present the RD1, which shows both that *AR(1)* has a very large p value = 0.7580 and *AR(3)* has a p value = 0.1214, which is smaller than the p value in the full model (2.55). So this RD1 is worse than the full model.

Table 2.10 Statistical results summary of three reduced models of the LV(2,3) Model in (2.55).

Dependent Variable: *CURR*									
Method: Least Squares									
Date: 06/23/14 Time: 05:36									
Sample (adjusted): 1918M01 2005M04									
Included observations: 1036 after adjustments									
Convergence achieved after 10 iterations									
C	−1.4171	−1.8449	0.0653	−1.5272	−6.2051	0.0000	−1.7307	−11.112	0.0000
CURR(−1)	1.1818	2.4799	0.0133	1.2215	9.9787	0.0000	1.3560	40.234	0.0000
CURR(−2)	−0.1791	−0.3747	0.7080	−0.2189	−1.7828	0.0749	−0.3539	−10.466	0.0000
T	−0.0015	−1.3138	0.1892	−0.0013	−1.9659	0.0496	−0.0011	−2.1591	0.0311
T^2	0.0000	1.4311	0.1527	0.0000	2.4357	0.0150	0.0000	2.7129	0.0068
@*MONTH* = 2	0.5865	0.4507	0.6523	0.7288	1.9581	0.0505	1.0890	5.7307	0.0000
@*MONTH* = 3	1.7369	1.3813	0.1675	1.8773	5.1188	0.0000	2.2313	10.8246	0.0000
@*MONTH* = 4	1.7928	2.5335	0.0114	1.8823	7.7395	0.0000	2.0786	11.4127	0.0000
@*MONTH* = 5	1.6877	2.9002	0.0038	1.7575	7.9201	0.0000	1.9274	11.3813	0.0000
@*MONTH* = 6	1.8662	3.0281	0.0025	1.9477	6.6458	0.0000	2.1202	12.3460	0.0000
@*MONTH* = 7	1.9912	3.7252	0.0002	2.0640	9.7912	0.0000	2.2053	12.7514	0.0000
@*MONTH* = 8	1.0247	2.2103	0.0273	1.0988	5.5253	0.0000	1.2121	7.1717	0.0000
@*MONTH* = 9	1.5570	1.7390	0.0823	1.6625	5.9973	0.0000	1.9098	10.9060	0.0000
@*MONTH* = 10	1.4717	2.0268	0.0429	1.5631	6.3081	0.0000	1.7643	9.6891	0.0000
@*MONTH* = 11	2.4604	3.3508	0.0008	2.5520	9.9296	0.0000	2.7549	14.2929	0.0000
@*MONTH* = 12	2.8066	9.6431	0.0000	2.8428	16.7524	0.0000	2.9213	17.3063	0.0000
AR(1)	0.1474	0.3032	0.7580	0.1417	1.1685	0.2429			
AR(2)				−0.2031	−5.7873	0.0000	−0.2259	−6.5700	0.0000
AR(3)	−0.0556	−1.5503	0.1214				−0.0859	−2.5630	0.0105

Dependent Variable: CURR
Method: Least Squares
Date: 06/23/14 Time: 13:27
Sample (adjusted): 1917M11 2005M04
Included observations: 1042 after adjustments
Convergence achieved after 21 iterations

Variable	Coefficient	Std. Error	t-Statistic	Prob.
C	-1.419502	2324.884	-0.000611	0.9995
CURR(-1)	1.173254	1503.300	0.000780	0.9994
CURR(-2)	-0.170586	1508.133	-0.000113	0.9999
T	-0.001415	2.572969	-0.000550	0.9996
T^2	2.37E-06	0.004294	0.000552	0.9996
@MONTH=2	0.571485	4060.933	0.000141	0.9999
@MONTH=3	1.715926	3892.337	0.000441	0.9996
@MONTH=4	1.767022	2143.238	0.000824	0.9993
@MONTH=5	1.658533	1769.003	0.000938	0.9993
@MONTH=6	1.849535	1868.458	0.000990	0.9992
@MONTH=7	1.974499	1598.237	0.001235	0.9990
@MONTH=8	1.021619	1364.431	0.000749	0.9994
@MONTH=9	1.540803	2757.138	0.000559	0.9996
@MONTH=10	1.457740	2213.464	0.000659	0.9995
@MONTH=11	2.434069	2245.887	0.001084	0.9991
@MONTH=12	2.789991	783.6857	0.003560	0.9972
AR(1)	0.170024	1503.302	0.000113	0.9999

R-squared	0.999965	Mean dependent var	125.0930
Adjusted R-squared	0.999964	S.D. dependent var	182.6421
S.E. of regression	1.095773	Akaike info criterion	3.036977
Sum squared resid	1230.736	Schwarz criterion	3.117717
Log likelihood	-1565.265	Hannan-Quinn criter.	3.067603
F-statistic	1807491.	Durbin-Watson stat	1.806678
Prob(F-statistic)	0.000000		
Inverted AR Roots	.17		

Forecast: CURRF_LVAR21
Actual: CURR
Forecast sample: 2001M08 2005M04
Included observations: 45

Root Mean Squared Error	9.616569
Mean Absolute Error	8.109949
Mean Abs. Percent Error	1.200584
Theil Inequality Coefficient	0.006991
Bias Proportion	0.463101
Variance Proportion	0.251191
Covariance Proportion	0.285708

Figure 2.65 Statistical results of a reduced model by deleting both AR(2) and AR(3).

- Then RD2 is obtained by deleting the term AR(3), or μ_{t-3}, which has the second greatest p value. The RD2 shows *AR(1)* has a p value = 0.2429, which much greater than its p value = 0.0596 in the full model.
- Finally, RD3 is obtained by deleting *AR(1)*, which shows that both *AR(2)* and *AR(3)* have very small p values of 0.0000 and 0.0105, respectively.
- As another comparison, by deleting both *AR(2)* and *AR(3)*, unexpected results are obtained as presented in Figure 2.65, where all parameters have p values > 0.99. I would say this LVAR(2,1) model is the worst fit model for the data.

2.7.2.4 LVAR(p,q) Year-Fixed-Effects Model of *CURR* with Polynomial Trend

Referring to the regression in Figure 2.55 with a very small DW statistic, the following examples present selected LV(p,q) year-fixed-effects models, or an ANCOVA model of *CURR* with a single factor *@Year*.

Dependent Variable: CURR
Method: Least Squares
Date: 06/13/14 Time: 15:44
Sample (adjusted): 1917M11 2005M04
Included observations: 1042 after adjustments
Convergence achieved after 21 iterations

Variable	Coefficient	Std. Error	t-Statistic	Prob.
C	0.569702	0.908821	0.626858	0.5309
CURR(-1)	1.024019	0.016170	63.32852	0.0000
T	-0.108955	0.024509	-4.445496	0.0000
T^2	0.000258	2.79E-05	9.259536	0.0000
@YEAR Dummies	Yes			
AR(1)	0.345182	0.035793	9.643908	0.0000
AR(2)	-0.149194	0.028519	-5.231314	0.0000

R-squared	0.999971	Mean dependent var	125.0930
Adjusted R-squared	0.999968	S.D. dependent var	182.6421
S.E. of regression	1.039212	Akaike info criterion	3.000683
Sum squared resid	1023.804	Schwarz criterion	3.447128
Log likelihood	-1469.356	Hannan-Quinn criter.	3.170024
F-statistic	345739.0	Durbin-Watson stat	1.913790
Prob(F-statistic)	0.000000		

Inverted AR Roots	.17+.35i	.17-.35i

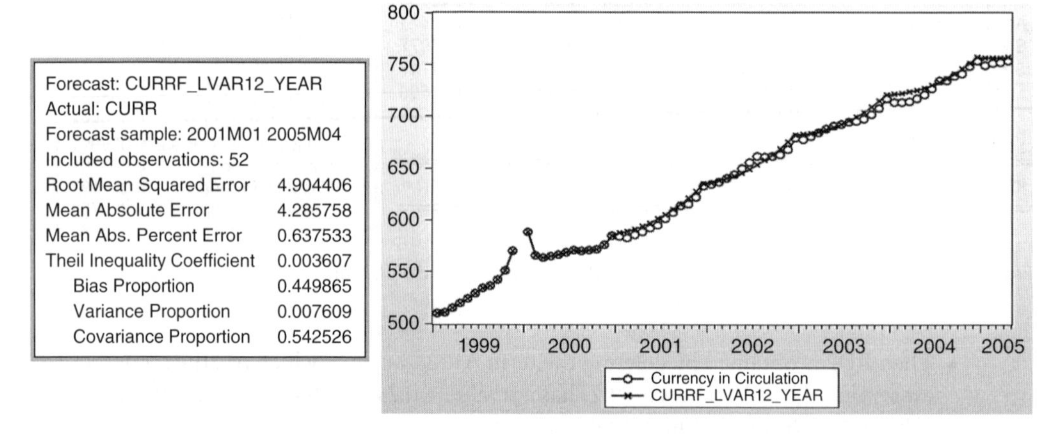

Figure 2.66 Statistical results based on the LVAR(1,2) model in (2.57).

Example 2.44 LVAR(1,2) Year-Fixed-Effects Model of *CURR* with Quadratic Trends
Figure 2.66 presents an unexpected forecasting of *CURR*, using the following equation, which shows the graphs of *CURR* and *CURRF* have a break point. In addition, this LVAR(1,2) has a very large R squared = 0.999971, as usual, and its IVs have significant joint effects, based on the *F*-statistic of $F_0 = 345729.0$, with a p value = 0.000000. Its forecast value has RMSE = 4.904406 and a very small TIC = 0.003607.

Compared to the forecast based on the LVAR(2,3) model (2.55) with RMSE = 2.754519 and TIC = 0.002031, this forecast worse, but this model has a relatively small RMSE compared to most of the others (Figure 2.66). See the following examples.

$$Curr\ C\ Curr(-1)\ t\ t\hat{}2\ @Expand(@Year, @Dropfirst)\ AR(1)\ AR(2) \tag{2.57}$$

Example 2.45 LVAR(2,3) Year-Fixed-Effects Models of *CURR* with Quadratic Trends
As two additional illustrations, Figure 2.67 presents the illustrative statistical results.

Based on these results, the following findings and notes are presented.

1) Figure 2.67a presents the statistical results of the following LVAR(2,3) ANCOVA models of *CURR*, by *@Year* with a quadratic trend of the time *t*, with its forecast has a large RMSE = 11.52437. So this forecast model is worse than the LVAR(1,2) model in (2.57).

$$\begin{aligned} &Curr\ C\ Curr(-1)\ Curr(-2)\ t\ t\hat{}2 \\ &@Expand(@Year, @Dropfirst)\ AR(1)\ AR(2)\ AR(3) \end{aligned} \tag{2.58}$$

2) As an alternative model, Figure 2.67b presents the statistical results of the following LVAR (2,3) model of *CURR*, by *@Year* with a quadratic trend of the numerical variable *@Month* with its forecast having a much greater RMSE = 97.52711 compared to the model (2.58). So, in statistical results, this forecast model is much worse than the LVAR(2,3) model in (2.55)

$$\begin{aligned} &Curr\ C\ Curr(-1)\ Curr(-2)\ @Month\ @Month\hat{}2 \\ &@Expand(@Year, @Dropfirst)\ AR(1)\ AR(2)\ AR(3) \end{aligned} \tag{2.59}$$

3) Note that both regression functions can be considered good fit ANCOVA or year-fixed-effects models, since each of the numerical independent variables (covariates), as well as each of the AR terms, has a significant effect at the 5% level of significance. These illustrations have shown that a good fit model does not directly give a good forecast model.
4) In addition, note that all models of *CURR* previously presented are additive ANCOVA or fixed-effects models. The following section presents selected interaction effects models with one of more interaction factors as their independent variables, as additional illustrations.

2.7.3 Interaction Forecast Models of *CURR*

It is recognized each of the forecast additive models of *CURR* presented previously can easily be modified or extended to several interaction model by inserting one or more interactions between their independent variables. The following subsections only present a few illustrative examples.

2.7.3.1 Interaction ANCOVA Models

Various LVAR(p,q) interaction ANCOVA models of any time series Y_t by *@Month* or by *@Year*, can easily be developed, for several or many alternative values of p and q. Those LVAR(p,q) interaction ANCOVA models of Y_t on a single time t would have at least one of the numerical interaction factors $t*@Curr(-p)$ for $p \geq 1$ as additional independent variable(s). However, we never can predict which one of those possible models would give the best forecasting. For this reason, the trial-and-error method should be applied. The following examples only present few models, which I obtain by doing analysis based on several alternative models.

Two of the simplest interaction ANCOVA models of Y_t by *@Month* are as follows:

1) A LV(1) interaction ANCOVA model Y_t with the TRE, by *@Month*, has the ES as follows:

$$Y\ C\ Y(-1)\ t\ t*Y(-1)\ @Expand(@Month, @Droplast) \tag{2.60}$$

Dependent Variable: CURR
Method: Least Squares
Date: 06/25/14 Time: 11:24
Sample (adjusted): 1918M01 2005M04
Included observations: 1036 after adjustments
Convergence achieved after 37 iterations

Variable	Coefficient	Std. Error	t-Statistic	Prob.
C	0.415375	0.394383	1.053228	0.2925
CURR(-1)	0.828179	0.051147	16.19219	0.0000
CURR(-2)	0.235093	0.053344	4.407134	0.0000
T	-0.056903	0.023060	-2.467537	0.0138
T^2	0.000160	2.68E-05	5.974155	0.0000
@YEAR=1919	0.547577	0.493106	1.110465	0.2671
...				
...				
@YEAR=2005	-165.2041	10.97869	-15.04771	0.0000
AR(1)	0.412097	0.062562	6.587067	0.0000
AR(2)	-0.360877	0.053893	-6.696129	0.0000
AR(3)	-0.113248	0.041970	-2.698294	0.0071

R-squared	0.999972	Mean dependent var	124.7080
Adjusted R-squared	0.999969	S.D. dependent var	181.9910
S.E. of regression	1.012845	Akaike info criterion	2.950621
Sum squared resid	965.3288	Schwarz criterion	3.403900
Log likelihood	-1433.422	Hannan-Quinn criter.	3.122603
F-statistic	355479.4	Durbin-Watson stat	1.925028
Prob(F-statistic)	0.000000		

Inverted AR Roots	.32+.63i	.32-.63i	-.22

Forecast: CURRF_EQ 258	
Actual: CURR	
Forecast sample: 2001M01 2005M04	
Included observations: 52	
Root Mean Squared Error	11.52347
Mean Absolute Error	9.506348
Mean Abs. Percent Error	1.357166
Theil Inequality Coefficient	0.008437
Bias Proportion	0.636286
Variance Proportion	0.219433
Covariance Proportion	0.144281

(a)

Dependent Variable: CURR
Method: Least Squares
Date: 06/25/14 Time: 11:31
Sample (adjusted): 1918M01 2005M04
Included observations: 1036 after adjustments
Convergence achieved after 16 iterations

Variable	Coefficient	Std. Error	t-Statistic	Prob.
C	-1.136473	0.279130	-4.071484	0.0001
CURR(-1)	0.842858	0.055355	15.22627	0.0000
CURR(-2)	0.287544	0.057735	4.980443	0.0000
@MONTH	0.195871	0.045496	4.305271	0.0000
@MONTH^2	-0.010919	0.003445	-3.169711	0.0016
@YEAR=1919	-0.149858	0.373624	-0.401093	0.6884
...				
...				
@YEAR=2005	-97.25221	7.549411	-12.88209	0.0000
AR(1)	0.390905	0.067145	5.821831	0.0000
AR(2)	-0.432592	0.052961	-8.168071	0.0000
AR(3)	-0.119713	0.047182	-2.537280	0.0113

R-squared	0.999971	Mean dependent var	124.7080
Adjusted R-squared	0.999968	S.D. dependent var	181.9910
S.E. of regression	1.025147	Akaike info criterion	2.974768
Sum squared resid	988.9222	Schwarz criterion	3.428047
Log likelihood	-1445.930	Hannan-Quinn criter.	3.146749
F-statistic	346998.2	Durbin-Watson stat	1.951331
Prob(F-statistic)	0.000000		

Inverted AR Roots	.30+.69i	.30-.69i	-.21

Forecast: CURRF_EQ 259	
Actual: CURR	
Forecast sample: 2001M01 2005M04	
Included observations: 52	
Root Mean Squared Error	97.52711
Mean Absolute Error	60.75160
Mean Abs. Percent Error	8.352198
Theil Inequality Coefficient	0.068436
Bias Proportion	0.388029
Variance Proportion	0.515603
Covariance Proportion	0.096363

(b)

Figure 2.67 Statistical results based on the LVAR(2,3) models in (2.58) and (2.59).

Dependent Variable: CURR
Method: Least Squares
Date: 06/26/14 Time: 16:05
Sample (adjusted): 1918M01 2005M04
Included observations: 1036 after adjustments
Convergence achieved after 18 iterations

Variable	Coefficient	Std. Error	t-Statistic	Prob.
C	1.541304	0.195736	7.874396	0.0000
CURR(-1)	1.737602	0.311036	5.586501	0.0000
CURR(-2)	-0.707690	0.313653	-2.256285	0.0243
T	-0.001073	0.000506	-2.121281	0.0341
T*CURR(-1)	-0.001068	0.000290	-3.688461	0.0002
T*CURR(-2)	0.001047	0.000292	3.592890	0.0003
@MONTH=1	-2.690802	0.156693	-17.17247	0.0000
....				
@MONTH=11	-0.727151	0.171533	-4.239128	0.0000
AR(1)	0.668308	0.083843	7.970989	0.0000
AR(2)	-0.457136	0.075427	-6.060678	0.0000
AR(3)	0.138389	0.050496	2.740599	0.0062

R-squared	0.999966	Mean dependent var	124.7080
Adjusted R-squared	0.999966	S.D. dependent var	181.9910
S.E. of regression	1.067457	Akaike info criterion	2.987552
Sum squared resid	1157.696	Schwarz criterion	3.082979
Log likelihood	-1527.552	Hannan-Quinn criter.	3.023759
F-statistic	1583328.	Durbin-Watson stat	1.852243
Prob(F-statistic)	0.000000		

Inverted AR Roots	.40	.14+.58i	.14-.58i

Forecast: CURRF	
Actual: CURR	
Forecast sample: 2001M01 2005M04	
Included observations: 52	
Root Mean Squared Error	7.558001
Mean Absolute Error	6.463748
Mean Abs. Percent Error	0.970135
Theil Inequality Coefficient	0.005589
Bias Proportion	0.281862
Variance Proportion	0.009546
Covariance Proportion	0.708592

(a)

Dependent Variable: CURR
Method: Least Squares
Date: 06/26/14 Time: 16:18
Sample: 1917M08 2005M04 IF @YEAR>1917 AND @YEAR<2005
Included observations: 1034
Convergence achieved after 10 iterations

Variable	Coefficient	Std. Error	t-Statistic	Prob.
C	-148.1952	10.46636	-14.15919	0.0000
CURR(-1)	0.733576	0.229680	3.193900	0.0015
CURR(-2)	-0.359175	0.233758	-1.536525	0.1247
T	0.102702	0.012225	8.400969	0.0000
T*CURR(-1)	0.000403	0.000246	1.638898	0.1016
T*CURR(-2)	0.000254	0.000250	1.013993	0.3108
@YEAR=1918	149.7707	10.31999	14.51268	0.0000
...				
@YEAR=2003	8.954468	0.647915	13.82043	0.0000
AR(1)	0.022849	0.009965	2.292824	0.0221
AR(2)	-0.013797	0.007416	-1.860359	0.0631

R-squared	0.999967	Mean dependent var	122.5943
Adjusted R-squared	0.999964	S.D. dependent var	178.4965
S.E. of regression	1.071543	Akaike info criterion	3.062584
Sum squared resid	1079.312	Schwarz criterion	3.511784
Log likelihood	-1489.356	Hannan-Quinn criter.	3.233034
F-statistic	308207.5	Durbin-Watson stat	1.589939
Prob(F-statistic)	0.000000		

Inverted AR Roots	.01-.12i	.01+.12i

Forecast: CURRF	
Actual: CURR	
Forecast sample: 2001M01 2005M04	
Included observations: 52	
Root Mean Squared Error	9.951670
Mean Absolute Error	7.673838
Mean Abs. Percent Error	1.094898
Theil Inequality Coefficient	0.007369
Bias Proportion	0.311496
Variance Proportion	0.118424
Covariance Proportion	0.570079

(b)

Figure 2.68 Statistical results based on the interaction ANCOVA models in (2.62) and (2.63).

2) A LVAR(1,1) interaction ANCOVA model Y_t with the TRE, by *@Month*, has the ES as follows:

$$Y\ C\ Y(-1)\ t\ t*Y(-1)\ @Expand(@Month, @Droplast)\ AR(1) \tag{2.61}$$

The data analysis can easily be done for all-time series Y_t. However, note the regression functions presented can be considered to be a good fit model, but its forecast can be worse with an unexpected large RMSE. Do this as an exercise for *Y* = *CURR* and *Y* = *log(CURR)*, respectively, with their forecasts. Similarly, for interaction, ANCOVA models of Y_t with the TRE by *@Year*. See the following examples.

Example 2.46 LVAR(*p,q*) Interaction ANCOVA Models of *CURR*
As an illustration, Figure 2.68 presents a part of the statistical results based on a LVAR(2,3) interaction ANCOVA model by *@Month*, and a LV(2,2) interaction ANCOVA model by *@Year*, using the following ESs, respectively.

$$\begin{aligned} &Curr\ C\ Curr(-1)\ Curr(-2)\ t\ t*Curr(-1)\ t*Curr(-2) \\ &@Expand(@Month, @Droplast)\ AR(1)\ AR(2)\ AR(3) \end{aligned} \tag{2.62}$$

and

$$\begin{aligned} &Curr\ C\ Curr(-1)\ Curr(-2)\ t\ t*Curr(-1)\ t*Curr(-2) \\ &@Expand(@Year, @Droplast)\ AR(1)\ AR(2) \end{aligned} \tag{2.63}$$

Based on these results, the following notes and comments are presented.

1) The regression in Figure 2.68a can be considered a good fit model, since each of the numerical independent variables, as well as each AR(*), has a p value < 0.01 or (0.05). Note that its forecast has RMSE = 97.52711, which is smaller than its forecast based on the LVAR(2,3) additive ANCOVA model in Figure 2.67a. So this LVAR(2,3) ANCOVA interaction model is better than the additive ANCOVA model to forecast *CURR*.

The regression in Figure 2.68b also can be considered a good fit model since only one of the numerical independent variables, namely *t*Curr(−2)*, has a p value = 0.3108. Note that this LVAR(2,2) interaction ANCOVA model of *CURR* by *@Year* gives a much better forecast with RMSE = 9.951670, compared to the LVAR(2,3) additive ANCOVA model of *CURR* by *@Year* in Figure 2.67b, with RMSE = 97.52711.

Example 2.47 Unexpected Forecast of *CURR*
EViews provides an option so that the forecast of *CURR* in previous example can also be obtained by using the LVAR(*p,q*) models of *log(CURR)*. As an illustration, Figure 2.69 presents a part of the statistical results based on a LVAR(2,3) interaction ANCOVA model of *log(Curr)* by *@Month*, and a LV(2,2) interaction ANCOVA model of *log(Curr)* by *@Year*, using the following ESs, respectively. Based on these results, the following findings and notes are presented.

$$\begin{aligned} &log(curr)\ C\ log(curr(-1))\ log(curr(-2))\ t\ t*log(curr(-1))\ t*log(curr(-2)) \\ &@expand(@Month, @droplast)\ ar(1)\ ar(2)\ ar(3) \end{aligned} \tag{2.64}$$

$$\begin{aligned} &log(curr)\ C\ log(curr(-1))\ log(curr(-2))\ t\ t*log(curr(-1))\ t*log(curr(-2)) \\ &@expand(@Year, @droplast)\ ar(1)\ ar(2) \end{aligned} \tag{2.65}$$

Dependent Variable: LOG(CURR)
Method: Least Squares
Date: 06/26/14 Time: 17:16
Sample (adjusted): 1918M01 2005M04
Included observations: 1036 after adjustments
Convergence achieved after 23 iterations

Variable	Coefficient	Std. Error	t-Statistic	Prob.
C	0.022496	0.004619	4.870434	0.0000
LOG(CURR(-1))	0.861013	0.115102	7.480457	0.0000
LOG(CURR(-2))	0.134231	0.114395	1.173401	0.2409
T	2.43E-05	1.42E-05	1.716553	0.0864
T*LOG(CURR(-1))	0.000696	8.52E-05	8.163200	0.0000
T*LOG(CURR(-2))	-0.000696	8.52E-05	-8.158735	0.0000
@MONTH=1	-0.038148	0.001441	-26.46515	0.0000
.....				
@MONTH=11	-0.007134	0.001108	-6.440413	0.0000
AR(1)	0.367159	0.116039	3.164095	0.0016
AR(2)	0.149873	0.056604	2.647720	0.0082
AR(3)	0.100018	0.036694	2.725739	0.0065

R-squared	0.999977	Mean dependent var	3.718876
Adjusted R-squared	0.999976	S.D. dependent var	1.594373
S.E. of regression	0.007790	Akaike info criterion	-6.852873
Sum squared resid	0.061653	Schwarz criterion	-6.757446
Log likelihood	3569.788	Hannan-Quinn criter.	-6.816667
F-statistic	2281906.	Durbin-Watson stat	1.971085
Prob(F-statistic)	0.000000		

Inverted AR Roots	.75	-.19+.31i	-.19-.31i

Dependent Variable: LOG(CURR)
Method: Least Squares
Date: 06/26/14 Time: 17:12
Sample: 1917M08 2005M04 IF @ YEAR>1917 AND @YEAR<2005
Included observations: 1034
Convergence achieved after 7 iterations

Variable	Coefficient	Std. Error	t-Statistic	Prob.
C	-2.051635	0.084211	-24.36290	0.0000
LOG(CURR(-1))	0.696355	0.038591	18.04449	0.0000
LOG(CURR(-2))	0.230691	0.039799	5.796400	0.0000
T	0.003865	0.000150	25.85244	0.0000
T*LOG(CURR(-1))	0.000286	8.49E-05	3.366392	0.0008
T*LOG(CURR(-2))	-0.000504	8.90E-05	-5.661243	0.0000
@YEAR=1918	2.140592	0.079791	26.82744	0.0000
.....				
@YEAR=2003	0.014378	0.003721	3.864529	0.0001
AR(1)	0.012129	0.005410	2.241775	0.0252
AR(2)	-0.013552	0.004035	-3.358496	0.0008

R-squared	0.999975	Mean dependent var	3.708200
Adjusted R-squared	0.999973	S.D. dependent var	1.588989
S.E. of regression	0.008248	Akaike info criterion	-6.671273
Sum squared resid	0.063942	Schwarz criterion	-6.222074
Log likelihood	3543.048	Hannan-Quinn criter.	-6.500824
F-statistic	412278.0	Durbin-Watson stat	1.516720
Prob(F-statistic)	0.000000		

Inverted AR Roots	.01-.12i	.01+.12i

Forecast: CURRF
Actual: LOG(CURR)
Forecast sample: 2001M01 2005M04
Included observations: 52

Root Mean Squared Error	0.052643
Mean Absolute Error	0.048448
Mean Abs. Percent Error	0.745613
Theil Inequality Coefficient	0.004056
Bias Proportion	0.846958
Variance Proportion	0.044614
Covariance Proportion	0.108428

Forecast: CURRF
Actual: CURR
Forecast sample: 2001M01 2005M04
Included observations: 52

Root Mean Squared Error	33.79819
Mean Absolute Error	31.34928
Mean Abs. Percent Error	4.709056
Theil Inequality Coefficient	0.025498
Bias Proportion	0.860336
Variance Proportion	0.027925
Covariance Proportion	0.111739

Forecast: CURRF
Actual: LOG(CURR)
Forecast sample: 2001M01 2005M04
Included observations: 52

Root Mean Squared Error	0.010620
Mean Absolute Error	0.007675
Mean Abs. Percent Error	0.117234
Theil Inequality Coefficient	0.000815
Bias Proportion	0.036936
Variance Proportion	0.123826
Covariance Proportion	0.839237

Forecast: CURRF
Actual: CURR
Forecast sample: 2001M01 2005M04
Included observations: 52

Root Mean Squared Error	7.789209
Mean Absolute Error	5.393085
Mean Abs. Percent Error	0.770875
Theil Inequality Coefficient	0.005735
Bias Proportion	0.041910
Variance Proportion	0.154612
Covariance Proportion	0.803477

Figure 2.69 Statistical results based on the interaction ANCOVA models in (2.64) and (2.65).

1) Figure 2.69a presents the forecast of *LOG(CURR)* and *CURR*, based on the LVAR(2,3) interaction ANCOVA model (2.64). Compared to the forecast of *CURR* in Figure 2.68a with RMSE = 7.558001, the forecast of *CURR* in Figure 2.69a, has an unexpected very large RMSE = 33.79819. So, in this case, the model of *LOG(CURR)* in (2.64) is not a good model to obtain the forecast of *CURR*.
2) On the other hand, Figure 2.69b presents a contradictory forecast of *CURR* using the LVAR (2,2) model of *LOG(CURR)* in (2.65) with RMSE = 7.789209, which is smaller than the forecast of *CURR* in Figure 2.68b with RMSE = 9.951670. So, this model can be considered a better model to forecast *CURR* than the model in (2.63).
3) Note both models present the forecast of l*og(CURR)* with a very small TIC, compared to the TIC = 0 as the perfect forecast. For instance, TIC = 0.004056 based on the model (2.64), and based on the model (2.65) with TIC = 0.000615. Would you consider these models are the best forecast models for *log(CURR)* or *CURR?* Compare to the following illustrative example.

2.7.3.2 Forecast Models of *CURR* with Heterogeneous Trend

As the extension of each of all LVAR(*p,q*) ANCOVA models by *@Month* or *@Year*, many alternative LVAR(*p,q*) models with heterogeneous trends, either linear, polynomial, logarithmic, or nonlinear trends can easily be developed.

Example 2.48 Heterogeneous Regressions Derived from the Reduced Model 3 in Table 2.10

In is recognized that several heterogeneous models of *CURR* can be developed based on each of the month-fixed-effects presented previously by inserting at least one set of the interactions *NIV*@Expand(@Month,@Dropfirst),* where *NIV* indicates a numerical independent variable. In order to check whether one or more interactions should be inserted as an additional independent variable, the OVT can be applied.

As an illustration, Figure 2.70 presents the statistical results of two OVTs based on the reduced model (RM3) in Table 2.10. Based on these results, the following notes and comments are presented.

1) Corresponding to the results in Figure 2.70a, 11 interaction independent variables: *t* (@month = 2)* up to *t*(@month = 12),* have significant joint effects on *CURR*. By using this heterogeneous regressions model, namely HRM1, it is found that *AR(2)* and *AR(3)* have insignificant effects, with *p* values of 0.1625, and 0.6941, respectively.

   A good fit reduced model is obtained by deleting both AR(2) and AR(3). Then we have a LV(2) HRM2, with the statistical results presented in Figure 2.71. And its forecast sample: 2001M01 2005M04, has RMSE = 7.845418 and TIC = 0.005792
2) The results in Figure 2.70b show that the 22 interaction independent variables, namely *Curr (−1)*(@month = m)* and *t*(@month = m),* for *m* = 2, ..., 12, have significant joint effects in the heterogeneous regression model. By using this heterogeneous regression model (HRM3), it is found that *AR(2)* and *AR(3)* have very large *p* values of 0.7975 and 0.5783, respectively, and only one of the numerical independent variables, namely *t*(@Month = 2),* has a very large *p* value = 0.9821.

   Furthermore, by deleting both AR(2) and AR(3), then we have the LV(2) HRM4, with only one of the numerical independent, namely *t*(@Month = 2),* has a negative coefficient with a *p* value = 0.3463. And its forecast sample of 2001M01 2005M04, has RMSE = 8.883632 and TIC = 0.006568.

Figure 2.70 Parts of the statistical results of two omitted variables tests based on model in (2.56a,b).

Omitted Variables Test
Equation: EQ16_LVAR23
Specification: CURR C CURR(-1) CURR(-2) T T^2 @EXPAND(@MONTH,@DROPFIRST) AR(1) AR(2) AR(3)
Omitted Variables: T*@EXPAND(@MONTH,@DROPFIRST)

	Value	df	Probability
F-statistic	2.097670	(11, 22)	0.0670
Likelihood ratio	37.29811	11	0.0001

(a)

Omitted Variables Test
Equation: EQ16
Specification: CURR C CURR(-2) T T^2 T^3 CURR(-1)*T CURR(-2)*T
Omitted Variables: CURR(-1)*@EXPAND(@MONTH,@DROPFIRST)T *@EXPAND(@MONTH,@DROPFIRST)

	Value	df	Probability
F-statistic	111.3540	(22, 1016)	0.0000
Likelihood ratio	1282.284	22	0.0000

(b)

Example 2.49 LVAR(*p,q*) Models of Log(CURR) with Heterogeneous Trend

As the extension of the LVAR(*p,q*) additive ANCOVA models, Figure 2.72 presents the statistical results of two LVAR(*p,q*) models of *log(CURR)* with heterogeneous linear trends, using the following ESs.

$$log(curr)\ C\ log(curr(-1))\ log(curr(-2))\ t\ t{*}@expand(@Month,@Droplast)\\ @expand(@Month,@droplast)\ ar(1)\ ar(2)\ ar(3) \quad (2.66)$$

and

$$log(curr)\ C\ log(curr(-1))\ log(curr(-2))\ t\ t{*}@expand(@Year,@Droplast)\\ @expand(@Year,@droplast)\ ar(1)\ ar(2) \quad (2.67)$$

Based on these results the following findings and notes are presented.

1) Based on the results of the model (2.66) in Figure 2.72a, the following notes are presented.
   1.1 The model is a LVAR(2,2) translog linear model with heterogeneous linear trend by *@Year.*
   1.2 The interaction *t∗@Expand(@Month,@Droplast)* has a significant effect based on the RVT, as presented on the right Figure 2.73.
   1.3 So, the set of 12 regressions have significant different linear time trends.

2) Based on the results of the model (2.67) in Figure 2.72b, the following notes are presented.
   2.1 The results are using the subsample {*@Year*>1917 and *@Year*<2006}.
   2.2 The model is a LVAR(2,2) translog linear model with heterogeneous linear trend by *@Year.*
   2.3 Unexpectedly, by doing the RVT, for the interaction *t∗@Expand(@Year,@Droplast)* the error message of "*Near Singular Matrix*" is shown on the screen. So we cannot test the hypothesis on the heterogeneous trends within all years.

Dependent Variable: CURR
Method: Least Squares
Date: 06/23/14 Time: 14:56
Sample (adjusted): 1917M10 2005M04
Included observations: 1045 after adjustments

Variable	Coefficient	Std. Error	t-Statistic	Prob.
C	1.239131	0.188453	6.575277	0.0000
CURR(-1)	1.359687	0.024522	55.44682	0.0000
CURR(-2)	-0.357433	0.024585	-14.53885	0.0000
T	-0.006667	0.000572	-11.65497	0.0000
T^2	1.70E-06	6.55E-07	2.590245	0.0097
T*(@MONTH=2)	0.003030	0.000456	6.647631	0.0000
T*(@MONTH=3)	0.006907	0.000470	14.70154	0.0000
T*(@MONTH=4)	0.006329	0.000423	14.96742	0.0000
T*(@MONTH=5)	0.005868	0.000418	14.03604	0.0000
T*(@MONTH=6)	0.006313	0.000418	15.11520	0.0000
T*(@MONTH=7)	0.006936	0.000416	16.68658	0.0000
T*(@MONTH=8)	0.003173	0.000412	7.700134	0.0000
T*(@MONTH=9)	0.005775	0.000435	13.27524	0.0000
T*(@MONTH=10)	0.005152	0.000421	12.24335	0.0000
T*(@MONTH=11)	0.009037	0.000422	21.42190	0.0000
T*(@MONTH=12)	0.009277	0.000408	22.71152	0.0000
@MONTH=2	-0.510493	0.248798	-2.051836	0.0404
@MONTH=3	-1.390214	0.252715	-5.501122	0.0000
@MONTH=4	-1.279951	0.248137	-5.158249	0.0000
@MONTH=5	-1.195615	0.248130	-4.818506	0.0000
@MONTH=6	-1.231197	0.246805	-4.988532	0.0000
@MONTH=7	-1.465196	0.246937	-5.933491	0.0000
@MONTH=8	-0.457804	0.246662	-1.856001	0.0637
@MONTH=9	-1.142652	0.249316	-4.583139	0.0000
@MONTH=10	-0.964690	0.245462	-3.930103	0.0001
@MONTH=11	-2.028061	0.245810	-8.250533	0.0000
@MONTH=12	-1.959097	0.244779	-8.003535	0.0000

R-squared	0.999981	Mean dependent var	125.2822
Adjusted R-squared	0.999980	S.D. dependent var	182.9596
S.E. of regression	0.808334	Akaike info criterion	2.437814
Sum squared resid	665.1644	Schwarz criterion	2.565754
Log likelihood	-1246.758	Hannan-Quinn criter.	2.486336
F-statistic	2057066.	Durbin-Watson stat	1.681824
Prob(F-statistic)	0.000000		

Figure 2.71 Statistical results of a LV(2) HRM2.

Dependent Variable: LOG(CURR)
Method: Least Squares
Date: 06/26/14 Time: 17:43
Sample (adjusted): 1918M01 2005M04
Included observations: 1036 after adjustments
Convergence achieved after 6 iterations

Variable	Coefficient	Std. Error	t-Statistic	Prob.
C	0.019208	0.001688	11.37879	0.0000
LOG(CURR(-1))	1.852673	0.020454	90.57818	0.0000
LOG(CURR(-2))	-0.853627	0.020432	-41.77929	0.0000
T	-9.19E-06	3.55E-06	-2.589743	0.0097
T*(@MONTH=1)	5.59E-05	4.27E-06	13.09563	0.0000
.....				
T*(@MONTH=11)	3.18E-05	4.26E-06	7.473208	0.0000
@MONTH=1	-0.074142	0.002603	-28.48506	0.0000
.....				
@MONTH=11	-0.022802	0.002591	-8.800867	0.0000
AR(1)	-0.394314	0.036812	-10.71152	0.0000
AR(2)	-0.328414	0.038021	-8.637757	0.0000
AR(3)	-0.143258	0.035576	-4.026761	0.0001

R-squared	0.999982	Mean dependent var	3.718876
Adjusted R-squared	0.999982	S.D. dependent var	1.594373
S.E. of regression	0.006778	Akaike info criterion	-7.122811
Sum squared resid	0.046257	Schwarz criterion	-6.984442
Log likelihood	3718.616	Hannan-Quinn criter.	-7.070312
F-statistic	2045548.	Durbin-Watson stat	1.974693
Prob(F-statistic)	0.000000		

Inverted AR Roots	.01-.58i	.01+.58i	-.42

Forecast: CURRF_EQ27	
Actual: LOG(CURR)	
Forecast sample: 2001M01 2005M04	
Included observations: 52	
Root Mean Squared Error	0.017515
Mean Absolute Error	0.014243
Mean Abs. Percent Error	0.218942
Theil Inequality Coefficient	0.001346
Bias Proportion	0.415904
Variance Proportion	0.089161
Covariance Proportion	0.494935

Forecast: CURRF	
Actual: CURR	
Forecast sample: 2001M01 2005M04	
Included observations: 52	
Root Mean Squared Error	11.52609
Mean Absolute Error	9.452882
Mean Abs. Percent Error	1.410503
Theil Inequality Coefficient	0.008542
Bias Proportion	0.400216
Variance Proportion	0.100267
Covariance Proportion	0.499518

Dependent Variable: LOG(CURR)
Method: Least Squares
Date: 07/22/14 Time: 09:42
Sample: 1917M08 2005M04 IF @YEAR>1917 AND @YEAR<2005
Included observations: 1034
Convergence achieved after 31 iterations

Variable	Coefficient	Std. Error	t-Statistic	Prob.
C	0.61135	0.04365	14.00720	0.00000
LOG(CURR(-1))	0.48053	0.03063	15.69046	0.00000
LOG(CURR(-2))	0.06748	0.02916	2.31398	0.02090
T	0.00227	0.00014	16.66661	0.00000
T*(@YEAR=1918)	0.00917	0.00071	12.87693	0.00000
T*(@YEAR=1919)	0.00287	0.00064	4.47185	0.00000
T*(@YEAR=1920)	0.00174	0.00062	2.78913	0.00540
.....	.....	.....	.....	.....
T*(@YEAR=2001)	0.00185	0.00062	2.96973	0.00310
T*(@YEAR=2002)	0.00061	0.00062	0.97482	0.32990
T*(@YEAR=2003)	0.00015	0.00062	0.23671	0.81290
"@YEAR dummies	Yes			
AR(1)	0.33207	0.04853	6.84324	0.00000
AR(2)	-0.27185	0.04637	-5.86279	0.00000
R-squared	0.99999	Mean dependent var		3.70820
Adjusted R-square	0.99998	S.D. dependent var		1.58899
S.E. of regression	0.00673	Akaike info criterion		-7.00928
Sum squared resid	0.03876	Schwarz criterion		-6.15867
Log likelihood	3801.80	Hannan-Quinn criter.		-6.68652
F-statistic	325393.20	Durbin-Watson stat		1.99218
Prob(F-statistic)	0.00000			
Inverted AR Roots	.17+.49i	.17-.49i		

Forecast: CURRF	
Actual: LOG(CURR)	
Forecast sample: 2001M01 2005M04	
Included observations: 52	
Root Mean Squared Error	0.004581
Mean Absolute Error	0.003555
Mean Abs. Percent Error	0.054406
Theil Inequality Coefficient	0.000352
Bias Proportion	0.032854
Variance Proportion	0.035539
Covariance Proportion	0.931607

Forecast: CURRF	
Actual: CURR	
Forecast sample: 2001M08 2005M04	
Included observations: 45	
Root Mean Squared Error	3.430184
Mean Absolute Error	2.624835
Mean Abs. Percent Error	0.374391
Theil Inequality Coefficient	0.002481
Bias Proportion	0.037395
Variance Proportion	0.053172
Covariance Proportion	0.909433

Figure 2.72 Statistical results based on the interaction ANCOVA models in (2.66) and (2.67).

Redundant Variables Test:
Equation: UNTITLED
Specification: LOG(CURR) C LOG(CURR(-1)) LOG(CURR(-2)) T T *@EXPAND(@MONTH.@DROPLAST)@EXPAND(@MONTH,@DROPLAST)AR(1) AR(2) AR(3)
Redundant Variables: T*@EXPAND(@MONTH,@DROPLAST)

	Value	df	Probability
F-statistic	37.92181	(11,1007)	0.0000
Likelihood ratio	359.0698	11	0.0000

Figure 2.73 An RVT of the model (2.67).

Wald Test:
Equation: Untitled

Test Statistic	Value	df	Probability
F-statistic	47.70893	(6,856)	0.0000
Chi-square	286.2536	6	0.0000

Null Hypothesis: C(5)=C(6)=C(7)=C(8)=C(9)=C(10)=0

Figure 2.74 A Wald test of the model (2.67).

2.4 However, with the 1% level of significance, the output in Figure 2.72b shows that the trend within each of the years 1918, 1919, 1920, and 2001 has a significant different with the *@Year=2004* as the reference year.

2.5 In addition, we can conduct the testing hypothesis on a set of trend differences using the Wald test. For instance, Figure 2.74 shows that the null hypothesis H_0: C(5) = ... = C(10) = 0 is rejected based on the *F*-statistic of $F_0 = 47.70893$, with $df = (6{,}856)$, and a *p value* = 0.0000. So it can be concluded that the six trends within the years 1918 to 2003 have a significant difference to the *@Year=2004*, as the reference group.

Example 2.50 LVAR(2,3) Interaction Models of Log(*CURR*) with Heterogeneous Trend
As an extension of the LVAR(2,3) interaction ANCOVA models (2.66), I am considering a LVAR(2,3) interaction models with heterogeneous linear trends, using the following ESs.

$$\begin{aligned}&log(curr)\ C\ log(curr(-1))\ log(curr(-2))\ t\ t*log(curr(-1))\\&\quad t*log(curr(-2))\ t*@expand(@Month,@droplast)\\&\quad @expand(@Month,@droplast)\ ar(1)\ ar(2)\ ar(3)\end{aligned}\tag{2.68}$$

Based on the results of this model, it is found that each of the interactions *t∗log(curr(−1))* and *t∗log(curr(−2))* have very large *p* values of 0.9265 and 0.9281, but each of the main variables *log(curr(−1)), log(curr(−2))* and *t* have *p* values of 0.0000, 0.0000, and 0.0098, respectively. These findings are the impacts of the multicollinearity between the independent variables, specifically the five main numerical variables.

So, in order to have the model with *t∗log(curr(−1))* and *t∗log(curr(−2))*, we have to delete at least one of the main variables using the trial-and-error method. Finally, I found the following two reduced models, which I consider the best possible reduced models.

$$\begin{gathered} log(curr)\ C\ log(curr(-2))\ t\ t{*}log(curr(-1))\ t{*}log(curr(-2)) \\ t{*}@expand(@month,@droplast)\ @expand(@month,@droplast) \\ ar(1)\ ar(2)\ ar(3) \end{gathered} \tag{2.69}$$

and

$$\begin{gathered} log(curr)\ C\ log(curr(-2))\ t{*}log(curr(-1))\ t{*}log(curr(-2)) \\ t{*}@expand(@month,@droplast)\ @expand(@month,@droplast) \\ ar(1)\ ar(2)\ ar(3) \end{gathered} \tag{2.70}$$

As an illustration, Figure 2.75 presents the statistical results based on the model (2.69). Based on these results, the following findings and notes are presented.

1) By using the Wald test, the null hypothesis H_0: C(4) = C(5) = 0 is rejected based on the *F*-statistic of $F_0 = 16.35$, with $df = (2,1006)$, and a *p* value = 0.0000. So the two interactions *t∗log(curr(−1))* and *t∗log(curr(−2))* have significant joint effects. In other words, the joint effects of *log(curr(−1))* and *log(curr(−2))* is significantly depend on the time *t*.
2) For the *@Month* = 12, as the reference group, we have the following regression.

$$\begin{aligned} LOG(CURR) &= 22.244 + 0.58186{*}LOG(CURR(-2)) - 0.00695{*}T \\ &+ 0.00056{*}T{*}LOG(CURR(-1)) - 0.00022{*}T{*}LOG(CURR(-2)) \end{aligned}$$

3) The forecasts of *log(CURR),* and *CURR* based on these models are worse than the forecasts based on the interaction ANCOVA model in Figure 2.74.
4) On the other hand, since the time *t* has a large *p* value, then we may delete this time variable from the model, and the statistical results of the reduced model (2.70) could be obtained easily, where each of the two interactions has a *p* value = 0.0000. But the forecasts based on this reduced model are also worse than the forecast in Figure 2.74. Do it as an exercise.
5) These findings show that the forecast based on a good fit model with a large number of independent variables is worse than a simpler model.

2.7.4 Forecast Models Based on Subsamples

In practice, specifically for policy analysis, we would never consider such long time observations for forecasting such as those presented in previous forecast models. As an illustration, Figure 2.76 presents three scatter graphs of *(t,log(Curr))* with their regression lines, based on the whole sample, and two selected subsamples. Based on these graphs, the following notes and comments are presented.

1) The scatter graph in Figure 2.76a clearly shows that a polynomial or nonlinear model of *log(Curr)* on the time *t* should be applied. Refer to the fourth degree polynomial model in (2.46), which can be considered the worst forecast model since it has such the greatest large RMSE compared to the other presented models.
2) On the other hand, by observing the scatter graphs based on the last couple years, such as presented in Figures 2.76b and c, I am very confident to say that the graphs support the application of a linear regression of *log(Curr)* on the time *t*, as the simplest possible model. Note this simplest linear regression in fact is the CGM, either geometric or exponential growth models, which can easily be extended to a lot of time series models – refer to Agung (2009a).

Dependent Variable: LOG(CURR)
Method: Least Squares
Date: 07/22/14 Time: 14:25
Sample (adjusted): 1918M01 2005M04
Included observations: 1036 after adjustments
Convergence achieved after 174 iterations

Variable	Coefficient	Std. Error	t-Statistic	Prob.
C	22.24399	142.3314	0.156283	0.8758
LOG(CURR(-2))	0.581861	0.043244	13.45522	0.0000
T	-0.006953	0.025724	-0.270304	0.7870
T*LOG(CURR(-1))	0.000555	9.71E-05	5.718373	0.0000
T*LOG(CURR(-2))	-0.000219	0.000115	-1.901071	0.0576
T*(@MONTH=1)	2.56E-05	2.79E-06	9.171506	0.0000
T*(@MONTH=2)	3.53E-05	4.46E-06	7.926683	0.0000
T*(@MONTH=3)	1.95E-05	4.65E-06	4.197689	0.0000
T*(@MONTH=4)	2.51E-05	4.61E-06	5.437524	0.0000
T*(@MONTH=5)	2.96E-05	4.63E-06	6.399672	0.0000
T*(@MONTH=6)	3.03E-05	4.71E-06	6.425694	0.0000
T*(@MONTH=7)	2.58E-05	4.65E-06	5.549382	0.0000
T*(@MONTH=8)	1.46E-05	4.57E-06	3.193671	0.0014
T*(@MONTH=9)	2.84E-08	4.60E-06	0.006179	0.9951
T*(@MONTH=10)	-9.65E-06	4.22E-06	-2.287914	0.0223
T*(@MONTH=11)	6.70E-06	2.87E-06	2.334431	0.0198
@MONTH=1	-0.041799	0.001625	-25.72612	0.0000
@MONTH=2	-0.061404	0.002645	-23.21398	0.0000
@MONTH=3	-0.038566	0.002677	-14.40821	0.0000
@MONTH=4	-0.035134	0.002788	-12.60079	0.0000
@MONTH=5	-0.038074	0.002838	-13.41583	0.0000
@MONTH=6	-0.035902	0.002911	-12.33379	0.0000
@MONTH=7	-0.030034	0.002894	-10.37913	0.0000
@MONTH=8	-0.028992	0.002835	-10.22618	0.0000
@MONTH=9	-0.019556	0.002775	-7.046667	0.0000
@MONTH=10	-0.011122	0.002560	-4.345146	0.0000
@MONTH=11	-0.015353	0.001673	-9.178525	0.0000
AR(1)	1.237804	0.035197	35.16760	0.0000
AR(2)	-0.611412	0.057241	-10.68132	0.0000
AR(3)	0.373168	0.039578	9.428562	0.0000

R-squared	0.999981	Mean dependent var	3.718876
Adjusted R-squared	0.999981	S.D. dependent var	1.594373
S.E. of regression	0.007032	Akaike info criterion	-7.048235
Sum squared resid	0.049742	Schwarz criterion	-6.905095
Log likelihood	3680.986	Hannan-Quinn criter.	-6.993926
F-statistic	1834798.	Durbin-Watson stat	1.855120
Prob(F-statistic)	0.000000		

Inverted AR Roots	1.00	.12-.60i	.12+.60i

Forecast: CURRF	
Actual: CURR	
Forecast sample: 2001M01 2005M04	
Included observations: 52	
Root Mean Squared Error	100.8318
Mean Absolute Error	71.60871
Mean Abs. Percent Error	9.916689
Theil Inequality Coefficient	0.070244
Bias Proportion	0.502920
Variance Proportion	0.460943
Covariance Proportion	0.036138

Figure 2.75 Statistical results based on the model (2.69) with heterogeneous trends.

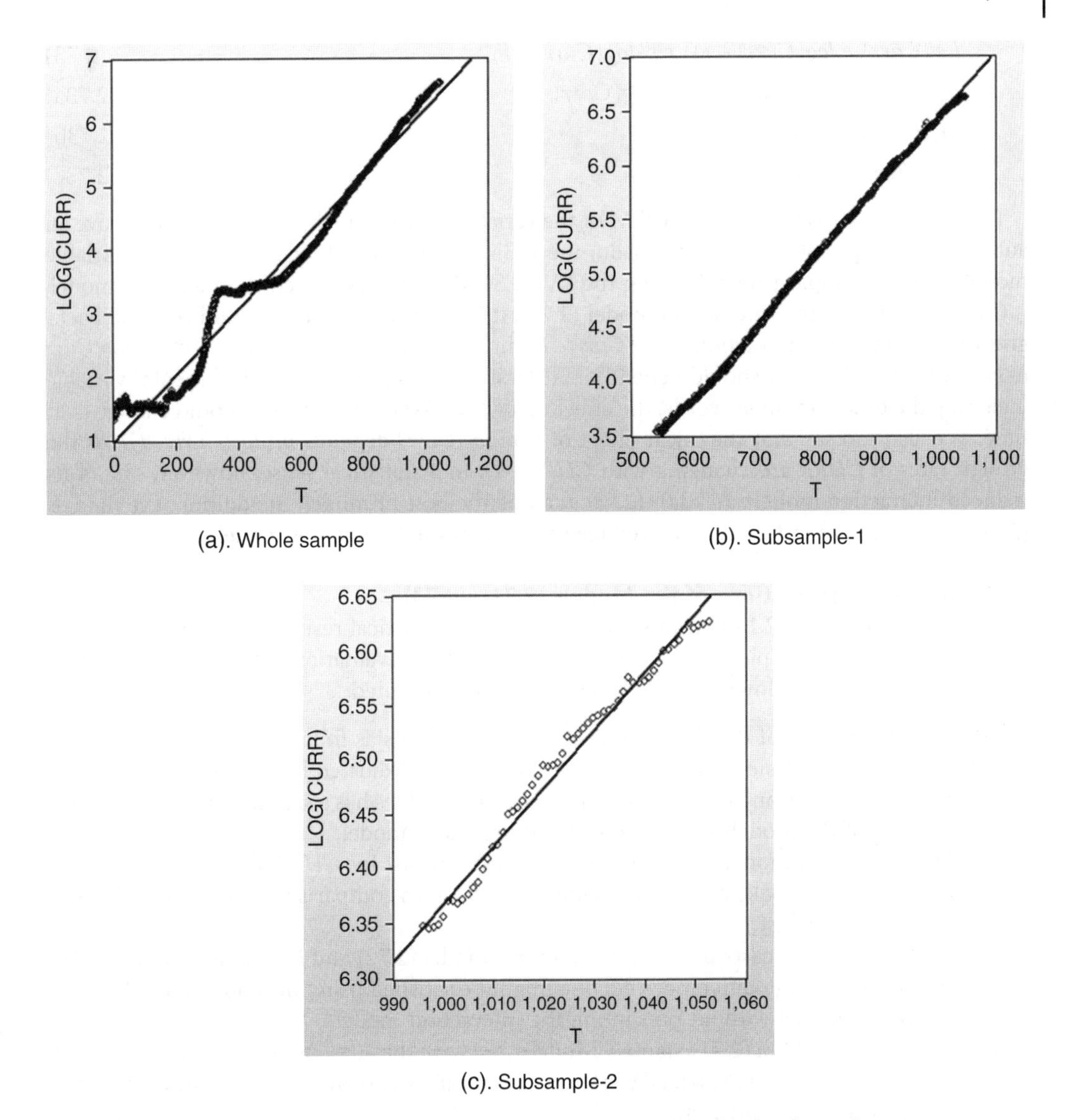

Figure 2.76 Scatter graphs of (*t*,log(*Curr*)) based on the whole sample and two subsamples.

3) However, in general, CGMs have autocorrelation problems. So it is recommended to apply an LVAR(*p*,*q*)_CGM, for non-negative integers *p* and *q*, which are highly dependent on the data, and they should be selected using the trial-and-error method.

2.7.4.1 Classical Growth Model of *CURR* and Alternative LV(1) Models

Referring to the last two scatter graphs with regression lines in Figure 2.76, we would have one of the following models, which can be the best fit model in the statistical sense.

$$log(Curr)\ c\ t \tag{2.71}$$

$$log(Curr)\ c\ log(Curr(-1))\ t \tag{2.72}$$

$$log(Curr)\ c\ log(Curr(-1))\ t\ t*log(Curr(-1) \tag{2.73}$$
$$log(Curr)\ c\ log(Curr(-1))\ t*log(Curr(-1) \tag{2.73a}$$
$$log(Curr)\ c\ t\ t*log(Curr(-1) \tag{2.73b}$$
$$log(Curr)\ c\ t*log(Curr(-1) \tag{2.73c}$$

The model (2.71) is the simplest CGM of the variable *CURR*, which in general would have an autocorrelation problem. For this reason, the CGM should be extended to a lagged variable model, with the simplest models being the LV(1) SGMs in (2.72), which is an additive model.

Corresponding to the interaction model (2.73), it is important to understand that the background in using the interaction *t*log(Curr(−1))* in the model is the linear effect of *log(Curr (−1))* on *log(Curr)*, which should depend on at least one exogenous or environmental variable including the time *t*. In other words, the effect of *log(Curr(−1))* on *log(Curr)* should change over time. We also can say that the linear effect of *t* on *log(Curr)* depends on *log(Curr(−1))*, or the growth rates of *CURR* are changing with *CURR(−1)*. In a statistical sense, however, one of its reduced interaction models: (2.73a)–(2.73c) can be the best fit model. In addition, the models (2.71) and (2.72) are in fact also the reduced models of the interaction model (2.73).

Example 2.51 Applications of the Models (2.71)–(2.73)

As an illustration, Table 2.11 presents a summary of the statistical results of the models (2.71)–(2.73), based on a subsample {*Year*>1990}, with their forecast evaluations RMSE and TIC. Based on this summary, the following findings and notes are presented.

1) Based on the results of the three models, we have findings as follows:
   1.1 Each of the regressions has very large R-squared and Adjusted R-squared values that are greater than 0.99, and each of their independent variables has a significant adjusted effect. So, each of the models can be considered a good fit model.
   1.2 The interaction model (2.73) presents the best forecast for *CURR* as well as *log(CURR)*, since it has the smallest values of RMSE and TIC. As a comparison, TIC = 0 is defined as the perfect forecast.
   1.3 Note that the forecasts based on the additive model (2.72) and interaction model (2.73), do not have a large differences. So one might be satisfied using the additive model as an acceptable model without presenting the interaction model.
2) Specific for the CGM in (2.71), namely Model 1, we have the autocorrelation problem, which can be identified by a very small DW statistic and it can be shown using the residual graph or the Serial Correlation LM test.
3) For this reason, the statistical results of two LV(1) models are presented. However, their residuals graphs still show they have autocorrelation problems, indicated by the systematic changing of the positive and negative signs of the residuals over times, but they are better than the CGM in (2.71). In addition, they also can be proven using the serial correlation LM test, as presented in Figure 2.77.
4) After doing analysis based on many alternative models of *log(Curr)*, I would conclude that it is not an easy task to develop a model without an autocorrelation. Furthermore, all researchers would present only a limited number of alternative models. Therefore, based on these three models, I would conclude that the interaction model (2.73) should be considered to be the best fit forecast model with a DW statistic = 1.1743. As a comparison, Baltagi (2009a,b) presents several models with very small DW statistics.
5) Since their residuals have significant serial correlations, then for the testing hypotheses on the model parameters, the HAC coefficient covariance matrix should be applied, which can easily be done – refer to Example 1.30.

Table 2.11 Summary of the statistical results of the models (2.71), (2.72), and (2.73), and their forecast evaluations, with forecast sample: 2001m01 2005m04.

Dependent Variable: LOG(*CURR*)

Method: Least Squares

Date: 07/23/14 Time: 13:52

Sample: 1917M08 2005M04 IF YEAR>1990

Included observations: 170

	Model 1			Model 2			Model 3		
Variable	**Coef.**	***t*-Stat.**	**Prob.**	**Coef.**	***t*-Stat.**	**Prob.**	**Coef.**	***t*-Stat.**	**Prob.**
C	0.50996	17.05463	0.00000	0.05278	3.22050	0.00150	−0.70631	−2.54467	0.01180
LOG(CURR(−1))				0.93135	35.26702	0.00000	1.00471	26.96496	0.00000
T	0.00585	189.4720	0.00000	0.00039	2.50549	0.01320	0.00152	3.45173	0.00070
T∗LOG(CURR(−1))							−0.00013	−2.73943	0.00680
R-squared	0.99531			0.99949			0.99951		
Adjusted R-squared	0.99529			0.99949			0.99951		
S.E. of regression	0.02009			0.00665			0.00652		
Sum squared residual	0.06821			0.00738			0.00706		
Log likelihood	426.56			612.605			616.364		
F-statistic	35,899.62			164,166			113,709.3		
Prob. (*F*-statistic)	0.00000			0.00000			0.00000		
DW statistic	0.11457			1.19089			1.17243		
R-squared	0.99531			0.99949			0.99951		
Adjusted R-squared	0.99529			0.99949			0.99951		
S.E. of regression	0.02009			0.00665			0.00652		
Sum squared residuals	0.06821			0.00738			0.00706		

(*Continued*)

Table 2.11 (Continued)

	Model 1			Model 2			Model 3		
Variable	**Coef.**	***t*-Stat.**	**Prob.**	**Coef.**	***t*-Stat.**	**Prob.**	**Coef.**	***t*-Stat.**	**Prob.**
Log likelihood	426.56			612.605			616.364		
F-statistic	35,899.62			164,166.00			113,709.3		
Prob. (*F*-statistic)	0.00000			0.00000			0.00000		
DW statistic	0.11457			1.19089			1.17243		
Forecast	*CURR*	*log(CURR)*		*CURR*	*log(CURR)*		*CURR*	*log(CURR)*	
RMSE	21.78505	0.020107		9.200528	0.013597		7.6868	0.011667	
TIC	0.01587	0.002231		0.006792	0.001044		0.005682	0.000896	

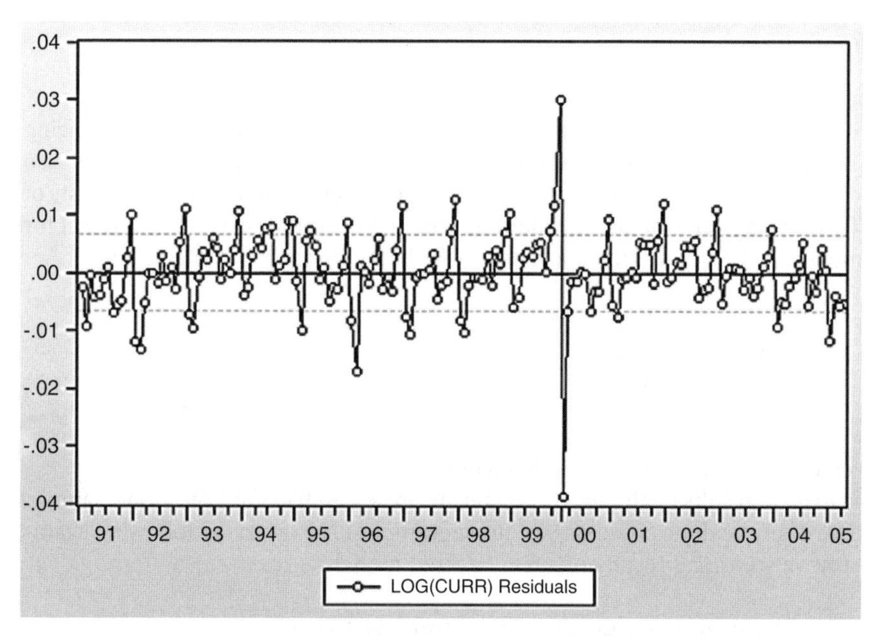

Breusch-Godfrey Serial Correlation LM Test:

F-statistic	8.777795	Prob. F(2,165)	0.0002
Obs*R-squared	16.34817	Prob. Chi-Square(2)	0.0003

(a). Residuals graph of the model (2.72), and
Its serial correlation LM test

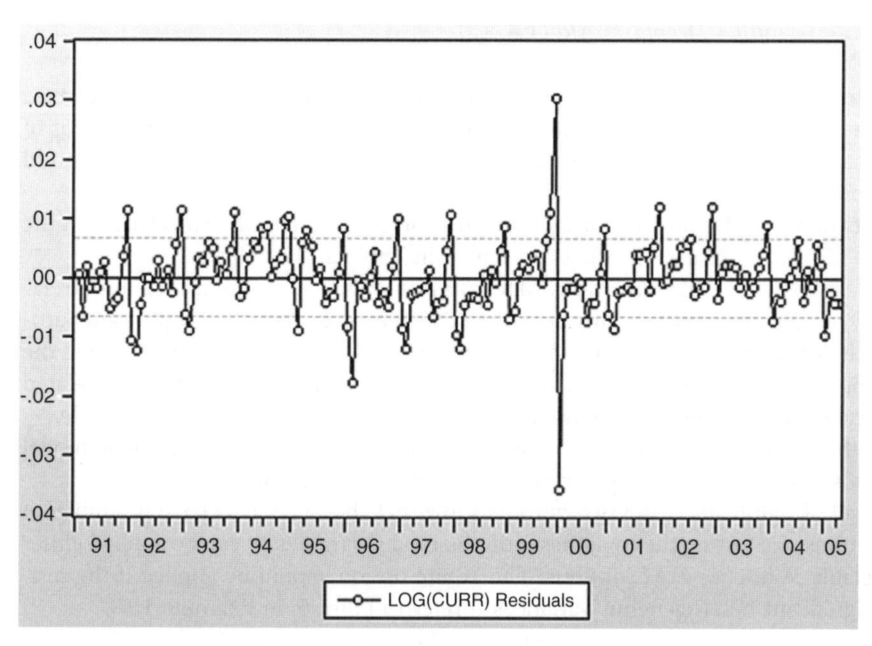

Breusch-Godfrey Serial Correlation LM Test:

F-statistic	9.651786	Prob. F(2,164)	0.0001
Obs*R-squared	17.90258	Prob. Chi-Square(2)	0.0001

(b). Residuals graph of the model (2.73), and
Its serial correlation LM test

Figure 2.77 Residuals graphs of the models (2.72) and (2.73), with their serial correlation LM tests.

2.7.4.2 AR(1) and LVAR(1,1) Additive Models

As an extension of each of the models (2.71)–(2.73), respectively, I am considering the following AR(1), and two LVAR(1,1) models with the time *t* as a predicator. It has been found that they also have autocorrelation problems. Do the data analyses as exercises, and compare their forecasts of *CURR* and *log(CURR)* to previous models. In addition, conduct the serial correlation LM test for any time series models presented in this book, as well as based on your own data sets.

$$log(Curr)\ c\ t\ ar(1) \quad (2.74)$$

$$log(Curr)\ c\ log(Curr(-1))\ t\ ar(1) \quad (2.75)$$

$$log(Curr)\ c\ log(Curr(-1))\ t*log(Curr(-1))\ ar(1) \quad (2.76)$$

2.7.4.3 Application of Interaction ANCOVA Models Based on a Subsample

Referring to the statistical results of all models previously presented based on the whole sample, the data analysis can easily be done based on various subsamples. However, the following examples present illustrative statistical results of a few selected models.

Example 2.52 LVAR(2,3) Interaction ANCOVA Model of *CURR*

As a comparison with the models (2.71) to (2.73), Figure 2.78 presents two statistical results based on the whole sample and a subsample {*Year*>2000}, with the following ES, which is presented again for easier reference to the readers. Based on these results, the following findings and notes are presented.

$$\begin{aligned} &Curr\ C\ Curr(-1)\ Curr(-2)\ t\ t*Curr(-1)\ t*Curr(-2) \\ &@Expand(@Month, @Droplast)\ AR(1)\ AR(2)\ AR(3) \end{aligned} \quad (2.77)$$

1) The forecast of *CURR* based on the subsample {*Year*>1990} is better than the forecast based on the whole sample, since it has a smaller RMSE and TIC. However, its standard error of *CURRF* is greater than the standard error based on the whole sample. Then which one would be a better forecast of *CURR*?
2) Both forecasts have smaller RMSE and TIC than the forecast models of *CURR* based on the three models (2.71), (2.72), and (2.73), as presented in Table 2.11.
3) In addition, based on the serial correlation LM test, its residuals have insignificant autocorrelation, at the 5% level of significance. However, note that the residuals have insignificant autocorrelations; it does not directly mean that the residuals does not have autocorrelation in the corresponding population. Compare this to a pair of a bivariate variables, which are correlated in a theoretical sense, but they might have an insignificant correlation based on a sample.
4) So, based on the subsample, I would conclude that the model (2.77) is the best forecast model of *CURR* in a statistical sense, compared to the models (2.71)–(2.73).
5) At the 5% level of significance, the two outputs show that their residuals have insignificant serial correlations. So, testing the hypotheses on the model parameters can be applied either using the default, White, or HAC options. The White option should be applied if the residuals have significant heterogeneous variances – refer to point 5, in Example 1.30.

Example 2.53 LVAR(2,3) ANCOVA Model of Log(*CURR*) by @Month

As a comparison with the models (2.71) to (2.73), Figure 2.79 presents a part of the statistical results of the LVAR(2,3) ANCOVA model of *log(CURR)* by *@Month* in (2.64), based on the subsample {*Year*>2000}, with the following ES, which is presented again for easier reference for the readers. Based on these results, the following findings and notes are presented.

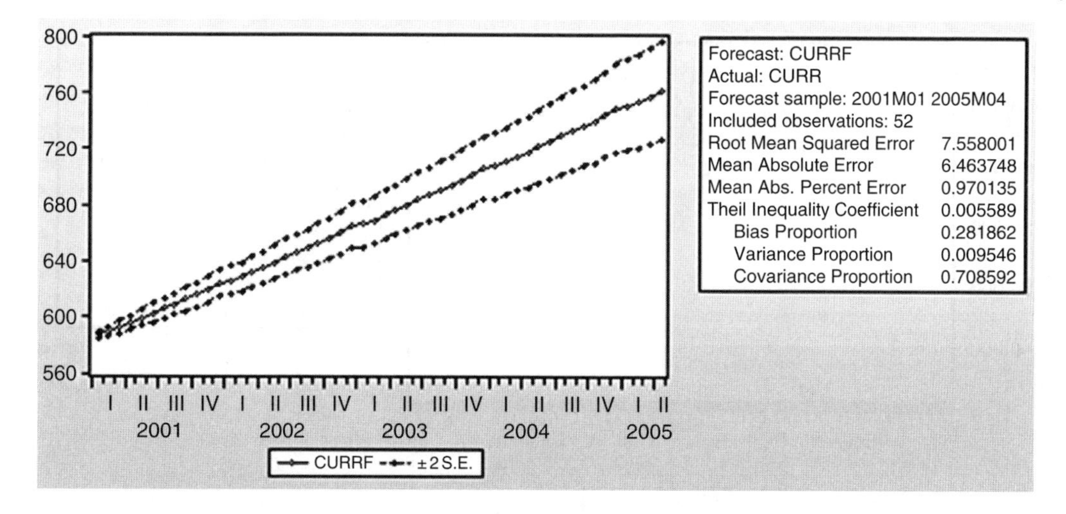

Breusch-Godfrey Serial Correlation LM Test:

F-statistic	1.941675	Prob. F(2,1014)	0.1440
Obs*R-squared	3.952467	Prob. Chi-Square(2)	0.1386

(a). Based on the whole sample

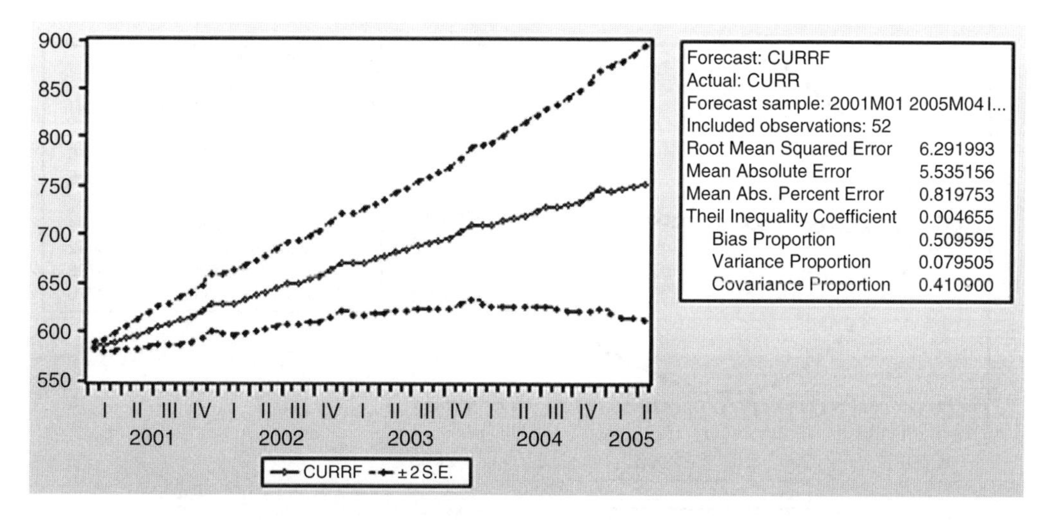

Breusch-Godfrey Serial Correlation LM Test:

F-statistic	2.361584	Prob. F(2,144)	0.0979
Obs*R-squared	5.271847	Prob. Chi-Square(2)	0.0717

(b). Based on the subsample {Year > 1990}

Figure 2.78 Two statistical results of the LVAR(2,3) model of *CURR* in (2.77).

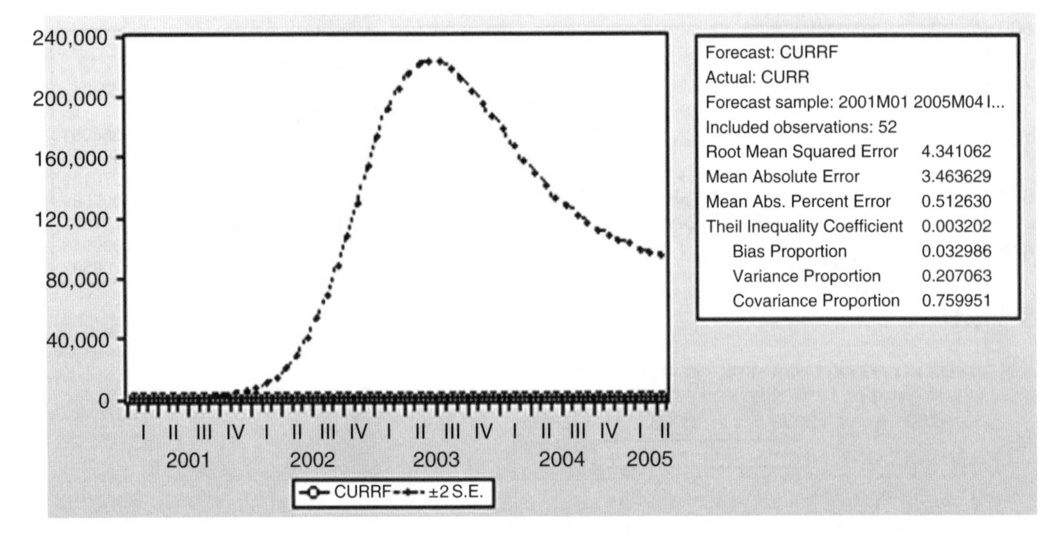

Breusch-Godfrey Serial Correlation LM Test:

F-statistic	1.804850	Prob. F(2,144)	0.1682
Obs*R-squared	4.059423	Prob. Chi-Square(2)	0.1314

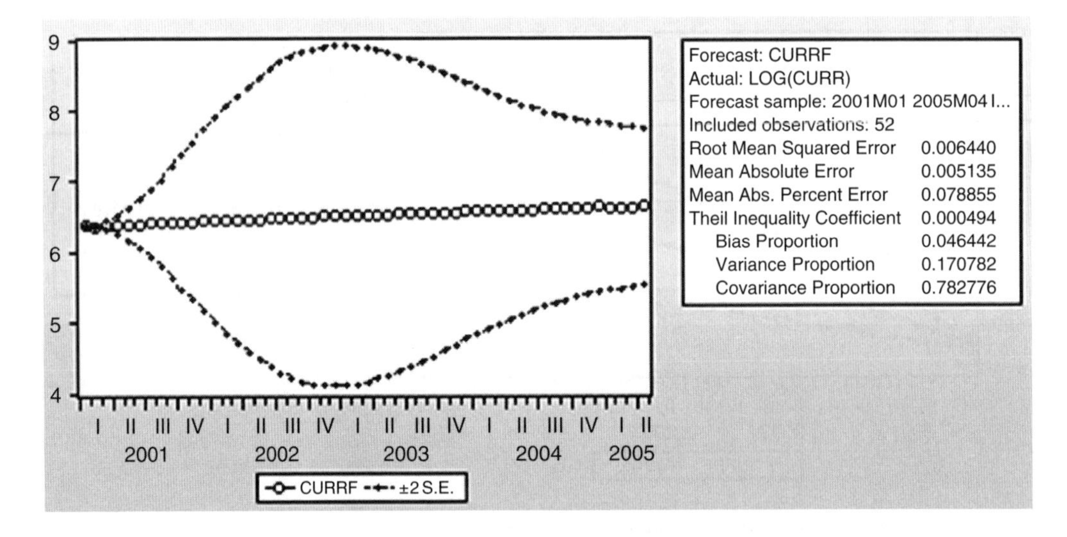

Heteroskedasticity Test: Breusch-Pagan-Godfrey

F-statistic	1.781849	Prob. F(16,149)	0.0384
Obs*R-squared	26.66105	Prob. Chi-Square(16)	0.0454
Scaled explained SS	66.59102	Prob. Chi-Square(16)	0.0000

Figure 2.79 Statistical results of the model log(*CURR*) in (2.78), based on a subsample {Year>1990}.

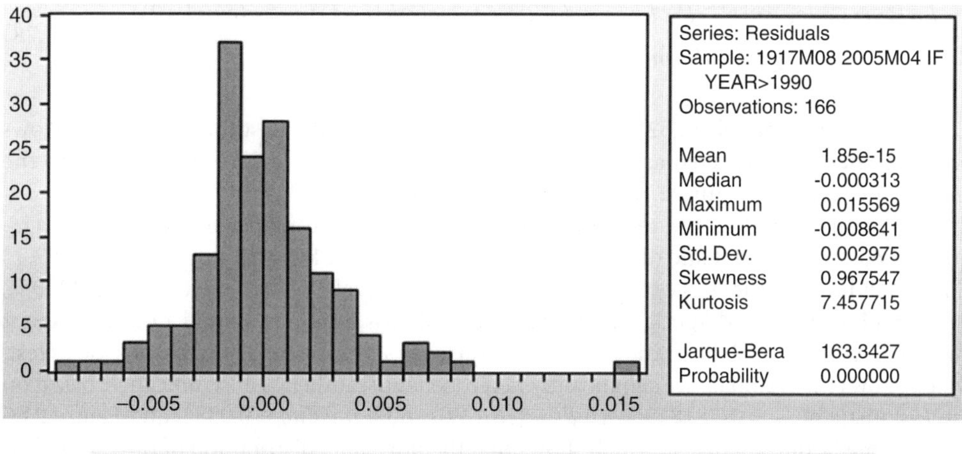

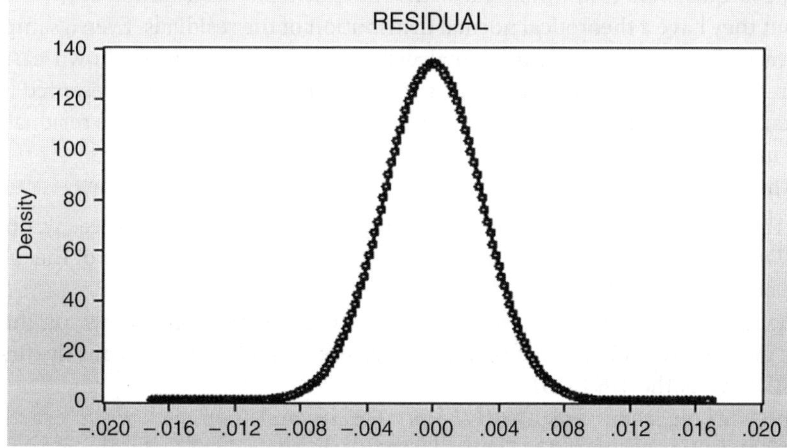

Figure 2.80 A residual's statistics and its theoretical distribution.

EViews - [Equation: EQ36 Workfile: CURR_16APRIL::Currcir\]

obs	Actual	Fitted	Residual
1991M01	5.65091	5.64517	0.00574
1991M02	5.64948	5.65254	-0.00306
1991M03	5.65741	5.65557	0.00184
1991M04	5.66136	5.66546	-0.00410
1991M05	5.66569	5.66688	-0.00119

(a). Residual table

EViews - [Group: UNTITLED Workfile: CURR_16APRIL::Currcir\]

1917M08
1917M09
1917M10
1917M11
1917M12
1918M01

(b). Empty Group

Figure 2.81 The residual table based on the subsample {*Year*>1990} and an empty group.

$$log(curr)\ c\ log(curr(-1))\ log(curr(-2))\ t\ t*log(curr(-1))\ t*log(curr(-2))$$
$$@expand(@month, @droplast)\ ar(1)\ ar(2)\ ar(3) \tag{2.78}$$

1) Compared to the models (2.62), (2.71)–(2.73), this model is the best forecast model. However, it presents unexpected and uncommon graphs of *CURRF* ± 2 S.E. Compare it to the graphs in Chapter 1, Figure 2.78, and the following example.
2) The BG serial correlation LM test shows that the residuals have an insignificant autocorrelation, based on the *F*-statistic of $F_0 = 1.804850$, with $df = (2{,}144)$, and a p value = 0.1682.
3) Corresponding to the basic assumptions of the residuals of a model, Agung (2009a, Section 14.3) presented special notes that the basic assumptions should not and cannot be tested. As additional illustration, Figure 2.79 also presents the BPG heteroskedasticity test, which shows the residuals are significantly heterogeneous, at the 5% of significance, based on the *F*-statistic of $F_0 = 1.781849$, with $df = (16{,}149)$, and a *p value* = 0.0384, and Figure 2.80 presents the Jarque–Bera test, which shows that the normality distribution of the residuals is rejected, but they have a theoretical normal distribution of the residuals. Even a sample of the zero-one variable has a theoretical normal distribution, which can be shown easily using EViews – refer to the histogram, empirical, and theoretical distributions, presented in Agung (2011a, Example 1.6). In addition, see the special notes and comments of the residual analysis presented in Chapter 1.
4) Note the theoretical distribution of the residuals, should be done as follows:
   - Having the statistical results of the regression on the screen, we can easily present its *Actual, Fitted, Residual Table* on the screen. Figure 2.81a presents a part of the table, with the first observation at 1991M01.
   - Block the table, and then click … *Copy/OK*, then the *Empty Group* shown on the screen, with the month 1917M08 in the first line, as presented in Figure 2.81b, but the first line of the RESIDUAL is the 1991M01.
   - Open the workfile, then click *Quick/Empty Group*, and then click *Paste* exactly at the month 1990M12, as presented in Figure 2.82, with the variable SER04 is the residual, which can be renamed RESIDUAL. Note that the other series are not important in this case, so they can be deleted. In other words, we can insert only the Residual.
   - Finally, the graphs in Figure 2.81 can be developed easily.

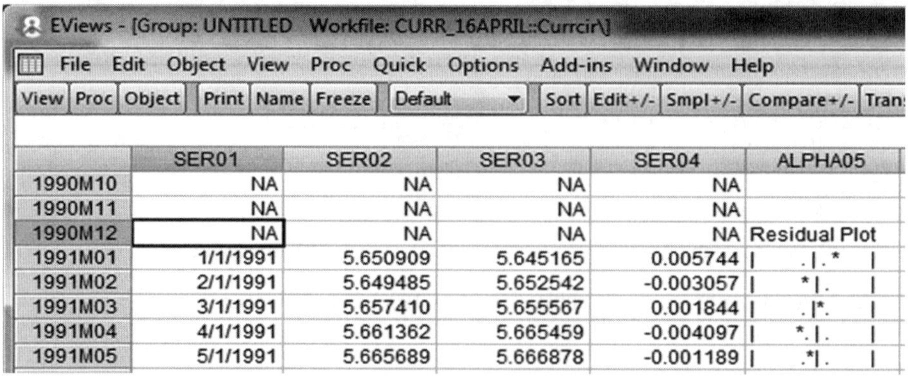

EViews - [Group: UNTITLED Workfile: CURR_16APRIL::Currcir\]

File Edit Object View Proc Quick Options Add-ins Window Help

View | Proc | Object | Print | Name | Freeze | Default | Sort | Edit+/- | Smpl+/- | Compare+/- | Tran:

	SER01	SER02	SER03	SER04	ALPHA05
1990M10	NA	NA	NA	NA	
1990M11	NA	NA	NA	NA	
1990M12	NA	NA	NA	NA	Residual Plot
1991M01	1/1/1991	5.650909	5.645165	0.005744	| . | . * |
1991M02	2/1/1991	5.649485	5.652542	-0.003057	| * | . |
1991M03	3/1/1991	5.657410	5.655567	0.001844	| . |* . |
1991M04	4/1/1991	5.661362	5.665459	-0.004097	| *. | . |
1991M05	5/1/1991	5.665689	5.666878	-0.001189	| .*| . |

Figure 2.82 Five series are inserted as additional new variables.

2.7.4.4 Application of Models with Heterogeneous Trend Based on Subsample

Example 2.54 LVAR(2,2) Models of *CURR* and Log(*CURR*) by @Year

As a comparison with the models (2.71) to (2.73), Figure 2.83 presents two sets of statistical results of the LVAR(2,2) ANCOVA models of *CURR* and *log(CURR)* by *@Year* in (2.79) and (2.80), based on a subsample {*Year*>2000 and *Year*<2005}, with the following ES.

$$\begin{aligned} &curr\ c\ curr(-1)\ curr(-2)\ t\ t{*}@expand(@year,@droplast) \\ &\qquad @expand(@year,@droplast) ar(1)\ ar(2) \end{aligned} \tag{2.79}$$

$$\begin{aligned} &log(curr)\ c\ log(curr(-1))\ log(curr(-2))\ t\ t{*}@expand(@year,@droplast) \\ &\qquad @expand(@year,@droplast)\ ar(1)\ ar(2) \end{aligned} \tag{2.80}$$

Based on the results in Figure 2.83, the following findings and notes are presented

1) Referring to the classification of forecasts presented in Example 1.8, then both models present perfect forecast, since their TIC < 0.01. Since the model (2.80) has a much smaller TIC = 0.000309 compared to the model (2.79) with a TIC = 0.002509, then the model (2.80) is a better forecasting model and it is the best compared to all previous models of *CURR*.
2) Both models have insignificant serial correlations. And at the 5% level of significance, both models also have insignificant heteroskedasticity.
3) However, at 10% level of significance they do have significant heteroskedasticity based on the *F*-statistic and *Obs**R-squared*. Based on this conclusion, for testing of hypotheses on the

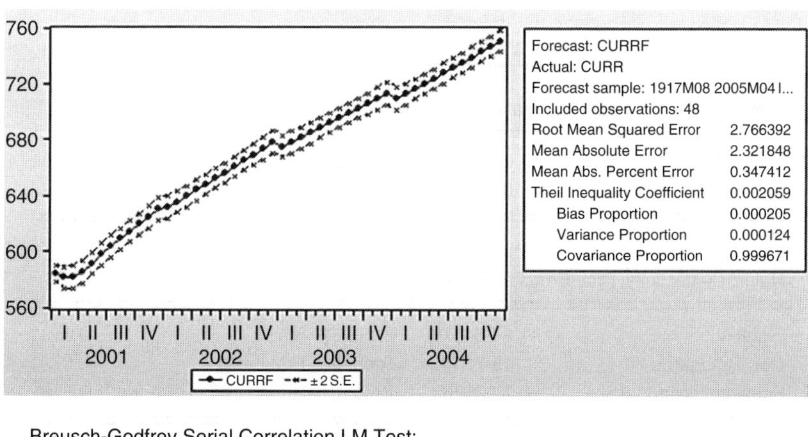

Breusch-Godfrey Serial Correlation LM Test:

F-statistic	0.107400	Prob. F(2,34)	0.8985
Obs*R-squared	0.301343	Prob. Chi-Square(2)	0.8601

Heteroskedasticity Test: Breusch-Pagan-Godfrey

F-statistic	2.030147	Prob. F(9,38)	0.0625
Obs*R-squared	15.58562	Prob. Chi-Square(9)	0.0761
Scaled explained SS	7.404505	Prob. Chi-Square(9)	0.5951

Figure 2.83 Statistical results based on the models (2.79) and (2.80) using a subsample {*Year>2000* and *Year<2005*}.

model parameters, it is recommended to apply the HAC coefficient covariance matrix for estimation. Refer to the special notes and comments presented in Section 1.8.

4) The estimation function of the model (2.80) is as follows.

   Then by using the HAC, all hypotheses on the model parameters can easily be tested using the Wald test, such as the null hypothesis H_0: C(5) = C(6) = C(7) = 0 is rejected based on the Wald (F-statistic) of 4.001439, with df = (3,36), and p value = 0.0018. Hence, it can be concluded that the model has significant heterogeneous time trends.

5) The forecast at several time-points beyond the sample point T = 1049 used to forecasting, can easily be computed using the forecast function. Figure 2.84 presents the forecast of four time-points for T > 1049, where *Obs_LCurr* = observed *log(Curr)* and *F_LCurr* = forecast of log(*Curr*).

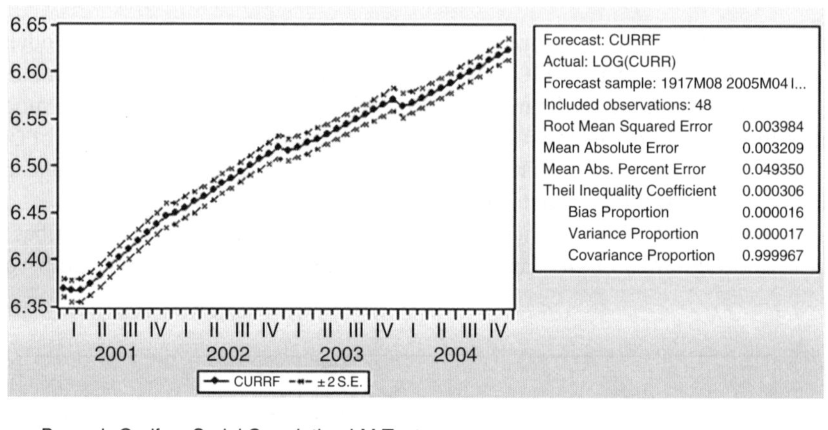

Breusch-Godfrey Serial Correlation LM Test:

F-statistic	0.188400	Prob. F(2,34)	0.8291
Obs*R-squared	0.526123	Prob. Chi-Square(2)	0.7687

Heteroskedasticity Test: Breusch-Pagan-Godfrey

F-statistic	2.064679	Prob. F(9,38)	0.0582
Obs*R-squared	15.76366	Prob. Chi-Square(9)	0.0720
Scaled explained SS	7.586962	Prob. Chi-Square(9)	0.5762

Figure 2.83 (Continued)

	t	Obs_Lcurr	F_Lcurr
2004M10	1047	6.607998	6.611576
2004M11	1048	6.61835	6.616852
2004M12	1049	6.624664	6.623227
2005M01	**1050**	**6.618998**	**6.629576**
2005M02	**1051**	**6.621874**	**6.633366**
2005M03	**1052**	**6.623058**	**6.63758**
2005M04	**1053**	**6.624856**	**6.642291**

Figure 2.84 Forecast beyond T = 1049.

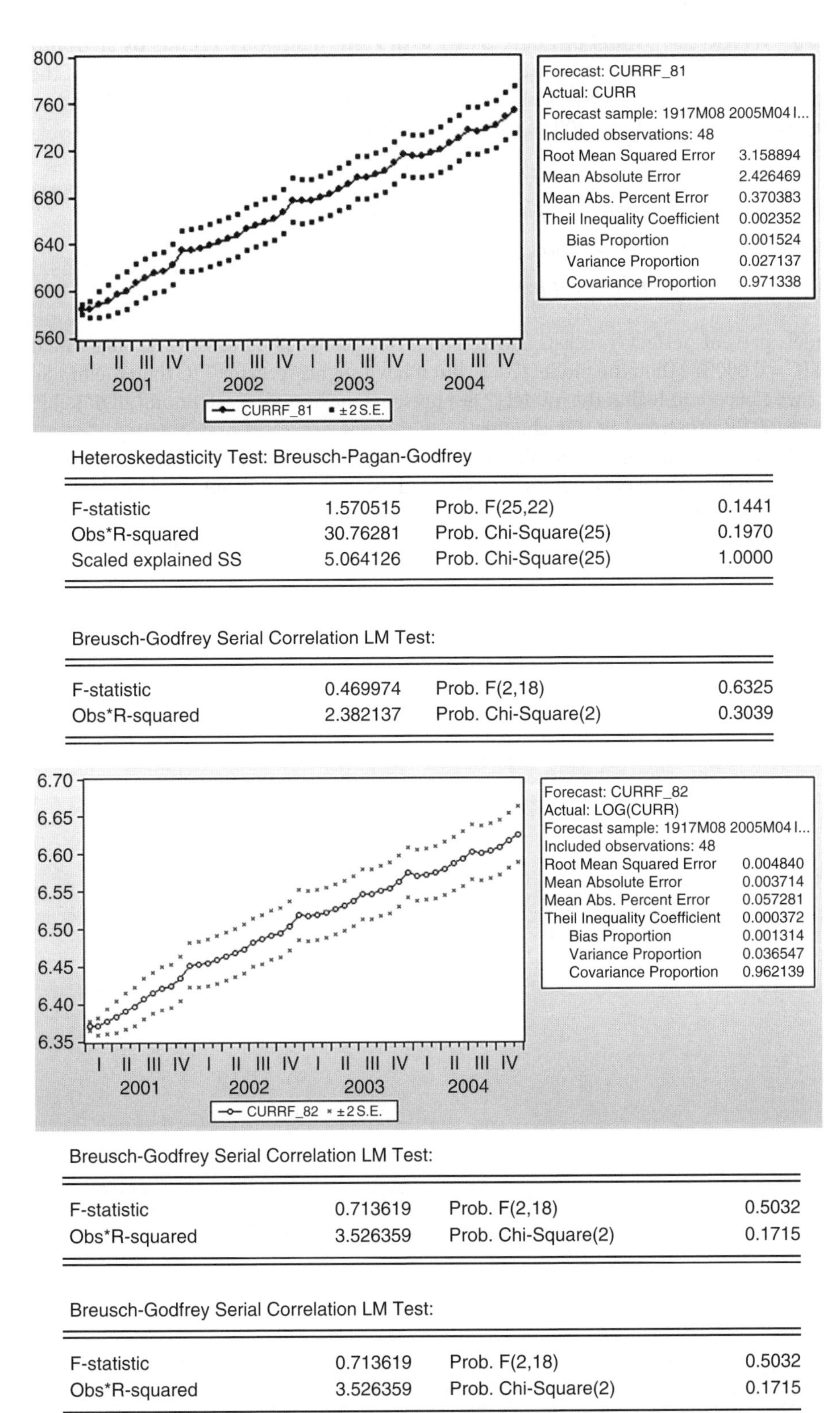

Heteroskedasticity Test: Breusch-Pagan-Godfrey

F-statistic	1.570515	Prob. F(25,22)	0.1441
Obs*R-squared	30.76281	Prob. Chi-Square(25)	0.1970
Scaled explained SS	5.064126	Prob. Chi-Square(25)	1.0000

Breusch-Godfrey Serial Correlation LM Test:

F-statistic	0.469974	Prob. F(2,18)	0.6325
Obs*R-squared	2.382137	Prob. Chi-Square(2)	0.3039

Breusch-Godfrey Serial Correlation LM Test:

F-statistic	0.713619	Prob. F(2,18)	0.5032
Obs*R-squared	3.526359	Prob. Chi-Square(2)	0.1715

Breusch-Godfrey Serial Correlation LM Test:

F-statistic	0.713619	Prob. F(2,18)	0.5032
Obs*R-squared	3.526359	Prob. Chi-Square(2)	0.1715

Figure 2.85 Statistical results based on models (2.81) and (2.82), using a subsample {*Year>2000* and *Year<2005*}.

Example 2.55 LVAR(2,3) Model of Log(*CURR*) with Heterogeneous Trends by *@Month*
As a comparison with the models (2.71)–(2.73), and (2.79), Figure 2.85 presents two sets of the statistical results of the LVAR(2,2) model of *log(CURR)* by *@Month* in (2.81 and (2.82)), based on a subsample {*Year*>2000 and *Year*<2005}, with the following ES.

$$curr\ c\ curr(-1)\ curr(-2)\ t\ t*@expand(@month,@droplast)\\ @expand(@month,@droplast)\ ar(1)\ ar(2) \tag{2.81}$$

$$log(curr)\ c\ log(curr(-1))\ log(curr(-2))\ t\ t*@expand(@month,@droplast)\\ @expand(@month,@droplast)\ ar(1)\ ar(2) \tag{2.82}$$

1) Both models present perfect forecasts, since their TIC < 0.001. The model (2.82) has much smaller TIC = 0.000372 than the model (2.81), but it has a slightly greater TIC than the model (2.80). So, we can conclude that the model (2.80) presents the best forecast among all possible forecasts of *CURR* presented in this chapter.
2) Both models have insignificant heteroskedasticity and serial correlations. For these reasons, all hypotheses on the model parameters can be tested using the results obtained by using the default coefficient covariance matrix.
3) The RVT of *t*@Expand(@Month,@Droplast)* gives the conclusion that the model (8.2) has significant heterogeneous trend by *@Month*.

3

Continuous Forecast Models

3.1 Introduction

This chapter presents a particular type of forecast model, namely continuous forecast models, models without dummy predictors, or models with numerical independent variables only, based on selected monthly time series. I have found that a monthly time series in EXER15_5.wf1, namely *FSPCOM*= S&P'S COMMON STOCK PRICE INDEX: COMPOSITE (1941–43 = 10), can be used to present illustration of various forecast models. Compared to the growth patterns of the two time series *HS* and *G* with dummy variables, *FSPCOM* has other types of growth curve, which can easily be identified using its growth curve on *@TREND* or the time t = *@TREND* + *1*, and *FSPCOM(−1)* as presented in Figure 3.1.

In fact, the work-file contains two monthly time series, namely *FSPCOM* and *RCAR6T* = RETAIL SALES: NEW PASSENGER CARS, TOTAL DOMESTICS + IMPORTS (THOUSANS). So I would say that all types of models presented in Chapters 1 and 2 for the monthly time series *HS* and *G*, could also be applied for *FSCOM* and *RCAR6T,* in addition to the continuous forecast models. Referring to all specific forecast models based on a single time series Y_t presented in Chapter 1, based on *FSCOM* and *log(FSPCOM)*, similar models also can easily be applied. Do these for some exercises.

However, based on the two scatter graphs of *FSPCOM* on the time t and *FSPCOM(−1)* with their regression lines and their NNF (Nearest Neighbor Fit) curves, I would present various alternative models of *FSPCOM* on the time t or *FSPCOM(−p)*; they present either good or worse forecasting.

Note that the scatter graphs in Figure 3.1 clearly show that a linear regression of *FSPCOM* on each of the time T and *FSPCOM(−1)* can be considered to be two acceptable models in the two-dimensional space, even though they might not be the best models, since the scatter graph of *FSPCOM* on the time T shows a waving curve. For this reason, alternative continuous models will also be considered. See the following subsections.

3.2 Forecasting of *FSPCOM*

3.2.1 Simple Continuous Models of FSPCOM

Based on the graph of *FSPCOM* in Figure 3.1, I would try to apply the simplest possible model with a linear time trend. In order to be general, then I present the variable *FSPCOM* as the time series Y_t. It is well-known that the lagged variable $Y_{t-1} = Y(-1)$ is a good predictor of Y_t. In general, Y_{t-1} has been found to have a significant linear effect on Y_t. for any endogenous time series Y_t. In addition, I would say that the linear effect of Y_{t-1} on Y_t, should depend on at least one

Advanced Time Series Data Analysis: Forecasting Using EViews, First Edition. I Gusti Ngurah Agung.

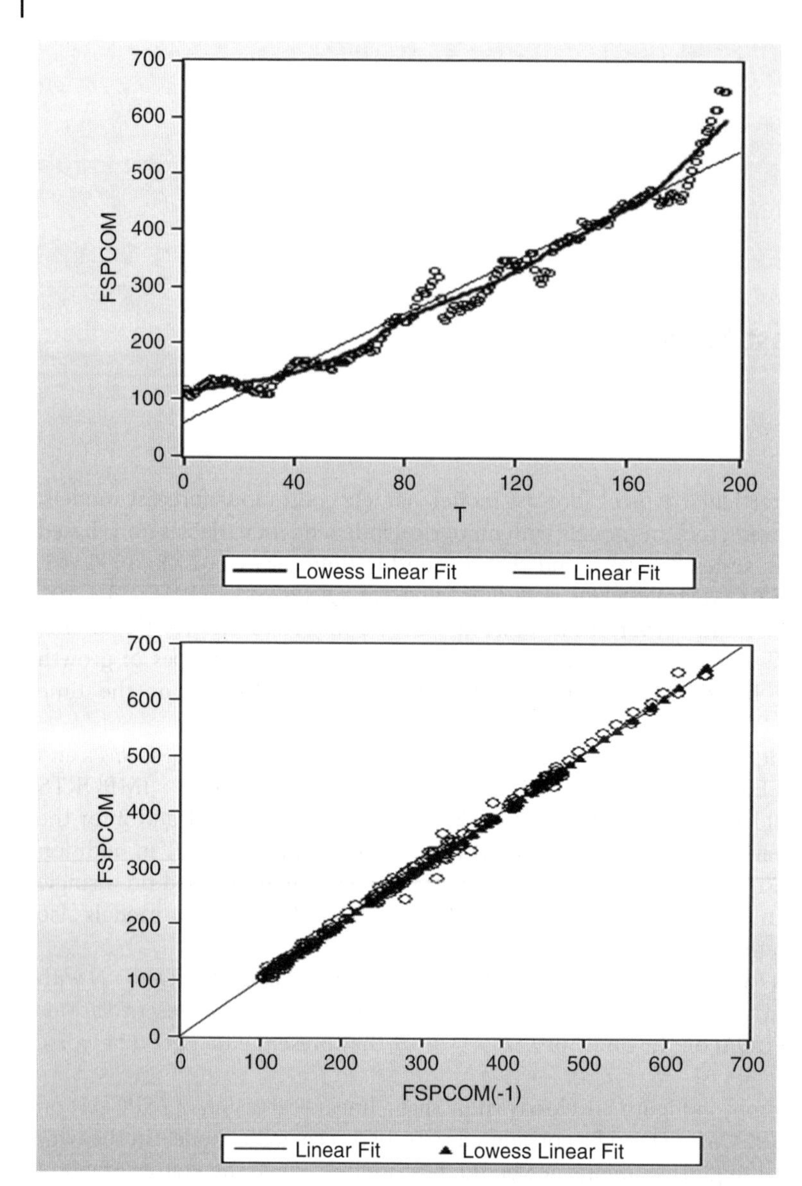

Figure 3.1 Scatter graphs of FSPCOM on the time *T* and *FSPCOM(−1)* with regression lines and NNFs.

exogenous or environmental variable, including the numerical time variable *@Trend,* or *t* = *@Trend* + *1*. So in a statistical sense, the interaction *t*∗*Y(−1)* should be used as an independent variable. Then the simplest continuous forecast model of Y_t can be presented using the ES (Equations Specification) as follows:

$$Y\ C\ Y(-1)\ t\ t{*}Y(-1) \tag{3.1}$$

This model is a hierarchical continuous model, where the time series Y_t can be the original observed/measured or transformed variables. Referring to all models with the dummies of the *@Month* and *@Year,* presented in Chapter 2, this model is a single multiple regression model for all observations. In fact, this model is the simplest *time-related-effect* (TRE) model of Y, which has been presented in Agung (2014, 2009a).

In practice, however, based on a data set, there are four alternative reduced models that can be obtained as good fits, such as follows:

$$Y\,C\,Y(-1)\;t*Y(-1) \tag{3.2}$$

$$Y\,C\,t\;t*Y(-1) \tag{3.3}$$

$$Y\,C\,t*Y(-1) \tag{3.4}$$

$$Y\,C\;Y(-1)\,t \tag{3.5}$$

where the first three models are nonhierarchical interaction reduced models and the last is an additive reduced model.

As more advanced models, each of the five models here can be extended to alternatives, such as: *lag-variable-autoregressive*: LVAR(p,q), *lag-variable-moving-average*: LVAMA(p,q), and *lag-variable-autoregressive-moving-average*: LVARMA(p,g,r) models for various integers $p \ge 1$, $q \ge 0$, and $r \ge 0$; which should be selected using the *trial-and-error* method or the manual stepwise/multistage selection method, with the results being highly dependent on the data used – refer to the many models based on time series variables presented in Agung (2009a). A special note should be presented for the LVAMA(p,q,r), since it was presented as *autoregressive-integrated-moving-average*: ARIMA(p,q) models in Hankle and Reitch (1992), and Wilson and Keating (1994).

Example 3.1 Application of the LV(1) Model (3.1)

Figure 3.2 presents the statistical results of the LV(1) model (3.1) of a monthly time series Y_t = *FSPCOM*. Based on these results the following findings and notes are presented.

1) At the $\alpha = 0.01$ level of significance, the interaction *@Trend***Y*(−1) has a positive significant adjusted effect on Y_t, at with a p-value = 0.0100/2 = 0.0050 < 0.01. So this finding shows that the linear effect of Y_{t-1} is significantly dependent on the numerical variable time: *@Trend*.
2) Even though the numerical time variable *@Trend* has an insignificant adjusted effect, at the 0.10 level of significance, this model can be considered to be a good fit, since *@Trend* has a positive significant adjusted effect on Y_t, with a p-value = 0.111 3/2 = 0.055 65 < 0.10. In general, if the interaction *@Trend***Y*(−1) has a significant effect, we always can keep the main variables to present the effect of *Y*(−1) on *Y* is significantly depends on the time t, since the variables *Y*(−1) and t**Y*(−1) have significant joint effects, based on the Wald test of F_0 = 1182.943 with df = (2,191), and a p-value = 0.0000.
3) Note the following regression function, which shows that the linear effect of *Y*(−1) on *Y* depends on [0.90186 + 0.00033*t].

   $$\hat{Y} = 11.26431 + 0.082116^{*}t + [0.90186 + 0.00033^{*}t]Y(-1)$$

   Furthermore, note that this regression function shows the linear effect of *Y*(−1) on *Y* increases with increasing scores or values of the time t, and it represents a set of heterogeneous simple linear regressions of $\hat{Y} = a_t + b_t \times Y(-1)$, for t = 0,1, …,T = 195.

4) Compared to the graphs of the forecasts of *HS* and *G* presented in Chapter 2, the forecast of *Y* = *FSPCOM* has a different pattern that is without a seasonal pattern. So, I have a good reason to present this continuous forecast model, even though this model might not be the best continuous forecast model. We have to apply the trial-and-error method using subjectively selected alternative models. Refer to alternative models presented in previous chapters, which also present unexpected forecast models, and see the following additional examples.

Equation: EQ01 Workfile: EXER15_5::Exer15_5\

View | Proc | Object | Print | Name | Freeze | Estimate | Forecast | Stats | Resids

Dependent Variable: Y
Method: Least Squares
Date: 05/13/14 Time: 12:52
Sample (adjusted): 1980M02 1996M04
Included observations: 195 after adjustments

Variable	Coefficient	Std. Error	t-Statistic	Prob.
C	11.26431	4.415493	2.551088	0.0115
Y(-1)	0.901862	0.039261	22.97077	0.0000
@TREND	0.082116	0.051331	1.599751	0.1113
@TREND*Y(-1)	0.000332	0.000127	2.603251	0.0100

R-squared	0.996086	Mean dependent var	293.5312
Adjusted R-squared	0.996024	S.D. dependent var	140.5846
S.E. of regression	8.864185	Akaike info criterion	7.222215
Sum squared resid	15007.59	Schwarz criterion	7.289353
Log likelihood	-700.1659	Hannan-Quinn criter.	7.249398
F-statistic	16202.25	Durbin-Watson stat	1.432110
Prob(F-statistic)	0.000000		

Forecast: YF_EQ01
Actual: Y
Forecast sample: 1990M01 1996M04
Included observations: 76

Root Mean Squared Error	33.64585
Mean Absolute Error	23.87421
Mean Abs. Percent Error	5.481669
Theil Inequality Coefficient	0.036628
Bias Proportion	0.395868
Variance Proportion	0.007537
Covariance Proportion	0.596595

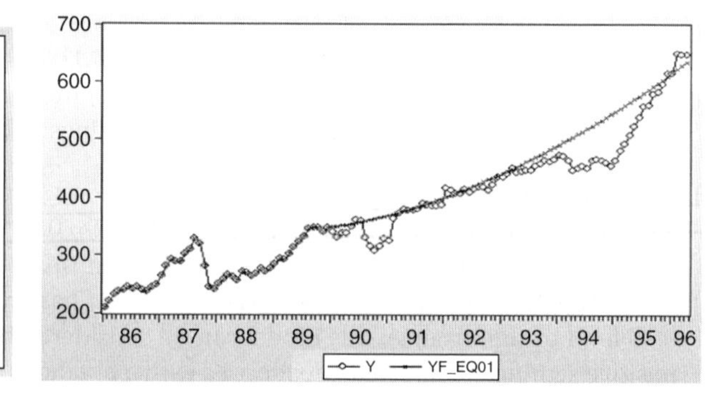

Figure 3.2 Statistical results of the LV(1) interaction model (3.1).

Example 3.2 Application of a LV(2) Model, as an Extension of the LV(1) Model (3.1)
As an extension of the LV(1) model (3.1), Figure 3.3 presents the statistical results of the LV(2) model of a monthly time series $Y_t = FSPCOM$. Based on these results the following findings and notes are presented.

1) Compared to the forecast based on the LV(1) model in Figure 3.2, this Figure 3.3 presents a better forecast model, since it has a smaller RMSE and TIC.
2) In a statistical sense, this regression function also is a better fit model, since each of independent variables has a p-value < 0.01, with a greater DW-statistic of 2.014195.
3) It is unexpected that *@Trend* has a significant adjusted effect with a p-value = 0.0075, compared to its insignificant effect based on the LV(1) regression in Figure 3.2, at the 10% level of significance with a p-value = 0.1113. These findings clearly show the unpredictable impacts of multicollinearity.

Dependent Variable: Y
Method: Least Squares
Date: 05/14/14 Time: 16:18
Sample (adjusted): 1980M03 1996M04
Included observations: 194 after adjustments

Variable	Coefficient	Std. Error	t-Statistic	Prob.
C	13.69624	4.261375	3.214044	0.0015
Y(-1)	1.183464	0.073339	16.13683	0.0000
Y(-2)	-0.316045	0.070600	-4.476573	0.0000
@TREND	0.137296	0.050833	2.700932	0.0075
@TREND*Y(-1)	0.000367	0.000122	3.005963	0.0030

R-squared	0.996435	Mean dependent var	294.4497
Adjusted R-squared	0.996360	S.D. dependent var	140.3605
S.E. of regression	8.468462	Akaike info criterion	7.136010
Sum squared resid	13554.11	Schwarz criterion	7.220233
Log likelihood	-687.1930	Hannan-Quinn criter.	7.170114
F-statistic	13207.71	Durbin-Watson stat	2.014195
Prob(F-statistic)	0.000000		

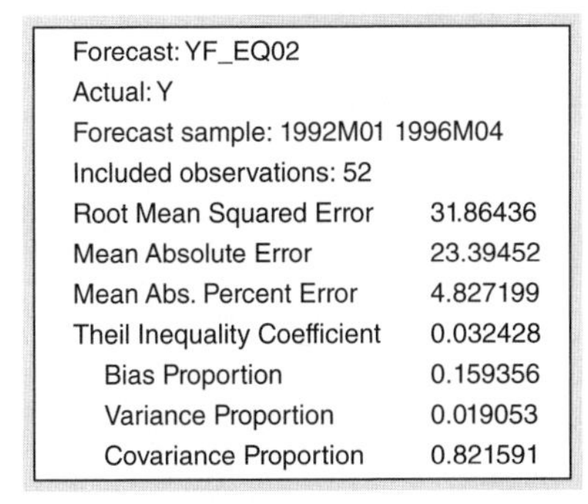
Forecast: YF_EQ02
Actual: Y
Forecast sample: 1992M01 1996M04
Included observations: 52

Root Mean Squared Error	31.86436
Mean Absolute Error	23.39452
Mean Abs. Percent Error	4.827199
Theil Inequality Coefficient	0.032428
Bias Proportion	0.159356
Variance Proportion	0.019053
Covariance Proportion	0.821591

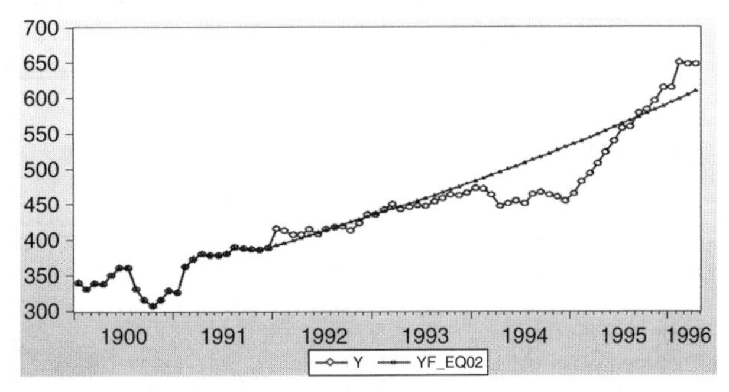

Figure 3.3 Statistical results of a LV(2) model, as an extension of the model (3.1).

Example 3.3 Application of a LV(1,1) Model, as an Extension of the LV(1) Model (3.1)
As an extension of the LV(1) model (3.1), Figure 3.4 presents the statistical results of the LVAR (1.1) model of a monthly time series $Y_t = FSPCOM$. Based on these results, the following findings and notes are presented.

1) Since each of the independent variables has a significant adjusted effect on $Y = FSPCOM$, then this model is an acceptable model in the statistical sense.
2) Compared to the forecasts based on the LV(1) and LV(2) models, in Figures 3.2 and 3.3, respectively, Figure 3.4 presents a better forecast model, since it has a smaller RMSE, and TIC. So this model is the best forecast model among the three defined models.
3) However, we cannot conclude that this model is the best possible forecast model of $Y = FSPCOM$, since we are presenting only three alternatives, and there are many other possible models, such as polynomial and nonlinear models.

Dependent Variable: Y
Method: Least Squares
Date: 07/08/14 Time: 09:17
Sample (adjusted): 1980M03 1996M04
Included observations: 194 after adjustments
Convergence achieved after 6 iterations

Variable	Coefficient	Std. Error	t-Statistic	Prob.
C	20.46945	7.976192	2.566318	0.0111
Y(-1)	0.800050	0.075966	10.53175	0.0000
@TREND	0.210687	0.096288	2.188105	0.0299
@TREND*Y(-1)	0.000547	0.000218	2.507557	0.0130
AR(1)	0.357615	0.096147	3.719469	0.0003

R-squared	0.996426	Mean dependent var	294.4497
Adjusted R-squared	0.996351	S.D. dependent var	140.3605
S.E. of regression	8.479040	Akaike info criterion	7.138507
Sum squared resid	13587.99	Schwarz criterion	7.222730
Log likelihood	-687.4352	Hannan-Quinn criter.	7.172611
F-statistic	13174.65	Durbin-Watson stat	2.007623
Prob(F-statistic)	0.000000		

Inverted AR Roots	.36

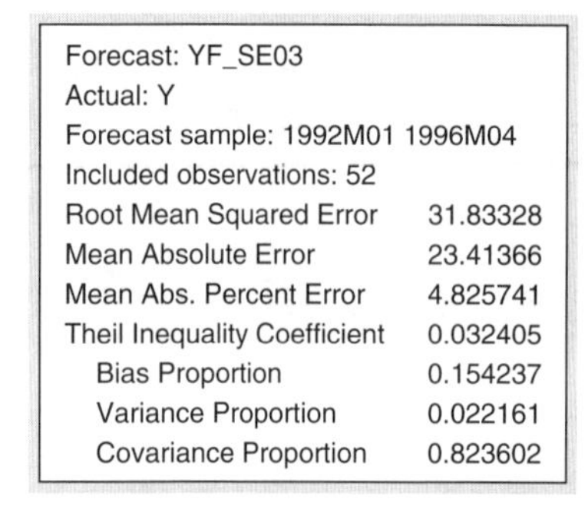
Forecast: YF_SE03
Actual: Y
Forecast sample: 1992M01 1996M04
Included observations: 52

Root Mean Squared Error	31.83328
Mean Absolute Error	23.41366
Mean Abs. Percent Error	4.825741
Theil Inequality Coefficient	0.032405
Bias Proportion	0.154237
Variance Proportion	0.022161
Covariance Proportion	0.823602

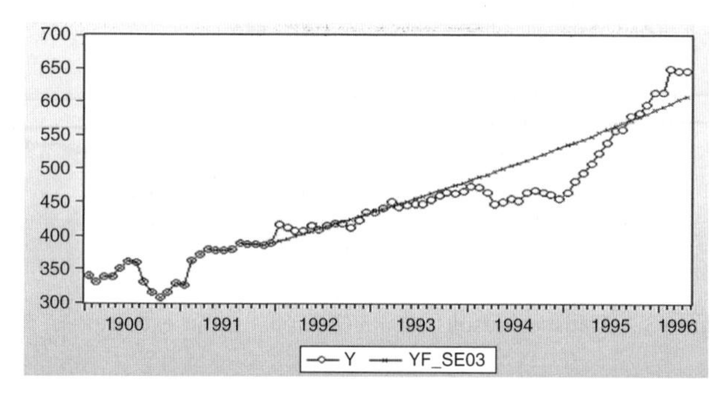

Figure 3.4 Statistical results of a LVAR(1,1) model, as an extension of the model (3.1).

3.2.2 LVAR(P,Q) Models of *Y* = *FSPCOM* with Polynomial Trend

The following examples present a few selected LVAR(*p*,*q*) models of *Y* = *FSPCOM* with polynomial trend, which are considered to be the best possible models among several LVAR(*p*,*q*) models, obtained by using the trial-and-error method.

Example 3.4 LV(2) and LVAR(1,1) Models with Quadratic Trend

Figures 3.5 presents the statistical results of LV(2) and LVAR(1,1) models of *Y* = *FSPCOM* with quadratic trend, with their forecast evaluations, using the following equation specifications. Based on these results, the following findings and notes need to be presented.

$$Y\,C\,Y(-1)\,Y(-2)\,T\,T\hat{}2 \tag{3.6}$$

Dependent Variable: Y
Method: Least Squares
Date: 07/08/14 Time: 18:53
Sample (adjusted): 1980M03 1996M04
Included observations: 194 after adjustments

Variable	Coefficient	Std. Error	t-Statistic	Prob.
C	9.342399	3.345527	2.792504	0.0058
Y(-1)	1.240126	0.069113	17.94338	0.0000
Y(-2)	-0.321776	0.071285	-4.513931	0.0000
T	0.065094	0.052106	1.249253	0.2131
T^2	0.000756	0.000292	2.590717	0.0103

R-squared	0.996393	Mean dependent var	294.4497
Adjusted R-squared	0.996317	S.D. dependent var	140.3605
S.E. of regression	8.518593	Akaike info criterion	7.147815
Sum squared resid	13715.06	Schwarz criterion	7.232038
Log likelihood	-688.3380	Hannan-Quinn criter.	7.181919
F-statistic	13052.15	Durbin-Watson stat	2.010470
Prob(F-statistic)	0.000000		

Forecast: YF_LV2	
Actual: Y	
Forecast sample: 1992M01 1996M04	
Included observations: 52	
Root Mean Squared Error	33.21535
Mean Absolute Error	25.20561
Mean Abs. Percent Error	5.149468
Theil Inequality Coefficient	0.033853
Bias Proportion	0.126482
Variance Proportion	0.074745
Covariance Proportion	0.798773

(a)

Dependent Variable: Y
Method: Least Squares
Date: 07/08/14 Time: 18:57
Sample (adjusted): 1980M03 1996M04
Included observations: 194 after adjustments
Convergence achieved after 5 iterations

Variable	Coefficient	Std. Error	t-Statistic	Prob.
C	14.76217	6.420091	2.299371	0.0226
Y(-1)	0.870466	0.055224	15.76242	0.0000
T	0.101862	0.088066	1.156661	0.2489
T^2	0.001199	0.000538	2.229685	0.0269
AR(1)	0.369660	0.097088	3.807475	0.0002

R-squared	0.996393	Mean dependent var	294.4497
Adjusted R-squared	0.996317	S.D. dependent var	140.3605
S.E. of regression	8.518593	Akaike info criterion	7.147815
Sum squared resid	13715.06	Schwarz criterion	7.232038
Log likelihood	-688.3380	Hannan-Quinn criter.	7.181919
F-statistic	13052.15	Durbin-Watson stat	2.010470
Prob(F-statistic)	0.000000		
Inverted AR Roots	.37		

Forecast: YF_LVAR11	
Actual: Y	
Forecast sample: 1992M01 1996M04	
Included observations: 52	
Root Mean Squared Error	33.21535
Mean Absolute Error	25.20561
Mean Abs. Percent Error	5.149468
Theil Inequality Coefficient	0.033853
Bias Proportion	0.126482
Variance Proportion	0.074745
Covariance Proportion	0.798773

(b)

Figure 3.5 Statistical results of LV(2) and LVAR(1,1) models of FSPCOM with quadratic trend.

and

$$Y\,C\,Y(-1)\,T\,T\hat{}2AR(1) \tag{3.7}$$

1) It is unexpected that both results present the same values of a set of statistics: R-Squared up to the Durbin–Watson statistic, as well as their forecast evaluations.
2) Compared to the forecast based on the LVAR(1,1) interaction model in Figure 3.4, with RMSE = 31.83328, these models have greater RMSE. So these two models give a worse forecast of *FSPCOM*.

Example 3.5 AR(2) and LVAR(1,3) Models with Third Degree Polynomial Trend

By using the trial-and-error method, I obtained the statistical results in Figure 3.6. This figure presents the statistical results of AR(2) and LVAR(1,2) models of $Y = FSPCOM$ with third degree polynomial trends and their forecast evaluations using the following equation specifications. Based on these results, the following findings and notes need to be presented.

$$Y\ C\ T\,T\hat{}3\ AR(1)\ \ AR(2) \tag{3.8}$$

and

$$Y\ C\ Y(-1)\ T\ T\hat{}3\ \ AR(1)AR(2) \tag{3.9}$$

1) The two models of $Y = FSPCOM$ in (3.7) and (3.8) in fact are the reduced models of Y on the third degree polynomial trend, since they do not have *T^2* as an independent variable.
2) Based on the results in Figure 3.6a, I made the following notes:
   2.1 Each of the time variables and each of the AR terms has a significant effect at the 1% level of significance. So the regression is a good fit model.
   2.2 Compared to the forecast based on the LVAR(1,1) interaction model in Figure 3.4, with RMSE = 31.83328, this AR(2) model has a slightly smaller RMSE = 31.80306. Hence, this AR(2) model should be considered a better forecast model, in a statistical sense.
3) Based on the results in Figure 3.6b, I have the findings and notes as follows:
   3.1 Even though the AR(2) has a large p-value > 0.50, I do not reduce the model in order to present its forecast with RMSE = 31.34839, which is smaller than the forecast based on the AR(2) model in Figure 3.6a with RMSE = 31.80306. On the other hand, I would say that a good fit model can have one of more independent variables with large p-values.
   3.2 Since the LVAR(1,2) forecast model is better than the LVAR(1,1) forecast model, then it can be concluded that this LVAR(1,2) model is the best forecast model of *FSPCOM*, compared to all previous models.
   3.3 As a comparison, Figure 3.7 presents the statistical results of the full model of the model (3.10), using the ES as follows:

$$Y\ C\ Y(-1)\ T\ T\hat{}2\ T\hat{}3\ \ AR(1)\ \ AR(2) \tag{3.10}$$

   Based on these results, the following findings and notes are presented.
   - Two of the independent variables, namely *T^2* and *AR(2),* have p-values > 0.50, and *T^3* has a p-value = 0.2690. Would you say this regression can be considered an acceptable model, in a statistical sense?
   - Note that the multicollinearity impacts are unpredictable, which are highly dependent on the data. So a large p-value of an independent variable does not directly mean that the variable can be deleted from the model.

Dependent Variable: Y
Method: Least Squares
Date: 07/08/14 Time: 19:28
Sample (adjusted): 1980M03 1996M04
Included observations: 194 after adjustments
Convergence achieved after 6 iterations

Variable	Coefficient	Std. Error	t-Statistic	Prob.
C	93.45082	22.03559	4.240904	0.0000
T	1.452864	0.338684	4.289729	0.0000
T^3	2.98E-05	8.09E-06	3.682837	0.0003
AR(1)	1.233090	0.069133	17.83660	0.0000
AR(2)	-0.318637	0.070777	-4.501982	0.0000

R-squared	0.996417	Mean dependent var	294.4497
Adjusted R-squared	0.996341	S.D. dependent var	140.3605
S.E. of regression	8.489971	Akaike info criterion	7.141083
Sum squared resid	13623.04	Schwarz criterion	7.225307
Log likelihood	-687.6851	Hannan-Quinn criter.	7.175188
F-statistic	13140.63	Durbin-Watson stat	2.009586
Prob(F-statistic)	0.000000		

Inverted AR Roots	.86	.37

Forecast: YF_AR2	
Actual: Y	
Forecast sample: 1992M01 1996M04	
Included observations: 52	
Root Mean Squared Error	31.80306
Mean Absolute Error	23.50123
Mean Abs. Percent Error	4.814647
Theil Inequality Coefficient	0.032423
Bias Proportion	0.124108
Variance Proportion	0.039291
Covariance Proportion	0.836601

(a)

Dependent Variable: Y
Method: Least Squares
Date: 07/08/14 Time: 19:36
Sample (adjusted): 1980M04 1996M04
Included observations: 193 after adjustments
Convergence achieved after 8 iterations

Variable	Coefficient	Std. Error	t-Statistic	Prob.
C	16.18317	8.143347	1.987287	0.0484
Y(-1)	0.846686	0.082037	10.32084	0.0000
T	0.219815	0.134032	1.640026	0.1027
T^3	4.95E-06	2.42E-06	2.041525	0.0426
AR(1)	0.376397	0.110631	3.402273	0.0008
AR(2)	0.048388	0.091809	0.527047	0.5988

R-squared	0.996425	Mean dependent var	295.4329
Adjusted R-squared	0.996330	S.D. dependent var	140.0542
S.E. of regression	8.484697	Akaike info criterion	7.145000
Sum squared resid	13462.15	Schwarz criterion	7.246431
Log likelihood	-683.4925	Hannan-Quinn criter.	7.186076
F-statistic	10425.47	Durbin-Watson stat	1.996655
Prob(F-statistic)	0.000000		

Inverted AR Roots	.48	-.10

Forecast: YF_LVAR12	
Actual: Y	
Forecast sample: 1992M01 1996M04	
Included observations: 52	
Root Mean Squared Error	31.34839
Mean Absolute Error	23.05532
Mean Abs. Percent Error	4.717976
Theil Inequality Coefficient	0.031987
Bias Proportion	0.108756
Variance Proportion	0.038164
Covariance Proportion	0.853080

(b)

Figure 3.6 Statistical results of AR(2) and LVAR(1,2) models of $Y = FSPCOM$ with third degree polynomial trends.

Dependent Variable: Y
Method: Least Squares
Date: 07/09/14 Time: 08:10
Sample (adjusted): 1980M04 1996M04
Included observations: 193 after adjustments
Convergence achieved after 9 iterations

Variable	Coefficient	Std. Error	t-Statistic	Prob.
C	13.47464	8.893838	1.515053	0.1315
Y(-1)	0.854717	0.077641	11.00853	0.0000
T	0.308415	0.228251	1.351212	0.1783
T^2	-0.001181	0.002338	-0.505052	0.6141
T^3	8.55E-06	7.71E-06	1.108699	0.2690
AR(1)	0.366217	0.107666	3.401410	0.0008
AR(2)	0.041413	0.091567	0.452274	0.6516

R-squared	0.996430	Mean dependent var	295.4329
Adjusted R-squared	0.996315	S.D. dependent var	140.0542
S.E. of regression	8.501878	Akaike info criterion	7.154047
Sum squared resid	13444.44	Schwarz criterion	7.272383
Log likelihood	-683.3655	Hannan-Quinn criter.	7.201969
F-statistic	8652.850	Durbin-Watson stat	1.995283
Prob(F-statistic)	0.000000		

Inverted AR Roots	.46	-.09

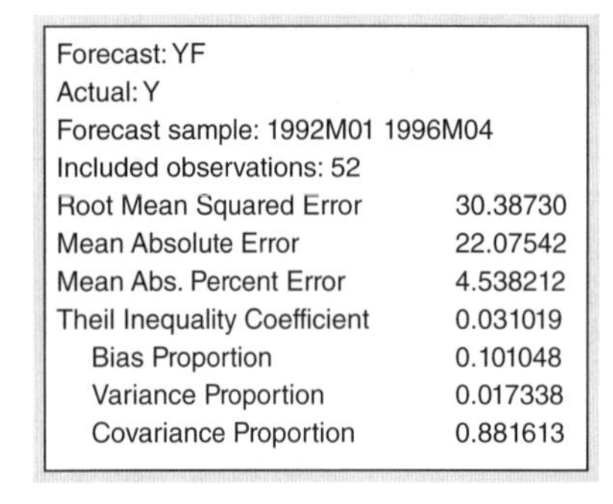

Forecast: YF
Actual: Y
Forecast sample: 1992M01 1996M04
Included observations: 52

Root Mean Squared Error	30.38730
Mean Absolute Error	22.07542
Mean Abs. Percent Error	4.538212
Theil Inequality Coefficient	0.031019
Bias Proportion	0.101048
Variance Proportion	0.017338
Covariance Proportion	0.881613

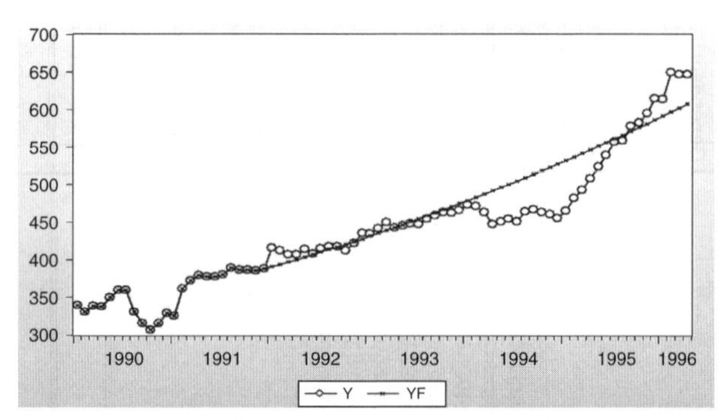

Figure 3.7 Statistical results of the LVAR(1,2) full model in (3.10).

- On the other hand, this full model can be considered the best possible model, in a theoretical sense, and it could give good parameter estimates for other data sets.
- It is common the forecast of a full model would have a smaller RMSE than its reduced model. In this case, the full model has RMSE = 30.38730 compared to its reduced model with RMSE = 31.34839. So this full model is the best forecast model compared to all previous models.

3.4 As an additional comparison, if the AR(2) is deleted from the model (3.10), then an LVAR (1,1) model of Y on a complete third degree polynomial trend would be obtained, with the statistical results presented in Figure 3.8. Based on these results, the following findings, and notes are presented.

- Only T^2 has a p-value > 0.30, and at the 15% level of significance (Lapin, 1973) T^3 has a positive significant effect with a p-value = 0.1535/2 = 0.07675. So I would consider this LVAR(1,1) model is a good fit model for the data set.

Dependent Variable: Y
Method: Least Squares
Date: 07/09/14 Time: 09:14
Sample (adjusted): 1980M03 1996M04
Included observations: 194 after adjustments
Convergence achieved after 9 iterations

Variable	Coefficient	Std. Error	t-Statistic	Prob.
C	10.28142	6.552857	1.568998	0.1183
Y(-1)	0.872020	0.053263	16.37193	0.0000
T	0.341847	0.195044	1.752672	0.0813
T^2	-0.001785	0.002087	-0.855208	0.3935
T^3	9.90E-06	6.91E-06	1.433228	0.1535
AR(1)	0.354530	0.095959	3.694605	0.0003

R-squared	0.996432	Mean dependent var	294.4497
Adjusted R-squared	0.996337	S.D. dependent var	140.3605
S.E. of regression	8.495001	Akaike info criterion	7.147272
Sum squared resid	13567.03	Schwarz criterion	7.248340
Log likelihood	-687.2854	Hannan-Quinn criter.	7.188197
F-statistic	10500.21	Durbin-Watson stat	2.003945
Prob(F-statistic)	0.000000		

Inverted AR Roots	.35

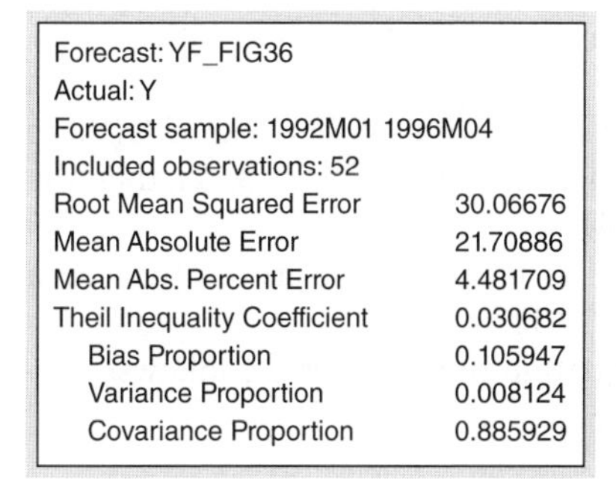

Forecast: YF_FIG36
Actual: Y
Forecast sample: 1992M01 1996M04
Included observations: 52

Root Mean Squared Error	30.06676
Mean Absolute Error	21.70886
Mean Abs. Percent Error	4.481709
Theil Inequality Coefficient	0.030682
Bias Proportion	0.105947
Variance Proportion	0.008124
Covariance Proportion	0.885929

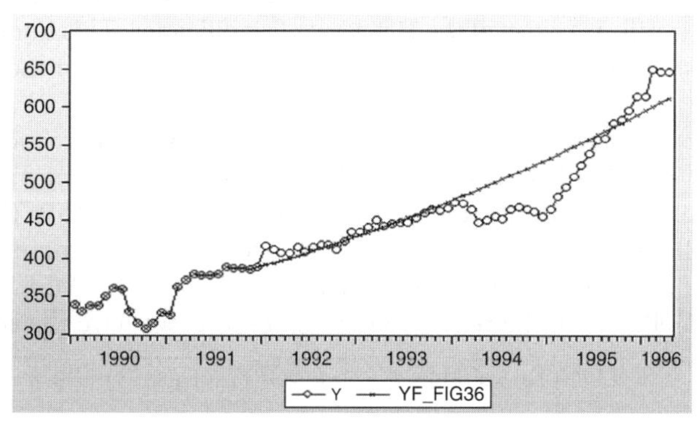

Figure 3.8 Statistical results of a LVAR(1,1) reduced model of the model in (3.10).

- It is surprising and unexpected that this reduced model has a smaller RMSE = 30.06676 compared to the full model in (3.10), with RMSE = 30.38730. So, in a statistical sense, this reduced model can be considered the best forecast model of *Y* = *FSPCOM*, compared to all previous models.

3.2.3 Translog Models with Time Predictor

Referring to all models of *Y* = *FSPCOM* have been previously presented, we can easily modify the models to the models of *LOG(Y)* with the time predictor. However, this section only presents some selected models. Note that using *LOG(Y)* as a dependent variable, EViews provides the option to forecast *Y* or *LOG(Y)*.

3.2.3.1 Simple Translog Linear Models

It is recognized there are two translog linear models can be considered to be the simplest translog linear models, with the equation specifications as follows;

1) *The LV(1) Translog Linear Model with Trend*

$$LOG(Y)\ C\ LOG(Y(-1))\ T \tag{3.11}$$

2) *The LV(1) Translog Linear Model with Logarithmic Trend*

$$LOG(Y)\ C\ LOG(Y(-1))\ LOG(T) \tag{3.12}$$

These models can easily be extended to various LVAR(*p,q*) translog linear models for alternative integers *p* and *q*. Furthermore, each of the models can be extended to ANCOVA or fixed-effect models by *@Month* or *@Year,* However, we have to apply the trial and error method, in order to obtain the best possible model.

Example 3.6 Application of the Models (3.11) and (3.12)

Figure 3.9 presents the statistical results of the two LV(1) translog linear models in (3.11) and (3.12) with $Y = FSPCOM$. Based on these results, the following findings and notes are presented.

1) Statistical results of both models present the forecasts of *Y*, as well as *LOG(Y)*, which show the forecast of the model (3.12) has smaller values of RMSE for the forecasts of *Y* and *LOG(Y)*. So the forecast model (3.12) is better than the model (2.11), even though *LOG(T)* has an insignificant effect at the 10% level of significance. This findings have demonstrated that a good or the best possible forecast model might have one or more independent variables with large *p*-values.
2) Compared to the forecast based on the LV(1) interaction model with RMSE = 33.64585, these models have smaller RMSEs. Referring to the three models (3.1), (3.11), and (3.12), then the model (3.12) is the best forecast model of $Y = FSPCOM$.
3) Referring to the forecast models for *Y*, with polynomial trends presented previously. These models are worse than the model (3.10) with RMSE = 30.38730.
4) Based on the regression in Figure 3.9a, we have the equation of the model as follows:

$$LOG(Y_t) = C(1) + C(2)*LOG(Y_{t-1}) + C(3)*T + \mu_t$$

with the following regression function.

$$LOG(Y) = 0.31346 + 0.93457^{*}LOG(Y(-1)) + 0.00059^{*}T$$

which can be presented as a nonlinear function as follows:

$$Y = Y(-1)^{C(2)*}Exp(C(1) + C(3)*T) = Y(-1)^{0.93457}Exp(0.31346 + 0.00059^{*}T)$$

5) Based on the regression in Figure 3.9b, we have the equation of the model as follows:

$$LOG(Y_t) = C(1) + C(2)*LOG(Y_{t-1}) + C(3)*LOG(T) + \mu_t$$

with the following regression function.

$$LOG(Y) = 0.04198 + 0.98920^{*}LOG(Y(-1)) + 0.00625^{*}LOG(T)$$

which can be presented as a nonlinear function as follows:

$$Y = Y(-1)^{C(2)}T^{C(3)}\cdot Exp(C(1)) = Y(-1)^{0.98920}T^{0.00625}Exp(0.94198)$$

Dependent Variable: LOG(Y)
Method: Least Squares
Date: 07/10/14 Time: 08:13
Sample (adjusted): 1980M02 1996M04
Included observations: 195 after adjustments

Variable	Coefficient	Std. Error	t-Statistic	Prob.
C	0.313460	0.120607	2.599020	0.0101
LOG(Y(-1))	0.934568	0.025807	36.21313	0.0000
T	0.000593	0.000233	2.542629	0.0118

R-squared	0.995844	Mean dependent var	5.558859
Adjusted R-squared	0.995800	S.D. dependent var	0.510774
S.E. of regression	0.033100	Akaike info criterion	-3.963299
Sum squared resid	0.210358	Schwarz criterion	-3.912945
Log likelihood	389.4216	Hannan-Quinn criter.	-3.942911
F-statistic	23001.88	Durbin-Watson stat	1.372106
Prob(F-statistic)	0.000000		

Forecast: YF
Actual: Y
Forecast sample: 1992M01 1996M04
Included observations: 52

Root Mean Squared Error	35.83002
Mean Absolute Error	26.19413
Mean Abs. Percent Error	5.485747
Theil Inequality Coefficient	0.036186
Bias Proportion	0.301169
Variance Proportion	0.000161
Covariance Proportion	0.698669

Forecast: YF
Actual: Y
Forecast sample: 1992M01 1996M04
Included observations: 52

Root Mean Squared Error	0.071569
Mean Absolute Error	0.052438
Mean Abs. Percent Error	0.850851
Theil Inequality Coefficient	0.005784
Bias Proportion	0.310194
Variance Proportion	0.004695
Covariance Proportion	0.685111

(a)

Dependent Variable: LOG(Y)
Method: Least Squares
Date: 07/10/14 Time: 08:17
Sample (adjusted): 1980M02 1996M04
Included observations: 195 after adjustments

Variable	Coefficient	Std. Error	t-Statistic	Prob.
C	0.041981	0.037383	1.122992	0.2628
LOG(Y(-1))	0.989202	0.010516	94.06257	0.0000
LOG(T)	0.006251	0.005928	1.054355	0.2930

R-squared	0.995729	Mean dependent var	5.558859
Adjusted R-squared	0.995684	S.D. dependent var	0.510774
S.E. of regression	0.033556	Akaike info criterion	-3.935955
Sum squared resid	0.216189	Schwarz criterion	-3.885601
Log likelihood	386.7556	Hannan-Quinn criter.	-3.915567
F-statistic	22378.85	Durbin-Watson stat	1.410284
Prob(F-statistic)	0.000000		

Forecast: YF
Actual: Y
Forecast sample: 1992M01 1996M04
Included observations: 52

Root Mean Squared Error	32.10326
Mean Absolute Error	23.01804
Mean Abs. Percent Error	4.517129
Theil Inequality Coefficient	0.033087
Bias Proportion	0.001989
Variance Proportion	0.193438
Covariance Proportion	0.804572

Forecast: YF
Actual: LOG(Y)
Forecast sample: 1992M01 1996M04
Included observations: 52

Root Mean Squared Error	0.060526
Mean Absolute Error	0.044899
Mean Abs. Percent Error	0.721454
Theil Inequality Coefficient	0.004905
Bias Proportion	0.009215
Variance Proportion	0.105775
Covariance Proportion	0.885010

(b)

Figure 3.9 Statistical results of the translog linear interaction model in (3.11) and (3.12).

3.2.3.2 Simple Nonlinear Models

Referring to the LS translog linear regression functions *log(Y)* presented in previous example, which can be presented as the nonlinear functions, the following example presents the corresponding nonlinear linear models, which should be estimated using the iteration computational method. So it is expected that their parameters' estimates would have different values.

Example 3.7 Application of Simple Nonlinear Models

Figure 3.10 presents the statistical results of two nonlinear models, using the following ESs, respectively. Based on these results, the following findings and notes are presented.

$$Y = Y(-1)\hat{}C(2)*Exp(C(1) + C(3)*T) \tag{3.13}$$

and

$$Y = Y(-1)\hat{}C(2)*T\hat{}C(3)*Exp(C(1)) \tag{3.14}$$

1) These nonlinear equation specifications are the alternative equations of the LS regression in (3.11) and (3.12), respectively.
2) Based on the results in Figure 3.10a, the following findings and notes are presented.
   2.1 The forecast of Y = *FSPCOM* has RMSE = 35.85100, which is greater than the forecast model based on the LS regression in Figure 3.8a. So the forecast model based on the nonlinear model (3.13) is worse than the LS regression model in (3.10).
   2.2 Each of the parameters has a significant positive effect at the 1% level of significance, since it has a p-value = Prob/2 < 0.01. So the regression can be considered to be a good fit model, in a statistical sense.
3) Based on the results in Figure 3.10b, the following findings and notes are presented.
   3.1 The forecast of Y = *FSPCOM* has an RMSE = 43.68615, which is greater than the forecast model based on the nonlinear model (3.13). So the forecast based on the nonlinear model (3.13) is better that the forecast based on the nonlinear model (3.14).
   3.2 However, two of the three model parameters of the model (3.14) have large probabilities greater the 0.50. Therefore, in a statistical sense, I would say that the nonlinear model (3.14) is an inappropriate forecast model for *FSPCOM*, in a statistical sense.

Example 3.8 An Extension of the Simple Nonlinear Model (3.13)

As an extension of the SNM (simple nonlinear model) (3.13), Figure 3.11 presents the statistical results a LV(2) nonlinear model, using the following ES.

$$Y = Y(-1)\hat{}C(1)*Y(-2)\hat{}C(2)*Exp(C(3) + C(4)*T) \tag{3.15}$$

Based on these results, the findings and notes presented are as follows:.

1) We have the following regression function,

   $$\hat{Y} = Y(-1)^{1.21550} \times Y(-2)^{-0.30190} \times Exp(0.00075T + 0.41232)$$

   which can be presented as a translog linear regression with a linear trend, with the equation as follows:

   $$\log(\hat{Y}) = 1.21550\log(Y(-1)) - 0.30190\log(Y(-2)) + 0.00075T + 0.41232$$

2) The forecast of *FSPCOM* has smaller RMSE = 31.96835.

(a)

Dependent Variable: Y
Method: Least Squares
Date: 07/11/14 Time: 06:29
Sample (adjusted): 1980M02 1996M04
Included observations: 195 after adjustments
Convergence achieved after 1 iteration
Y = Y(-1)^C(2)*EXP(C(1)+C(3)*T)

	Coefficient	Std. Error	t-Statistic	Prob.
C(2)	0.935227	0.025731	36.34643	0.0000
C(1)	0.311320	0.122573	2.539872	0.0119
C(3)	0.000582	0.000218	2.666048	0.0083

R-squared	0.996074	Mean dependent var	293.5312
Adjusted R-squared	0.996033	S.D. dependent var	140.5846
S.E. of regression	8.854845	Akaike info criterion	7.215072
Sum squared resid	15054.39	Schwarz criterion	7.265425
Log likelihood	-700.4695	Hannan-Quinn criter.	7.235459
Durbin-Watson stat	1.426779		

Forecast: YF_NL1
Actual: Y
Forecast sample: 1992M01 1996M04
Included observations: 52

Root Mean Squared Error	35.85100
Mean Absolute Error	26.26590
Mean Abs. Percent Error	5.500028
Theil Inequality Coefficient	0.036210
Bias Proportion	0.300393
Variance Proportion	0.000018
Covariance Proportion	0.699590

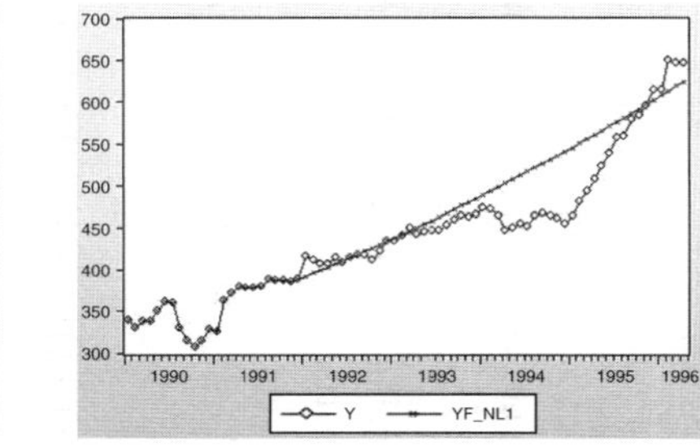

(a)

(b)

Dependent Variable: Y
Method: Least Squares
Date: 07/11/14 Time: 06:22
Sample (adjusted): 1980M02 1996M04
Included observations: 195 after adjustments
Convergence achieved after 1 iteration
Y=Y(-1)^C(2)*T^C(3)*EXP(C(1))

	Coefficient	Std. Error	t-Statistic	Prob.
C(2)	0.995196	0.013396	74.28903	0.0000
C(3)	0.005996	0.010203	0.587692	0.5574
C(1)	0.008839	0.039575	0.223340	0.8235

R-squared	0.995935	Mean dependent var	293.5312
Adjusted R-squared	0.995893	S.D. dependent var	140.5846
S.E. of regression	9.009958	Akaike info criterion	7.249803
Sum squared resid	15586.43	Schwarz criterion	7.300157
Log likelihood	-703.8558	Hannan-Quinn criter.	7.270191
Durbin-Watson stat	1.463605		

Forecast: YF_NL2
Actual: Y
Forecast sample: 1992M01 1996M04
Included observations: 52

Root Mean Squared Error	43.68615
Mean Absolute Error	32.73061
Mean Abs. Percent Error	6.880284
Theil Inequality Coefficient	0.043647
Bias Proportion	0.460337
Variance Proportion	0.023319
Covariance Proportion	0.516345

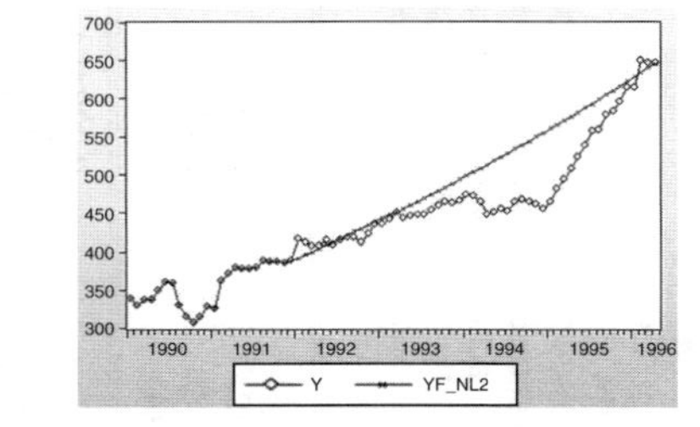

(b)

Figure 3.10 Statistical results of the nonlinear models in (3.13) and (3.14).

Dependent Variable: Y
Method: Least Squares
Date: 07/12/14 Time: 09:21
Sample (adjusted): 1980M03 1996M04
Included observations: 194 after adjustments
Convergence achieved after 16 iterations
Y = Y(-1)^C(1)*Y(-2)^C(2)*EXP(C(3)+C(4)*T)

	Coefficient	Std. Error	t-Statistic	Prob.
C(1)	1.215497	0.069557	17.47481	0.0000
C(2)	-0.301902	0.070222	-4.299223	0.0000
C(3)	0.412322	0.119908	3.438651	0.0007
C(4)	0.000751	0.000213	3.527151	0.0005

R-squared	0.996401	Mean dependent var	294.4497
Adjusted R-squared	0.996344	S.D. dependent var	140.3605
S.E. of regression	8.486595	Akaike info criterion	7.135256
Sum squared resid	13684.24	Schwarz criterion	7.202634
Log likelihood	-688.1198	Hannan-Quinn criter.	7.162539
Durbin-Watson stat	1.991670		

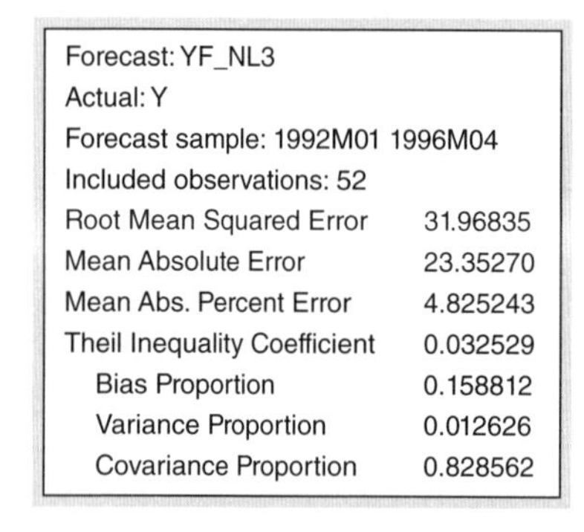
Forecast: YF_NL3
Actual: Y
Forecast sample: 1992M01 1996M04
Included observations: 52

Root Mean Squared Error	31.96835
Mean Absolute Error	23.35270
Mean Abs. Percent Error	4.825243
Theil Inequality Coefficient	0.032529
Bias Proportion	0.158812
Variance Proportion	0.012626
Covariance Proportion	0.828562

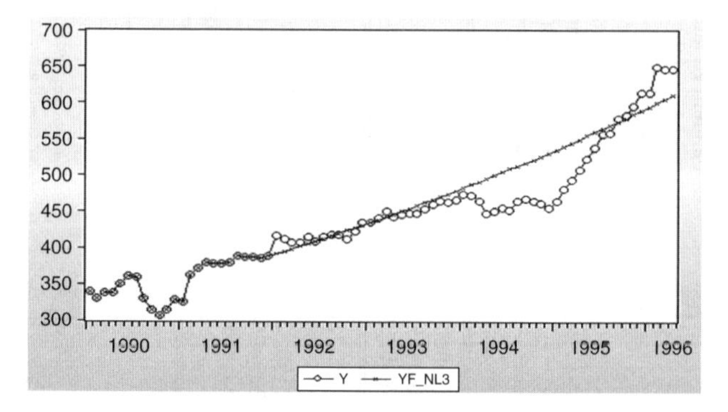

Figure 3.11 Statistical results of the nonlinear LV(2) model in (3.15).

3) An alternative translog linear, with the following ES, is recommended as an exercise for the readers, as a comparison.

$$log(Y)\ C\ log(Y(-1))\ log(Y(-2))\ log(T) \tag{3.16a}$$

or

$$log(Y)\ C\ log(Y(-1))\ log(Y(-2))\ log(@Trend+1) \tag{3.16b}$$

Example 3.9 A Modification of the Model (3.15)

As a comparison, Figure 3.12 presents the statistical results of a nonlinear LS regression with the time predictor, using the ES as follows:

$$Y = C(1)*Y(-1)\hat{}C(2)*Y(-2)\hat{}C(3)*Exp(C(4)*T) \tag{3.17}$$

Based on these results the following findings and notes are presented.

1) Compared to the forecast of the model (3.15) with RMSE = 31.96835, unexpectedly this forecast has exactly the same RMSE = 31.96835. However, note that based on the results,

Dependent Variable: Y
Method: Least Squares
Date: 05/18/14 Time: 14:47
Sample (adjusted): 1980M03 1996M04
Included observations: 194 after adjustments
Convergence achieved after 23 iterations
Y =C(1)*(Y(-1)^C(2))*(Y(-2)^C(3))*(EXP(C(4)*T))

	Coefficient	Std. Error	t-Statistic	Prob.
C(1)	1.511456	0.181547	8.325420	0.0000
C(2)	1.215497	0.069557	17.47481	0.0000
C(3)	-0.301902	0.070222	-4.299223	0.0000
C(4)	0.000751	0.000213	3.527151	0.0005

R-squared	0.996401	Mean dependent var	294.4497
Adjusted R-squared	0.996344	S.D. dependent var	140.3605
S.E. of regression	8.486595	Akaike info criterion	7.135256
Sum squared resid	13684.24	Schwarz criterion	7.202634
Log likelihood	-688.1198	Hannan-Quinn criter.	7.162539
Durbin-Watson stat	1.991670		

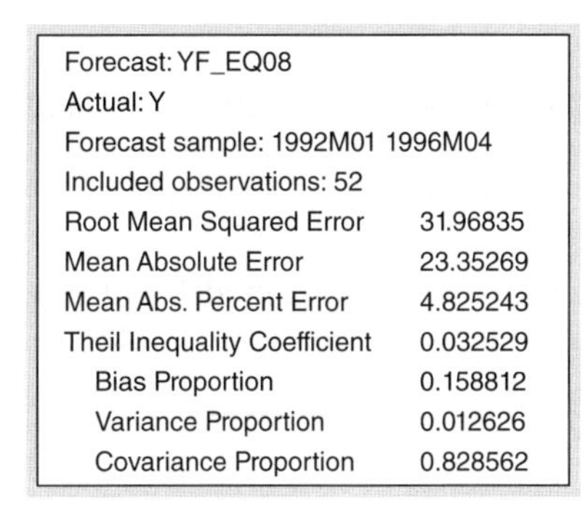

Forecast: YF_EQ08
Actual: Y
Forecast sample: 1992M01 1996M04
Included observations: 52

Root Mean Squared Error	31.96835
Mean Absolute Error	23.35269
Mean Abs. Percent Error	4.825243
Theil Inequality Coefficient	0.032529
Bias Proportion	0.158812
Variance Proportion	0.012626
Covariance Proportion	0.828562

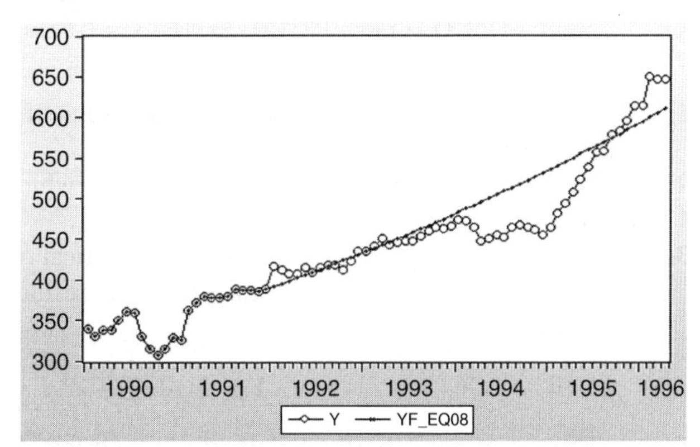

Figure 3.12 Statistical results of the nonlinear LV(2) model (3.17).

we have the following regression function, which is the same as the regression function in Figure 3.11.

$$\hat{Y} = 1.511456Y(-1)^{1.215497} \times Y(-2)^{-0.301902} \times Exp(0.000751T)$$

2) The nonlinear function above can be presented as the translog linear regression with a linear trend T, as follows:

$$\log(\hat{Y}) = \log(1.511456) + 1.215497\log(Y(-1))^{1.215497}$$
$$-0.301902\log(Y(-2)) + 0.000751T$$

Example 3.10 Application of a Nonlinear LV(2) Models with a Linear Trend
As a modification of the NLS model (3.15), Figure 3.13 presents the statistical results of two nonlinear LV(2) models of *Y* with a linear trend, using the ESs as follows:

$$Y = C(1)*Y(-1)\hat{}C(2)*Y(-2)\hat{}2 + C(4)*T \tag{3.18}$$

Based on these results, the following findings and notes are presented.

1) The model (3.18) is a model without an intercept parameter, and its forecast of *Y* has greatest RMSE than the previous two models (3.15) and (3.17).
2) As an upper model of the model (3.18), the model (3.19) is a model with an intercept parameter C(5). And it is found that this model has the smallest RMSE = 30.96426 compared to most of the previous models, aside from the LVAR(1,2) model in Figure 3.7, and the LVAR (1,1) model in Figure 3.8, with RMSE = 30.38730 and RMSE = 30.06676, respectively.

$$Y = C(1)*Y(-1)\hat{}C(2)*Y(-2)\hat{}2 + C(4)*T + C(5) \tag{3.19}$$

Example 3.11 An Extension of the LV(2) Nonlinear Model (3.15), with a Polynomial Trend
As an extension of the LV(2) nonlinear model (3.15), Figure 3.14 presents the statistical results a LV(2) nonlinear model, using the following ES, which is a LV(2) model of *Y* on a third degree polynomial of the time *T*. Based on these results, the following findings and notes are presented.

$$Y = Y(-1)\hat{}C(1)*Y(-2)\hat{}C(2)*Exp(C(3) + C(4)*T + C(5)*T\hat{}2 + C(6)*T\hat{}3) \tag{3.20}$$

1) The Eq. (3.20) present a nonlinear model, and each of the variables *Y(−1), Y(−2), T, T ^2,* and *T ^3* has significant adjusted effect on *Y.*
2) Its forecast evaluation presents TIC = 0.026591, which can be considered to be a very good forecast.
3) As a comparison, Figure 3.15 presents the statistical results of the following model, with one of its possible reduced models. Based on these results the following findings and notes are presented.

$$LOG(Y)\ LOG(Y(-1))\ LOG(Y(-2))\ C\ T\ T\hat{}2\ T\hat{}3 \tag{3.21}$$

3.1 The estimate of both models present perfect forecasts with TICs of 0.004868 and 0.006873 less than 0.01, respectively, and they are better forecast models compared to previous ones.
3.2 Since each of the *T* ^2 and *T* ^3 has insignificant effect with a large *p*-value, then a reduced model should be explored. And it is found to be an acceptable reduced model by deleting the time *T*, which is unexpected, since it has the smallest *p*-value = 0.0125. The regression function of the reduced model can be written in another form as follows:

$$Y = (Y(-1))^{1.266}(Y(-2))^{-0.332} \times \text{Exp}(0.325 + 7506e-06*T\hat{}2 - 2.554e-08*T\hat{}3$$

4) As an additional comparison, Figure 3.16 presents the statistical results based on the following LV(2) model of *Y* and one of its possible reduced models. Based on these results, the following findings and notes are presented.

$$Y\ Y(-1)\ Y(-2)\ C\ T\ T\hat{}2\ T\hat{}3 \tag{3.22}$$

Dependent Variable: Y
Method: Least Squares
Date: 05/20/14 Time: 12:36
Sample (adjusted): 1980M03 1996M04
Included observations: 194 after adjustments
Convergence achieved after 9 iterations
Y=C(1)*Y(-1)^C(2)*Y(-2)^C(3) +C(4)*T

	Coefficient	Std. Error	t-Statistic	Prob.
C(1)	0.994091	0.032096	30.97214	0.0000
C(2)	1.286007	0.074271	17.31498	0.0000
C(3)	-0.288775	0.075047	-3.847901	0.0002
C(4)	0.082277	0.046588	1.766055	0.0790

R-squared	0.996226	Mean dependent var	294.4497
Adjusted R-squared	0.996166	S.D. dependent var	140.3605
S.E. of regression	8.690991	Akaike info criterion	7.182854
Sum squared resid	14351.33	Schwarz criterion	7.250232
Log likelihood	-692.7368	Hannan-Quinn criter.	7.210137
Durbin-Watson stat	1.966460		

Forecast: YF_EQ10
Actual: Y
Forecast sample: 1992M01 1996M04
Included observations: 52

Root Mean Squared Error	36.28025
Mean Absolute Error	28.44821
Mean Abs. Percent Error	5.822469
Theil Inequality Coefficient	0.036907
Bias Proportion	0.146577
Variance Proportion	0.088674
Covariance Proportion	0.764749

(a). ES (3.18)

Dependent Variable: Y
Method: Least Squares
Date: 07/17/14 Time: 15:12
Sample (adjusted): 1980M03 1996M04
Included observations: 194 after adjustments
Convergence achieved after 1 iteration
Y=C(1)*Y(-1)^C(2)*Y(-2)^C(3) + C(4)*T +C(5)

	Coefficient	Std. Error	t-Statistic	Prob.
C(1)	0.531207	0.120791	4.397736	0.0000
C(2)	1.428498	0.100208	14.25539	0.0000
C(3)	-0.346950	0.088188	-3.934189	0.0001
C(4)	0.228572	0.067671	3.377708	0.0009
C(5)	22.22296	7.007600	3.171265	0.0018

R-squared	0.996391	Mean dependent var	294.4497
Adjusted R-squared	0.996315	S.D. dependent var	140.3605
S.E. of regression	8.520764	Akaike info criterion	7.148324
Sum squared resid	13722.05	Schwarz criterion	7.232547
Log likelihood	-688.3875	Hannan-Quinn criter.	7.182429
F-statistic	13045.48	Durbin-Watson stat	1.990999
Prob(F-statistic)	0.000000		

Forecast: YF
Actual: Y
Forecast sample: 1992M01 1996M04
Included observations: 52

Root Mean Squared Error	30.96426
Mean Absolute Error	22.59914
Mean Abs. Percent Error	4.658580
Theil Inequality Coefficient	0.031545
Bias Proportion	0.141382
Variance Proportion	0.019080
Covariance Proportion	0.839538

(b). ES (3.19)

Figure 3.13 Statistical results of the nonlinear LV(2) models in (3.18) and (3.19).

Dependent Variable: Y
Method: Least Squares
Date: 07/12/14 Time: 11:24
Sample (adjusted): 1980M03 1996M04
Included observations: 194 after adjustments
Convergence achieved after 1 iteration
Y = Y(-1)^C(1)*Y(-2)^C(2)*EXP(C(3)+C(4)*T+C(5)*T^2+C(6)*T^3)

	Coefficient	Std. Error	t-Statistic	Prob.
C(1)	1.190471	0.070322	16.92876	0.0000
C(2)	-0.291188	0.070917	-4.106017	0.0001
C(3)	0.449423	0.122695	3.662933	0.0003
C(4)	0.002088	0.000737	2.831714	0.0051
C(5)	-1.22E-05	6.01E-06	-2.030666	0.0437
C(6)	3.54E-08	1.72E-08	2.050824	0.0417

R-squared	0.996480	Mean dependent var	294.4497
Adjusted R-squared	0.996386	S.D. dependent var	140.3605
S.E. of regression	8.437708	Akaike info criterion	7.133738
Sum squared resid	13384.64	Schwarz criterion	7.234806
Log likelihood	-685.9726	Hannan-Quinn criter.	7.174663
Durbin-Watson stat	1.986108		

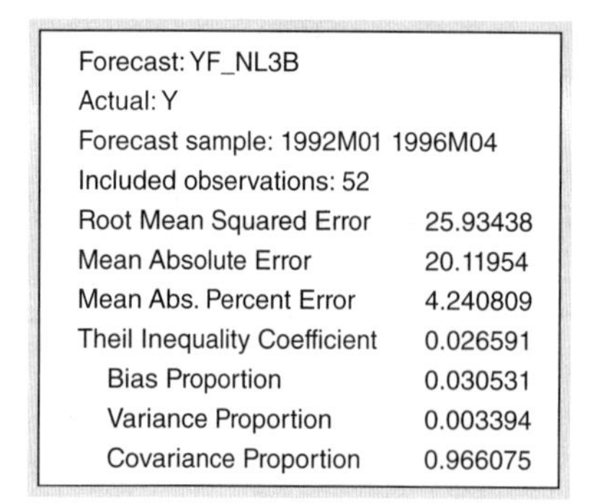

Forecast: YF_NL3B
Actual: Y
Forecast sample: 1992M01 1996M04
Included observations: 52

Root Mean Squared Error	25.93438
Mean Absolute Error	20.11954
Mean Abs. Percent Error	4.240809
Theil Inequality Coefficient	0.026591
Bias Proportion	0.030531
Variance Proportion	0.003394
Covariance Proportion	0.966075

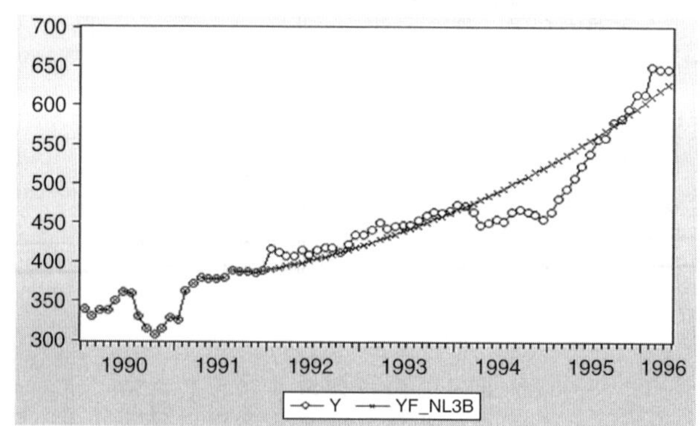

Figure 3.14 Statistical results of the nonlinear LV(2) model in (3.20).

4.1 Both of these models have greater TICs than the two models in Figure 3.15. So they can be considered to be worse forecast models than the model (3.21) and its reduced model.

4.2 Note that, at the 10% level of significance, T^3 has a positive significant adjusted effect on Y with a p-value = 0.1537/2 = 0.07685 < 0.10. So the model (3.22) can be considered to be an acceptable model, even though T ^2 has an insignificant adjusted effect.

4.3 However, the statistical results of a reduced model also is presented in Figure 3.16, which is obtained by deleting the variable T ^2.

5) Note that in order to have greater coefficients of T ^2 and T ^3, we can easily use a scale time variable, namely *ST*, such as *ST* = *T/10* or ST = T/100. At around a polynomial of degree 3, Basilevsky (1994) states that polynomial models of degree higher than 3 are rare in practice because of interpretational difficulties.

Dependent Variable: LOG(Y)
Method: Least Squares
Date: 07/18/14 Time: 13:36
Sample (adjusted): 1980M03 1996M04
Included observations: 194 after adjustments

Variable	Coefficient	Std. Error	t-Statistic	Prob.
LOG(Y(-1))	1.241126	0.068398	18.14557	0.0000
LOG(Y(-2))	-0.338249	0.069051	-4.898529	0.0000
C	0.448996	0.124524	3.605682	0.0004
T	0.001199	0.000476	2.520993	0.0125
T^2	-2.48E-06	5.00E-06	-0.496210	0.6203
T^3	4.80E-09	1.67E-08	0.287934	0.7737

R-squared	0.996313	Mean dependent var	5.563039
Adjusted R-squared	0.996215	S.D. dependent var	0.508740
S.E. of regression	0.031299	Akaike info criterion	-4.060022
Sum squared resid	0.184170	Schwarz criterion	-3.958954
Log likelihood	399.8221	Hannan-Quinn criter.	-4.019097
F-statistic	10160.49	Durbin-Watson stat	1.880049
Prob(F-statistic)	0.000000		

Dependent Variable: LOG(Y)
Method: Least Squares
Date: 07/18/14 Time: 13:37
Sample (adjusted): 1980M03 1996M04
Included observations: 194 after adjustments

Variable	Coefficient	Std. Error	t-Statistic	Prob.
LOG(Y(-1))	1.265719	0.068652	18.43685	0.0000
LOG(Y(-2))	-0.332901	0.069990	-4.756429	0.0000
C	0.329557	0.116780	2.822039	0.0053
T^2	7.56E-06	3.06E-06	2.467412	0.0145
T^3	-2.55E-08	1.17E-08	-2.179201	0.0306

R-squared	0.996188	Mean dependent var	5.563039
Adjusted R-squared	0.996108	S.D. dependent var	0.508740
S.E. of regression	0.031739	Akaike info criterion	-4.037085
Sum squared resid	0.190396	Schwarz criterion	-3.952862
Log likelihood	396.5972	Hannan-Quinn criter.	-4.002980
F-statistic	12349.10	Durbin-Watson stat	1.867487
Prob(F-statistic)	0.000000		

Forecast: YF
Actual: LOG (Y)
Forecast sample: 1992M01 1996M04
Included observations: 52

Root Mean Squared Error	0.060131
Mean Absolute Error	0.044780
Mean Abs. Percent Error	0.722980
Theil Inequality Coefficient	0.004868
Bias Proportion	0.090667
Variance Proportion	0.030514
Covariance Proportion	0.878819

Forecast: YF
Actual: LOG (Y)
Forecast sample: 1992M01 1996M04
Included observations: 52

Root Mean Squared Error	0.085053
Mean Absolute Error	0.072125
Mean Abs. Percent Error	1.166952
Theil Inequality Coefficient	0.006873
Bias Proportion	0.238376
Variance Proportion	0.076980
Covariance Proportion	0.684644

Figure 3.15 Statistical results based on the LV(2) model in (3.21) and its reduced model.

Dependent Variable: Y
Method: Least Squares
Date: 07/18/14 Time: 14:30
Sample (adjusted): 1980M03 1996M04
Included observations: 194 after adjustments

Variable	Coefficient	Std. Error	t-Statistic	Prob.
Y(-1)	1.226549	0.069571	17.63024	0.0000
Y(-2)	-0.309157	0.071632	-4.315915	0.0000
C	6.758184	3.792927	1.781786	0.0764
T	0.219376	0.119600	1.834248	0.0682
T^2	-0.001142	0.001356	-0.841722	0.4010
T^3	6.39E-06	4.46E-06	1.432219	0.1537

R-squared	0.996432	Mean dependent var	294.4497
Adjusted R-squared	0.996337	S.D. dependent var	140.3605
S.E. of regression	8.495001	Akaike info criterion	7.147272
Sum squared resid	13567.03	Schwarz criterion	7.248340
Log likelihood	-687.2854	Hannan-Quinn criter.	7.188197
F-statistic	10500.21	Durbin-Watson stat	2.003945
Prob(F-statistic)	0.000000		

Forecast: YF
Actual: Y
Forecast sample: 1992M01 1996M04
Included observations: 52

Root Mean Squared Error	30.06676
Mean Absolute Error	21.70886
Mean Abs. Percent Error	4.481709
Theil Inequality Coefficient	0.030682
Bias Proportion	0.105947
Variance Proportion	0.008124
Covariance Proportion	0.885929

(a)

Dependent Variable: Y
Method: Least Squares
Date: 07/18/14 Time: 14:34
Sample (adjusted): 1980M03 1996M04
Included observations: 194 after adjustments

Variable	Coefficient	Std. Error	t-Statistic	Prob.
Y(-1)	1.232527	0.069154	17.82295	0.0000
Y(-2)	-0.317823	0.070833	-4.486907	0.0000
C	8.766208	2.946478	2.975148	0.0033
T	0.128114	0.050444	2.539726	0.0119
T^3	2.72E-06	9.56E-07	2.846823	0.0049

R-squared	0.996418	Mean dependent var	294.4497
Adjusted R-squared	0.996343	S.D. dependent var	140.3605
S.E. of regression	8.488447	Akaike info criterion	7.140724
Sum squared resid	13618.16	Schwarz criterion	7.224948
Log likelihood	-687.6503	Hannan-Quinn criter.	7.174829
F-statistic	13145.36	Durbin-Watson stat	2.009109
Prob(F-statistic)	0.000000		

Forecast: YF
Actual: Y
Forecast sample: 1992M01 1996M04
Included observations: 52

Root Mean Squared Error	31.72745
Mean Absolute Error	23.41450
Mean Abs. Percent Error	4.798886
Theil Inequality Coefficient	0.032345
Bias Proportion	0.124380
Variance Proportion	0.036867
Covariance Proportion	0.838752

(b)

Figure 3.16 Statistical results of the LV(2) model of Y in (3.22), and its reduced model.

3.2.3.3 Translog Interaction Models

Referring to the basic interaction model (3.1), the following example presents selected translog interaction models.

Example 3.12 Translog Interaction Models

As a modification of the LV(1) interaction model (3.1), Figures 3.17 presents the statistical results of two LV(1) translog interaction models of $Y = FSPCOM$ on the time $T = @Trend + 1$, using the following equation specifications, respectively. Note that the function $T = @Trend + 1$ is applied so that *log(T)* is valid for all time points. Based on these results, the following findings and notes are presented.

$$log(Y)\ C\ log(Y(-1))\ T\ T*log(Y(-1)) \tag{3.23}$$

and

$$log(Y)\ C\ log(Y(-1))\ log(T)\ log(T)*log(Y(-1)) \tag{3.24}$$

1) By comparing their forecasting graphs, the regression in Figure 3.17a can be considered to be a better forecast model than the regression in Figure 3.17b since its graph is very close to the observed scores of *Y*. However, *T*log(Y(−1))* has a large *p*-value.
2) In addition, the forecast in Figure 3.17a based on the model (3.23) has smaller RMSE and TIC than the model (3.24). So the model (3.23) is a better forecasting model of $Y = SPCOM$ than the model (3.24).
3) Even though *T*log(Y(−1))* has insignificant effect, but the H_0: C(2) = C(4) = 0 is rejected based on the *F*-statistic of $F_0 = 654.1197$ with $df = (2,191)$, and *p*-value = 0.0000. So it can be concluded that the linear effect of *log(Y(−1))* on *log(Y)* is significantly dependent on the time *t*, specifically the function {C(2) + C(4)**t*}, since the equation of the regression can be written as follows:

$$\log(Y) = \hat{C}(1) + \hat{C}(3) + \{\hat{C}(3) + \hat{C}(4)*t\}*\log(Y(-1))$$

4) As an additional illustration, Figure 3.18 presents the statistical results based on an acceptable reduced model of the model (3.23). Based on these results, the findings and notes presented are as follows:
   4.1 The reduced model in Figure 3.18 shows the interaction *T*log*(*Y*(−1)) has a positive significant adjusted effect on *log(Y)* with a *p*-value = 0.0322/2 = 0.0161.
   4.2 However, the forecast of *Y* based on this reduced model has the greatest RMSE = 42.48046 compared to both forecasts in Figure 3.17. So this reduced model presents the worst forecast, which also can be derived based on their forecasting graphs.

3.3 Forecasting Based on Subsamples

By observing the graph of *FPSCOM* in Figure 3.1 with a break point, I would experiment to forecast by using a subsample after the break point, the subsample selected is $\{t > 131\}$ or {@Year>1990}. Figure 3.19 presents the scatter graph *FPSCOM* on the time *T* with its regression line and *NNF* (*Nearest Neighbor Fit*).

I am very confident that all previous models can be applied for this subsample data set and the data analysis can be done easily. Hence, the following examples will present alternative lower and upper bound dynamic models, since any time series would never increase without an upper bound. However, by doing experiment with the lower and upper bound models, I found the

Dependent Variable: LOG(Y)
Method: Least Squares
Date: 07/09/14 Time: 20:33
Sample (adjusted): 1980M02 1996M04
Included observations: 195 after adjustments

Variable	Coefficient	Std. Error	t-Statistic	Prob.
C	0.309698	0.120898	2.561644	0.0112
LOG(Y(-1))	0.934520	0.025843	36.16109	0.0000
T	0.001045	0.000700	1.493482	0.1370
T*LOG(Y(-1))	-7.06E-05	0.000103	-0.685471	0.4939

R-squared	0.995854	Mean dependent var	5.558859
Adjusted R-squared	0.995789	S.D. dependent var	0.510774
S.E. of regression	0.033146	Akaike info criterion	-3.955499
Sum squared resid	0.209842	Schwarz criterion	-3.888361
Log likelihood	389.6612	Hannan-Quinn criter.	-3.928316
F-statistic	15292.40	Durbin-Watson stat	1.368033
Prob(F-statistic)	0.000000		

Forecast: YF_323
Actual: Y
Forecast sample: 1980M01 1996M04 I...
Included observations: 52

Root Mean Squared Error	33.08001
Mean Absolute Error	24.77772
Mean Abs. Percent Error	5.088015
Theil Inequality Coefficient	0.033668
Bias Proportion	0.152214
Variance Proportion	0.041066
Covariance Proportion	0.806720

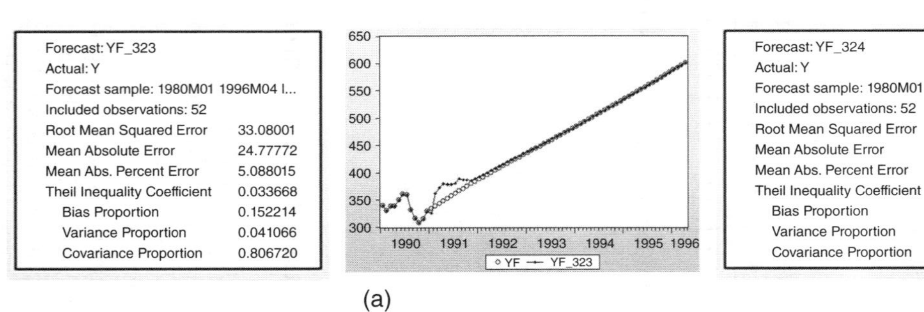

(a)

Dependent Variable: LOG(Y)
Method: Least Squares
Date: 07/09/14 Time: 20:54
Sample (adjusted): 1980M02 1996M04
Included observations: 195 after adjustments

Variable	Coefficient	Std. Error	t-Statistic	Prob.
C	0.657193	0.325631	2.018216	0.0450
LOG(Y(-1))	0.857262	0.070162	12.21831	0.0000
LOG(T)	-0.097527	0.054888	-1.776842	0.0772
LOG(T)*LOG(Y(-1))	0.023186	0.012192	1.901698	0.0587

R-squared	0.995808	Mean dependent var	5.558859
Adjusted R-squared	0.995742	S.D. dependent var	0.510774
S.E. of regression	0.033329	Akaike info criterion	-3.944456
Sum squared resid	0.212172	Schwarz criterion	-3.877317
Log likelihood	388.5844	Hannan-Quinn criter.	-3.917272
F-statistic	15123.75	Durbin-Watson stat	1.383164
Prob(F-statistic)	0.000000		

Forecast: YF_324
Actual: Y
Forecast sample: 1980M01 1996M04 I...
Included observations: 52

Root Mean Squared Error	39.41009
Mean Absolute Error	28.58314
Mean Abs. Percent Error	5.988613
Theil Inequality Coefficient	0.039547
Bias Proportion	0.404516
Variance Proportion	0.039724
Covariance Proportion	0.555760

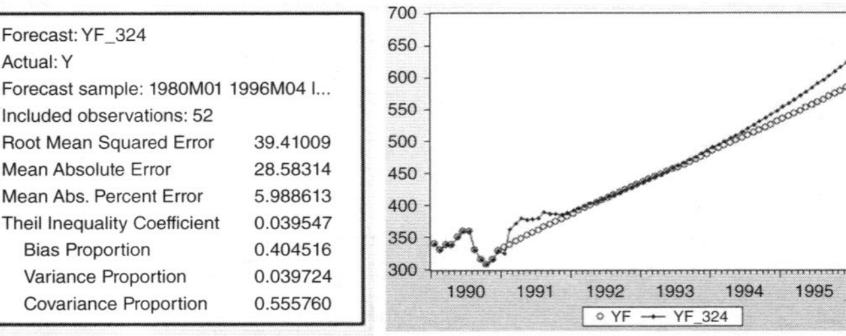

(b)

Figure 3.17 Statistical results of the translog interaction model in (3.23) and (3.24).

Dependent Variable: LOG(Y)
Method: Least Squares
Date: 11/04/16 Time: 20:52
Sample (adjusted): 1980M02 1996M04
Included observations: 195 after adjustments

Variable	Coefficient	Std. Error	t-Statistic	Prob.
C	0.258644	0.116336	2.223257	0.0274
LOG(Y(-1))	0.947288	0.024466	38.71795	0.0000
T*LOG(Y(-1))	7.44E-05	3.45E-05	2.157928	0.0322

R-squared	0.995806	Mean dependent var	5.558859
Adjusted R-squared	0.995762	S.D. dependent var	0.510774
S.E. of regression	0.033252	Akaike info criterion	-3.954146
Sum squared resid	0.212292	Schwarz criterion	-3.903792
Log likelihood	388.5292	Hannan-Quinn criter.	-3.933758
F-statistic	22791.42	Durbin-Watson stat	1.384516
Prob(F-statistic)	0.000000		

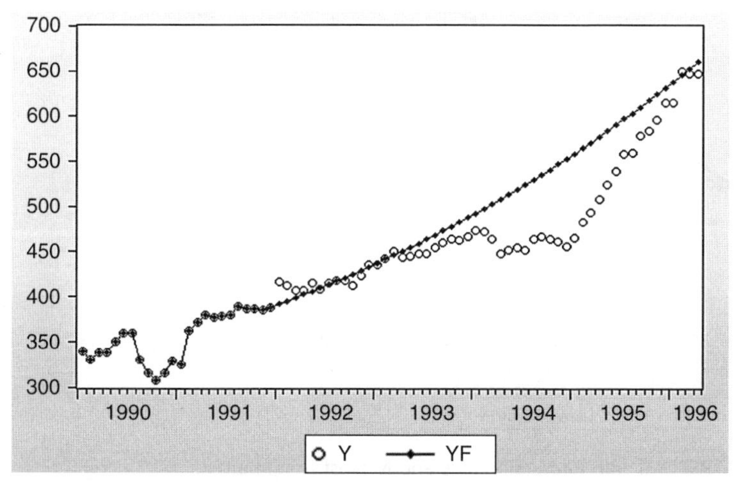

With RMSE = 42.48046 and TIC = 0.042471

Figure 3.18 Statistical results of a reduced model of the model in (3.23).

forecast values were decreasing after time points $t = 193$. To support those findings, I present the second graph in Figure 3.19, which shows *FSPCOM* has a maximum score for $t = 193$. So, I have been doing experiments using alternative models with a maximum at several time-points around $t = 193$.

3.3.1 Lag Variable Models With Lower and Upper Bounds

It has been well known that the lower and upper bounds for a variable have to be defined based on expert judgment and they can be very subjective, since they are highly dependent on the firm or region measured or observed. However, in practice, we can subjectively select two numbers outside of the interval of the observed scores.

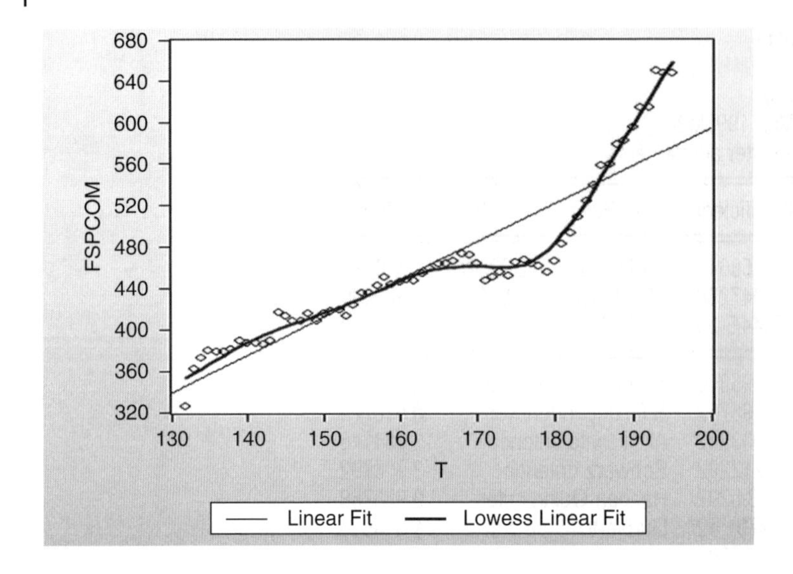

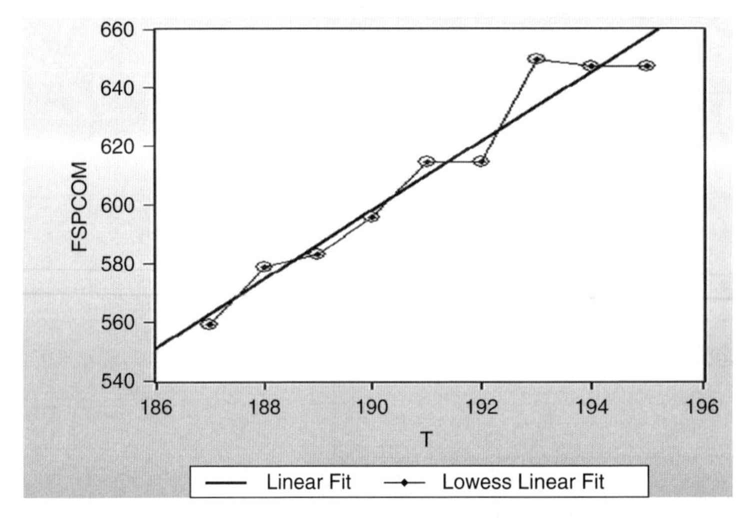

Figure 3.19 Scatter graph of (T,FPSCOM) with Regression Line and Nearest Neighbor Fit.

3.3.1.1 LV(p) Models *Y* with Upper and Lower Bounds

The equation of a *lag-variable*: LV(*p*) model of *Y* = *FSPCOM* with upper and lower bound can be represented using the general equation specification as follows:

$$LNYul\ C\ Y(-1) \ldots Y(-p) \tag{3.25}$$

where *LNYul* = *log((Y−L)/(U−Y))*, with *U* and *L* are selected fixed upper and lower bounds of the dependent variable *Y*.

Example 3.13 The Simplest Model of (3.25)

As an illustration, Figure 3.20 presents the statistical results of a specific selected model (3.25), namely LV(1), with the following equation specification using a subsample {@*Year*>1990} with

Dependent Variable: LNY
Method: Least Squares
Date: 11/11/16 Time: 21:29
Sample: 1980M01 1996M04 IF @YEAR>1990
Included observations: 64

Variable	Coefficient	Std. Error	t-Statistic	Prob.
C	-5.936380	0.117104	-50.69302	0.0000
Y(-1)	0.012091	0.000254	47.66127	0.0000

R-squared	0.973432	Mean dependent var	-0.423116
Adjusted R-squared	0.973003	S.D. dependent var	0.887861
S.E. of regression	0.145882	Akaike info criterion	-0.981282
Sum squared resid	1.319462	Schwarz criterion	-0.913817
Log likelihood	33.40102	Hannan-Quinn criter.	-0.954704
F-statistic	2271.596	Durbin-Watson stat	2.019951
Prob(F-statistic)	0.000000		

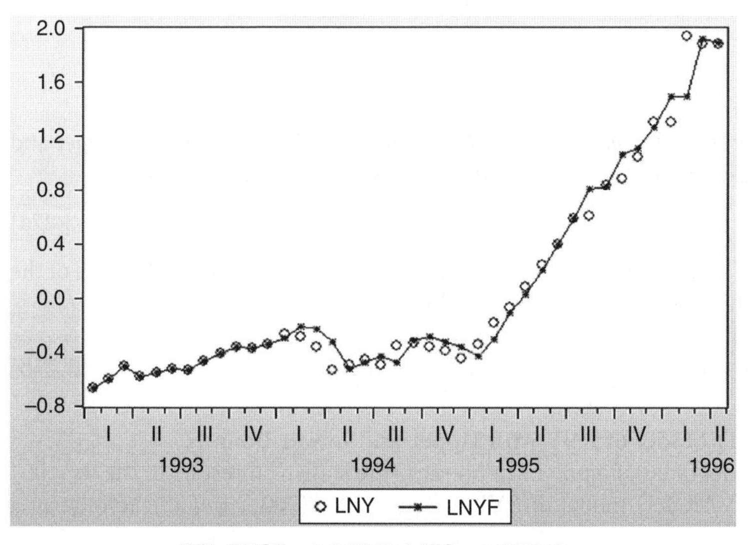

Figure 3.20 Statistical results of the LV(1)_ Model in (3.26).

its forecast sample {@*Year*>1993}. Based on these results, the following findings and notes are presented:

$$LNY\ C\ Y(-1) \tag{3.26}$$

where $LNY = log((Y\text{-}300)/(700\text{-}Y))$.

1) Compared to the forecasts presented previously, this forecast has the smallest TIC = 0.075219 < 0.10. So the forecast can be considered to be a good one – refer to the classification proposed in Chapter 1.
2) Note that the predicted scores beyond the sample period, namely for $t > 196$, tend to increase, as presented on the right side, which can be computed based on the following equation, starting with t = 197, using Excel.

	T	Y
1996M01	193	614.42
1996M02	194	649.542
1996M03	195	647.07
1996M04	196	647.17
	197	***647.429***
	198	***647.572***
	199	***647.651***
	200	***647.694***

$$Y(t) = (300 + 700 * Exp(*))/(1 + Exp(*))$$

where

$$\text{Exp}(*) = \text{Exp}(-5.93637985545 + 0.0120907695556 * \text{Y}(\text{t}-1))$$

For t = 197, we obtain $Y(197)$ = 647.429 by using the observed score $Y(t-1)$ = 647.17, then we can compute $Y(198)$ using this equation for $Y(t-1) = Y(197)$ obtained at the first step. Similarly, we can compute $Y(199)$ using $Y(t-1) = Y(198)$, and $Y(200)$, by using $Y(t-1) = (199)$.
3) Note that the forecast scores are increasing after t = 196, since we are using a model with an upper bound. Alternative upper bounds, as it is expected to increase up to 5 or 10% in the long run, could be applied. However, note that the forecasting should be done again whenever a new observed score is available.

3.3.1.2 LVAR(p,q) Model of *LNY*

The equation of a *lag-variable-autoregressive:* LVAR(p,q) model of $Y = FSPCOM$ with upper and lower bound can be represented using the following general equation specification:

$$LNYul\ C\ Y(-1)...Y(-p)\ AR(1)...AR(q) \tag{3.27a}$$

where $LNYul = log((Y-L)/(U-Y))$, with U and L are selected fixed upper and lower bounds of the dependent variable Y, with the model equation as follows:

$$LNYul_t = \beta_o + \sum_{i=1}^{p} \beta_i \times Y(-i) + \sum_{j=1}^{q} \rho_j \times \mu_{t-j} + \varepsilon_t \tag{3.27b}$$

Example 3.14 LVAR(1,1) Model of LNY with Upper and Lower Bounds

As an extension of the model (3.26), Figure 3.21a presents the statistical results of the simplest model in (3.28), namely LVAR(1,1) model of $LNY = log((Y-300)/(700-Y))$ with the following equation specifications.

$$LNY\ C\ Y(-1)\ AR(1) \tag{3.28}$$

As a comparative study, I present the statistical results of an alternative LVAR(1,1) model with the following equation specification, in Figure 3.21b. Based on these results, it is very easy to see that the model (3.28) presents a better forecast than the model (3.29).

$$LNY\ C\ LNY(-1)\ AR(1) \tag{3.29}$$

3.3.1.3 LVAMA(*p,r*) Model of LNY

The equation of a *lag-variable-moving-average*: LVAMA(p,r) model of $Y = FSPCOM$ with upper and lower bound can be represented using the following general equation specification:

$$LNYul\ C\ Y(-1)...Y(-p)\ MA(1)...MA(r) \tag{3.30a}$$

Dependent Variable: LNY
Method: Least Squares
Date: 11/10/16 Time: 06:16
Sample: 1980M01 1996M04 IF @YEAR>1990
Included observations: 63
Convergence achieved after 4 iterations

Variable	Coefficient	Std. Error	t-Statistic	Prob.
C	-5.826629	0.076466	-76.19861	0.0000
Y(-1)	0.011872	0.000165	71.92600	0.0000
AR(1)	-0.199801	0.095466	-2.092900	0.0406

R-squared	0.984140	Mean dependent var	-0.387176
Adjusted R-squared	0.983611	S.D. dependent var	0.846765
S.E. of regression	0.108402	Akaike info criterion	-1.559502
Sum squared resid	0.705053	Schwarz criterion	-1.457448
Log likelihood	52.12430	Hannan-Quinn criter.	-1.519363
F-statistic	1861.547	Durbin-Watson stat	1.747463
Prob(F-statistic)	0.000000		

Inverted AR Roots	-.20

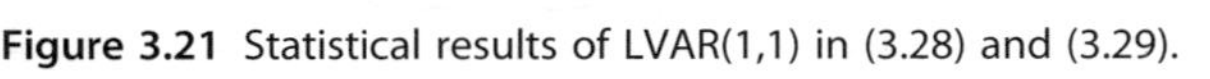

With RMSE = 127135 & TIC = 0.076453

Dependent Variable: LNY
Method: Least Squares
Date: 11/10/16 Time: 06:19
Sample: 1980M01 1996M04 IF @YEAR>1990
Included observations: 62
Convergence achieved after 3 iterations

Variable	Coefficient	Std. Error	t-Statistic	Prob.
C	0.071163	0.017869	3.982535	0.0002
LNY(-1)	1.033562	0.019752	52.32776	0.0000
AR(1)	0.036458	0.088129	0.413686	0.6806

R-squared	0.980342	Mean dependent var	-0.366147
Adjusted R-squared	0.979676	S.D. dependent var	0.836929
S.E. of regression	0.119315	Akaike info criterion	-1.366918
Sum squared resid	0.839931	Schwarz criterion	-1.263992
Log likelihood	45.37446	Hannan-Quinn criter.	-1.326507
F-statistic	1471.170	Durbin-Watson stat	2.280342
Prob(F-statistic)	0.000000		

Inverted AR Roots	.04

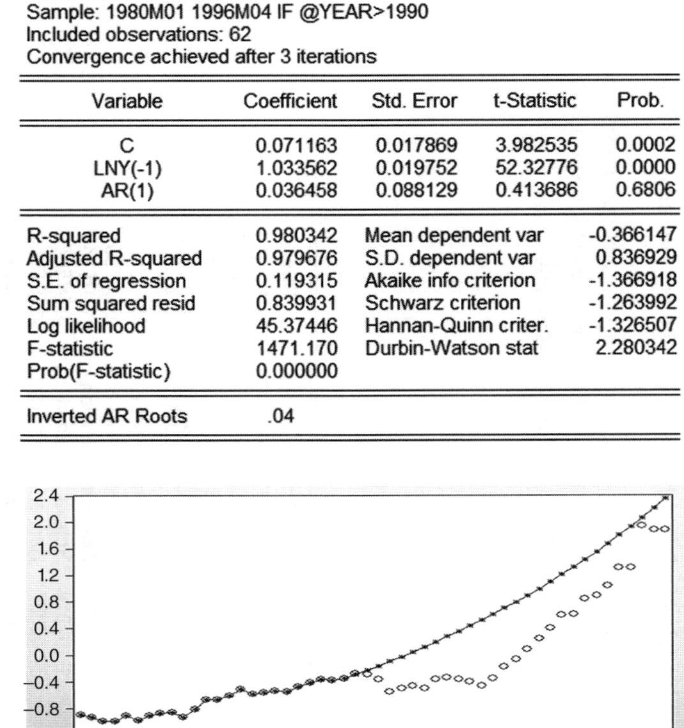

With RMSE = 0.630577 & TIC = 0.315370

Figure 3.21 Statistical results of LVAR(1,1) in (3.28) and (3.29).

where *LNYul* = *log*((*Y*–*L*)/(*U*–*Y*)), with *U* and *L* are selected fixed upper and lower bounds of the dependent variable *Y*, with the model equation as follows:

$$LNYul_t = \beta_o + \sum_{i=1}^{p} \beta_i \times Y(-i) + \sum_{j=1}^{r} \theta_j \times \nu_{t-j} + \varepsilon_t \tag{3.30b}$$

Example 3.15 LVAMA(1,1) Model with Upper and Lower Bounds and Alternatives

1) As an extension of the model (3.26), Figure 3.22 presents the statistical results of the simplest model in (3.30), namely LVAMA(1,1), with the following equation specification. Based on these results, the following findings and notes are presented.

Dependent Variable: LNY
Method: Least Squares
Date: 11/06/16 Time: 14:01
Sample: 1980M01 1996M04 IF @YEAR>1990
Included observations: 64
Convergence achieved after 9 iterations
MA Backcast: 1979M12

Variable	Coefficient	Std. Error	t-Statistic	Prob.
C	-5.887128	0.065666	-89.65215	0.0000
Y(-1)	0.011993	0.000143	84.08787	0.0000
MA(1)	-0.451756	0.105574	-4.279055	0.0001

R-squared	0.976596	Mean dependent var	-0.423116
Adjusted R-squared	0.975828	S.D. dependent var	0.887861
S.E. of regression	0.138038	Akaike info criterion	-1.076840
Sum squared resid	1.162318	Schwarz criterion	-0.975642
Log likelihood	37.45887	Hannan-Quinn criter.	-1.036973
F-statistic	1272.684	Durbin-Watson stat	1.202727
Prob(F-statistic)	0.000000		

Inverted MA Roots	.45

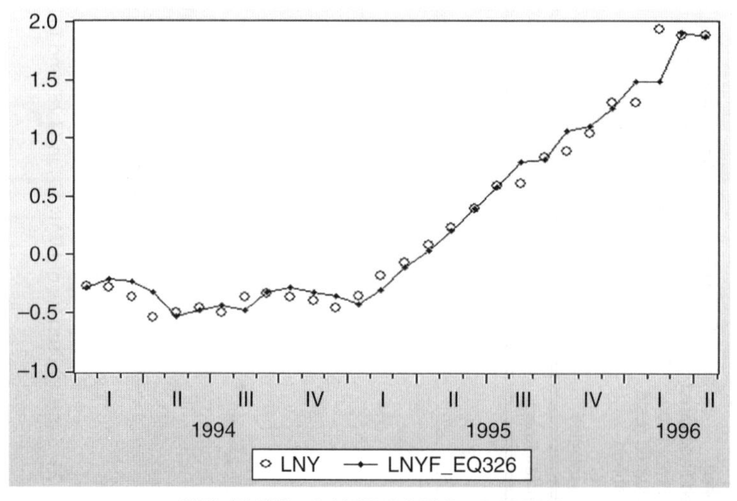

With RMSE = 0.126374 & TIC = 0.075632

Figure 3.22 Statistical results of the LVAMA(1,1)_ Model (3.31).

$$LNY\ C\ Y(-1)\ MA(1) \tag{3.31}$$

2) Even though this model has a slightly greater RMSE and TIC than the LV(1)_model (3.25), I would say this model presents a very good forecast, since its TIC < 0.10 – refer to the classification defined in Chapter 1.
3) By doing further experiments, I have found two good forecast models with TIC < 0.10, namely LVAMA(1,3) and LVAMA(2,2), with the following equation specifications. Based on the statistical results in Figure 3.23, findings and notes presented are as follows:

$$LNY\ C\ Y(-1)\ MA(1)\ MA(2)\ MA(3) \tag{3.32}$$

and

$$LNY\ C\ Y(-1)\ Y(-2)\ MA(1)\ MA(2) \tag{3.33}$$

	T	LNY	LNYf_331	LNYF_332	LNYF_333
1995M10	190	0.88229	1.05425	1.04892	0.95105
1995M11	191	1.03987	1.10400	1.09849	1.15557
1995M12	192	1.30351	1.25526	1.2492	1.2272
1996M01	193	1.30128	1.48361	1.47672	1.39814
1996M02	194	**1.93548**	1.48181	1.47493	1.58611
1996M03	195	1.88056	**1.90304**	**1.89464**	1.66983
1996M04	195	1.88273	1.87339	1.86510	**2.01124**

3.1 Since each of the independent variables has adjusted significant effect on *LNY*, then both models are acceptable models, in a statistical sense.

3.2 In addition, both models present the forecasts with their TIC < 0.10. By referring to the TIC interval of classifications presented in Example 1.8, then both models can be considered to be very good forecasting models. However, by observing the maximum values of *LNY*, *LNYF*_331, *LNYF*_332, and *LNYF*_333 in the last five months on the right, I would say the model (3.31) is the worst forecast model among the three, since the observed scores tend to decrease from $t = 194$.

3.3.1.4 LVARMA(*p,q,r*) Model of *LNY*

As a combination of the models LVAMA(*p,r*) in (3.27) and LVAR(*p,q*) in (3.31), I define a *lag-variable-autoregressive-moving-average*: LVARMA(*p,q,r*) model, with the following equation specification. The integers $p > 0$, $q \geq 0$, and $r \geq 0$ should be selected using the trial-and-error-method to obtain a good fit model, since they are highly dependent on the data used.

$$LNYul\ C\ Y(-1)...Y(-p)\ AR(1)...AR(q)\ MA(1)...MA(r) \tag{3.34a}$$

where $LNYul = log\,((Y-L)/(U-Y))$, with U and L are selected fixed upper and lower bounds of the dependent variable Y, with the model equation as follows:

$$LNYul_t = \beta_o + \sum_{i=1}^{p} \beta_i \times Y(-i) + \sum_{j=1}^{q} \rho_j \times \mu_{t-j} + \sum_{k=1}^{r} \theta_k \times \nu_{t-k} + \varepsilon_t \tag{3.34b}$$

For $q = 0$ or $r = 0$, we would have LVARMA(*p*,0,0) = LV(*p*), LVARMA(*p,q*,0) = LVAR(*p,q*), and LVARMA(*p*,0,*r*) = LVAMA(*p,r*). Furthermore, the basic assumptions of the error terms or the residuals have identical independent normal distributions with zero means, and contain variances that could be taken for granted; they should not and cannot be tested – refer to special notes presented in Chapter 1 and Agung (2009a, Section 2.14.3)

Example 3.16 LVARMA(1,1,1) Model of *LNY*

As an illustration, Figure 3.24a presents the statistical results of the simplest model in (3.34), namely LVARMA(1,1,1) model with the following equation specifications. Based on these results, the following findings and notes are presented.

$$LNY\,C\,Y(-1)\,AR(1)\,MA(1) \tag{3.35a}$$

Dependent Variable: LNY
Method: Least Squares
Date: 11/06/16 Time: 13:48
Sample: 1980M01 1996M04 IF @YEAR>1990
Included observations: 64
Convergence achieved after 15 iterations
MA Backcast: 1979M10 1979M12

Variable	Coefficient	Std. Error	t-Statistic	Prob.
C	-5.867355	0.061339	-95.65441	0.0000
Y(-1)	0.011950	0.000134	89.30723	0.0000
MA(1)	-0.529507	0.104880	-5.048682	0.0000
MA(2)	0.302438	0.125716	2.405725	0.0193
MA(3)	-0.299598	0.120026	-2.496103	0.0154

R-squared	0.977499	Mean dependent var	-0.423116
Adjusted R-squared	0.975973	S.D. dependent var	0.887861
S.E. of regression	0.137624	Akaike info criterion	-1.053676
Sum squared resid	1.117484	Schwarz criterion	-0.885014
Log likelihood	38.71764	Hannan-Quinn criter.	-0.987232
F-statistic	640.7635	Durbin-Watson stat	1.236907
Prob(F-statistic)	0.000000		

Inverted MA Roots	.70	-.09-.65i	-.09+.65i

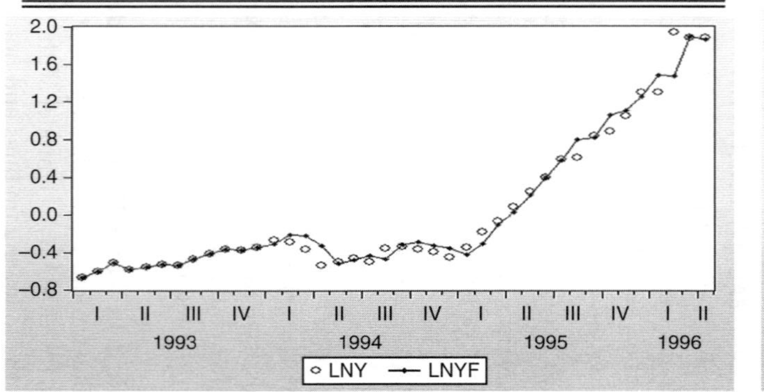

With RMSE=0.126962 & TIC=0.076135

Dependent Variable: LNY
Method: Least Squares
Date: 11/10/16 Time: 06:58
Sample: 1980M01 1996M04 IF @YEAR>1990
Included observations: 64
Convergence achieved after 19 iterations
MA Backcast: 1979M11 1979M12

Variable	Coefficient	Std. Error	t-Statistic	Prob.
C	-5.983511	0.134830	-44.37820	0.0000
Y(-1)	0.002426	0.001124	2.157891	0.0350
Y(-2)	0.009892	0.001151	8.596461	0.0000
MA(1)	0.666767	0.125443	5.315282	0.0000
MA(2)	-0.309716	0.123363	-2.510609	0.0148

R-squared	0.983213	Mean dependent var	-0.423116
Adjusted R-squared	0.982074	S.D. dependent var	0.887861
S.E. of regression	0.118873	Akaike info criterion	-1.346622
Sum squared resid	0.833713	Schwarz criterion	-1.177960
Log likelihood	48.09191	Hannan-Quinn criter.	-1.280178
F-statistic	863.8808	Durbin-Watson stat	2.118384
Prob(F-statistic)	0.000000		

Inverted MA Roots	.32	-.98

2.5
2.0
1.5
1.0
0.5
0.0
−0.5
−1.0
I II III IV I II III IV I II III IV I II
1993
1994
1995
1996
LNY
LNYF

With RMSE=0.146348 & TIC =0.088768

Figure 3.23 Statistical results of LVAMA in (3.32) and (3.33).

Dependent Variable: LNY
Method: Least Squares
Date: 11/10/16 Time: 16:51
Sample: 1980M01 1996M04 IF @YEAR>1990
Included observations: 63
Convergence achieved after 8 iterations
MA Backcast: 1979M12

Variable	Coefficient	Std. Error	t-Statistic	Prob.
C	-5.830807	0.084362	-69.11641	0.0000
Y(-1)	0.011879	0.000182	65.26969	0.0000
AR(1)	-0.299549	0.129860	-2.306714	0.0246
MA(1)	0.215764	0.182442	1.182640	0.2417

R-squared	0.984542	Mean dependent var	-0.387176
Adjusted R-squared	0.983756	S.D. dependent var	0.846765
S.E. of regression	0.107921	Akaike info criterion	-1.553446
Sum squared resid	0.687171	Schwarz criterion	-1.417374
Log likelihood	52.93354	Hannan-Quinn criter.	-1.499928
F-statistic	1252.616	Durbin-Watson stat	2.004305
Prob(F-statistic)	0.000000		

Inverted AR Roots	-.30
Inverted MA Roots	-.22

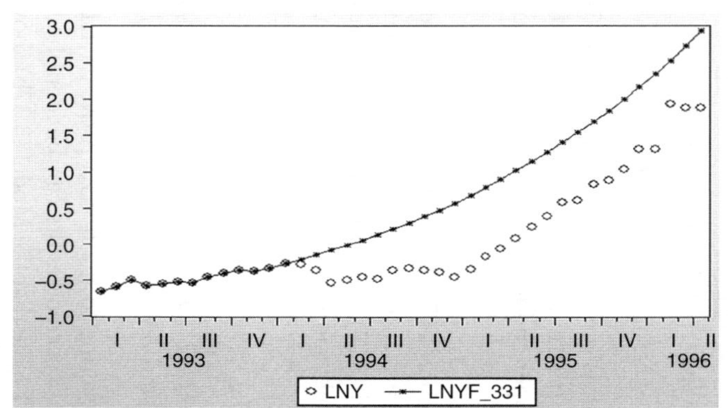

With RMSE = 0.127069 & TIC = 0.076395

Dependent Variable: LNY
Method: Least Squares
Date: 11/10/16 Time: 16:53
Sample: 1980M01 1996M04 IF @YEAR>1990
Included observations: 62
Convergence achieved after 7 iterations
MA Backcast: 1979M12

Variable	Coefficient	Std. Error	t-Statistic	Prob.
C	0.074104	0.013919	5.323904	0.0000
LNY(-1)	1.047489	0.016923	61.89786	0.0000
AR(1)	0.200485	0.110638	1.812080	0.0752
MA(1)	-0.355010	0.171733	-2.067226	0.0432

R-squared	0.981562	Mean dependent var	-0.366147
Adjusted R-squared	0.980608	S.D. dependent var	0.836929
S.E. of regression	0.116546	Akaike info criterion	-1.398723
Sum squared resid	0.787810	Schwarz criterion	-1.261488
Log likelihood	47.36041	Hannan-Quinn criter.	-1.344841
F-statistic	1029.223	Durbin-Watson stat	2.005132
Prob(F-statistic)	0.000000		

Inverted AR Roots	.20
Inverted MA Roots	.36

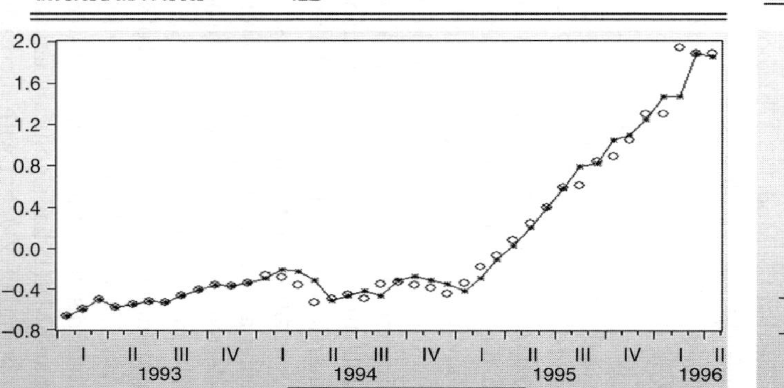

With RMSE = 0.797038 & TIC = 0.356902

Figure 3.24 Statistical results of LVARMA(1,1,1) in (3.35) and (3.36).

1) At the 15% level of significance, MA(1) has a positive significant effect based on the t-statistic of $t_0 = 1.18264$ with p-value $= 0.2417/2 = 0.12085 < 0.15$.
2) In addition, its forecast is a good forecast, since its TIC = 0.076395 < 0.10, and the forecast values are relatively very close to the observed scores. Hence the model (3.35) is an acceptable forecast model in the statistical sense.
3) As a comparative study, in Figure 3.24b, I present the statistical results of an alternative model of LVARMA(1,1,1) with the following equation specification.

$$LNY\ C\ LNY(-1)\ AR(1)\ MA(1) \tag{3.36}$$

   It is very clear that the model (3.35) presents a much better forecast than the model (3.36).
4) Furthermore, I have found unexpected statistical results were obtained by using different upper and lower bounds for the model (3.35). See the following example.

Example 3.17 LVARMA(1,1,1) Models with Alternative Upper and Lower Bounds
By doing an experiment using alternative upper and lower bounds, as a modification of the model (3.35), I have found various unexpected statistical results. As an illustration, Table 3.1 presents summary the results of four alternative models of *LNYk*, for k = 1, 2, 3, and 4. Based in this summary, the following findings and notes are presented.

$$LNYk\ C\ Y(-1)\ AR(1)\ MA(1) \tag{3.37}$$

with *LNY1 = log((Y−300)/(800−Y)), LNY2 = log((Y−300)/(1000−Y)), LNY3 = log((Y−200)/(800 −Y)),* and *LNY4 = log((Y−300)/(800−Y)),* which are presented as the models (3.37a)–(3.37d).

1) Revering to the five LVARMA(1,1,1) models in (3.32), (3.37a) up to (3.37d), the following unexpected results can be identified.
   1.1 Note that the Model (3.37b) shows AR(1) has insignificant effect with the largest p-value of 0.614, the smallest Adj. R-squared, and the second largest RMSE compared to the models (3.37a), (3.37c), and (3.37d).
   1.2 On the other hand, the Model (3.37b) presents the forecast with the smallest TIC = 0.053008, compared to the other four models with TICs of 0.076, 0.077, 0.071, and 0.070, respectively. Could we conclude that the model (3.37b) is the best forecast model among the five models? Since the five models have TICs within the interval (0.05,0.10), then the five models could be considered to be good forecast models – refer to the classification of TICs presented in previous chapter. However, we would never know which one will give the best forecast values beyond the sample period.
   1.3 On the other hand, note that the five models are five distinct models, since they have different dependent variables, namely *LNY, LNY1, LNY2, LNY3,* and *LNY4.* So their forecasts are not comparable, but they are acceptable models in the statistical sense, even though AR(1) in the model (3.37b) has a large p-value > 0.50.
   1.4 As additional illustrative forecast models, many acceptable models can be obtained by using alternative upper and lower bounds of Y = *FSPOM*, which can be subjectively selected, or by using a percentage of changes or increases from the observed scores interval. Do these for some exercises.
2) It is also unexpected, at the 5% level of significance, the statistical results of model (3.35) show that AR(1) has significant adjusted effect and MA(1) has insignificant adjusted effect, but the statistical results of the two models (3.37a) and (3.37b) show AR(1) has insignificant adjusted effect and MA(1) has significant adjusted effect. Note that the three models using the same lower bound of 300, but with increasing upper bounds of 700, 800, and 1000, respectively.

Table 3.1 Summary of the statistical results of the four models in (3.37).

Dep. Variable	LNY1; Model (3.37a)			LNY2; Model(3.37b)			LNY3; Model(3.37c)			LNY4; Model(3.37d)		
Variable	Coef.	*t*-Stat.	Prob.	Coef.	*t*-Stat.	Prob.	Coef.	*t*-Stat.	Prob.	Coef.	*t*-Stat.	Prob.
C	−5.078	−54.637	0.000	−4.792	−41.743	0.000	−3.549	−67.310	0.000	−3.373	−71.657	0.000
Y(−1)	0.009	46.855	0.000	0.008	31.221	0.000	0.007	63.204	0.000	0.006	61.160	0.000
AR(1)	−0.106	−1.007	0.318	0.050	0.507	0.614	−0.547	−2.470	0.016	−0.486	−2.454	0.017
MA(1)	0.280	1.772	0.082	0.354	2.477	0.016	0.569	2.324	0.024	0.543	2.462	0.017
R-squared	0.981			0.974			0.986			0.985		
Adj. R-squared	0.980			0.973			0.985			0.985		
F-statistic	1026.2			734.4			1345.5			1326.2		
Prob (*F*-statistic)	0.000			0.000			0.000			0.000		
DW-Stat	1.911			1.838			2.070			2.049		
RMSE		0.098562			0.100682			0.070078			0.06138	
TIC		0.076671			0.053008			0.070845			0.07004	

Other unexpected statistical results also would be obtained by using alternative upper bounds U > 700. Do these for some exercises.

3) On the other hand, based on the two models (3.37c) and (3.37d), each of AR(1) and MA(1) has significant adjusted effects, by using the same lower bound of 200, and upper bounds of 800, and 900, respectively. Do this as an exercise for alternative lower bound <300 and upper bounds >700.

3.3.1.5 LVARMA(*p,q,r*) Model of *LNYul* with *TIME* Predictor

Referring to the model of $Y = FSPCOM$ in (3.1) with predictors $Y(-1)$, T, and $T*Y(-1)$, based on each of the four models (3.25), (3.27), (3.30), and (3.34), we can have a more advanced models by using the variables T and $T*Y(-1)$ as two additional predictors. The most general model could be presented as LVARMA(*p,q,r*) model with the following equation specification.

$$LNYul\ C\ Y(-1)...Y(-p)\ T\ T*Y(-1)...T*Y(-p)$$
$$AR(1)...AR(q)\ MA(1)...MA(r) \tag{3.37a}$$

where $LNYul = log((Y-L)/(U-Y))$, with U and L are selected fixed upper and lower bounds of the dependent variable Y, with the model equation as follows:

$$LNYul_t = \beta_o + \sum_{i=1}^{p} \beta_i \times Y(-i) + \sum_{i=0}^{p} \delta_i \times T \times Y(-i) + \sum_{j=1}^{q} \rho_j \times \mu_{t-j} + \sum_{k=1}^{r} \theta_k \times \nu_{t-k} + \varepsilon_t \tag{3.37b}$$

with $Y(0) = 1$, for alternative integers $p > 0$, $q \geq 0$, and $r \geq 0$ that should be selected using the trial-and-error-method to obtain a good fit model, since they are highly dependent on the data used. Furthermore, for $q = 0$ or $r = 0$, we would have LVARMA(p,0,0) = LV(p), LVARMA(p,-q,0) = LVAR(p,q), and LVARMA(p,0,r) = LVAMA(p,r).

Example 3.18 An Application of LVAR(1,1,1) Model of *LNYul* with the Time Predictor

As an illustration, Figure 3.25 presents the statistical results of a simple LVARMA(1,1,1)-Model of $LNYul = log((Y-300)/(1000-Y))$, $Y = FSPCOM$, based on the subsample {@*Year*>1990} with the following equation. Based on these results, the following findings and notes are presented.

$$LNYul\ C\ Y(-1)\, t\, t*Y(-1)\ AR(1)\ MA(1) \tag{3.38}$$

1) Compared to the LVARMA(1,1,1)-Model (3.37b), this model has exactly the same upper and lower bounds: U = 1000 and L = 300, or *LNYul* = *LNY2*. But this model is a better forecast model, since each of its independent variables, including AR(1) and MA(1), has a significant adjusted effect on *LNYul* = *LNY2*, and it has smaller RMSE and TIC.
2) Furthermore, compared to the five models in (3.35), (3.37a) up to (3.37d) without the time predictor, which are classified as good forecast models, this model is classified as a very good forecast model with TIC = 0.038477 < 0.05.
3) As an additional comparative study, Figure 3.26 presents the statistical results of the model, $LNYul1 = log((Y-200)/(800-Y)) = LNY3$ in Table 3.1, with the following ES. Based on these results, findings, and notes presented are as follows:

$$LNYull\ C\ Y(-1)\ t\, t*Y(-1)\ AR(1)\ MA(1) \tag{3.39}$$

Dependent Variable: LNYUL
Method: Least Squares
Date: 11/14/16 Time: 20:03
Sample: 1980M01 1996M04 IF @YEAR>1990
Included observations: 64
Convergence achieved after 14 iterations
MA Backcast: 1979M12

Variable	Coefficient	Std. Error	t-Statistic	Prob.
C	-9.059368	0.646317	-14.01691	0.0000
Y(-1)	0.016037	0.001871	8.571460	0.0000
T	0.026086	0.003249	8.029337	0.0000
T*Y(-1)	-5.04E-05	9.04E-06	-5.571582	0.0000
AR(1)	-0.349392	0.032883	-10.62540	0.0000
MA(1)	0.427478	0.128085	3.337447	0.0015

R-squared	0.987485	Mean dependent var	-1.301887
Adjusted R-squared	0.986406	S.D. dependent var	0.615330
S.E. of regression	0.071743	Akaike info criterion	-2.342385
Sum squared resid	0.298532	Schwarz criterion	-2.139990
Log likelihood	80.95633	Hannan-Quinn criter.	-2.262651
F-statistic	915.2810	Durbin-Watson stat	2.055502
Prob(F-statistic)	0.000000		

Inverted AR Roots	-.35
Inverted MA Roots	-.43

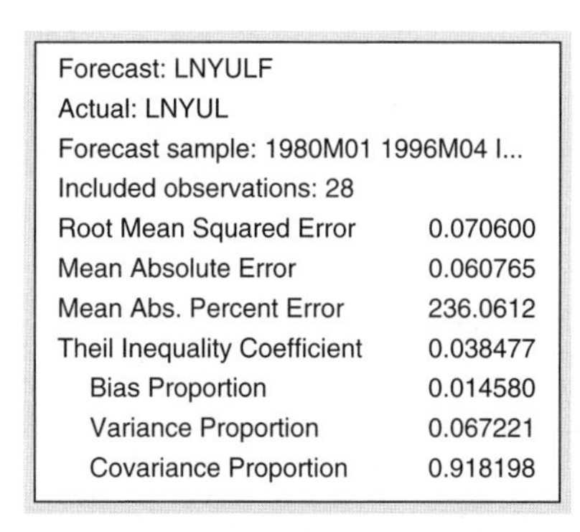

Forecast: LNYULF
Actual: LNYUL
Forecast sample: 1980M01 1996M04 I...
Included observations: 28

Root Mean Squared Error	0.070600
Mean Absolute Error	0.060765
Mean Abs. Percent Error	236.0612
Theil Inequality Coefficient	0.038477
Bias Proportion	0.014580
Variance Proportion	0.067221
Covariance Proportion	0.918198

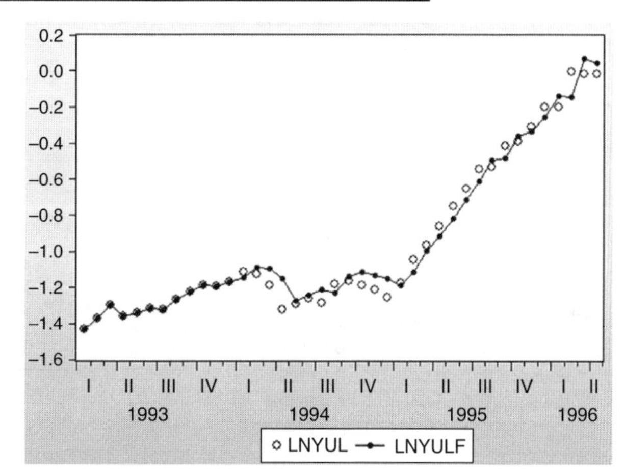

Figure 3.25 Statistical results of LVARMA(1,1,1) in (3.38).

3.1 This model has a slightly smaller RMSE = 0.067344 and TIC = 0.067076 compared to the model (3.37c) with RMSE = 0.070083 and TIC = 0.070845. Note that each of *Y(−1)* and MA(1) has insignificant effect, at the 10% level of significance, however, at the 15% level of significance each of the two variables has a positive significant effect with the p-values of 0.2515/2 = 0.12575 and 0.2450/2 = 0.1225.

3.2 Referring to these findings, one may argue about which one should be considered the better forecast model. I would choose the model (3.39) to be better, since the time t is one of the best predictors of any time series, and the results show each of t and $t*Y(-1)$ has significant effect, at the 1% level of significance.

Dependent Variable: LNYUL1
Method: Least Squares
Date: 11/17/16 Time: 15:39
Sample: 1980M01 1996M04 IF @YEAR>1990
Included observations: 63
Convergence achieved after 14 iterations
MA Backcast: 1979M12

Variable	Coefficient	Std. Error	t-Statistic	Prob.
C	-1.795926	0.601609	-2.985202	0.0042
Y(-1)	0.002010	0.001736	1.158421	0.2515
T	-0.007789	0.002906	-2.680087	0.0096
T*Y(-1)	2.49E-05	8.33E-06	2.984346	0.0042
AR(1)	-0.352898	0.230779	-1.529159	0.1318
MA(1)	0.302785	0.257750	1.174723	0.2450

R-squared	0.987731	Mean dependent var	-0.251357
Adjusted R-squared	0.986654	S.D. dependent var	0.514877
S.E. of regression	0.059480	Akaike info criterion	-2.715958
Sum squared resid	0.201659	Schwarz criterion	-2.511850
Log likelihood	91.55268	Hannan-Quinn criter.	-2.635681
F-statistic	917.7484	Durbin-Watson stat	1.998683
Prob(F-statistic)	0.000000		

Inverted AR Roots	-.35
Inverted MA Roots	-.30

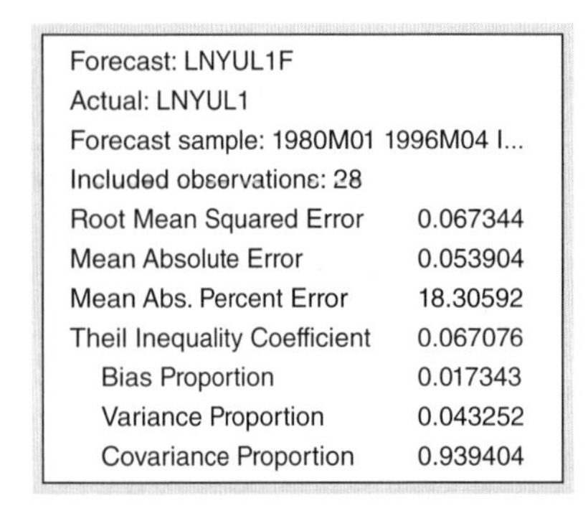

Forecast: LNYUL1F	
Actual: LNYUL1	
Forecast sample: 1980M01 1996M04 I...	
Included observations: 28	
Root Mean Squared Error	0.067344
Mean Absolute Error	0.053904
Mean Abs. Percent Error	18.30592
Theil Inequality Coefficient	0.067076
Bias Proportion	0.017343
Variance Proportion	0.043252
Covariance Proportion	0.939404

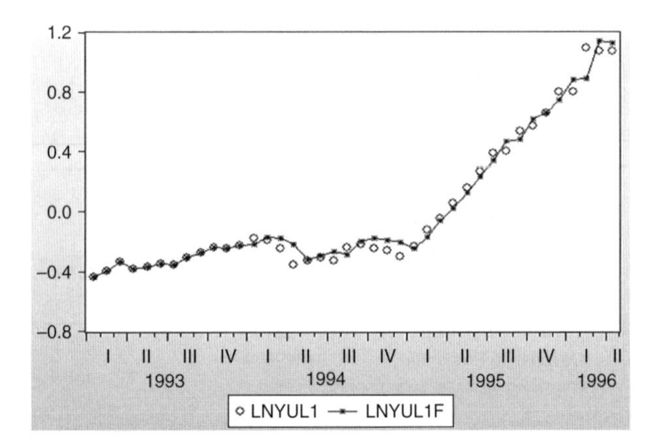

Figure 3.26 Statistical results of LVARMA(1,1,1) in (3.39).

3.4 Special LV(12) Models of *HS* with Upper and Lower Bounds

Referring to the special LV(12) model of *HS* in (1.7), as modification of the LVARMA(p,q,r) models of *LNYul* in (3.34) and (3.37), two groups of special models for $p = 12$ are presented in the following subsections.

3.4.1 Special LVARMA(12,q,r) Model of LNYul Without Time Predictor

Referring to the LVARMA(p,q,r) model in (3.34), special LVARMA(12,*q.r*) models of *LNYul* for various $q \ge 0$ and $r \ge 0$ can be presented using the following equation.

$$LNYul\ C\ Y(-12)\ AR(1)...AR(q)\ MA(1)...MA(r) \tag{3.40a}$$

with the general equation as follows:

$$LNYul_t = \beta_o + \beta_1 \times Y(-12) + \sum_{j=1}^{q} \rho_j \times \mu_{t-j} + \sum_{k=1}^{r} \theta_k \times \nu_{t-k} + \varepsilon_t \tag{3.40b}$$

For $q = 0$ or $r = 0$, we would have LVARMA(*12*,0,0) = LV(12), LVARMA(12,*q*,0) = LVAR(12,*q*), and LVARMA(12,0,*r*) = LVAMA(12,*r*).

Example 3.19 Application of Specific Models in (3.40)
Table 3.2 presents the statistical results summary of four specific models, based on the subsample {@*Year*>1990}, with the following equation specifications. Based on this summary, the following findings and comments are presented.

$$LNYul\ C\ Y(-12) \tag{3.41}$$

$$LNYul\ C\ Y(-12)\ AR(1)\ AR(2) \tag{3.42}$$

$$LNYul\ C\ Y(-12)\ MA(1)\ MA(2) \tag{3.43}$$

$$LNYul\ C\ Y(-12)\ AR(1)\ AR(2)\ MA(1)\ MA(2) \tag{3.44}$$

1) By using the trial-and-error method, I have obtained a set of four acceptable special LVARMA(12,*q*,*r*) models of *LNYul* = *log((y−300)/(1000−Y)* in (3.41) up to (3.44), for a comparative study.
2) Note that only the LVARMA(12,2,0) = LVAR(12,2) in (3.42) has a term AR(2) with a very large Prob. = 0.682. However, since it has the smallest values of RMSE and TIC among the four models, then it should be considered to be the best forecasting, even though they have small or very small differences of RMSE and TIC.
3) On the other hand, note that the LVARMA(12,2,2) in (3.44) has the largest R-squared = 0.962. Hence, this model is the best fit model of *LNYul* = *log((Y−300)/(1000−Y),* based on the subsample {@*Year*>1990}. However, its forecast *LNYULF_EQ344* presents unexpected large deviations from the observed values *LNYUL*, especially during the last two years, as presented in Figure 3.27.
4) Hence, based on this comparative study, it can be concluded that a good fit model would not directly be a good forecast model.

3.4.2 Special LVARMA(12,q,r) of LNYul With Time Predictor

Referring to the LVARMA(*p,q,r*) in (3.37), a special LVARMA(12,*q.r*) of *LNYul* with the time-*t* as predictor, can be presented using the following equation.

$$LNYul\ C\ Y(-12)\ T\ T{*}Y(-12) AR(1)...AR(q)\ MA(1)...MA(r) \tag{3.45a}$$

and the model has the general equation as follows:

Table 3.2 Statistical results summary of the four models (3.41) up to (3.44).

Dependent Variable: LNYUL

Method: Least Squares

Date: 11/20/16 Time: 15:30

Sample: 1980M01 1996M04 IF "YEAR>1990

Included observations: 64

	Model (3.41)			Model (3.42)			Model (3.43)			Model (3.44)		
Variable	**Coef.**	***t*-Stat.**	**Prob.**	**Coef.**	***t*-Stat.**	**Prob.**	**Coef.**	***t*-Stat.**	**Prob.**	**Coef.**	***t*-Stat.**	**Prob.**
C	−5.505	−16.636	0.000	−5.557	−11.056	0.000	−4.317	−10.126	0.000	1.078	1.163	0.250
Y(−12)	0.010	12.798	0.000	0.010	8.561	0.000	0.007	7.247	0.000	−0.004	−3.413	0.001
AR(1)				0.362	2.924	0.005				0.458	9.469	0.000
AR(2)				0.047	0.411	0.682				0.440	12.150	0.000
MA(1)							1.055	64.143	0.000	1.523	19.246	0.000
MA(2)							0.962	82.971	0.000	0.669	3.783	0.000
R-squared	0.725			0.777			0.937			0.962		
Adj. R-squared	0.721			0.765			0.934			0.958		
F-statistic	163.79			69.53			298.60			291.48		
Prob (*F*-statistic)	0.000			0.000			0.000			0.000		
DW Stat	0.338			0.952			1.637			3.005		
Inverted AR Roots				0.460	−0.100					0.93	−0.47	
Inverted MA Roots							−0.53 + 0.83i	−0.53 − 0.83i		−0.76 − 0.30i	−0.76 + 0.30i	
RMSE		0.373548			0.372637			0.438265			0.38813	
TIC		0.212682			0.211031			0.227972			0.21569	

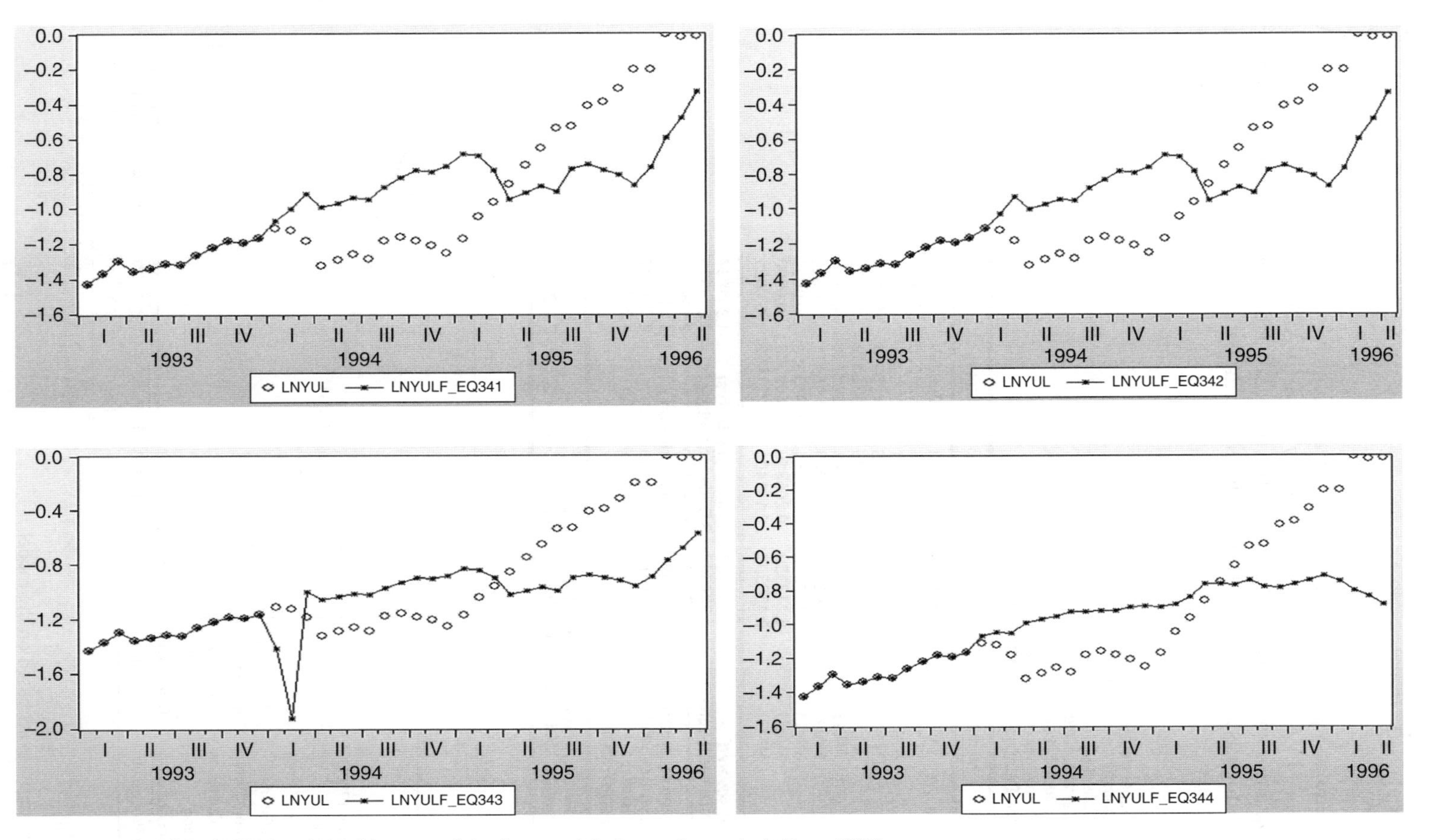

Figure 3.27 Graphs of LNYul and the forecasts of the four models for a subsample {@Year>1993}.

$$LNYul_t = \beta_o + \beta_1 \times Y(-12) + \delta_0 \times T + \delta_1 \times T \times Y(-12) + \sum_{j=1}^{q} \rho_j \times \mu_{t-j} + \sum_{k=1}^{r} \theta_k \times \nu_{t-k} + \varepsilon_t \tag{3.45b}$$

for alternative integers $p > 0$, $q \geq 0$, and $r \geq 0$. For $q = 0$ or $r = 0$, we would have LVARMA (12,0,0) = LV(12), LVARMA(12,q,0) = LVAR(12,q), and LVARMA(12,0,r) = LVAMA(12,r). The integers q, and r should be selected using the trial-and-error method, in order to obtain the best possible forecast model, which is highly depend on the data as well as the subsample used.

Dependent Variable: LNYUL
Method: Least Squares
Date: 11/21/16 Time: 15:26
Sample: 1980M01 1996M04 IF @YEAR>1990
Included observations: 64
Convergence achieved after 41 iterations
MA Backcast: 1979M09 1979M12

Variable	Coefficient	Std. Error	t-Statistic	Prob.
C	0.887145	3.000272	0.295688	0.7686
Y(-12)	-0.019954	0.006806	-2.931830	0.0049
T	-0.007422	0.019991	-0.371256	0.7119
T*Y(-12)	0.000106	4.40E-05	2.410334	0.0194
AR(1)	-0.252110	0.048805	-5.165612	0.0000
AR(2)	-0.051582	0.037604	-1.371724	0.1758
MA(1)	1.506827	0.134963	11.16474	0.0000
MA(2)	1.407345	0.203551	6.913959	0.0000
MA(3)	1.113352	0.205819	5.409378	0.0000
MA(4)	0.456391	0.141404	3.227569	0.0021

R-squared	0.986602	Mean dependent var	-1.301887
Adjusted R-squared	0.984368	S.D. dependent var	0.615330
S.E. of regression	0.076932	Akaike info criterion	-2.149179
Sum squared resid	0.319604	Schwarz criterion	-1.811853
Log likelihood	78.77372	Hannan-Quinn criter.	-2.016289
F-statistic	441.8118	Durbin-Watson stat	1.911131
Prob(F-statistic)	0.000000		

Inverted AR Roots	-.13+.19i	-.13-.19i		
Inverted MA Roots	-.02-.86i	-.02+.86i	-.74+.27i	-.74-.27i

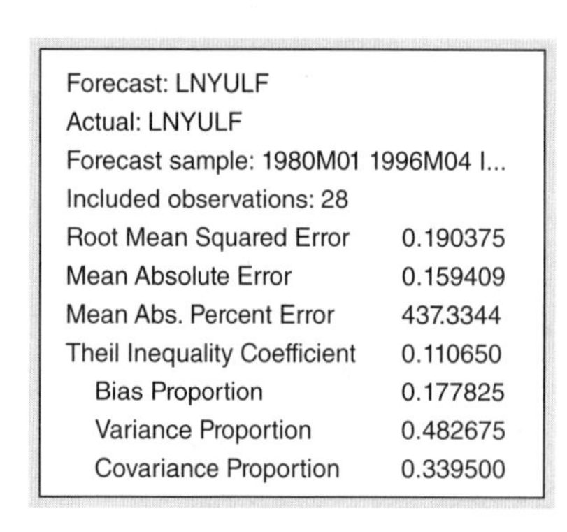

Forecast: LNYULF
Actual: LNYULF
Forecast sample: 1980M01 1996M04 I...
Included observations: 28

Root Mean Squared Error	0.190375
Mean Absolute Error	0.159409
Mean Abs. Percent Error	437.3344
Theil Inequality Coefficient	0.110650
Bias Proportion	0.177825
Variance Proportion	0.482675
Covariance Proportion	0.339500

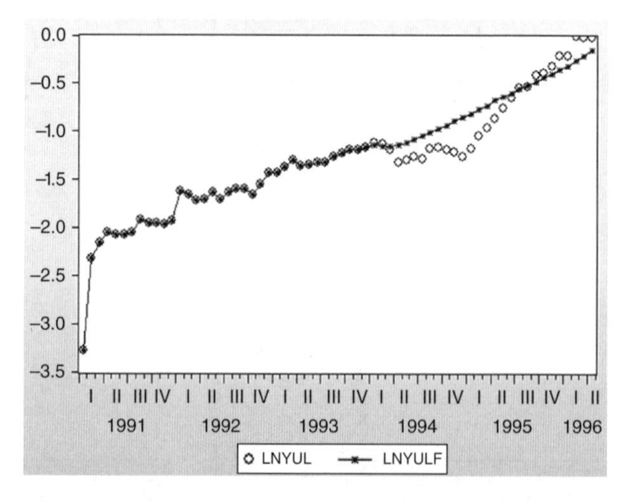

Figure 3.28 Statistical results based on the model.

Example 3.20 An Application of the Model (3.45)

The statistical results presented in Figure 3.28 are obtained by using the trial-and-error method based on the equation specification (3.45). Based on these results, the following findings and notes are presented.

1) The estimation equation of the model could be written as follows:

$$\begin{aligned} &LNYUL = C(1) + C(2)*Y(-12) + C(3)*T + C(4)*T*Y(-12) + \\ &[AR(1) = C(5), AR(2) = C(6), MA(1) = C(7), MA(2) = C(8), \\ &MA(3) = C(9), MA(4) = C(10), BACKCAST = 1980M01, \\ &ESTSMPL = "1980M01\ 1996M04\ IF@YEAR > 1990"] \end{aligned} \tag{3.46}$$

2) Even though the time T has a large p-value, it can be kept as an independent variable, because its interaction $T*Y(-12)$ has a significant adjusted effect, at the 5% level of significance. In addition, T and $T*Y(-12)$ have a joint significant effects, based on the *Wald* (F-statistic) of $F_0 = 60.79859$ with $df = (2,54)$, and p-value $= 0.0000$.
3) However, AR(2) could be deleted from the model, since it has an insignificant effect at the 10% level of significance. Do it as an exercise. However, it is found AR(1) and AR(2) have a joint significant effects, based on the *Wald* (F-statistic) of $F_0 = 21.41494$ with $df = (2,54)$, and p-value $= 0.0000$.
4) Compared to the four models with the statistical results presented in Table 3.2, this model is the best forecast model, since it has the smallest RMSE and TIC.

4

Forecasting Based on (X_t, Y_t)

4.1 Introduction

As the extension of all forecast models presented in previous chapters, this chapter presents various models based on bivariate time series (X_t, Y_t). It is well-known that the $Y(-1)$ is one of the best predictor for any time series Y, in addition to a cause (exogenous, upstream, or source) factor or variable X. So the simplest model Y considered should use possible independent variables X or $X(-1)$, and $Y(-1)$ with the following alternative form of forecast models. Note that these models can be considered to be the modifications of all cross-section GLMs (General Linear Models) based on a bivariate time series (X_i, Y_i) presented in Agung (2011a).

4.2 Forecast Models Based on (X_t, Y_t)

Figure 4.1 presents the graphs of three alternative simple relationships or up-and-down relationships between the variables X or $X(-1)$, Y, and $Y(-1)$. Based on these graphs, referring to the two-way interaction model (3.1) with its possible reduced models, I propose the basic full models as follows:

i) Based on the graph in Figure 4.1a, the simplest LV(1) two-way interaction full model can be presented using the following equation specification.

$$Y\ C\ Y(-1)\ X(-1)\ Y(-1)*X(-1) \quad (4.1)$$

ii) Based on the graph in Figure 4.1b, the simplest LV(1) two-way interaction full model can be presented using the following equation specification.

$$Y\ C\ Y(-1)\ X\ Y(-1)*X \quad (4.2)$$

iii) The graph in Figure 4.1c is presented to forecast beyond the sample period $t > 591$. Hence, a pair of LV(1) and LV(p) models should be applied using the following pairs of specific equations. However. the estimations should be done one by one, and the integer p should be selected using trial-and-error in order to have the best possible forecasting of X.

$$Y\ C\ \ Y(-1)\ X(-1)\ Y(-1)*X(-1) \quad (4.3a)$$

$$X\ C\ X(-1)...X(-p) \quad (4.3b)$$

However, an alternative estimation can be done once by using the system equation as follows:

$$Y = C(10) + C(11)*Y(-1) + C(12)*X(1) + C(13)*Y(-1)*X(-1) \quad (4.4a)$$

Advanced Time Series Data Analysis: Forecasting Using EViews, First Edition. I Gusti Ngurah Agung.

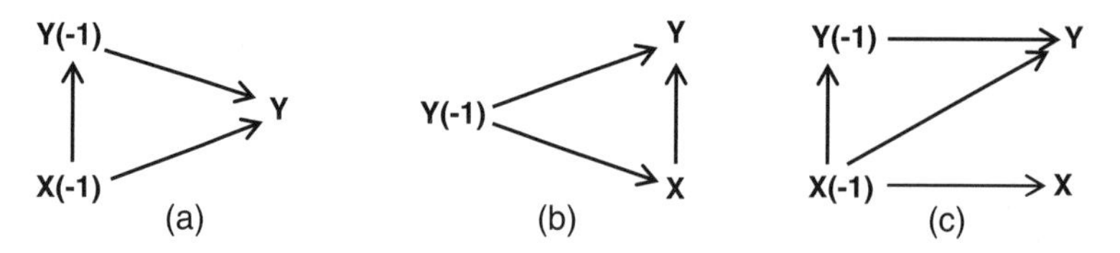

Figure 4.1 Three alternative up-and-down simple relationships based on bivariate (X_t,Y_t).

$$X = C(20) + C(21)*X(-1) + \cdots + C(2p)*X(-p) \tag{4.4b}$$

Note that the interaction *Y(−1)*X(−1)* is valid if both variables *X* and *Y* are positive variables. If both time series have negative observed scores then their interaction would be misleading. Furthermore, each of these model can easily be modified to more advanced models, such as *Lag-Variable-Autogressive-Moving Average*, namely LVARMA(*p,q,r*) models of *LNY* = *log(Y)* or *LNYul* = *log((Y−L)/(U−Y)*, and heterogeneous regression models by *@Year* or *@Month*, as presented in previous chapters. In addition, forecasting based on the data of the last few years should be considered or experimented in order to obtain better forecasting, which have been illustrated in Chapter 3. However, this chapter presents only some selected models as illustrative data analyses.

4.3 Data Analysis Based on a Monthly Time Series

For illustrative examples, the work file used is the GARCH.wf1, which contains six time series *FSCOM* = S&PS common stock price index composites, *IP* = industrial production total, *PW* = producer price index all composite (82 = 100,NSA), *R3* = R3/100 = interest rate, *RAAA* = RAAA/100 bond yield Moody's AAA corporate (% per annum), and *FSDXP* = S&PS common stock price dividend yield index composite (% per annum), with their graphs presented in Figure 4.2.

Based on the growth patterns of these six variables, I select *Y* = *FSCOM* as the predicted or dependent variable, then we have to select either one of *IP* and *PW* as an predictor or the *X*-variable, since the other three variables should be inappropriate predictors for *FSCOM*.

By looking at the growth of *FSCOM*, it might be better to conduct the data analysis based on the last few years, such as the subsample {*Year* > 1990}, since it has more stable observed scores.

4.4 Forecast Models without a Time Predictor

4.4.1 Two-Way Interaction Models

In general, either the linear effect of a cause factor or an exogenous variable *X* or *X(−1)* on *Y* depends on *Y(−1)*, or the linear effect of a *Y(−1)* on *Y* depends on a cause factor or an exogenous variable *X* or *X(−1)*. So that the models of *Y* on *Y(−1)*, and *X or X(−1)*, should be two-way interaction models with an independent variable *X∗Y(−1)* or *X(−1)∗Y(−1)*, similar to the model (3.1) and its possible reduced models. As a comparison, refer to full *IxJ* factorial ANOVA models, which should have the interaction categorical independent variables. For this reason, I consider that a first lagged variable interaction model of *Y* on *X* or LV(1) interaction model is the simplest forecast model based on (X_t, Y_t). Two alternative models can be presented using the following equation specifications, which are the *first-order lagged variables interaction models*, namely LV(1)_IM.

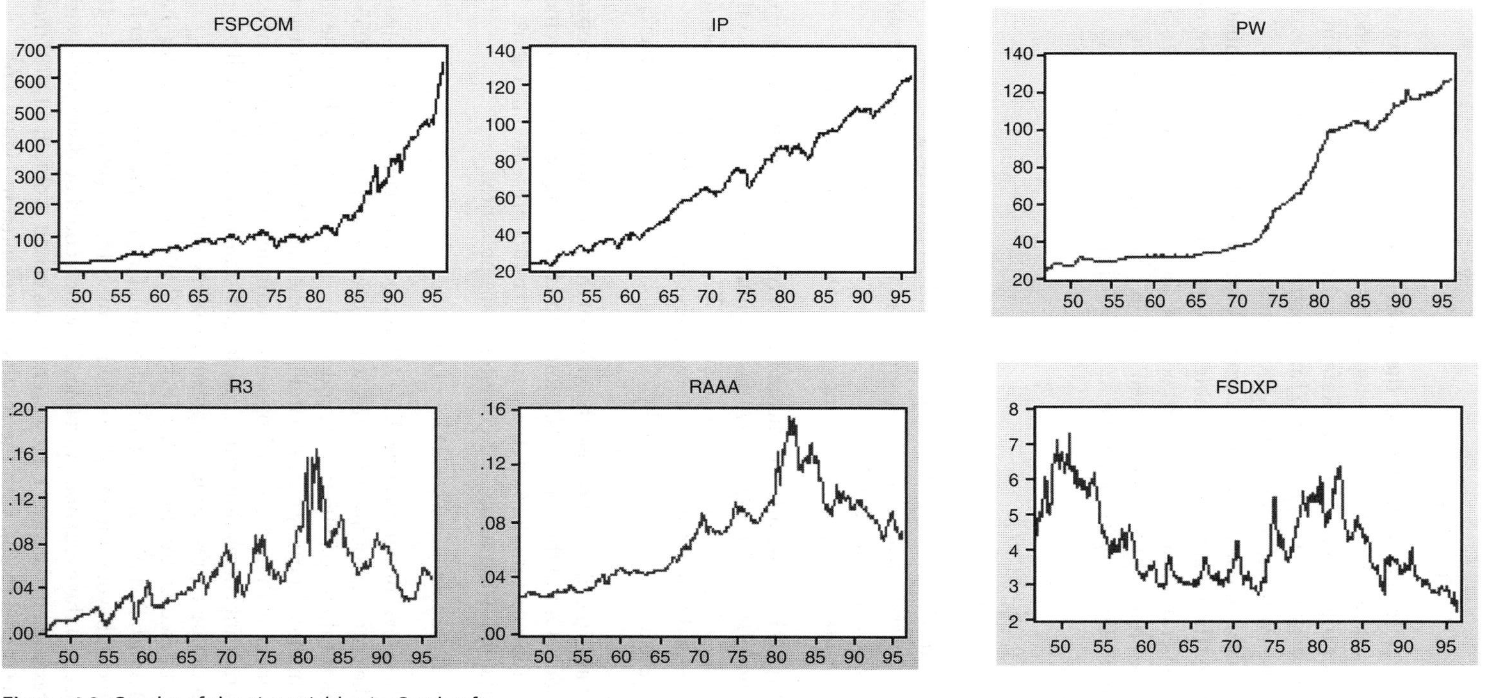

Figure 4.2 Graphs of the six variables in Garch.wf1.

$$Y\ C\ Y(-1)\ X\ X*Y(-1) \tag{4.5}$$

and

$$Y\ C\ Y(-1)\ X(-1)\ X(-1)*Y(-1) \tag{4.6}$$

However, I prefer or recommend applying the model (4.6), since everyone can be very confident that the two variables *X(−1)* and *Y(−1)* should have effects on the endogenous variable *Y*, either linear effects or other types of effects. Refer to various models presented in Agung (2014, 2011a, 2009a). In other words, *X(−1)* and *Y(−1)* are upstream (cause, source, or exogenous) variables of the downstream (endogenous, or impact) variable *Y*.

Note that this model is hierarchical model, where *X* and *Y* could be the original measured or observed variables, or their transformed variables. In a statistical sense, however, one of the following reduced nonhierarchical models might become the best fit model, in the statistical sense, which is highly dependent on the data set used, and in fact is unpredictable or uncontrollable because of the multicollinearity between the *X(−1), Y(−1)* and *X(−1)*Y(−1).*

$$Y\ C\ Y(-1)\ X(-1)*Y(-1) \tag{4.6a}$$

$$Y\ C\ X(-1)\ X(-1)*Y(-1) \tag{4.6b}$$

$$Y\ C\ X(-1)*Y(-1) \tag{4.6c}$$

Example 4.1 Application of the Model in (4.6) and Alternatives

Figure 4.3 presents the statistical results of the model in (4.6), with its forecast evaluation, using the variables *Y1* = *FSCOM* and *X1* = *IP*.

Based on these results, the following findings and notes are presented.

1) Even though *X1(−1)* has insignificant adjusted effect on *Y* with a large *p*-value = 0.7205, it can be kept in the model, since *X1(−1)*Y1(−1)* has a significant effect, which shows that the linear effect of *Y1(−1)* on *Y* is significantly dependent on *X1(−1)*. In other words, the data supports the hypothesis stated "the effect of *Y1(−1)* on *Y* depends on *X1(−1)*."
2) A reduced model can easily be obtained by deleting *X1(−1)*. Do this as an exercise, and compare its forecast.
3) On the other hand, in order to increase the value of the DW statistic, I have observed several alternative models. Out of those models, I found the following equation specification represents the best fit forecast model with the statistical results presented in Figure 4.4. Based on these results, the following notes and comments are presented.

$$Y\ C\ Y(-1)\ Y(-2)\ X(-1)\ X(-1)*Y(-1) \tag{4.7}$$

   3.1 At the $\alpha = 0.15$ level of significance, *X1(−1)* has a positive significant adjusted effect on *Y1*, based on the *t*-statistic of $t_0 = 1.078924$, with a *p*-value = 0.2811/2 = 0.14055 < 0.15, compared to its *p*-value = 0.7905 in the model (4.1). These findings show the unpredictable impacts of multicollinearity between independent variables on the parameters' estimates.

   3.2 As an additional analysis, Figure 4.5 presents each pairs of the independent variables has a positive significant correlation with a very large *t*-statistic or a very small *p*-value = 0.0000, which also can be representing a high multicollinearity.

   3.3 The forecast based on this model has RMSE = 27.80623, which is smaller than the model in Figure 4.3 with RMSE = 36.98385. So the model in (4.6) presents a better forecast of *Y1*.
4) In order to forecast beyond the sample period, namely for *t* = *@trend* > 591 = 1996 M04, we have to forecast also the exogenous variable *X1*, which could be done using various alternative models presented in Chapter 1. However, in this case it is obtain a LV(3) model as a good fit model of the model (4.3b), with the results as presented in Figure 4.6.

Dependent Variable: Y1
Method: Least Squares
Date: 05/09/14 Time: 12:47
Sample (adjusted): 1947M02 1996M04
Included observations: 591 after adjustments

Variable	Coefficient	Std. Error	t-Statistic	Prob.
C	0.990549	0.775505	1.277294	0.2020
Y1(-1)	0.964193	0.020342	47.40016	0.0000
X1(-1)	0.006091	0.017018	0.357916	0.7205
X1(-1)*Y1(-1)	0.000370	0.000153	2.412533	0.0161

R-squared	0.998391	Mean dependent var	140.9656
Adjusted R-squared	0.998383	S.D. dependent var	136.5858
S.E. of regression	5.492871	Akaike info criterion	6.251524
Sum squared resid	17710.75	Schwarz criterion	6.281181
Log likelihood	-1843.325	Hannan-Quinn criter.	6.263077
F-statistic	121406.9	Durbin-Watson stat	1.479404
Prob(F-statistic)	0.000000		

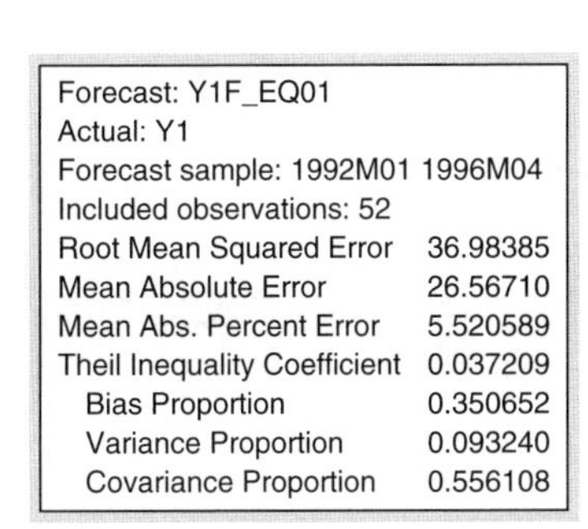

Forecast: Y1F_EQ01
Actual: Y1
Forecast sample: 1992M01 1996M04
Included observations: 52

Root Mean Squared Error	36.98385
Mean Absolute Error	26.56710
Mean Abs. Percent Error	5.520589
Theil Inequality Coefficient	0.037209
Bias Proportion	0.350652
Variance Proportion	0.093240
Covariance Proportion	0.556108

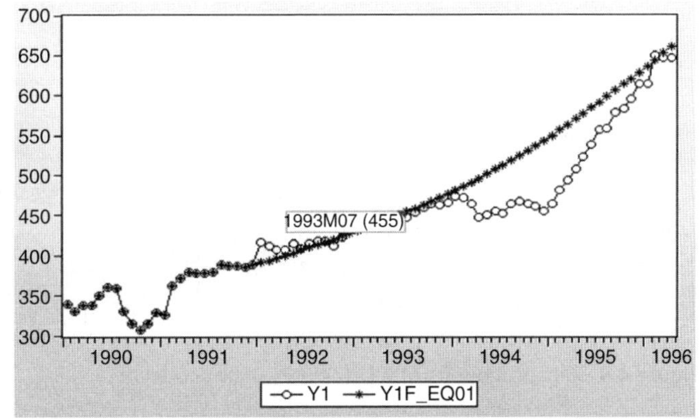

Figure 4.3 Statistical results of the LV(1) interaction model in (4.2).

$$XI\ C\ XI(-1)\ XI(-2)\ XI(-3) \tag{4.8}$$

Based on these results the following findings and notes are presented

4.1 The regression function could be considered as the best fit model, since by using the LV(4) model it was found X1(−3) has an insignificant adjusted effect with a p-value = 0.6100.

4.2 Based on the following pairs of regression functions, namely the regressions in Figures 4.4 and 4.6, the forecast of Y1, for t > 591, can be computed easily using Excel. Do it as an exercise.

$$\begin{aligned} YI &= 0.777888343584 + 1.22340268193*YI(-1) - 0.266926868745*YI(-2) \\ &\quad + 0.017853628051*XI(-1) + 0.000388288010249*XI(-1)*YI(-1) \end{aligned} \tag{4.9a}$$

and

$$\begin{aligned} XI &= 0.0597624879954 + 1.31077002912*XI(-1) \\ &\quad - 0.180611831503*XI(-2) - 0.129585330106*XI(-3) \end{aligned} \tag{4.9b}$$

Dependent Variable: Y1
Method: Least Squares
Date: 05/10/14 Time: 19:21
Sample (adjusted): 1947M03 1996M04
Included observations: 590 after adjustments

Variable	Coefficient	Std. Error	t-Statistic	Prob.
C	0.777888	0.753192	1.032790	0.3021
Y1(-1)	1.223403	0.043913	27.85962	0.0000
Y1(-2)	-0.266927	0.040423	-6.603402	0.0000
X1(-1)	0.017854	0.016548	1.078924	0.2811
X1(-1)*Y1(-1)	0.000388	0.000148	2.616700	0.0091

R-squared	0.998500	Mean dependent var	141.1778
Adjusted R-squared	0.998490	S.D. dependent var	136.6042
S.E. of regression	5.307981	Akaike info criterion	6.184739
Sum squared resid	16482.18	Schwarz criterion	6.221858
Log likelihood	-1819.498	Hannan-Quinn criter.	6.199199
F-statistic	97380.67	Durbin-Watson stat	1.985873
Prob(F-statistic)	0.000000		

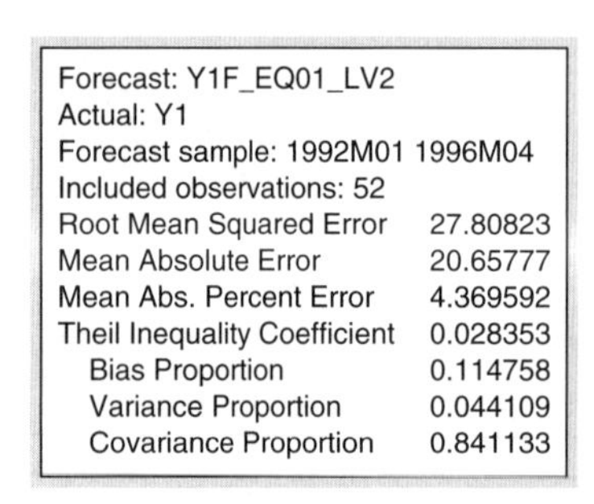

Forecast: Y1F_EQ01_LV2
Actual: Y1
Forecast sample: 1992M01 1996M04
Included observations: 52

Root Mean Squared Error	27.80823
Mean Absolute Error	20.65777
Mean Abs. Percent Error	4.369592
Theil Inequality Coefficient	0.028353
Bias Proportion	0.114758
Variance Proportion	0.044109
Covariance Proportion	0.841133

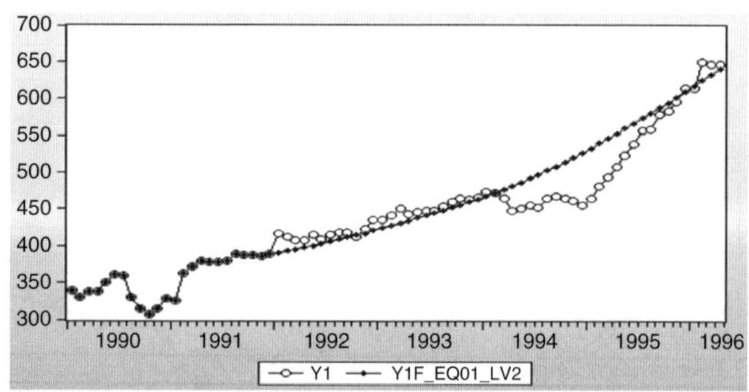

Figure 4.4 Statistical results of a LV(2) interaction model in (4.7).

Covariance Analysis: Ordinary
Date: 05/10/14 Time: 21:32
Sample (adjusted): 1947M03 1996M04
Included observations: 590 after adjustments
Balanced sample (listwise missing value deletion)

Correlation t-Statistic Probability	Y1	Y1(-1)	Y1(-2)	X1(-1)	X1(-1)*Y1(-1)
Y1	1.000000 ----- -----				
Y1(-1)	0.999185 600.3136 0.0000	1.000000 ----- -----			
Y1(-2)	0.997923 375.6907 0.0000	0.999169 594.3305 0.0000	1.000000 ----- -----		
X1(-1)	0.865330 41.86549 0.0000	0.866678 42.12705 0.0000	0.868211 42.42878 0.0000	1.000000 ----- -----	
X1(-1)*Y1(-1)	0.994655 233.5904 0.0000	0.995053 242.8806 0.0000	0.994097 222.1760 0.0000	0.840680 37.64390 0.0000	1.000000 ----- -----

Figure 4.5 Correlation matrix of the variables of the LV(2) model in Figure 4.2.

Dependent Variable: X1
Method: Least Squares
Date: 12/12/16 Time: 07:26
Sample (adjusted): 1947M04 1996M04
Included observations: 589 after adjustments

Variable	Coefficient	Std. Error	t-Statistic	Prob.
C	0.059762	0.054139	1.103878	0.2701
X1(-1)	1.310770	0.041154	31.85011	0.0000
X1(-2)	-0.180612	0.067959	-2.657677	0.0081
X1(-3)	-0.129585	0.041274	-3.139670	0.0018

R-squared	0.999674	Mean dependent var	66.10910
Adjusted R-squared	0.999672	S.D. dependent var	29.66055
S.E. of regression	0.537224	Akaike info criterion	1.601964
Sum squared resid	168.8365	Schwarz criterion	1.631699
Log likelihood	-467.7784	Hannan-Quinn criter.	1.613549
F-statistic	597258.5	Durbin-Watson stat	2.012617
Prob(F-statistic)	0.000000		

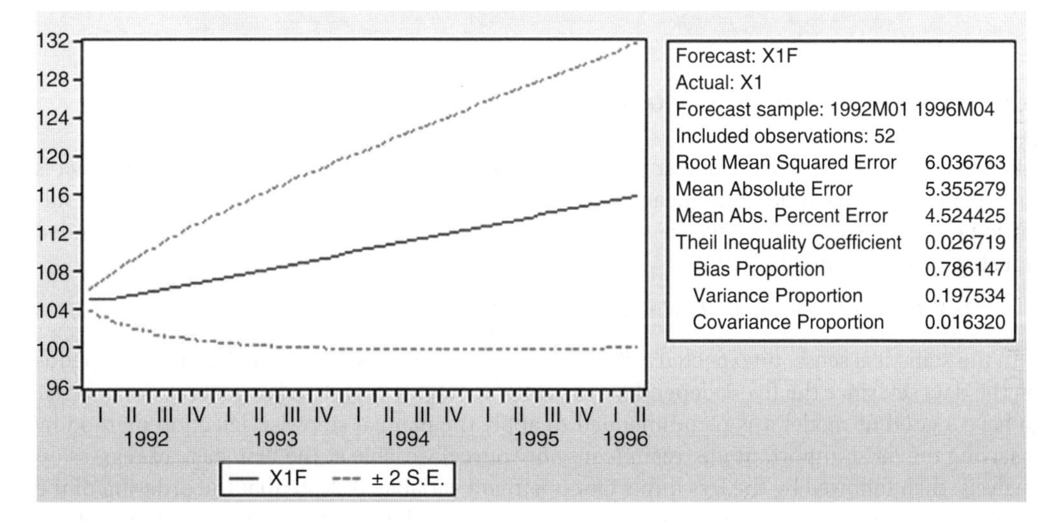

Figure 4.6 Statistical results of the LV(3) model (4.8).

4.4.2 Cobb–Douglass Model and Alternatives

Another type of the reduced model of the model in (4.6) is an additive model with the following ES, which shows the linear effect of *Y(−1)* on *Y* is constant for all values of *X(−1)*.

$$Y\ C\ Y(-1)\ X(-1) \tag{4.10}$$

For most cases, if both variables *X* and *Y* are the original observed or measured variables, I would say that additive model is an inappropriate model, in a theoretical sense, since the effect of *Y(−1)* on *Y*, in general, should depend on *X1(−1)*. One well known additive model is a translog linear model, which is known as a Cobb–Douglass production function, with the ES as follows:

$$log(Y)\ C\ log(Y(-1))\ log(X(-1)) \tag{4.11}$$

which in fact is derived from a nonlinear model with the general equation specification as follows:

$$Y = C(1)*Y(1)\hat{}C(2)*X(-1)\hat{}C(3) \tag{4.11a}$$

Note that the translog linear model in (4.11) can be extended to a bounded translog linear model with the following alternative general equations, where *U* and *L* are fixed upper and lower bounds of Y_t, which have been been illustrated in Section 3.3; refer also to various alternative translog linear models presented in Agung (2009a).

$$Yul\ C\ \log(Y(-1))\ \log(X(-1)),\ Yul = \log((Y-L)/(U-Y)) \tag{4.12}$$

4.5 Translog Quadratic Model

As an extension of the translog linear model, we have a translog quadratic model, which in fact is the Taylor's expansion of a nonlinear model called the CES production function, with the following equation specification where *ln(∗) = log(∗)*. Then it can be easily modified to a bounded translog quadratic model.

$$lnY\ C\ ln\ Y(-1)*lnX(-1)\ (lnY(-1))\hat{}2\ (lnX(-1))\hat{}2\ lnY(-1)\ lnX(-1) \tag{4.13}$$

In the statistical sense, unexpected reduced model would be obtained, which is highly depends on the data set, since the five independent variables are highly correlated in general. So, in order to have a good fit model it is recommended to apply the manual stepwise selection method by inserting the most important upstream (cause or source) variable at the first stage of regression analysis, then followed by the less important upstream variable(s). Note that the ordering of the five independent variables presented in ES (4.13) presents the five stages of regression analyses that should be done in order to obtain an acceptable reduced model. Furthermore, the first three independent variables are the most important variables to represent the CES (Constant Elasticity of Substitution) production function. See the following example.

Example 4.2 Manual Stepwise Selection Method Based on the ES (4.13)
Corresponding to the ordering of the five independent variables in (4.13), the stages of regression analyses are as follows:

1) At the first stage, it is obtain an OLS regression of *lnY1* on *lnY1(−1)∗lnX1(−1)*, but the statistical results are not presented. Since it is found that *lnY1(−1)∗lnX1(−1)* has a significant effect on *lnY1* based on the *t*-statistic of $t_0 = 198.2593$ with a *p*-value 0.000, then it can continue to the second stage.
2) At the second stage, a stepwise regression method with the default options is applied using the equation specification *"lnY1 C lnY1(−1)∗lnX1(−1)"* and the search regressors *"lnY1(−1)*

Dependent Variable: LNY1
Method: Stepwise Regression
Date: 05/12/14 Time: 09:17
Sample (adjusted): 1947M02 1996M04
Included observations: 591 after adjustments
Number of always included regressors: 2
Number of search regressors: 2
Selection method: Stepwise forwards
Stopping criterion: p-value forwards/backwards = 0.5/0.5

Variable	Coefficient	Std. Error	t-Statistic	Prob.*
C	2.128402	0.012291	173.1612	0.0000
LNY1(-1)*LNX1(-1)	0.965600	0.018833	51.27262	0.0000
LNX1(-1)^2	-0.528800	0.011258	-46.97145	0.0000
LNY1(-1)^2	-0.324089	0.008098	-40.01991	0.0000

R-squared	0.997686	Mean dependent var	4.526843
Adjusted R-squared	0.997674	S.D. dependent var	0.942446
S.E. of regression	0.045450	Akaike info criterion	-3.337649
Sum squared resid	1.212584	Schwarz criterion	-3.307992
Log likelihood	990.2752	Hannan-Quinn criter.	-3.326096
F-statistic	84365.33	Durbin-Watson stat	0.878836
Prob(F-statistic)	0.000000		

Selection Summary

Added LNX1(-1)^2
Added LNY1(-1)^2

*Note: p-values and subsequent tests do not account for stepwise selection.

(a)

Dependent Variable: LNY1
Method: Stepwise Regression
Date: 05/11/14 Time: 11:28
Sample (adjusted): 1947M02 1996M04
Included observations: 591 after adjustments
Number of always included regressors: 4
Number of search regressors: 2
Selection method: Stepwise forwards
Stopping criterion: p-value forwards/backwards = 0.5/0.5

Variable	Coefficient	Std. Error	t-Statistic	Prob.*
C	0.320596	0.281626	1.138376	0.2554
LNY1(-1)*LNX1(-1)	0.009548	0.058060	0.164457	0.8694
LNY1(-1)^2	-0.006251	0.016845	-0.371065	0.7107
LNX1(-1)^2	0.016091	0.055280	0.291078	0.7711
LNY1(-1)	1.022512	0.099448	10.28185	0.0000
LNX1(-1)	-0.180196	0.232182	-0.776094	0.4380

R-squared	0.998776	Mean dependent var	4.526843
Adjusted R-squared	0.998766	S.D. dependent var	0.942446
S.E. of regression	0.033110	Akaike info criterion	-3.967863
Sum squared resid	0.641319	Schwarz criterion	-3.923378
Log likelihood	1178.504	Hannan-Quinn criter.	-3.950534
F-statistic	95487.28	Durbin-Watson stat	1.530392
Prob(F-statistic)	0.000000		

Selection Summary

Added LNY1(-1)
Added LNX1(-1)

*Note: p-values and subsequent tests do not account for stepwise selection.

(b)

Figure 4.7 Statistical results of the second and third manual stepwise selection methods.

Dependent Variable: LNY1
Method: Least Squares
Date: 05/12/14 Time: 09:24
Sample (adjusted): 1947M02 1996M04
Included observations: 591 after adjustments

Variable	Coefficient	Std. Error	t-Statistic	Prob.
C	0.113173	0.088754	1.275134	0.2028
LNY1(-1)*LNX1(-1)	0.039972	0.042814	0.933611	0.3509
LNX1(-1)^2	-0.022687	0.023643	-0.959544	0.3377
LNY1(-1)^2	-0.012376	0.014877	-0.831850	0.4058
LNY1(-1)	0.952461	0.041734	22.82215	0.0000

R-squared	0.998775	Mean dependent var	4.526843
Adjusted R-squared	0.998767	S.D. dependent var	0.942446
S.E. of regression	0.033099	Akaike info criterion	-3.970218
Sum squared resid	0.641979	Schwarz criterion	-3.933147
Log likelihood	1178.200	Hannan-Quinn criter.	-3.955778
F-statistic	119440.0	Durbin-Watson stat	1.528413
Prob(F-statistic)	0.000000		

(a)

Dependent Variable: LNY1
Method: Least Squares
Date: 05/11/14 Time: 11:36
Sample (adjusted): 1947M02 1996M04
Included observations: 591 after adjustments

Variable	Coefficient	Std. Error	t-Statistic	Prob.
C	-1.792247	0.209076	-8.572215	0.0000
LNY1(-1)*LNX1(-1)	0.550491	0.026661	20.64766	0.0000
LNY1(-1)^2	-0.138872	0.011763	-11.80581	0.0000
LNX1(-1)^2	-0.545936	0.008951	-60.99448	0.0000
LNX1(-1)	1.986528	0.105821	18.77254	0.0000

R-squared	0.998555	Mean dependent var	4.526843
Adjusted R-squared	0.998545	S.D. dependent var	0.942446
S.E. of regression	0.035947	Akaike info criterion	-3.805130
Sum squared resid	0.757212	Schwarz criterion	-3.768059
Log likelihood	1129.416	Hannan-Quinn criter.	-3.790689
F-statistic	101241.2	Durbin-Watson stat	1.314046
Prob(F-statistic)	0.000000		

(b)

Figure 4.8 Statistical results of two alternative OLS regressions based on the model in Figure 4.5a.

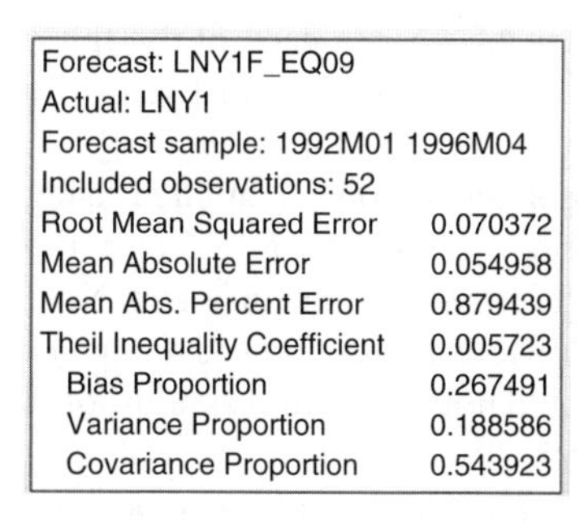

Forecast: LNY1F_EQ09	
Actual: LNY1	
Forecast sample: 1992M01 1996M04	
Included observations: 52	
Root Mean Squared Error	0.070372
Mean Absolute Error	0.054958
Mean Abs. Percent Error	0.879439
Theil Inequality Coefficient	0.005723
Bias Proportion	0.267491
Variance Proportion	0.188586
Covariance Proportion	0.543923

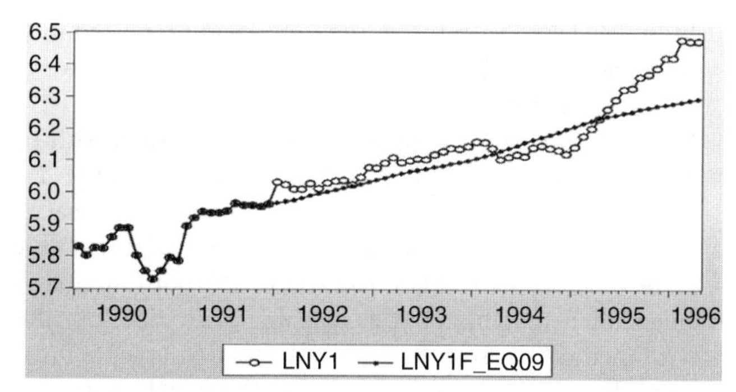

Figure 4.9 Forecast evaluation based on the model in Figure 4.6b.

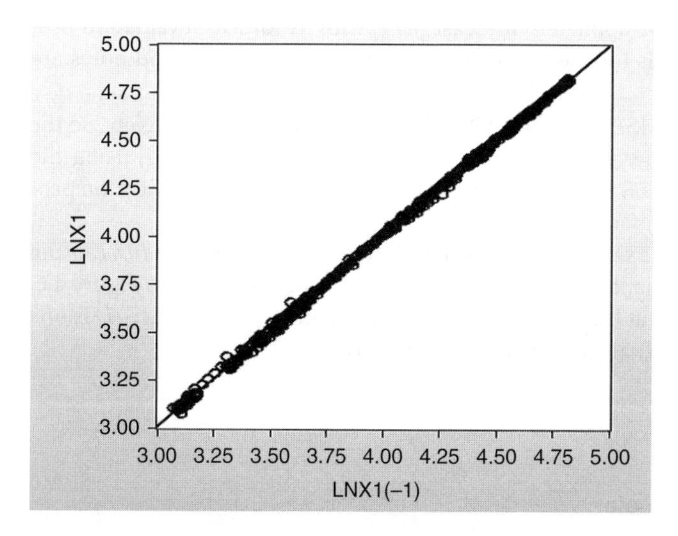

Figure 4.10 Scatter plot of (*lnX1(−1),lnX1*) with regression line.

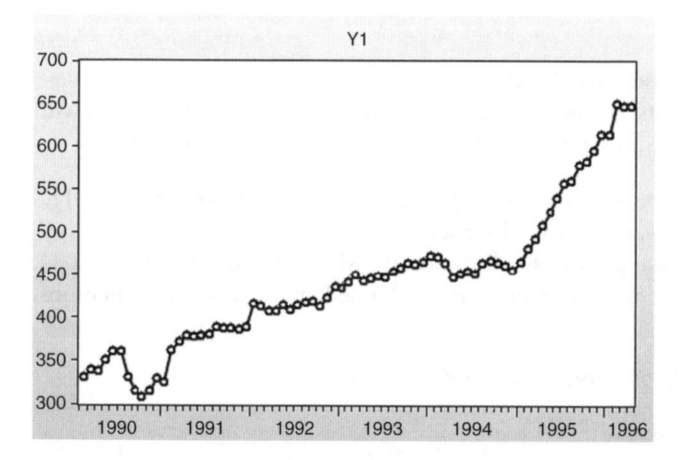

Figure 4.11 The growth of Y1 based on a subsample.

^2 lnX1(−1)^2," with the statistical results presented in Figure 4.7a. Since each of the three independent variables has a significant effect, then we can continue to the third stage, by inserting additional independent variables.

3) At the third stage, a stepwise regression method, with the default options, is applied using the equation specification *"lnY1 C lnY1(−1)*lnX1(−1 lnY1(−1)^2 lnX1(−1)^2,"* and the search regressors *"lnY1(−1) lnX1(−1)"* with the statistical results presented in Figure 4.7b. Note that the results show that each of the three variables of the previous regression has a large *p*-value, by using both additional variables. So both variables should not be inserted as additional independent variables. In this case we have to develop alternative models, using the trial-and-error method, to use either one of the two variables *lnY1(−1)* and *lnX1(−1)*, which can be done using the OLS regression analysis, as the fourth stage of analysis.
4) Then at the fourth stage, we will have the two OLS regressions in Figure 4.8, which show that the regression in Figure 4.8b is a good fit model, in the statistical sense, since each of the independent variables has a significant adjusted effect on *lnY1*.
5) So the model in Figure 4.8b can be applied to forecast *lnY1*, with its forecast evaluation presented in Figure 4.9. Based on this forecast evaluation, the following findings and notes are presented.
   5.1 The forecast has very small RMSE, and TIC of 0.070372 and 0.005723, respectively. So the forecast model can be considered to be giving a perfect forecasting of *lnY1*, using the quadratic translog model or CES model, since its TIC < 0.001 – refer to classification proposed in Chapter 1.
   5.2 In order to forecast beyond the sample period, say $t > 591$, we have to forecast *lnX1*, using either one of the models in Chapters 1 and 2. Referring to the forecast of *X1* in Figure 4.6, then we have to forecast *lnX1* at least using *lnX1(−1)* as a predictor, since *lnX1(−1)* is one of the best predictors of *lnX1*, as presented in Figure 4.10.

4.5.1 Forecasting Using a Subsample

By looking at the graph of *Y1* = *FSCOM* for the last several years, as presented in Figure 4.11, I would say that it is better to use an upper and lower translog linier or quadratic model, based on a subsample {*Year*>1992}, which should be done by using the trial-and-error method, as illustrated in Chapter 3.

Example 4.3 Application of a Model in (4.12)

As an illustration, Figure 4.12 presents the results of a model *Yul* = *log((Y1–400)/(800–Y1))* with the Eq. (4.12). Based on these results, the following findings and notes are presented.

1) At the 10% level of significance, *log(X1(−1))* has a significant negative adjusted effect on *Yul*. So the model is an acceptable model, in statistical sense.
2) Its forecast evaluation of *Yul* presents very small values of RMSE = 0.002 and TIC = 0.044. The forecast values of *Y1* = *FSCOM* could easily be computed based on the following equations, using the Excel.

$$\begin{aligned} Yul = & -0.329819518905 + 0.0666599660894*LOG(Y1(-1)) \\ & -0.0140305991754*LOG(X1(-1)) \end{aligned} \tag{4.14a}$$

and

Dependent Variable: YUL
Method: Least Squares
Date: 12/12/16 Time: 09:14
Sample: 1947M01 1996M04 IF YEAR>1992
Included observations: 40

Variable	Coefficient	Std. Error	t-Statistic	Prob.
C	-0.329820	0.032346	-10.19653	0.0000
LOG(Y1(-1))	0.066660	0.003159	21.10170	0.0000
LOG(X1(-1))	-0.014031	0.009226	-1.520757	0.1368

R-squared	0.959590	Mean dependent var	0.016383
Adjusted R-squared	0.957406	S.D. dependent var	0.007716
S.E. of regression	0.001592	Akaike info criterion	-9.975095
Sum squared resid	9.38E-05	Schwarz criterion	-9.848429
Log likelihood	202.5019	Hannan-Quinn criter.	-9.929297
F-statistic	439.3112	Durbin-Watson stat	1.372328
Prob(F-statistic)	0.000000		

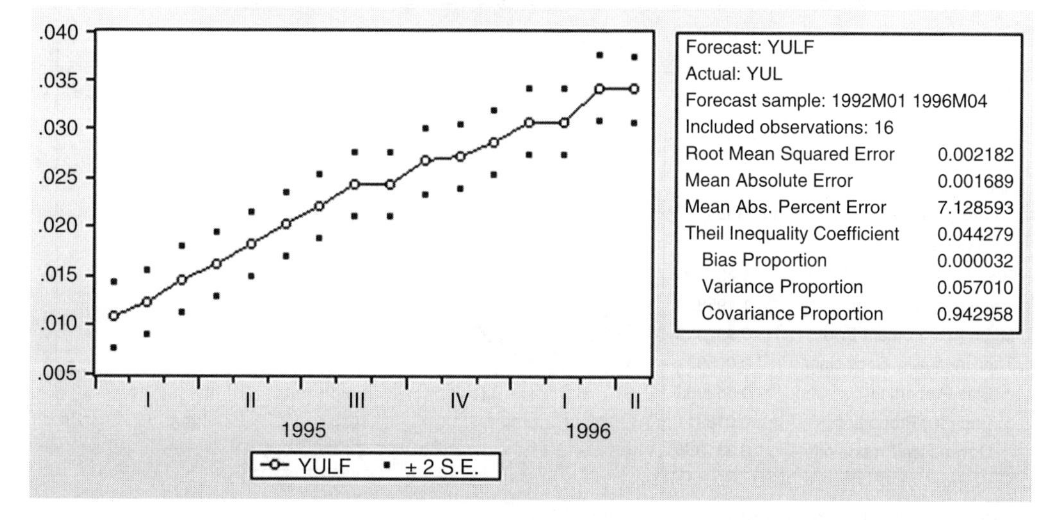

Figure 4.12 Statistical results based on the model in (4.12).

$$Y1 = (400 + 800{*}Exp(Yul))/(1 + EXp(Yul) \tag{4.14b}$$

3) In addition, note that the negative coefficient of *log(X1(−1))* indicates the unexpected impact of the correlation between *log(X1(−1))* and *log(Y1(−1))*, since *log(X1(−1))* has a significant positive correlation with each of the variables *Yul* and *log(Y1(−1))*, which can easily be presented.
4) Thence, to forecast beyond the sample period, we also should to forecast *log(X1)* using the subsample {*Year*>1992}, with a very small TIC = 0.005981 as presented in Figure 4.13, based on a simple LV(1) translog linear model with the following equation specification.

$$log(X1)\ C\ log(X1(-1)) \tag{4.15}$$

In general, we may apply LV(*p*) model, where *p* should be selected using the trial-and-error method. In this case, however, by using the LV(2) model, it was found *log(X1(−2))* has an insignificant adjusted effect with a *p*-value = 0.5990. For this reason, I decide to present the LV(1) model (4.15). Thence, the forecast values of *Yul(t)* beyond the sample period can easily be computed using the following pairs of equations.

Dependent Variable: LOG(X1)
Method: Least Squares
Date: 12/12/16 Time: 15:37
Sample: 1992M11 1996M04 IF YEAR>1992
Included observations: 40

Variable	Coefficient	Std. Error	t-Statistic	Prob.
C	0.070224	0.077251	0.909043	0.3691
LOG(X1(-1))	0.985914	0.016209	60.82371	0.0000

R-squared	0.989833	Mean dependent var	4.768739
Adjusted R-squared	0.989565	S.D. dependent var	0.040580
S.E. of regression	0.004145	Akaike info criterion	-8.085013
Sum squared resid	0.000653	Schwarz criterion	-8.000569
Log likelihood	163.7003	Hannan-Quinn criter.	-8.054480
F-statistic	3699.523	Durbin-Watson stat	2.108660
Prob(F-statistic)	0.000000		

Forecast: X1F
Actual: X1
Forecast sample: 1992M11 1996M04 I...
Included observations: 28

Root Mean Squared Error	1.435409
Mean Absolute Error	1.195838
Mean Abs. Percent Error	0.992653
Theil Inequality Coefficient	0.005981
Bias Proportion	0.652035
Variance Proportion	0.016911
Covariance Proportion	0.331054

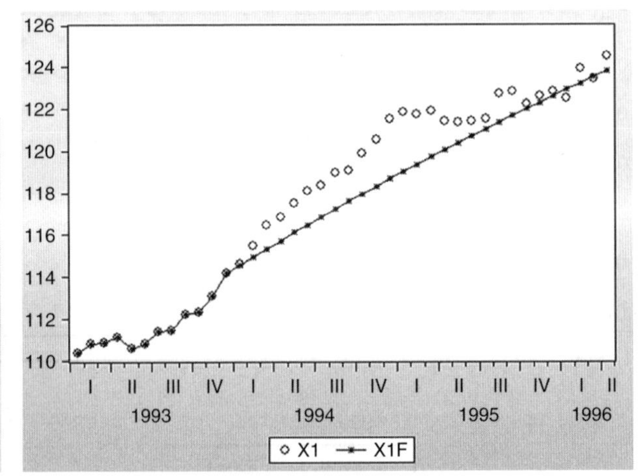

Figure 4.13 Statistical results of the LV(1) model (4.15).

$$Yul(t) = -0.329819518905 + 0.0666599660894*LOG(Y1(t-1)) - 0.0140305991754*LOG(X1(t-1)) \quad (4.16)$$

and

$$LOG(X1(t-1)) = 0.0702243839576 + 0.985913552418*LOG(X1(t-2)) \quad (4.17)$$

6) Furthermore, we could be doing experiments using alternative upper and lower bounds of *Y1* in order to obtain the best possible forecasting, in the statistical sense. However, we should always be thinking about the uncertainty of the future.

Dependent Variable: Y1
Method: Least Squares
Date: 05/12/14 Time: 20:51
Sample (adjusted): 1947M03 1996M04
Included observations: 590 after adjustments

Variable	Coefficient	Std. Error	t-Statistic	Prob.
C	5.175351	1.633930	3.167425	0.0016
Y1(-1)	1.169387	0.047119	24.81760	0.0000
Y1(-2)	-0.257745	0.040258	-6.402337	0.0000
X1(-1)	-0.205205	0.075496	-2.718097	0.0068
X1(-1)*Y1(-1)	0.000752	0.000190	3.954919	0.0001
@TREND	0.040372	0.013337	3.027168	0.0026

R-squared	0.998524	Mean dependent var	141.1778
Adjusted R-squared	0.998511	S.D. dependent var	136.6042
S.E. of regression	5.271328	Akaike info criterion	6.172559
Sum squared resid	16227.55	Schwarz criterion	6.217103
Log likelihood	-1814.905	Hannan-Quinn criter.	6.189912
F-statistic	78993.54	Durbin-Watson stat	1.982415
Prob(F-statistic)	0.000000		

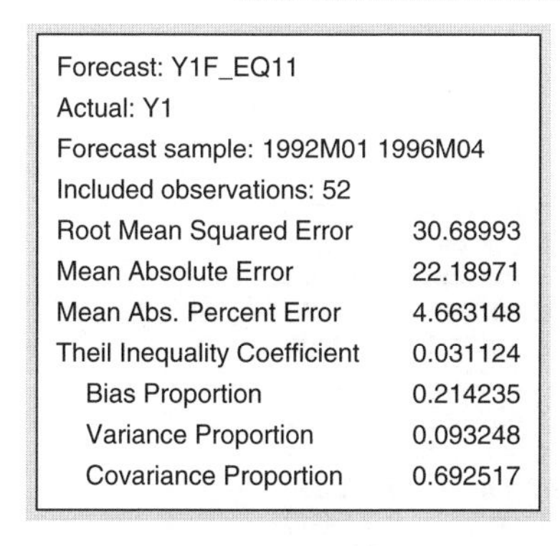
Forecast: Y1F_EQ11
Actual: Y1
Forecast sample: 1992M01 1996M04
Included observations: 52

Root Mean Squared Error	30.68993
Mean Absolute Error	22.18971
Mean Abs. Percent Error	4.663148
Theil Inequality Coefficient	0.031124
Bias Proportion	0.214235
Variance Proportion	0.093248
Covariance Proportion	0.692517

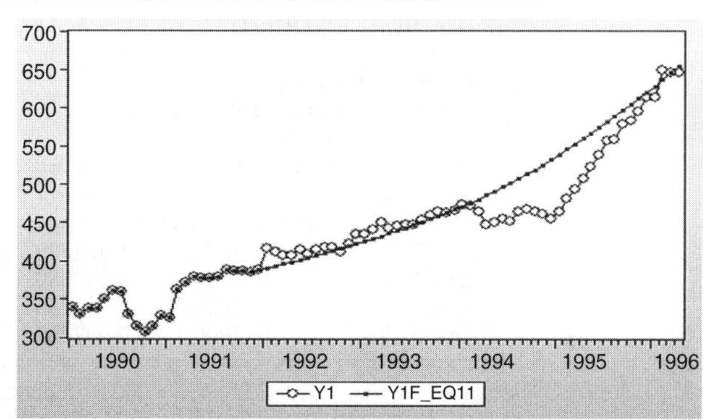

Figure 4.14 Statistical results of the LV(1) model with trend in (4.18).

4.5.2 Forecast Model with Trend

It is recognized that all models without the time predictors presented here can be modified easily to models with trend. So this section will present only some models as an extension of those previously presented.

Example 4.4 An Extension of the LV(2) Model in Figure 4.2 with Trend
As an extension of the model (4.6) with the results in Figure 4.4., Figure 4.14 presents the statistical results by using the following equation specification, based on the whole sample with the forecast sample {*Year*>1991}. Based on these results and the results in Figure 4.2, the following findings and notes are presented.

$$Y1\ C\ \ Y1(-1)\ X1(-1)\ X1(-1)*Y1(-1)\ @Trend \tag{4.18}$$

1) In the statistical sense, the regression function is a good fit function, since each of the independent variables has significant adjusted effects at the 1% level of significance, with an Adjusted R-squared of 0.999.

2) However, compared to the model (4.6) without trend, this model has greater RMSE and TIC. Hence, in the statistical sense, this model is a worse forecast model.

Example 4.5 An Extension of the Translog Linear Model (4.11) with Trend
Figure 4.15 presents the statistical using the following equation specification, based on the sub-sample {*Year*>1992}, with the forecast sample {*Year*>1993}. Based on these results the following findings are presented.

$$log(Y1)\ C\ log(Y(-1))\ log(X(-1))\ @Trend \tag{4.19}$$

1) The coefficient of *@Trend* indicates that *Y1* has an adjusted growth rate 0,43% over the last 40 months of observations, adjusted for or by taking into account both variables *Y1(−1)* and *X1(−1)*.
2) At the 10% level of significance, *log(X1(−1))* has a negative adjusted effect with a *p*-value = 0.1790/2 = 0.0895, but *@Trend* has a positive adjusted effect with a *p*-value = 0.1101/2 = 0.05505
3) The TIC = 0.019 < 0.05 indicates the model is a very good forecasting model.

Dependent Variable: LOG(Y1)
Method: Least Squares
Date: 12/18/16 Time: 11:19
Sample: 1992M11 1996M04 IF YEAR>1992
Included observations: 40

Variable	Coefficient	Std. Error	t-Statistic	Prob.
C	1.980107	1.652889	1.197967	0.2388
LOG(Y1(-1))	0.865519	0.092680	9.338775	0.0000
LOG(X1(-1))	-0.755587	0.551255	-1.370668	0.1790
@TREND	0.004311	0.002632	1.637928	0.1101

R-squared	0.983465	Mean dependent var	6.206530
Adjusted R-squared	0.982087	S.D. dependent var	0.125649
S.E. of regression	0.016817	Akaike info criterion	-5.238253
Sum squared resid	0.010181	Schwarz criterion	-5.069365
Log likelihood	108.7651	Hannan-Quinn criter.	-5.177188
F-statistic	713.7400	Durbin-Watson stat	1.683257
Prob(F-statistic)	0.000000		

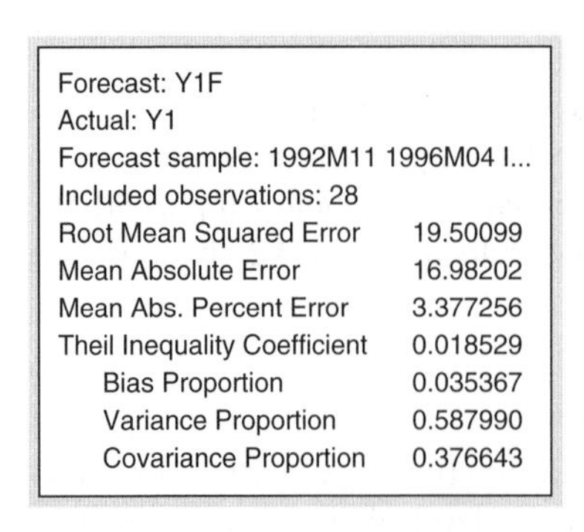
Forecast: Y1F
Actual: Y1
Forecast sample: 1992M11 1996M04 I...
Included observations: 28

Root Mean Squared Error	19.50099
Mean Absolute Error	16.98202
Mean Abs. Percent Error	3.377256
Theil Inequality Coefficient	0.018529
Bias Proportion	0.035367
Variance Proportion	0.587990
Covariance Proportion	0.376643

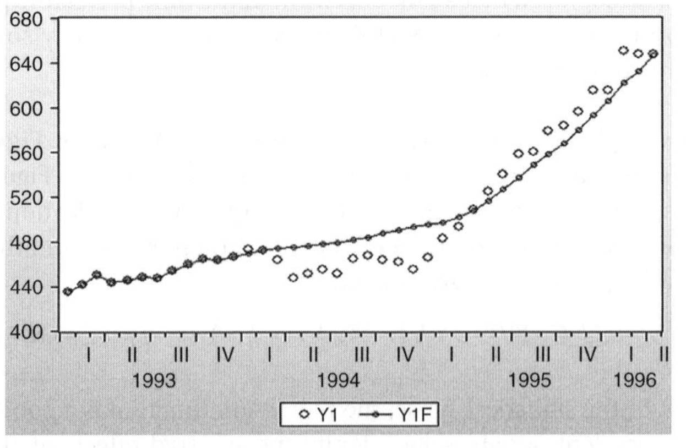

Figure 4.15 Statistical results of the LV(1) model with trend in (4.19).

4) The regression function in Figure 4.15 can be presented as a nonlinear function as follows:

$$Y1 = C(1)*Y1(-1)\hat{}C(2)*X1(-1)\hat{}C(3)*Exp(C(4)*@Trend) \quad (4.20a)$$

$$Y1 = 1.980*Y1(-1)\hat{}0.866*X1(-1)\hat{}(0.756)*Exp(0.004*@TREND) \quad (4.20b)$$

Example 4.6 An Extension of the Translog Linear Model in Figure 4.12 with Trend
Figure 4.16 presents the statistical results using the following equation specification (SE), as an extension of the model in (4.19). Based on these results, the following comments are presented.

$$Yul\ C\ log(Y1(-1))\ log(X1(-1))\ @Trend \quad (4.21)$$

1) Compared to the results in Figure 4.15, where *log(X1(−1))* has a significant negative adjusted effect on *log(Y1)*. In this model, however, it has an insignificant effect, which is the impact of *@Trend* as an additional independent variable. It is found *log(X1(−1))* and *@Trend* have significant correlation with $r = 0.969999$. Hence the model should be modified.

Dependent Variable: YUL
Method: Least Squares
Date: 12/19/16 Time: 06:30
Sample: 1992M11 1996M04 IF YEAR>1992
Included observations: 40

Variable	Coefficient	Std. Error	t-Statistic	Prob.
C	-0.253978	0.158148	-1.605948	0.1170
LOG(Y1(-1))	0.062605	0.008868	7.059941	0.0000
LOG(X1(-1))	-0.039476	0.052744	-0.748437	0.4591
@TREND	0.000123	0.000252	0.490140	0.6270

R-squared	0.959858	Mean dependent var	0.016383
Adjusted R-squared	0.956513	S.D. dependent var	0.007716
S.E. of regression	0.001609	Akaike info criterion	-9.931746
Sum squared resid	9.32E-05	Schwarz criterion	-9.762858
Log likelihood	202.6349	Hannan-Quinn criter.	-9.870682
F-statistic	286.9403	Durbin-Watson stat	1.301709
Prob(F-statistic)	0.000000		

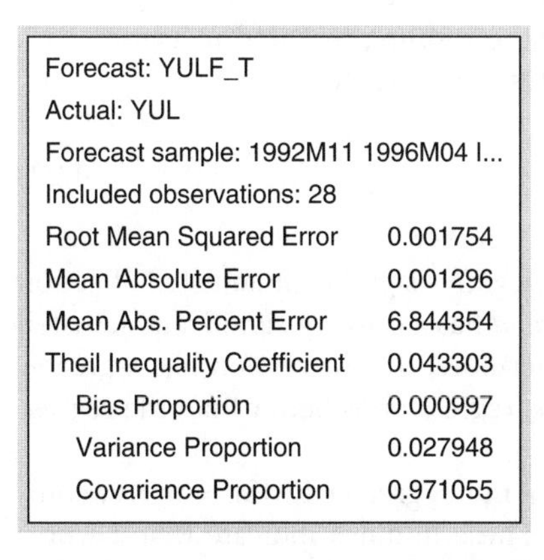

Forecast: YULF_T
Actual: YUL
Forecast sample: 1992M11 1996M04 I...
Included observations: 28

Root Mean Squared Error	0.001754
Mean Absolute Error	0.001296
Mean Abs. Percent Error	6.844354
Theil Inequality Coefficient	0.043303
Bias Proportion	0.000997
Variance Proportion	0.027948
Covariance Proportion	0.971055

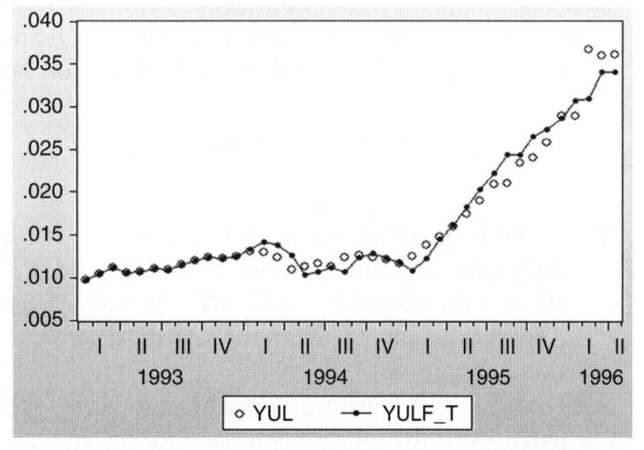

Figure 4.16 Statistical results of the LV(1) model with trend in (4.21).

Dependent Variable: YUL1
Method: Least Squares
Date: 12/19/16 Time: 06:42
Sample: 1992M11 1996M04 IF YEAR>1992
Included observations: 40

Variable	Coefficient	Std. Error	t-Statistic	Prob.
C	-12.49808	4.481949	-2.788536	0.0084
LOG(Y1(-1))	2.479997	0.251310	9.868282	0.0000
LOG(X1(-1))	-2.013230	1.494774	-1.346846	0.1864
@TREND	0.010852	0.007138	1.520463	0.1371

R-squared	0.984241	Mean dependent var	-0.522641
Adjusted R-squared	0.982927	S.D. dependent var	0.348989
S.E. of regression	0.045600	Akaike info criterion	-3.243187
Sum squared resid	0.074856	Schwarz criterion	-3.074299
Log likelihood	68.86373	Hannan-Quinn criter.	-3.182122
F-statistic	749.4493	Durbin-Watson stat	1.873553
Prob(F-statistic)	0.000000		

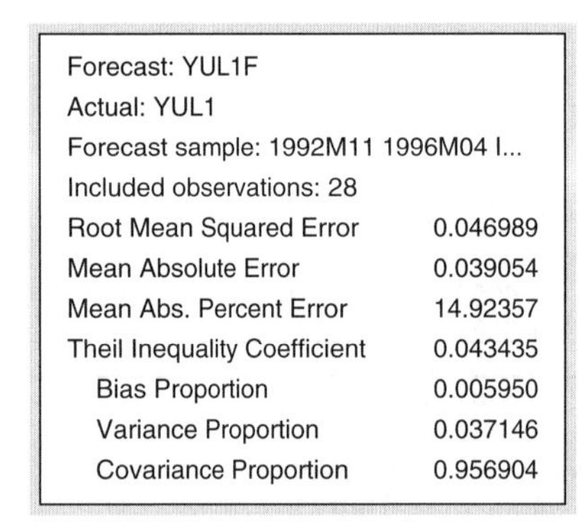

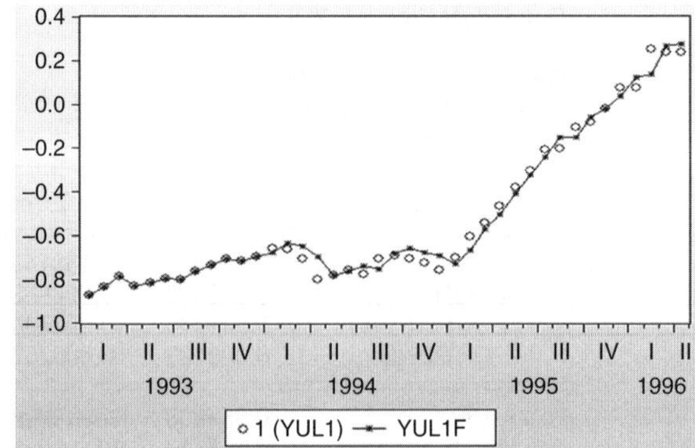

Figure 4.17 Statistical results of the LV(1) model with trend in (4.22).

2) By looking at the deviations of the last three observed values and their forecasts, and selecting alternative upper or lower bounds for *Y1*, I present an alternative regressions using the following SE, with *Yul1* = *log((Y1−200)/(1000−Y1))* and the results presented in Figure 4.17. Based on these results, the following findings and notes are presented.

$$Yul1\ C\ log(Y1(-1))\ \ log(X1(-1))\ @Trend \tag{4.22}$$

2.1 At the 10% level of significance, *Log(X1(−1))* has a negative significant adjusted effect on *Yul1*, with a *p*-value = 0.1864/2 = 0.0932, and *@Trend* has a positive significant adjusted effect, with a *p*-value = 0.1371/2 = 0.06805. Furthermore, this regression has a greater Adjusted R-squared. So, in the statistical sense, this regression is better than the regression in Figure 4.16.

2.2 However, this regression presents a forecast with a much greater RMSE = 0.046, compared to 0.0018 of the regression in Figure 4.16. These findings indicate that a worse regression can give a better forecasting within the sample period.

2.3 Thence, these findings present a type of contradictory result. Since, in general, we expect a better fit regression would present a better forecast, at least within the sample period.

4.6 Forecasting of *FSXDP*

By observing the graph of *FSXDP* in Figure 4.1, it seems to be very difficult to forecast *FSXDP* beyond the sample period, say for t = *@trend* > 591. However, by observing the scatter graph of *(Y2(−1),Y2)* with regression line, where *Y2* = *FSXDP*, as presented in Figure 4.18, then *Y2(−1)* can be considered to be one of the best predictors of *Y2*. For this reason, then alternative *LVARMA (p,q,r)* model of *Y2*, such as presented Section 3.2, could easily be applied. However, we can never predict the best possible forecast model. Do this as an exercise. In the following subsections, illustrative forecasting will be done using a subsample in order to present simple possible forecast models based on the bivariate *(X1,Y2)*.

4.6.1 Forecasting of Y2 Based on a Subsample

By observing the graphs of *Y2* = *FSXDP* and *X1* = *IP* based on the whole sample as presented in Figure 4.1, at the first stage I try to observe the graphs of *(Y2(−1),Y2)* and *(X1(−1),Y2*) for the subsample {*@year*>1986}, as presented in Figure 4.19, which show that *Y2* has a positive trend with respect to *Y2(−1)*, and has a negative trend with respect to *X1(−1)* in a two-dimensional space (orthogonal coordinates).

Even though *X1(−1)* has a nonlinear effect on *Y2* in a two-dimensional space, in a three-dimensional space with the axes *X1(−1)*, *Y2(−1)*, and *Y2*, in general, the linear effect of *X1 (−1)* on *Y2* would be accepted for each score of *Y2(−1)*, or the linear effect of *Y2(−1)* on *Y2* would be accepted for each score of *X1(−1)*. This statement is supported by the results in Figure 4.20, which shows that there are at most three observed scores of *X1(−1)* for each of the scores of *Y2 (−1)* based on a subsample *@Year*>1994 as an illustration. Then the set of regression lines of *Y2* on *X1(−1)* in the three-dimensional space are acceptable and show the correct specification model. For these reasons, the model applied should have the interaction *Y2(−1)*X1(−1)* as an independent variable (IV) with the following general equation.

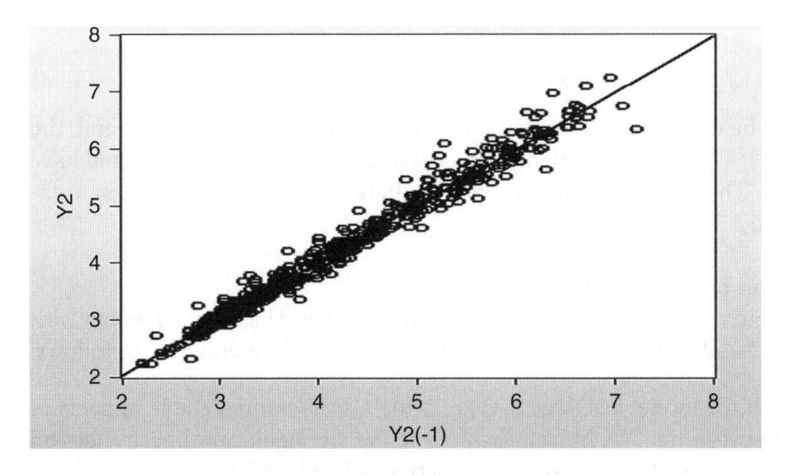

Figure 4.18 Scatter graph of (*Y2(−1),Y2*) with regression line.

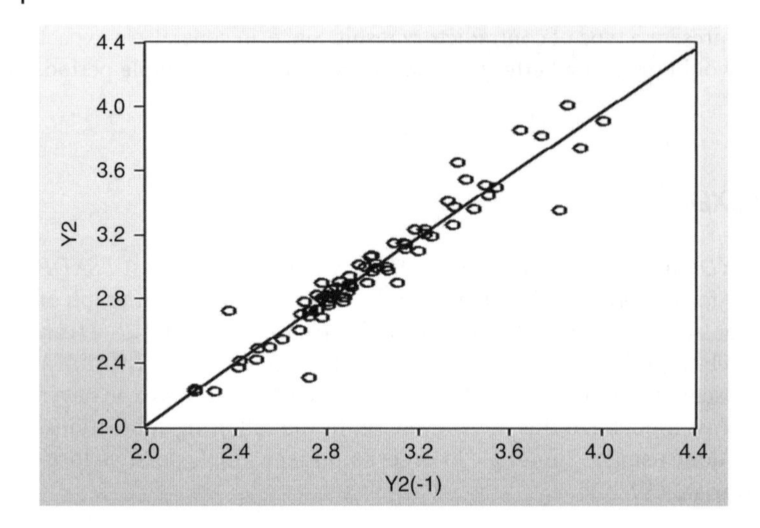

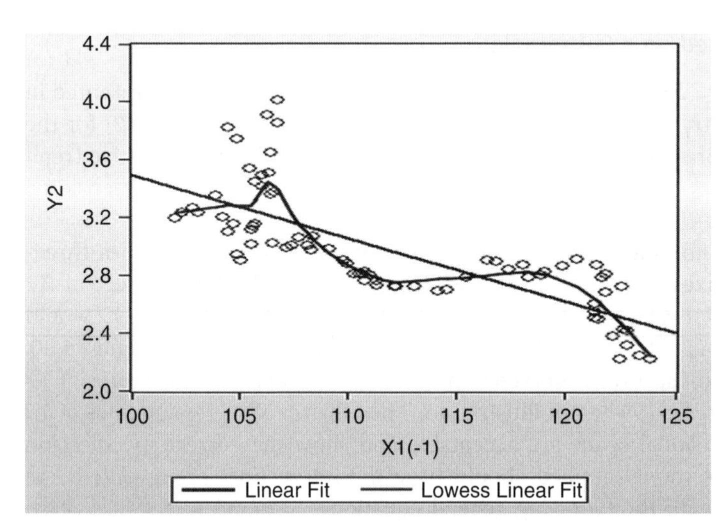

Figure 4.19 Scatter graphs of (*Y2(−1),Y2*) and (*X1(−1), Y2*) with regression lines.

$$Y2 = \beta_0 + \beta_1 * Y2(-1) + \beta_2 * X1(-1) + \beta_3 * Y2(-1) * X1(-1) + \varepsilon \tag{4.23}$$

The data analysis could be done using the equation specification (4.1). Since, in general, the IVs: *Y2(−1), X1(−1),* and *Y2(−1)∗X1(−1)* are highly correlated, at least one of the IVs would have large or very large p-value. Therefore, a reduced model would be an acceptable model, in a statistical sense. See Example 4.7.

Example 4.7 Application of the Model (4.23)

Figure 4.21 presents the statistical results of the LV(1) model (4.23) and its reduced model with their forecast evaluations. Based on these results, the findings and notes presented are as follows:

1) Note that each of IVs of the model has large p-value, which is the unpredictable impacts of the multicollinearity between the IVs. In fact, each of the IVs has significant linear effect on Y2, as presented in the results of their correlation matrix in Figure 4.20.

Covariance Analysis: Ordinary
Date: 12/27/16 Time: 15:51
Sample (adjusted): 1990M01 1996M04
Included observations: 76 after adjustments

Correlation t-Statistic Probability	Y2	Y2(-1)	X1(-1)	Y2(-1)*X1(-1)
Y2	1.000000 ----- -----			
Y2(-1)	0.959375 29.25152 0.0000	1.000000 ----- -----		
X1(-1)	-0.763946 -10.18427 0.0000	-0.771496 -10.43115 0.0000	1.000000 ----- -----	
Y2(-1)*X1(-1)	0.862577 14.66651 0.0000	0.913983 19.37723 0.0000	-0.450729 -4.343547 0.0000	1.000000 ----- -----

Figure 4.20 Bivariate correlation of the four variables in the model (4.23).

2) Even though $Y2(-1)*X1(-1)$ has the greatest p-value, it cannot be deleted in order to have a reduced model to show the effect of $Y2(-1)$ on $Y2$ depends on $X1(-1)$. Since, $X1(-1)$ has the second greatest p-value, then I decide to delete it from the model, with the statistical results presented in Figure 4.21, with the regression function as follows:

$$Y2 = 0.19195 + 1.05748*Y2(-1) - 0.00114*Y2(-1)*X1(-1) \tag{4.24}$$

3) Even though $X1(-1)$ and $Y2(-1)*X1(-1)$ have large p-values, the full model (4.23) has smaller RMSE and TIC than its reduced model. Hence, the full model can be considered to be a better forecast model of $Y2$. This finding indicates that a model with all significant independent variables does not directly become a good forecast model.
4) As a comparison, by deleting $Y2(-1)$ from both models in Figure 4.21, the statistical results in Figure 4.22 are obtained. Based on the four models of Y2 in these two figures, the following findings and notes are presented.

4.1 Since the third model,

$$Y2 = \beta_0 + \beta_1*X1(-1) + \beta_2*Y2(-1)*X1(-1) + \epsilon \tag{4.25}$$

has the smallest RMSE and TIC, then it is the best forecasting model in the statistical sense.

4.2 And the forth model,

$$Y2 = \delta_0 + \delta_1*Y2(-1)*X1(-1) + \mu \tag{4.26}$$

is the worst forecast model of $Y2 = FSXDP$, with TIC close to one.

4.3 Note that the IV $Y2(-1)$ of the full model (4.23) has the smallest p-value, which is deleted from the full model. So, unexpectedly the reduced model would have given the best forecasting.

4.4 Another possible reduced model is an additive model $Y2 = \gamma_0 + \gamma_1*Y2(-1) + \gamma_2*X1(-1) + \nu$. However, I would consider this model as inappropriate to demonstrate that the linear effect of $Y2(-1)$ on $Y2$ depends on $X1(-1)$ or the linear effect of $X1(-1)$ on $Y2$ depends on $Y2(-1)$, also for the additive model with multivariate independent variables.

Dependent Variable: Y2
Method: Least Squares
Date: 12/27/16 Time: 13:27
Sample: 1947M01 1996M04 IF @YEAR>1989
Included observations: 76

Variable	Coefficient	Std. Error	t-Statistic	Prob.
C	1.128094	1.928907	0.584836	0.5605
Y2(-1)	0.727324	0.682855	1.065122	0.2904
X1(-1)	-0.008493	0.017438	-0.487071	0.6277
Y2(-1)*X1(-1)	0.001873	0.006274	0.298608	0.7661

R-squared	0.921895	Mean dependent var	2.966053
Adjusted R-squared	0.918641	S.D. dependent var	0.393464
S.E. of regression	0.112230	Akaike info criterion	-1.485345
Sum squared resid	0.906874	Schwarz criterion	-1.362675
Log likelihood	60.44312	Hannan-Quinn criter.	-1.436320
F-statistic	283.2803	Durbin-Watson stat	1.951783
Prob(F-statistic)	0.000000		

4.4
4.0
3.6
3.2
2.8
2.4
2.0
1990 1991 1992 1993 1994 1995 1996
o Y2 —— Y2F_EQ01

RMSE=0.170500 & TIC=0.032829

Dependent Variable: Y2
Method: Least Squares
Date: 12/27/16 Time: 13:29
Sample: 1947M01 1996M04 IF @YEAR>1989
Included observations: 76

Variable	Coefficient	Std. Error	t-Statistic	Prob.
C	0.191948	0.162337	1.182405	0.2409
Y2(-1)	1.057477	0.082213	12.86261	0.0000
Y2(-1)*X1(-1)	-0.001138	0.001060	-1.073808	0.2864

R-squared	0.921638	Mean dependent var	2.966053
Adjusted R-squared	0.919491	S.D. dependent var	0.393464
S.E. of regression	0.111642	Akaike info criterion	-1.508371
Sum squared resid	0.909862	Schwarz criterion	-1.416369
Log likelihood	60.31811	Hannan-Quinn criter.	-1.471603
F-statistic	429.2873	Durbin-Watson stat	1.922279
Prob(F-statistic)	0.000000		

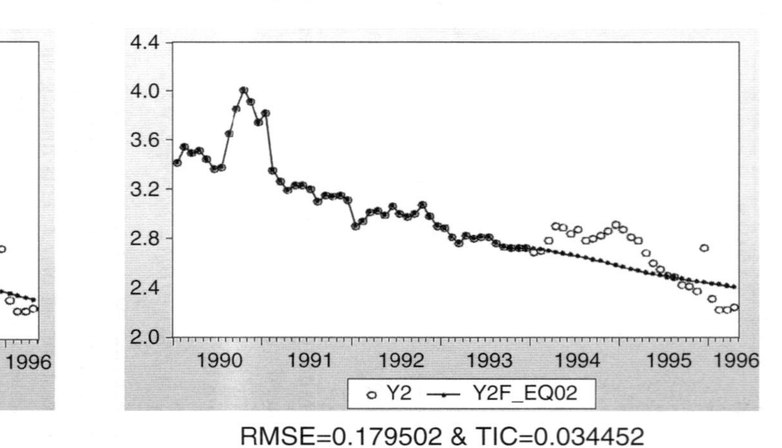

RMSE=0.179502 & TIC=0.034452

Figure 4.21 Statistical results of the LV(1) model (4.23) and its reduced model.

Dependent Variable: Y2
Method: Least Squares
Date: 12/27/16 Time: 16:26
Sample: 1947M01 1996M04 IF @YEAR>1989
Included observations: 76

Variable	Coefficient	Std. Error	t-Statistic	Prob.
C	3.150305	0.341063	9.236722	0.0000
X1(-1)	-0.026930	0.002112	-12.74839	0.0000
Y2(-1)*X1(-1)	0.008536	0.000485	17.61070	0.0000

R-squared	0.920665	Mean dependent var	2.966053
Adjusted R-squared	0.918491	S.D. dependent var	0.393464
S.E. of regression	0.112333	Akaike info criterion	-1.496027
Sum squared resid	0.921163	Schwarz criterion	-1.404024
Log likelihood	59.84903	Hannan-Quinn criter.	-1.459258
F-statistic	423.5728	Durbin-Watson stat	1.983187
Prob(F-statistic)	0.000000		

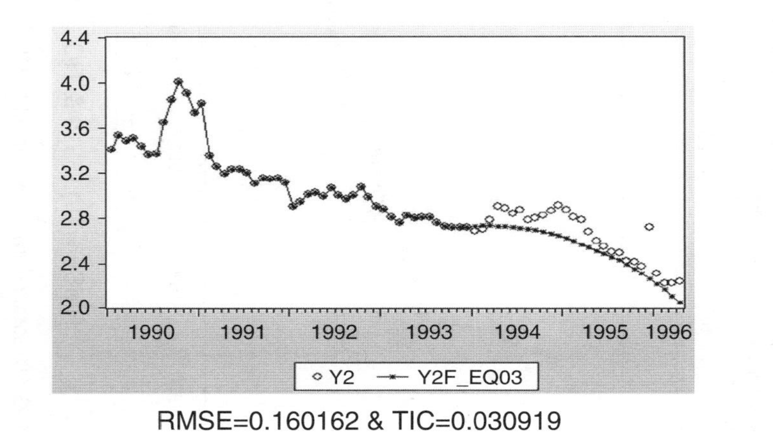

RMSE=0.160162 & TIC=0.030919

Dependent Variable: Y2
Method: Least Squares
Date: 12/27/16 Time: 16:54
Sample: 1947M01 1996M04 IF @YEAR>1989
Included observations: 76

Variable	Coefficient	Std. Error	t-Statistic	Prob.
C	-0.790322	0.257149	-3.073403	0.0030
Y2(-1)*X1(-1)	0.011321	0.000772	14.66651	0.0000

R-squared	0.744039	Mean dependent var	2.966053
Adjusted R-squared	0.740580	S.D. dependent var	0.393464
S.E. of regression	0.200404	Akaike info criterion	-0.351000
Sum squared resid	2.971968	Schwarz criterion	-0.289665
Log likelihood	15.33801	Hannan-Quinn criter.	-0.326488
F-statistic	215.1065	Durbin-Watson stat	0.859327
Prob(F-statistic)	0.000000		

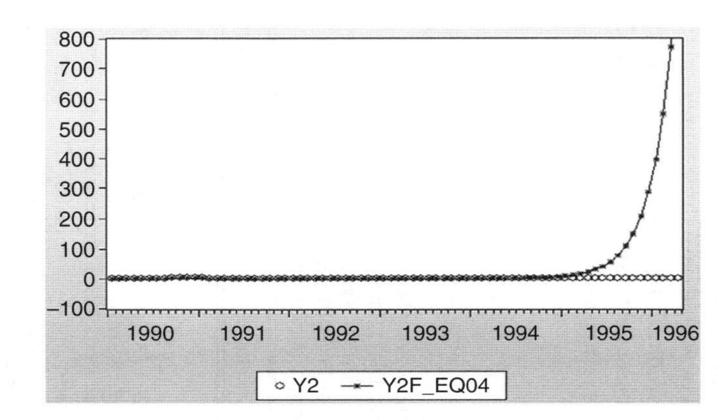

RMSE=290.5311 & TIC=0.987273

Figure 4.22 Statistical results by deleting *Y2(−1)* from both models in Figure 4.27.

4.6.2 Extension of the Model (4.25) with Time Variables

It is well-known that the growth of a time series depends on the time variable. For the monthly time series we would have three numerical time variables, such as *@Trend* or $t = @Trend/100$, *@Month* = *Month*, and *@Year* = *Year*. The following examples present alternative models with time variables as extensions of the model (4.25). For the extensions of the other models, please do these as exercises.

Dependent Variable: Y2
Method: Least Squares
Date: 01/02/17 Time: 05:57
Sample: 1947M01 1996M04 IF @YEAR>1989
Included observations: 76

Variable	Coefficient	Std. Error	t-Statistic	Prob.
C	6.289776	1.452300	4.330908	0.0000
Y2(-1)	0.042476	0.251671	0.168775	0.8664
Y2(-1)*X1(-1)	0.006104	0.001964	3.107838	0.0027
@TREND	-0.009893	0.002344	-4.220224	0.0001

R-squared	0.937178	Mean dependent var	2.966053
Adjusted R-squared	0.934560	S.D. dependent var	0.393464
S.E. of regression	0.100653	Akaike info criterion	-1.703089
Sum squared resid	0.729427	Schwarz criterion	-1.580419
Log likelihood	68.71739	Hannan-Quinn criter.	-1.654064
F-statistic	358.0319	Durbin-Watson stat	1.994596
Prob(F-statistic)	0.000000		

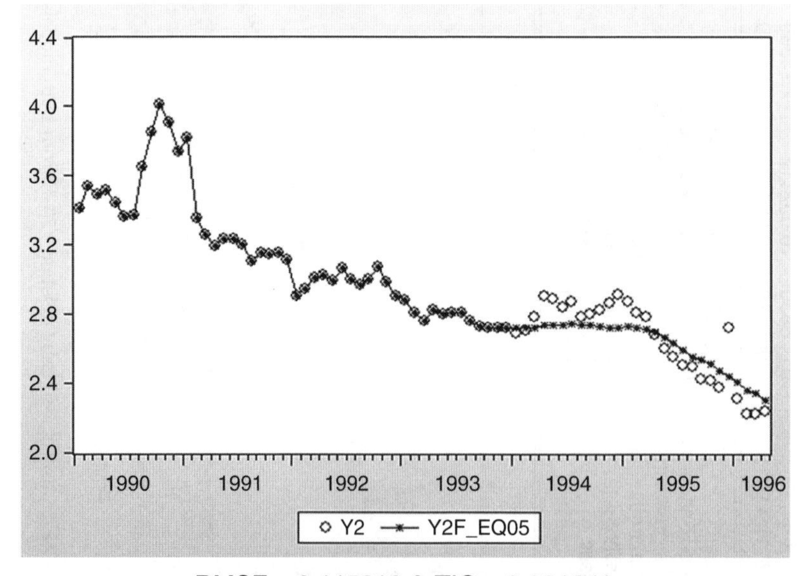

RMSE = 0.115310 & TIC = 0.021849

Figure 4.23 Statistical results of the model (4.27).

4.6.2.1 Extension of the Model (4.25) with Trend

As the simplest extension of the model (4.24) is the model with trend, use the following equation specification where *@Trend* is a numerical variable that can be replaced by t = *@Trend* or t = *@Trend*/100.

$$Y2\ C\ Y2(-1)\ Y2(-1)*X1(-1)\ @Trend \tag{4.27}$$

Example 4.8 Application of the Model (4.27)

Figure 4.23 presents the statistical results of the model (4.27). Based on these results, the following findings and notes are presented.

1) Compared to the model (4.24), this model has a smaller RMSE and TIC. Hence, this is a better forecast model.
2) Even though *Y2(−1)* has a very large p-value = 0.8664, it should not be deleted from the model, because if it is, then the reduced model would have been the extension of (4.26), which is the worst forecast model. Furthermore, since *Y2(−1)*X1(−1)* has a significant adjusted effect, then this is a suitable model to show that the effect of *X1(−1)* on *Y2* is significantly dependent on *Y2(−1)* or the effect of *Y2(−1)* on *Y2* is significantly dependent on *X1(−1).*

4.6.2.2 Extension of the Model (4.24) with Trend by @Year

A more advanced extension of the model (4.24) is a model with polygonal or heterogeneous trend by *@Year,* with the equation specification as follows:

$$Y2\ Y2(-1)\ Y2(-2)*X1(-1)\ @Trend*@Expand(@Year)\ @Expand(@Year) \tag{4.28a}$$

Note that in this equation *@Trend* is a numerical variable and *@Expand(@Year)* is there to present a set of dummy variables of the selected *@Year* > 1989. It is found that exactly the same statistical results would be obtained using the following equation specification, where *@Month* is the numerical trend variable.

$$Y2\ \ Y2(-1)\ Y2(-1)*X1(-1)\ @Month*@Expand(@Year)\ @Expand(@Year) \tag{4.28b}$$

Example 4.9 Application of the Model (4.28b)

Figure 4.24 presents the statistical results of the model (4.28b). Based on these results, the following findings and notes are presented.

1) Compared to the model (4.27), this model has smaller RMSE and TIC. Hence this model is a better forecast model.
2) In addition, note that each of the variables *Y2(−1)* and *Y2(−1)*X1(−1)* has significant adjusted effect on *Y2*. Hence, in the statistical sense, this model presents a better estimate compared to the model (4.27).
3) Furthermore, note that the regression function in fact presents a set of seven regression functions. The first and seventh regressions have the following equations.

$$Y2(1990) = -1.360*Y2(-1) + 0.016*Y2(-1)*X1(-1) + 0.025*@Month + 0.048$$

$$Y2(1996) = -1.360*Y2(-1) + 0.016*Y2(-1)*X1(-1) + 0.073*@\mathrm{Month} + 0.480$$

4) To compute the predicted scores of *Y2* beyond the sample period, the regressions of *X1* in (4.9b), (4.15) or other alternative models could be applied. Do these as exercises.
5) Compared to the model (4.27), which shows a negative trend within the forecast period, this model presents a positive trend within the *Year* = 1996. Hence, I consider this model a better forecasting, since within the year 1996 the observed scores of *Y2* tend to increase.

Dependent Variable: Y2
Method: Least Squares
Date: 01/02/17 Time: 07:04
Sample: 1947M01 1996M04 IF @YEAR>1989
Included observations: 76

Variable	Coefficient	Std. Error	t-Statistic	Prob.
Y2(-1)	-1.360235	0.775426	-1.754177	0.0845
Y2(-1)*X1(-1)	0.016488	0.006743	2.445264	0.0174
@MONTH*(@YEAR=1990)	0.025067	0.009640	2.600175	0.0117
@MONTH*(@YEAR=1991)	-0.032694	0.012704	-2.573613	0.0126
@MONTH*(@YEAR=1992)	-0.019415	0.011291	-1.719520	0.0907
@MONTH*(@YEAR=1993)	-0.015799	0.008845	-1.786218	0.0791
@MONTH*(@YEAR=1994)	-0.020638	0.013183	-1.565533	0.1227
@MONTH*(@YEAR=1995)	-0.004878	0.009599	-0.508153	0.6132
@MONTH*(@YEAR=1996)	0.072888	0.044011	1.656143	0.1029
@YEAR=1990	2.048260	0.412008	4.971405	0.0000
@YEAR=1991	2.281211	0.498590	4.575327	0.0000
@YEAR=1992	1.879425	0.393442	4.776884	0.0000
@YEAR=1993	1.558140	0.343929	4.530412	0.0000
@YEAR=1994	1.335374	0.317813	4.201765	0.0001
@YEAR=1995	0.933213	0.375015	2.488466	0.0156
@YEAR=1996	0.479524	0.372403	1.287647	0.2028

R-squared	0.959167	Mean dependent var	2.966053
Adjusted R-squared	0.948958	S.D. dependent var	0.393464
S.E. of regression	0.088893	Akaike info criterion	-1.818108
Sum squared resid	0.474116	Schwarz criterion	-1.327427
Log likelihood	85.08810	Hannan-Quinn criter.	-1.622008
Durbin-Watson stat	1.935723		

Forecast: Y2F_EQ08	
Actual: Y2	
Forecast sample: 1947M01 1996M04 I...	
Included observations: 28	
Root Mean Squared Error	0.084101
Mean Absolute Error	0.066281
Mean Abs. Percent Error	2.579692
Theil Inequality Coefficient	0.015846
Bias Proportion	0.000209
Variance Proportion	0.003973
Covariance Proportion	0.995818

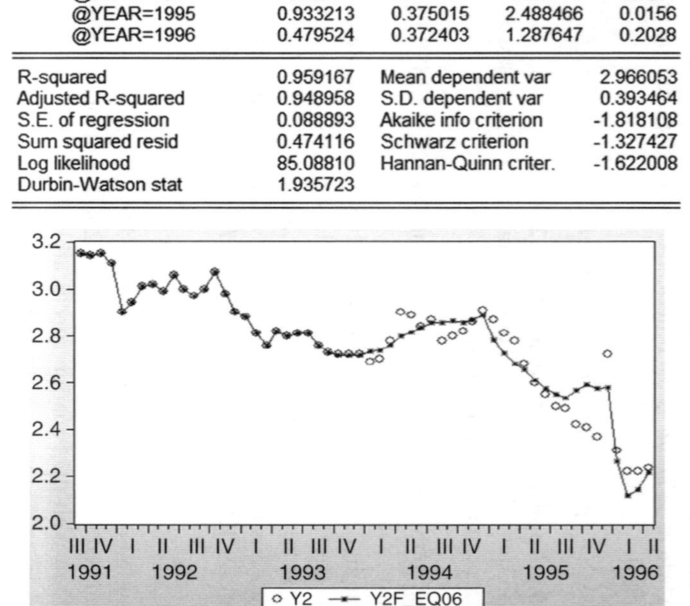

Figure 4.24 Statistical results based on the model (4.28b).

4.6.2.3 Extension of the Model (4.25) with Trend by Time Period

By observing the coefficients of the variable *@Month* in Figure 4.24, which are positive for the *@Year* = 1990, negative for the *@Years* = 1991 up to 1995, and positive for the *@Year* = 1996, then the model (4.28) can be modified by using the Time-Period (TP) with three levels; TP = 1 for *@Year* = 1990, TP = 2 for *@Years* = 1991 up to 1995, and TP = 3 for *@Year* = 1996, which can easily be generated using the following equation.

Dependent Variable: Y2
Method: Least Squares
Date: 01/02/17 Time: 11:30
Sample: 1989M05 1996M04 IF @YEAR>1989
Included observations: 76

Variable	Coefficient	Std. Error	t-Statistic	Prob.
Y2(-1)	-0.524110	0.507929	-1.031856	0.3058
Y2(-1)*X1(-1)	0.009803	0.004093	2.394875	0.0194
@TREND*(TP=1)	0.019066	0.008919	2.137667	0.0361
@TREND*(TP=2)	-0.015331	0.005221	-2.936212	0.0045
@TREND*(TP=3)	0.080080	0.041705	1.920156	0.0590
TP=1	-8.182575	4.464853	-1.832664	0.0712
TP=2	9.768375	3.082318	3.169165	0.0023
TP=3	-46.57649	24.63504	-1.890661	0.0629

R-squared	0.953376	Mean dependent var	2.966053
Adjusted R-squared	0.948577	S.D. dependent var	0.393464
S.E. of regression	0.089224	Akaike info criterion	-1.896025
Sum squared resid	0.541347	Schwarz criterion	-1.650685
Log likelihood	80.04896	Hannan-Quinn criter.	-1.797975
Durbin-Watson stat	2.020821		

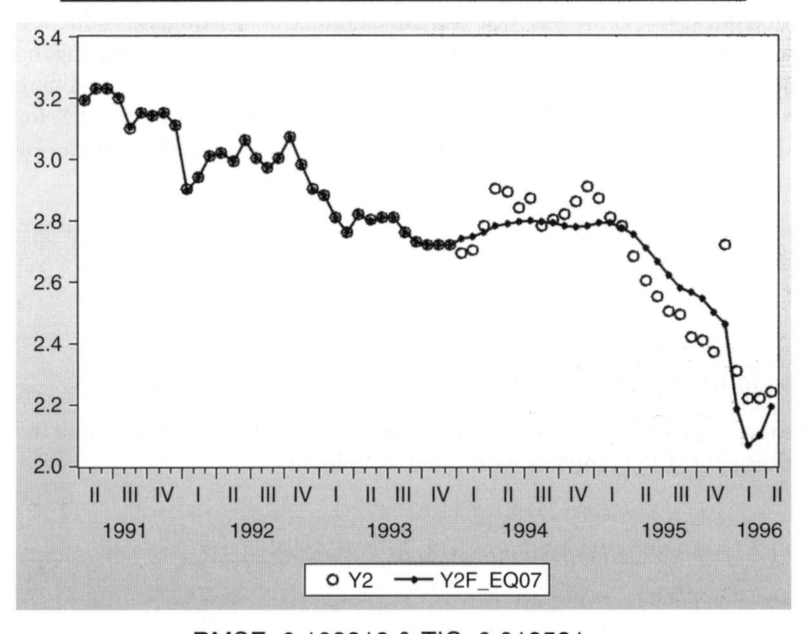

RMSE=0.103318 & TIC=0.019521

Figure 4.25 Statistical results of the model (4.29).

$$TP = 1 + 1*(@Year > 1990) + 1*(@Year > 1995)$$

Then as a modification or a simplification of the model (4.28a), we have a model with the following equation specification.

$$Y2\ Y2(-1)\ Y2(-1)*X1(-1)\ @Trend*@Expand(TP))@Expand(TP) \tag{4.29}$$

Example 4.10 Application of the Model (4.29)

Figure 4.25 presents the statistical results of the model (4.29). Based on these results, the findings and notes presented are as follows:

1) Compared to the model (4.28a), (4.28b), which presents a set of seven multiple regressions, this model presents only three multiple regressions. So this model is a far simpler model with relatively small values of RMSE and TIC.
2) Similar to the model (4.29), this model also has a positive trend within the *Year* = 1996, but with a greater slope of 0.080 compared to the 0.073 of (4.28a), (4.28b). So both models can be used to predict the scores beyond the sample period.
3) However, to compute the predicted scores of *Y2* beyond the sample period, the regression of *X1* in (4.9b) or (4.15) should be applied. Otherwise, the readers also could easily apply alternative models for *X1*. Do these for exercises.

4.7 Translog Linear Models

4.7.1 Basic Translog Linear Model

Based on the bivariate time series (X_t, Y_t), I propose the following two equation specifications to present two basic translog linear LV(1) models, which are in fact the Cobb–Douglas production functions. Note that the models are additive models. However, I would recommend applying the first model, since we can be very confident that *log(Y(−1))* and *log(X(−1))* are the causes or source factors for *Y*. These LV(1) models could easily be extended to LVARMA(*p,q,r*) models of *Y*, for various $p > 0$, $q \geq 0$, and $r \geq 0$. Refer to Subsection 3.2.1 for illustrative examples. By using the *log (Y)* as a dependent variable, we could forecast either *Y* or *log(Y)*

$$log(Y)\ C\ log(Y(-1))\ log(X1(-1)) \tag{4.30}$$

$$log(Y)C\ log(Y(-1))\ log(X1) \tag{4.31}$$

4.7.2 Tanslog Linear Model with Trend

As the extension of the model (4.30), based on the monthly time series, I propose to conduct the forecasting based on alternative LV(1) models with trend as follows:

$$log(Y)\ C\ log(Y(-1))\ log(X(-1))@Trend \tag{4.32}$$

$$log(Y)\ C\ log(Y(-1))\ log(X(-1))\ @Month*@Expand(@Year)\ @Expand(@Year,@Droplast) \tag{4.33a}$$

$$log(Y)\ C\ log(Y(-1))\ log(X(-1))\ @Trend*@Expand(@Year)\ @Expand(@Year,@Droplast) \tag{4.33b}$$

Example 4.11 An Alternative LV(1) Model with Trend

As a modification of the model (4.29), this example presents an alternative additive translog linear model with heterogeneous trends, with the following equation specification based on the subsample {@*Year*>1989}. Its estimate and the forecast evaluation of *Y2* are presented in Figure 4.26. In fact, we can also forecast directly the variable *log(Y)* by selecting the suitable option. Based on the results in this figure, the following findings and notes are presented.

Dependent Variable: LOG(Y2)
Method: Least Squares
Date: 01/05/17 Time: 12:16
Sample: 1989M05 1996M04 IF @YEAR>1989
Included observations: 76

Variable	Coefficient	Std. Error	t-Statistic	Prob.
C	-17.33955	8.121345	-2.135059	0.0364
LOG(Y2(-1))	0.470787	0.108433	4.341717	0.0000
LOG(X1(-1))	1.576214	0.573819	2.746885	0.0077
@TREND*(TP=1)	0.005687	0.002764	2.057189	0.0435
@TREND*(TP=2)	-0.007344	0.002316	-3.171270	0.0023
@TREND*(TP=3)	0.017232	0.015117	1.139930	0.2583
TP=1	7.703015	9.342209	0.824539	0.4125
TP=2	14.54984	8.344205	1.743706	0.0857

R-squared	0.957192	Mean dependent var	1.078739
Adjusted R-squared	0.952786	S.D. dependent var	0.130741
S.E. of regression	0.028408	Akaike info criterion	-4.184962
Sum squared resid	0.054879	Schwarz criterion	-3.939622
Log likelihood	167.0286	Hannan-Quinn criter.	-4.086912
F-statistic	217.2149	Durbin-Watson stat	1.819953
Prob(F-statistic)	0.000000		

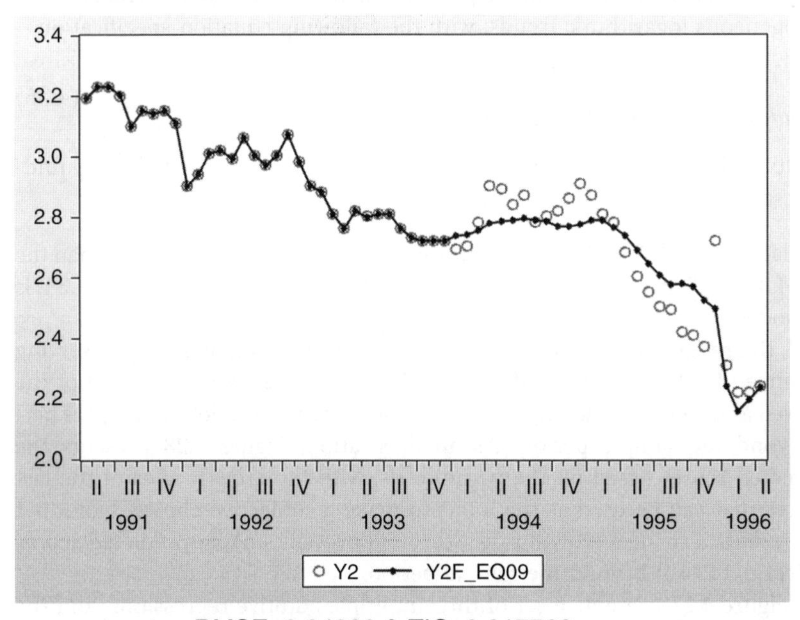

RMSE=0.04090 & TIC=0.017762

Figure 4.26 Statistical results based on the model (4.34).

$$log(Y2)\ C\ log(Y2(-1))\ log(XI(-1))$$
$$@Trend*@Expand(TP)\ @Expand(TP,@Drop(3)) \tag{4.34}$$

1) Compared to the forecast of *Y2* in Figure 4.24, the forecast based on this model has smaller RMSE and TIC. So, in the statistical sense, this model is a better forecast model.
2) Each of *log(Y2(−1))* and *log(X1(−1))* has a significant adjusted effect at the 1% level of significance, with positive coefficients of 0.470787 and 1.576214, respectively.
3) For the last time period (TP = 3), the following multiple regression is obtained, where *@Trend* has insignificant adjusted effect with a *p*-value = 0.2583.

$$LOG(Y2) = -17.33955 + 0.47079*LOG(Y2(-1))$$
$$+ 1.57621*LOG(X1(-1)) + 0.017232*@TREND$$

3. The additive function here can be written as a multiplicative function as follows:

$$Y2 = \left\{Exp(-17.33955) \times (Y2(-1))^{0.47079} \times (X2(-1))^{1.57621}\right\} \times Exp(0.017232*@Trend)$$

Note that the coefficient of *@Trend* in fact presents an exponential growth rate of *Y2*, adjusted for or by taking into account *log(Y2(−1)) & log(X1(−1))* over the last four months: 1996M01 up to 1996M04.

4) The null hypothesis H_0:C(2) = C(3) = ...= C(7) = 0 is rejected based on the *F*-statistic of $F_0 = 217.2149$ with a *p*-value 0.000000. Hence, it can be concluded that all independent variables have significant joint effects.

Example 4.12 An Alternative LV(1) Model with Logarithmic Trend

As a modification of the model (4.34), this example presents an alternative additive translog linear model with heterogeneous logarithmic trends, with the following equation specification.

$$log(Y2)\ C\ log(Y2(-1))\ log(X1(-1))$$
$$log(t)*@Expand(TP)\ @Expand(TP,@Drop(3)) \tag{4.35}$$

Based on these results and its forecast evaluation presented in Figure 4.27, the following findings and notes are presented.

1) Its forecast has a relatively small RMSE and TIC, however, they are a little bit greater than the RMSE and TIC of (4.34) in Figure 4.25. Despite this, in the statistical sense, the model (4.34) is a better forecast model.
2) On the other hand, the researcher and users should be using their judgments by observing the growths difference of *Y2f_Eq09* = *Y2f_434* and *Y2F_Eq10* = *Y2f_435* at the end of the sample period, as well as un-observable cause factors, to select which model would gives better predictions beyond the sample period. As an illustration, Figure 4.28 presents the observed and predicted scores based on the two models, with the growth rates of the last two predicted scores that can be used as the input to make a subjective choice. Note that the model (4.34) presents a smaller growth rate. So, it will present smaller predicted scores than the model (4.41a), (4.41b) beyond the sample period.
3) The estimation in Figure 4.27 presents a set of three multiple additive regressions, and the third regression for TP = 3 has the equation is as follows:

Dependent Variable: LOG(Y2)
Method: Least Squares
Date: 01/07/17 Time: 11:37
Sample: 1989M05 1996M04 IF @YEAR>1989
Included observations: 76

Variable	Coefficient	Std. Error	t-Statistic	Prob.
C	-74.75330	55.75172	-1.340825	0.1844
LOG(Y2(-1))	0.479709	0.107958	4.443469	0.0000
LOG(X1(-1))	1.461935	0.551042	2.653036	0.0099
LOG(T)*(TP=1)	2.907202	1.442997	2.014698	0.0479
LOG(T)*(TP=2)	-3.855350	1.245268	-3.096000	0.0028
LOG(T)*(TP=3)	10.67746	8.903005	1.199309	0.2346
TP=1	50.41468	59.79705	0.843096	0.4021
TP=2	92.77634	53.50341	1.734027	0.0874

R-squared	0.956936	Mean dependent var	1.078739
Adjusted R-squared	0.952503	S.D. dependent var	0.130741
S.E. of regression	0.028493	Akaike info criterion	-4.179001
Sum squared resid	0.055207	Schwarz criterion	-3.933660
Log likelihood	166.8020	Hannan-Quinn criter.	-4.080951
F-statistic	215.8661	Durbin-Watson stat	1.792784
Prob(F-statistic)	0.000000		

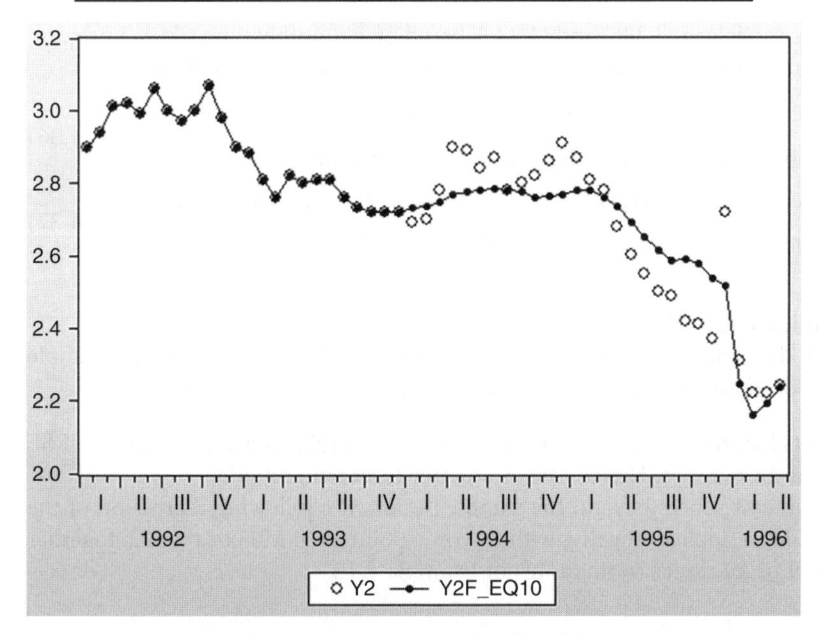

RMSE=0.098138 & TIC=0.018526

Figure 4.27 Statistical results based on the model (4.35).

Figure 4.28 Observed and predicted scores of *Y2* with their selected growth rates.

	T	Y2	Y2f_Eq09	Y2f_Eq10
1996M01	588	2.31	2.238444	2.245996
1996M02	589	2.22	2.155565	2.158428
1996M03	590	2.22	2.19336	2.192473
1996M04	591	2.24	2.235525	2.236036
Growth rate of the last two points			0.019041	0.019674

$$LOG(Y2) = -74.75330 + 0.47969*LOG(Y2(-1))$$
$$+ 1.46218*LOG(X1(-1)) + 10.69428*LOG(T)$$

which can be written as a multiplicative function as follows:

$$Y2 = Exp(-74.75330) \times (Y2(-1))^{0.47969} \times (X1(-1))^{1.46218} \times (T)^{10.69428}$$

4) The predicted scores of *Y2* beyond the sample period can easily be computed based on this equation using Excel, by using the regression of *X1* in (4.9b) or (4.15), step by step. Otherwise, the readers also could easily apply alternative models for *X1*. Do these as exercises.
5) Note that Figure 4.28 presents the minimum observed score of *Y2* = 2.22 based on the whole sample, and the observed score increases to 2.24 at the last time sample period. Based on these conditions, I expect greater increasing predicted scores beyond the sample period. For these reasons, I would select the model (4.26) as a better forecasting model, since its forecast scores have greater growth rates as presented in Figure 4.28.

4.7.3 Heterogeneous Tanslog Linear Model

As the extension of the models in (4.33b) and (4.33c), based on a monthly time series, I propose two heterogeneous translog linear models with the equation specifications as follows.

$$log(Y2)\ C\ log(Y2(-1))*@Expand(@Year)\ log(X1(-1))*@Expand(@Year)$$
$$log(@Month)*@Expand(@Year)\ @Expand(@Year, @Droplast) \quad (4.36)$$

$$log(Y2)\ C\ log(Y2(-1))*@Expand(@Year)\ log(X1(-1))*@Expand(@Year)$$
$$log(t)*@Expand(@Year)\ @Expand(@Year, @Droplast) \quad (4.37)$$

Example 4.13 Application of the Model (4.36)
Figure 4.29 presents the statistical results based on the model (4.36), for the subsample {*@Year*>1989}. Based on these results, findings and notes presented are as follows:

1) Compared to results based on the model (4.33b), it was found this model has smaller MRSE and TIC. Do this as an exercise. Hence, this is a better forecasting model.
2) To compute the forecast scores beyond the sample period, the following regression of the *@Year* = 1996 should be applied, starting with the time point 1964 M05 or $t = 592$, together with a LV(p) model of *X1*. Refer to the notes in Example 4.12.

$$LOG(Y2) = -0.26306 + 0.38216*LOG(Y2(-1))$$
$$+ 0.14924*LOG(X1(-1)) + 0.03327*LOG(@MONTH) \quad (4.38)$$

Example 4.14 Unexpected "Error Message" Based on the Model (4.37)
Figure 4.30 shows a part of the estimates of model (4.37) based on a subsample {*@Year*>1989} as the background of the error message *"Log of non positive number."*

The error message was obtained in the process of computing its forecast evaluation or in using the object *"Forecast."* It is a very unexpected error message, since a forecast evaluation is obtained based on the model (4.36), as presented in Figure 4.29. I cannot find an explanation yet.

Dependent Variable: LOG(Y2)
Method: Least Squares
Date: 01/08/17 Time: 10:36
Sample: 1989M05 1996M04 IF @YEAR>1989
Included observations: 76

Variable	Coefficient	Std. Error	t-Statistic	Prob.
C	-0.263062	19.75129	-0.013319	0.9894
LOG(Y2(-1))*(@YEAR=1990)	0.829025	0.181196	4.575302	0.0000
LOG(Y2(-1))*(@YEAR=1991)	-0.541041	0.348841	-1.550971	0.1275
LOG(Y2(-1))*(@YEAR=1992)	0.285783	0.516587	0.553213	0.5827
LOG(Y2(-1))*(@YEAR=1993)	0.050376	0.924457	0.054493	0.9568
LOG(Y2(-1))*(@YEAR=1994)	0.421828	0.419698	1.005076	0.3199
LOG(Y2(-1))*(@YEAR=1995)	0.191820	0.387676	0.494794	0.6230
LOG(Y2(-1))*(@YEAR=1996)	0.382157	0.522583	0.731285	0.4682
LOG(X1(-1))*(@YEAR=1990)	3.207271	1.679624	1.909517	0.0622
LOG(X1(-1))*(@YEAR=1991)	2.403109	0.922073	2.606203	0.0122
LOG(X1(-1))*(@YEAR=1992)	-2.881680	2.088200	-1.379983	0.1740
LOG(X1(-1))*(@YEAR=1993)	-1.315133	2.077848	-0.632930	0.5298
LOG(X1(-1))*(@YEAR=1994)	-0.368469	1.839028	-0.200361	0.8420
LOG(X1(-1))*(@YEAR=1995)	1.743636	2.236307	0.779694	0.4394
LOG(X1(-1))*(@YEAR=1996)	0.149236	4.136923	0.036074	0.9714
LOG(@MONTH)*(@YEAR=1990)	-0.001393	0.015419	-0.090323	0.9284
LOG(@MONTH)*(@YEAR=1991)	-0.127467	0.034488	-3.695946	0.0006
LOG(@MONTH)*(@YEAR=1992)	0.061429	0.039941	1.537993	0.1306
LOG(@MONTH)*(@YEAR=1993)	-0.007260	0.022506	-0.322598	0.7484
LOG(@MONTH)*(@YEAR=1994)	0.022695	0.043980	0.516020	0.6082
LOG(@MONTH)*(@YEAR=1995)	-0.057133	0.035245	-1.621033	0.1116
LOG(@MONTH)*(@YEAR=1996)	0.033273	0.092477	0.359801	0.7206
@YEAR=1990	-14.47037	21.27706	-0.680093	0.4997
@YEAR=1991	-8.869097	20.15444	-0.440057	0.6619
@YEAR=1992	14.41648	21.86904	0.659219	0.5129
@YEAR=1993	7.441563	22.28404	0.333941	0.7399
@YEAR=1994	2.583047	21.59321	0.119623	0.9053
@YEAR=1995	-7.247258	22.54800	-0.321415	0.7493

R-squared	0.973291	Mean dependent var	1.078739
Adjusted R-squared	0.958267	S.D. dependent var	0.130741
S.E. of regression	0.026709	Akaike info criterion	-4.130351
Sum squared resid	0.034241	Schwarz criterion	-3.271659
Log likelihood	184.9533	Hannan-Quinn criter.	-3.787176
F-statistic	64.78255	Durbin-Watson stat	1.839731
Prob(F-statistic)	0.000000		

Forecast: Y2F_EQ11	
Actual: Y2	
Forecast sample: 1989M01 1996M04 I...	
Included observations: 28	
Root Mean Squared Error	0.071269
Mean Absolute Error	0.049416
Mean Abs. Percent Error	1.883930
Theil Inequality Coefficient	0.013435
Bias Proportion	0.000800
Variance Proportion	0.002018
Covariance Proportion	0.997182

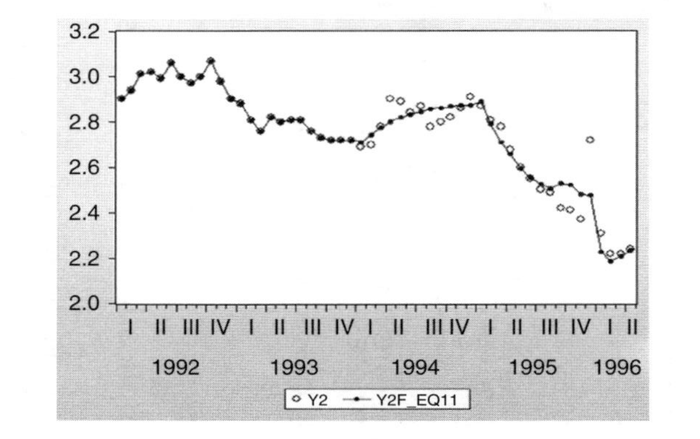

Figure 4.29 Statistical results based on the model (4.36).

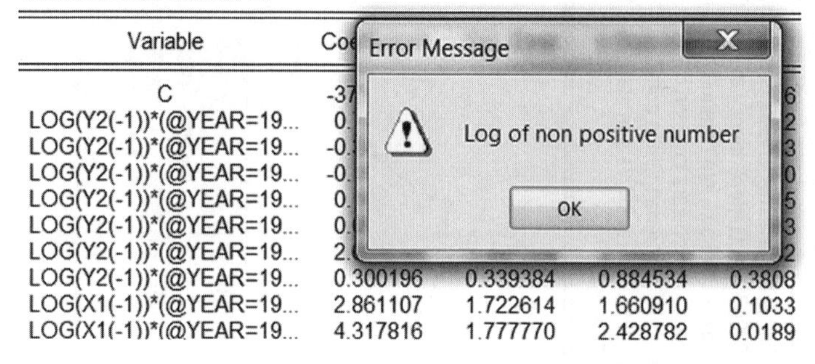

Dependent Variable: LOG(Y2)
Method: Least Squares
Date: 01/08/17 Time: 10:41
Sample: 1989M05 1996M04 IF @YEAR>1989
Included observations: 76
Variable
Error Message
Log of non positive number
OK
C
LOG(Y2(-1))*(@YEAR=19...
LOG(Y2(-1))*(@YEAR=19... 0.300196 0.339384 0.884534 0.3808
LOG(X1(-1))*(@YEAR=19... 2.861107 1.722614 1.660910 0.1033
LOG(X1(-1))*(@YEAR=19... 4.317816 1.777770 2.428782 0.0189

Figure 4.30 Unexpected error message.

4.8 Application of VAR Models

It should already be well-known that all multivariate autoregressive models (MAR) are valid models for forecasting. So, all MAR models presented in Agung (2009a) can be easily be applied for the bivariate (X_t,Y_t). However, the object/option *"Forecast"* is not applicable for the MAR models. But the forecast values beyond the sample period can easily be computed using Excel. This section only presents additional illustrative examples of the VAR models. If computing the in-sample forecast values is needed, then each regression of a VAR model should be estimated individually.

It is known that the statistical results are obtained by selecting *Object/New Object /VAR ... OK*, so then we can insert the endogenous variables with their lag intervals, and additional exogenous variables. The default options are Unrestricted VAR, with lag intervals 1, 2, and C as an exogenous variable. Refer to all VAR models presented in Agung (2009a), and the following additional examples.

4.8.1 Unstructured VAR Models Based on $(X1_t,Y1_t)$

4.8.1.1 The Simplest VAR Model

The simplest VAR model with the lag intervals "1 1," namely VAR(1,1) model, has the following equations, which can be obtained from the output of the statistical results.

$$Y1 = C(1,1)*Y1(-1) + C(1,2)*X1(-1) + C(1,3) \quad (4.39a)$$

$$X1 = C(2,1)*Y1(-1) + C(2,2)*X1(-1) + C(2,3) \quad (4.39b)$$

with the following estimates, which can be used to compute the forecast values of *Y1* and *X1* by using Excel starting from the time period 1996M5 or t = *@Trend* + 1 = 593.

$$\mathrm{Y1} = 1.0125945685*\mathrm{Y1}(-1) - 0.0119825113166*\mathrm{X1}(-1) + 0.0956888223815 \quad (4.40a)$$

$$\mathrm{X1} = 0.000777170250331*\mathrm{Y1}(-1) + 0.998232195589*\mathrm{X1}(-1) + 0.180336730783 \quad (4.40b)$$

4.8.1.2 The Simplest Translog Linear VAR Model

The simplest translog linear VAR model, namely TL_VAR(1,1) model, has the equation as follows:

$$LOG(Y1) = C(1,1)*LOG(Y1(-1)) + C(1,2)*LOG(X1(-1)) + C(1,3) \quad (4.41a)$$

$$LOG(X1) = C(2,1)*LOG(Y1(-1)) + C(2,2)*LOG(X1(-1)) + C(2,3) \quad (4.41b)$$

with the following estimates, which can be used to compute the forecast values of *Y1* and *X1* by using Excel starting from the time period 1996M5 or t = *@Trend* + 1 = 593. However, remember the symbol LOG should be replaced by LN in Excel.

$$\begin{aligned} LOG(Y1) = {} & 1.00458442882*LOG(Y1(-1)) \\ & -0.00914295994341*LOG(X1(-1)) + 0.0228263420192 \end{aligned} \quad (4.42a)$$

$$\begin{aligned} LOG(X1) = {} & 0.00533279530197*LOG(Y1(-1)) \\ & +0.989126808748*LOG(X1(-1)) + 0.0230398887276 \end{aligned} \quad (4.42b)$$

4.8.1.3 The Simplest Interaction VAR Model

The simplest interaction VAR model, namely I_VAR(1,1) model, has the following equation, which can be obtained by inserting *Y1(−1)*X1(−1)* as an additional exogenous variable for the VAR(1,1).

$$Y1 = C(1,1)*Y1(-1) + C(1,2)*X1(-1) + C(1,3) + C(1,4)*Y1(-1)*X1(-1) \quad (4.43a)$$

$$X1 = C(2,1)*Y1(-1) + C(2,2)*X1(-1) + C(2,3) + C(2,4)*Y1(-1)*X1(-1) \quad (4.43b)$$

with the following estimates, which can be used to compute the forecast values of *Y1* and *X1* by using Excel starting from the time period 1996M5 or t = *@Trend* + 1 = 593. Note that the results

$$\begin{aligned} Y1 = {} & 0.964192730615*Y1(-1) + 0.00609117947255*X1(-1) \\ & +0.9905485058420 + 0.000369892797659*Y1(-1)*X1(-1) \end{aligned} \quad (4.44a)$$

$$\begin{aligned} X1 = {} & 0.00900941152166*Y1(-1) + 0.995158201094*X1(-1) \\ & +0.0281379478455 - 6.29118002091e-05*Y1(-1)*X1(-1) \end{aligned} \quad (4.44b)$$

4.8.1.4 Forecast Values Beyond the Sample Period

Figure 4.31 shows the spreadsheet for computing the forecast values beyond the sample period, based on the VAR(1,1), TL_VAR(1,1), and I_VAR(1,1) models. The steps of the computation are as follows:

1) Copy the last four observed scores of the three variables *t*, *Y1*, and *X1*. In fact, it is sufficient to use the last observed score, namely t = 592.
2) The six forecast values of the variables *Y1* and *X1* based on the three models VAR(1,1), TL_VAR(1,1), and I_VAR(1,1) for t = 593 can easily be computed by using each of the equations in Figure 4.32, which are derived from the Eqs. (4.40), (4.42a), (4.42b), and (4.44a), (4.44b). Note that the symbol "LOG" in the Eq. (4.42a), (4.42b) should be replaced by "LN."
3) Then the other forecast values for $t > 593$ can be obtained by using the block-copy-paste method of the six forecast values for t = 593.

	A	B	C	D	E	F	G	H
1			VAR(1,1)		TL_VAR(1,1)		I_VAR(1,1)	
2		T	Y1	X1				
3	1996M01	589	614.42	122.5				
4	1996M02	590	649.542	123.9				
5	1996M03	591	647.07	123.4				
6	1996M04	592	647.17	124.5				
7	1996M05	593	653.9247	124.9632	652.6233	125.1367	655.5487	124.687
8			660.7589	125.4308	659.4439	125.6042	662.4875	125.1368
9			667.6736	125.903	666.3448	126.0761	669.5100	125.5906

Figure 4.31 Spreadsheet for computing the forecast values beyond the sample period.

THE EQUATIONS USED TO COMPUTE FORECAST VALUES VAR(1,1)
*=1.0125945685*C6 - 0.0119825113166*D6 + 0.0956888223815*
*= 0.000777170250331*C6 + 0.998232195589*D6 + 0.180336730783*

THE EQUATIONS USED TO COMPUTE FORECAST VALUES TL_VAR(1,1)
*=EXP(1.00458442882*LN(C6)-0.00914295994341*LN(D6)+0.0228263420192)*
*=EXP(0.00533279530197*LN(C6)+ 0.989126808748*LN(D6) + 0.0230398887276)*

THE EQUATIONS USED TO COMPUTE FORECAST VALUES I_VAR(1,1)
*=0.964192730615*C6+0.00609117947255*D6+0.990548505842+0.000369892797659*C6*D6*
*=0.00900941152166*C6 0.995158201094*D6+ 0.028137947845-0.0000629118002091*C6*D6*

Figure 4.32 The equations used to compute the forecast values for t = 593.

Table 4.1 Forecast models with alternative *Trends* for the simplest V(1.1) models.

Additional Time IV to insert	Forecast Model (FM) with
1. @Trend	Linear Trend
2. @Trend^2 @trend	Quadratic Trend
3. @Trend^3 @Trend^2 @Trend	Cubic Trend
4. log(@Trend+1) = log(t)	Logarithmic Trend
5. @Month*@Expand(@Year) @Expand(@Year,@Droplast)	Heterogeneous Linear Trend by Years
6. @Month*@Expand(@Year) @Month^2*@Expand(@Year) @Expand(@Year,@Droplast)	Heterogeneous Quadratic Trend by Years
7. @Month*@Expand(@Year) @Month^2*@Expand(@Year) @Month^3*@Expand(@Year) @Expand(@Year,@Droplast)	Heterogeneous Third Degree Polynomial Trend by Years

4.8.2 The Simplest VAR Models with Alternative Trends

As an extension of each of the simplest VAR(1,1), TL_VA(1,1), and I_VAR(1,1) models, I propose the VAR models with alternative specific Trends, as presented in Table 4.1, which can easily be analyzed by inserting additional exogenous variables. However, I only presenta few selected examples. Please do the others as exercises.

Example 4.15 TL_Var(1,1) with Logarithmic Trend
Since, by using the whole sample, I found that the exogenous variable *log(t)* has insignificant effects in both regressions, in the statistical sense *log(t)* can be deleted from both regressions. For this reason, I tried to use a subsample {@*Year*>1990}, and the results show *log(t)* has an insignificant effect on *log(Y1)* but it has significant effects on *log(X1)* with the equations as follows:

$$LOG(Y1) = 0.957569004238*LOG(Y1(-1)) + 0.139932650233*LOG(X1(1)) - 0.0879712079845 - 0.0479656498258*LOG(T) \quad (4.45a)$$

$$LOG(X1) = -0.029663194585*LOG(Y1(-1)) + 0.690917747766*LOG(X1(-1)) - 2.83893754567 + 0.708531223739*LOG(T) \quad (4.45b)$$

Example 4.16 I_VAR(1,1) with Linear Trend
As an extension of the I_VAR(1,1) in (4.43a), (4.43b), based on the whole sample, the following regressions are obtained, where the time variable $T = @Trend + 1$ has significant effect on *Y1* but an insignificant effect on *X1*.

$$Y1 = 0.911913227142*Y1(-1) - 0.251016919854*X1(-1) + 6.02782894958 + 0.000792148877291*Y1(-1)*X1(-1) + 0.0466007154522*T \quad (4.46a)$$

$$X1 = 0.00882067058953*Y1(-1) + 0.994229982211*X1(-1) + 0.0463236792151 - 6.1387359448e-05*Y1(-1)*X1(-1) + 0.00016823921207*T \quad (4.46b)$$

4.8.2.1 Unrestricted VAR Models with Heterogeneous Linear Trends by @Year
As a more complex extension of the I_VAR(1,1) models in (4.43a), (4.43b), I_VAR(1,1) models with heterogeneous linear trends (HLT) by @*Year* are proposed. The models can be developed by inserting the following exogenous independent variables into the VAR(1,1) models, respectively.

$$Y1(-1)*X1(-1)\ @Month*@Expand(@Year)\ @Expand(@Year) \quad (4.47a)$$

$$Y1(-1)*X1(-1)\ (T/100)*@Expand(@Year)\ @Expand(@Year) \quad (4.47b)$$

without the intercept or symbol *C*, so that the regression for each @*Year* can be written directly from the output. Note that both equations give the same results, and the numerical variable (*T*/100) is used, instead of *T*, especially for a large number of observations.

Example 4.17 I_VAR(1,1) with Heterogeneous Linear Trends by Year
The results of I_VAR(1,1) in (4.43a), (4.43b) with the exogenous variables (4.47a), present a set of 50 regressions of *Y1* and *X1* from @*Year* = 1947 up to @*Year* = 1996, with the last pairs of regressions having the following equations, which can be used to compute the forecast values beyond the sample period starting with the time period 1996M05.

$$Y1(1996) = 0.985547435679*Y1(-1) - 0.0414882698125*X1(-1) - 0.00341290621092*Y1(-1)*X1(-1) + 2.78307187127*@MONTH + 280.83359939 \quad (4.48a)$$

$$X1(1996) = 0.0411595809069*Y1(-1) + 0.757971201171*X1(-1) \\ - 0.000375146945489*Y1(-1)*X1(-1) + 0.449959800356*@MONTH \\ + 32.2912885364 \tag{4.48b}$$

As an illustration, Figure 4.33 presents a spreadsheet for computing three forecast values beyond the sample period. Refer to Figures 4.30 and 4.31.

4.8.2.2 Unrestricted VAR Models with Heterogeneous Linear Trends by @Month

As another complex extension of the I_VAR(1,1) in (4.43a), (4.43b), for the I_VAR(1,1) with HLT, namely I_VAR(1,1) with HLT, by @Month, the exogenous variables inserted are as follows:

$$Y1(-1)*X1(-1)\ @Year*@Expand(@Month)\ @Expand(@Month) \tag{4.49a}$$

$$Y1(-1)*X1(-1)\ (T/100)@Expand(@Month)\ @Expand(@Month) \tag{4.49b}$$

without the intercept or symbol *C*, so that the regression for each *@Month* can be written directly from the output. Note that both equations give the same results, the numerical variable $(T/100)$ instead of T is used, especially for a large number of observations. The results of the data analysis would present a set of 12 regressions, with the first pairs of the equations for *@Month* = 1, using the EVs (4.49b) as follows:

$$Y1(@Month = 1) = 0.914001950873*Y1(-1) - 0.243925082067*X1(-1) \\ + 0.000776379422291*Y1(-1)*X1(-1) + 4.93115038389*(T/100) \\ + 5.7635831232 \tag{4.50a}$$

$$X1(@Month = 1) = 0.00873058035265*Y1(-1) + 0.994367807856*X1(-1) \\ - 6.07476331286e-05*Y1(-1)*X1(-1) - 0.0268623808604*(T/100) \\ + 0.105765324676 \tag{4.50b}$$

4.8.2.3 Unrestricted VAR Models with Heterogeneous Trends by Time Period

As another complex extension of the I_VAR(1,1) in (4.43a), (4.43b), for the I_VAR(1,1) with HLT namely I_VAR(1,1) with HLT by TP, the exogenous variables inserted are as follows:

	A	B	C	D	E
1			**I_VAR(1,1) with HLT**		
2		T	Y1	X1	Month
3	1996M01	589	614.42	122.5	1
4	1996M02	590	649.542	123.9	2
5	1996M03	591	647.07	123.4	3
6	1996M04	592	647.17	124.5	4
7	1996M05	593	**652.4134**	**125.3192**	5
8			653.4952	125.7105	6
9			655.9927	126.3548	7

Figure 4.33 Forecast values beyond the sample period.

```
Estimation Proc:
===============================
LS(NOCONST) 1 1 Y1 X1  @ Y1(-1)*X1(-1) ( T/100)*@EXPAND(TP)
@EXPAND(TP)

VAR Model:
===============================
Y1 = C(1,1)*Y1(-1) + C(1,2)*X1(-1) + C(1,3)*Y1(-1)*X1(-1) + C(1,4)*(T/100)*
(TP=1) + C(1,5)*(T/100)*(TP=2) + C(1,6)*(T/100)*(TP=3) + C(1,7)*(TP=1)
+ C(1,8)*(TP=2) + C(1,9)*(TP=3)

X1 = C(2,1)*Y1(-1) + C(2,2)*X1(-1) + C(2,3)*Y1(-1)*X1(-1) + C(2,4)*(T/100)*
(TP=1) + C(2,5)*(T/100)*(TP=2) + C(2,6)*(T/100)*(TP=3) + C(2,7)*(TP=1)
+ C(2,8)*(TP=2) + C(2,9)*(TP=3)

VAR Model - Substituted Coefficients:
===============================
Y1 = 0.856414800484*Y1(-1) - 0.27926817066*X1(-1) + 0.00106284974883
*Y1(-1)*X1(-1) + 6.76440963262*(T/100)*(TP=1) + 5.30866851895*(T/100)*
(TP=2) + 9.87361986624*(T/100)*(TP=3) + 6.597828608*(TP=1) +
9.01436178472*(TP=2) - 8.81553007487*(TP=3)

X1 = 0.0151125321765*Y1(-1) + 0.993840865292*X1(-1) - 6.64492766177e-
05*Y1(-1)*X1(-1) - 0.22151165239*(T/100)*(TP=1) - 0.0495066687625*
(T/100)*(TP=2) - 1.5375645218*(T/100)*(TP=3) + 0.0627844018927*(TP=
1) - 0.267024005738*(TP=2) + 6.16142233124*(TP=3)


Estimation Command:
====================
LS(DERIV=AA)

Estimated Equations:
====================
Y1 = C(11)*Y1(-1) + C(12)*X1(-1) + C(13)*Y1(-1)*X1(-1) + C(14)*(T/100)*(TP=1)
+ C(15)*(T/100)*(TP=2) + C(16)*(T/100)*(TP=3) + C(17)*(TP=1) + C(18)*(TP=
2) + C(19)*(TP=3)

X1 = C(21)*Y1(-1) + C(22)*X1(-1) + C(23)*Y1(-1)*X1(-1) + C(24)*(T/100)*(TP=1)
+ C(25)*(T/100)*(TP=2) + C(26)*(T/100)*(TP=3) + C(27)*(TP=1) + C(28)*(TP=
2) + C(29)*(TP=3)

Substituted Coefficients:
====================
Y1 = 0.856414800483*Y1(-1) - 0.27926817066*X1(-1) + 0.00106284974883*Y1(-
1)*X1(-1) + 6.76440963263*(T/100)*(TP=1) + 5.30866851896*(T/100)*(TP=2) +
9.87361986629*(T/100)*(TP=3) + 6.597828608*(TP=1) + 9.01436178475*(TP=
2) - 8.81553007503*(TP=3)

X1 = 0.0151125321764*Y1(-1) + 0.993840865292*X1(-1) - 6.64492766167e-05
*Y1(-1)*X1(-1) - 0.22151165238*(T/100)*(TP=1) - 0.0495066687547*(T/100)*
(TP=2) - 1.53756452179*(T/100)*(TP=3) + 0.0627844019033*(TP=1) -
0.267024005722*(TP=2) + 6.16142233125*(TP=3)
```

Figure 4.34 Equations of the I_VAR(1,1) and their corresponding SCM.

Vector Autoregression Estimates
Date: 03/16/17 Time: 20:23
Sample (adjusted): 1947M02 1996M04
Included observations: 591 after adjustments
Standard errors in ( ) & t-statistics in [ ]

	Y1	X1
Y1(-1)	0.856415	0.015113
	(0.03406)	(0.00353)
	[25.1447]	[4.28666]
X1(-1)	-0.279268	0.993841
	(0.08189)	(0.00848)
	[-3.41013]	[117.243]
Y1(-1)*X1(-1)	0.001063	-6.64E-05
	(0.00027)	(2.8E-05)
	[3.88989]	[-2.34950]
(T/100)*(TP=1)	6.764410	-0.221512
	(1.62213)	(0.16791)
	[4.17008]	[-1.31926]
(T/100)*(TP=2)	5.308669	-0.049507
	(1.52599)	(0.15795)
	[3.47884]	[-0.31342]
(T/100)*(TP=3)	9.873620	-1.537565
	(4.09820)	(0.42420)
	[2.40926]	[-3.62460]

System: SYS01
Estimation Method: Least Squares
Date: 03/16/17 Time: 20:25
Sample: 1947M02 1996M04
Included observations: 591
Total system (balanced) observations 1182

	Coefficient	Std. Error	t-Statistic	Prob.
C(11)	0.856415	0.034060	25.14466	0.0000
C(12)	-0.279268	0.081894	-3.410134	0.0007
C(13)	0.001063	0.000273	3.889890	0.0001
C(14)	6.764410	1.622128	4.170084	0.0000
C(15)	5.308669	1.525988	3.478841	0.0005
C(16)	9.873620	4.098197	2.409259	0.0161
C(21)	0.015113	0.003525	4.286657	0.0000
C(22)	0.993841	0.008477	117.2429	0.0000
C(23)	-6.64E-05	2.83E-05	-2.349500	0.0190
C(24)	-0.221512	0.167906	-1.319263	0.1873
C(25)	-0.049507	0.157954	-0.313424	0.7540
C(26)	-1.537565	0.424202	-3.624602	0.0003
Determinant residual covariance		9.001602		

Figure 4.35 The first six parameters' estimates of the I_VAR(1,1) and their corresponding SCMs.

$$Y1(-1)*X1(-1)\ (T/100)*@Expand(TP)\ @Expand(TP) \tag{4.51}$$

without the intercept or symbol *C*, so that the regression for each TP can be written directly from the output. Note that the variable TP should be defined using real critical events, such as before and after the World Trade Center (September 11, 2001) terror attack in New York; the three time periods of the first and second Bali Attacks/Explosions; before, during, and after the Global Financial Crisis; or by looking the growth patterns of the endogenous variables.

Example 4.18 The VAR Model (4.51) and a System Equations Model

As an illustration, in this case, TP with three-levels is generated using the following equation.

$$TP = 1 + 1*(@Year >= 1970) + 1*(@Year >= 1984) \tag{4.52}$$

Figure 4.34 presents exactly the same pairs of equations of the VAR Model and an Estimated Functions using the object *"System"*, but by using different symbols of their parameters. Based on these results, the following findings and notes are presented.

1) The results of the system equations model (SEM), which was presented as SCM (Seemingly Casual Model) in Agung (2014, 2011a, 2009a), because the name "SEM" could be misleading, since it is also used for the Structural Equation Model.
2) Note that the VAR models are using the symbol C(i,j) for their parameters, but the SCM should be using the symbol C(ij). An error message will be obtained if the symbol C(i,j) is used for the SCM.
3) Figure 4.35 presents the first six parameters' estimates of the I_VAR(1,1) and their corresponding SCM, in order to show the equality of the assumptions of the error terms of both models, which are presented by the equality of the standard errors of each parameters' estimates. For instance, both C(1,1) and C(11) have S.E. = 0.03406. Those assumptions are the theoretical or mathematical concepts of statistics, which should be taken for granted. They should not be proven by using a sample data: refer to special notes and comments on the residual analysis presented in Agung (2011a, 2009a), with additional notes presented in Chapter 1.

Dependent Variable: Y1
Method: Least Squares
Date: 03/16/17 Time: 06:41
Sample (adjusted): 1947M02 1996M04
Included observations: 591 after adjustments
Y1 = C(11)*Y1(-1) + C(12)*X1(-1) + C(13)*Y1(-1)*X1(-1) + C(14)*(T/100)
(TP=1) + C(15)(T/100)*(TP=2) + C(16)*(T/100)*(TP=3) + C(17)
(TP=1) + C(18)(TP=2) + C(19)*(TP=3)

	Coefficient	Std. Error	t-Statistic	Prob.
C(11)	0.856415	0.034060	25.14466	0.0000
C(12)	-0.279268	0.081894	-3.410134	0.0007
C(13)	0.001063	0.000273	3.889890	0.0001
C(14)	6.764410	1.622128	4.170084	0.0000
C(15)	5.308669	1.525988	3.478841	0.0005
C(16)	9.873620	4.098197	2.409259	0.0163
C(17)	6.597829	1.794990	3.675691	0.0003
C(18)	9.014362	4.219250	2.136485	0.0331
C(19)	-8.815530	17.57144	-0.501697	0.6161

R-squared	0.998443	Mean dependent var	140.9656
Adjusted R-squared	0.998421	S.D. dependent var	136.5858
S.E. of regression	5.427042	Akaike info criterion	6.235777
Sum squared resid	17141.52	Schwarz criterion	6.302505
Log likelihood	-1833.672	Hannan-Quinn criter.	6.261770
Durbin-Watson stat	1.472890		

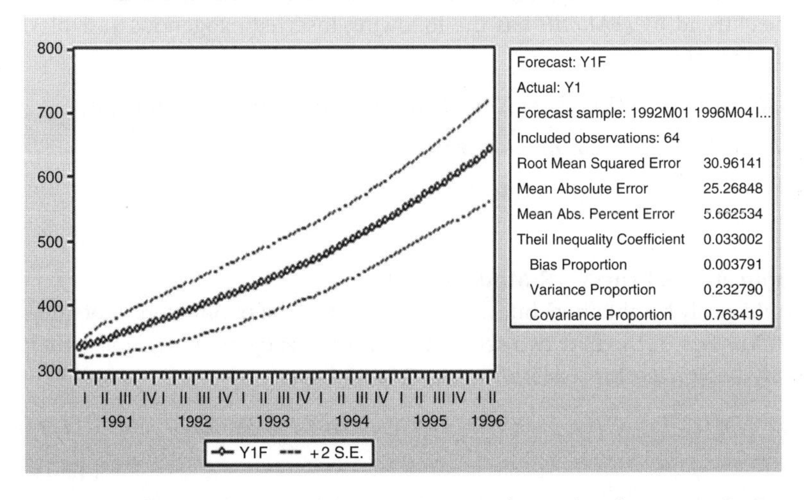

Figure 4.36 Statistical results of the LS regression of *Y1* and its forecast evaluation.

4) The object/option *"Forecast"* cannot be used to compute the in-sample forecast values, based on both VAR Model and SCM.

   The advantages of using the SCM are (i) the Wald-test can be applied; (ii) each of the models can be reduced; and (iii) each of the estimated function can be estimated as a LS Regression, then the object *"Forecast"* can be applied. As an illustration, Figure 4.26 presents the LS Regression of Y1, using the specific equation as presented in the output, and its forecast evaluation for @Year>1990.

5) Referring to the statistical results of the I_VAR(1,1) model and the SCM in Figures 4.34 and 4.35, and the LS Regression in Figure 4.36, I would say that the data analysis based each of the

models can be done easily. However, if a reduced model of the VAR model is considered important, then the object *"System"* should be applied.

4.8.3 Complete Heterogeneous VAR Models by @Month

This section will only present complete heterogeneous VAR models of *(X1,Y1)* as the extensions of the simplest I_VAR(1,1) in (4.43). Similar models could be developed easily, based on the VAR (1,1) model in (4.39a), (4.39b), and TL_VAR(1,1) model in (4.41). The results of each pairs of the VAR models would present a set of 12 multiple regressions with numerical independent variables.

4.8.3.1 A Complete Heterogeneous I_VAR(1,1) Model by @Month

As an extension of the I_VAR(1,1) in (4.43), the complete heterogeneous I_VAR(1,1) of *(X1,Y1)* by *@Month* has the following inserted exogenous variables, without the intercept or symbol C.

$$\begin{gathered} Y1(-1)*@Expand(@Month,@Dropfirst)\ X1(-1)*@Expand(@Month,@Dropfirst) \\ Y1(-1)*X1(-1)*@Expand\ (@Month)\ t*@Expand(@Month) \end{gathered} \tag{4.53}$$

4.8.3.1.1 A Complete Heterogeneous I_VAR(1,1) Model with Trend by @Month

As another complex extension of the I_VAR(1,1) in (4.43), the complete heterogeneous I_VAR (1,1) of *(X1,Y1)* with linear trend by *@Month* has the following inserted exogenous variables, without the intercept or symbol C.

$$\begin{gathered} Y1(-1)*@Expand(@Month,@Dropfirst)\ X1(-1)*@Expand(@Month,@Dropfirst) \\ Y1(-1)*X1(-1)*@Expand(@Month)\ t*@Expand(@Month) \\ @Expand(@Month) \end{gathered} \tag{4.54}$$

Example 4.19 An Application of the VAR Model (4.53)

The statistical results could easily be obtained, but with a large number of parameters. Note that by using the function *@Dropfirst* in the first two sets of the inserted exogenous variables, then the VAR model can easily be written for *@Month* = m = 1, as follows:

$$Y1(m=1) = C(1,1,)*Y1(-1) + C1(1,1)*X1(-1) + C(1,25)*Y1(-1)*X1(-1) + C1(1,37) \tag{4.55a}$$

$$X1(m=1) = C(2,1,)*Y1(-1) + C(2,2)*X1(-1) + C(2,25)*Y1(-1)*X1(-1) + C(2,37) \tag{4.55b}$$

with the following regression functions, which can be applied easily to compute the forecast values beyond the sample period using Excel – refer to the illustration presented in Figure 4.31. Do the others as exercises.

$$\begin{aligned} Y1(m=1) &= 1.04415664598*Y1(-1) + 0.0258635260698*X1(-1) \\ &\quad - 0.000296595635504*Y1(-1)*X1(-1) - 1.88937236243 \end{aligned} \tag{4.56a}$$

$$\begin{aligned} X1(m=1) &= 0.0131292498917*Y1(-1) + 0.990867801189*X1(-1) \\ &\quad - 9.39625786858e-05*Y1(-1)*X1(-1) + 0.0637372385822 \end{aligned} \tag{4.56b}$$

Since each regression of the VAR output is an OLS regression, then in order to have the in-sample forecast values with its forecast evaluation, each of the models (4.55a) and (4.55b) should be run as an OLS Regression, which can easily be done by selecting the sample {@*Month* = 1}, and then we can apply each of the following equation specifications. Do these as exercises and also for the other months.

$$Y1\ Y1(-1)\ X1(-1)\ Y1(-1)*X1(-1)\ C \tag{4.57a}$$

and

$$X1\ Y1(-1)\ X1(-1)\ Y1(-1)*X1(-1)\ C \tag{4.57b}$$

4.8.4 Bayesian VAR Models

All VAR models presented here can be applied as the Bayesian VAR (BVAR) models. For this reason, only one of the models will be presented as an illustrative example to present its differences with the basic VAR model, with special notes and comments

Example 4.20 The Simplest VAR(1,1) and BVAR(1,1)
Figure 4.37 presents the statistical results of the simplest VAR(1,1) in (4.39a), (4.39b), and BVAR (1,1). Based on these results, the following findings and notes are presented.

1) The differences between their parameters' estimates indicate that each of the BVAR output is not an OLS regression. On the other hand, the different values of the *t*-statistic do not directly indicate both models are using different assumptions of residuals.
2) Both results are acceptable results, in the statistical sense. However, we never know the true VAR model for the corresponding population, since a sample data is a set of scores of selected variables, which happen to be selected or available to researchers, as stated in Agung (2011a).

4.8.5 VEC Models

As a more advanced VAR model, each equation specification of these VAR models can be applied for the VEC model. However, the VEC model has three types of VAR Specifications: namely Basic as the default, and alternative options for Cointergration, and VEC Restrictions, as presented in Figure 4.38.

Example 4.21 The Simplest Basic VEC Model
By using the same input as the simplest VAR(1,1) model based on $(X1_t, Y1_t)$, the results are obtained of the simplest Basic VEC(1,1) with Deterministic Trend Specification: 3) *Intercept (no trend) in CE and VAR*, as presented in Figure 4.39. Based on these results, the following findings and notes are presented.

1) The output presents two kinds of statistical results: a *Cointegrating Eq (CointEq1)* and the output of the VAR model of the first differences $DY1 = Y1 - Y1(-1)$ and $DX1 = X1\text{-}X(-1)$, with exogenous variables "C CointEq1."
2) The forecast values of *Y1* and *X1*, beyond the sample period for $t = 593$ can be computed using Excel, with the results presented in Figure 4.40. The forecast values of *Y1* and X1 for $t = 593$ are obtained by using the Eqs. (4.58a) and (4.58b), respectively.

Vector Autoregression Estimates
Date: 03/19/17 Time: 10:33
Sample (adjusted): 1947M02 1996M04
Included observations: 591 after adjustments
Standard errors in ( ) & t-statistics in [ ]

	Y1	X1
Y1(-1)	1.012595	0.000777
	(0.00337)	(0.00035)
	[300.423]	[2.20896]
X1(-1)	-0.011983	0.998232
	(0.01534)	(0.00160)
	[-0.78096]	[623.283]
C	0.095689	0.180337
	(0.68383)	(0.07138)
	[0.13993]	[2.52644]
R-squared	0.998375	0.999626
Adj. R-squared	0.998369	0.999625
Sum sq. resids	17886.35	194.8825
S.E. equation	5.515339	0.575702
F-statistic	180626.7	785799.2
Log likelihood	-1846.241	-510.7593
Akaike AIC	6.258007	1.738610
Schwarz SC	6.280249	1.760853
Mean dependent	140.9656	65.96169
S.D. dependent	136.5858	29.71828

Bayesian VAR Estimates
Date: 03/19/17 Time: 10:27
Sample (adjusted): 1947M02 1996M04
Included observations: 591 after adjustments
Prior type: Litterman/Minnesota
Initial residual covariance: Univariate AR
Hyper-parameters: Mu: 0, L1: 0.1, L2: 0.99, L3: 1
Standard errors in ( ) & t-statistics in [ ]

	Y1	X1
Y1(-1)	1.011446	0.000825
	(0.00337)	(0.00035)
	[300.381]	[2.33959]
X1(-1)	-0.007450	0.997978
	(0.01533)	(0.00161)
	[-0.48596]	[621.433]
C	-0.041824	0.190310
	(0.68340)	(0.07159)
	[-0.06120]	[2.65823]
R-squared	0.998375	0.999626
Adj. R-squared	0.998369	0.999625
Sum sq. resids	17889.89	194.8908
S.E. equation	5.515884	0.575714
F-statistic	180591.0	785765.6
Mean dependent	140.9656	65.96169
S.D. dependent	136.5858	29.71828

Figure 4.37 The results of the simplest VAR model in (4.39) and BVAR(1,1).

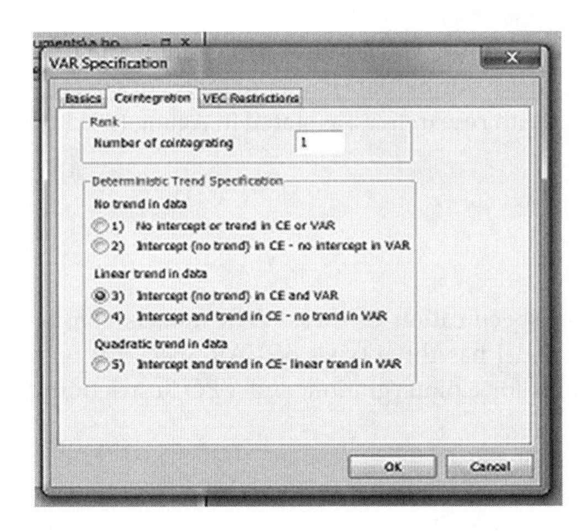

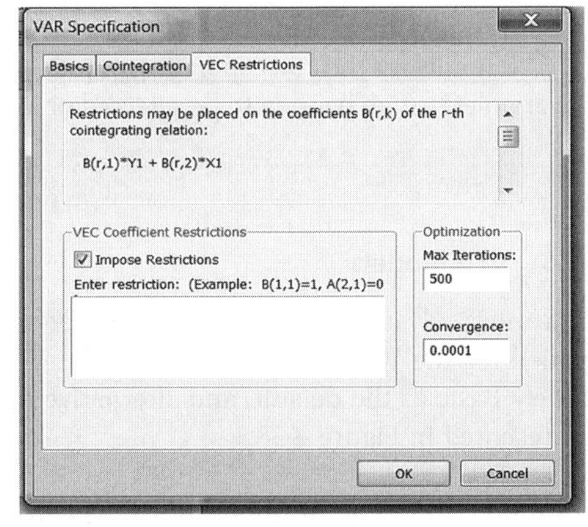

Figure 4.38 VAR specification for the VEC models.

$$= C6 + 0.00912049178045*(C6 - 0.984798926916*D6 - 75.2463799458) + 0.257529092398*(C6 - C5) - 0.844017374182*(C6 - C5) + 0.938802381225 \tag{4.58a}$$

$$= D6 + 0.000458062536887*(C6 - 0.984798926916*D6 - 75.2463799458) - 0.00383349721226*(C6 - C5) + 0.350301627152*(C6 - C5) + 0.117019946722 \tag{4.58b}$$

Vector Error Correction Estimates
Date: 03/19/17 Time: 13:10
Sample (adjusted): 1947M03 1996M04
Included observations: 590 after adjustments
Standard errors in ( ) & t-statistics in [ ]

Cointegrating Eq:	CointEq1	
Y1(-1)	1.000000	
X1(-1)	-0.984799	
	(0.73670)	
	[-1.33677]	
C	-75.24638	
Error Correction:	**D(Y1)**	**D(X1)**
CointEq1	0.009120	0.000458
	(0.00208)	(0.00021)
	[4.37669]	[2.16481]
D(Y1(-1))	0.257529	-0.003833
	(0.04045)	(0.00411)
	[6.36704]	[-0.93341]
D(X1(-1))	-0.844017	0.350302
	(0.38144)	(0.03873)
	[-2.21273]	[9.04457]
C	0.938802	0.117020
	(0.23317)	(0.02368)
	[4.02619]	[4.94251]

R-squared	0.131535	0.135785
Adj. R-squared	0.127089	0.131360
Sum sq. resids	16543.99	170.5707
S.E. equation	5.313385	0.539515
F-statistic	29.58470	30.69060
Log likelihood	-1820.602	-471.0867
Akaike AIC	6.185092	1.610464
Schwarz SC	6.214788	1.640159
Mean dependent	1.070119	0.172881
S.D. dependent	5.687042	0.578873

Determinant resid covariance (dof adj.)	8.213637
Determinant resid covariance	8.102643
Log likelihood	-2291.544
Akaike information criterion	7.801843
Schwarz criterion	7.876082

```
VAR Model:
===============================
D(Y1) = A(1,1)*(B(1,1)*Y1(-1) + B(1,2)*X1(-1) + B(1,3)) + C
(1,1)*D(Y1(-1)) + C(1,2)*D(X1(-1)) + C(1,3)

D(X1) = A(2,1)*(B(1,1)*Y1(-1) + B(1,2)*X1(-1) + B(1,3)) + C
(2,1)*D(Y1(-1)) + C(2,2)*D(X1(-1)) + C(2,3)

VAR Model - Substituted Coefficients:
===============================
D(Y1) = 0.00912049178045*( Y1(-1) - 0.984798926916*X1(-1)
- 75.2463799458 ) + 0.257529092398*D(Y1(-1)) -
0.844017374182*D(X1(-1)) + 0.938802381225

D(X1) = 0.000458062536887*( Y1(-1) - 0.984798926916*X1(-1)
- 75.2463799458 ) - 0.00383349721226*D(Y1(-1)) +
0.350301627152*D(X1(-1)) + 0.117019946722
```

Figure 4.39 Statistical results of the simplest basic VEC(1,1) model based on $(X1_t, Y1_t)$, with its regression models and functions.

	A	B	C	D
1			**VEC(1,1)**	
2		T	Y1	X1
3	1996M01	589	614.42	122.5
4	1996M02	590	649.542	123.9
5	1996M03	591	647.07	123.4
6	1996M04	592	647.17	124.5
7	1996M05	593	***651.304***	***125.208***
8			*656.840*	*125.764*
9			*662.909*	*126.265*

Figure 4.40 Forecast values beyond the sample period based on the VEC(1,1).

3) In order to have the in-sample forecast evaluation, the data analysis should be done based on each of the regressions of the VEC(1,1) model, with the steps as follows:

3.1 Generate the three series *ContEq1* = *Y1(−1)*−0.984798926916∗*X1(−1)*−75.2463799458,

DY1 = *Y1*-*Y1(−1)*, and *DX1* = *X1*-*X1(−1)*.

Dependent Variable: DY1
Method: Least Squares
Date: 03/20/17 Time: 17:50
Sample (adjusted): 1947M03 1996M04
Included observations: 590 after adjustments

Variable	Coefficient	Std. Error	t-Statistic	Prob.
COINTEQ1	0.009120	0.002084	4.376692	0.0000
DY1(-1)	0.257529	0.040447	6.367038	0.0000
DX1(-1)	-0.844017	0.381437	-2.212733	0.0273
C	0.938802	0.233174	4.026187	0.0001

R-squared	0.131535	Mean dependent var	1.070119
Adjusted R-squared	0.127089	S.D. dependent var	5.687042
S.E. of regression	5.313385	Akaike info criterion	6.185092
Sum squared resid	16543.99	Schwarz criterion	6.214788
Log likelihood	-1820.602	Hannan-Quinn criter.	6.196661
F-statistic	29.58470	Durbin-Watson stat	1.978690
Prob(F-statistic)	0.000000		

Dependent Variable: DX1
Method: Least Squares
Date: 03/20/17 Time: 17:51
Sample (adjusted): 1947M03 1996M04
Included observations: 590 after adjustments

Variable	Coefficient	Std. Error	t-Statistic	Prob.
COINTEQ1	0.000458	0.000212	2.164813	0.0308
DY1(-1)	-0.003833	0.004107	-0.933414	0.3510
DX1(-1)	0.350302	0.038731	9.044566	0.0000
C	0.117020	0.023676	4.942510	0.0000

R-squared	0.135785	Mean dependent var	0.172881
Adjusted R-squared	0.131360	S.D. dependent var	0.578873
S.E. of regression	0.539515	Akaike info criterion	1.610464
Sum squared resid	170.5707	Schwarz criterion	1.640159
Log likelihood	-471.0867	Hannan-Quinn criter.	1.622032
F-statistic	30.69060	Durbin-Watson stat	2.069966
Prob(F-statistic)	0.000000		

Figure 4.41 The statistical results of the models (4.59a) and (4.59b).

3.2 Then the outputs in Figure 4.41 are obtained by using each of the following equation specifications.

$$DY1\ CointEq1\ DY1(-1)\ DX1(-1)\ \ C \tag{4.59a}$$

$$DX1\ CointEq1\ DY1(-1)\ DX1(-1)\ C \tag{4.59b}$$

Further analysis could be done to obtain the forecast evaluation, based on each of the output in Figure 4.41. Do these as exercises.

4.9 Forecast Models Based on $(Y1_t,Y2_t)$

It should be understood that all models presented based on the bivariate $(X1_t,Y1_t)$ are applicable to the models based on $(Y1_t,Y2_t)$ by using *Y2* to replace *X1*. First of all, it should be known that *Y1 (−1)* and *Y2(−1)* could be upper (source or cause) variables of both *Y1* and *Y2*. Referring to the models (4.1) and (4.2), Figures 4.42a and b present specific up-and-down or causal relationships based on time series $(Y1_t,Y2_t)$ and their first lag-variables. As an extension, Figure 4.42c presents a path diagram under the assumption that *Y1* and *Y2* have reciprocal causal effects. Based on these path diagrams, the following three pairs of general forecasting models are presented.

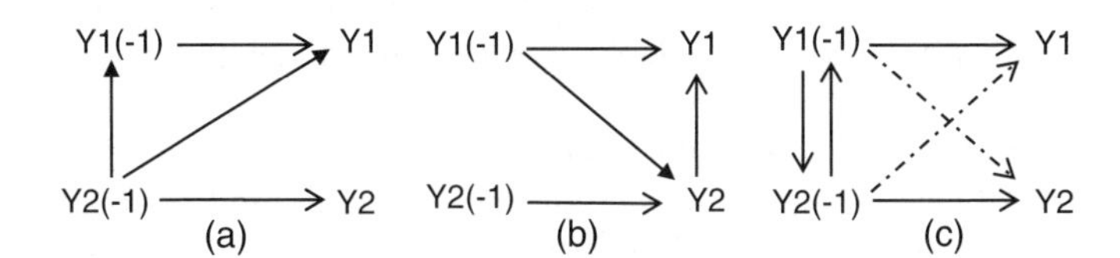

Figure 4.42 Alternative path diagrams based on bivariate (*Y1,Y2*).

4.9.1 Forecast Models Based on Figures 4.42a and b

These two path diagrams are presented under the assumption that *Y2* has a direct effect on *Y1*. However, they represent different type of *two-way-interaction lag-variables models,* namely TWI_LVM (*p1,p2*), as follows:

4.9.1.1 TWI_LVM(*p1,p2*) Based on Figure 4.42a

Note that in this figure the direct effect of *Y2* on *Y1* is represented by the direct effect of *Y2(−1)* on *Y1(−1)*. In the statistical sense, the indirect effect of *Y2(−1)* on *Y1* through *Y1(−1)* indicates that the effect of *Y2(−1)* on *Y1* depends on *Y1(−1)*, since *Y1(−1)* is one of the best predictor of *Y1*. Then, in the statistical sense, the model of *Y1* should have the two-way interaction *Y1(−1)∗Y2 (−1)* as an independent variable with the following pairs general equations, or SCM (Seemingly Causal Model).

$$Y1\ C\ Y1(-1)*Y2(-1)\ Y1(-1)...\ Y1(-p1)\ Y2(-1) \tag{4.60a}$$

$$Y2\ C\ Y2(-1)\ ...\ Y2(-p2) \tag{4.60b}$$

4.9.1.2 TWI_LVM(*p1,p2*) Based on Figure 4.42b

Based on this figure, the TWI_LVM(*p1,p2*) has the following pairs of general equation specifications or SCM, with the interaction *Y1(−1)∗Y2* as an independent variable of the model of *Y1*.

$$Y1\ C\ Y1(-1)*Y2(-1)\ Y1(-1)...\ Y1(-p1)\ Y2 \tag{4.61a}$$

$$Y2\ C\ Y2(-1)\ ...\ Y2(-p2) \tag{4.61b}$$

Note that the model (4.61b) is exactly the same as the model (4.60b), which should be applied to compute the first forecast values of *Y1* beyond the sample period, especially based on the model (4.61a). See the following example.

Example 4.22 Application of the Models (4.60a) and (4.61a)

Figure 4.43 presents the statistical results of specific models in (4.60a) and (4.61a) for $p1 = 2$, which are obtained by using the trial-and-error method, with their forecast evaluations: RMSE & TIC for the sample {@*Year*>1990}. Based these results, the following findings and notes are presented.

1) Even though *Y1(−1)∗Y2(−1)* in Figure 4.43a has a large *p*-value, the reduced model is not presented, in order to have similar set of independent variables for both models. Note that the model (4.60a) has smaller RMSE than the model (4.61a), even though they do not have a large deviation. So it could be concluded that the model (4.60a) is a better forecast model.
2) However, a reduced model of the model (4.60a) can be explored. Do this as an exercise.
3) Based on the model (4.61a), the following special notes are presented.
   3.1 Compared to the model (4.60a), the first forecast values of *Y1* beyond the sample period, namely for the time $(T+1) = 593$, cannot be computed based on the model (4.61a) because the value/score of $Y2(T+1)$ is not available. Hence, the model (4.60b) or (4.61b) should be applied.
   3.2 Figure 4.44 presents two alternative estimates of the model (4.60b), which can be used to compute the forecast value of *Y1* beyond the sample period, based on both model (4.60a) and (4.61a). However, to compute the forecast values of *Y1* beyond the sample period, based on the model (4.61a), first we have to compute the forecast values of *Y2* beyond the sample period, and then we can compute the forecast values of *Y1*.
4) Figure 4.45 presents the spreadsheet for computing the forecast values of *Y1* and *Y2* beyond the sample period, based on the pairs of model (4.60a) and (4.61a), (4.61b). These results show that the model (4.60a), (4.60b) presents greater forecast values beyond the sample period than the model (4.60a), (4.60b). However, we cannot say which one is a better predicted value. The steps of the computation are as follows:
   4.1 Copy the last five observed scores of the variables *T* = *@Trend, Y2* = *Y2f_460b*, and *Y1* = *Y1f_460a* = *Y1f_461a*, as presented in Figure 4.45.
   4.2 The forecast values for $T = 592$ can be computed using the following three equations, which are obtained from the outputs by clicking the object *"Representations,"* with the ordering (1), (2), and (3); or (2), (1), and (3).

      1) **Y2F_460B** = 0.0360838984878 + 1.21127332164∗Y2(−1)
      − 0.297691175521∗Y2(−2) + 0.0766910758572∗Y2(−3)
      C7 = 0.0360838984878 + 1.21127332164∗C6
      − 0.297691175521∗C5 + 0.0766910758572∗C4
      2) **Y1_460A** = −1.70895185414 + 0.00149610912258∗Y1(−1)∗Y2(−1)
      + 1.27321355653∗Y1(−1) − 0.269061205299∗Y1(−2) + 0.287645463594∗Y2(−1)

D7 = −1.70895185414 + 0.00149610912258∗C6∗D6 + 1.27321355653∗C6 − 0.269061205299∗C5 + 0.287645463594∗D6

3) **Y1_461A** = −0.374705158588 − 0.00530880228601∗Y1(−1)∗Y2 + 1.27146869698∗Y1(−1) − 0.248975733297∗Y1(−2) + 0.157465038132∗Y2

E7 = −0.374705158588 − 0.00530880228601∗E6∗C7 + 1.27146869698∗E6 − 0.248975733297∗E5 + 0.157465038132∗C7

Dependent Variable: Y1
Method: Least Squares
Date: 10/04/17 Time: 16:35
Sample (adjusted): 1947M03 1996M04
Included observations: 590 after adjustments

Variable	Coefficient	Std. Error	t-Statistic	Prob.
C	-1.708952	1.156921	-1.477155	0.1402
Y1(-1)*Y2(-1)	0.001496	0.002296	0.651496	0.5150
Y1(-1)	1.273214	0.039915	31.89827	0.0000
Y1(-2)	-0.269061	0.040585	-6.629603	0.0000
Y2(-1)	0.287645	0.260146	1.105706	0.2693

R-squared	0.998491	Mean dependent var	141.1778
Adjusted R-squared	0.998481	S.D. dependent var	136.6042
S.E. of regression	5.324798	Akaike info criterion	6.191065
Sum squared resid	16586.78	Schwarz criterion	6.228185
Log likelihood	-1821.364	Hannan-Quinn criter.	6.205526
F-statistic	96765.62	Durbin-Watson stat	1.986097
Prob(F-statistic)	0.000000		

RMSE=29.425 & TIC=0.0319

(a). Model (4.60a)

Dependent Variable: Y1
Method: Least Squares
Date: 10/04/17 Time: 16:16
Sample (adjusted): 1947M03 1996M04
Included observations: 590 after adjustments

Variable	Coefficient	Std. Error	t-Statistic	Prob.
C	-0.374705	1.151012	-0.325544	0.7449
Y1(-1)*Y2	-0.005309	0.002279	-2.329930	0.0201
Y1(-1)	1.271469	0.039825	31.92653	0.0000
Y1(-2)	-0.248976	0.040770	-6.106815	0.0000
Y2	0.157465	0.258639	0.608821	0.5429

R-squared	0.998498	Mean dependent var	141.1778
Adjusted R-squared	0.998487	S.D. dependent var	136.6042
S.E. of regression	5.312725	Akaike info criterion	6.186525
Sum squared resid	16511.65	Schwarz criterion	6.223645
Log likelihood	-1820.025	Hannan-Quinn criter.	6.200986
F-statistic	97206.58	Durbin-Watson stat	1.945630
Prob(F-statistic)	0.000000		

RMSE=29.570 & TIC=0.0318

(b). Model (4.61a)

Figure 4.43 The statistical results of specific models in (4.60a) and (4.61a) for $p1 = 2$, with their forecast evaluations for the sample {@*Year*>*1990*}.

Dependent Variable: Y2
Method: Least Squares
Date: 10/05/17 Time: 06:28
Sample (adjusted): 1947M05 1996M04
Included observations: 588 after adjustments

Variable	Coefficient	Std. Error	t-Statistic	Prob.
C	0.039780	0.024704	1.610282	0.1079
Y2(-1)	1.214516	0.041322	29.39141	0.0000
Y2(-2)	-0.313790	0.064769	-4.844787	0.0000
Y2(-3)	0.151577	0.064646	2.344715	0.0194
Y2(-4)	-0.062925	0.041310	-1.523242	0.1282

R-squared	0.980667	Mean dependent var	4.065306
Adjusted R-squared	0.980535	S.D. dependent var	1.129409
S.E. of regression	0.157573	Akaike info criterion	-0.849393
Sum squared resid	14.47540	Schwarz criterion	-0.812176
Log likelihood	254.7214	Hannan-Quinn criter.	-0.834892
F-statistic	7393.341	Durbin-Watson stat	1.999484
Prob(F-statistic)	0.000000		

Dependent Variable: Y2
Method: Least Squares
Date: 10/05/17 Time: 06:29
Sample (adjusted): 1947M04 1996M04
Included observations: 589 after adjustments

Variable	Coefficient	Std. Error	t-Statistic	Prob.
C	0.036084	0.024604	1.466601	0.1430
Y2(-1)	1.211273	0.041129	29.45069	0.0000
Y2(-2)	-0.297691	0.063491	-4.688690	0.0000
Y2(-3)	0.076691	0.041206	1.861174	0.0632

R-squared	0.980592	Mean dependent var	4.066469
Adjusted R-squared	0.980492	S.D. dependent var	1.128801
S.E. of regression	0.157660	Akaike info criterion	-0.849990
Sum squared resid	14.54107	Schwarz criterion	-0.820255
Log likelihood	254.3219	Hannan-Quinn criter.	-0.838405
F-statistic	9852.325	Durbin-Watson stat	1.992965
Prob(F-statistic)	0.000000		

Figure 4.44 Two alternative LS estimates of the model (4.60b).

	A	B	C	D	E
1	**Year**	**T**	**Y2F_460B**	**Y1F_460A**	**Y1F_461A**
2	1995M12	587	2.720	614.570	614.570
3	1996M01	588	2.310	614.420	614.420
4	1996M02	589	2.220	649.542	649.542
5	1996M03	590	2.220	647.070	647.070
6	1996M04	591	2.240	647.170	647.170
7	**1996M05**	**592**	**2.259**	**650.961**	**653.972**
8	1996M06	593	2.275	654.805	662.459
9	1996M07	594	2.292	658.699	671.400

Figure 4.45 Spreadsheet for computing the forecast values beyond the sample period.

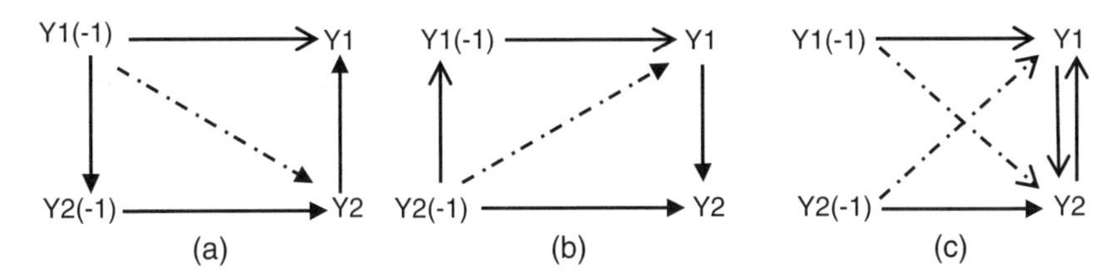

Figure 4.46 Alternative path diagrams based on bivariate (*Y1,Y2*) with reciprocal effects.

4.9.2 Reciprocal Causal Effects Models

Under the assumption that *Y1* and *Y2* have reciprocal causal effects, then we have three alternative path diagrams as presented in Figure 4.46, in addition to the path diagram in Figure 4.42c. Hence, we have the following four passible pairs of general models, namely the *Reciprocal-Causal-Effects* (RCE) TWI_LVM(*p1,p2*), with the simplest models for $p1 = p2 = 1$.

4.9.2.1 RCE TWI_LVM(*p1,p2*) Based on Figure 4.42c

As an extension of the path diagram in Figure 4.42c, the double arrows between *Y1(−1)* and *Y2 (−1)* represent the assumption that *Y1* and *Y2* have reciprocal causal effects, Hence, we have the following pairs of general equation specifications, which are called full lag-variables models, because all independent variables are the lag-variables.

$$Y1\ C\ Y1(-1)*Y2(-1)\ Y1(-1)\ \ldots\ Y1(-p1)\ Y2(-1) \tag{4.62a}$$

$$Y2\ C\ Y1(-1)*Y2(-1)\ Y2(-1)\ \ldots\ Y2(-p2)\ Y1(-1) \tag{4.62b}$$

Note that based on these models the forecast values of *Y1* and *Y2* beyond the sample period can be computed directly. Refer to the notes presented in Example 4.22.

4.9.2.2 RCE TWI_LVM(*p1,p2*) Based on Figure 4.46a

In this figure, the reciprocal causal effects between *Y1* and *Y2* are represented by the direct effect of *Y1* on *Y2* and the direct effect of *Y2(−1)* on *Y1(−1)*. Then the pairs of TWI_LVM(*p1,p2*) have the following equation specifications

$$Y1\ C\ Y1(-1)*Y2(-1)\ Y1(-1)\ \ldots\ Y1(-p1)\ Y2 \tag{4.63a}$$

$$Y2\ C\ Y1(-1)*Y2(-1)\ Y2(-1)\ \ldots\ Y2(-p2)\ Y1(-1) \tag{4.63b}$$

Note that the model (4.63a) is developed based on the triple time series *Y1*, *Y1(−1)*, and *Y2*, and the model (4.63b) is developed based on the triple time series *Y2*, *Y2(−1)*, and *Y1(−1)*.

4.9.2.3 RCE TWI_LVM(*p1,p2*) Based on Figure 4.46b

In this figure, the reciprocal causal effects between *Y1* and *Y2* are represented by the direct effect of *Y1* on *Y2* and the direct effect of *Y2(−1)* on *Y1(−1)*. Then the pairs of TWI_LVM(*p1,p2*) have the following equation specifications

$$Y1\ C\ Y1(-1)*Y2(-1)\ Y1(-1)\ \ldots\ Y1(--p1)\ Y2(-1) \tag{4.64a}$$

$$Y2\ C\ Y1(-1)*Y2(-1)\ Y2(-1)\ \ldots\ Y2(-p2)\ Y1 \tag{4.64b}$$

Note that the model (4.64a) is developed based on the triple time series *Y1*, *Y1(−1)*, and *Y2(−1)*, and the model (4.64b) is developed based on the triple time series *Y2*, *Y2(−1)*, and *Y1*.

4.9.2.4 RCE TWI_LVM(*p1,p2*) Based on Figure 4.46c

In this figure, the reciprocal causal effects between *Y1* and *Y2* are represented by the double arrows between *Y1* on *Y2*. Then the pairs of TWI_LVM(*p1,p2*) have the following equation specifications

$$Y1\ C\ Y1(-1)*Y2(-1)\ Y1(-1)\ \ldots\ Y1(-p1)\ Y2 \tag{4.65a}$$

$$Y2\ C\ Y1(-1)*Y2(-1)\ Y2(-1)\ \ldots\ Y2(-p2)\ Y1 \tag{4.65b}$$

Note that the model (4.65a) is developed based the triple time series *Y1*, *Y1(−1)*, and *Y2*, and the model (4.65b) is developed based the triple time series *Y2*, *Y2(−1)*, and *Y1*.

4.9.3 Models with the Time Independent Variables

The models based on the endogenous variables *Y1* and *Y2* can easily be extended to models with the time independent variable, which have been presented in previous chapters. However, this section only presents two of the models, as follows.

4.9.3.1 Models with Alternative Trends

It should be well accepted that a model with linear trend can always be applied for all of the time series *Y1* and *Y2*, without looking at their growth curves, in order to identify only whether each of them has a significant positive or negative trend. However, for having a good fit model, their growth curve should be taken into account. In mathematical sense, a polynomial growth model can always be used to fit the growth curve of a time series. Refer to all models with alternative trends presented in Chapter 2.

By looking at the growth curves of *Y1* = *FSPCOM* and *Y2* = *FSDXP* in Figure 4.2, it is clear that a model with a polynomial trend would be a good fit model for each of *Y1* and *Y2*. In the mathematical sense, a model with a polynomial trend can be applied for each of *Y1* and *Y2*, as presented in the following example.

Example 4.23 Models with Polynomial Trends

As an extension of the models in (4.63a and 4.63b), Figure 4.47a and b present the statistical results of the models with polynomial trends, which use *(T/100)* instead of *T* as the time independent variable, in order to have sufficiently large parameter estimates. The statistical results are obtained by using the trial-and-error method.

Dependent Variable: Y1
Method: Least Squares
Date: 10/12/17 Time: 12:39
Sample (adjusted): 1947M03 1996M04
Included observations: 590 after adjustments

Variable	Coefficient	Std. Error	t-Statistic	Prob.
C	0.231946	2.889249	0.080279	0.9360
Y1(-1)*Y2(-1)	-0.007416	0.005691	-1.303191	0.1930
Y1(-1)	1.242004	0.040301	30.81800	0.0000
Y1(-2)	-0.287397	0.040220	-7.145632	0.0000
Y2(-1)	-0.158015	0.414871	-0.380877	0.7034
T/100	7.273783	2.146577	3.388549	0.0008
(T/100)^2	-3.421101	0.870855	-3.928439	0.0001
(T/100)^3	0.571156	0.131042	4.358578	0.0000

R-squared	0.998541	Mean dependent var	141.1778
Adjusted R-squared	0.998524	S.D. dependent var	136.6042
S.E. of regression	5.248807	Akaike info criterion	6.167345
Sum squared resid	16034.09	Schwarz criterion	6.226737
Log likelihood	-1811.367	Hannan-Quinn criter.	6.190483
F-statistic	56910.17	Durbin-Watson stat	1.989294
Prob(F-statistic)	0.000000		

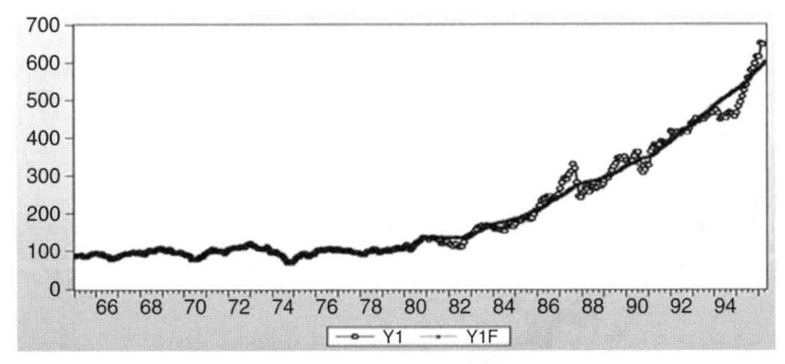

RMSE=23.255 & TIC=0.0347

Figure 4.47 (a) The statistical results of a model in (4.63a) with a polynomial trend. (b) The statistical results of a model in (4.63b) with a polynomial trend.

1) Based on the results in Figure 4.47, the findings and notes presented are as follows:
   1.1 At the 10% level, the interaction *Y1(−1)∗Y2(−1)* has a negative significant effect on *Y1* with a p-value = 0.1930/2 = 0.0965. For this reason, the main variable *Y2(−1)* can be kept in the model, even though it has a large p-value.
   1.2 The graphs of *Y1* and *Y1F* show that the model with a third degree polynomial trend and a small value of TIC show the model is a good forecast model for *Y1*. Its forecast values beyond the sample period, specifically at the time $(T + 1) = 592$, can be computed directly using its regression function.
   1.3 As a comparison, the model of *Y1* with polynomial trend has been presented in Chapter 3, but without using *Y2(−1)* as an independent variable.
2) Based on the results in Figure 4.47b, the findings and notes presented are as follows:
   2.1 Compared to *Y1*, the growth curve of *Y2* clearly shows that the model of *Y2* should have a higher degree polynomial trend. However, by using the trial-and-error method, it is found the regression with a fourth degree polynomial trend is a good forecast model, with TIC < 0.10.
   2.2 The same as the forecast values of *Y1* at $(T + 1) = 592$, the forecast values of *Y2* beyond the sample period at $(T + 1) = 592$ also can be computed directly using its regression function.
3) The forecast values beyond the sample period for both *Y1* and *Y2* could easily be computed using the Excel – refer to Figure 4.42.

Dependent Variable: Y2
Method: Least Squares
Date: 10/12/17 Time: 13:35
Sample (adjusted): 1947M04 1996M04
Included observations: 589 after adjustments

Variable	Coefficient	Std. Error	t-Statistic	Prob.
C	0.135443	0.077095	1.756838	0.0795
Y1(-1)*Y2(-1)	-0.000409	0.000162	-2.522299	0.0119
Y2(-1)	1.210337	0.041772	28.97481	0.0000
Y2(-2)	-0.291761	0.063022	-4.629498	0.0000
Y2(-3)	0.064421	0.041141	1.565841	0.1179
Y1(-1)	0.000840	0.000499	1.684322	0.0927
(T/100)^2	-0.052551	0.020803	-2.526070	0.0118
(T/100)^3	0.025791	0.008804	2.929559	0.0035
(T/100)^4	-0.002888	0.001022	-2.826109	0.0049

R-squared	0.981074	Mean dependent var	4.066469
Adjusted R-squared	0.980813	S.D. dependent var	1.128801
S.E. of regression	0.156356	Akaike info criterion	-0.858194
Sum squared resid	14.17946	Schwarz criterion	-0.791292
Log likelihood	261.7383	Hannan-Quinn criter.	-0.832129
F-statistic	3758.310	Durbin-Watson stat	1.988469
Prob(F-statistic)	0.000000		

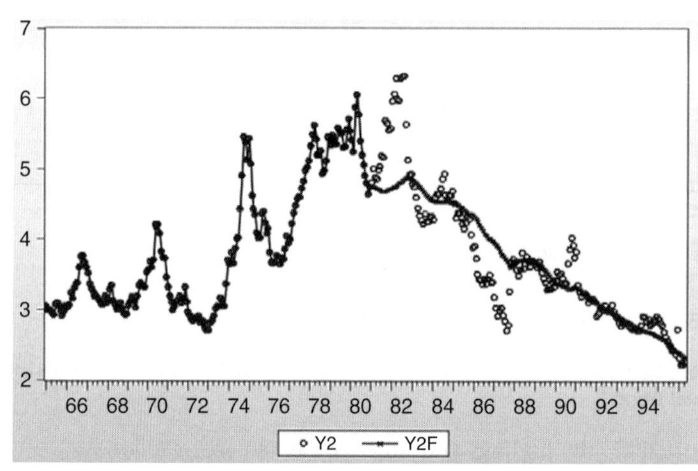

Figure 4.47 (Continued)

4.9.3.2 The Time-Related-Effects Models

Referring to various time-related-effects (TRE) models presented in Agung (2014, 2009a), and the simplest TRE model (3.1), all pairs of models of *Y1* and *Y2* also can be extended to TRE models. As an illustration, the following general equation specifications are considered as the extension of the models (4.62a), (4.62b).

$$\begin{aligned} &Y1\ C\ Y1(-1)*Y2(-1)*T\ \ Y1(-1)*T\ \ Y1(-1)*Y2(-1)\ T\ T\hat{}2\ \ldots\ T\hat{}k \\ &\qquad Y1(-1)\ \ldots\ Y1(-p1)\ Y2(-1) \end{aligned} \tag{4.66a}$$

$$Y2\ C\ Y1(-1)*Y2(-1)*T\ Y2(-1)*T\ Y1(-1)*Y2(-1)\ T\ T\hat{}2\ \ldots\ T\hat{}m \\ Y2(-1)\ \ldots\ Y1(-p2)\ Y1(-1) \tag{4.66b}$$

Note that these models are three-way-interaction lag-variables models, namely 3WI_LVM (*p1,p2*), which also can be considered to be an extension of the two-way-interaction lag-variables models: 2WI_LVM(2,3), in Figures 4.47a and b. Having models with such a large number of independent variables, the manual multistage selection method, which is the combination of the STEPLS-Combinatorial and trial-and-error selection methods, should be used to obtain the best possible forecast model based on each of the equation specifications. For the TRE models, the most important predictors are the set of time interaction variables. See the following examples.

Example 4.24 Alternative Reduced Models Based on the General Model (4.66a)
Figure 4.48 presents the results of four reduced models of the general model (4.66a), namely RM-1 to RM-4, which are obtained using the manual multistage selection method. Based on these results, the following findings and notes are presented.

1) By using four stages of regression analyses, I obtain RM-1, which is a good fit 3WI_TRE-model, which could be the best fit model with its adjusted R-squared = 0.9981. However, it obtained an unbelievable worst forecast model with a value of TIC close to one.
2) For this reason, I should be using the trial-and-error method to reduce the model in order to obtain a better forecasting. Then I obtain RM-2 and RM-3, which are even worse forecast models. Finally, I found RM-4 is a good forecast of *Y1* with TIC = 0.0481, and it is the best among the four reduced models.
3) As a graphical comparison of forecast values of *Y1F_RM-1* and *Y1F_RM-4*, Figure 4.49 presents the scatter graphs of *(Y1,Y1_Fit,Y1_RM1*), and *(Y1,Y1F_RM4)*. It is very clear the deviation of *Y1* and *Y1F_RM-1* is very large.

Variable	RM-1			RM-2			RM-3			RM-4		
	Coef.	t-Stat.	Prob.	Coef.	t-Stat.	Prob.	Coef.	t-Stat.	Prob.	Coef.	t-Stat.	Prob.
C	49.745	13.692	0.000	-5.776	-1.969	0.049	46.138	14.078	0.000	12.216	3.248	0.001
Y1(-1)*Y2(-1)*(T/100)	-0.037	-7.824	0.000	-0.024	-3.717	0.000	-0.033	-10.948	0.000	0.002	3.390	0.001
Y1(-1)*(T/100)	0.173	34.946	0.000	0.188	46.085	0.000	0.168	35.238	0.000	0.020	3.657	0.000
Y1(-1)*Y2(-1)	0.240	10.997	0.000	0.170	5.871	0.000	0.205	14.891	0.000			
T/100	-8.778	-3.377	0.001	17.644	5.724	0.000						
(T/100)^2	3.125	2.986	0.003	-3.062	-2.278	0.023						
(T/100)^3	-0.426	-2.441	0.015	-0.350	-1.495	0.135						
Y1(-1)										1.122	19.979	0.000
Y1(-2)										-0.280	-6.952	0.000
Y1(-3)	-0.042	-1.515	0.130				-0.041	-1.463	0.144			
Y2(-1)	-9.208	-19.544	0.000				-8.597	-20.997	0.000	-1.527	-2.878	0.004
R-squared	0.9982			0.9966			0.9981			0.9985		
Adj. R-squared	0.9981			0.9966			0.9981			0.9985		
F-statistic	39615			28476			62326			79286		
Prob(F-stat.)	0.0000			0.0000			0.0000			0.0000		
DW-Stat.	1.3307			0.6863			1.2479			2.0113		
RMSE/TIC	10562.65/0.9415			9.43E+10/1.0000			62973980/0.9999991			***32.8845/0.0481***		

Figure 4.48 Four alternative reduced models of the model (4.66a), with their forecast evaluations, RMSE and TIC, for the sample {@Year>1980}.

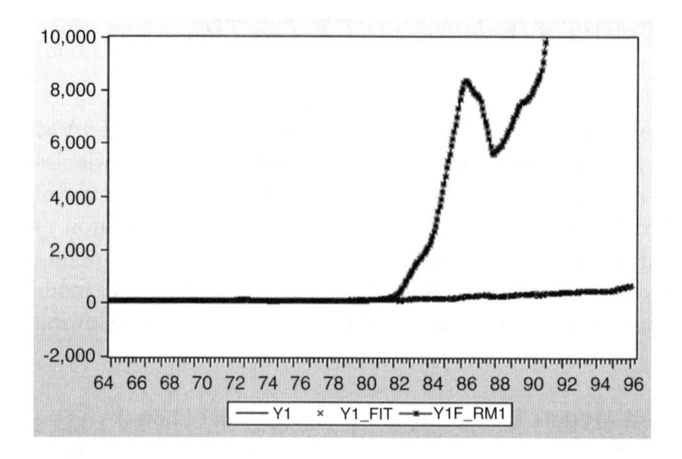

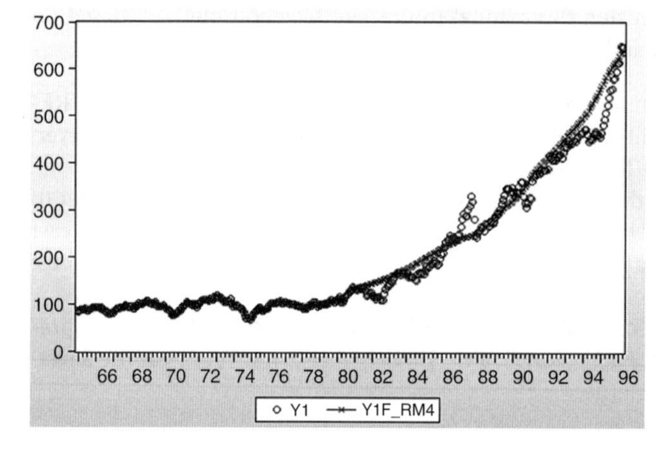

Figure 4.49 Scatter graphs of (*Y1,Y1_Fit,Y1f_RM1*) and (*Y1,Y1F_RM4*).

4) As an additional illustration, Figure 4.50 presents a better forecast model of *Y1* compared to RM-4, with a slightly smaller TIC.

Example 4.25 Application of the Model (4.66b)

By using the multistage trial-and-error selection method, Figure 4.51 presents three intentionally selected OLS regressions of *Y2* with the general equation specification (4.46b) out of many possible regressions. Based on these results, the following findings and notes are presented.

1) Compared to RM-2 of *Y1* in Figure 4.48, RM-1 of *Y2* is obtained after two stages of regression analysis with very small values of RMSE and TIC. By looking at the graph of *Y2F_RM1* in Figure 4.52, RM-1 of *Y2* could be considered a good forecast model, at least for the last several sample-points, where six of the last in-sample forecast values are presented in Figure 4.53. However, compared to RM-2 and RM_3, the RM-1 of *Y2* is the worst forecast model, since it has the greatest values of RMSE and TIC.

Dependent Variable: Y1
Method: Least Squares
Date: 10/22/17 Time: 06:55
Sample (adjusted): 1947M03 1996M04
Included observations: 590 after adjustments

Variable	Coefficient	Std. Error	t-Statistic	Prob.
C	12.97409	3.880719	3.343219	0.0009
Y1(-1)*Y2(-1)	0.012322	0.003553	3.468380	0.0006
Y1(-1)*(T/100)	0.026786	0.006766	3.959108	0.0001
Y1(-1)	1.083199	0.062110	17.43999	0.0000
Y1(-2)	-0.278488	0.040156	-6.935179	0.0000
Y2(-1)	-1.784644	0.583088	-3.060675	0.0023

R-squared	0.998530	Mean dependent var	141.1778
Adjusted R-squared	0.998518	S.D. dependent var	136.6042
S.E. of regression	5.259243	Akaike info criterion	6.167969
Sum squared resid	16153.23	Schwarz criterion	6.212512
Log likelihood	-1813.551	Hannan-Quinn criter.	6.185322
F-statistic	79357.50	Durbin-Watson stat	2.008909
Prob(F-statistic)	0.000000		

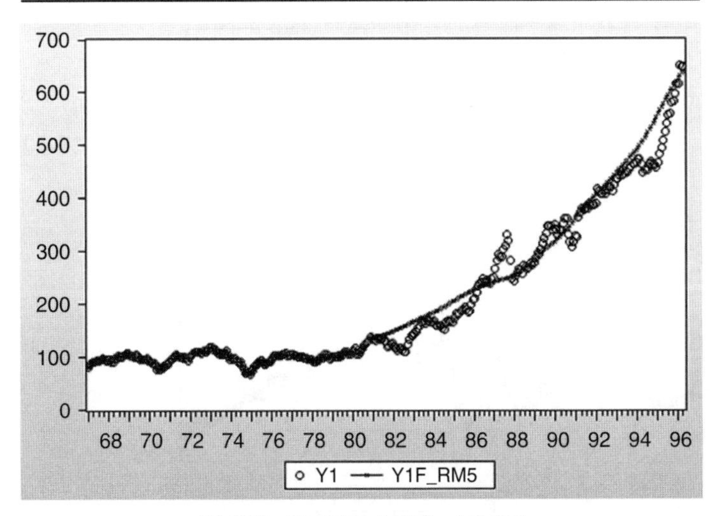

RMSE= 31.8979 & TIC= 0.0468

Figure 4.50 A better forecasting *2WI_TRE-LVM(2)* of *Y1* compared to *RM-4*.

2) As an additional illustration, Figure 4.53 presents the forecast values of *Y2* for $t = 592$ based on the three models, which shows that RM-2 and RM-3 have better forecast values compared to RM-1
3) Even though *Y2(−1)∗(T/100)* has a large p-value = 0.389, and *Y1(−1)* has a p-value = 0.289, they can be used in the model because the interaction *Y1(−1)∗Y2(−1)∗(T/100)* has a negative significant adjusted effect on *Y2* at the 5% level, with a p-value = 0.085/2 = 0.0475. In addition, I also want to show that a good forecasting model can have one or more independent variables with large p-values or insignificant effects.
4) Similarly, for RM-3 of *Y2*, at the 10% level, the interaction *Y1(−1)∗Y2(−1)∗(T/100)* has a negative significant adjusted effect on *Y2* with a p-value = 0.112/2 = 0.056.

Variable	RM-1			RM-2			RM-3		
	Coef.	t-Stat.	Prob.	Coef.	t-Stat.	Prob.	Coef.	t-Stat.	Prob.
C	6.072	39.307	0.000	0.269	2.072	0.039	0.344	2.596	0.010
Y1(-1)*Y2(-1)*(T/100)	-0.001	-3.115	0.002	0.000	-1.723	0.085	0.000	-1.590	0.112
Y2(-1)*(T/100)	0.266	32.704	0.000	0.007	0.863	**0.389**	0.013	1.628	0.104
Y1(-1)*Y2(-1)	0.004	2.103	0.036	0.001	1.389	0.165	0.000	0.920	**0.358**
T/100	-4.080	-13.392	0.000	-0.487	-3.370	0.001	-0.376	-3.773	0.000
(T/100)^2	1.162	5.875	0.000	0.218	2.418	0.016	0.110	3.728	0.000
(T/100)^3	-0.198	-3.772	0.000	-0.042	-1.677	0.094	-0.010	-2.061	0.040
(T/100)^4	0.018	3.497	0.001	0.003	1.263	0.207			
Y2(-1)				0.955	51.724	0.000	1.171	26.507	0.000
Y1(-1)				0.001	1.187	0.236	0.001	2.325	0.020
Y2(-2)							-0.289	-4.623	0.000
Y2(-3)							0.066	1.608	0.108
R-squared	0.8872			0.9804			0.9814		
Adj. R-squared	0.8859			0.9801			0.9811		
F-statistic	655.12			3226.4			3050.3		
Prob(F-statistic)	0.0000			0.0000			0.0000		
DW-Stat.	0.2361			1.5949			1.9842		
RMSE/TIC	1.6210/0.1851			0.5028/0.0658			0.4767/0.0626		

Figure 4.51 Three selected OLS regressions of the general equation specification (4.46b).

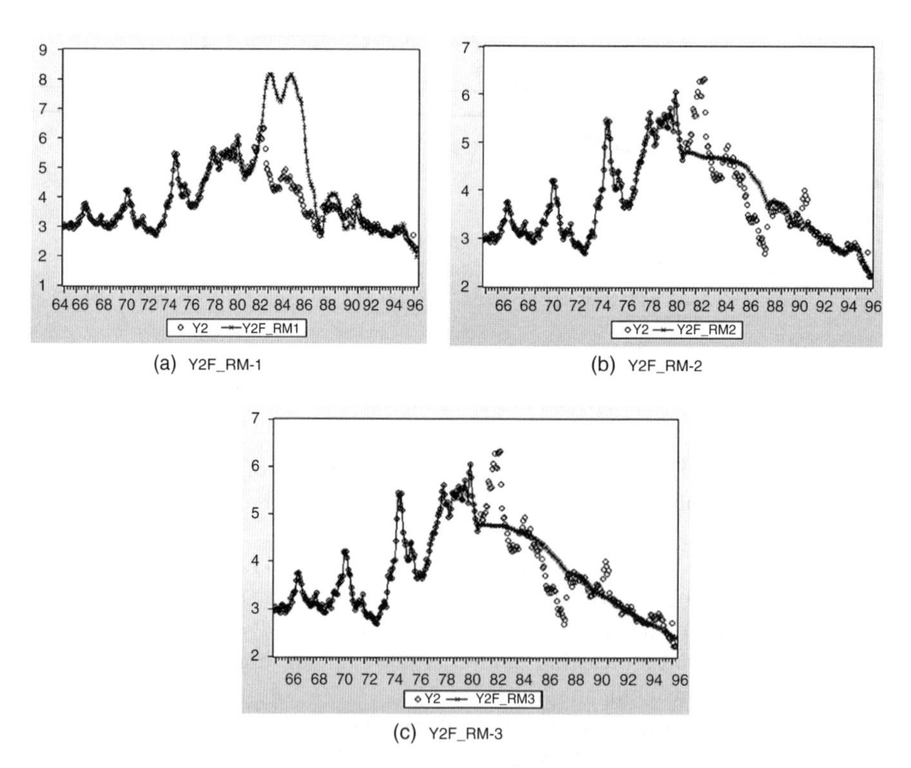

(a) Y2F_RM-1 (b) Y2F_RM-2 (c) Y2F_RM-3

Figure 4.52 The graphs of *Y2* and the forecast values *Y2F_RM-1*, *Y2F_RM-2*, and *Y2F-RM-3*.

	A	B	C	D	E	F	G
1		t	Y1	Y2	Y2F_RM1	Y2F_RM2	Y2F_RM3
2	1995M11	586	595.53	2.334	2.415	2.466	2.466
3	1995M12	587	614.57	2.720	2.242	2.380	2.453
4	1996M01	588	614.42	2.310	2.162	2.334	2.440
5	1996M02	589	649.542	2.220	2.227	2.306	2.429
6	1996M03	590	647.07	2.220	1.954	2.251	2.416
7	1996M04	591	647.17	2.240	2.167	2.215	2.404
8		**592**			**1.987**	**2.223**	**2.302**

Figure 4.53 The forecast values of *Y2*, based on the three models in Figure 4.51.

4.10 Special Notes and Comments

Referring to the RM-1 of *Y1* in Figure 4.48, having each of the eight independent variables has either a positive or negative significant adjusted effect on *Y1* at the 1% level and at the 10% level; only *Y3(−1)* has a negative significant adjusted effect with a p-value = 0.130/2 = 0.065. So, RM-1 is a good fit model, but it is the worst forecast model with a very large TIC = 0.9415, which is very close to one (refer to the graph in Figure 4.49) and similarly for the models of RM-2 and RM-3.

On the other hand, alternative worse fit models *Y2* presented in Figure 4.51 have one or more insignificant independent variables, compared to the models of *Y1*. Note that each of the models of *Y2* has a small or very small values of RMSE and TIC. Hence, based on these findings, I made the following notes.

The best possible fit regression model could be the worse forecasting model with a large or very large values of RMSE and TIC.
On the other hand, a worse fit regression model could be a better forecasting model with a small or very small values of RMSE and TIC.
Hence, based on these findings, complete statistical results should be presented and evaluated in all studies on forecasting.

For the data analysis based on each model having a large number of independent variables, the manual multistage selection method, which is a combination of the STEPLS: Combinatorial and trial-and-error selection methods, should be applied.

5

Forecasting Based on $(X1_t, X2_t, Y_t)$

5.1 Introduction

As the extension of all forecast models presented in previous chapters, especially the models based on the bivariate time series (X_t, Y_t) presented in Chapter 4, this chapter presents various models based on trivariate or triple time series $(X1_t, X2_t, Y_t)$, and/or $(Y1_t, Y2_t, Y3_t)$. However, I will start with various translog linear models as the extension of those presented in Section 4.7. The examples will be presented based on the GARCH.wf1. In order to be general, I select $X1 = IP$, $X2 = PW$, $Y1 = FSPCOM$, and $Y2 = FSDXP$.

5.2 Translog Linear Models Based on $(X1, X2, Y1)$

5.2.1 Basic Translog Linear Model

Based on the trivariate time series $(X1_t, X2_t, Y1_t)$, I propose the following equation specifications (ESs) to present three basic translog linear LV(1) models, which are in fact the Cobb–Douglass production functions. Note that the models are additive models. However, I would recommend applying the first model, since we can be very confident that *log*(*Y*1(−1)), *log*(*X*(−1)), and *log*(*X*2 (−1) are the cause, source, or upstream factors for *Y*1. These LV(1) models could easily be extended to LVARMA(*p*,*q*,*r*) models of *Y*1 for various $p > 0$, $q \geq 0$, and $r \geq 0$. Refer to Subsection 3.2.1 for illustrative examples. By using the *log*(*Y*1) as a dependent variable, we can forecast either *Y*1 or *log*(*Y*1).

$$log(Y1)\,C\;log(Y1(-1))\;log(X1(-1))\;log(X2(-1)) \tag{5.1}$$

$$log(Y1)\,C\;log(Y1(-1))\;log(X1)\;log(X2(-1)) \tag{5.2}$$

$$log(Y1)\,C\;log(Y1(-1))\;log(X1)\;log(X2) \tag{5.3}$$

In addition, to compute the predicted values/scores of *Y1* or *log(Y1)* beyond the sample period, forecasting models also should be applied for each of the time series *X1* and *X2* using LV(*p*) models; the simplest possible models are as follows:

$$log(X1)\,C\;log(X1(-1)) \tag{5.4}$$

$$log(X2)\,C\;log(X2(-1)) \tag{5.5}$$

In practice, we might need to apply the LV(*p*) model, where $p > 1$ should be selected using the trial-and-error method, in order to have the best possible fit for each of *log(X1)* and *log(X2)*, with suitable values of Durbin–Watson statistics.

Advanced Time Series Data Analysis: Forecasting Using EViews, First Edition. I Gusti Ngurah Agung.

Example 5.1 An Application of the Model (5.1)

Figure 5.1 presents statistical results of the model (5.1). Based on these results, the following findings and notes are presented.

1) The estimation function has the equation as follows

$$\begin{aligned} LOG(Y1) &= C(1) + C(2) * LOG(Y1(-1)) \\ &\quad + C(3) * LOG(X1(-1)) + C(4) * LOG(X2(-1)) \end{aligned} \tag{5.6a}$$

which is an additive linear function, and can be presented as the following Cobb–Douglas multiplicative or nonlinear function.

Dependent Variable: LOG(Y1)
Method: Least Squares
Date: 01/10/17 Time: 14:45
Sample (adjusted): 1947M02 1996M04
Included observations: 591 after adjustments

Variable	Coefficient	Std. Error	t-Statistic	Prob.
C	0.025457	0.016888	1.507401	0.1322
LOG(Y1(-1))	1.004157	0.004730	212.2902	0.0000
LOG(X1(-1))	-0.023989	0.010466	-2.292011	0.0223
LOG(X2(-1))	0.015122	0.005755	2.627461	0.0088

R-squared	0.998786	Mean dependent var	4.526843
Adjusted R-squared	0.998779	S.D. dependent var	0.942446
S.E. of regression	0.032926	Akaike info criterion	-3.982345
Sum squared resid	0.636391	Schwarz criterion	-3.952688
Log likelihood	1180.783	Hannan-Quinn criter.	-3.970792
F-statistic	160927.6	Durbin-Watson stat	1.539248
Prob(F-statistic)	0.000000		

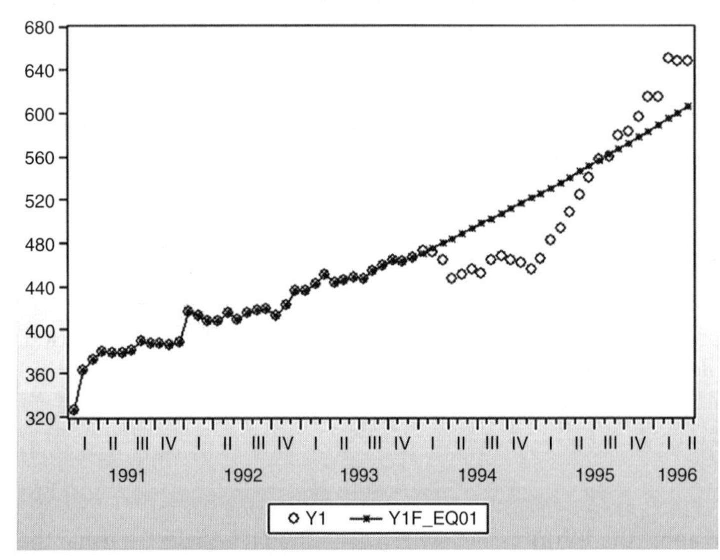

RMSE=36.78286 & 0.034642

Figure 5.1 Statistical results of the model (5.1) with a forecast sample {@*Year*>1993}.

$$Y1 = Exp(C(1)) * (Y1(-1))^{C(2)} * (X1(-1))^{C(3)} * (X1(-1))^{C(4)} \tag{5.6b}$$

or $$Y1 = \alpha * (Y1(-1))^{\beta 1} * (X1(-1)^{\beta 2}) * (X2(-1)^{\beta 4} \tag{5.6c}$$

Note that this multiplicative function indicates that the effect of *Y1(−1)* depends on both exogenous variables *X1(−1)* and *X2(−1)*. However, the additive model (5.6a) indicates that the effect of *log(Y1(−1))* on *log(Y(1))* is an adjusted effect, that is the effect of *log(Y1(−1))* on *log(Y1)* is adjusted for or by taking into account of the other independent variables.

2) As a comparison, Figure 5.2 presents the statistical results of the model (5.1) using a subsample {@*Year*>1990}, with the same forecast sample {@*Year*>1993}. Based on these results, the findings and notes presented are as follows:

Dependent Variable: LOG(Y1)
Method: Least Squares
Date: 01/10/17 Time: 14:55
Sample: 1947M01 1996M04 IF @YEAR>1990
Included observations: 64

Variable	Coefficient	Std. Error	t-Statistic	Prob.
C	-3.686073	0.811879	-4.540175	0.0000
LOG(Y1(-1))	0.894916	0.037013	24.17836	0.0000
LOG(X1(-1))	-0.196487	0.111653	-1.759796	0.0835
LOG(X2(-1))	1.100607	0.251240	4.380695	0.0000

R-squared	0.985008	Mean dependent var	6.121370
Adjusted R-squared	0.984259	S.D. dependent var	0.153347
S.E. of regression	0.019240	Akaike info criterion	-5.003240
Sum squared resid	0.022210	Schwarz criterion	-4.868310
Log likelihood	164.1037	Hannan-Quinn criter.	-4.950084
F-statistic	1314.085	Durbin-Watson stat	2.225089
Prob(F-statistic)	0.000000		

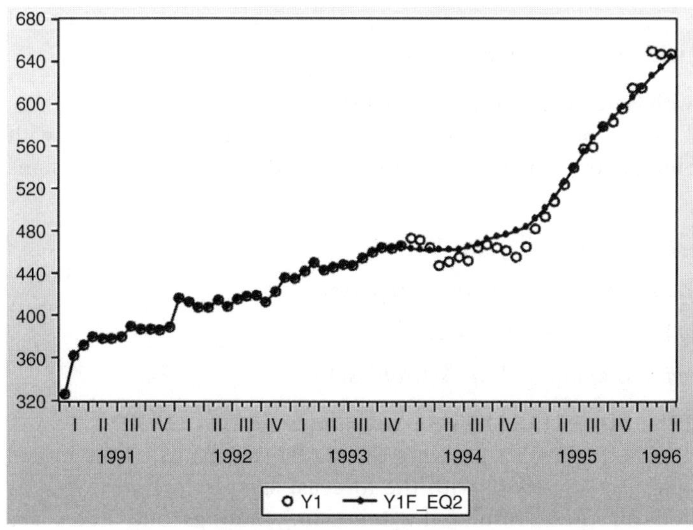

Figure 5.2 Statistical results of the model (5.1) using a subsample {@*Year*>1990}.

2.1 Compared to the forecast using the whole sample period in Figure 5.1, Figure 5.2 presents a much smaller RMSE (Root Mean Squared Error), TIC (Theil Inequality Coefficient), and very small values deviations of the forecasting values from the observed values of *Y1*.

2.2 Hence, better predicted values of *Y1* or *log(Y1)* beyond the sample period would be obtained by using the following three regression functions, where the last two are the regressions of the models (5.4) and (5.5), which could easily be done using Excel. Do this as an exercise. Note that instead of *log(V),* the symbol *ln(V)* should be used for the computation using Excel. Refer to Table 5.4, in Example 5.6 later, to see if it is found to be difficult.

$$ln(Y1) = -3.68607272725 + 0.89491557008 * ln(Y1(-1)) - 0.196486932785 * ln(X1(-1)) + 1.10060726755 * ln(X2(-1)) \quad (5.7a)$$

$$ln(X1) = -0.0147162870192 + 1.00369133779 * ln(X1(-1)) \quad (5.7b)$$

$$ln(X2) = 2.72021415565 + 0.437537824669 * ln(X1(-2)) \quad (5.7c)$$

5.2.2 Tanslog Linear Model with Trend

As the extension of the model (5.1) based on the monthly time series, I propose to conduct the following two forecast models with a linear trend and logarithmic trend. Similar forecast models can easily be defined as the extensions of the models in Eqs. (5.2) or (5.3).

$$log(Y1)\ C\ log(Y1(-1))\ log(X1(-1))\ log(X2(-1))\ @Trend \quad (5.8)$$

$$log(Y1)\ C\ log(Y1(-1))\ log(X1(-1))\ log(X2(-1))\ log(@Trend + 1) \quad (5.9)$$

Note that *log(@Trend + 1)* should be used in (5.9), because *@Trend* has a score of zero based on the whole sample.

5.2.3 Tanslog Linear Model with Heterogeneous Trends

5.2.3.1 Translog Linear Model with Heterogeneous Linear Trends

As the extension of the model (5.1), I propose the following alternative forecast models with heterogeneous linear trends. Similar forecast models can easily be defined as the extensions of the models (5.2) or (5.3).

$$log(Y1)\ C\ log(Y1(-1))\ log(X1(-1))\ log(X2(-1)\ @Month * @Expand(@Year) @Expand(@Year, @Droplast) \quad (5.10)$$

$$log(Y1)\ C\ log(Y1(-1))\ log(X1(-1))\ log(X2(-1))\ @Trend * @Expand(@Year) @Expand(@Year, @Droplast) \quad (5.11)$$

Example 5.2 Application of the Model (5.11) Using Subsample {@*Year*>1990}

As an extension of the model (5.1), Figure 5.3 presents the statistical results of the model (5.11) using the same subsample {@*Year*>1990} with the forecast sample {@*Year*>1993} as presented in Figure 5.2. Based on these results the following findings and notes are presented.

1) Compared to the forecasting in previous figures, this forecast has the smallest RMSE and TIC. Hence, in the statistical sense, this model can be expected to give the best possible predicted values of *Y1* beyond the sample period.
2) Note that the regression in fact is presenting six different regressions for the *@Year = 1991* up to *@Year* = 1996. So to compute the predicted value of *Y1* or *log(Y1)*, the following regression of *@Year* = 1996 can be applied using Excel, together with the regressions Eqs. (5.6a) and (5.6b).

$$ln(Y1) = 5.93089570278 + 0.108524819492 * ln(Y1(-1)) - 1.82104937866 * ln(X1(-1))$$
$$- 0.458150533337 * ln(X2(-1)) + 0.0183395190273 * @TREND$$

Note that by using the function *@Expand(@Year,@Droplast)* in (5.11), this regression could be written directly based on the output of the object *"Representations."* Do this as an exercise.

5.2.3.2 Translog Linear Model with Heterogeneous Logarithmic Trends

As an extension of the models Eqs. (5.10) and (5.11), I propose to conduct the following alternative forecast models with heterogeneous logarithmic trends, where t = *@trend* + 1, because *@Trend* can have a zero value.

$$log(Y1)\ C\ log(Y1(-1))\ log(X1(-1))\ log(X2(-1))$$
$$log@Month * @Expand(@Year)@Expand(@Year,@Droplast) \quad (5.12)$$

$$log(Y1)\ C\ log(Y1(-1))\ log(X1(-1))\ log(X2(-1))$$
$$log(t)*@Expand(@Year)@Expand(@Year,@Droplast) \quad (5.13)$$

Example 5.3 An Application of the Model (5.13) Using Subsample {@*Year*>1990}

Figure 5.4 presents the statistical results of the ES (5.13) using the subsample {@*Year*>1990} and the forecast sample {@*Year*>1993}. Based on these results, the findings and notes presented are as follows:

1) Compared to the forecast evaluation in Figure 5.3, this model has slightly smaller RMSE and TIC. So, in the statistical sense, this model is a better forecast model.
2) Similar to the note presented in Example 5.2, the following regression for the @*Year*=1996, together with the regressions (5.6a) and (5.6b), can be used to compute the forecast values beyond the sample period, using Excel.

$$ln(Y1) = -52.7409003864 + 0.103714008271 * ln(Y1(-1))$$
$$- 1.81653538871 * ln(X1(-1)) - 0.477431902664 * ln(X2(-1))$$
$$+ 10.9051618301 * ln(t)$$

5.3 Translog Linear Models Based on *(X1,X2,Y2)*

By looking at the three graphs in Figure 5.5, it is very clear that the model of *Y2* = *FSXDP* for the whole sample should be presented using a polynomial trend, such as a fifth degree polynomial. However, based on a subsample, we could present a model with a lower degree polynomial trend, such as a second degree polynomial for the subsample {@*Year*>1975} and a linear trend for the subsample {@*Year*>1990}.

Dependent Variable: LOG(Y1)
Method: Least Squares
Date: 01/14/17 Time: 17:14
Sample: 1947M01 1996M04 IF @YEAR>1990
Included observations: 64

Variable	Coefficient	Std. Error	t-Statistic	Prob.
C	5.930896	4.900816	1.210185	0.2320
LOG(Y1(-1))	0.108525	0.149237	0.727196	0.4706
LOG(X1(-1))	-1.821049	0.550869	-3.305777	0.0018
LOG(X2(-1))	-0.458151	0.588061	-0.779087	0.4397
@TREND*(@YEAR=1991)	0.012275	0.003163	3.881336	0.0003
@TREND*(@YEAR=1992)	0.011168	0.003208	3.481078	0.0011
@TREND*(@YEAR=1993)	0.009145	0.002307	3.964750	0.0002
@TREND*(@YEAR=1994)	0.008654	0.003089	2.801903	0.0073
@TREND*(@YEAR=1995)	0.024296	0.004877	4.981945	0.0000
@TREND*(@YEAR=1996)	0.018340	0.008490	2.160211	0.0357
@YEAR=1991	3.448255	4.551793	0.757560	0.4523
@YEAR=1992	4.051744	4.491697	0.902052	0.3714
@YEAR=1993	5.189315	4.591385	1.130229	0.2639
@YEAR=1994	5.480042	4.598750	1.191637	0.2391
@YEAR=1995	-3.491055	4.380382	-0.796975	0.4293

R-squared	0.991351	Mean dependent var	6.121370
Adjusted R-squared	0.988879	S.D. dependent var	0.153347
S.E. of regression	0.016171	Akaike info criterion	-5.209487
Sum squared resid	0.012814	Schwarz criterion	-4.703499
Log likelihood	181.7036	Hannan-Quinn criter.	-5.010152
F-statistic	401.1524	Durbin-Watson stat	1.887991
Prob(F-statistic)	0.000000		

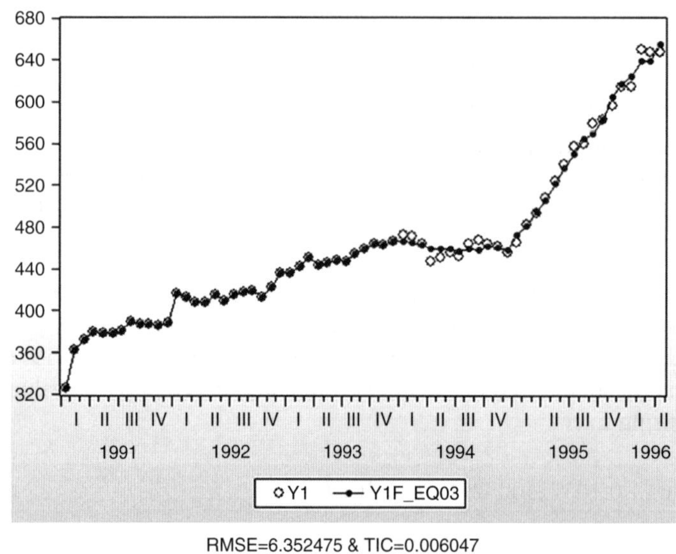

Figure 5.3 Statistical results of the model (5.11) using a subsample {@*Year*>1990} and forecast sample {@*Year*>1993}.

Even though, based on my experiment, I have found that alternative models with heterogeneous trends were presenting the best possible forecast values beyond the sample period, which have been presented in previous chapters. Hence, the following examples also show that the models with heterogeneous trends are the best forecast models, based on the trivariate $(X1_t, X2_t, Y2_t)$.

Dependent Variable: LOG(Y1)
Method: Least Squares
Date: 01/15/17 Time: 05:31
Sample: 1947M01 1996M04 IF @YEAR>1990
Included observations: 64

Variable	Coefficient	Std. Error	t-Statistic	Prob.
C	-52.74090	30.08228	-1.753222	0.0858
LOG(Y1(-1))	0.103714	0.149856	0.692093	0.4921
LOG(X1(-1))	-1.816535	0.549866	-3.303597	0.0018
LOG(X2(-1))	-0.477432	0.590348	-0.808729	0.4226
LOG(T)*(@YEAR=1991)	6.577452	1.690552	3.890712	0.0003
LOG(T)*(@YEAR=1992)	6.122938	1.756390	3.486093	0.0010
LOG(T)*(@YEAR=1993)	5.124072	1.289561	3.973501	0.0002
LOG(T)*(@YEAR=1994)	4.942564	1.762455	2.804363	0.0072
LOG(T)*(@YEAR=1995)	14.24845	2.854410	4.991731	0.0000
LOG(T)*(@YEAR=1996)	10.90516	5.015497	2.174293	0.0345
@YEAR=1991	27.45303	29.33442	0.935864	0.3539
@YEAR=1992	30.31913	28.86654	1.050321	0.2987
@YEAR=1993	36.64864	29.54029	1.240632	0.2206
@YEAR=1994	37.81285	29.50568	1.281545	0.2060
@YEAR=1995	-21.31403	27.96973	-0.762039	0.4497

R-squared	0.991364	Mean dependent var	6.121370
Adjusted R-squared	0.988896	S.D. dependent var	0.153347
S.E. of regression	0.016159	Akaike info criterion	-5.210984
Sum squared resid	0.012795	Schwarz criterion	-4.704996
Log likelihood	181.7515	Hannan-Quinn criter.	-5.011649
F-statistic	401.7586	Durbin-Watson stat	1.884538
Prob(F-statistic)	0.000000		

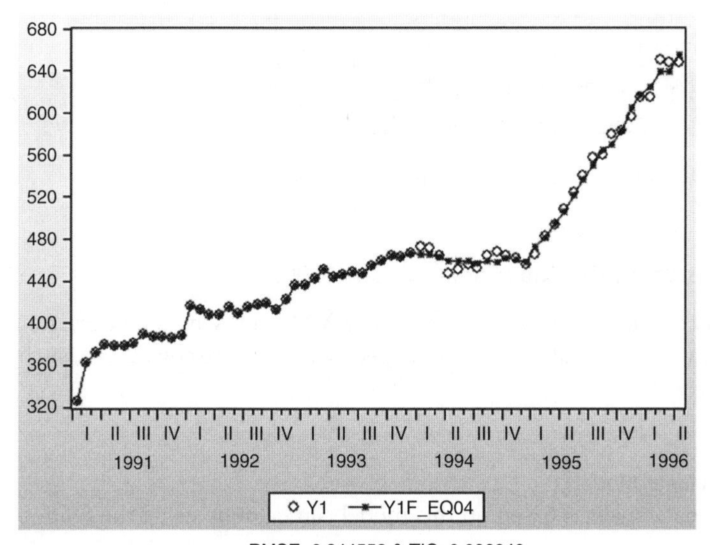

RMSE=6.344558 & TIC=0.006040

Figure 5.4 Statistical results of the model (5.13) using a subsample {@*Year*>1990) and forecast sample {@*Year*>1993}.

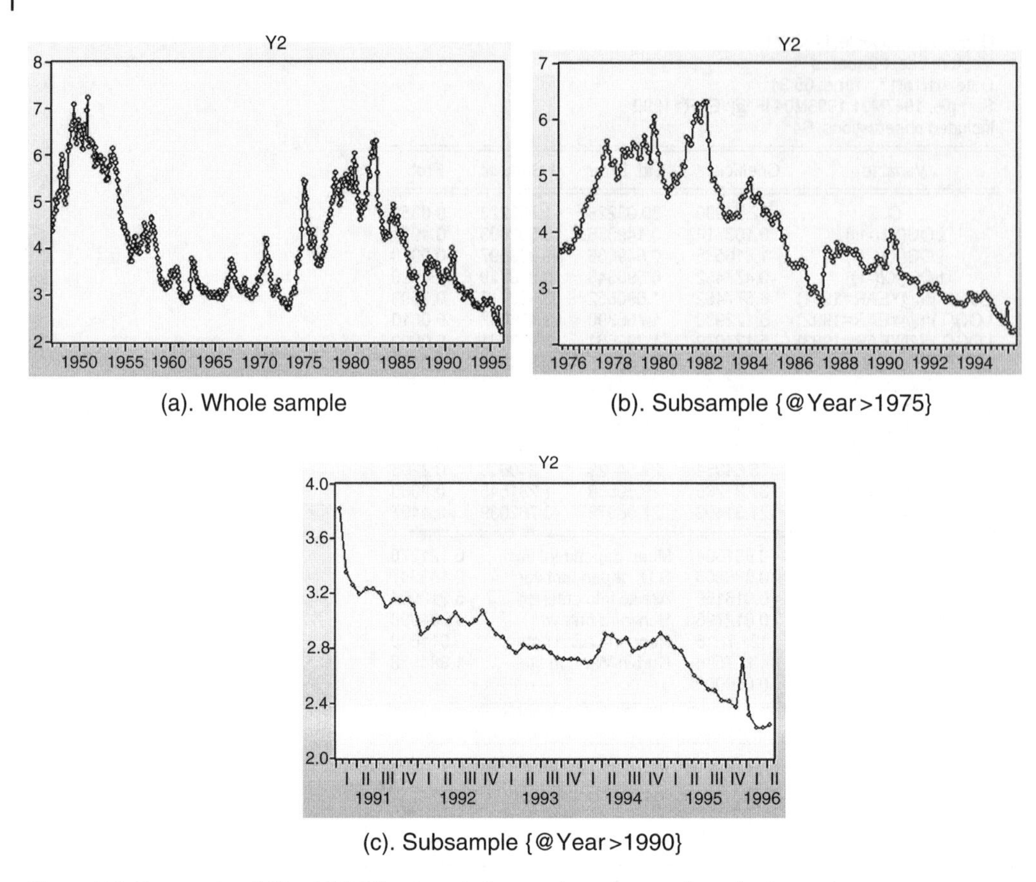

Figure 5.5 The graphs of $Y2 = FSXDP$ for the whole sample and two selected subsamples.

5.3.1 Translog Linear Models Using the Subsample {@Year>1990}

Using this subsample, I am very confident that all models for *(X1,X2,Y1)* presented here are suitable for *(X1,X2,Y2)*. However, as the illustration, only four specific models are considered, such as follows

5.3.1.1 The Simplest Translog Linear Model

Similar to the model (5.1), the data analysis based on this model can be done using the following ES.

$$log(Y2)\ C\ log(Y2(-1))\ log(X1(-1))\ log(X2(-1)) \tag{5.14}$$

5.3.1.2 Translog Linear Model with a Linear Trend

Similar to the model (5.4), the data analysis based on this model can be done using the following ES.

$$log(Y2)\ C\ log(Y2(-1))\ log(X2(-1))\ log(X2(-1))@Trend \tag{5.15}$$

5.3.1.3 Translog Linear Model with Logarithmic Trend

Similar to the model (5.5), the data analysis based on this model can be done using the following ES.

$$log(Y2)\ C\ log(Y2(-1))\ log(X1(-1))\ log(X2(-1))\ log(t) \qquad (5.16)$$

5.3.1.4 Translog Linear Model with Heterogeneous Logarithmic Trend by Year

Similar to the model in Eq. (5.13), the data analysis based on this model can be done using the following ES.

$$\begin{aligned} &log(Y2)\ C\ log(Y2(-1))\ log(X1(-1))\ log(X1(-1)) \\ &\qquad log(t)*@Expand(@Year) \end{aligned} \qquad (5.17)$$

Example 5.4 Application of the Models Here Using Subsample {@*Year*>1990}

Table 5.1 presents the summary of the forecast evaluations based on the four models (5.10) up to (5.13). Based on this summary, the following findings and notes are presented.

1) The four models are acceptable forecast models, since their TIC < 0.02, and we never can predict which model presents the best possible predicted values beyond the sample period. Subjective judgments of the users and researchers should be taken into consideration in selecting alternative models.
2) However, in the statistical sense, the model (5.13) can be considered to be the best forecast, since it has the smallest RMSE and TIC. However, the model has 14 independent variables, which is a very large number compared to the other models, and their TICs only have small differences.
3) Furthermore, note that the regression of the model (5.13) presents six additive multiple regressions. Even though this model has 14 independent variables, to compute the predicted values of *Y2* or *log(Y2)* beyond the sample period, only the sixth regression, that is, for the *@Year=1996*, with the following equation can be applied together with both regressions in (5.6ab) and (5.6ac) by using Excel.

$$\begin{aligned} ln(Y2) = &-24.5383910246 + 0.282089025732*ln(Y2(-1)) \\ &+ 1.94765530028*ln(X1(-1)) - 0.705890428848*ln(X2(-1)) \\ &+ 3.00035556195*ln(t) \end{aligned}$$

Table 5.1 Summary of the forecast evaluation based on the four models before.

	Models			
Statistics	**(5.10)**	**(5.11)**	**(5.12)**	**(5.13)**
RMSE	0.104996	0.082487	0.082536	0.074140
TIC	0.019949	0.015566	0.015573	0.013969
Bias proportion	0.146566	0.003519	0.002382	0.000142
Number of IVs	3	4	4	14

5.3.2 Translog Linear Models Using the Subsample {@Year>1975}

Referring to the graph in Figure 5.5b, I apply a model with quadratic trend by using the time variable $(t/100)$ and a model with quadratic logarithmic trend with the following ESs in addition to the model (5.14), (5.15), (5.16), and (5.17). Note that the time variable $(t/100)$ is used in order to have a large coefficient of the time variable.

$$log(Y2)\ C\ log(Y2(-1))\ log(X1(-1))\ log(X2(-1))\ (t/100)\ (t/100)\hat{\ }2 \tag{5.18}$$

$$log(Y2)\ C\ log(Y2(-1))\ log(X1(-1))\ log(X2(-1))\ log(t))\ log(t)\hat{\ }2 \tag{5.19}$$

Example 5.5 Application of the Six Models Using Subsample {@*Year*>1975}
Table 5.2 presents the summary of the forecast evaluations based on the six models (5.14) up to (5.19). Based on this summary, the following findings and notes are presented.

1) Refer to the notes presented in previous example. Based on this summary, also it is found that the model (5.17) is the best forecasting model, since it has the smallest RMSE and TIC.
2) Compared to the results of the model (5.17) in Table 5.1, it can be concluded that the forecast using the subsample {@*Year*>1990} can be expected to present better forecast values, since it has smaller RMSE and TIC.
3) Looking at the results of the models (5.18) and (5.19), both are worse forecast models than (5.17).
4) Even though it has 44 independent variables, there is no problem in doing the analysis and to compute the forecast values beyond the sample period the following simple multiple regression for the *@Year=1996*, together with both regressions in (5.7b) and (5.7c), can be applied easily using Excel.

$$\begin{aligned} ln(Y2) = &-116.482469768 + 0.630534910542* > ln(Y2(-1)) \\ &+ 0.590556159789* ln(X1(-1)) - 0.263926185349* lnX2(-1) \\ &+ 18.0513225042* ln(t) \end{aligned}$$

5.3.3 Translog Linear Models Using the Whole Sample

Based on the statistical results and notes presented in the last examples, I am very confident that models with heterogeneous trends would be the best forecasting models, compared to those with the k-th degree polynomial trend for any $k > 2$. Refer to the results in Table 5.2, and see the following illustrative example.

Table 5.2 Summary of the forecast evaluations based on the six models before.

	Models					
Statistics	**(5.14)**	**(5.15)**	**(5.16)**	**(5.17)**	**(5.18)**	**(5.19)**
RMSE	0.191069	0.195898	0.229089	0.083242	0.206181	0.207986
TIC	0.036050	0.037369	0.043035	0.015687	0.039066	0.039373
Bias proportion	0.000161	0.080595	0.12855	0.000040	0.009101	0.039373
Number of IVs	3	4	4	44	5	5

Example 5.6 Application of the Four Models Using the Whole Sample
Table 5.3 presents a summary of the forecast evaluations based on the four models (5.14), (5.15), (5.16), and (5.17), using the whole sample. Based on this summary, the following findings and notes are presented.

1) This summary shows that the model (5.17) also is the best forecast model using the whole sample, with the forecast sample {@*Year*>1993}. However, compared to the results using the subsamples {@*Year*>1990} and {@*Year*>1975}, this forecasting is the worst, even though the three forecasts have very small differences between RMSEs and TICs.
2) Although it has 102 independent variables, it can be estimated in a short time, and to compute the forecast values beyond the sample period, the following simple multiple regression for the @*Year=1996*, together with both regressions in (5.7b) and (5.7c), can be applied easily using Excel.

$$\begin{aligned} ln(Y2) &= -114.071656686 + 0.614260429515 * ln(Y2(-1)) \\ &\quad + 0.142694403437 * ln(X1(-1)) + 0.239245918032 * ln(X2(-1)) \\ &\quad + 17.6368400965 * ln(t) \end{aligned}$$

3) As an illustration, Table 5.4 presents the forecast values computed for t = 592 = 1996M05 using the following three equations, by using Excel. Then for t = 593 = 1996M06, and so on, the forecast values can be obtained easily using the block-copy-paste method.

$$\begin{aligned} ln(Y2(592)) &= -114.071656686 + 0.614260429515 * \text{D5} \\ &\quad + 0.142694403437 * \text{E5} + 0.239245918032 * \text{F5} \\ &\quad + 17.6368400965 * \text{G5} \end{aligned}$$

$$ln(X1(592)) = -0.0147162870192 + 1.00369133779 * \text{E5}$$
$$ln(X2(592)) = 2.72021415565 + 0.437537824669 * \text{F5}$$

where D5 = $ln(Y2_{591}(-1))$, E5 = $ln(X1_{591}(-1))$, F5 = $ln(X2_{591}(-1))$, and G5 = ln(591).

Table 5.3 Summary of the forecast evaluations based on four models.

	Models			
Statistics	**(5.14)**	**(5.15)**	**(5.16)**	**(5.17)**
RMSE	0.187841	0.235698	0.269180	0.087809
TIC	0.035534	0.045987	0.049793	0.016548
Bias proportion	0.003792	0.547578	0.162669	0.000023
Number of IVs	3	4	4	102

Table 5.4 A spreadsheet for computing forecast values beyond the sample period.

A	B	C	D	E	F	G
1		T	LOG(Y2)	LOG(X1)	LOG(X2)	LOG(T)
2	1996M01	588	0.837247	4.808111	4.837075	6.376727
3	1996M02	589	0.797507	4.819475	4.835488	6.378426
4	1996M03	590	0.797507	4.815431	4.839451	6.380123
5	1996M04	591	0.806476	4.824306	4.848116	6.381816
		592	**0.82709**	**4.8274**	**4.84145**	**6.38351**

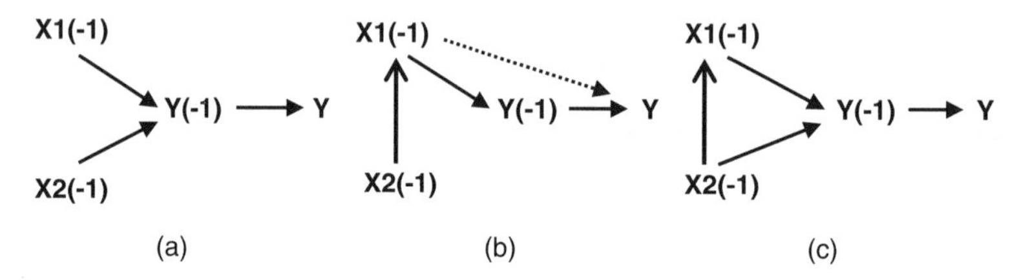

Figure 5.6 Alternative up-and-down relationships between four selected variables.

5.4 Forecast Models Using Original *(X1,X2,Y)*

Compared to the additive translog linear models (5.1), (5.2), and (5.3), the models using the observed scores time series *X1*, *X2*, and *Y* will be more complex, as are extensions of all models presented in previous chapters, specifically the models using the bivariate *(X,Y)* presented in Chapter 4. Similar to all examples presented here, I will use *X1(−1)*, *X2(−1)*, and *Y(−1)* as the exogenous or independent variables of the first set of forecast models. For this reason, I present three alternative up-and-down or causal relationships between the three independent variables and an endogenous or dependent variable *Y* in Figure 5.6. Note that the three graphs presenting *X1(−1)* and *X2(−1)* are the source/causal factors of *Y1(−1)* and the effect of *Y(−1)* on *Y* depends on *X1(−1)* and *X2(−1)*. As the extension of the model (4.1), I propose the following alternative models.

5.4.1 Model Based on Figure 5.6a

Figure 5.6a presents the up-and-down stream relationship between the four variables *X1(−1)*, *X2 (−1)*, *Y(−1)*, and *Y* under the assumption that *X1(−1)* and *X2(−1)* do not have up-and-down stream or causal relationship. Note that the arrows from *X1(−1)* and *X2(−1)* to *Y(−1)*, together with the arrow from *Y(−1)* to *Y*, indicate that each *X1(−1)* and *X(2(−1)* has an indirect effect on *Y*, through *Y(−1)*. In other words, the effect of *Y(−1)* on *Y* depends on both *X1(−1)* and *X2(−1)*, or the effect of both *X1(−1) and X2(−1)* on *Y* depending on *Y1(−1)*. Furthermore, even though, the arrows from *X1(−1)* and *X2(−1)* to *Y* are not presented, it should be understood that both variables are the source, upper, or causal variables of *Y*. For these reasons, based on this graph, I define a two-way or two-factor interactions model based on the following ES, or modeling the first stage of the analysis in order to obtain a good fit model in the statistical sense.

$$Y\ C\ Y(-1)*X1(-1)\ Y(-1)*X2(-1) \tag{5.20}$$

Since the impacts of multicollinearity or bivariate correlations of the independent variables on the parameter estimates are unpredictable, then it is recommended to use *Y(−1)*, *X1(−1)*, and *X2 (−1)* as the search variables by using either the stepwise selection method, specifically the combinatorial selection method or the manual selection method – refer to Agung (2009a, 2011a, 2014). The best possible fit model would be highly dependent on the data used and could be unexpected. However, for the forecasting, we may use the full model with the five independent variables, even though some of them have large or very large p-values. So we would apply the following equation of the full model, which is a hierarchical model, at least as a comparative result. See the following example.

$$Y\ C\ Y(-1)*X1(-1)\ Y(-1)*X2(-1)\ Y(-1)\ X1(-1)\ X2(-1) \tag{5.21}$$

Example 5.7 Application of the Models (5.20) and (5.21)
Figure 5.7 presents the statistical results based on (5.20) and (5.21) for *Y1 = FSPCOM*, using the subsample {@*Year*>1990} with the forecast sample {@*Year*>1993}. Based on these results, the findings and notes presented are as follows:

1) For the model (5.20), the interaction *Y1(−1)*X2(−1)* has a positive significant adjusted effect at the 1% level. However, *Y1(−1)*X1(−1)* has a negative significant adjusted effect on *Y1* at the 10% level of significance. So the model is acceptable for showing the effect of *Y1(−1)* on *Y1* is significantly dependent on *X1(−1)* and *X2(−1)*. Note that both interactions have significant joint effects based on the F-statistic of $F_0 = 2462697$ with a p-value = 0.000000.
2) As expected, the forecast based on the model (5.21) has smaller RMSE and TIC, because it has five independent variables, compared to the model (5.20) having only two predictors. However, they have small differences.
3) Note that each independent variables of the model (5.21) has an insignificant adjusted effect at the 10% level, which are the unexpected impacts of all the highly correlated pairs of independent variables with correlations greater than 0.869 and they are significant based on the t-statistic with p-values = 0.0000.
4) Compared to previous forecast models in Figures 5.1 and 5.2, these two models are better forecast models since they have smaller RMSEs and TICs. Therefore, among the four models, which one would you use to forecast beyond the sample period?
5) As a comparison, do the analysis based on the same models for *Y2 = FSDXP*.

5.4.2 Model Based on Figure 5.6b

Referring to the notes presented based on Figure 5.6a, then Figure 5.6b shows that *X1(−1)* has an indirect effect on *Y* through *Y(−1)*, and *X2(−1)* has an indirect effect on *Y* through *X1(−1)* and *Y(−1)*, which should be presented as a two-way interaction *Y(−1)*X1(−1)*, and a three-way interaction *Y(−1)*X1(−1)*X2(−1)*, respectively. Therefore, I propose the following two alternative based forecasting models.

5.4.2.1 The Simplest Model

The simplest based model is a two-way interaction model with the ES as follows:

$$Y\ C\ Y(-1)*X1(-1)\ X1(-1)*X2(-1) \tag{5.22}$$

In addition, note that it could be accepted that both *X1(−1) and X2(−1)* are the upper, source, or causal factors/variables of *Y*, even though there are no arrows presented. Note that the arrow,

Dependent Variable: Y1
Method: Least Squares
Date: 01/22/17 Time: 18:12
Sample: 1947M01 1996M04 IF @YEAR>1990
Included observations: 64

Variable	Coefficient	Std. Error	t-Statistic	Prob.
C	53.66400	9.869583	5.437312	0.0000
Y1(-1)*X1(-1)	-0.001377	0.000990	-1.391014	0.1693
Y1(-1)*X2(-1)	0.008729	0.001089	8.012831	0.0000

R-squared	0.987767	Mean dependent var	460.9648
Adjusted R-squared	0.987366	S.D. dependent var	74.47930
S.E. of regression	8.371678	Akaike info criterion	7.133327
Sum squared resid	4275.185	Schwarz criterion	7.234524
Log likelihood	-225.2665	Hannan-Quinn criter.	7.173193
F-statistic	2462.697	Durbin-Watson stat	2.264135
Prob(F-statistic)	0.000000		

(a)

Dependent Variable: Y1
Method: Least Squares
Date: 01/22/17 Time: 18:23
Sample: 1947M01 1996M04 IF @YEAR>1990
Included observations: 64

Variable	Coefficient	Std. Error	t-Statistic	Prob.
C	617.1897	692.3408	0.891454	0.3764
Y1(-1)*X1(-1)	-0.014949	0.010587	-1.411961	0.1633
Y1(-1)*X2(-1)	0.033664	0.021777	1.545893	0.1276
Y1(-1)	-1.508230	1.616507	-0.933018	0.3547
X1(-1)	5.897208	4.667263	1.263526	0.2115
X2(-1)	-10.05360	9.427925	-1.066364	0.2907

R-squared	0.988112	Mean dependent var	460.9648
Adjusted R-squared	0.987087	S.D. dependent var	74.47930
S.E. of regression	8.463579	Akaike info criterion	7.198481
Sum squared resid	4154.666	Schwarz criterion	7.400877
Log likelihood	-224.3514	Hannan-Quinn criter.	7.278215
F-statistic	964.1388	Durbin-Watson stat	2.233397
Prob(F-statistic)	0.000000		

(b)

Figure 5.7 Statistical results based on the models (5.20) and (5.21).

with a dotted line from *X1(−1)* to *Y* to indicate that *X2(−1)* can be considered, has an indirect effect on *Y* through *X1(−1)*. For this reason, *X1(−1)∗X2(−1)* is used in the based model (5.23). For these reasons, two sets of search variables should be considered; the first set is a single interaction *Y(−1)∗X2(−1)*, which indicates that *X2(−1)* has an indirect effect on *Y* through *Y(−1)*, even though there is no arrow from *X2(−1)* to *Y(−1)*, and the second set is the set of the main variables *Y(−1)*, *X1(−1)*, and *X2(−1)* that should be selected using the manual multistage selection method, depending on the subjective judgment of the researchers.

5.4.2.2 A More Complex Model

A more complex based model is a three-way interaction model based on the model that follows:

$$Y\ C\ Y(-1)*X1(-1)*X2(-1) \tag{5.23}$$

where the interaction *Y(−1)∗X1(−1)∗X2(−1)* indicates the indirect effect of *X2(−1)* on *Y* through *X1(−1)* and *Y(−1)*. In this case, the interactions *Y(−1)∗X(1(−1)* and *X1(−1)∗X2(−1)* should be used as the search variables at the second stage of analysis to represent the arrow from *X1 (−1)* to *Y(−1)*, and the arrow from *X2(−1)* to *X1(−1)*, respectively. Then, the main variables/factors *Y(−1)*, *X1(−1)*, and *X2(−1)* can be used as the search variables for the third stage.

Example 5.8 Application of the Model (5.23) for *Y2 = FSDXP*

Figure 5.8a presents the final regression of the second stage of the analysis, which shows that both interactions have small *p*-values of 0.0000. In fact, even though the interaction *X1(−1)∗*

Dependent Variable: Y2
Method: Least Squares
Date: 01/26/17 Time: 12:57
Sample: 1947M01 1996M04 IF @YEAR>1990
Included observations: 64

Variable	Coefficient	Std. Error	t-Statistic	Prob.
C	0.516519	0.209712	2.462989	0.0166
Y2(-1)*X1(-1)*X2(-1)	-0.000175	1.35E-05	-12.94909	0.0000
Y2(-1)*X1(-1)	0.028127	0.001549	18.15853	0.0000

R-squared	0.872032	Mean dependent var	2.845937
Adjusted R-squared	0.867836	S.D. dependent var	0.287815
S.E. of regression	0.104633	Akaike info criterion	-1.630975
Sum squared resid	0.667832	Schwarz criterion	-1.529777
Log likelihood	55.19119	Hannan-Quinn criter.	-1.591108
F-statistic	207.8409	Durbin-Watson stat	1.858447
Prob(F-statistic)	0.000000		

(a)

Dependent Variable: Y2
Method: Stepwise Regression
Date: 01/26/17 Time: 13:30
Sample: 1947M01 1996M04 IF @YEAR>1990
Included observations: 64
Number of always included regressors: 3
Number of search regressors: 3
Selection method: Combinatorial
Number of search regressors: 2

Variable	Coefficient	Std. Error	t-Statistic	Prob.*
C	-12.92164	8.758947	-1.475250	0.1455
Y2(-1)*X1(-1)*X2(-1)	-0.000457	0.000248	-1.842373	0.0704
Y2(-1)*X1(-1)	0.058159	0.031271	1.859859	0.0679
Y2(-1)	0.455097	0.195954	2.322474	0.0237
X2(-1)	0.111343	0.072069	1.544954	0.1277

R-squared	0.903357	Mean dependent var	2.845937
Adjusted R-squared	0.896805	S.D. dependent var	0.287815
S.E. of regression	0.092458	Akaike info criterion	-1.849229
Sum squared resid	0.504356	Schwarz criterion	-1.680566
Log likelihood	64.17532	Hannan-Quinn criter.	-1.782784
F-statistic	137.8733	Durbin-Watson stat	2.282784
Prob(F-statistic)	0.000000		

(b)

Figure 5.8 Statistical results using the model (5.22) at the first stage of analysis.

X2(−1) has a large *p*-value, as long as the interaction *Y2(−1)*X1(−1)*X2(−1)* has a significant adjusted effect on *Y2*, the regression would be accepted as the final regression at the second stage of analysis. The steps of analyses are as follows:

At the second stage of analysis, first, I try to insert both interactions *Y2(−1)*X1(−1)* and *X1(−1)*X2(−1)* as additional variables for the based model (5.23), and it is found that *Y2(−1)*X1(−1)*X2(−1)* has a very large *p*-value = 0.9851. So, both interactions cannot be inserted as additional independent variables to present a three-way interaction model.

For this reason, I have to select one of them. Since *Y2(−1)* is the best predictor for *Y2*, then I decide to insert *Y2(−1)*X1(−1)* as an additional independent variable with the results presented in Figure 5.8a, which is a good fit model, since each independent variable has a *p*-value = 0.0000.

At the third stage of analysis, I insert the three main variables *Y1(−1)*, *X1(−1)*, and *X2(−1)* as additional independent variables, but it was found that each of the five independent variables has a large or very large *p*-value with a maximum of 0.9545 for the main variable *X1(−1)*. In this case, I have tried to apply three different methods with exactly the same results. First, I used only using *Y2(−1)* and *X2(−1)* as additional independent variables for the LS regression, since *X1(−1)* has the greatest *p*-value, so it can be omitted. Then to reconfirm the result, I applied the default/basic stepwise selection method and the combinatorial selection method with the result presented in Figure 5.8b, which is a good fit model, since each of the interactions has a significant adjusted effect at the 10% level, even though *X2(−1)* has an insignificant adjusted effect.

The steps of the combinatorial selection methods are as follows:

1) Click *Quick/Estimate Equation.../STEPLS-Stepwise Least Squares*
2) Insert the Equation Specification: the set of *Y2 C Y2(−1)*X1(−1)*X2(−1) Y2(−1*X2(−1)* and the list of search regressors; *Y2(−1) X1(−1) X2(−1)*.
3) Click *Options/Selection Method: Combinatorial/Number of regressors to select*: 2 ... *OK*, the result are obtained.

Example 5.9 Differential Alternative Selection Methods

Figure 5.9a presents exactly the same result as in Figure 5.8b using the combinatorial selection method with the based model (5.23) and the list of search regressors *Y2(−1)*X1(−1) X1(−1)*X2(−1) Y2(−1) X1(−1) X2(−1)*. However, different multiple regressions are obtained by using the default/basic stepwise selection method, as presented in Figure 5.9b.

In general, different results would be obtained by using the combinatorial and the default stepwise selection methods, since they have different criteria. In this case, the combinatorial selection method is used to select the best subset of three regressors out of 10 possible subsets, with the results presented in Figure 5.9a. And Figure 5.9b presents the results by using the default stepwise forward selection method with the stopping criterion forward/backward = 0.5/05. Based on these results, the following findings and notes are presented.

1) Looking at their forecast evaluation, the results in Figure 5.9b have a smaller RMSE and TIC. So, in the statistical sense, this model is a better forecast model, even they have small differences.
2) However, in the theoretical or substantial sense, the graph in Figure 5.6b is presenting that *Y2(−1)*X1(−1)* is a more important predictor than *X1(−1)*X2(−1)*, since it is defined that *Y2(−1)* has direct effect on *Y2*. So I prefer to apply this model to forecast beyond the sample period even though each of the interactions *Y2(−1)*X1(−1)*X2(−1)* and *X1(−1)*X2(−1)* has an insignificant adjusted effect on *Y2* with *p*-values of 0.1082 and 0.2587. In fact, at the 10% level, the interaction *Y2(−1)*X1(−1)*X2(−1)* has a positive significant adjusted effect on *Y2*, with a *p*-value = 0.1802/2 = 0.0901 < 0.10.

Dependent Variable: Y2
Method: Stepwise Regression
Date: 01/26/17 Time: 14:54
Sample: 1947M01 1996M04 IF @YEAR>1990
Included observations: 64
Number of always included regressors: 2
Number of search regressors: 5
Selection method: Combinatorial
Number of search regressors: 3

Variable	Coefficient	Std. Error	t-Statistic	Prob.*
C	-12.92164	8.758947	-1.475250	0.1455
Y2(-1)*X1(-1)*X2(-1)	-0.000457	0.000248	-1.842373	0.0704
Y2(-1)	0.455097	0.195954	2.322474	0.0237
X2(-1)	0.111343	0.072069	1.544954	0.1277
Y2(-1)*X1(-1)	0.058159	0.031271	1.859859	0.0679

R-squared	0.903357	Mean dependent var	2.845937
Adjusted R-squared	0.896805	S.D. dependent var	0.287815
S.E. of regression	0.092458	Akaike info criterion	-1.849229
Sum squared resid	0.504356	Schwarz criterion	-1.680566
Log likelihood	64.17532	Hannan-Quinn criter.	-1.782784
F-statistic	137.8733	Durbin-Watson stat	2.282784
Prob(F-statistic)	0.000000		

RMSE=0.100144 & TIC=0.018932

(a)

Dependent Variable: Y2
Method: Stepwise Regression
Date: 01/26/17 Time: 16:21
Sample: 1947M01 1996M04 IF @YEAR>1990
Included observations: 64
Number of always included regressors: 2
Number of search regressors: 5
Selection method: Stepwise forwards
Stopping criterion: p-value forwards/backwards = 0.5/0.5

Variable	Coefficient	Std. Error	t-Statistic	Prob.*
C	-4.220500	2.880304	-1.465297	0.1481
Y2(-1)*X1(-1)*X2(-1)	-9.28E-05	5.69E-05	-1.631272	0.1082
Y2(-1)	2.024655	0.771037	2.625885	0.0110
X1(-1)	0.055507	0.022930	2.420652	0.0186
X1(-1)*X2(-1)	-0.000106	9.25E-05	-1.140471	0.2587

R-squared	0.902568	Mean dependent var	2.845937
Adjusted R-squared	0.895963	S.D. dependent var	0.287815
S.E. of regression	0.092834	Akaike info criterion	-1.841101
Sum squared resid	0.508472	Schwarz criterion	-1.672438
Log likelihood	63.91522	Hannan-Quinn criter.	-1.774656
F-statistic	136.6377	Durbin-Watson stat	2.307079
Prob(F-statistic)	0.000000		

RMSE=0.096223 & TIC=0.018178

(b)

Figure 5.9 Differential results between combinatorial and default stepwise selection methods.

Dependent Variable: Y2
Method: Stepwise Regression
Date: 01/27/17 Time: 10:06
Sample: 1947M01 1996M04 IF @YEAR>1990
Included observations: 64
Number of always included regressors: 2
Number of search regressors: 5
Selection method: Stepwise forwards
Stopping criterion: p-value forwards/backwards = 0.25/0.5

Variable	Coefficient	Std. Error	t-Statistic	Prob.*
C	-5.585204	2.626530	-2.126457	0.0376
Y2(-1)*X1(-1)*X2(-1)	-0.000129	4.75E-05	-2.710596	0.0087
Y2(-1)	2.510614	0.644208	3.897210	0.0002
X1(-1)	0.054859	0.022981	2.387159	0.0201

R-squared	0.900420	Mean dependent var	2.845937
Adjusted R-squared	0.895441	S.D. dependent var	0.287815
S.E. of regression	0.093066	Akaike info criterion	-1.850545
Sum squared resid	0.519682	Schwarz criterion	-1.715615
Log likelihood	63.21743	Hannan-Quinn criter.	-1.797389
F-statistic	180.8439	Durbin-Watson stat	2.216971
Prob(F-statistic)	0.000000		

(a)

Dependent Variable: Y2
Method: Stepwise Regression
Date: 01/27/17 Time: 10:08
Sample: 1947M01 1996M04 IF @YEAR>1990
Included observations: 64
Number of always included regressors: 2
Number of search regressors: 5
Selection method: Stepwise forwards
Stopping criterion: p-value forwards/backwards = 0.5/0.25

Variable	Coefficient	Std. Error	t-Statistic	Prob.*
C	-5.585204	2.626530	-2.126457	0.0376
Y2(-1)*X1(-1)*X2(-1)	-0.000129	4.75E-05	-2.710596	0.0087
Y2(-1)	2.510614	0.644208	3.897210	0.0002
X1(-1)	0.054859	0.022981	2.387159	0.0201

R-squared	0.900420	Mean dependent var	2.845937
Adjusted R-squared	0.895441	S.D. dependent var	0.287815
S.E. of regression	0.093066	Akaike info criterion	-1.850545
Sum squared resid	0.519682	Schwarz criterion	-1.715615
Log likelihood	63.21743	Hannan-Quinn criter.	-1.797389
F-statistic	180.8439	Durbin-Watson stat	2.216971
Prob(F-statistic)	0.000000		

(b)

Figure 5.10 Unexpected results of the stepwise selection method using two different *p*-values.

3) It is well known that the stepwise selection method could present other forms of multiple regressions by using different stopping criteria. In fact I have been doing several analyses using alternative stopping criteria of *p*-values. Figure 5.10 presents two of the same results, which are unexpected, by using different *p*-values forward/backward of 0.25/0.50 and 0.50/0.25, respectively. In addition, by using the *p*-values of 0.3/0.5 and 0.4/05, the same multiple regressions as in Figure 5.9b were obtained.

5.4.3 Model Based on Figure 5.6c

This graph presents complete up-and-down or causal relationships between the three variables *Y(−1), X1(−1),* and *X2(−1).* In fact, for the graph in Figure 5.6b, in a theoretical sense, it could be said that *X2(−1)* is an upper or source variable of *Y(−1)*, but the arrow is not presented. So the graph in Figure 5.6c can be considered to be the same graph as Figure 5.6b. However, in Figure 5.6c, it is defined that *Y(−1), X1(−1),* and *X2(−1)* have a complete up-and-down or causal relationships. Hence, I propose a full model with the following ES, which is a hierarchical three-way interaction model.

$$\begin{aligned} &Y\,C\,Y(-1)*X1(-1)*X2(-1)\;Y(-1)*X1(-1)\;Y(-1)*X2(-1) \\ &\qquad X1(-1)*X2(-1)\;Y(-1)\;X1(-1)\;X2(-1) \end{aligned} \tag{5.24}$$

Since, in general, all pairs of independent variables are highly or significantly correlated, then we would explore obtaining a reduced model as an acceptable or good fit model, in the statistical sense, and it could be one of the models presented based on the graphs in Figure 5.6a and b. In this case, I again would consider the following two alternative based models.

5.4.3.1 A Two-Way Interaction Based Model

As an extension of the graph in Figure 5.6b based on the model (5.22), I propose the following ES for the first stage of the regression analysis, based on the graph in Figure 5.6c. However, note that the results of the first stage analysis could be unexpected, because the three interaction would be highly correlated, in general. See the notes presented in Example 5.10. Then, at the second stage of analysis, we only have the main variables *Y(−1), X1(−1),* and *X2(−1)* as the search regressors.

$$Y\,C\,Y(-1)*X1(-1)\;Y(-1)*X2(-1)\;X1(-1)*X2(-1) \tag{5.25}$$

Example 5.10 Four Alternative Models (5.25)

As illustrative regressions, Figure 5.11 presents a short summary of the statistical results using a subsample {@Year>1990}} based on the following four alternative ESs of *Y1 = FSPCOM*, where the exogenous variables *V1 and V2* are the four intentionally selected combination of two variables from the four variables *X1 = IP, X2 = PW, X3 = R3,* and *X4 = RAAA* out of six possible combinations. Based on these results, the following findings and notes are presented.

$$Y1\,C\,Y1(-1)*X1(-1)\;Y1(-1)*X2(-1)\;X1(-1)*X2(-1) \tag{5.26}$$

$$Y1\,C\,Y1(-1)*X1(-1)\;Y1(-1)*X3(-1)\;X1(-1)*X3(-1) \tag{5.27}$$

$$Y1\,C\,Y1(-1)*X2(-1)\;Y1(-1)*X3(-1)\;X2(-1)*X3(-1) \tag{5.28}$$

$$Y1\,C\,Y1(-1)*X3(-1)\;Y1(-1)*X4(-1)\;X3(-1)*X4(-1) \tag{5.29}$$

1) Under the assumption that the graph in Figure 5.6c presents the true population model for each of the pairs of *(X1,X2), (X1,X3), (X2,X3),* and *(X(3,X4),* then each of the models in Figure 5.11 is an acceptable model, in the statistical sense, even though the second model has insignificant independent variables at the 10% level.
2) Even though *X1(−1)*X2(−1)* in the model (5.27) has a large p-value = 0.282, it should not be deleted, since it is defined that *Y1(−1), X1(−1),* and *X2(−1)* has complete up-and-down relationships. In other words, the based model should have three two-way interactions as independent variables.
3) It has been found that all pairs of the three independent variables of each model are significantly correlated with p-values = 0.0000, and they have unexpected impacts on the parameter

Dependent Variable: Y1
Method: Least Squares
Date: 11/14/17 Time: 13:09
Sample (adjusted): 1947M02 1996M04
Included observations: 591 after adjustments

	Model (5.27)			Model (5.28)			Model (5.29)			Model (5.30)		
Variable	Coef.	t-Stat.	Prob.	Coef.	t-Stat.	Prob.	Coef.	t-Stat.	Prob.	Coef.	t-Stat.	Prob.
C	29.811	28.09	0.000	33.274	35.21	0.000	42.994	42.63	0.000	32.103	22.61	0.000
Y1(-1)*V1(-1)	0.009	21.89	0.000	0.008	76.98	0.000	0.007	61.08	0.000	1.314	2.95	0.003
Y1(-1)*V2(-1)	-0.002	-4.20	0.000	0.345	1.08	0.282	3.476	9.60	0.000	11.488	42.66	0.000
V1(-1)*V2(-1)	0.003	9.12	0.000	0.881	1.88	0.061	-4.045	-8.51	0.000	-7845.2	-26.22	0.000
R-squared	0.9906			0.9901			0.98514			0.9732		
Adj. R-squared	0.9906			0.9900			0.98506			0.9731		
F-statistic	20688.3			19505.2			12971.1			7108.4		
Prob(F-stat.)	0.000			0.000			0.000			0.000		

Figure 5.11 Statistical results summary of the four models (5.26) to (5.29).

Dependent Variable: Y1
Method: Stepwise Regression
Date: 01/31/17 Time: 12:58
Sample: 1947M01 1996M04 IF @YEAR>1990
Included observations: 64
Number of always included regressors: 4

	Full Model			Reduced Model-1			Reduced Model-2		
Variable	Coef.	t-Stat.	Prob.	Coef.	t-Stat.	Prob.	Coef.	t-Stat.	Prob.
C	-2.518	-2.335	0.020	232.8	1.942	0.057	144.07	2.7617	0.0077
Y1(-1)*X3(-1)	0.797	4.847	0.000	9.808	1.789	0.079	5.5794	2.9284	0.0048
Y1(-1)*X4(-1)	-1.051	-4.426	0.000	-3.222	-1.084	**0.283**	-0.8667	-1.0856	**0.2821**
X3(-1)*X4(-1)	278.31	1.146	**0.252**	7428.8	0.206	**0.838**	-21320.2	-2.3624	0.0215
Y1(-1)	1.052	78.468	0.000	0.672	4.520	0.000	0.6810	4.6015	0.0000
X3(-1)	-179.70	-4.130	0.000	-4250.1	-0.823	**0.414**			
X4(-1)	142.87	4.033	0.000						
R-squared	0.9984			0.9875			0.9874		
Adj. R-squared	0.9984			0.9865			0.9865		
F-statistic	62468.6			918.3			1154.0		
Prob(F-stat)	0.0000			0.0000			0.0000		
DW-Stat	1.5422			2.2633			2.2794		

Figure 5.12 Alternative regressions for the second stage of analysis based on the model (5.29).

estimates of each model. In other words, the parameter estimates of the model can be unexpected, because of the multicollinearity of the independent variables of each model.

4) The second stage of the regression analysis can be done easily using *Y1(−1)*, *V1(−1)*, and *V2(−1)* as the search regressors. It is suggested to apply the combinatorial selection method, because of the unpredicted impacts of the multicollinearity by inserting at least one additional independent variable.
5) As an illustrative result, Figure 5.12 presents three alternative regressions as the second stage of analysis based on the model (5.29) using the combinatorial selection method. The reasons to present three alternative regressions are as follows:

Dependent Variable: Y1
Method: Stepwise Regression
Date: 02/02/17 Time: 16:25
Sample: 1947M01 1996M04 IF @YEAR>1990
Included observations: 64
Number of always included regressors: 2
Number of search regressors: 6
Selection method: Combinatorial
Number of search regressors: 6

Variable	Coefficient	Std. Error	t-Statistic	Prob.*
C	1375.601	959.0048	1.434405	0.1570
Y1(-1)*X3(-1)*X4(-1)	-590.6758	466.1457	-1.267148	0.2103
Y1(-1)	-1.816020	2.033455	-0.893071	0.3756
Y1(-1)*X3(-1)	59.39600	39.81289	1.491879	0.1413
X3(-1)	-27376.47	19462.02	-1.406661	0.1651
Y1(-1)*X4(-1)	26.70166	24.41420	1.093694	0.2788
X3(-1)*X4(-1)	281672.2	224032.4	1.257283	0.2139
X4(-1)	-13687.61	11345.98	-1.206384	0.2327

R-squared	0.987897	Mean dependent var	460.9648
Adjusted R-squared	0.986384	S.D. dependent var	74.47930
S.E. of regression	8.690701	Akaike info criterion	7.278853
Sum squared resid	4229.584	Schwarz criterion	7.548713
Log likelihood	-224.9233	Hannan-Quinn criter.	7.385164
F-statistic	653.0040	Durbin-Watson stat	2.329006
Prob(F-statistic)	0.000000		

(a). RMSE=13.78230 & TIC=0.013235

Dependent Variable: Y1
Method: Stepwise Regression
Date: 02/02/17 Time: 16:07
Sample: 1947M01 1996M04 IF @YEAR>1990
Included observations: 64
Number of always included regressors: 2
Number of search regressors: 6
Selection method: Combinatorial
Number of search regressors: 5

Variable	Coefficient	Std. Error	t-Statistic	Prob.*
C	525.6841	118.0952	4.451359	0.0000
Y1(-1)*X3(-1)*X4(-1)	-177.5630	57.47950	-3.089154	0.0031
Y1(-1)*X3(-1)	24.22514	5.831666	4.154069	0.0001
X3(-1)	-10644.37	5258.694	-2.024147	0.0477
Y1(-1)*X4(-1)	5.110712	3.395500	1.505143	0.1378
X3(-1)*X4(-1)	85261.45	42614.60	2.000757	0.0502
X4(-1)	-3591.169	958.0906	-3.748256	0.0004

R-squared	0.987725	Mean dependent var	460.9648
Adjusted R-squared	0.986433	S.D. dependent var	74.47930
S.E. of regression	8.675255	Akaike info criterion	7.261745
Sum squared resid	4289.823	Schwarz criterion	7.497873
Log likelihood	-225.3758	Hannan-Quinn criter.	7.354767
F-statistic	764.4197	Durbin-Watson stat	2.295545
Prob(F-statistic)	0.000000		

(b). RMSE=12.56734 & TIC=0.012046

Figure 5.13 Statistical results of the second stage of analyses.

5.1 The full model has an insignificant interaction variable with a p-value = 0.252. Even though it is not very large, as a comparative study, I try to observe two reduced models using the combinatorial selection method by selecting two and one of the three search regressors *Y1(−1), X3(−1),* and *X4(−1).* Unexpectedly, three out of five independent variables of the Reduced Model have large p-values. It is worse than the full model and the worst of the three models.

5.2 Hence, we have a choice between the full model and the Reduced Model 2. Note that the Full Model has a DW Statistic = 1.5422 compared to 2.2794 for the Reduced Model. And we can easily compute their forecast evaluations. Do this as an exercise.

5.4.3.2 A Three-Way Interaction Based Model

In order to generate a three-way interaction model based on the graph in Figure 5.6c, we should definitely start doing the analysis using the following based model. Then we have three two-way interactions and three main variables as the two sets of search regressors. However, in Example 5.11 I present an alternative method that is to use both sets as a single set of search regressors using the STEPLS, a combinatorial selection method.

$$Y1\ C\ Y1(-1)*X1(-1)*X2(-1) \tag{5.30}$$

Example 5.11 The Second Stage of Analyses

For the model (5.30), it was found that *Y1(−1)*X3(−1)*X4(−1)* has a significant effect on *Y1* based on the t-statistic of $t_0 = 48.94303$ with $df = 62$ and p-value = 0.0000. Hence, as the second stage of analysis, Figure 5.13 presents the statistical results by using the six variables as search regressors for the combinatorial selection method. Note that Figure 5.13 shows two statistical results, which are obtained by inserting six and five regressors to be selected, respectively. Based on these results, the following findings and notes are presented.

1) Each variable of the full model in Figure 5.13a has insignificant adjusted effects on *Y1* at the 10% level of significance.
2) However, at the 15% level of significance, six out of seven independent variables either have positive or negative adjusted with p-values = Prob./2 < 0.15 – refer to Lapin (1973). For instance, *Y1(−1)*X4(−1)* has a positive adjusted effect on *Y1*, with a p-value = 0.2788/2 = 0.1394 < 0.15. So there is a reason to accept the full model as a good fit.
3) As a comparison, Figure 5.13b presents much better estimates than the full model and it has smaller values of RMSE and TIC. Even though their values have small differences, the reduced model should be considered to be a better forecast model in the statistical sense.

5.5 Alternative Forecast Models Using Original *(X1,X2,Y)*

Referring to Figure 5.6, which presents three up-and-down or causal relationships between the four time series *X1(−1), X2(−1), Y(−1),* and *Y,* Figure 5.14 presents three alternative up-and down relationships based on the triple time series *(X1,X2,Y),* where the arrows represent the true up-and-down stream or causal relationships between the variables. Note that the arrow from *Y(−1)* to *Y* is omitted because the two variables have an up-and-down relationship. Also note that, even though the three graphs present only three or four pairs of up-and-down stream

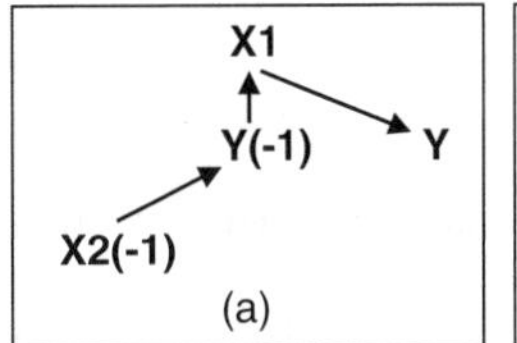

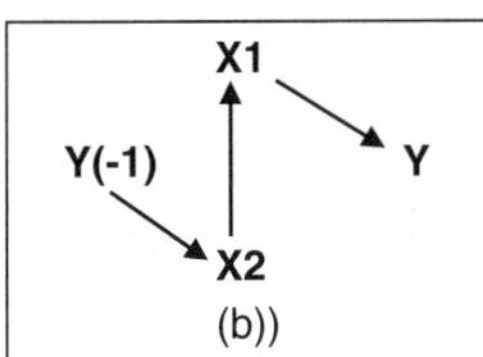

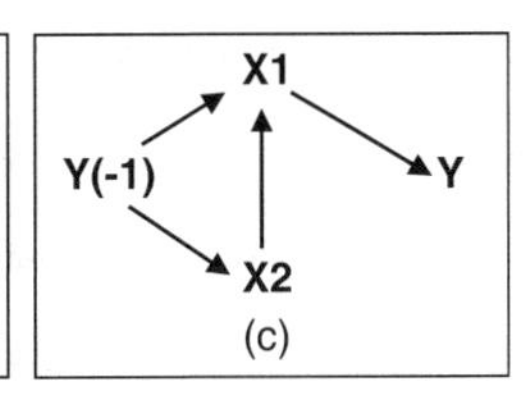

Figure 5.14 Alternative up-and-down relationships based on a triple time series (*X1,X2,Y*).

relationships, in fact, in the theoretical or substantial sense, they can be considered to be presenting six pairs of up-and-down-stream relationships. For instance, in Figure 5.14a, $X2(-1)$ is an upper variable of both $X1$ and Y, and $Y(-1)$ is the natural upper variable of Y.

Referring to the three interaction models presented based on the graphs in Figures 5.6b and c, then each of the graphs could represent a reduced acceptable model of the full three-way interaction models as follows, as well as the two-way interaction models. However, this section does not present an example of the statistical results. It is recommended to readers to apply all alternative models in the previous section, based on the time series $X1$, $X2$, $X3$, $X4$, $Y1$, and $Y2$.

5.5.1 Three-Way Interaction Based on Figure 5.14a

The up-and-down-stream relationship of $X2(-1) \rightarrow Y(-1) \rightarrow X1 \rightarrow Y$ in Figure 5.14a can be presented as an effect of the interaction $X2(-1)*Y(-1)*X1$ on Y in the statistical sense. Hence, the graph in Figure 5.14a presents one of all possible reduced models of an hierarchical three-way interaction model with the ES as follows:

$$\begin{aligned} &Y\,C\,Y(-1)*X1*X2(-1)\;Y(-1)*X1\;Y(-1)*X2(-1) \\ &\qquad X1*X2(-1)\;Y(-1)\;X1\;X2(-1) \end{aligned} \tag{5.31}$$

5.5.2 Three-Way Interaction Based on Figure 5.14b and c

The up-and-down-stream relationship of $Y(-1) \rightarrow X2 \rightarrow X1 \rightarrow Y$ in Figure 5.14b can be presented as an effect of the interaction $Y(-1)*X2*X1$ on Y in the statistical sense. On the other hand, the complete pair-wise causal relationships between $Y(-1)$, $X1$, and $X2$ in Figure 5.14c can also be presented as the effect of $Y(-1)*X2*X1$ on Y in the statistical sense – refer to the model based on Figure 5.6c. Hence, the diagrams in both Figure 5.14b and c in fact present two of all possible reduced models of an hierarchical three-way interaction model with the ES as follows:

$$Y\,C\,Y(-1)*X1*X2\;Y(-1)*X1\;Y(-1)*X2\;X1*X2\;Y(-1)\;X1\;X2 \tag{5.32}$$

5.6 Forecasting Models with Trends Using Original *(X1,X2,Y)*

As the extension of all acceptable models based on the three sets of four time series, namely $\{X1(-1),X2(-1),Y(-1),Y\}$, $\{X1,X2(-1),Y(-1),Y\}$, and $\{X1,X2,Y(-1),Y\}$, this section presents seven alternative forecast models with specific trends by inserting additional selected special time independent variables. Refer to Table 4.1, which presents 11 alternative specific trends that can be used as additional independent variables of the forecast models using triple as well as multiple time series, either as univariate or multivariate autoregressive models.

Table 5.5 The results summary of the model (5.33) and its reduced models.

	Model (5.33)			Reduced Model 1			Reduced Model 2		
Variable	**Coef.**	***t*-Stat.**	**Prob.**	**Coef.**	***t*-Stat.**	**Prob.**	**Coef.**	***t*-Stat.**	**Prob.**
C	176.486	1.162	0.250	94.063	0.798	0.428	178.625	1.179	0.243
*Y1(−1)*X3(−1)*X4(−1)*	−313.775	−4.670	0.000	−333.587	−5.294	0.000	−304.138	−4.600	0.000
*Y1(−1)*X3(−1)*	44.700	5.445	0.000	48.734	7.236	0.000	38.996	7.950	0.000
*Y1(−1)*X4(−1)*	−3.544	−0.868	**0.389**	−5.741	−1.800	0.077			
*X3(−1)*X4(−1)*	185 180.9	3.732	0.000	192 730.0	3.955	0.000	156 997.6	4.195	0.000
X3(−1)	−23 287.0	−3.768	0.000	−25 016.2	−4.288	0.000	−18 777.0	−5.627	0.000
X4(−1)	−1017.9	−0.864	**0.391**				−1654.4	−1.798	0.078
@TREND	0.830	3.304	0.002	0.973	5.175	0.000	0.690	3.588	0.001
R-squared	0.9897			0.9896			0.9896		
Adj. R-squared	0.9884			0.9885			0.9885		
F-statistic	770.7351			903.0900			902.9809		
Prob. (F-stat)	0.0000			0.0000			0.0000		
DW-Stat	2.1355			2.0136			2.2205		
	RMSE = 8.8348 & TIC = 0.0084			RMSE = 8.9468 & TIC = 0.0085			RMSE = 9.3670 & TIC = 0.0092		

Example 5.12 Application of Forecast Model with a Linear Trend

As an extension of the model in Figure 5.13b, Table 5.5 presents the statistical results of the model with the following ES and two of its possible reduced models, namely Reduced Model 1 and Reduced Model 2. Based on these results, findings and notes are presented as follows:

$$\begin{gathered} Y1\,C\,Y1(-1)*X3(-1)*X4(-1)\;Y1(-1)*X3(-1)\;Y1(-1)*X4(-1) \\ X3(-1)*X4(-1)\;Y1(-1)\;X3(-1)\;X4(-1)@Trend \end{gathered} \tag{5.33}$$

1) Even though the estimate of the model (5.33) has two IVs with large p-values, I would say that the model is an acceptable forecast model of *Y1*. However, its forecast evaluation should be compared with other forecast models of *Y1* using exactly the same data set and the same set of main variables.
2) Note that the two reduced models RM-1 and RM-2 with all IVs (independent variables) having p-values <0.10 are obtained by deleting a single *IV*, namely *X4(−1)* and *Y1(−1)*X4(−1)*, respectively, which are the unexpected impacts of the multicollinearity of the *IVs*. So, in the statistical sense, RM-1 and RM-2 are better regressions than the model (5.33).
3) Since RM-1 has the smallest RMSE and TIC, compared to all previous forecast models of *Y1*, then RM-1 is the best forecast model of *Y1*.

Example 5.13 Application of Forecast Model with Heterogeneous Trend by Year
As a more advanced extension of the model in Figure 5.13b, the following two EQs (ESs) represent the same forecast model for $Y1 = FSPCOM$.

$$\begin{gathered} Y1\;C\;Y1(-1)*X3(-1)*X4(-1)\;Y1(-1)*X3(-1)\;Y1(-1)*X4(-1) \\ X3(-1)*X4(-1)\;Y1(-1)\;X3(-1)\;X4(-1)\;@Month^{*}@Expand(@Year) \\ @Expand(@Year,@Droplast) \end{gathered} \tag{5.34a}$$

Note that homogeneity (equality) of trends can be tested using the Wald-test. However, specifically for a large number of years, it is recommended to apply the following ES, since the set of *IVs* in *@Month*@Expand(@Year,@Droplast)* can be inserted to conduct the *Redundant Variables Test* (RVT) instead of using the Wald-test.

$$\begin{gathered} Y1\;C\;Y1(-1)*X3(-1)*X4(-1)\;Y1(-1)*X3(-1)\;Y1(-1)*X4(-1) \\ X3(-1)*X4(-1)Y1(-1)\;X3(-1)\;X4(-1)\;@Month \\ @Month^{*}@Expand(@Year,@Droplast)\;@Expand(@Year,@Droplast) \end{gathered} \tag{5.34b}$$

Figure 5.15 presents the statistical results based on the ES (5.34b). Based on these results, the following findings and special notes are presented.

1) Compared to the Reduced Model 1 in Table 5.5, this model has smaller RMSE and TIC. So this model is a better forecast model than the Reduced Model 1.
2) The estimate of the model presents six multiple regressions for *@Year* = 1991 up to 1996. The regression for *@Year* = 1996 can easily be copied from the output by clicking the object *"Representations."* The following estimation function is obtained and the equation of the regression for *@Year* = 1996 – refer to the first two lines of the ES (5.34b).

$$\begin{gathered} Y1(1996) = C(1) + C(2)*Y1(-1)*X3(-1)*X4(-1) + C(3)*Y1(-1)*X3(-1) \\ + C(4)*Y1(-1)*X4(-1) + C(5)*X3(-1)*X4(-1) + C(6)*Y1(-1) \\ + C(7)*X3(-1) + C(8)*X4(-1) + C(9)*@Month \end{gathered} \tag{5.35}$$

$$\begin{gathered} Y1(1996) = -4156.035 + 2669.403*Y1(-1)*X3(-1)*X4(-1) \\ -200.393*Y1(-1)*X3(-1) \\ -146.474*Y1(-1)*X4(-1) - 1177669.981*X3(-1)*X4(-1) + 10.924*Y1(-1) \\ + 88953.745*X3(-1) + 63390.885*X4(-1) + 14.601*@Month \end{gathered} \tag{5.36}$$

3) In order to compute the forecast values of $Y1(t)$ for $t > 591$, we have to forecast $X3$ and $X4$. It is found that the LV(2) Models of $X3$ and $X4$ are acceptable models, since they have the Durbin–Watson statistics of 2.057 396 and 1.857 360, as presented in Figure 5.16. Then the forecast values of $X3(t)$ and $X4(t)$ for $t > 591$, can be computed easily using the following equations. For instance, Table 5.6 presents their forecast values for *@Month* = 5, 6, and 7.

$$X3 = 0.00151223274018 + 1.48153811785*E5 - 0.518469535409*E4$$

$$X4 = 0.00418380323598 + 1.42642824425*F5 - 0.480516741509*F4$$

4) Note that the regression (5.36) can easily be used to compute a forecast value at the first time point beyond the sample period, using Excel, as presented in Table 5.7. The result $Y1$ (1996M05) = $Y1(t = 592)$ = 656.8672 is obtained by using the following formula.

Dependent Variable: Y1
Method: Least Squares
Date: 02/09/17 Time: 06:28
Sample: 1947M01 1996M04 IF @YEAR>1990
Included observations: 64

Variable	Coefficient	Std. Error	t-Statistic	Prob.
C	-4156.035	1431.079	-2.904127	0.0057
Y1(-1)*X3(-1)*X4(-1)	2669.403	756.2450	3.529813	0.0010
Y1(-1)*X3(-1)	-200.3929	66.25877	-3.024397	0.0041
Y1(-1)*X4(-1)	-146.4739	37.02204	-3.956397	0.0003
X3(-1)*X4(-1)	-1177670.	352800.8	-3.338059	0.0017
Y1(-1)	10.92416	3.110790	3.511701	0.0010
X3(-1)	88953.75	31020.02	2.867623	0.0063
X4(-1)	63390.89	17064.43	3.714797	0.0006
@MONTH	14.60097	4.345498	3.360022	0.0016
@MONTH*(@YEAR=1991)	-15.08194	4.434716	-3.400881	0.0014
@MONTH*(@YEAR=1992)	-12.37424	4.284675	-2.888022	0.0059
@MONTH*(@YEAR=1993)	-14.30591	4.635142	-3.086402	0.0035
@MONTH*(@YEAR=1994)	-13.01196	4.671216	-2.785561	0.0078
@MONTH*(@YEAR=1995)	-1.922458	5.577563	-0.344677	0.7319
@YEAR=1991	-202.2685	43.43722	-4.656572	0.0000
@YEAR=1992	-192.7276	41.97961	-4.590982	0.0000
@YEAR=1993	-143.5256	42.79866	-3.353507	0.0016
@YEAR=1994	-138.5793	42.97916	-3.224337	0.0024
@YEAR=1995	-134.6129	39.05157	-3.447054	0.0012

R-squared	0.995210	Mean dependent var	460.9648
Adjusted R-squared	0.993294	S.D. dependent var	74.47930
S.E. of regression	6.099241	Akaike info criterion	6.695735
Sum squared resid	1674.033	Schwarz criterion	7.336653
Log likelihood	-195.2635	Hannan-Quinn criter.	6.948225
F-statistic	519.4005	Durbin-Watson stat	1.998823
Prob(F-statistic)	0.000000		

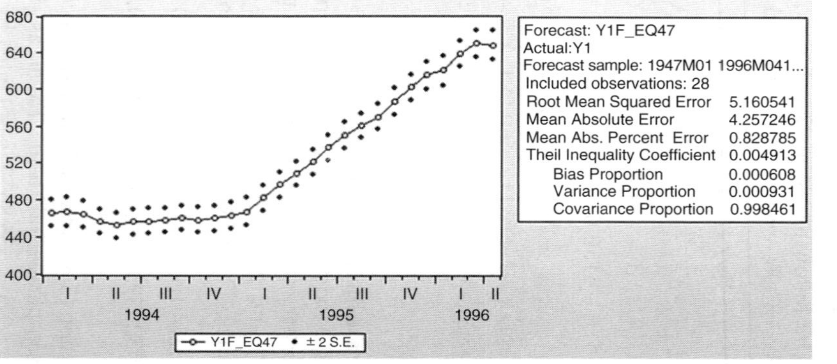

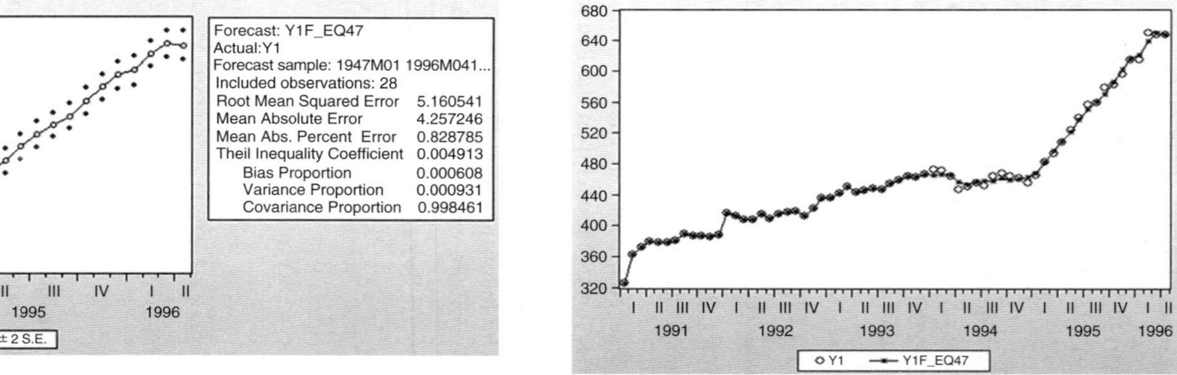

Figure 5.15 Statistical results based on (5.34b).

Dependent Variable: X3
Method: Least Squares
Date: 02/09/17 Time: 12:37
Sample: 1947M01 1996M04 IF @YEAR>1990
Included observations: 64

Variable	Coefficient	Std. Error	t-Statistic	Prob.
C	0.001512	0.000846	1.787868	0.0788
X3(-1)	1.481538	0.107777	13.74635	0.0000
X3(-2)	-0.518470	0.103551	-5.006891	0.0000

R-squared	0.977522	Mean dependent var	0.043734
Adjusted R-squared	0.976785	S.D. dependent var	0.010940
S.E. of regression	0.001667	Akaike info criterion	-9.910010
Sum squared resid	0.000169	Schwarz criterion	-9.808812
Log likelihood	320.1203	Hannan-Quinn criter.	-9.870143
F-statistic	1326.394	Durbin-Watson stat	2.057396
Prob(F-statistic)	0.000000		

Dependent Variable: X4
Method: Least Squares
Date: 02/09/17 Time: 12:40
Sample: 1947M01 1996M04 IF @YEAR>1990
Included observations: 64

Variable	Coefficient	Std. Error	t-Statistic	Prob.
C	0.004184	0.002179	1.920103	0.0595
X4(-1)	1.426428	0.111357	12.80949	0.0000
X4(-2)	-0.480517	0.108524	-4.427762	0.0000

R-squared	0.952242	Mean dependent var	0.078880
Adjusted R-squared	0.950676	S.D. dependent var	0.006684
S.E. of regression	0.001485	Akaike info criterion	-10.14172
Sum squared resid	0.000134	Schwarz criterion	-10.04052
Log likelihood	327.5349	Hannan-Quinn criter.	-10.10185
F-statistic	608.1330	Durbin-Watson stat	1.857350
Prob(F-statistic)	0.000000		

Figure 5.16 Statistical results of acceptable LV(2) Models of *X3* and *X4*.

Table 5.6 An Excel file to compute the forecast values of X3 and X4.

1		*T*	*Y1*	*X3*	*X4*	*@MONTH*
2	1996 M01	588	614.42	0.0502	0.0681	1
3	1996 M02	589	649.542	0.0487	0.0699	2
4	1996 M03	590	647.07	0.0496	0.0735	3
5	1996 M04	591	647.17	0.0499	0.075	4
6		**592**		**0.049725**	**0.075848**	5
7				**0.04931**	**0.076337**	6
8				**0.048786**	**0.076626**	7

Table 5.7 Spreadsheet for computing the forecast values of *Y1(t)* for $t > 591$.

A	B	C	D	E	F	G
1		*T*	*Y1*	*X3*	*X4*	*@MONTH*
2	1996 M01	588	614.42	0.0502	0.0681	1
3	1996 M02	589	649.542	0.0487	0.0699	2
4	1996 M03	590	647.07	0.0496	0.0735	3
5	1996 M04	591	647.17	0.0499	0.075	4
6		**592**	**656.8672**	**0.049725**	**0.075848**	5
7				**0.04931**	**0.076337**	6
8				**0.048786**	**0.076626**	7

$$Y1(t=592) = -4156.03532907 + 2669.40331942*D5*E5*F5$$
$$-200.392862135*D5*E5 - 146.473875354*D5*F5 - 1177669.98196*E5*F5$$
$$+10.924162729*D5 + 88953.745109*E5 + 63390.8850472*F5 + 14.6009701863*G6$$

5) Finally, additional forecast values of *Y1(t)* for the @Month = 6 and 7 can be obtained by using the block-copy-paste method of D6.
6) Unexpectedly, an error message *"Near singular matrix error. Regressors may be perfectly collinear"* appears for testing the equality of the slopes of *@Month* using the Redundant Variables Test (RVT) by inserting the function

   @Month@Expand(@Year, @Droplast)*.

   So, I try to apply the Wald-test, and it is found that H_0: C(10) = C(11) = C(12) = C(13) = C(14) = 0 is rejected based on the *F*-statistic of $F_0 = 7.577004$ with $df = (5.45)$ and a *p*-value = 0.0000. I think the RVT and the Wald-test should give the same results. Nothing wrong with the RVT, but the data is not suitable for the test. See the following findings as a comparison.

7) Both RVT of *@Month*@Expand(@Year,@Droplast) @Expand(@Year,@Droplast)* and the Wald-test present exactly the same value of the *F*-statistic, as presented in Figure 5.17. However, in addition to the *F*-statistic, RVT presents the Likelihood Ratio Statistic, which has different assumptions compared to the Chi-square statistic presented in the results of the Wald-test.

5.7 Application of VAR Models Based on $(X1_t, X2_t, Y1_t)$

5.7.1 Unrestricted VAR Models

5.7.1.1 The Simplest VAR Model

The simplest VAR model with the lag intervals "1,1," namely VAR(1,1) model, has the following equation, which can be obtained from the output of the statistical results.

$$Y1 = C(1,1)*Y1(-1) + C(1,2)*X1(-1) + C(1,3)*X2(-1) + C(1,4) \quad (5.37a)$$
$$X1 = C(2,1)*Y(1(-1) + C(2,2)*X1(-1) + C(2,3)*X2(-1) + C(2,4) \quad (5.37b)$$

Redundant Variables Test
Equation: EQ47
Specification: Y1 C Y1(-1)*X3(-1)*X4(-1) Y1(-1)*X3(-1) Y1(-1)*X4(-1) X3(-1)*X4(-1) Y1(-1) X3(-1) X4(-1) @MONTH @MONTH *@EXPAND(@YEAR,@DROPLAST) @EXPAND(@YEAR,@DROPLAST)
Redundant Variables: @MONTH*@EXPAND(@YEAR,@DROPLAST) @EXPAND(@YEAR,@DROPLAST)

	Value	df	Probability
F-statistic	6.321746	(10, 45)	0.0000
Likelihood ratio	56.15874	10	0.0000

Wald Test:
Equation: EQ47

Test Statistic	Value	df	Probability
F-statistic	6.321746	(10, 45)	0.0000
Chi-square	63.21746	10	0.0000

Null Hypothesis: C(10)=C(11)=C(12)=C(13)=C(14)=C(15) =C(16)=C(17)=C(18)=C(19)=0

Figure 5.17 Two alternative tests of the equality of the slopes *@Month* based on the model (5.35b).

Table 5.8 The statistical results summary of the VAR(1,1) model (5.37a), (5.37b), (5.38c).

	Y1(−1)	*X1(−1)*	*X2(−1)*	*C*
Y1	1.0111	−0.0414	0.0316	0.3082
	289.3389	−1.7562	1.6423	0.4434
	0.0000	0.0796	0.1011	0.6576
X1	0.0009	1.0014	−0.0034	0.1577
	2.5810	407.2542	−1.6741	2.1744
	0.0101	0.0000	0.0946	0.0301
X2	−0.0012	0.0116	0.9968	−0.2242
	−4.5845	6.5113	685.0844	−4.2681
	0.0000	0.0000	0.0000	0.0000

$$X2 = C(3,1)*Y1(-1) + C(3,2)*X1(-1) + C(3,3)*X2(-1) + C(3,4) \tag{5.37c}$$

with the summary of its regressions presented in Table 5.8, which can be used to compute the forecast values of *Y1*, *X1*, and *X2* by using Excel starting from the time point 1996M5 or $t = @Trend + 1 = 593$, Do this as an exercise – refer to Figures 4.29 and 4.35.

5.7.2 The Simplest Two-Way Interaction VAR Model

The simplest interaction VAR model, namely 2WI_VAR(1,1), has the following equation, which can be obtained by inserting *Y1(−1)∗X1(−1)*, *Y1(−1)∗X2(−1)*, and *X1(−1)∗X2(−1)* as an additional exogenous variable for the VAR(1,1) model.

$$\begin{aligned} Y1 &= C(1,1)*Y1(-1) + C(1,2)*X1(-1) + C(1,3) + C(1,4)*Y1(-1)*X1(-1) \\ &+ C(1,5)*Y1(-1)*X2(-1) + C(1,6)*X1(-1)*X2(-1) \end{aligned} \tag{5.38a}$$

$$\begin{aligned} X1 &= C(2,1)*Y1(-1) + C(2,2)*X1(-1) + C(2,3) + C(2,4)*Y1(-1)*X1(-1) \\ &+ C(2,5)*Y1(-1)*X2(-1) + C(2,6)*X1(-1)*X2(-1) \end{aligned} \tag{5.38b}$$

Table 5.9 The statistical results summary of the *2WI_VAR(1,1)* model (5.38a), (5.38b), and (5.38c).

	Y1(−1)	X1(−1)	X2(−1)	Y1(−1)*X1(−1)	Y1(−1)*X2(−1)	X1(−1)*X2(−1)	C
Y1	0.9492	0.0705	−0.1043	−0.0003	0.0007	0.0005	2.2878
	32.5961	0.8364	−0.8029	−0.5541	1.2122	0.3758	0.8250
	0.0000	***0.4033***	***0.4224***	***0.5797***	0.2259	***0.7072***	***0.4097***
X1	0.0170	0.9604	0.0603	0.0002	−0.0003	−0.0003	−0.7037
	5.7915	112.87	4.5977	4.2129	−5.2805	−2.2724	−2.5139
	0.0000	0.0000	0.0000	0.0000	0.0000	0.0234	0.0122
X2	−0.0083	0.0081	1.0134	0.0002	−0.0001	−0.0001	−0.3260
	−3.8973	1.3104	106.3792	4.0584	−1.9923	−1.1517	−1.6033
	0.0001	0.1906	0.0000	0.0001	0.0468	0.2499	0.1094

$$\begin{aligned} X1 = {} & C(2,1)*Y1(-1) + C(2,2)*X1(-1) + C(2,3) + C(2,4)*Y1(-1)*X1(-1) \\ & + C(2,5)*Y1(-1)*X2(-1) + C(2,6)*X1(-1)*X2(-1) \end{aligned} \quad (5.38c)$$

with the summary of its regressions presented in Table 5.9, which can be used to compute the forecast values of *Y1*, *X1*, and *X2* by using Excel starting from the time point 1996 M5 or $t = @Trend + 1 = 593$. Do this as an exercise – refer to Figure 4.32 and Figure 4.38.

Based on the results in Table 5.9, the following notes and comments are presented.

1) Even though a VAR model have many insignificant independent variables. The VAR model is an acceptable model to compute the forecast values of each of the dependent variables beyond the sample period, since the VAR model is an acceptable model in the theoretical sense.
2) However, in the statistical sense, since four out of the six independent variables of the *Y1*-regression have *p*-values >0.30, then the best possible reduced model should be explored. However, as the object VAR cannot be applied anymore, we should use the Equation system Model or the Seemingly Causal Model (SCM), as presented in Agung (2014, 2011a, 2009a).
3) Since each regression of the VAR model in fact is an OLS regression, in order to have a three-way interaction model of *Y1*, I am using the trial-and-error method, and the combinatorial selection method, to obtain an OLS regression of *Y1* in Figure 5.18 as an acceptable model.

5.7.3 The Simplest Three-Way Interaction VAR Model

The simplest three-way interaction VAR model, namely 3WI_VAR(1,1), has the following equation, which can be obtained by inserting the interactions *Y1(−1)*X1(−1)*, *Y1(−1)*X2(−1)*, *X1 (−1)*X2(−1)*, and *Y1(−1)*X1(−1*X2(−1)* as additional exogenous variables for the VAR(1,1).

Dependent Variable: Y1
Method: Stepwise Regression
Date: 03/26/17 Time: 18:36
Sample (adjusted): 1947M02 1996M04
Included observations: 591 after adjustments
Number of always included regressors: 5
Number of search regressors: 3
Selection method: Combinatorial
Number of search regressors: 1

Variable	Coefficient	Std. Error	t-Statistic	Prob.*
Y1(-1)*X1(-1)*X2(-1)	1.88E-05	4.85E-06	3.878980	0.0001
Y1(-1)*X1(-1)	-0.002011	0.000612	-3.286810	0.0011
Y1(-1)*X2(-1)	-0.001373	0.000437	-3.143171	0.0018
X1(-1)*X2(-1)	0.000818	0.000244	3.350001	0.0009
C	-1.558278	0.995460	-1.565385	0.1180
Y1(-1)	1.124946	0.044568	25.24135	0.0000

R-squared	0.998435	Mean dependent var	140.9656
Adjusted R-squared	0.998422	S.D. dependent var	136.5858
S.E. of regression	5.425610	Akaike info criterion	6.230238
Sum squared resid	17220.79	Schwarz criterion	6.274723
Log likelihood	-1835.035	Hannan-Quinn criter.	6.247567
F-statistic	74664.77	Durbin-Watson stat	1.491800
Prob(F-statistic)	0.000000		

Figure 5.18 An OLS regression of *Y1*.

$$\begin{aligned} Y1 = {} & C(1,1)*Y1(-1) + C(1,2)*X1(-1) + C(1,3) + C(1,4)*Y1(-1)*X1(-1) \\ & + C(1,5)*Y1(-1)*X2(-1) + C(1,6)*X1(-1)*X2(-1) \\ & + C(1,7)*Y1(-1)*X1(-1)*X2(-1) \end{aligned} \tag{5.39a}$$

$$\begin{aligned} X1 = {} & C(2,1)*Y1(-1) + C(2,2)*X1(-1) + C(2,3) + C(2,4)*Y1(-1)*X1(-1) \\ & + C(2,5)*Y1(-1)*X2(-1) + C(2,6)*X1(-1)*X2(-1) \\ & + C(2,7)*Y1(-1)*X1(-1)*X2(-1) \end{aligned} \tag{5.39b}$$

$$\begin{aligned} X2 = {} & C(3,1)*Y1(-1) + C(3,2)*X1(-1) + C(3,3) + C(3,4)*Y1(-1)*X1(-1) \\ & + C(3,5)*Y1(-1)*X2(-1) + C(3,6)*X1(-1)*X2(-1) \\ & + C(3,7)*Y1(-1)*X1(-1)*X2(-1) \end{aligned} \tag{5.39c}$$

with the summary of its regressions presented in Table 5.10, which can be used to compute the forecast values of *Y1*, *X1*, and *X2* by using Excel starting from the time point 1996M5 or t = *@Trend* + 1 = 593. Do this as an exercise – refer to Figure 4.31 and Figure 4.32.

Corresponding to these results, refer to the notes and comment presented based on the results of the VAR model (5.38a), (5.38b) in Table 5.9.

Table 5.10 Summary of the statistical results of the *3WI_VAR(1,1)* model (5.39a), (5.39b), and (5.39c).

	Y1(−1)	*X1(−1)*	*X2(−1)*	*Y1(−1)±X1(−1) ±X2(−1)*	*Y1 (−1)±X1 (−1)*	*Y1 (−1)±X2 (−1)*	*X1 (−1)±X2 (−1)*	*C*
Y1	1.104	0.099	−0.122	0.000	-0.003	-0.001	0.001	-0.701
	22.777	1.189	−0.949	3.971	-3.408	-1.050	0.968	-0.247
	0.000	0.235	0.343	0.000	0.001	0.294	0.334	0.805
XI	0.010	0.959	0.061	0.000	0.000	0.000	0.000	−0.572
	2.063	112.481	4.661	−1.712	4.019	−3.593	−2.506	−1.974
	0.040	0.000	0.000	0.088	0.000	0.000	0.013	0.049
X2	−0.016	0.007	1.014	0.000	0.000	0.000	0.000	−0.187
	−4.337	1.093	106.88	−2.502	4.534	−0.382	−1.517	−0.890
	0.000	0.275	0.000	0.013	0.000	0.703	0.130	0.374

5.8 Applications of the Object "System"

Note that the Basic VAR models are in fact vector lag-variables models. For this reason, I present the *Multivariate Lag-Variable-Autoregressive,* namely MLVAR(***p***), models in Agung (2014, 2009a). In fact, the VAR models are special MLVAR(***p***) models where all multiple regressions of a MLVAR model have the same set of independent variables. The object *"System"* can be applied easily for all types of MLVAR(***p***) models. Refer to the various MLVAR(*p,q*) models presented in Agung (2014, 2009a).

5.8.1 The MLV(1,1,1) Models of (*Y1,Y2,Y3*) on (*Y1(−1),Y2(−1),Y3(−1)*)

Note that, based on the six variables *Y1(−1), Y2(1), Y3(−1), Y1, Y2,* and *Y3,* we would have C (2,6) = 15 pairs of up-and-down or causal relationships. Each dependent variable of *Yk* would have 25 possible independent variables, namely five main variables, $C_5^2 = C(2,5) = 10$ possible two-way interactions, and C(2,5) = 10 possible three-way interactions, which of course should not all be used as independent variables of the model of *Yk.* For this reason, a researcher should use his/her subjective expert judgment to select a set of independent variables for each *Yk,* which are valid in the theoretical sense.

The simplest possible model of *(Y1,Y2,Y3)* on *(Y1(−1),Y2(−1),Y3(−1))* has the following ES, which is an additive MLV(1,1,1) = MLV(**1**) model. The vector or symbol (1,1,1) = (**1**) is used to indicate each of the three regressions is a LV(1) model, which is the same set of the VAR (1,1,1) model of *(Y1,Y2,Y3).* In fact, all VAR models presented based on *(X1,X2,Y)* can be presented for the models based on *(Y1,Y2,Y3).*

$$
\begin{aligned}
Y1 &= C(10)*Y1(-1) + C(12)*Y2(-1) + C(13)*Y3(-1) + C(10) \\
Y2 &= C(20)*Y1(-1) + C(22)*Y2(-1) + C(23)*Y3(-1) + C(20) \\
Y3 &= C(30)*Y1(-1) + C(32)*Y2(-1) + C(33)*Y3(-1) + C(30)
\end{aligned}
\tag{5.40}
$$

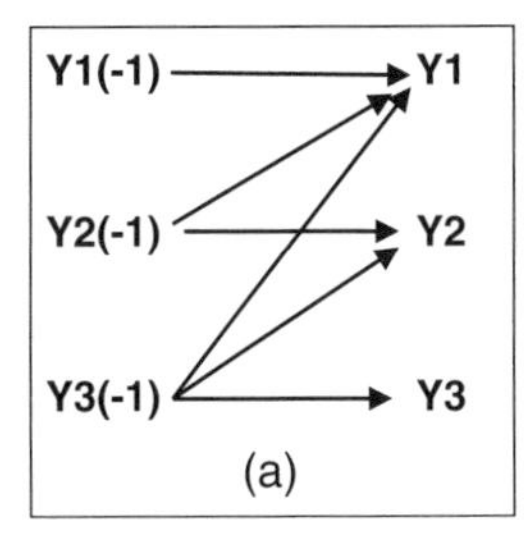

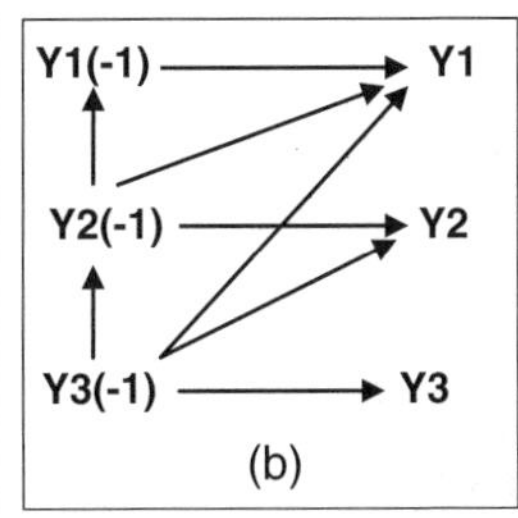

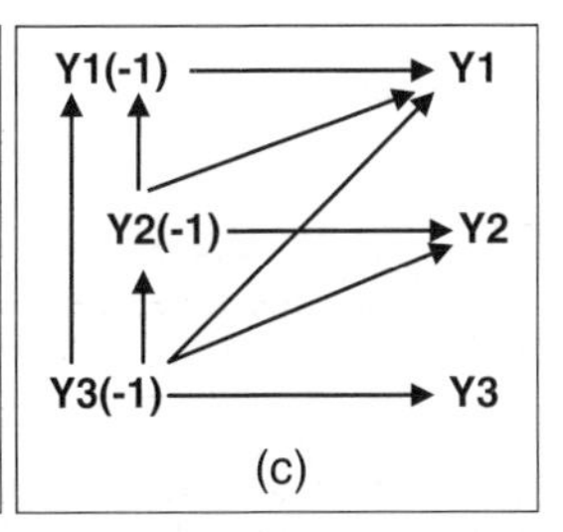

Figure 5.19 Alternative up-and-down relationships based on a triple time series (*Y1,Y2,Y3*).

The readers are advised to use the 250 examples of various time series models presented in Agung (2009a) as their reference to conduct a forecast based on their own data.

As alternative causal effects or up-and-down relationship presented by the object VAR in EViews, Figure 5.19 presents three alternative up-and-down relationships between *(Y1,Y2,Y3)* and their first lags. Under the assumption that each of these path diagrams is well defined to represent specific causal effects of *(Y1(−1),Y2(−1),Y3(−1))* on *(Y1,Y2,Y3)*, then the following statistical models are acceptable models in the theoretical sense. However, note that only one out of the three path diagrams would be valid for a specific set of variables. The data analysis should be done using the object *"System"*. As illustrative examples, the data used is the data in GARCH. wf1, with *Y1* = *FSPCOM*, *Y2* = *FSDXP*, and *Y3* = *100*∗*RAAA*.

5.8.1.1 Additive Triangular-Effects MLV(1,1,1) Model

Based on the path diagram in Figure 5.19a, the following equation system is considered as an *additive-triangular-effects* multivariate-lag-variables model, namely ATE_MLV(**1**) = ATE_MLV(1,1,1), in the theoretical sense.

$$
\begin{aligned}
Y1 &= C(11)*Y1(-1) + C(12)*Y2(-1) + C(13)*Y3(-1) + C(14) \\
Y2 &= C(12)*Y2(-1) + C(22)*Y3(-1) + C(24) \\
Y3 &= C(31)*Y3(-1) + C(34)
\end{aligned}
\tag{5.41}
$$

Example 5.14 Application of the ES Model (5.41)

Figure 5.20 presents the OLS estimate of the model (5.41). Based on this result, the following findings and notes are presented.

1) At the 5% level of significance, *Y1(−1)* has a positive significant adjusted effect on *Y1*, with a *p*-value = 0.0785/2 = 0.03925 and at the 10% level of significance each of *Y2(−1)* and *Y3(−1)* has a negative significant adjusted effect on *Y1*, with *p*-values of 0.1942/2 = 0.0971, and 0.06335, respectively.
2) For additional discussion, Figure 5.20 presents the bivariate correlation tests between *Y1* and each of *Y1(−1)*, *Y2(−1)*, and *Y3(−1)*. Note that these correlation tests show that each of *Y1 (−1)*, *Y2(−1)*, and *Y3(−1)* has a significant *unadjusted* linear effect on *Y1*. They are quite different than the *adjusted* linear effect of each lags in the multiple regression of *Y1* on *Y1(−1)*, *Y2(−1)*, and *Y3(−1)*, which should be accepted because of the multicollinearity between the independent variables. Note that *Y3(−1)* has a positive significant correlation with *Y1* but the regression of *Y1* shows *Y3(−1)* has a negative significant adjusted effect on *Y1* at the 10% level of significance, which has been indicated before (Figure 5.21).

System: SYS03
Estimation Method: Least Squares
Date: 03/26/17 Time: 20.12
Sample: 1947M02 1996M04
Included observations: 591
Total system (balanced) observations 1773

	Coefficient	Std. Error	t-Statistic	Prob.
C(11)	1.013192	0.002214	457.6108	0.0000
C(12)	0.420408	0.238823	1.760332	0.0785
C(13)	-0.105057	0.080891	-1.298751	0.1942
C(14)	-1.767761	1.156824	-1.528116	0.1267
C(21)	0.991796	0.005926	167.3564	0.0000
C(22)	-8.65E-05	0.002087	-0.041420	0.9670
C(23)	0.030189	0.029629	1.018909	0.3084
C(31)	0.996286	0.002741	363.4912	0.0000
C(32)	0.033803	0.020724	1.631079	0.1031

Determinant residual covariance 0.020837

Equation: Y1 = C(11)*Y1(-1) + C(12)*Y2(-1) + C(13)*Y3(-1) + C(14)
Observations: 591

R-squared	0.993384	Mean dependent var	140.9657
Adjusted R-squared	0.998376	S.D. dependent var	136.5858
S.E of regression	5.504084	Sum squared resid	17783.13
Durbin-Watson stat	1.480006		

Equation: Y2 = C(21)*Y2(-1) + C(22)*Y3(-1) + C(23)
Observations: 591

R-squared	0.979531	Mean dependent var	4.067919
Adjusted R-squared	0.979461	S.D. dependent var	1.127182
S.E. of regression	0.161542	Sum squared resid	15.34425
Durbin-Watson stat	0.595101		

Equation: Y3 = C(31)*Y3(-1) + C(32)
Observations: 591

R-squared	0.995562	Mean dependent var	6.863469
Adjusted R-squared	0.995554	S.D. dependent var	3.188500
S.E. of regression	0.212595	Sum squared resid	26.62078
Durbin-Watson stat	1.233485		

Figure 5.20 The OLS estimate of the model (5.41).

Correlation t-Statistic Probability	Y1(-1)	Y2(-1)	Y3(-1)	Y1
Y1	0.999186	-0.504846	0.442036	1.000000
	601.2423	-14.19387	11.95980	-----
	0.0000	0.0000	0.0000	-----

Figure 5.21 Correlation tests between *Y1* and each of *Y1(−1)*, *Y2(−1)*, and *Y3(−1)*.

3) Even though *Y3(−1)* has an insignificant effect on *Y2* with a very large p-value = 0.9670, it should not be deleted from the model because *Y2(−1)* and *Y3(−1)* have significant joint effects on *Y2* based on the Wald Chi-square test of 284 137.74 with df = 2 and p-value = 0.0000. In addition, if *Y3(−1)* is deleted from the model then it would not be a triangular effects model anymore.

4) Hence, it can be concluded that the data supports the defined causal or up-and-down relationships between the six variables presented in Figure 5.19a.

5.8.1.2 Partial Two-Way Interaction MLV(1,1,1) Based on Figure 5.19b

Based on the path diagram in Figure 5.19b, which is well defined or assumed to be true, the following equation system is a set of *partial two-way interaction* MLV(1,1,1) models in the statistical sense.

$$\begin{aligned} Y1 &= C(11)*Y1(-1)*Y2(-1) + C(12)*Y2(-1)*Y3(-1) \\ &\quad + C(13)*Y1(-1) + C(14)*Y2(-1) + C(15)*Y3(-1) + C(16) \\ Y2 &= C(21)*Y2(-1)*Y3(-1) + C(22)*Y2(-1) + C(23)*Y3(-1) + C(24) \\ Y3 &= C(31)*Y3(-1) + C(32) \end{aligned} \tag{5.42}$$

System: SYS04
Estimation Method: Least Squares
Date: 03/27/17 Time: 15:37
Sample: 1947M02 1996M04
Included observations: 591
Total system (balanced) observations 1773

	Coefficient	Std. Error	t-Statistic	Prob.
C(11)	0.006598	0.004262	1.548027	0.1218
C(12)	0.173261	0.081845	2.116947	0.0344
C(13)	0.998739	0.011547	86.49233	0.0000
C(14)	-0.725368	0.499661	-1.451720	0.1468
C(15)	-1.181993	0.435156	-2.716252	0.0067
C(16)	4.257403	2.551193	1.668789	0.0953
C(21)	-0.001097	0.002063	-0.531710	0.5950
C(22)	0.998455	0.013856	72.05908	0.0000
C(23)	0.005226	0.010207	0.511988	0.6087
C(24)	-0.002986	0.069078	-0.043221	0.9655
C(31)	0.996286	0.002741	363.4912	0.0000
C(32)	0.033803	0.020724	1.631079	0.1031

Equation: Y1 = C(11)*Y1(-1)*Y2(-1) + C(12)*Y2(-1)*Y3(-1) + C(13)*Y1(-1)+C(14)*Y2(-1)+C(15)*Y3(-1)+C(16)
Observations: 591

R-squared	0.998405	Mean dependent var	140.9657
Adjusted R-squared	0.998391	S.D. dependent var	136.5858
S.E. of regression	5.478787	Sum squared resid	17560.01
Durbin-Watson stat	1.478983		

Equation: Y2 = C(21)*Y2(-1)*Y3(-1) + C(22)*Y2(-1) + C(23)*Y3(-1) +C(24)
Observations: 591

R-squared	0.979540	Mean dependent var	4.067919
Adjusted R-squared	0.979436	S.D. dependent var	1.127182
S.E. of regression	0.161640	Sum squared resid	15.33687
Durbin-Watson stat	1.594906		

Equation: Y3 = C(31)*Y3(-1) + C(32)
Observations: 591

R-squared	0.995562	Mean dependent var	6.863469
Adjusted R-squared	0.995554	S.D. dependent var	3.188500
S.E. of regression	0.212595	Sum squared resid	26.62078
Durbin-Watson stat	1.233485		

Figure 5.22 Statistical results of the MLV(1,1,1) model (5.42).

Note that the interaction *Y1(−1)∗Y2(−1)* is used as an IV to present the indirect effect of *Y2(−1)* on *Y1* through *Y1(−1)*, and the interaction *Y2(−1)∗Y3(−1)* is used to present the indirect effect of *Y3(−1)* on *Y1* through *Y2(−1)*. And the model is called a *partial two-way interaction* model, because the interaction *Y1(−1)∗Y3(−1)* is omitted or not defined in Figure 5.18b in the theoretical sense, compared to Figure 5.18c.

Example 5.15 Application of the Model (5.42)

Figure 5.22 presents the OLS estimate of the model (5.41). Based on this result, the following findings and notes are presented.

1) The results show that *Y2(−1)∗Y3(−1)* has insignificant effects on *Y2* with a large *p*-value = 0.5950 ≥ 0.30. However, the results are acceptable for computing the forecast values of *Y1*, *X1*, and *X2* beyond the sample period.
2) In statistical sense, the regression of *Y2* can be reduced by deleting the main independent variables *Y2(−1)* or *Y3(−1)* in order still to have a two-way interaction model. Figure 5.23

Dependent Variable: Y2
Method: Least Squares
Date: 03/28/17 Time: 06:59
Sample (adjusted): 1947M02 1996M04
Included observations: 591 after adjustments
Y2 = C(21)*Y2(-1)*Y3(-1) + C(23)*Y3(-1)+C(24)

	Coefficient	Std. Error	t-Statistic	Prob.
C(21)	0.133273	0.002768	48.14340	0.0000
C(23)	-0.649823	0.014554	-44.64882	0.0000
C(24)	4.834912	0.050972	94.85473	0.0000

R-squared	0.798558	Mean dependent var	4.067919
Adjusted R-squared	0.797873	S.D. dependent var	1.127182
S.E. of regression	0.506765	Akaike info criterion	1.483523
Sum squared resid	151.0044	Schwarz criterion	1.505765
Log likelihood	-435.3809	Hannan-Quinn criter.	1.492187
F-statistic	1165.477	Durbin-Watson stat	0.177898
Prob(F-statistic)	0.000000		

Dependent Variable: Y2
Method: Least Squares
Date: 03/28/17 Time: 07:08
Sample (adjusted): 1947M02 1996M04
Included observations: 591 after adjustments
Y2 = C(21)*Y2(-1)*Y3(-1) + C(22)*Y1(-1)+C(24)

	Coefficient	Std. Error	t-Statistic	Prob.
C(21)	0.027637	0.002114	13.07270	0.0000
C(22)	-0.004691	0.000263	-17.85134	0.0000
C(24)	3.959389	0.073907	53.57235	0.0000

R-squared	0.426444	Mean dependent var	4.067919
Adjusted R-squared	0.424493	S.D. dependent var	1.127182
S.E. of regression	0.855104	Akaike info criterion	2.529876
Sum squared resid	429.9474	Schwarz criterion	2.552119
Log likelihood	-744.5784	Hannan-Quinn criter.	2.538541
F-statistic	218.5919	Durbin-Watson stat	0.034464
Prob(F-statistic)	0.000000		

Figure 5.23 Statistical results of two alternative reduced models of the *Y2* regression in (5.42).

presents the results of two alternative reduced models. Based on these results, findings and notes presented are as follows:

2.1 Both regressions are acceptable models to show that the interaction *Y2(−1)∗Y3(−1)* has a significant effect on *Y2*. The first regression is obtained by deleting *Y2(−1)*, even though it has a significant adjusted effect. This reduced model is a better predictor for *Y2*, since it has a greater R-squared than the second reduced model.

2.2 Compared to the full regression of *Y2* in Figure 5.22, both reduced regressions have smaller R-squared values. Hence, it can be said that the full regression is the best fit model of *Y2* to compute the forecast values of *Y2* within the sample period. However, we never know which one gives the best forecast values beyond the sample period. For this reason, I myself would choose the reduced regression in Figure 5.23a together with the *Y2* and *Y3* regressions as the final forecast two-way interaction model.

3) On the other hand, it is not wrong if one would prefer to use the full two-way interaction MLV(1,1,1) in (5.42) as the forecast model.

5.8.1.3 Nonhierarchical 3-Way Interaction MLV(1,1,1) Based on Figure 5.18b

As an extension of the model (5.42), based on the path diagram in Figure 5.19b, the following equation system is considered an acceptable three-way interaction MLV(1,1,1) model, in the theoretical sense, because *Y3(−1)* has an indirect effect on *Y1* through *Y2(−1)* and *Y1(−1)*, which can be presented by the three-way interaction *Y1(−1)∗Y2(−1)∗Y3(−1)* in the statistical sense.

$$\begin{aligned} Y1 &= C(11)*Y1(-1)*Y2(-1)*Y3(-1) + C(12)*Y1(-1)*Y2(-1) \\ &\quad + C(13)*Y2(-1)Y3(-1) \\ &\quad + C(14)*Y1(-1) + C(15)*Y2(-1) + C(16)*Y3(-1) + C(17) \\ Y2 &= C(21)*Y2(-1)*Y3(-1) + C(22)*Y2(-1) + C(23)*Y3(-1) + C(24) \\ Y3 &= C(31)*Y3(-1) + C(32) \end{aligned} \tag{5.43}$$

Note that the model of *Y1* is a nonhierarchical three-way interaction model, because the two-way interaction *Y1(−1)∗Y3(−1)* is not in the model.

Example 5.16 Application of the Model (5.43)

Figure 5.24 presents the OLS estimate of the model (5.43). Based on this estimate, especially for the regression of *Y1*, the following findings and notes are presented. Note that the regression of *Y2* has been discussed in Example 5.15.

1) At the 10% level of significance, *Y1(−1)∗Y2(−1)∗Y3(−1)* has a insignificant adjusted effect on *Y1* with a p-value = 0.1053 > 0.10, however, it has a negative significant adjusted effect on *Y1* with a p-value = 0.1053/2 = 0.05265 < 0.10. This contradicted conclusion could lead to a different opinion on the validity of the regression to present a three-way interaction model. My own opinion is to accept the three-way interaction – refer also to a 15% level of significance used by Lapin (1973).
2) On the other hand, if the regression of *Y1* should be reduced, then it is recommended to delete one or two of the other independent variables by using the manual stepwise or the combinatorial selection method. I am very confident about a reduced model where the

System: SYS04
Estimation Method: Least Squares
Date: 03/28/17 Time: 06:20
Sample: 1947M02 1996M04
Included observations: 591
Total system (balanced) observations 1773

	Coefficient	Std. Error	t-Statistic	Prob.
C(11)	-0.000800	0.000494	-1.620479	0.1053
C(12)	0.015840	0.007116	2.225851	0.0262
C(13)	0.172682	0.081732	2.112781	0.0348
C(14)	0.990647	0.012566	78.83676	0.0000
C(15)	-0.583204	0.506622	-1.151162	0.2498
C(16)	-0.936460	0.460210	-2.034853	0.0420
C(17)	1.919018	2.927944	0.655415	0.5123
C(21)	-0.001097	0.002063	-0.531710	0.5950
C(22)	0.998455	0.013856	72.05908	0.0000
C(23)	0.005226	0.010207	0.511988	0.6087
C(24)	-0.002986	0.069078	-0.043221	0.9655
C(31)	0.996286	0.002741	363.4912	0.0000
C(32)	0.033803	0.020724	1.631079	0.1031
Determinant residual covariance		0.020941		

Equation: Y1 = C(11)*Y1(-1)*Y2(-1)*Y3(-1) + C(12)*Y1(-1)*Y2(-1) + C(13)*Y2(-1)*Y3(-1) +C(14)*Y1(-1)+C(15)*Y2(-1)+C(16)*Y3(-1) +C(17)
Observations: 591

R-squared	0.998412	Mean dependent var	140.9657
Adjusted R-squared	0.998395	S.D. dependent var	136.5858
S.E. of regression	5.471189	Sum squared resid	17481.40
Durbin-Watson stat	1.479718		

Equation: Y2 = C(21)*Y2(-1)*Y3(-1) + C(22)*Y2(-1) + C(23)*Y3(-1) +C(24)
Observations: 591

R-squared	0.979540	Mean dependent var	4.067919
Adjusted R-squared	0.979436	S.D. dependent var	1.127182
S.E. of regression	0.161640	Sum squared resid	15.33687
Durbin-Watson stat	1.594906		

Equation: Y3 = C(31)*Y3(-1) + C(32)
Observations: 591

R-squared	0.995562	Mean dependent var	6.863469
Adjusted R-squared	0.995554	S.D. dependent var	3.188500
S.E. of regression	0.212595	Sum squared resid	26.62078
Durbin-Watson stat	1.233485		

Figure 5.24 Statistical results of the model (5.43).

interaction $Y1(-1)*Y2(-1)*Y3(-1)$ has an insignificant adjusted effect on $Y1$, since $Y1$ and the interaction have a significant correlation based on the t-statistic of $t_0 = 36.443$, with $df = 589$, and p-value = 0.000. Do this as an exercise.

5.8.1.4 Full 2-Way Interaction MLV(1,1,1) Based on Figure 5.18c

Based on the path diagram in Figure 5.19c, which is well-defined theoretically, the following equation system is considered an acceptable full two-way interaction MLV(1,1,1) model in the statistical sense.

$$
\begin{aligned}
YI &= C(111)*YI(-1)*Y2(-1) + C(12)*YI(-1)*Y3(-1) \\
&\quad + C(13)*Y2(-1)*Y3(-1) \\
&\quad + C(14)*YI(-1) + C(15)*Y2(-1) + C(16)*Y3(-1) + C(17) \\
Y2 &= C(21)*Y2(-1)*Y3(-1) + C(22)*Y2(-1) + C(23)*Y3(-1) + C(24) \\
Y3 &= C(31)*Y3(-1) + C(32)
\end{aligned}
\tag{5.44}
$$

Example 5.17 Application of the Model (5.44)

Figure 5.25 presents the OLS estimate of the model (5.44). Based on this result, specifically the regression of *Y1*, the following findings and notes are presented.

System: SYS06
Estimation Method: Least Squares
Date: 03/26/17 Time: 21:01
Sample: 1947M02 1996M04
Included observations: 591
Total system (balanced) observations 1773

	Coefficient	Std. Error	t-Statistic	Prob.
C(11)	0.013706	0.005095	2.689952	0.0072
C(12)	-0.005060	0.002008	-2.519321	0.0118
C(13)	0.063376	0.092414	0.685782	0.4929
C(14)	1.018384	0.013890	73.31690	0.0000
C(15)	-0.010700	0.572601	-0.018686	0.9851
C(16)	-0.382599	0.536962	-0.712526	0.4762
C(17)	-1.522688	3.422495	-0.444906	0.6564
C(21)	-0.001097	0.002063	-0.531710	0.5950
C(22)	0.998455	0.013856	72.05908	0.0000
C(23)	0.005226	0.010207	0.511988	0.6087
C(24)	-0.002986	0.069078	-0.043221	0.9655
C(31)	0.996286	0.002741	363.4912	0.0000
C(32)	0.033803	0.020724	1.631079	0.1031
Determinant residual covariance		0.020878		

Equation: Y1 = C(11)*Y1(-1)*Y2(-1) + C(12)*Y1(-1)*Y3(-1)+ C(13)*Y2(-1)
*Y3(-1)+C(14)*Y1(-1)+C(15)*Y2(-1)+C(16)*Y3(-1)+C(17)
Observations: 591

R-squared	0.998422	Mean dependent var	140.9657
Adjusted R-squared	0.998406	S.D. dependent var	136.5858
S.E. of regression	5.453919	Sum squared resid	17371.21
Durbin-Watson stat	1.481534		

Equation: Y2 = C(21)*Y2(-1)*Y3(-1) + C(22)*Y2(-1) + C(23)*Y3(-1)
+C(24)
Observations: 591

R-squared	0.979540	Mean dependent var	4.067919
Adjusted R-squared	0.979436	S.D. dependent var	1.127182
S.E. of regression	0.161640	Sum squared resid	15.33687
Durbin-Watson stat	1.594906		

Equation: Y3 = C(31)*Y3(-1) + C(32)
Observations: 591

R-squared	0.995562	Mean dependent var	6.863469
Adjusted R-squared	0.995554	S.D. dependent var	3.188500
S.E. of regression	0.212595	Sum squared resid	26.62078
Durbin-Watson stat	1.233485		

Figure 5.25 Statistical results of the 2WI_MLV(1,1,1) model in (5.44).

1) Note that one out of the three two-way interactions of the *Y1*-regression has a large *p*-value > 0.30. However, their joint (adjusted) effects on *Y1* is significant, based on the Wald Chi-square statistic of 13.84809 with $df = 3$ and a *p*-value = 0.0031. So, in the statistical sense, we do not have to reduce the model.
2) If the model should be reduced, then it is recommended to delete one or two of the main variables *Y1(−1)*, *Y2(−1)*, and *Y3(−1)*. Do this as an exercise.

5.8.1.5 Hierarchical three-way Interaction MLV(1,1,1) Based on Figure 5.18c

As a more advanced model, a hierarchical three-way interaction MLV(1,1,1) is a valid statistical model for the path diagram in Figure 5.19c, with the following equation.

$$\begin{aligned}
Y1 &= C(11)*Y1(-1)*Y2(-1)*Y3(-1) + C(12)*Y1(-1)*Y2(-1) \\
&\quad + C(13)*Y1(-1)*Y3(-1) \\
&\quad + C(14)*Y2(-1)*Y3(-1) + C(15)*Y1(-1) + C(16)*Y2(-1) \\
&\quad + C(17)*Y3(-1) + C(18) \\
Y2 &= C(12)*Y2(-1)*Y3(-1) + C(22)*Y2(-1) + C(23)*Y3(-1) + C(24) \\
Y3 &= C(13)*Y3(-1) + C(32)
\end{aligned} \tag{5.45}$$

Example 5.18 Application of the Model (5.45)

Figure 5.26 presents the OLS estimates of the model (5.45). Based on this result, the following findings and notes are presented.

1) Since the interaction *Y1(−1)*Y2(−1)*Y3(−1)* has significant effect on *Y1* with a *p*-value = 0.0058, then the hierarchical three-way interaction regression of *Y1* is acceptable, regardless of the status of the other independent variables. In this case, it happens that each of the other independent variables has either positive or negative adjusted effects on *Y1* based on the *t*-statistic at the 10% level of significance. For instance, *Y1(−1)*Y2(−1)* has a negative significant adjusted effect on *Y1* based on the *t*-statistic of $t_0 = -1.291585$ with a *p*-value = 0.1967/2 = 0.09835 < 0.10.
2) The problems of the *Y2*-regression have been presented in Example 5.15.
3) There is no problem with the *Y3*-regression, since *Y3(−1)* has a positive significant effect on *Y3* at the 10% level of significance based on the *t*-statistic of $t_0 = 1.631079$ with a *p*-value = 0.1031/2 = 0.050155 < 0.10.

5.8.2 Circular Effects MLV(1,1,1) Models

Referring to the circular AR(1)_SCM with Trend presented in Agung (2014, Section 2.4.1.2), Figure 5.27 presents two alternative up-and-down relationships between the six variables *Y1*, *Y2*, *Y3*, *Y1(−1)*, *Y2(−1)*, and *Y3(−1)*, which are called *circular effects multivariate lag-variable:* CE_MLV(1,1,1) models. Note that their circularities are better represented in Figure 5.27b between the first-lag-variables.

System: SYS05
Estimation Method: Least Squares
Date: 03/26/17 Time: 20:45
Sample: 1947M02 1996M04
Included observations: 591
Total system (balanced) observations 1773

	Coefficient	Std. Error	t-Statistic	Prob.
C(11)	0.004718	0.001707	2.763617	0.0058
C(12)	-0.014874	0.011516	-1.291585	0.1967
C(13)	-0.023501	0.006965	-3.374026	0.0008
C(14)	-0.333694	0.170551	-1.956566	0.0506
C(15)	1.137690	0.045326	25.10033	0.0000
C(16)	1.755789	0.856012	2.051127	0.0404
C(17)	1.083239	0.752607	1.439315	0.1502
C(18)	-8.802011	4.303453	-2.045337	0.0410
C(21)	-0.001097	0.002063	-0.531710	0.5950
C(22)	0.998455	0.013856	72.05908	0.0000
C(23)	0.005226	0.010207	0.511988	0.6087
C(24)	-0.002986	0.069078	-0.043221	0.9655
C(31)	0.996286	0.002741	363.4912	0.0000
C(32)	0.033803	0.020724	1.631079	0.1031
Determinant residual covariance		0.020658		

Equation: Y1 = C(11)*Y1(-1)*Y2(-1)*Y3(-1) + C(12)*Y1(-1)*Y2(-1) + C(13)*Y1(-1)*Y3(-1)+C(14)*Y2(-1)*Y3(-1)+C(15)*Y1(-1)+C(16)*Y2(-1)+C(17)*Y3(-1)+C(18)
Observations: 591

R-squared	0.998442	Mean dependent var	140.9657
Adjusted R-squared	0.998423	S.D. dependent var	136.5858
S.E. of regression	5.423187	Sum squared resid	17146.59
Durbin-Watson stat	1.487514		

Equation: Y2 = C(21)*Y2(-1)*Y3(-1) + C(22)*Y2(-1) + C(23)*Y3(-1) +C(24)
Observations: 591

R-squared	0.979540	Mean dependent var	4.067919
Adjusted R-squared	0.979436	S.D. dependent var	1.127182
S.E. of regression	0.161640	Sum squared resid	15.33687
Durbin-Watson stat	1.594906		

Equation: Y3 = C(31)*Y3(-1) + C(32)
Observations: 591

R-squared	0.995562	Mean dependent var	6.863469
Adjusted R-squared	0.995554	S.D. dependent var	3.188500
S.E. of regression	0.212595	Sum squared resid	26.62078
Durbin-Watson stat	1.233485		

Figure 5.26 Statistical results of the 3WI_MLV(1,1,1) model in (5.45).

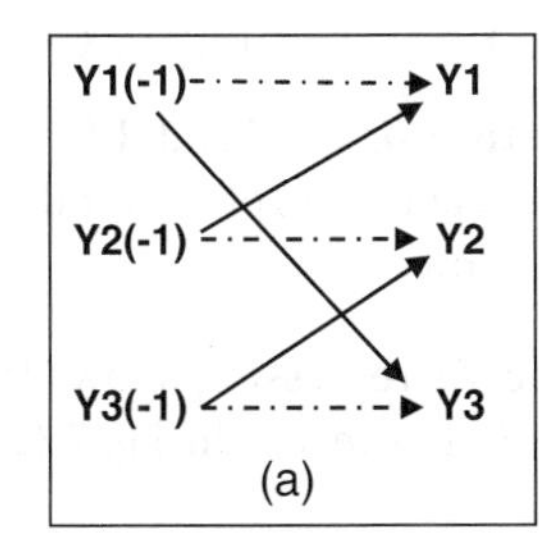

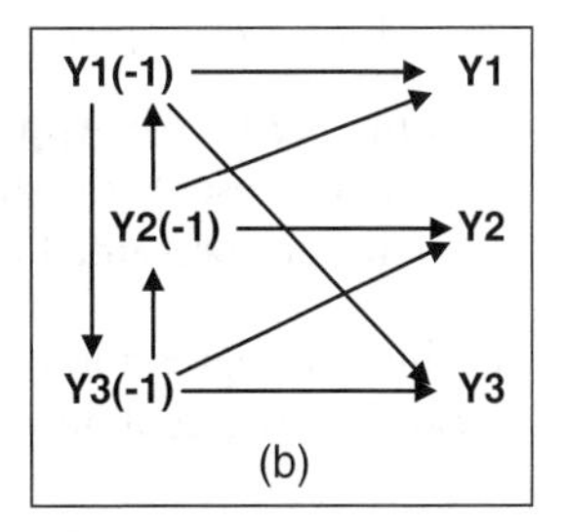

Figure 5.27 Alternative theoretical CE_MLV (1,1,1) based on (*Y1,Y2,Y3*).

5.8.2.1 Additive CE_MLV(1,1,1) Model

The path diagram in Figure 5.27a and b, which is accepted in theoretical sense, or assumed to be a true relationship, can be represented as an additive multivariate lag-variables model, namely ACE_MLV(1,1,1) with the following equation system.

$$\begin{aligned} Y1 &= C(11)*\mathbf{Y2}(-\mathbf{1}) + C(12)*Y1(-1) + C(10) \\ Y2 &= C(21)*\mathbf{Y3}(-\mathbf{1}) + C(22)*Y2(-1) + C(20) \\ Y3 &= C(31)*\mathbf{Y1}(-\mathbf{1}) + C(32)*Y3(-1) + C(30) \end{aligned} \tag{5.46}$$

Note that their circularity is represented by *Y2(−1)* that has a direct effect on *Y1*, *Y3(−1)* has a direct effect on *Y2*, and *Y1(−1)* has a direct effect on *Y3*.

Example 5.19 An Application of the Model (5.46)

Figure 5.28 presents the statistical results of the model (5.46). Based on these results, the following findings and notes are presented.

1) Note that *Y2(−1)* has insignificant *adjusted* effect on *Y(1)* with a very large *p*-value. In fact *Y1* and *Y2(−1)* have a negative significant correlation of −0.504846 based on the *t*-statistic of $t_0 = -14.19387$ with a *p*-value = 0.0000. Should we omit or delete *Y2(−1)* from the model? The answer is no, because by deleting *Y2(−1)*, we are suggesting that *Y2(−1)* has no effect on *Y1*, while it is defined that it does, in the theoretical sense. A similar explanation can be presented for the *Y2*-regression and the *Y3*-regression. The insignificance of the effects of the independent variables *Y2(−1)*, *Y3(−1)*, and *Y1(−1)*, is demonstrated by the uncontrollable and unexpected impacts of bivariate correlations between the independent variables.
2) It is found that the joint effects of independent variables in each regression in Figure 5.28 has a significant effect on the corresponding dependent variable, based on the Chi-square statistic as presented in Figure 5.29.
3) Furthermore, the null hypothesis H_0: C(11) = C(12) = C(21) = C(22) = C(31) = C(32) = 0 is rejected, based on the Chi-square statistic of 478 186.1 with *df* = 6 and *p*-value = 0.0000. So it can be concluded that the data supports the causal or up-and-down relationships between the six variables presented in Figure 5.27a.

5.8.2.2 Two-Way Interaction CE_MLV(1,1,1) Model

The path diagram in Figure 5.27b, which is assumed to be a true relationships, can be represented as a two-way interaction circular effects, *2WI_CE_MLV(1,1,1)* model, with the following equation system.

$$\begin{aligned} Y1 &= \mathbf{Y2}(-\mathbf{1})*(C(11)*Y1(-1) + C(12)) + C(13)*Y1(-1) + C(10) \\ Y2 &= \mathbf{Y3}(-\mathbf{1})*(C(21)*Y2(-1) + C(22)) + C(22)*Y2(-1) + C(20) \\ Y3 &= \mathbf{Y1}(-\mathbf{1})*(C(31)*Y3(-1) + C(32)) + C(33)*Y3(-1) + C(30) \end{aligned} \tag{5.47}$$

Example 5.20 An Application of the Model (5.47)

Figure 5.30 presents the statistical results of the model (5.47). Based on these results, the findings and notes are presented as follows:

1) The first regression or the *Y1*-regression shows that the effect of *Y2(−1)* on *Y1* is insignificantly dependent on *Y1(−1)*. However, the H_0: C(11) = C(12) = C(13) = 0 is rejected based on

System: SYS08
Estimation Method: Full Information Maximum Likelihood (BFGS / Marquardt steps)
Date: 03/30/17 Time: 10:44
Sample: 1947M02 1996M04
Included observations: 591
Total system (balanced) observations 1773
Convergence achieved after 26 iterations
Coefficient covariance computed using outer product of gradients

	Coefficient	Std. Error	z-Statistic	Prob.
C(11)	0.184865	0.610756	0.302683	0.7621
C(12)	1.010240	0.002367	426.8046	0.0000
C(10)	-1.115936	2.566929	-0.434736	0.6638
C(21)	-0.000978	0.002349	-0.416372	0.6771
C(22)	0.993852	0.009006	110.3605	0.0000
C(20)	0.027930	0.037032	0.754227	0.4507
C(31)	-1.42E-05	8.67E-05	-0.164029	0.8697
C(32)	0.995917	0.003361	296.3231	0.0000
C(30)	0.038319	0.043638	0.878099	0.3799

Log likelihood	-1371.246	Schwarz criterion	4.737611
Avg. log likelihood	-0.773404	Hannan-Quinn criter.	4.696876
Akaike info criterion	4.670882		
Determinant residual covariance		0.020792	

Equation: Y1 = C(11)*Y2(-1)+C(12)*Y1(-1)+C(10)
Observations: 591

R-squared	0.998378	Mean dependent var	140.9657
Adjusted R-squared	0.998372	S.D. dependent var	136.5858
S.E. of regression	5.511047	Sum squared resid	17858.53
Durbin-Watson stat	1.474365		

Equation: Y2 = C(21)*Y3(-1)+C(22)*Y2(-1) +C(20)
Observations: 591

R-squared	0.979519	Mean dependent var	4.067919
Adjusted R-squared	0.979450	S.D. dependent var	1.127182
S.E. of regression	0.161586	Sum squared resid	15.35272
Durbin-Watson stat	1.597398		

Equation: Y3 = C(31)*Y1(-1) + C(32)*Y3(-1)+C(30)
Observations: 591

R-squared	0.995563	Mean dependent var	6.863469
Adjusted R-squared	0.995548	S.D. dependent var	3.188500
S.E. of regression	0.212749	Sum squared resid	26.61423
Durbin-Watson stat	1.233556		

Figure 5.28 Statistical results of the equation system (5.46).

Figure 5.29 Summary of testing hypotheses on the joint effects of the independent variables of each regressions in Figure 5.27.

H_0	Chi-square	df	Prob.
C(11)=C(12)=0	240253.7	2	0.0000
C(21)=C(22)=0	15610.19	2	0.0000
C(31)=C(32)=0	15610.19	2	0.0000

the Wald Chi-square test of 381 706.9 with $df = 3$ and p-value = 0.0000. Hence, its independent variables have significant joint effects.

2) Similarly, the *Y3*-regression shows that the effect of *Y1(−1)* on *Y3* is insignificantly dependent on *Y3(−1)*, and its independent variables have significant joint effects, based on the Wald Chi-square test of 1 311 770.3 with $df = 3$ and p-value = 0.0000.

System: SYS09
Estimation Method: Least Squares
Date: 03/31/17 Time: 11:48
Sample: 1947M02 1996M04
Included observations: 591
Total system (balanced) observations 1773

	Coefficient	Std. Error	t-Statistic	Prob.
C(11)	-0.000328	0.002359	-0.139219	0.8893
C(12)	0.375521	0.268922	1.396392	0.1628
C(13)	1.012804	0.007318	138.3906	0.0000
C(10)	-2.089144	1.196030	-1.746733	0.0809
C(21)	0.116124	0.011900	9.758052	0.0000
C(22)	-0.483318	0.060304	-8.014690	0.0000
C(20)	6.135931	0.346206	17.72334	0.0000
C(31)	-1.34E-07	6.00E-07	-0.223975	0.8228
C(32)	2.99E-05	0.000336	0.089027	0.9291
C(33)	0.996465	0.004190	237.8386	0.0000
C(30)	0.033469	0.021091	1.586894	0.1127
Determinant residual covariance		1.239414		

Equation: Y1 = Y2(-1)*(C(11)*Y1(-1)+C(12))+C(13)*Y1(-1)+C(10)
Observations: 591

R-squared	0.998380	Mean dependent var	140.9657
Adjusted R-squared	0.998371	S.D. dependent var	136.5858
S.E. of regression	5.511895	Sum squared resid	17833.64
Durbin-Watson stat	1.475309		

Equation: Y2 = Y3(-1)*(C(21)*Y2(-1)+C(22))+C(22)*Y2(-1)+C(20)
Observations: 591

R-squared	0.202702	Mean dependent var	4.067919
Adjusted R-squared	0.199990	S.D. dependent var	1.127182
S.E. of regression	1.008188	Sum squared resid	597.6686
Durbin-Watson stat	0.029861		

Equation: Y3 = Y1(-1)*(C(31)*Y1(-1)+ C(32))+C(33)*Y3(-1)+C(30)
Observations: 591

R-squared	0.995565	Mean dependent var	6.863469
Adjusted R-squared	0.995542	S.D. dependent var	3.188500
S.E. of regression	0.212882	Sum squared resid	26.60205
Durbin-Watson stat	1.235127		

Figure 5.30 Statistical results of the model (5.47).

3) Based on the findings presented before, it can be concluded that the data supports the proposed model (5.47).
4) However, in the statistical sense, the regressions of *Y1* and *Y3* can be reduced by deleting their main independent variables in order to show the two-way interactions independent variables have significant effects. By using the trial-and-error, I obtain the estimate of the reduced model as presented in Figure 5.31, which shows that the effect of *Y2(−1)* on *Y1* is significantly dependent on *Y1(−1)*, and the effect of *Y1(−1)* on *Y3* is significantly dependent on *Y3(−1)*.

5.8.2.3 A Three-Way Interaction CE_MLV(1,1,1) Model

The path diagram in Figure 5.27b, also could be represented as a three-way interaction circular effects, *3WI_CE_MLV(1,1,1)* model, with the following equation system.

System: SYS10
Estimation Method: Least Squares
Date: 03/31/17 Time: 11:44
Sample: 1947M02 1996M04
Included observations: 591
Total system (balanced) observations 1773

	Coefficient	Std. Error	t-Statistic	Prob.
C(11)	0.314335	0.003641	86.33463	0.0000
C(12)	-21.65191	1.255890	-17.24029	0.0000
C(10)	74.26934	6.148253	12.07975	0.0000
C(21)	0.116124	0.011900	9.758052	0.0000
C(22)	-0.483318	0.060304	-8.014690	0.0000
C(20)	6.135931	0.346206	17.72334	0.0000
C(31)	-9.76E-05	4.33E-06	-22.54860	0.0000
C(32)	0.058929	0.002245	26.25500	0.0000
C(30)	2.306041	0.185371	12.44012	0.0000
Determinant residual covariance		3908.093		

Equation: Y1 = Y2(-1)*(C(11)*Y1(-1)+C(12))+C(10)
Observations: 591

R-squared	0.945517	Mean dependent var	140.9657
Adjusted R-squared	0.945331	S.D. dependent var	136.5858
S.E. of regression	31.93555	Sum squared resid	599689.1
Durbin-Watson stat	0.062196		

Equation: Y2 = Y3(-1)*(C(21)*Y2(-1)+C(22))+C(22)*Y2(-1)+C(20)
Observations: 591

R-squared	0.202702	Mean dependent var	4.067919
Adjusted R-squared	0.199990	S.D. dependent var	1.127182
S.E. of regression	1.008188	Sum squared resid	597.6686
Durbin-Watson stat	0.029861		

Equation: Y3 = Y1(-1)*(C(31)*Y1(-1)+ C(32))+C(30)
Observations: 591

R-squared	0.568183	Mean dependent var	6.863469
Adjusted R-squared	0.566714	S.D. dependent var	3.188500
S.E. of regression	2.098814	Sum squared resid	2590.151
Durbin-Watson stat	0.017261		

Figure 5.31 Statistical results of a reduced model of (5.47).

$$
\begin{aligned}
Y1 &= \mathbf{Y2(-1)}*(C(11)*Y1(-1)*Y3(-1) + C(12)*Y1(-1) \\
&\quad + C(13)) + C(14)*Y1(-1) + C(10) \\
Y2 &= \mathbf{Y3(-1)}*(C(21)*Y1(-1)*Y2(-1) + C(22)*Y2(-1) \\
&\quad + C(23)) + C(24)*Y2(-1) + C(20) \\
Y3 &= \mathbf{Y1(-1)}*(C(31)*Y2(-1)*Y3(-1) + C(32)*Y3(-1) \\
&\quad + C(33)) + C(34)*Y3(-1) + C(30)
\end{aligned}
\tag{5.48}
$$

Example 5.21 An Application of the Model (5.48)
Figure 5.32 presents the statistical results of the model (5.48). Based on these results, the following findings and notes are presented.

System: SYS11
Estimation Method: Least Squares
Date: 04/04/17 Time: 16:22
Sample: 1947M02 1996M04
Included observations: 591
Total system (balanced) observations 1773

	Coefficient	Std. Error	t-Statistic	Prob.
C(11)	-0.000924	0.000345	-2.675684	0.0075
C(12)	0.017135	0.006936	2.470529	0.0136
C(13)	0.323598	0.268226	1.206442	0.2278
C(14)	0.986799	0.012144	81.26142	0.0000
C(10)	-3.001708	1.237721	-2.425189	0.0154
C(21)	-7.85E-06	3.41E-06	-2.300592	0.0215
C(22)	-0.003583	0.002322	-1.542668	0.1231
C(23)	0.024982	0.013311	1.876851	0.0607
C(24)	1.009331	0.014593	69.16708	0.0000
C(20)	-0.081295	0.076784	-1.058749	0.2899
C(31)	-5.03E-05	1.83E-05	-2.755774	0.0059
C(32)	0.000146	0.000107	1.364258	0.1727
C(33)	-0.000276	0.000555	-0.497653	0.6188
C(34)	1.017691	0.006857	148.4061	0.0000
C(30)	-0.032233	0.034886	-0.923951	0.3556

Determinant residual covariance 0.020579

Equation: Y1 = Y2(-1)*(C(11)*Y1(-1)*Y3(-1)+C(12)*Y1(-1)+C(13))+C(14)
*Y1(-1)+C(10)
Observations: 591

R-squared	0.998399	Mean dependent var	140.9657
Adjusted R-squared	0.998388	S.D. dependent var	136.5858
S.E. of regression	5.483203	Sum squared resid	17618.39
Durbin-Watson stat	1.478386		

Equation: Y2 = Y3(-1)*(C(21)*Y1(-1)*Y2(-1)+C(22)*Y2(-1)+C(23))+C(24)
*Y2(-1)+C(20)
Observations: 591

R-squared	0.979724	Mean dependent var	4.067919
Adjusted R-squared	0.979585	S.D. dependent var	1.127182
S.E. of regression	0.161052	Sum squared resid	15.19959
Durbin-Watson stat	1.598220		

Equation: Y3 = Y1(-1)*(C(31)*Y2(-1)*Y3(-1)+C(32)*Y3(-1)+C(33))+C(34)
*Y3(-1)+C(30)
Observations: 591

R-squared	0.995654	Mean dependent var	6.863469
Adjusted R-squared	0.995624	S.D. dependent var	3.188500
S.E. of regression	0.210922	Sum squared resid	26.06993
Durbin-Watson stat	1.251404		

Figure 5.32 Statistical results of the model (5.48).

1) The results are really unexpected, which show that each of the three-way interactions has significant effect on the corresponding dependent variable at the 1 or 5% levels. In this case, only the effect of *Y1(−1)* on *Y3* has a very large *p*-value = 0.6188., so it can be concluded the data supports the *3WI_CE_MLV(1,1,1)* model, whatever happens with the two-way interaction and the main independent variables.
2) In fact, even though most or all two-way interactions and the main independent variables have large *p*-values, the model is still an acceptable *3WI_CE_MLV(1,1,1)* model. However,

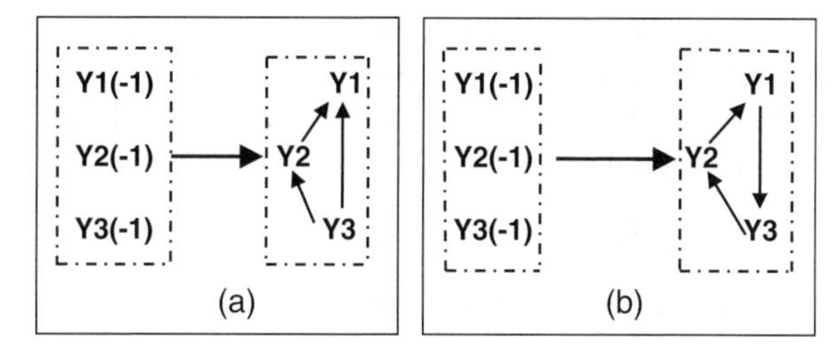

Figure 5.33 Alternative causal relationships between *Y1, Y2,* and *Y3.*

if a three-way interaction has a large p-value > 0.30, then the regression should be reduced by deleting one or two of the main variables or the two-way interactions by using the trial-and-error or combinatorial selection method.

5.9 Models Presenting Causal Relationships *Y1,Y2,* and *Y3*

Referring to the five path diagrams based on the variables *Y1, Y2, Y3, Y1(−1), Y2(−1),* and *Y3(−1)* presented in Figures 5.19, 5.27, and 5.33, I present two additional specific up-and-down or causal relationships between the six variables out of many possible relationships.

5.9.1 Triangular Effects Models

Referring to the causal or up-and-down relationships between *Y1(−1), Y2(−1),* and *Y3(−1)* in Figure 5.19c; in fact, the variables *Y1, Y2,* and *Y3* should have the same relationships, in the theoretical sense, even though there are no arrows between the last three variables. The arrows are not presented because their causal relations are not taken into account in the model. In this section, I present alternative models showing the causal Triangular Effects between the variables *Y1, Y2,* and *Y3* with the path diagram presented in Figure 5.33a, such as follows.

5.9.1.1 Additive Triangular Effects Model

An additive triangular effects model, called the *additive triangular effects multivariate lag-variables model,* namely ATE_MLV(1,1,1) = ATE_MLV(**1**) model of *(Y1,Y2,Y3),* has the following equation system.

$$\begin{aligned} Y1 &= C(11)*\mathbf{Y2} + C(12)*\mathbf{Y3} + C(13)*Y1(-1) + C(14)*Y2(-1) \\ &\quad + C(15)*Y3(-1) + C(10) \\ Y2 &= C(21)*\mathbf{Y3} + C(22)*Y1(-1) + C(23)*Y2(-1) + C(24)*Y3(-1) + C(20) \\ Y3 &= C(31)*Y1(-1) + C(32)*Y2(-1) + C(33)*Y3(-1) + C(30) \end{aligned} \tag{5.49}$$

Note the first equation in (5.49) shows that *Y2* and *Y3* have direct effects on *Y1* and the second shows that *Y3* has a direct effect on *Y2*. These types of causal relations are defined as triangular effects models of *Y1, Y2,* and *Y3.*

5.9.1.2 Two-Way Interaction Triangular Effects Model

It has been found and noted that variable *Yk(−1)* is one of the best predictors of *Yk*. For this reason, I propose a *two-way interaction triangular effect multivariate autoregressive* model, namely 2WTE_MLV(1,1,1) = 2WTE_MLV(**1**) model of *(Y1,Y2,Y3)*, with the following equation system as an extension of the model (5.49), which shows that the effects of *Y2* and *Y3* on *Y1* depend on *Y1(−1)*, and the effect of *Y3* on *Y2* depends on *Y2(−1)*. Note that the effect of each *Y2* and *Y3* on *Y1*, respectively, depends on the functions of *Y1(−1)*.

$$\begin{aligned} Y1 &= \mathbf{Y2}^*(C(11)*\mathbf{Y1}(-\mathbf{1}) + C(12)) + \mathbf{Y3}*(C(13)*\mathbf{Y1}(-\mathbf{1}) + C(14)) \\ &\quad + C(15)*Y1(-1) + C(16)*Y2(-1) + C(17)*Y3(-1) + C(10) \\ Y2 &= \mathbf{Y3}^*(C(21)*\mathbf{Y2}(-\mathbf{1}) + C(22)) + C(23)*Y1(-1) + C(23)*Y2(-1) \\ &\quad + C(24)*Y3(-1) + C(20) \\ Y3 &= C(31)*Y1(-1) + C(32)*Y2(-1) + C(33)*Y3(-1) + C(30) \end{aligned} \tag{5.50}$$

5.9.1.3 Three-Way Interaction Triangular Effects Model

As an extension of the model (5.50), I propose the following 3WTE_MLV(**1**) model to test the hypothesis that joint effects of *Y2∗Y3*, *Y2*, and *Y3* on *Y1* depend on *Y1(−1)*.

$$\begin{aligned} Y1 &= C(11)*\mathbf{Y2}*\mathbf{Y3}*\mathbf{Y}(-\mathbf{1}) + \mathbf{Y2}*(C(12)*\mathbf{Y1}(-\mathbf{1}) + C(13)) \\ &\quad + \mathbf{Y3}*(C(14)*\mathbf{Y1}(-\mathbf{1}) + C(15)) \\ &\quad + C(16)*Y1(-1) + C(17)*Y2(-1) + C(18)*Y3(-1) + C(10) \\ Y2 &= \mathbf{Y3}*(C(21)*\mathbf{Y2}(-\mathbf{1}) + C(22)) + C(23)*Y1(-1) + C(23)*Y2(-1) \\ &\quad + C(24)*Y3(-1) + C(20) \\ Y3 &= C(31)*Y1(-1) + C(32)*Y2(-1) + C(33)*Y3(-1) + C(30) \end{aligned} \tag{5.51}$$

Example 5.22 Application of the Model (5.51)

Figure 5.34 presents the LS estimate of the model (5.51). Based on this estimate, the findings and notes presented are as follows:

1) Even though the interaction *Y2∗Y3∗Y1(−1)* has a large *p*-value, it is found that the null hypothesis H_0: C(11) = C(12) = C(14) = 0 is rejected based on the Wald Chi-square statistic of 346.2443 with *df* = 3 and *p*-value = 0.0000. Hence, it can be concluded that the set of interactions *(C11)∗Y2∗Y3 + C(12)∗Y2 + C(14)∗Y3)∗Y1(−1)* has a significant effect on *Y1*. In other words, the joint effects of the three variables *Y2∗Y3*, *Y2*, and *Y3* on *Y1* are significantly dependent on *Y1(−1)*.
2) In the statistical sense, it is well known that the impacts of multicollinearity between all independent variables would give the interaction *Y2∗Y3∗Y1(−1)* a large *p*-value. In fact, *Y1* and *Y2∗Y3∗Y1(−1)* have a correlation of *r* = 0.827394, and it is significant based on the *t*-statistic of t_0 = 35.75417 with a *p*-value = 0.0000. Hence, in the statistical sense, a reduced model can be explored to obtain a model with the interaction *Y2∗Y3∗Y1(−1)* that a significant effect by deleting one or two of the main variables or two-way interactions using the trial-and-error method. My first choice is to delete *Y1(−1)* from the regression because *Y1(−1)* has been used in three interactions. A reduced model is found that presents a significant interaction *Y2∗Y3∗Y1(−1)*, as presented in Figure 5.35.

System: SYS20
Estimation Method: Least Squares
Date: 04/21/17 Time: 22:45
Sample: 1947M02 1996M04
Included observations: 591
Total system (balanced) observations 1773

	Coefficient	Std. Error	t-Statistic	Prob.
C(11)	-0.000182	0.000730	-0.249000	0.8034
C(12)	-0.010814	0.006683	-1.618133	0.1058
C(13)	-17.40642	1.200326	-14.50140	0.0000
C(14)	-0.001032	0.002664	-0.387447	0.6985
C(15)	-3.360272	0.911171	-3.687861	0.0002
C(16)	1.050669	0.020577	51.06120	0.0000
C(17)	18.11769	1.184819	15.29152	0.0000
C(18)	3.799235	0.902000	4.212014	0.0000
C(10)	-4.704257	1.669190	-2.818287	0.0049
C(21)	0.146277	0.003609	40.53158	0.0000
C(22)	-0.383125	0.095374	-4.017084	0.0001
C(23)	0.001123	0.000225	4.991674	0.0000
C(24)	-0.347326	0.095917	-3.621121	0.0003
C(20)	4.865888	0.049922	97.47014	0.0000
C(31)	-0.000123	8.54E-05	-1.438412	0.1505
C(32)	-0.016006	0.009214	-1.737153	0.0825
C(33)	0.998211	0.003121	319.8641	0.0000
C(30)	0.102963	0.044630	2.307031	0.0212

Determinant residual covariance 0.200623

Equation: Y1 = C(11)*Y2*Y3*Y1(-1)+Y2*(C(12)*Y1(-1)+C(13))+Y3 *(C(14)*Y1(-1)+C(15)) +C(16)*Y1(-1) +C(17)*Y2(-1) +C(18)*Y3(-1)+C(10)
Observations: 591

R-squared	0.998987	Mean dependent var	140.9657
Adjusted R-squared	0.998973	S.D. dependent var	136.5858
S.E. of regression	4.376964	Sum squared resid	11149.85
Durbin-Watson stat	1.629249		

Equation: Y2 = Y3*(C(21)*Y2(-1)+C(22))+C(23)*Y1(-1)+C(23)*Y2(-1) +C(24)*Y3(-1)+C(20)
Observations: 591

R-squared	0.811899	Mean dependent var	4.067919
Adjusted R-squared	0.810615	S.D. dependent var	1.127182
S.E. of regression	0.490532	Sum squared resid	141.0040
Durbin-Watson stat	0.183192		

Equation: Y3 = C(31) *Y1(-1)+C(32)*Y2(-1)+C(33)*Y3(-1)+C(30)
Observations: 591

R-squared	0.995587	Mean dependent var	6.863469
Adjusted R-squared	0.995565	S.D. dependent var	3.188500
S.E. of regression	0.212346	Sum squared resid	26.46825
Durbin-Watson stat	1.237501		

Figure 5.34 The LS estimate of the model (5.52).

3) I am very confident that other alternative reduced models of *Y1* with a significant interaction *Y2∗Y3∗Y1(−1)* can be obtained by deleting one or two other independent variables. However, using only the trial-and-error method would be difficult and time consuming. For this reason, it is suggested to apply the combination of the *STEPLS: Combinatorial* and the trial-and-error methods to select a list of always-included regressors and a list of search regressors. Of course,

System: SYS21
Estimation Method: Least Squares
Date: 04/22/17 Time: 21:48
Sample: 1947M02 1996M04
Included observations: 591
Total system (balanced) observations 1773

	Coefficient	Std. Error	t-Statistic	Prob.
C(11)	-0.031719	0.000912	-34.78253	0.0000
C(12)	0.267815	0.009024	29.67877	0.0000
C(13)	-17.90203	2.807339	-6.376868	0.0000
C(14)	0.118974	0.002935	40.54270	0.0000
C(15)	-1.713017	2.129793	-0.804311	0.4213
C(17)	16.89802	2.770599	6.099048	0.0000
C(18)	3.856603	2.109676	1.828054	0.0677
C(10)	-6.497837	3.903187	-1.664752	0.0961
C(21)	0.146277	0.003609	40.53158	0.0000
C(22)	-0.383125	0.095374	-4.017084	0.0001
C(23)	0.001123	0.000225	4.991674	0.0000
C(24)	-0.347326	0.095917	-3.621121	0.0003
C(20)	4.865888	0.049922	97.47014	0.0000
C(31)	-0.000123	8.54E-05	-1.438412	0.1505
C(32)	-0.016006	0.009214	-1.737153	0.0825
C(33)	0.998211	0.003121	319.8641	0.0000
C(30)	0.102963	0.044630	2.307031	0.0212
Determinant residual covariance		0.780622		

Equation: Y1 = C(11)*Y2*Y3*Y1(-1)+Y2*(C(12)*Y1(-1)+C(13))+Y3
*(C(14)*Y1(-1)+C(15))+C(17)*Y2(-1)+C(18)*Y3(-1)+C(10)

				c(11)=c(12
	c(11)=c(12)=	c(11)=c(12)=	c(11)=c(12)=	)=c(13)=c(
	c(13)=c(14)=	c(13)=c(14)=	c(13)=c(14)=	14)=c(15)=
Observations: 591	c(15)=0	c(15)=0	c(15)=0	0

R-squared	0.994449	Mean dependent var	140.9657
Adjusted R-squared	0.994382	S.D. dependent var	136.5858
S.E. of regression	10.23724	Sum squared resid	61098.98
Durbin-Watson stat	0.343704		

Equation: Y2 = Y3*(C(21)*Y2(-1)+C(22))+C(23)*Y1(-1)+C(23)*Y2(-1)
+C(24)*Y3(-1)+C(20)

				c(11)=c(12
	c(11)=c(12)=	c(11)=c(12)=	c(11)=c(12)=	)=c(13)=c(
	c(13)=c(14)=	c(13)=c(14)=	c(13)=c(14)=	14)=c(15)=
Observations: 591	c(15)=0	c(15)=0	c(15)=0	0

R-squared	0.811899	Mean dependent var	4.067919
Adjusted R-squared	0.810615	S.D. dependent var	1.127182
S.E. of regression	0.490532	Sum squared resid	141.0040
Durbin-Watson stat	0.183192		

Equation: Y3 = C(31) *Y1(-1)+C(32)*Y2(-1)+C(33)*Y3(-1)+C(30)

				c(11)=c(12
	c(11)=c(12)=	c(11)=c(12)=	c(11)=c(12)=	)=c(13)=c(
	c(13)=c(14)=	c(13)=c(14)=	c(13)=c(14)=	14)=c(15)=
Observations: 591	c(15)=0	c(15)=0	c(15)=0	0

R-squared	0.995587	Mean dependent var	6.863469
Adjusted R-squared	0.995565	S.D. dependent var	3.188500
S.E. of regression	0.212346	Sum squared resid	26.46825
Durbin-Watson stat	1.237501		

Figure 5.35 The LS estimate of a reduced model of (5.52).

Dependent Variable: Y1
Method: Stepwise Regression
Date: 04/23/17 Time: 20:55
Sample (adjusted): 1947M02 1996M04
Included observations: 591 after adjustments
Number of always included regressors: 2
Number of search regressors: 7
Selection method: Combinatorial
Number of search regressors: 5

Variable	Coefficient	Std. Error	t-Statistic	Prob.*
C	-5.579398	1.588650	-3.512038	0.0005
Y2*Y3*Y1(-1)	-0.000948	0.000234	-4.050876	0.0001
Y1(-1)	1.029728	0.004996	206.1102	0.0000
Y3(-1)	3.618475	0.892341	4.055034	0.0001
Y3	-3.095093	0.897812	-3.447373	0.0006
Y2(-1)	18.26952	1.177363	15.51732	0.0000
Y2	-17.58297	1.189653	-14.77991	0.0000

R-squared	0.998982	Mean dependent var	140.9656
Adjusted R-squared	0.998971	S.D. dependent var	136.5858
S.E. of regression	4.380538	Akaike info criterion	5.803994
Sum squared resid	11206.44	Schwarz criterion	5.855893
Log likelihood	-1708.080	Hannan-Quinn criter.	5.824211
F-statistic	95502.42	Durbin-Watson stat	1.623884
Prob(F-statistic)	0.000000		

Selection Summary

Number of combinations compar...	21

*Note: p-values and subsequent tests do not account for stepwise selection.

Figure 5.36 The LS estimate of *Y1* using Combinatorial STEPLS Regression.

the list of always-included regressors should contain at least *C* and *Y2**Y3**Y1(−1)*. The LS estimate in Figure 5.36 is obtained by using *C* and *Y2**Y3**Y1(−1)* and *Y2**Y1(−1)*, *Y3**Y1(−1)*. *Y2*, *Y3*, *Y1(−1)*, *Y2(−1)*, and *Y3(−1)* as the list of search regressors. Then by using the trial-and-error method to select 1, 2, 3, 4, and 5 of the search regressors, respectively, we can obtain five reduced models with a significant interaction *Y2**Y3**Y1(−1)*, and the fifth is presented in Figure 5.36. It is also found that by selecting six out of the seven search regressors a model was obtained that had a negative insignificant interaction *Y2**Y3**Y1(−1)*, with a *p*-value = 0.2835.

4) By using other list of always-included regressors and selected subsets of the search regressors, we could obtained other alternative reduced models with a significant interaction. For instance, such as by using *C*, *Y2**Y3**Y1(−1)*, and *Y2**Y1(−1)* as the list of always-included regressors, with any alternative subsets of the six variables *Y3**Y1(−1)*. *Y2*, *Y3*, *Y1(−1)*, *Y2 (−1)*, and *Y3(−1)* as one of the search regressors. Do these as exercises.

5.9.1.4 Computing the Forecast Values Beyond the Sample Period

Based on each of the triangular effect models presented before, the first forecast value beyond the sample period can easily be computed using the steps as follows:

1) *The first step*. The forecast values of $Y3(T+k)$ for $k = 1,2, \ldots$ can easily be compute based on the regression of *Y3* using Excel. Do this as an exercise – refer to spreadsheets presented before.

2) *The second step*. By using the forecast values of *Y3(T + k)* computed at the first step, the forecast values of *Y2(T + k)* can be computed.
3) Finally, the forecast values of *Y1(T + k)* can be computed.

5.9.2 Circular Effects Models

Referring the causal or up-and-down relationships between *Y1(−1)*, *Y2(−1)*, and *Y3(−1)* in Figure 5.26b, in fact the variables *Y1*, *Y2*, and *Y3* should have the same relationships, in the theoretical sense, even though there are no arrows between the variables. The arrows are not presented because their causal relations are not taken into account in the model. In this section, I present alternative models showing the circular causal relationships between the variables *Y1*, *Y2*, and *Y3* with the path diagram presented in Figure 5.31b. However, it is found that each regression in a circular-effects model (CE_M) cannot be directly used to compute the forecast values beyond the sample period. See the following explanation.

5.9.2.1 Additive Circular Effects Model

The *additive circular effects* MLV model, namely ACE_MLV(1,1,1) = ACE_MLV(**1**) has the equation system as follows:

$$\begin{aligned}
Y1 &= C(11)*\mathbf{Y2} + C(12)*Y1(-1) + C(13)*Y2(-1) + C(14)*Y3(-1) + C(10) \\
Y2 &= C(21)*\mathbf{Y3} + C(22)*Y1(-1) + C(23)*Y2(-1) + C(24)*Y3(-1) + C(20) \\
Y3 &= C(31)*\mathbf{Y1} + C(32)*Y1(-1) + C(33)*Y2(-1) + C(34)*Y3(-1) + C(30)
\end{aligned} \quad (5.52)$$

Compared to the triangular effects models presented before, based on each of these regressions we can compute the in-sample forecast values, but we cannot compute directly the forecast values beyond the sample period, because each regression has an independent variable of the same level. For instance, the forecast value of *Y1(T + 1)* cannot be computed because there is no score for *Y2(T + 1)*. Similarly, the forecast value of *Y3(T + 1)* cannot be computed because there is no data for *Y1(T + 1)*. See Section 5.8.2.1 that presents how to compute the forecast values beyond the sample period.

In order to be able to compute the forecast values beyond the sample period of *Y1*, the model of *Y3* should be replaced by a new model, which can be used to computing its forecast values beyond the sample period. The simplest method is by deleting *Y1* from the model of *Y3*. Note that *Y1* and *Y1(−1)* are highly or significantly correlated, so by deleting one of them *Y1* from the model of *Y3*, the model (5.52) still could be considered as presenting a circular effects model. Hence, we would have the same model as *Y3* in (5.49).

Then the forecast value *Y3(T + 1)* can be computed, which can be used to compute *Y2(T + 1)*, and finally *Y1(T + 1)* can be computed by using *Y2(T + 1)*, which can be done using Excel. Similarly, it can be done to compute additional *Y1(T + k)* for $k > 1$.

5.9.2.2 A Two-Way Interaction Circular Effects Model

The *two-way interaction circular effects* MLV model, namely 2WICE_MLV(1,1,1) = 2WICE_MLV(**1**) has the following equation system, which shows that the effect *of Y2* on *Y1* depends on *Y1(−1)*, the effect of *Y3* on *Y2* depends on *Y2(−1)*, and a the effect of *Y1* on *Y3* depends on *Y3(−1)*.

$$
\begin{aligned}
Y1 &= \mathbf{Y2}*(C(11)*\mathbf{Y1}(-\mathbf{1}) + C(12)) + C(13)*Y1(-1) \\
&\quad + C(14)*Y2(-1) + C(15)*Y3(-1) + C(10) \\
Y2 &= \mathbf{Y3}*(C(21)*\mathbf{Y2}(-\mathbf{1}) + C(22)) + C(23)*Y1(-1) \\
&\quad + C(24)*Y2(-1) + C(25)*Y3(-1) + C(20) \\
Y3 &= \mathbf{Y1}*(C(31)*\mathbf{Y3}(-\mathbf{1}) + C(32)) + C(33)*Y1(-1) \\
&\quad + C(34)*Y2(-1) + C(35)*Y3(-1) + C(30)
\end{aligned}
\tag{5.53}
$$

Similar to the model (5.52), based on this model we can not compute the forecast values of each *Yk*. So in order to be able to compute the forecast values beyond the sample period for *Y1*, the independent variable *Y1* in model of *Y3* should be replaced by *Y1(−1)*.

5.9.3 Reciprocal Effects Models

Under the assumption that each pair of the three time series *Y1*, *Y2*, and *Y3* have reciprocal causal effects or up-and-down relationships. Then we would have three alternative models, as presented in the following subsections

5.9.3.1 Additive Reciprocal Causal Effects Model

The *additive reciprocal causal effects* MLV model, namely ASCE_MLV(1,1,1) = ACE_MLV(**1**) has the equation system as follows:

$$
\begin{aligned}
Y1 &= C(11)*\mathbf{Y2} + C(12)*\mathbf{Y3} + C(13)*Y1(-1) + C(14)*Y2(-1) \\
&\quad + C(15)*Y3(-1) + C(10) \\
Y2 &= C(21)*\mathbf{Y1} + C(22)*\mathbf{Y3} + C(23)*Y1(-1) + C(24)*Y2(-1) \\
&\quad + C(25)*Y3(-1) + C(20) \\
Y3 &= C(31)*\mathbf{Y1} + C(32)*\mathbf{Y2} + C(33)*Y1(-1) + C(34)*Y2(-1) \\
&\quad + C(35)*Y3(-1) + C(30)
\end{aligned}
\tag{5.54}
$$

Note that by doing analysis based on each of the three regressions, we can compute its forecast evaluation, but we cannot compute its forecast beyond the sample period.

5.9.3.2 A Two-Way Interaction Reciprocal Effects Model

The *two-way interaction reciprocal effects* MLV model: 2WISE_MLV(1,1,1) = 2WISE_MLV(**1**) has the following equation system, which shows that the joint effects of *Y2* and *Y3* on *Y1* depend on *Y1(−1)*, the joint effects of *Y1 and Y3* on *Y2* depend on *Y2(−1)*, and the joint effects of *Y1* and *Y2* on *Y3* depend on *Y3(−1)*.

$$
\begin{aligned}
Y1 &= \mathbf{Y2}^*(C(11)*\mathbf{Y1(-1)} + C(12)) + \mathbf{Y3}^*(C(13)*Y1(-1) + C(14)) \\
&\quad + C(15)*Y1(-1) + C(16)*Y2(-1) + C(17)*Y3(-1) + C(10) \\
Y2 &= \mathbf{Y1}^*(C(21)*\mathbf{Y2(-1)} + C(22)) + \mathbf{Y3}^*(C(23)*Y2(-1) + C(24)) \\
&\quad + C(25)*Y1(-1) + C(26)*Y2(-1) + C(27)*Y3(-1) + C(20) \\
Y3 &= \mathbf{Y1}^*(C(31)*\mathbf{Y3(-1)} + C(32)) + \mathbf{Y2}^*(C(33)*Y3(-1) + C(34)) \\
&\quad + C(35)*Y1(-1) + C(36)*Y2(-1) + C(37)*Y3(-1) + C(30)
\end{aligned}
\tag{5.55}
$$

Note that the *Y1*-regression of this equation system is exactly the same as the *Y1*-regression of the equation system in (5.50). However, its forecast values beyond the sample period cannot be computed. In order to be able to compute its forecast values beyond the sample period, the independent variables *Y1* and *Y2* of the model of *Y3* can be replaced by *Y1(−1)* and *Y2(−1)*, respectively.

5.9.3.3 A Three-Way Interaction Reciprocal Effects Model

The *three-way interaction Reciprocal effects* MLV model: 3WISE_MLV(1,1,1) = 3WISE_MLV (**1**) has the following equation system, which shows that the joint effects of *Y2∗Y3*, *Y2*, and *Y3* on *Y1* depend on *Y1(−1)*, the joint effects of *Y1∗Y3*, *Y1*, and *Y3* on *Y2* depends on *Y2 (−1)*, and the joint effects of *Y1∗Y2*, *Y1*, and *Y2* on *Y3* depends on *Y3(−1)*.

$$
\begin{aligned}
Y1 &= C(11)*Y > \mathbf{2*Y3*Y1(-1)} + \mathbf{Y2}*(C(12)*\mathbf{Y1(-1)} + C(13)) \\
&\quad + \mathbf{Y3}*(C(14)*Y1(-1) + C(15)) \\
&\quad + C(16)*Y1(-1) + C(17)*Y2(-1) + C(18)*Y3(-1) + C(10) \\
Y2 &= C(21)*\mathbf{Y1*Y3*Y2(-1)} + \mathbf{Y1}*(C(22)*\mathbf{Y2(-1)} + C(13)) \\
&\quad + \mathbf{Y3}^*(C(24)*Y2(-1) + C(25)) \\
&\quad + C(26)*Y1(-1) + C(27)*Y2(-1) + C(28)*Y3(-1) + C(20) \\
Y3 &= C(31)*\mathbf{Y1*Y2*Y3(-1)} + \mathbf{Y1}*(C(32)*\mathbf{Y3(-1)} + C(33)) \\
&\quad + \mathbf{Y2}*(C(34)*Y3(-1) + C(35)) \\
&\quad + C(36)*Y1(-1) + C(37)*Y2(-1) + C(38)*Y3(-1) + C(30)
\end{aligned}
\tag{5.56}
$$

Note that the *Y1*-regression of this equation system is exactly the same as the *Y1*-regression of the equation system in Eq. (5.51).

Example 5.23 An Application of the Model (5.56)

It is found the LS estimate of the model (5.56) shows that each of the 3-way interaction has insignificant adjusted effect on the corresponding dependent variable. Since the *Y1*-regression of this model is exactly the same as the *Y1*-regression of the system equation (5.51), then its LS estimate would be exactly the same as presented in Figure 5.54, in the Example 5.22, with a conclusion that the joint effects of *Y2***Y3*, *Y2* and *Y3* on *Y1* is significantly depends on *Y1*(−*1*). In addition, special notes and comments are presented specific for the *Y1*-regression in order to obtain possible reduced models having significant *3*-way interaction.

System: SYS23_B
Estimation Method: Least Squares
Date: 04/23/17 Time: 16:36
Sample: 1947M02 1996M04
Included observations: 591
Total system (balanced) observations 1773

	Coefficient	Std. Error	t-Statistic	Prob.
C(11)	-0.031719	0.000912	-34.78253	0.0000
C(12)	0.267815	0.009024	29.67877	0.0000
C(13)	-17.90203	2.807339	-6.376868	0.0000
C(14)	0.118974	0.002935	40.54270	0.0000
C(15)	-1.713017	2.129793	-0.804311	0.4213
C(17)	16.89802	2.770599	6.099048	0.0000
C(18)	3.856603	2.109676	1.828054	0.0677
C(10)	-6.497837	3.903187	-1.664752	0.0961
C(21)	0.000195	4.11E-05	4.738295	0.0000
C(22)	-0.000426	0.000594	-0.716614	0.4737
C(23)	-0.023188	0.003914	-5.924070	0.0000
C(24)	0.131736	0.004121	31.96337	0.0000
C(25)	-0.517561	0.096220	-5.378940	0.0000
C(26)	0.021628	0.003714	5.823063	0.0000
C(28)	-0.253772	0.095281	-2.663410	0.0078
C(20)	5.350406	0.097243	55.02110	0.0000
C(31)	-0.000682	5.10E-05	-13.37226	0.0000
C(32)	0.004853	0.000252	19.26455	0.0000
C(33)	-0.034618	0.004169	-8.303826	0.0000
C(34)	0.193342	0.003250	59.49844	0.0000
C(35)	-0.759341	0.134022	-5.665783	0.0000
C(36)	0.013257	0.003987	3.325480	0.0009
C(37)	-0.218560	0.134164	-1.629048	0.1035
C(30)	5.748687	0.103033	55.79476	0.0000

Determinant residual covariance 2.388695

Equation: Y1=C(11)*Y2*Y3*Y1(-1)+Y2*(C(12)*Y1(-1)+C(13))+Y3*(C(14)*Y1(-1)+C(15))+C(17)*Y2(-1)+C(18)*Y3(-1)+C(10)
Observations: 591

R-squared	0.994449	Mean dependent var	140.9657
Adjusted R-squared	0.994382	S.D. dependent var	136.5858
S.E. of regression	10.23724	Sum squared resid	61098.98
Durbin-Watson stat	0.343704		

Equation: Y2=C(21)*Y1*Y3*Y2(-1)+Y1*(C(22)*Y2(-1)+C(23))+Y3*(C(24)*Y2(-1)+C(25))+C(26)*Y1(-1)+C(28)*Y3(-1)+C(20)
Observations: 591

R-squared	0.835331	Mean dependent var	4.067919
Adjusted R-squared	0.833354	S.D. dependent var	1.127182
S.E. of regression	0.460142	Sum squared resid	123.4389
Durbin-Watson stat	0.160999		

Equation: Y3=C(31)*Y1*Y2*Y3(-1)+Y1*(C(32)*Y3(-1)+C(33))+Y2*(C(34)*Y3(-1)+C(35))+C(36)*Y1(-1)+C(37)*Y2(-1)+C(30)
Observations: 591

R-squared	0.982461	Mean dependent var	6.863469
Adjusted R-squared	0.982250	S.D. dependent var	3.188500
S.E. of regression	0.424797	Sum squared resid	105.2038
Durbin-Watson stat	0.430240		

Figure 5.37 The LS estimate of an acceptable reduced model of (5.57).

Since *Y2*-regression and *Y3*-regression of the system equation (5.56) have similar problems to the *Y1*-regression, then the LS estimate of the model (5.56) is not presented. Do the analysis as exercises, similar to *Y1*-regression presented in Figure 5.34.

However, Figure 5.37 presents the LS estimate of an acceptable reduced model of the system equation (5.56), which is obtained by deleting *Y1(−1)*, *Y2(−1)*, and *Y3(−1)*, respectively, from the *Y1*, *Y2* and *Y3*-regressions.

On the other hand, if the 3-way interactions would be deleted from the system, then the LS estimate of the 2WISE-MLV(1) in (5.56) would be obtained.

5.10 Extended Models

With exactly the same extensions as presented before, we can easily develop various models based on *(Y1,Y2,Y3)* as the extension of all models based on *(Y1,Y2)* presented in Section 5.8. However, this section only presents the extension of all possible multiple regressions, so that the object *"Forecast"* can be applied directly in order to have their forecast evaluation, as well as the graphs of their in-sample forecast values over times compared to the observed scores.

5.10.1 Extension to the Models with Additional Exogenous Variables

All of the multiple regressions of *Yk* can be extended to various models with additional exogenous or environmental variables. The simplest models would be adding only the main variables, which should be selected using a combination of the STEPLS: Combinatorial and trial-and-error methods, which has been illustrated in Example 5.22. More advanced models can be obtained by using very few selected two- or three-way interaction exogenous variables. Refer to specific equation systems or SCM presented based on (*Y1,Y2,Y3*).

Example 5.24 Alternative Extensions of the *Y1*-Regression in (5.56)
As the first extension of *Y1*-regression in (5.56), Figure 5.38(a) presents the results of the *Y1*-regression as presented in Figure 5.37 with three additional exogenous variables: *X1(−1) = IP (−1)*, *X2(−1) = PW(−1)*, and *X3(−1) = R3(−1)* in GARCH.wf1. Based on these results, the following findings and notes are presented.

$$\begin{aligned} &Y1\; C\; \boldsymbol{Y2*Y3*Y1(-1)}\; \boldsymbol{Y2*Y1(-1)}\; \boldsymbol{Y3*}Y1(-1)\; Y2\; Y3 \\ &\quad Y1(-1)\, Y2(-1)\, Y3(-1)\, X1(-1)\, X2(-1)\, X3(-1) \end{aligned} \tag{5.57}$$

1) This model can be considered to be an extension with the simplest additional set of three additive exogenous variables. Other alternative models with three additive exogenous variables could be defined by using a combination of three out of the six variables *X1(−1) X2(−1) X3(−1)*, *X1*, *X2*, and *X3*.
2) Since the three-way interaction has insignificant effects with a very large *p*-value, then the results of an acceptable interaction reduced model are presented in Figure 5.38(b), which is obtained by deleting the main variable *Y1(−1)*. Refer to the notes presented in the Example 5.22.
3) However, the reduced model has unexpectedly large values of RMSE and TIC compared to the full model. Hence, I try to explore alternative three-way interaction reduced models by

Dependent Variable: Y1
Method: Least Squares
Date: 04/26/17 Time: 06:30
Sample (adjusted): 1947M02 1996M04
Included observations: 591 after adjustments

Variable	Coefficient	Std. Error	t-Statistic	Prob.
C	-3.774250	1.581615	-2.386327	0.0173
Y2*Y3*Y1(-1)	-0.000108	0.001013	-0.106804	0.9150
Y2*Y1(-1)	-0.040971	0.010054	-4.074989	0.0001
Y3*Y1(-1)	-0.002104	0.002882	-0.730072	0.4656
Y2	-15.32254	1.161953	-13.18689	0.0000
Y3	-3.999771	0.909871	-4.395976	0.0000
Y1(-1)	1.063013	0.021914	48.50841	0.0000
Y2(-1)	15.18726	1.177651	12.89623	0.0000
Y3(-1)	2.716102	0.854729	3.177733	0.0016
X1(-1)	0.077141	0.047973	1.608001	0.1084
X2(-1)	0.324724	0.039100	8.304976	0.0000
X3(-1)	68.28238	16.56585	4.121877	0.0000

R-squared	0.999113	Mean dependent var	140.9656
Adjusted R-squared	0.999096	S.D. dependent var	136.5858
S.E. of regression	4.106415	Akaike info criterion	5.683074
Sum squared resid	9763.473	Schwarz criterion	5.772045
Log likelihood	-1667.348	Hannan-Quinn criter.	5.717732
F-statistic	59286.94	Durbin-Watson stat	1.629761
Prob(F-statistic)	0.000000		

RMSE=21.882 & TIC=0.024

(a)

Dependent Variable: Y1
Method: Least Squares
Date: 04/26/17 Time: 06:32
Sample (adjusted): 1947M02 1996M04
Included observations: 591 after adjustments

Variable	Coefficient	Std. Error	t-Statistic	Prob.
C	-4.064449	3.556071	-1.142960	0.2535
Y2*Y3*Y1(-1)	-0.041321	0.001241	-33.28436	0.0000
Y2*Y1(-1)	0.345518	0.013788	25.05879	0.0000
Y3*Y1(-1)	0.124835	0.002714	45.98914	0.0000
Y2	-17.21285	2.611061	-6.592281	0.0000
Y3	2.378125	2.024278	1.174801	0.2406
Y2(-1)	14.70554	2.647732	5.554014	0.0000
Y3(-1)	2.875229	1.921756	1.496147	0.1352
X1(-1)	-0.943382	0.096938	-9.731850	0.0000
X2(-1)	0.413706	0.087815	4.711099	0.0000
X3(-1)	280.7851	35.92064	7.816818	0.0000

R-squared	0.995508	Mean dependent var	140.9656
Adjusted R-squared	0.995431	S.D. dependent var	136.5858
S.E. of regression	9.232849	Akaike info criterion	7.301849
Sum squared resid	49442.39	Schwarz criterion	7.383406
Log likelihood	-2146.697	Hannan-Quinn criter.	7.333619
F-statistic	12853.94	Durbin-Watson stat	0.451665
Prob(F-statistic)	0.000000		

RMSE=1061.886 & TIC=0.551

(b)

Figure 5.38 The LS estimate of the model (5.58) and one of its reduced models,with their forecast evaluation, RMSE and TIC.

Dependent Variable: Y1
Method: Stepwise Regression
Date: 01/27/18 Time: 19:24
Sample (adjusted): 1947M02 1996M04
Included observations: 591 after adjustments
Number of always included regressors: 2
Number of search regressors: 10
Selection method: Combinatorial
Number of search regressors: 9

Variable	Coefficient	Std. Error	t-Statistic	Prob.*
C	-3.865164	1.576070	-2.452406	0.0145
Y2*Y3*Y1(-1)	-0.000727	0.000554	-1.312251	0.1900
Y1(-1)	1.048486	0.009176	114.2584	0.0000
X2(-1)	0.326274	0.039026	8.360328	0.0000
Y2*Y1(-1)	-0.036291	0.007742	-4.687272	0.0000
X3(-1)	72.17678	15.67720	4.603932	0.0000
Y3	-3.931703	0.904717	-4.345782	0.0000
Y3(-1)	2.752307	0.852946	3.226824	0.0013
Y2	-15.28623	1.160420	-13.17301	0.0000
Y2(-1)	15.14542	1.175782	12.88114	0.0000
X1(-1)	0.061812	0.043117	1.433587	0.1522

R-squared	0.999112	Mean dependent var	140.9656
Adjusted R-squared	0.999097	S.D. dependent var	136.5858
S.E. of regression	4.104762	Akaike info criterion	5.680610
Sum squared resid	9772.461	Schwarz criterion	5.762166
Log likelihood	-1667.620	Hannan-Quinn criter.	5.712379
F-statistic	65268.13	Durbin-Watson stat	1.628554
Prob(F-statistic)	0.000000		

Selection Summary

Number of combinations compared:	10

*Note: p-values and subsequent tests do not account for stepwise selection.

RMSE=21.280 & TIC=0.0238

(a)

Dependent Variable: Y1
Method: Stepwise Regression
Date: 01/27/18 Time: 20:45
Sample (adjusted): 1947M02 1996M04
Included observations: 591 after adjustments
Number of always included regressors: 2
Number of search regressors: 10
Selection method: Combinatorial
Number of search regressors: 8

Variable	Coefficient	Std. Error	t-Statistic	Prob.*
C	-3.826473	1.577269	-2.426011	0.0156
Y2*Y3*Y1(-1)	-0.001299	0.000385	-3.369674	0.0008
Y1(-1)	1.047049	0.009130	114.6839	0.0000
X2(-1)	0.334991	0.038585	8.681955	0.0000
Y2*Y1(-1)	-0.028786	0.005709	-5.041941	0.0000
X3(-1)	75.09962	15.55817	4.827021	0.0000
Y3	-3.538438	0.862908	-4.100598	0.0000
Y3(-1)	2.731810	0.853600	3.200339	0.0014
Y2	-15.33858	1.160898	-13.21268	0.0000
Y2(-1)	15.11749	1.176688	12.84749	0.0000

R-squared	0.999109	Mean dependent var	140.9656
Adjusted R-squared	0.999095	S.D. dependent var	136.5858
S.E. of regression	4.108488	Akaike info criterion	5.680763
Sum squared resid	9807.089	Schwarz criterion	5.754905
Log likelihood	-1668.665	Hannan-Quinn criter.	5.709644
F-statistic	72388.45	Durbin-Watson stat	1.628298
Prob(F-statistic)	0.000000		

Selection Summary

Number of combinations compared:	45

*Note: p-values and subsequent tests do not account for stepwise selection.

RMSE=21.790 & TIC=0.0244

(b)

Figure 5.39 Statistical results based on two additional reduced models of (5.58).

using the STEPLS – *Combinatorial Selection*, with the results shown in Figure 5.37. Based on these results, the following findings and notes are presented.

3.1 The result in Figure 5.39(a) is obtained by using *"Y1 C Y2*Y3*Y1(−1)"* as the specific equation and the other 10 variables as the search regressors. Nine out of the 10 search regressors are selected to present the best possible regression. This regression is an acceptable three-way interaction model, since at the 10% level, *Y2*Y3*Y1(−1)* has a negative significant effect on *Y1* with a p-value = 0.1900/2 = 0.0950 < 0.10.

3.2 As an alternative reduced model, eight out of the 10 search regressors are selected, with the result presented in Figure 5.39(b). This result shows that *Y2*Y3*Y1(−1)* has a negative significant effect with a p-value = 0.0008/2. However, it has slightly greater values of RMSE and TIC compared to the regression in Figure 5.39(a).

4) As alternative reduced model, I have tried to use *"Y1 C"* as the specific equation with 11 search regressors. Then by selecting 10 out of 11 regressors, the best possible regression obtained does not have the interaction *Y2*Y3*Y1(−1)*. In other words, the model is a two-way interaction model with slightly greater values of RMSE = 21.921 and TIC = 0.0245, compared to the regression in Figure 5.27(b). However, unexpectedly, by selecting 9 out of 11, the best regression out of 55 possible regressions was obtained, with the smallest values of RMSE = 20.628 and TIC = 0.0231, compared to previous regressions.
5) In order to be able to compute the forecast values from beyond the sample period, the forecast models of *Y2* and *Y3* should be developed or defined so that the forecast values beyond the sample period of *Y2* and *Y3* are available for computing the forecast value beyond the sample period of *Y1*. The simplest possible model is the LV(p) model for each of *Y2* and *Y3*, where the integer p should be selected using the trial-and-error method. Refer to other alternative forecast models presented in Chapter 4 based on a bivariate time series. Do these as exercises.
6) More complex models can be obtained by inserting selected two- or three-way interactions of the variables *X1(−1)*, *X2(−1)*, and *X3(−1)*, or *X1*, *X2*, and *X3*. In this case, the data analysis should be done using the *manual multistage – selection* (MMS) method, by making subsets or subgroups of the search regressors and inserting the most important subset at the first stage. And at each stage, a combination of STEPLS: Combinatorial and the trial-and-error selection methods should be applied. Do these as exercises – refer to the many examples presented in Agung (2014, 2011a, 2009a) and the following examples.

5.10.2 Extension to the Models with Alternative Trends

All of the multiple regression of *Yk*, including all proposed or defined multiple regressions with additional exogenous or environmental variables, as mentioned in Subsection 5.10.1, can easily be extended to the models with alternative trends, as presented in Table 4.1.

Example 5.25 An Application of a Manual-Two-Stage-Selection (M2SS) Method
Figure 5.40 presents an extension of a reduced *Y1*-regression in Figure 5.36 with the ES as follows:

$$Y1\,C\,Y2*Y3*Y1(-1)\,Y2*Y1(-1)\,Y3*Y1(-1)\,Y2\,Y3\,Y2(-1)\,Y3(-1) \tag{5.58}$$

The steps of the regression analysis are as follows:

1) At the first stage, the search regressors are the variables T and $T\,^\wedge\,2$. However, the output is not presented. I am using the quadratic trend, namely $Y1 = a*T\,^\wedge\,2 + b*T + c$, with $a > 0$

Dependent Variable: Y1
Method: Stepwise Regression
Date: 04/27/17 Time: 07:11
Sample (adjusted): 1947M02 1996M04
Included observations: 591 after adjustments
Number of always included regressors: 10
Number of search regressors: 6
Selection method: Combinatorial
Number of search regressors: 5

Variable	Coefficient	Std. Error	t-Statistic	Prob.*
C	45.37438	5.712540	7.942942	0.0000
Y2*Y3*Y1(-1)	-0.043187	0.001398	-30.90133	0.0000
Y2*Y1(-1)	0.332984	0.016954	19.63991	0.0000
Y3*Y1(-1)	0.095262	0.002733	34.85471	0.0000
Y2	-15.19403	1.965107	-7.731907	0.0000
Y3	12.06153	1.599679	7.539968	0.0000
Y2(-1)	11.77791	1.882988	6.254904	0.0000
Y3(-1)	-3.288207	1.448001	-2.270860	0.0235
T	-0.139549	0.024690	-5.652056	0.0000
T^2	0.000388	7.91E-05	4.911944	0.0000
X1(-1)*X2(-1)	0.021906	0.001644	13.32370	0.0000
X2(-1)	-2.752892	0.163089	-16.87971	0.0000
X2(-1)*X3(-1)	22.19238	1.102145	20.13563	0.0000
X1(-1)*X3(-1)	-26.38870	1.478461	-17.84877	0.0000
X3(-1)	596.7518	103.2969	5.777054	0.0000

R-squared	0.997785	Mean dependent var	140.9656
Adjusted R-squared	0.997731	S.D. dependent var	136.5858
S.E. of regression	6.506392	Akaike info criterion	6.608500
Sum squared resid	24383.89	Schwarz criterion	6.719714
Log likelihood	-1937.812	Hannan-Quinn criter.	6.651823
F-statistic	18530.67	Durbin-Watson stat	0.616653
Prob(F-statistic)	0.000000		

Selection Summary

Number of combinations compar...	6

*Note: p-values and subsequent tests do not account for stepwise selection.

Figure 5.40 The LS estimate of an extension of the *Y1*-regression in (5.59).

because the growth curve of the time series *Y1* looks like a part of a parabolic curve,. The data supports the results of the first stage regression analysis, which shows *T ^ 2* has a significant positive adjusted effect on *Y1*.

2) At the second stage, six variables: *X1(−1)∗X2(−1)*, *X1(−1)∗X3(−1)*, *X2(−1)∗X3(−1)*, *X1(−1)*, *X2(−1)*, and *X3(−1)* are used as the search regressors. By using a combination of the STEPLS: Combinatorial and trial-and-error methods, acceptable results are obtained as presented in Figure 5.40, which is the best out of the six possible multiple regressions.
3) As an additional analysis, its forecast evaluation can easily be obtained. However, the forecast values of *Y1* beyond the sample period cannot be computed. Refer to the notes presented in Example 5.24.

Example 5.26 An Application of a M5SMS Method for a Special Model

Example 5.25 presents the model by only inserting the time and exogenous variables as additional independent variables of the model (5.58). This example presents special search

Dependent Variable: Y1
Method: Stepwise Regression
Date: 04/27/17 Time: 12:52
Sample (adjusted): 1947M02 1996M04
Included observations: 591 after adjustments
Number of always included regressors: 8
Number of search regressors: 3
Selection method: Combinatorial
Number of search regressors: 3

Variable	Coefficient	Std. Error	t-Statistic	Prob.*
C	37.43233	2.536722	14.75618	0.0000
Y2*Y3*Y1(-1)	-0.015279	0.000905	-16.88765	0.0000
Y2*Y1(-1)	0.132157	0.006591	20.05048	0.0000
Y3*Y1(-1)	0.035140	0.002842	12.36526	0.0000
Y2	-19.41966	1.563923	-12.41728	0.0000
Y3	1.582418	1.192542	1.326928	0.1851
Y2(-1)	11.87277	1.550294	7.658399	0.0000
Y3(-1)	2.413688	1.148455	2.101683	0.0360
Y1(-1)*X2(-1)	0.003490	0.000297	11.73180	0.0000
Y1(-1)*X1(-1)	0.001438	0.000321	4.477796	0.0000
Y1(-1)*X3(-1)	0.293450	0.107139	2.738956	0.0064

R-squared	0.998380	Mean dependent var	140.9656
Adjusted R-squared	0.998353	S.D. dependent var	136.5858
S.E. of regression	5.543826	Akaike info criterion	6.281684
Sum squared resid	17825.73	Schwarz criterion	6.363241
Log likelihood	-1845.238	Hannan-Quinn criter.	6.313454
F-statistic	35755.24	Durbin-Watson stat	0.879408
Prob(F-statistic)	0.000000		

Selection Summary

Number of combinations compared: 1

*Note: p-values and subsequent tests do not account for stepwise selection.

(a)

Dependent Variable: Y1
Method: Stepwise Regression
Date: 04/27/17 Time: 12:57
Sample (adjusted): 1947M02 1996M04
Included observations: 591 after adjustments
Number of always included regressors: 11
Number of search regressors: 1
Selection method: Combinatorial
Number of search regressors: 1

Variable	Coefficient	Std. Error	t-Statistic	Prob.*
C	37.89998	2.528919	14.98663	0.0000
Y2*Y3*Y1(-1)	-0.017346	0.001180	-14.69715	0.0000
Y2*Y1(-1)	0.140845	0.007300	19.29499	0.0000
Y3*Y1(-1)	0.041455	0.003665	11.30966	0.0000
Y2	-18.88993	1.567737	-12.04917	0.0000
Y3	1.750283	1.187716	1.473655	0.1411
Y2(-1)	11.25268	1.558846	7.218598	0.0000
Y3(-1)	2.037494	1.150674	1.770696	0.0771
Y1(-1)*X2(-1)	0.003359	0.000300	11.20250	0.0000
Y1(-1)*X1(-1)	0.001387	0.000320	4.337101	0.0000
Y1(-1)*X3(-1)	-0.247293	0.226463	-1.091977	0.2753
X1(-1)*X2(-1)*X3(-1)	0.015529	0.005739	2.706064	0.0070

R-squared	0.998401	Mean dependent var	140.9656
Adjusted R-squared	0.998370	S.D. dependent var	136.5858
S.E. of regression	5.513854	Akaike info criterion	6.272500
Sum squared resid	17603.09	Schwarz criterion	6.361471
Log likelihood	-1841.524	Hannan-Quinn criter.	6.307158
F-statistic	32859.78	Durbin-Watson stat	0.861897
Prob(F-statistic)	0.000000		

Selection Summary

Number of combinations compared: 1

*Note: p-values and subsequent tests do not account for stepwise selection.

(b)

Figure 5.41 The statistical results of the first two stages of LS regressions.

Dependent Variable: Y1
Method: Stepwise Regression
Date: 04/27/17 Time: 13:12
Sample (adjusted): 1947M02 1996M04
Included observations: 591 after adjustments
Number of always included regressors: 14
Number of search regressors: 2
Selection method: Combinatorial
Number of search regressors: 3

Variable	Coefficient	Std. Error	t-Statistic	Prob.*
C	26.92067	2.908065	9.257245	0.0000
Y2*Y3*Y1(-1)	-0.012164	0.001655	–7.349133	0.0000
Y2*Y1(-1)	0.063189	0.015768	4.007540	0.0001
Y3*Y1(-1)	0.039418	0.003974	9.919715	0.0000
Y2	-15.93598	1.474642	-10.80668	0.0000
Y3	-1.405969	1.280011	-1.098404	0.2725
Y2(-1)	10.64287	1.453419	7.322640	0.0000
Y3(-1)	1.090598	1.048012	1.040635	0.2985
Y1(-1)*X2(-1)	0.006824	0.000661	10.31776	0.0000
Y1(-1)*X1(-1)	-0.001741	0.000611	-2.849725	0.0045
Y1(-1)*X3(-1)	0.339273	0.262034	1.294765	0.1959
X1(-1)*X2(-1)*X3(-1)	-0.049286	0.016617	-2966093	0.0031
X1(-1)*X3(-1)	-9.612192	1.070958	-8.975322	0.0000
X2(-1)*X3(-1)	14.34272	1.569435	9.138778	0.0000
X1(-1)	1.529128	0.146428	10.44289	0.0000
X2(-1)	-0.913095	0.102971	-8.867456	0.0000

R-squared	0.998722	Mean dependent var	140.9656
Adjusted R-squared	0.998688	S.D. dependent var	135.5858
S.E. of regression	4.946412	Akaike info criterion	6.061902
Sum squared resid	14068.52	Schwarz criterion	6.180530
Log likelihood	-1775.292	Hannan-Quinn criter.	6.108112
F-statistic	29952.67	Durbin-Watson stat	1.024549
Prob(F-statistic)	0.000000		

Selection Summary	
Number of combinations compared:	3

*Note: p-values and subsequent tests do not account for stepwise selection.

(a)

Dependent Variable: Y1
Method: Least Squares
Date: 04/27/17 Time: 13:51
Sample (adjusted): 1947M02 1996M04
Included observations: 591 after adjustments

Variable	Coefficient	Std. Error	t-Statistic	Prob.
C	45.22707	5.371562	8.419725	0.0000
Y2*Y3*Y1(-1)	-0.010512	0.001840	-5.714155	0.0000
Y2*Y1(-1)	0.042874	0.018591	2.306204	0.0215
Y3*Y1(-1)	0.036410	0.003991	9.123401	0.0000
Y2	-15.61660	1.513756	-10.31646	0.0000
Y3	-1.415118	1.361621	-1.039289	0.2991
Y2(-1)	10.24205	1.446963	7.078310	0.0000
Y3(-1)	**-0.283188**	**1.087229**	**-0.260468**	**0.7946**
Y1(-1)*X2(-1)	0.006799	0.000653	10.41150	0.0000
Y1(-1)*X1(-1)	-0.001785	-0.000606	-2.943657	0.0034
Y1(-1)*X3(-1)	0.435314	0.262740	1.656828	0.0981
X1(-1)*X2(-1)*X3(-1)	-0.062475	0.016760	-3727552	0.0002
X1(-1)*X3(-1)	-9.817009	1.175245	-8.353163	0.0000
X2(-1)*X3(-1)	16.39306	1.707624	9.599921	0.0000
X1(-1)	1.318741	0.169790	7.766909	0.0000
X2(-1)	-1.176533	0.121344	-9.695838	0.0000
T	**0.001962**	**0.024614**	**0.079722**	**0.9365**
T^2	0.000214	5.49E-05	3.894726	0.0001

R-squared	0.998759	Mean dependent var	140.9656
Adjusted R-squared	0.998722	S.D. dependent var	136.5858
S.E. of regression	4.883242	Akaike info criterion	6.039479
Sum squared resid	13663.79	Schwarz criterion	6.172936
Log likelihood	-1766.666	Hannan-Quinn criter.	6.091466
F-statistic	27118.02	Durbin-Watson stat	1.017643
Prob(F-statistic)	0.000000		

(b)

Figure 5.42 The statistical results of the last three stages of the LS regressions.

regressors *Y1(−1)∗X1(−1), Y1(−1)∗X2(−1),* and *Y1(−1)∗X3(−1)* to show that the effects of *X1(−1), X2(−1),* and *X3(−1)* on *Y1* depend on *Y1(−1),* since *Y1(−1)* is one of the best predictors for *Y1*. In this case, a M5SS method is applied by using five sets of search regressors as follows:

(1) *Y1(−1)∗X1(−1), Y1(−1)∗X2(−1),* and *Y1(−1)∗X3(−1)*;

(2) *X1(−1)∗X2(−1)∗X3(−1)*;

(3) *X1(−1)∗X2(−1), X1(−1)∗X3(−1),* and *X2(−1)∗X3(−1)*;

(4) *X1(−1), X2(−1),* and *X3(−1)*;

(5) *T, and T^2.*

The results of the first stage analysis in Figure 5.41a clearly show that each of the interactions of three- and two-way interaction independent variables has a significant effect on *Y1*. Hence, it can be concluded that the adjusted joint effects of the variables *Y2∗Y3, Y2∗Y1, Y3∗Y1, X1(−1), X2(−1),* and *X3(−1)* on *Y1* is significantly dependent *Y1(−1).*

By inserting the interaction *X1(−1)∗X2(−1)∗X3(−1),* the results of the second stage analysis in Figure 5.41(b) show the two-way interaction *Y1(−1)∗X3(−1)* is insignificant at the 10% level. However, the joint effects of the six interactions have significant effects on *Y1* based on the Wald (*F*-statistic) of $F_0 = 17\,127.03$, with *df* = (6579), and a *p*-value = 0.0000. So, these results also show that the adjusted joint effects of the variables *Y2∗Y3, Y2∗Y1), Y3∗Y1, X1(−1), X2(−1),* and *X3(−1)* on *Y1* re significantly dependent *Y1(−1).*

Furthermore, Figure 5.42 presents the statistical results of the last three stages of the analysis. By inserting the two sets of three two-way interactions, and three main variables, respectively, an acceptable estimate in Figure 5.42(a) is obtained and it could easily be tested whether the joint effects of the variables *Y2∗Y3, Y2∗Y1), Y3∗Y1, X1(−1), X2(−1),* and *X3(−1)* on *Y1* are significantly dependent on *Y1(−1).* On the other hand, note that *Y3* and *Y3(−1)* have a Prob. < 0.30. However, this is not a problem in statistical sense, since at the 15% level (applied by Lapin 1973) *Y3,* and *Y3(−1),* respectively, have negative and positive significant effects on *Y1* with *p*-values of 0.2725/2 = 0.13625 and 0.2985/2 = 0.14925.

Finally, Figure 5.42b presents the fifth stage of the analysis by inserting both the variables *T* and *T* ^ 2. This results show the variables *Y3(−1)* and *T* have very large *p*-values, which are the impacts of the multicollinearity of the independent variables. In fact, each has positive correlations of 0.44 and 0.85 with *Y1,* and they are significant based on the *t*-statistics of 11.95 and 38.53, respectively, with *df* = 589 and *p*-values = 0.0000.

The variable *T* can be kept in the model, since *T* ^ *2* has a positive significant effect on *Y1* with a *p*-value = 0.0001. However, a reduced model could be explored with respect to the variable *Y3(−1).* Since, both *Y3* and *Y3(−1)* are in the model, then we have a choice to delete either one of them at the first trial. In general, one might delete *Y3(−1)* from the model because it has a much greater probability. However, my choice is to delete *Y3,* because I can be very sure that *Y3(−1)* is an upper, source, or cause factor/variable of *Y1.* Then I obtain a reduced model with *Y3(−1)* that has a negative significant adjusted effect on *Y1,* with a *p*-value = 0.1246/2 = 0.0623 at the 10% level of significance. As a comparison, while by deleting *Y3(−1),* the reduced model shows that *Y3* has a negative significant adjusted effect on *Y1,* with a *p*-value = 0.0664/2 = 0.0332 < 0.05. What is your choice? Do these as exercises to see what happens with the other independent variables, specifically for the variables *T* and *T* ^ *2.*

5.10.3 Extension to LVARMA(*p,q,r*)

All of the multiple regressions of *Yk*, presented in Subsections 5.10.1 and 5.10.2 with illustrative examples can easily be extended to the *lag-variable-autoregressive-moving average models*, namely LVARMA(*p,q,r*) models for $p \geq 1$, $q \geq 0$, and $r \geq 0$, which should be selected using the trial-and-error method. So, the model should be presented using the list of dependent and independent variables, with the general ES, as follows:

$$Yk\ C\ Yk(-1) \cdots Yk(-p)\ X1\ X2 \cdots Xn \cdots AR(1) \ldots AR(q)\ MA(1) \cdots MA(r) \tag{5.59}$$

where *Xn* can be a main variable, two-way or three-way interaction of the independent variables of all possible models, which has been presented before based on *(Y1,Y2,Y3)*, and new defined models in Subsections 5.10.1 and 5.10.2.

Example 5.27 An Extension of the Model (5.58) to LVARMA(*p,q,r*) Model
Figure 5.43 presents the LS estimate of the following LVARMA(1,3,2) model with its forecast evaluation for a forecast sample {@*Year*>1990}, which shows the graph of the in-sample forecast values have large deviations from the observed scores, corresponding to a very large RMSE.

$$\begin{aligned} &Y1\ C\ Y2*Y3*Y1(-1)\ Y2*Y1(-1)\ Y3*Y1(-1)\ Y2\ Y3\ Y2(-1)\ Y3(-1) \\ &AR(1)\ AR(2)\ AR(3)\ MA(1)\ MA(2) \end{aligned} \tag{5.60}$$

So it can be concluded that the LVARMA(1,3,2) model (5.60) is not a good forecast model for *Y1*, even though its TIC is relatively small. For this reason, and referring to the statement "The model must follow the data, and not the other way around" (Bezzecri 1973, in Gifi 1991, p. 25), I try to obtain alternative better forecast models to fit the data, which are presented in the following examples.

On the other hand, unexpectedly, the results show an additional independent variable, namely SIGMASQ. Why, and what is the idea? There is no explanation yet!

Example 5.28 An Extension of the Model (5.60) with a Quadratic Trend
Referring to the growth pattern of *Y1* over time, which looks like a part of parabolic curve, I try to apply an extension of the model (5.60) to the following LVARMA(1,3,2) model with a quadratic trend. Figure 5.44 presents its LS estimate with the forecast evaluation. Based on these results, the following findings and notes are presented.

$$\begin{aligned} &Y1\ C\ Y2*Y3*Y1(-1)\ Y2*Y1(-1)\ Y3*Y1(-1)\ Y2\ Y3\ Y2(-1)\ y3(-1) \\ &T\ T\hat{}2 AR(1)\ AR(2)\ AR(3)\ MA(1)\ MA(2) \end{aligned} \tag{5.61}$$

1) This model is a better forecast model than (5.60) because it has smaller values of RMSE and TIC.
2) On the other hand, the results present one of the inverted AR roots = 1.00, but without any statement. So the LS estimate is an acceptable statistical result. In other words, the estimated AR process is stationary. Compared to results in Figure 5.46 with an inverted AR root = 1.00 ad the statement "*Estimated AR proses is nonstationary.*"
3) However, the in-sample forecast values and the observed scores of *Y1* have different trends during the time period @*Year*>1994. So I try to apply another model with piece-wise or heterogeneous quadratic trends, which is presented in the following example.

Dependent Variable: Y1
Method: ARMA Maximum Likelihood (OPG - BHHH)
Date: 04/28/17 Time: 16:07
Sample: 1947M02 1996M04
Included observations: 591
Convergence achieved after 186 iterations
Coefficient covariance computed using outer product of gradients

Variable	Coefficient	Std. Error	t-Statistic	Prob.
C	228.8820	201.5709	1.135491	0.2566
Y2*Y3*Y1(-1)	-0.026857	0.001691	-15.88365	0.0000
Y2*Y1(-1)	0.108676	0.012659	8.584610	0.0000
Y3*Y1(-1)	0.061990	0.003034	20.43225	0.0000
Y2	-6.445940	1.915310	-3.365482	0.0008
Y3	8.388418	1.645697	5.097182	0.0000
Y2(-1)	-2.513111	1.533862	-1.638420	0.1019
Y3(-1)	-0.141922	0.628817	-0.225697	0.8215
AR(1)	-0.682829	0.075149	-9.086274	0.0000
AR(2)	0.878453	0.026878	32.68262	0.0000
AR(3)	0.803880	0.080171	10.02710	0.0000
MA(1)	1.543759	0.084790	18.20675	0.0000
MA(2)	0.631761	0.091366	6.914580	0.0000
SIGMASQ	13.90569	0.516469	26.92455	0.0000

R-squared	0.999253	Mean dependent var	140.9656
Adjusted R-squared	0.999237	S.D. dependent var	136.5858
S.E. of regression	3.774001	Akaike info criterion	5.531683
Sum squared resid	8218.261	Schwarz criterion	5.635482
Log likelihood	-1620.612	Hannan-Quinn criter.	5.572117
F-statistic	59400.62	Durbin-Watson stat	1.850731
Prob(F-statistic)	0.000000		

Inverted AR Roots	1.00	-.84+.31i	-.84-.31i
Inverted MA Roots	-.77-.19i	-.77+.19i	

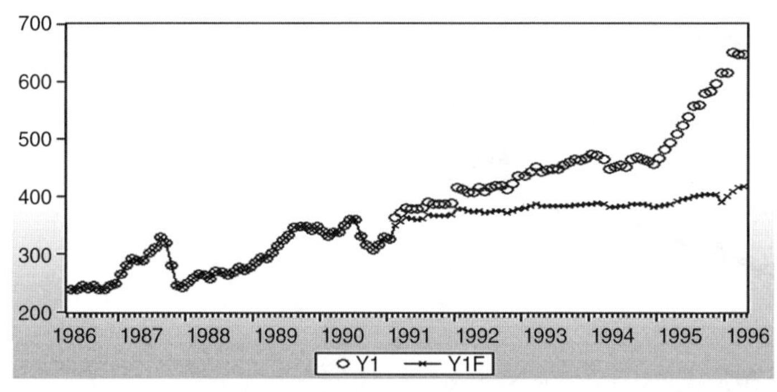

RMSE=99.61873 & TIC=0.117376

Figure 5.43 The LS estimate of the LVARMA(1,3,2) in (5.61), and its forecast evaluations.

Example 5.29 An Extension of the Model (5.61) with Two-Piece Quadratic-Trends

As a better fit model, an extension of the model (5.61) is presented by inserting additional two interactions *t*(@Year=<1994}* and *t^2*(@Year=<1994* as additional independent variables.

$$
\begin{aligned}
&Y1\,C\,Y2*Y3*Y1(-1)\,Y2*Y1(-1)\,Y3*Y1\\
&(-1)\,Y2\,Y3\,Y2(-1)\,Y3(-1)\,T\,T\hat{}2\\
&T*(@YEAR=<1994)\\
&T\hat{}2*(@YEAR=<1994\,AR(1)\,AR(2)\,AR(3)\,MA(1)\,MA(2)
\end{aligned}
\tag{5.62}
$$

Dependent Variable: Y1
Method: ARMA Maximum Likelihood (OPG - BHHH)
Date: 04/28/17 Time: 16:55
Sample: 1947M02 1996M04
Included observations: 591
Convergence achieved after 177 iterations
Coefficient covariance computed using outer product of gradients

Variable	Coefficient	Std. Error	t-Statistic	Prob.
C	47.85768	110.9599	0.431306	0.6664
Y2*Y3*Y1(-1)	-0.024784	0.001608	-15.41237	0.0000
Y2*Y1(-1)	0.085240	0.012284	6.939389	0.0000
Y3*Y1(-1)	0.057325	0.002610	21.96776	0.0000
Y2	-5.844752	1.734982	-3.368769	0.0008
Y3	7.770745	1.548348	5.018734	0.0000
Y2(-1)	-3.383901	1.353617	-2.499896	0.0127
Y3(-1)	-0.437431	0.596611	-0.733193	0.4637
T	-0.651958	0.447662	-1.456362	0.1458
T^2	0.002298	0.000522	4.406112	0.0000
AR(1)	-0.648105	0.074234	-8.730575	0.0000
AR(2)	0.853332	0.032019	26.65109	0.0000
AR(3)	0.780357	0.081495	9.575496	0.0000
MA(1)	1.474045	0.087296	16.88561	0.0000
MA(2)	0.576009	0.095184	6.051529	0.0000
SIGMASQ	12.49682	0.467468	26.73301	0.0000

R-squared	0.999329	Mean dependent var	140.9656
Adjusted R-squared	0.999311	S.D. dependent var	136.5858
S.E. of regression	3.583930	Akaike info criterion	5.425874
Sum squared resid	7385.620	Schwarz criterion	5.544502
Log likelihood	-1587.346	Hannan-Quinn criter.	5.472085
F-statistic	57090.12	Durbin-Watson stat	1.894558
Prob(F-statistic)	0.000000		

Inverted AR Roots	1.00	-.82-.33i	-.82+.33i
Inverted MA Roots	-.74-.18i	-.74+.18i	

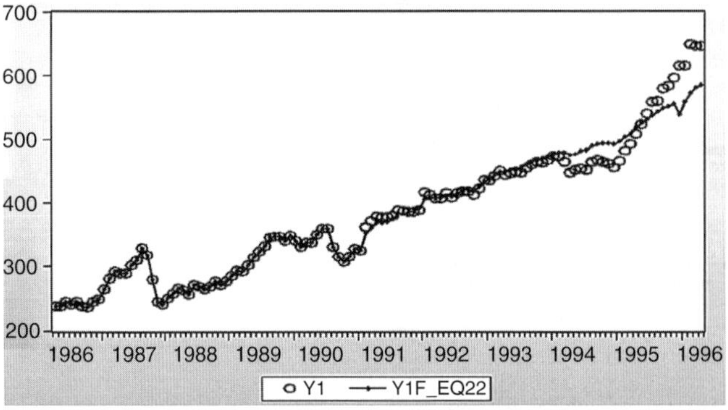

RMSE=24.46816 & TIC=0.026306

Figure 5.44 The LS estimate of the LVARMA(1,3,2) in (5.62) and its forecast evaluations.

where (@*Year*=<1994), is a dummy variable with (@*Year*=<1994) = 1 for the time period @*Year*≤1994, and (@*Year*=<1994) = 0, for the time period @*Year*>1994.

The results are presented in Figure 5.45, with its forecast evaluation for the sample {@*Year*>1990}. Based on these results, findings and notes are presented as follows

Dependent Variable: Y1
Method: Least Squares
Date: 04/29/17 Time: 20:00
Sample (adjusted): 1947M05 1996M04
Included observations: 588 after adjustments
Convergence achieved after 55 iterations
MA Backcast: 1947M03 1947M04

Variable	Coefficient	Std. Error	t-Statistic	Prob.
C	170.7271	51.74693	3.299269	0.0010
Y2*Y3*Y1(-1)	-0.022129	0.001951	-11.34221	0.0000
Y2*Y1(-1)	0.063978	0.018724	3.416933	0.0007
Y3*Y1(-1)	0.052105	0.003923	13.28137	0.0000
Y2	-6.232413	1.214194	-5.132961	0.0000
Y3	6.826332	1.538422	4.437231	0.0000
Y2(-1)	-4.050877	1.170511	-3.460777	0.0006
Y3(-1)	-0.336437	0.740909	-0.454087	0.6499
T	-6.783390	0.772649	-8.779390	0.0000
T^2	0.012458	0.001326	9.397983	0.0000
T*(@YEAR=<1994)	5.707133	0.794391	7.184289	0.0000
T^2*(@YEAR=<1994)	-0.009914	0.001376	-7.203559	0.0000
AR(1)	-0.627457	0.096608	-6.494896	0.0000
AR(2)	0.780567	0.041370	18.86782	0.0000
AR(3)	0.787808	0.097007	8.121145	0.0000
MA(1)	1.407623	0.122565	11.48468	0.0000
MA(2)	0.552193	0.121715	4.536770	0.0000

R-squared	0.999400	Mean dependent var	141.6074
Adjusted R-squared	0.999383	S.D. dependent var	136.6373
S.E. of regression	3.393442	Akaike info criterion	5.310052
Sum squared resid	6575.323	Schwarz criterion	5.436590
Log likelihood	-1544.155	Hannan-Quinn criter.	5.359355
F-statistic	59444.94	Durbin-Watson stat	1.929286
Prob(F-statistic)	0.000000		

Inverted AR Roots	.98	-.81-.39i	-.81+.39i
Inverted MA Roots	-.70-.24i	-.70+.24i	

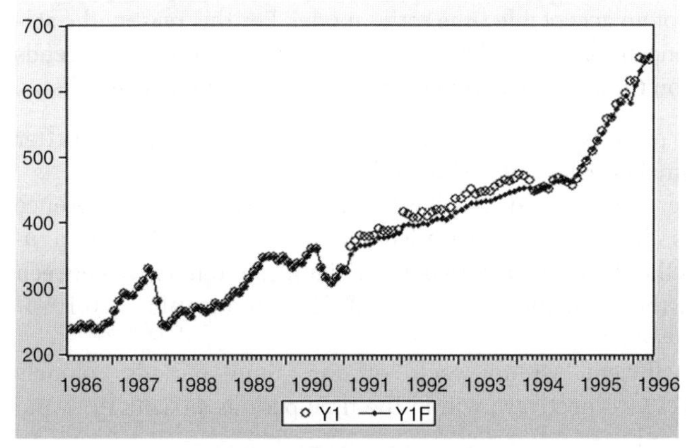

RMSE=12.92098 & TIC=0.013974

Figure 5.45 The LS estimate of the model (5.62) with its forecast evaluation.

1) This model is the best forecast model, compared to the models (5.61) and (5.62), since it has the smallest RMSE = 12.921 and TIC = 0.014.
2) In addition, the graphs of the in-sample forecast values and observed scores of *Y1* are very close.
3) On the other hand, compared to the results in Figures (5.41) and (5.42), these results do not present the independent variables SIGMASQ. Why?
4) The in-sample forecast values can easily be obtained. The forecast values of Y1 beyond the sample period can be computed using the following function, which is obtained as the output of the option *View/Representations* using Excel, supported by the forecasting for *Y2* and *Y3*. Refer to special notes presented in Example 5.24, point 4. Do this as an exercise – refer to Figures 4.30 and 4.32.

$$\begin{aligned} Y1 = {} & 170.727063938 - 0.0221288827275*Y2*Y3*Y1(-1) \\ & + 0.0639775543524*Y2*Y1(-1) + 0.0521045663176*Y3*Y1(-1) \\ & - 6.23241296051*Y2 + 6.82633228078*Y3 - 4.05087651345*\mathrm{Y2}(-1) \\ & - 0336437383382*Y3(-1) - 678339039191*T + 0.012458145644*T\hat{\ }2 \end{aligned}$$

Example 5.30 An Extension of the Model in Figure 5.41(a)
Figure 5.46 presents the statistical results based on a LVARMA(1,3,2) model with the following ES.

$$\begin{gathered} Y1\ C\ Y2*Y3*Y1(-1)\ Y2*Y1(-1)\ Y3*Y1 \\ (-1)\ Y2\ Y3\ Y2(-1)\ Y3(-1)\ Y1(-1)*X2(-1) \\ Y1(-1)*X1(-1)\ Y1(-1)*X3(-1)\ AR(1)\ AR(2)\ AR(2)\ MA(1)\ MA(2) \end{gathered} \tag{5.63}$$

Since one of the inverted AR roots = 1.00, with the statement "*Estimated AR process is non stationary,*" then the regression is not an acceptable time series model. For this reason, the statistical results of an acceptable extension of the model (5.65) with piece-wise quadratic trends are presented in Figure 5.45. Based on these results, findings and notes are presented as follows:

1) Compared to the model with the piece-wise quadratic trends in (5.63), this model is a better forecast model, because it has smaller values of RMSE and TIC.
2) At the 10% level of significance, the slopes of t^2 have a negative significant difference between the two time periods, based on the t-statistic of $t_0 = -1.613261$ with a p-value = 0.1073/2 = 0.05365. And the slopes of t have a positive significant difference between the two time periods, based on the t-statistic of $t_0 = -1.603261$ with a p-value = 0.1094/2 = 0.0547.
3) The t^2 has a positive coefficient within each of the time periods, namely (0.003713–0.002335) and 0.003713, respectively, within the time periods @*Year*≤1994 and @*Year*>1994.

5.10.4 Extension to Heterogeneous Regressions by Months

Referring to the heterogeneous regressions LVAR(p,q) models by *MONTH* presented in Chapter 2, Section 2.5. As their extensions, various heterogeneous regressions by *MONTH* of

Dependent Variable: Y1
Method: Least Squares
Date: 04/30/17 Time: 11:27
Sample (adjusted): 1947M05 1996M04
Included observations: 588 after adjustments
Convergence achieved after 21 iterations
MA Backcast: 1947M03 1947M04

Variable	Coefficient	Std. Error	t-Statistic	Prob.
C	5.236660	47.30819	0.110692	0.9119
Y2*Y3*Y1(-1)	-0.013672	0.001925	-7.102004	0.0000
Y2*Y1(-1)	-0.035593	0.018578	-1.915841	0.0559
Y3*Y1(-1)	0.006535	0.005430	1.203671	0.2292
Y2	-4.728477	1.089373	-4.340549	0.0000
Y3	8.263517	1.350592	6.118439	0.0000
Y2(-1)	-3.064561	1.048787	-2.922006	0.0036
Y3(-1)	-1.958223	0.762454	-2.568317	0.0105
Y1(-1)*X2(-1)	0.004746	0.000795	5.971770	0.0000
Y1(-1)*X1(-1)	0.000965	0.000899	1.072530	0.2839
Y1(-1)*X3(-1)	0.830428	0.219086	3.790419	0.0002
AR(1)	-0.528309	0.075257	-7.020051	0.0000
AR(2)	0.797314	0.046133	17.28307	0.0000
AR(3)	0.743862	0.082023	9.068902	0.0000
MA(1)	1.079190	0.098924	10.90933	0.0000
MA(2)	0.259288	0.099748	2.599431	0.0096

R-squared	0.999468	Mean dependent var	141.6074
Adjusted R-squared	0.999454	S.D. dependent var	136.6373
S.E. of regression	3.193166	Akaike info criterion	5.186737
Sum squared resid	5832.289	Schwarz criterion	5.305831
Log likelihood	-1508.901	Hannan-Quinn criter.	5.233140
F-statistic	71616.17	Durbin-Watson stat	2.013236
Prob(F-statistic)	0.000000		

Inverted AR Roots	1.00	-.77-.39i	-.77+.39i
	Estimated AR process is nonstationary		
Inverted MA Roots	-.36	-.72	

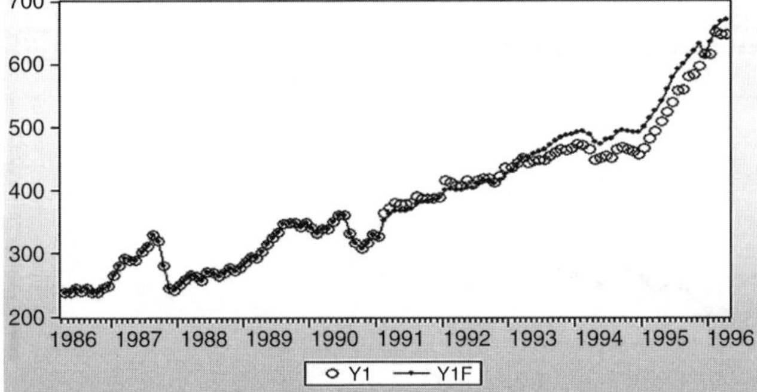

RMSE=21.00107 & TIC=0.022178

Figure 5.46 The statistical results based on the LVARMA(1,3,2) model in (5.63).

Dependent Variable: Y1
Method: Least Squares
Date: 04/30/17 Time: 15:50
Sample (adjusted): 1947M05 1996M04
Included observations: 588 after adjustments
Convergence achieved after 31 iterations
MA Backcast: 1947M03 1947M04

Variable	Coefficient	Std. Error	t-Statistic	Prob.
C	107.2627	27.79809	3.858636	0.0001
Y2*Y3*Y1(-1)	-0.014242	0.001937	-7.352414	0.0000
Y2*Y1(-1)	-0.026778	0.018718	-1.430586	0.1531
Y3*Y1(-1)	0.010893	0.005834	1.867015	0.0624
Y2	-5.117537	1.092671	-4.683511	0.0000
Y3	7.920484	1.367937	5.790096	0.0000
Y2(-1)	-3.418619	1.060248	-3.224357	0.0013
Y3(-1)	-1.785566	0.783480	-2.279019	0.0230
Y1(-1)*X2(-1)	0.004430	0.000815	5.433617	0.0000
Y1(-1)*X1(-1)	0.000566	0.000896	0.631877	0.5277
Y1(-1)*X3(-1)	0.809707	0.230549	3.512085	0.0005
T	-1.813701	0.882235	-2.055802	0.0403
T^2	0.003713	0.001538	2.414305	0.0161
T*(@YEAR=<1994)	1.339502	0.835486	1.603261	0.1094
T^2*(@YEAR=<1994)	-0.002335	0.001447	-1.613088	0.1073
AR(1)	-0.576315	0.076555	-7.528107	0.0000
AR(2)	0.733141	0.047532	15.42427	0.0000
AR(3)	0.772255	0.080755	9.562924	0.0000
MA(1)	1.139938	0.102449	11.12689	0.0000
MA(2)	0.342644	0.102859	3.331193	0.0009

R-squared	0.999483	Mean dependent var	141.6074
Adjusted R-squared	0.999466	S.D. dependent var	136.6373
S.E. of regression	3.157862	Akaike info criterion	5.171089
Sum squared resid	5664.147	Schwarz criterion	5.319957
Log likelihood	-1500.300	Hannan-Quinn criter.	5.229093
F-statistic	57811.23	Durbin-Watson stat	2.011632
Prob(F-statistic)	0.000000		

Inverted AR Roots	.98	-.78+.43i	-.78-.43i
Inverted MA Roots	-.57-.13i	-.57+.13i	

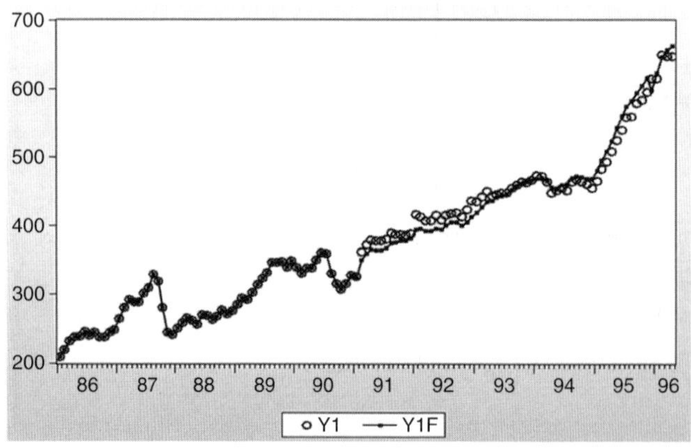

RMSE=12.82141 & TIC=0.013740

Figure 5.47 The statistical results of the model (5.63) with piece-wise quadratic trends.

Dependent Variable: Y1
Method: Least Squares
Date: 05/01/17 Time: 13:05
Sample: 1947M01 1996M04 IF M=2
Included observations: 49
Convergence achieved after 20 iterations

Variable	Coefficient	Std. Error	t-Statistic	Prob.
C	129.1833	29.54569	4.372322	0.0001
Y2*Y3*Y1(-1)	**-0.004218**	**0.004582**	**-0.920605**	**0.3637**
Y2*Y1(-1)	-0.183455	0.046996	-3.903681	0.0004
Y3*Y1(-1)	-0.056549	0.010936	5.170954	0.0000
Y2	-3.972840	2.757661	-1.440656	0.1588
Y3	-5.037554	3.336403	-1.509876	0.1403
Y2(-1)	-4.445710	1.949240	-2.280740	0.0290
Y3(-1)	-4.728537	1.421349	-3.326795	0.0021
T	-33.96624	11.15248	-3.045622	0.0045
T*2	0.060490	0.019395	3.118784	0.0037
T*(@YEAR=<1994)	33.76997	11.15964	3.026080	0.0047
T*2*(@YEAR=<1994)	-0.058667	0.019372	-3.028404	0.0047
AR(1)	1.456856	0.138364	10.52914	0.0000
AR(2)	-0.407256	0.170988	-2.381785	0.0230
AR(3)	-0.112656	0.102183	-1.102499	0.2780

R-squared	0.999928	Mean dependent var	146.2566
Adjusted R-squared	0.999898	S.D. dependent var	144.9028
S.E. of regression	1.464631	Akaike info criterion	3.847869
Sum squared resid	72.93490	Schwarz criterion	4.426998
Log likelihood	-79.27279	Hannan-Quinn criter.	4.067590
F-statistic	33556.67	Prob(F-statistic)	0.000000

Inverted AR Roots	.81-. 14i	.81+. 14i	-.17

Dependent Variable: Y1
Method: Least Squares
Date: 05/01/17 Time: 13.16
Sample: 1947Mo1 1996M04 IF M=2
Included observations: 49
Convergence achieved after 35 iterations

Variable	Coefficient	Std. Error	t-Statistic	Prob.
C	127.2414	29.52331	4.309864	0.0001
Y2*Y3*Y1(-1)	-0.010286	0.002299	-4.474502	0.0001
Y2*Y1(-1)	-0.119307	0.018884	-6.317987	0.0000
Y3*Y1(-1)	0.052425	0.010170	5.154678	0.0000
Y2	-5.383630	2.616418	-2.057634	0.0471
Y2(-1)	-3.647404	1.930432	-1.889424	0.0671
Y3(-1)	-6.135785	1.114879	-5.503545	0.0000
T	-30.72261	8.367642	-3.671597	0.0008
T^2	0.054801	0.014552	3.765983	0.0006
T*(@YEAR=<1994)	30.39642	8.362756	3.634736	0.0009
T^2*(@YEAR=<1994)	-0.052821	0.014513	-3.639581	0.0009
AR(1)	1.396158	0.128399	10.87363	0.0000
AR(2)	-0.378901	0.150834	-2.512040	0.0168
AR(3)	-0.080859	0.094609	-0.854666	0.3985

R-squared	0.999923	Mean dependent var	146.2566
Adjusted R-squared	0.999895	S.D. dependent var	144.9028
S.E. of regression	1.487523	Akaike info criterion	3.867058
Sum squared resid	77.44534	Schwarz criterion	4.407578
Log likelihood	-80.74291	Hannan-Quinn criter.	4.072130
F-statistic	35034.09	Prob(F-statistic)	0.000000

Inverted AR Roots	.77-04i	.77+.04i	-.14

Figure 5.48 The LS estimate of the model (5.64) for M = 2 and one of its reduced models.

each *Yk* can be done based on all multiple regression presented in this chapter, by using the following alternative methods. However, examples are presented only for a few selected models.

5.10.4.1 The Simplest Method

The data analysis based each multiple regression of *Yk*, including the LVARM*(p,q,r)* model (5.60a,b) can easily be conducted using each of the 12 subsamples of the months. The first step is to select a sample of the *@Month* = *M*, then run the model so that the general ES (5.60a,b) can be applied.

Example 5.31 An Application of the LVARMA(1,3,2) Model (5.63) for M = 2
By using the ES (5.63), an unexpected the error message on the right is obtained.

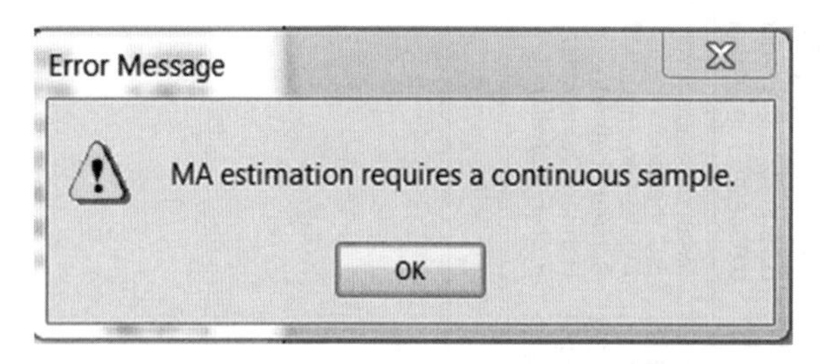

For this reason, I do analysis based on a model without the MA term, namely the LVARMA (1,3,0) = LVAR(1,3), with ES as follows:

$$\begin{aligned} &Y1\ C\ Y2*Y3*Y1(-1)\ Y2*Y1(-1)\ Y3*Y1(-1)\ Y2\ Y3\ Y2(-1)\ Y3(-1)\ T\ T\hat{}2 \\ &T^*(@YEAR=<1994)\ T\hat{}2*(@YEAR=<1994)\ AR(1)AR(2)AR(3) \end{aligned} \quad (5.64)$$

Figure 5.48 presents the LS estimates of the LVAR(1,3) model in (5.64) and one of its possible reduced models where the three-way interaction *Y2∗Y3∗Y1(−1)* has a significant effect, which is obtained by deleting *Y3* from the full model.

Example 5.32 An Application of a LVARMA(1,*q.r*) for M = 2
In order to apply a LVARMA(*1,q,r*) for $r > 0$ and $M = 2$, I have found that the data should be sorted by *M=@Month and Year=@Year,* which can be done by selecting *Proc/Sort Current Page* ... after the data GARCH.wf1 shown on the screen, then I save as Sort_GARCH.wf1. However, by using the ES (5.63) for $M = 2$, an error message *"Near singular matrix"* is obtained. Then by using the trial-and-error method to select the independent variables, as well as the integers q and r, I have obtained several regressions with a significant *Y2∗Y3∗Y1(−1)*; one of the results is LVARMA(1,1,2) model as presented in Figure 5.49. Based on these results, the findings and notes presented are as follows:

1) The results are obtained by deleting *t∗@Expand(M) and t^2∗@Expand(M),* and two of the AR terms, namely AR(2) and AR(3), from the LVARMA(1,3,2) model in (5.63) because they have large p-values.
2) At the 10% level of significance, each of the time variables *T* and *T^2* has significant negative and positive adjusted effects, respectively, with p-values of 0.1510/2 = 0.0755 and 0.0771/2 = 0.03855.
3) There is a problem with the interaction *Y2∗Y3∗Y1(−1),* which has an insignificant adjusted effect. There are three alternative options. First, keep the model as it is, because an acceptable

Dependent Variable: Y1
Method: Least Squares
Date: 05/02/17 Time: 15:23
Sample: 1 592 IF M=2
Included observations: 50
Convergence achieved after 35 iterations
MA Backcast: -1 0

Variable	Coefficient	Std. Error	t-Statistic	Prob.
C	4792.978	3840.323	1.248066	0.2198
Y2*Y3*Y1(-1)	0.005149	0.005818	0.885016	0.3819
Y2*Y1(-1)	-0.314424	0.074730	-4.207469	0.0002
Y3*Y1(-1)	0.033645	0.020181	1.667201	0.1039
Y2	7.620309	5.078233	1.500583	0.1419
Y3	-11.97730	5.048687	-2.372360	0.0230
Y2(-1)	-21.78424	3.510582	-6.205306	0.0000
Y3(-1)	3.335279	2.220521	1.502025	0.1416
T	-128.2474	87.45130	-1.466501	0.1510
T^2	0.903245	0.496680	1.818565	0.0771
AR(1)	0.873418	0.116297	7.510240	0.0000
MA(1)	0.114433	0.013901	8.231907	0.0000
MA(2)	0.980010	0.005351	183.1556	0.0000

R-squared	0.993581	Mean dependent var	143.6474
Adjusted R-squared	0.991499	S.D. dependent var	144.5984
S.E. of regression	13.33177	Akaike info criterion	8.237072
Sum squared resid	6576.240	Schwarz criterion	8.734198
Log likelihood	-192.9268	Hannan-Quinn criter.	8.426381
F-statistic	477.2755	Durbin-Watson stat	1.915787
Prob(F-statistic)	0.000000		

Inverted AR Roots	.87	
Inverted MA Roots	-.06+.99i	-.06-.99i

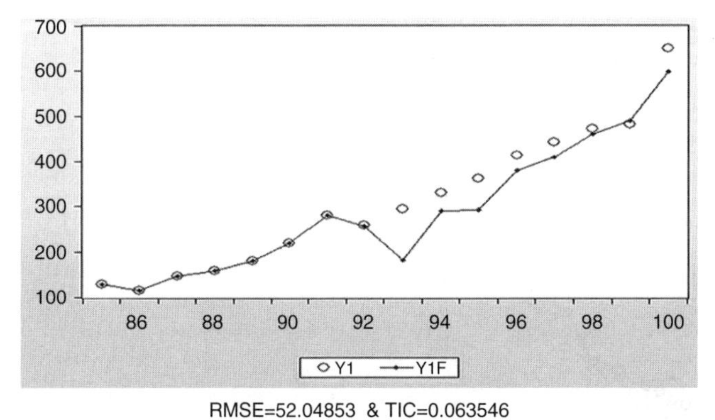

Figure 5.49 The statistical results of a LVARMA(1,1,2) model with its forecast evaluation for forecast sample {M = 2 and @*Year*>1990}.

model can have one or more insignificant adjusted effects. The second option is to delete one or two other independent variables to obtain a reduced model with the interaction *Y2*∗*Y3*∗*Y1 (−1)* having a significant effect. Refer to alternative models presented in the Example 5.24. The last option is to delete the interaction so we would have a two-way interaction LVARMA(1,1,2) model. Do these as exercises.

Dependent Variable: Y1
Method: Least Squares
Date: 05/03/17 Time: 14:09
Sample: 1 592 IF M<3
Included observations: 98
Convergence achieved after 76 iterations
MA Backcast: -1 0

Variable	Coefficient	Std. Error	t-Statistic	Prob.
C	2.575701	10.07534	0.255644	0.7989
Y2*Y3*Y1(-1)	0.031784	0.004035	7.877763	0.0000
Y2*Y1(-1)	-0.501070	0.029016	-17.26852	0.0000
Y3*Y1(-1)	-0.083100	0.017190	-4.834186	0.0000
Y2	-0.404469	2.542473	-0.159085	0.8740
Y3	-14.02724	1.970438	-7.118844	0.0000
Y1(-1)	2.029501	0.122021	16.63242	0.0000
Y2(-1)	10.23805	2.349946	4.356717	0.0000
Y3(-1)	2.846299	1.618938	1.758127	0.0825
T	2.292272	0.664163	3.451369	0.0009
T^2	0.063894	0.012544	5.093436	0.0000
T*(M=2)	-5.230971	0.684213	-7.645242	0.0000
T^2*(M=2)	-0.007290	0.009877	-0.738081	0.4626
AR(1)	0.525897	0.127825	4.114187	0.0001
MA(1)	-0.403277	0.125744	-3.207129	0.0019
MA(2)	-0.567339	0.128226	-4.424520	0.0000

R-squared	0.995297	Mean dependent var	144.9908
Adjusted R-squared	0.994436	S.D. dependent var	141.7413
S.E. of regression	10.57251	Akaike info criterion	7.702675
Sum squared resid	9165.801	Schwarz criterion	8.124710
Log likelihood	-361.4311	Hannan-Quinn criter.	7.873379
F-statistic	1156.829	Durbin-Watson stat	1.695770
Prob(F-statistic)	0.000000		

Inverted AR Roots	.55	
Inverted MA Roots	.98	-.58

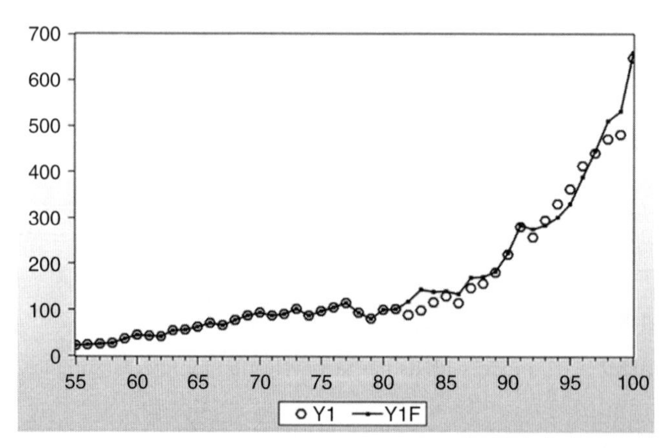

RMSE=24.472162 & TIC=0.038595

Figure 5.50 The statistical results of a LtVARMA(1,1,2) model with its forecast evaluation, for forecast sample {M < 3 and $t > 80$}.

Example 5.33 An Application of a LVARMA(1,1,2) for M < 3

Figure 5.50 presents the results of a LVARMA(1,1,2) model of *Y1* with heterogeneous trends using the following ES for the subsample {M < 3}. Based on these results, the following findings and notes are presented.

$$\begin{gathered} Y1\ C\ Y2*Y3*Y1(-1)\ Y2*Y1(-1)\ Y3*Y1(-1) \\ Y2\ Y3\ Y1(-1)\ Y2(-1)\ Y3(-1)\ T\ T\hat{}2 \\ T^{*}@EXPAND(M,@DROPFIRST)\ T\hat{}2*@EXPAND(M,@DROPFIRST) \\ AR(1)\ MA(1)\ MA(2) \end{gathered} \quad (5.65)$$

1) The model is a three-way interaction model for $M = 1$ and $M = 2$, with all IVs having the same slopes, except the time variables T and T^2.
2) The main objective of this model is to test the effects differences of the time variables T and T^2 between $M = 1$ and $M = 2$. The t-statistic of $t_0 = 7.272258$ with a p-value = 0.0000 of the IV (Independent Variable): *T*(M = 2)* indicate that T has a significantly greater effect on *Y1* for $M = 1$ than $M = 2$. But, at the 10% level, the IV: *T^2*(M = 2)* has insignificantly different effects, based on the t-statistic of $t_0 = 1.090412$ with a p-value = 0.2787.
3) By using the SE (5.65), the following LS regression, specific for $M = 1$, can easily be obtained or copied from the output of the *"Representations,"* which is the estimate of the first and the last lines of the ES (5.65).

$$\begin{aligned} Y1 = {} & 2575701 + 0.031784*Y2*Y3*Y1(-1) - 0.501070*Y2*Y1(-1) \\ & -0.083100*Y3*Y1(-1) \\ & -0.404469*Y2 - 14.027239*Y3 + 2.029501*Y1(-1) + 10.238048*Y2(-1) \\ & +2.846299*Y3(-1) + 2.292272*T + 0.0638937*T\hat{}2 \\ & +[AR(1) = 4.114187, MA(1) = -3.207129, MA(2) = -4.424520] \end{aligned}$$

The model uses an assumption that all of the other IVs have the same slopes within the two months.

5.10.4.2 Alternative Equation Specifications

The first alternative method is to do the analysis using the following general ES as an extension of the model (5.65) by using the function *@Expand(M)*, where *M = @Month* should be generated before doing the regression analysis. Compared to the simplest method, this method has a limitation since the terms AR(p) and MA(r) are valid for the whole 12 months. In other words, the function *AR(p)*@Expand(M)* cannot be applied. This model should be applied if and only if we want to test the effects differences of one or more numerical independent variables on *Yk* between the months.

$$\begin{gathered} Yk\ Yk(-1)...Yk(-p)\ X1...Xnef\ C \\ Yk(-1)*@Expand(M,@Dropfirst)...Yk(-p)*@Expand(M,@Dropfirst) \\ X1*@Expand(M,@Dropfirst)...Xn*@Expand(M,@Dropfirst \\ @Expand(M,@Dropfirst)\ AR(1)...AR(q)\ MA(1)...MA(r) \end{gathered} \quad (5.66a)$$

Dependent Variable: Y1
Method: Least Squares
Date: 05/04/17 Time: 06:57
Sample (adjusted): 1947M04 1996M04
Included observations: 589 after adjustments
Convergence achieved after 40 iterations
MA Backcast: 1947M03

Variable	Coefficient	Std. Error	t-Statistic	Prob.
Y1(-1)	1.011817	0.006120	165.3331	0.0000
Y2(-1)	0.212054	0.372594	0.569129	0.5695
Y3(-1)	0.166921	0.253459	0.658574	0.5104
C	-1.549247	0.518942	-2.985396	0.0030
Y1(-1)*(M=2)	0.025563	0.007723	3.309873	0.0010
Y1(-1)*(M=3)	-0.004915	0.008612	-0.570741	0.5684
Y1(-1)*(M=4)	-0.012687	0.008886	-1.427749	0.1539
Y1(-1)*(M=5)	0.007533	0.009787	0.769709	0.4418
Y1(-1)*(M=6)	0.007187	0.009750	0.737185	0.4613
Y1(-1)*(M=7)	0.004529	0.009609	0.471297	0.6376
Y1(-1)*(M=8)	-0.000499	0.009497	-0.052584	0.9581
Y1(-1)*(M=9)	-0.002207	0.009424	-0.234172	0.8149
Y1(-1)*(M=10)	-0.018944	0.009279	-2.041663	0.0417
Y1(-1)*(M=11)	-0.007918	0.009005	-0.879283	0.3796
Y1(-1)*(M=12)	0.009362	0.007915	1.182813	0.2374
Y2(-1)*(M=2)	0.170420	0.457103	0.372826	0.7094
Y2(-1)*(M=3)	0.050064	0.514575	0.097292	0.9225
Y2(-1)*(M=4)	0.086749	0.529402	0.163863	0.8699
Y2(-1)*(M=5)	0.170815	0.538091	0.317447	0.7510
Y2(-1)*(M=6)	0.022247	0.540990	0.041123	0.9672
Y2(-1)*(M=7)	0.334825	0.541792	0.617994	0.5368
Y2(-1)*(M=8)	0.103516	0.546499	0.189417	0.8498
Y2(-1)*(M=9)	0.290215	0.547399	0.530170	0.5962
Y2(-1)*(M=10)	0.244761	0.540176	0.453113	0.6506
Y2(-1)*(M=11)	0.369831	0.525500	0.703770	0.4819
Y2(-1)*(M=12)	0.239088	0.452036	0.528914	0.5971
Y3(-1)*(M=2)	-0.563499	0.326267	-1.727110	0.0847
Y3(-1)*(M=3)	-0.065798	0.366402	-0.179578	0.8576
Y3(-1)*(M=4)	0.034972	0.379338	0.092191	0.9266
Y3(-1)*(M=5)	-0.379395	0.391806	-0.968324	0.3333
Y3(-1)*(M=6)	-0.258734	0.392780	-0.658725	0.5103
Y3(-1)*(M=7)	-0.470966	0.392416	-1.200170	0.2306
Y3(-1)*(M=8)	-0.221062	0.391085	-0.565253	0.5721
Y3(-1)*(M=9)	-0.411744	0.391155	-1.052637	0.2930
Y3(-1)*(M=10)	-0.131114	0.385166	-0.340410	0.7337
Y3(-1)*(M=11)	-0.312504	0.373608	-0.836448	0.4033
Y3(-1)*(M=12)	-0.393073	0.327096	-1.201705	0.2300
AR(1)	1.241281	0.040974	30.29462	0.0000
AR(2)	-0.282143	0.041217	-6.845266	0.0000
MA(1)	-0.995898	0.002945	-338.1086	0.0000

R-squared	0.998667	Mean dependent var	141.3917
Adjusted R-squared	0.998573	S.D. dependent var	136.6213
S.E. of regression	5.161289	Akaike info criterion	6.185745
Sum squared resid	14624.76	Schwarz criterion	6.483092
Log likelihood	-1781.702	Hannan-Quinn criter.	6.301592
F-statistic	10550.04	Durbin-Watson stat	2.005282
Prob(F-statistic)	0.000000		

Inverted AR Roots	.94	.30
Inverted MA Roots	1.00	

Figure 5.51 The statistical results of the model (5.67) with its forecast evaluation for a forecast sample {@YEAR>1990}.

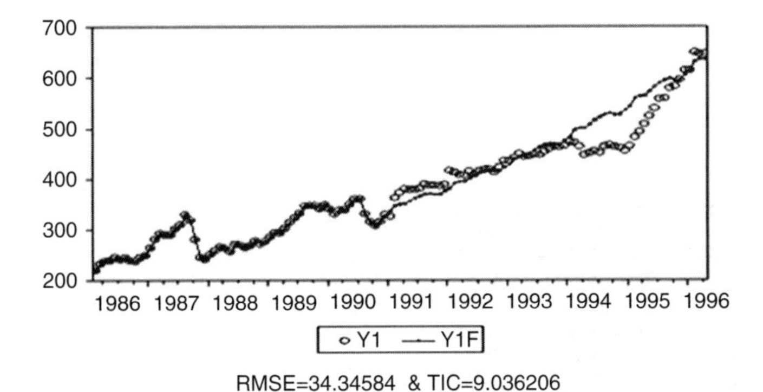

Figure 5.51 (Continued)

$$Yk\ Yk(-1)*@Expand(M)...Yk(-p)*@Expand(M)$$
$$X1^{*}@Expand(M)...Xn*@Expand(M)@Expand(M) \qquad (5.66b)$$
$$AR(1)...AR(q)\ MA(1)...MA(r)$$

where *Xn* can be a main exogenous, environmental and time variable, and two- or three-way interaction between the main variables. However, it is recommended to apply the ES (5.66a), since it has the following advantages.

a) The equation of multiple regression for $M = 1$ can be written directly based on the output of the *"Representations."*
b) The slopes difference of each of the numerical IVs between $M = 1$ and another level of *M* can be tested directly using the *t*-statistic presented in the statistical results.
c) Testing the hypotheses on the effects differences of two or more IVs on *Y1* between the 12 months is more easy to do by using the RVT.

Example 5.34 An Extension of the LV(1) Model of *Y1* in (5.41)

As an extension if the LV(1) model of *Y1* in (5.41), the results of a LVARMA(1,2,1) model of *Y1* by *M (Month)* with the following ES are presented in Figure 5.51. Based on these results, the following findings and notes are presented.

$$Y1\ Y1(-1)\ Y2(-1)\ Y3(-1)\ C\ Y1(-1)*@Expand(M,@Dropfirst)$$
$$Y2(-1)*@Expand(M,@Dropfirst)\ Y3(-1)*@Expand(M,@Dropfirst) \qquad (5.67)$$
$$AR(1)\ AR(2)\ MA(1)$$

1) The model is an additive model by *M*. Specific for $M = 1$, we have the following multiple regression, which can easily be copied from the output of the *"Representations."*

$$Y1 = 1.012*Y1(-1) + 0.212*Y2(-1) + 0.167*Y3(-1) - 1.549$$
$$[AR(1) = 1.241, AR(2) = -0.282, MA(1) = -0.996]$$

2) The results show that each of *Y2(−1)* and *Y3(−1)* has an insignificant adjusted effect on *Y1* with a large Prob. > 0.50. However, they have significant joint effects on *Y1*, based on the *F*-statistic of $F_0 = 20\,676.60$, with $df = (3549)$, and *p*-value = 0.0000.

Dependent Variable:Y1
Method: Least Squares
Date: 04/30/17 Time: 16:44
Sample (adjusted): 1947M05 1996M04
Included observations: 588 after adjustments
Convergence achieved after 18 iterations
MA Backcast: 1947M03 1947M04

Variable	Coefficient	Std. Error	t-Statistic	Prob.
Y2*Y3*Y1(-1)*(M=1)	3.54E-05	0.002813	0.012569	0.9900
Y2*Y3*Y1(-1)*(M=2)	0.001751	0.002879	0.608281	0.5433
Y2*Y3*Y1(-1)*(M=3)	-0.001505	0.003029	-0.496858	0.6195
Y2*Y3*Y1(-1)*(M=4)	-0.003637	0.003314	-1.097543	0.2730
Y2*Y3*Y1(-1)*(M=5)	-0.004396	0.003111	-1.413070	0.1583
Y2*Y3*Y1(-1)*(M=6)	-0.000776	0.002690	-0.288511	0.7731
Y2*Y3*Y1(-1)*(M=7)	-0.002654	0.002741	-0.968314	0.3334
Y2*Y3*Y1(-1)*(M=8)	-0.002726	0.003004	-0.907459	0.3646
Y2*Y3*Y1(-1)*(M=9)	-0.000954	0.002701	-0.353099	0.7242
Y2*Y3*Y1(-1)*(M=10)	-0.002112	0.002916	-0.724138	0.4694
Y2*Y3*Y1(-1)*(M=11)	0.000127	0.003090	0.041132	0.9672
Y2*Y3*Y1(-1)*(M=12)	-0.004842	0.003146	-1.539250	0.1244
*				
*				
*				
T^2*(M=1)	0.001170	0.000700	1.672055	0.0952
T^2*(M=2)	0.001014	0.000703	1.442046	0.1500
T^2*(M=3)	0.000963	0.000706	1.365163	0.1729
T^2*(M=4)	0.000794	0.000708	1.120428	0.2631
T^2*(M=5)	0.000911	0.000703	1.296443	0.1955
T^2*(M=6)	0.000806	0.000704	1.145695	0.2525
T^2*(M=7)	0.000830	0.000700	1.185429	0.2365
T^2*(M=8)	0.000981	0.000698	1.404410	0.1609
T^2*(M=9)	0.000873	0.000699	1.250216	0.2119
T^2*(M=10)	0.000963	0.000698	1.380350	0.1682
T^2*(M=11)	0.001029	0.000698	1.475063	0.1409
T^2*(M=12)	0.000884	0.000700	1.262759	0.2073
*				
*				
*				
M=12	-5.393787	249.9672	-0.021578	0.9828
AR(1)	-0.330100	0.103733	-3.182220	0.0016
AR(2)	0.977041	0.023026	42.43285	0.0000
AR(3)	0.342645	0.103643	3.306020	0.0010
MA(1)	0.784396	0.117306	6.686724	0.0000
MA(2)	–0.215495	0.112892	-1.908858	0.0569

R-squared	0.999746	Mean dependen var	141.6074
Adjusted R-squared	0.999669	S.D dependent var	136.6373
S.E of regression	2.485516	Akaike info criterion	4.859565
Sum squared resid	2786.184	Schwarz criterion	5.879312
Log likelihood	-1291 . 712	Hannan-Quinn criter.	5.256890
Durbin-Watson stst	2.012167		

Inverted AR Roots	1.00	-.35	-.97
Inverted MA Roots	.22	-1.00	

Figure 5.52 The LS estimate of the LVARMA(1,3,2) model in (5.68).

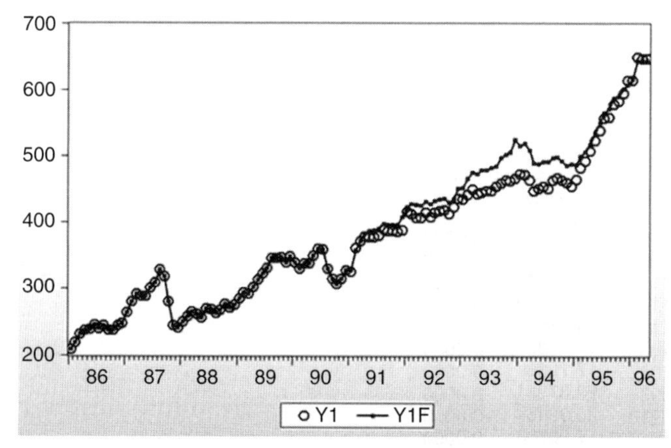

RMSE=24.47575 & TIC=0.025688

Figure 5.52 (Continued)

3) Based on the RVT above, it can be concluded that *Y1(−1)* has significantly different effects on *Y1* between the 12 months. However, the results present a WARNING, which shows a limitation of the RVT based on a model with a MA term.

Example 5.35 An Extension of the *Y1*-Regression in (5.67)

Figure 5.48 presents a part of the LS estimate of the following LVARMA(1,3,1) model by *Months* as an extension of the *Y1*-regression in the 3WISE_MLV(**1**) model (5.67).

$$\begin{gathered} Y1\ Y2*Y3*Y1(-1)*@Expand(M)\ Y2*Y1(-1)*@Expand(M)\ Y>3*Y1 \\ (-1)*@Expand(M) \\ Y2*@Expand(M)\ Y3*@Expand(M)\ Y1(-1)*@Expand(M)\ Y2 \\ (-1)*@Expand(M) \\ Y3(-1)*@Expand(M)\,T*@Expand(M)\,T\hat{}2*@\,expand(M) \\ @Expand(M)\,AR(1)\,AR(2)\,AR(3)\,MA(1)\,MA(2) \end{gathered} \tag{5.68}$$

Based on these results, the following findings and notes are presented.

1) By doing LS regression analysis based on the model without the time variables *T* and *T^2*, the results present one of the inverted AR roots = 1.00 with the note *"Estimated AR proses is nonstationary."* However, the output of the analysis is not presented.
2) The results in Figure 5.52 also show an inverted AR root = 1.00, but without the note. However, each of the three-way interactions has an insignificant effect on *Y1* at the 10% level, which are the unexpected impacts of the multicollinearity. I am very sure that by deleting one or more of the two-way interactions or the main variables, a reduced model with three-way interactions having significant effects would be obtained.
3) By using the trial-and-error method, I obtain an intended reduced model by deleting the interaction *Y2*Y1(−1)*@Expand(M)*, with the results presented in Figure 5.53, which shows each of the three-way interaction has a significant adjusted effect on *Y1*.
4) In fact, I have found alternative intended reduced models also can be obtained by deleting *Y1 (−1)*@Expand(M)* or both *Y2*@Expand(M)* and *Y3*@Expand(M)*. Do this as an exercise.

Dependent Variable: Y1
Method: Least Squares
Date: 04/30/17 Time: 20:33
Sample (adjusted): 1947M05 1996M04
Included observations: 588 after adjustments
Convergence achieved after 29 iterations
MA Backcast: 1947M03 1947M04

Variable	Coefficient	Std. Error	t-Statistic	Prob.
Y2*Y3*Y1(-1)*(M=1)	-0.018002	0.001168	-15.41885	0.0000
Y2*Y3*Y1(-1)*(M=2)	-0.018662	0.001187	-15.72666	0.0000
Y2*Y3*Y1(-1)*(M=3)	-0.017425	0.001183	-14.73040	0.0000
Y2*Y3*Y1(-1)*(M=4)	-0.019440	0.001180	-16.47042	0.0000
Y2*Y3*Y1(-1)*(M=5)	-0.017519	0.001191	-14.71553	0.0000
Y2*Y3*Y1(-1)*(M=6)	-0.018007	0.001038	-17.35267	0.0000
Y2*Y3*Y1(-1)*(M=7)	-0.018560	0.001047	-17.72867	0.0000
Y2*Y3*Y1(-1)*(M=8)	-0.020936	0.000936	-22.37911	0.0000
Y2*Y3*Y1(-1)*(M=9)	-0.019255	0.000918	-20.96464	0.0000
Y2*Y3*Y1(-1)*(M=10)	-0.018508	0.001072	-17.26784	0.0000
Y2*Y3*Y1(-1)*(M=11)	-0.019263	0.001144	-16.83435	0.0000
Y2*Y3*Y1(-1)*(M=12)	-0.016626	0.001153	-14.41600	0.0000
Y3*Y1(-1)*(M=1)	0.005450	0.007367	0.739712	0.4598
Y3*Y1(-1)*(M=2)	0.003001	0.007558	0.397042	0.6915
Y3*Y1(-1)*(M=3)	0.005635	0.007878	0.715268	0.4748
Y3*Y1(-1)*(M=4)	0.025026	0.007264	3.445041	0.0006
Y3*Y1(-1)*(M=5)	0.015126	0.007474	2.023820	0.0436
Y3*Y1(-1)*(M=6)	0.019335	0.006305	3.066474	0.0023
Y3*Y1(-1)*(M=7)	0.019930	0.006100	3.267147	0.0012
Y3*Y1(-1)*(M=8)	0.029420	0.005786	5.085100	0.0000
Y3*Y1(-1)*(M=9)	0.014298	0.005081	2.814174	0.0051
Y3*Y1(-1)*(M=10)	0.013712	0.005481	2.501643	0.0127
Y3*Y1(-1)*(M=11)	0.012744	0.006634	1.920946	0.0554
Y3*Y1(-1)*(M=12)	-0.012653	0.006406	-1.975229	0.0488
T^2*(M=1)	0.002591	0.000634	4.084874	0.0001
T^2*(M=2)	0.002444	0.000638	3.828654	0.0001
T^2*(M=3)	0.002491	0.000640	3.890901	0.0001
T^2*(M=4)	0.002366	0.000643	3.681454	0.0003
T^2*(M=5)	0.002458	0.000635	3.869851	0.0001
T^2*(M=6)	0.002457	0.000635	3.869029	0.0001
T^2*(M=7)	0.002355	0.000633	3.721251	0.0002
T^2*(M=8)	0.002457	0.000631	3.864052	0.0001
T^2*(M=9)	0.002370	0.000632	3.747967	0.0002
T^2*(M=10)	0.002443	0.000631	3.870681	0.0001
T^2*(M=11)	0.002478	0.000632	3.922955	0.0001
T^2*(M=12)	0.002316	0.000634	3.654934	0.0003
*				
*				
M=12	115. 1933	101. 5513	1.134336	0.2572
AR(1)	1. 598244	0.099420	16.07564	0.0000
AR(2)	-0.356022	0.176892	-2.012657	0.0447
AR(3)	-0.244378	0.095050	-2.571043	0.0105
MA(1)	-1. 121777	0.102132	-10.98364	0.0000
MA(2)	0.298105	0.094478	3.155286	0.0017

R-squared	0.999729	Mean dependent var	141.6074
Adjusted R-squared	0.999656	S.D. dependent var	136.6373
S.E. of regression	2.533364	Akaike info criterion	4.883143
Sum squared resid	2971.503	Schwarz criterion	5.813570
Log likelihood	-1310.644	Hannan-Quinn criter.	5.245667
Durbin-Watson stat	1.995084		

Inverted AR Roots	.98	.89	-.28
Inverted MR Roots	.69	.43	

Figure 5.53 The LS estimate of a reduced model of the LVARMA(1,3,2) model in (5.68).

5.11 Special Notes and Comments

Referring to a lot of possible models, specifically the alternative models and special notes presented in the Example 5.24 as well as to be presented in the following chapters, it is important to present the following notes and comments.

1) About the true population model, I would say that researchers never know which one of all possible models would be the true population model with perfect forecast values beyond the sample period, since the results are highly dependent on the data used, which is defined as a set of scores of variables that happen to be selected by or available to researchers, as stated in Agung (2011a). And the estimates of the model parameters could be unexpected because of the multicollinearity between the independent variables.
2) Talking about data analysis, Tukey (1962, in Gifi 1991, p. 22) presented the statements on the statistical results of data analysis as follows:

   1) In data analysis we must look to every emphasis on judgment. At least three different sorts or sources in almost every instance:
   (a1) judgment based upon the experience ofa particular field of subject matter from which the data come
   (a2) judgment based upon a broad experience of how particular techniques of data analysis work out in a variety of fields of applications.
   (a3) judgment based on abstract results about the properties of particular techniques, whether obtained by mathematical proof of empirical sampling.
   2) Often in statistics one in using a parametric model, such as the common model of normality distributed errors, or that of exponential distributed dobservations. Classical (parametric) statistics derives results unde the assumption that theses models are strictly true. However, apart from some simple discrete models perhaps, such model never exactly true (Hample 1973, quoted by Gifi, 1990, p. 27)
   3) The mathematical foundation of statistical analysis are more algebraic and geometric (and if many dimension are involved geometric ideas merge with algebraic calculations) then probabilistic; it is better speak of the aveage and principle axes, etc,... which are defined interma of a finite number of theactual data points, than to speak of expected value, etc,... defined in a potentially infinite universe. But probabilistic concepts and ideas can suggest algebraic operations and sometimes can be used to evaluate their usefulness (Benzecri, 1973, quoted by Gifi, p.26)

3) Hampel (1973, in Gifi 1991, p. 27) presented notes on statistical models as follows:

   Classical (parametric) statistics derives results under the assumption these models are strictly true. However apart from simple discrete models perhaps, such models are never exactly true. After all a statistical model has to be simple (where "simple," of course as a relative meaning, depending on state and standards of the subject matter).

4) Bezzecri (1973, in Gifi 1991, p. 25), presented notes on the relationship between the data set and the model as follows:

   The model must follow the data, and not the other way around. This is another error in the application of mathematics to the human sciences: to an abundance of models, which are built a priori and then confronted with the data by what one calls a "test." Often the "test" is used to justify a model in which the parameters to be fitted are larger than the number of the data points often is used, on the contrary, to reject as invalid the most judicial remarks of the experimenter. But what we need is a rigorous method to extract structure, starting from the data.

5.11 Special Notes and Comments

6

Forecasting Quarterly Time Series

6.1 Introduction

I have found that all forecast models for the monthly time series, presented in the five previous chapters, can be used for the quarterly time series by replacing the time variable *@Month* = *M* with *@Quarter* = *Q* for all relevant models. For this reason, this chapter only presents selected illustrative examples. Referring to all possible models based on a single time series Y_t, bivariate time series (X_t,Y_t) or $(Y1_t,Y2_t)$, and triple time series $(X1_t,X2_t,Y_t)$ or $(Y1_t,Y2_t,Y3_t)$, which have been presented in Chapters 1–5 with a summary of alternative trends presented in Table 4.1, then based on a quarterly time series we have the summary of alternative trends as presented in Table 6.1.

Illustrative examples of the forecasting are presented based on CONS1.wf1, which contains five quarterly time series *GC* = Personal Consumption Expenditures, *GCDAN* = Personal Consumption Expenditures, New Autos, *GWY* = National Income Wages and Salaries, *GYD* = Personal Income: Disposable personal Income, and *R* = Interest Rate.

6.2 Alternative LVARMA(*p*,*q*,*r*) of a Single Time Series

This section presents examples of statistical results based on alternative LVARMA(*p*,*q*,*r*) models of single time series in CONS1.wf1, specifically $GCDAN_t$ for some acceptable *p*, *q*, and *r* that are selected using the trial-and-error method. Even though we are using a single quarterly time series, we also have the time variables *@Trend*, *@Year*, and *@Quarter* that can be used as additional independent variables, either numerical or categorical. See the following examples.

Various forecast models could be presented, similar to the models presented in previous chapters. In this chapter, I only present selected forecast models of the time series *GCDAN*. Note that, even though the growth pattern of *GCDAN* is not smooth as presented in Figure 6.1a, *GCDAN (−1)* can be considered as one of the best predictors of *GCDAN*, as presented in Figure 6.1b. For this reason, the variable *GCDAN(−1)* should be used as an independent variable for any forecast models of *GCDAN*, similar to all those presented based on the monthly time series.

6.2.1 LV(P) Forecast Model of $GCDAN_t$

Example 6.1 An Acceptable LV(p) Model of $GCDAN_t$

Figure 6.2 presents the statistical results of an LV(2) = LVARMA(2,0,0) model of $GCDAN_t$, the graph of the observed scores together with its forecast values *GCDANF*, and two statistics of its forecast evaluation, using the following equation specification (ES). Based on these results, the following findings and notes are presented.

Advanced Time Series Data Analysis: Forecasting Using EViews, First Edition. I Gusti Ngurah Agung.

Table 6.1 Forecast models with specific trends for all models based on a single quarterly time series, bivariate, and triple time series.

Continuous Regressions with Trend	
Additional Time IV to insert	**Forecast Model (FM) with**
1. *@Trend*	Linear Trend
2. *@Trend ^ 2 @trend*	Quadratic Trend
3. *@Trend ^ 3 @Trend ^ 2 @Trend*	Cubic Trend
4. *log(@Trend + 1) = log(t)*	Logarithmic Trend
5. *Exp(rt)*	Exponential Trend, for a fixed selected *r*
Regressions with Heterogeneous Trends by @Year	
6. [a]*Q*@Expand(@Year) @Expand(@Year,@Droplast)*	
7. [a]*Q*@Expand(@Year)* [a]*Q ^ 2*@Expand(@Year) @Expand(@Year,@Droplast)*	
8. [a]*Q*@Expand(@Year)* [a]*Q ^ 2*@Expand(@Year) Q ^ 3*[a]*@Expand(@Year) @Expand(@Year,@Droplast)*	
Regressions with Heterogeneous Trends by TP([b])	
9. *t*@Expand(TP) @Expand(@TP,@Droplast)*	
10. *t*@Expand(TP) t ^ 2*[a]*@Expand(TP) @Expand(TP,@Droplast)*	
11. *t*@Expand(TP) t ^ 2*[a]*@Expand(TP) t ^ 3*[a]*@Expand(TP) @Expand(TP,@Droplast)*	

[a] Q = *@Quarter.*
[b] Replace *@Year* in 6–7 by a defined *TP* (time period).

$$GCDAN\ GCDAN(-1)\ GCDAN(-2)\ C \tag{6.1}$$

1) LV(2) model of quarterly time series can be considered to be the simplest additive forecast model of *GCDAN*, since by using a LV(3) model, *GCDAN(−3)* has an insignificant adjusted effect on *GCDAN*, with Prob. = 0.2274.
2) Based on the DW-statistic value, it can be concluded that the model does not have an autocorrelation problem.

6.2.2 LVARMA(*p,q.r*) Forecast Models of GCDN

This section presents alternative LVARMA(*p,q,r*) models of *GCDAN*, where the integers $p \geq 0$, $q \geq 0$, and $r \geq 0$ should be selected using the trial-and-error method in order to obtain the best possible model.

Example 6.2 Four Alternative LVARMA(*p,q,r*) Models of *GCDAN*
Figure 6.3 presents the summary of the statistical results of four models, with their forecast evaluations; namely the LVARMA(2,2,2), LVARMA(1,3,2), LVARMA(1,2,3), and a reduced model of LVARMA(0,3,3) = ARMA(3,3) or ARIMA(3,3) models, with the following ESs, respectively. Based on these results, the findings and notes presented are as follows:

$$GCDAN(-1)\ GCDAN(-2)\ C\ AR(1)\ AR(2)\ MA(1)\ MA(2) \tag{6.2}$$

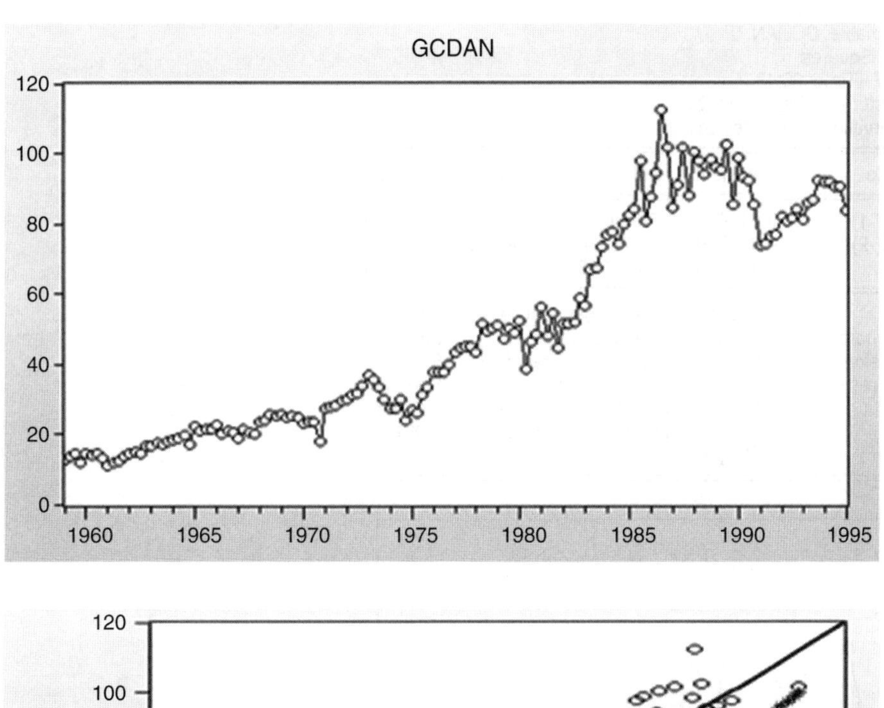

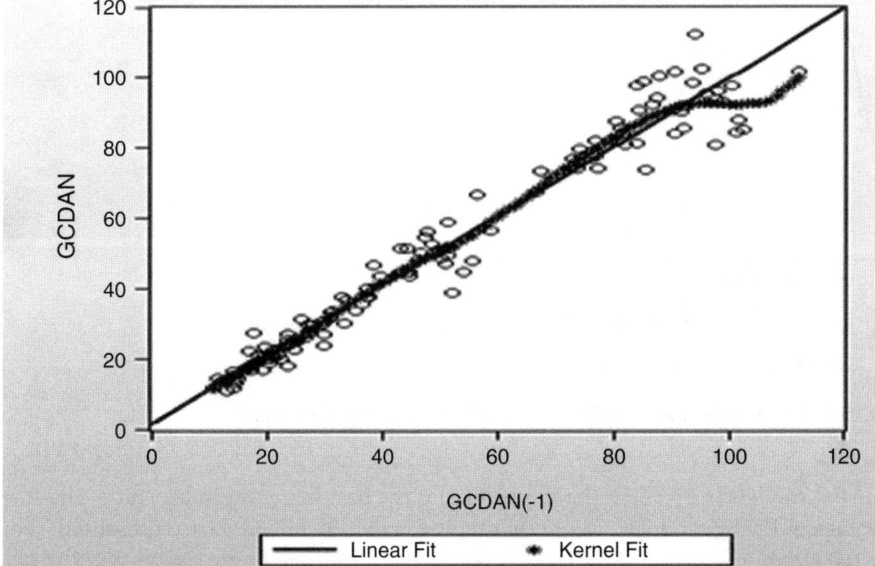

Figure 6.1 The growth pattern of *GCDAN* over time, and the scatter graph of *GCDAN* on *GCDAN(−1)*, with regression line and kernel fit curve.

$$GCDAN\ GCDAN(-1)\ C\ AR(1)\ AR(2)\ MA(1)\ MA(2) \tag{6.3}$$

$$GCDAN\ GCDAN(-1)\ C\ AR(1)\ AR(2)\ MA(1)\ MA(2)\ MA(3) \tag{6.4}$$

$$GCDAN\ C\ AR(2)\ AR(3)\ MA(1)\ MA(2)\ MA(3) \tag{6.5}$$

1) Unexpectedly, the forecast models (6.2) and (6.3) present exactly the same values of RMSE (Root Mean Squared Error) and TIC (Theil Inequality Coefficient) in three decimal points.

Dependent Variable: GCDAN
Method: Least Squares
Date: 05/06/17 Time: 05:48
Sample (adjusted): 1959Q3 1995Q1
Included observations: 143 after adjustments

Variable	Coefficient	Std. Error	t-Statistic	Prob.
GCDAN(-1)	0.618661	0.079011	7.830048	0.0000
GCDAN(-2)	0.371716	0.079178	4.694709	0.0000
C	1.162005	0.803648	1.445913	0.1504

R-squared	0.972212	Mean dependent var	49.19538
Adjusted R-squared	0.971815	S.D. dependent var	29.74958
S.E. of regression	4.994464	Akaike info criterion	6.075293
Sum squared resid	3492.255	Schwarz criterion	6.137451
Log likelihood	-431.3835	Hannan-Quinn criter.	6.100551
F-statistic	2449.082	Durbin-Watson stat	2.061007
Prob(F-statistic)	0.000000		

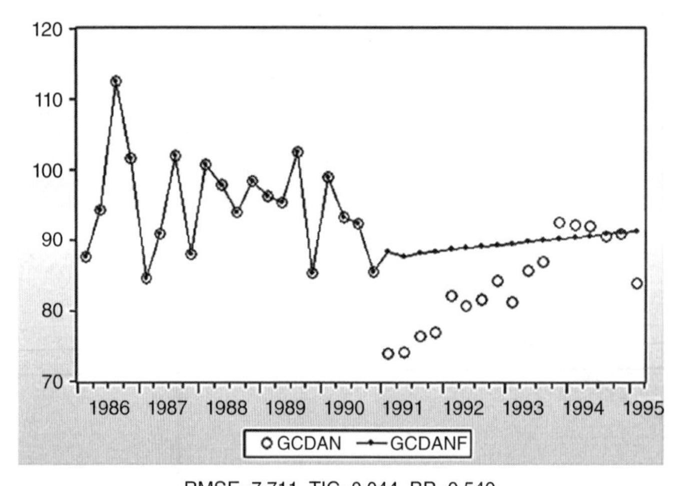

RMSE=7.711, TIC=0.044, BP=0.540

Figure 6.2 The LS estimate of the LV(2) model in (6.1), with its forecast evaluation.

2) Among the five models (6.1)–(6.5), the model (6.2) is the best forecast model, since it has the smallest values of RMSE, and TIC. Note that typing errors of TIC=0.39 are presented, they should be 0.039.

6.2.3 Forecast Models of GCDAN with Time Variables

As the extension of the model (6.1), this section only presents examples of selected LV(2) models of *GCDAN* with alternative trends, as follows:

Example 6.3 A LV(2) Model of GCDAN with a Linear Trend

I would say that a forecast model with *@Trend* or linear trend is always acceptable, even though the growth pattern of the dependent variable is non-linear over time, since the trend indicates that either the dependent variable tends to increase or decrease over time. However, a good forecast model is highly dependent on the growth pattern of the dependent variable. As an

	Model (6.2)			Model (6.3)			Model (6.4)			Model (6.5)		
Variable	Coef.	t-Stat.	Prob.	Coef.	t-Stat.	Prob.	Coef.	t-Stat.	Prob.	Coef.	t-Stat.	Prob.
GCDAN(-1)	0.621	7.453	0.000	0.994	105.326	0.000	0.995	109.00	0.000			
GCDAN(-2)	0.370	4.420	0.000									
C	1.130	1.534	0.127	0.823	1.529	0.129	0.749	1.450	0.149	176.28	0.620	0.536
AR(1)	0.815	30.53	0.000	0.443	5.255	0.000	-1.275	-32.58	0.000			
AR(2)	-0.969	-37.80	0.000	-0.666	-9.554	0.000	-0.928	-25.50	0.000	0.227	1.753	0.082
AR(3)				-0.361	-4.328	0.000				0.762	5.897	0.000
MA(1)	-0.899	-34.83	0.000	-0.899	-34.860	0.000	0.874	9.744	0.000	0.588	7.191	0.000
MA(2)	0.971	57.742	0.000	0.971	57.816	0.000	0.529	4.959	0.000	0.443	3.351	0.001
MA(3)							0.085	-4.055	0.000	0.094	-3.769	0.000
R-squared	0.974			0.974			0.974			0.973		
Adj.R-squared	0.973			0.973			0.973			0.972		
F-statistic	853.37			853.37			843.70			985.25		
Prob(F-stat.)	0.000			0.000			0.000			0.000		
DW-stat.	1.988			1.988			1.943			1.982		
Inv. AR Roots	.41+.90i	.41-.90i		.41+.90i	.41-.90i	-0.370	-.64-.72i	-.64+.72i		1.000	-.50-.72i	-.50+.72i
Inv. MA Roots	.45+.88i	.45-.88i		.45+.88i	.45-.88i		0.360	-.61+.77i	-.61-.77i	0.410	-.50-.78i	-.50+.78i
	Rmse=6.662 & TIC=0.39			Rmse=6.662 & TIC=0.39			Rmse=7.611 & TiC=0.044			Rmse=8.673 & TIC=0.050		

Figure 6.3 Statistical results summary of the four models (6.2) up to (6.5).

Dependent Variable: GCDAN
Method: Least Squares
Date: 05/06/17 Time: 20:13
Sample (adjusted): 1959Q3 1995Q1
Included observations: 143 after adjustments

Variable	Coefficient	Std. Error	t-Statistic	Prob.
GCDAN(-1)	0.575577	0.080721	7.130466	0.0000
GCDAN(-2)	0.332877	0.080390	4.140757	0.0001
C	0.590655	0.839408	0.703656	0.4828
@TREND	0.062196	0.029631	2.098998	0.0376

R-squared	0.973066	Mean dependent var	49.19538
Adjusted R-squared	0.972485	S.D. dependent var	29.74958
S.E. of regression	4.934800	Akaike info criterion	6.058075
Sum squared resid	3384.964	Schwarz criterion	6.140952
Log likelihood	-429.1524	Hannan-Quinn criter.	6.091752
F-statistic	1673.909	Durbin-Watson stat	2.033273
Prob(F-statistic)	0.000000		

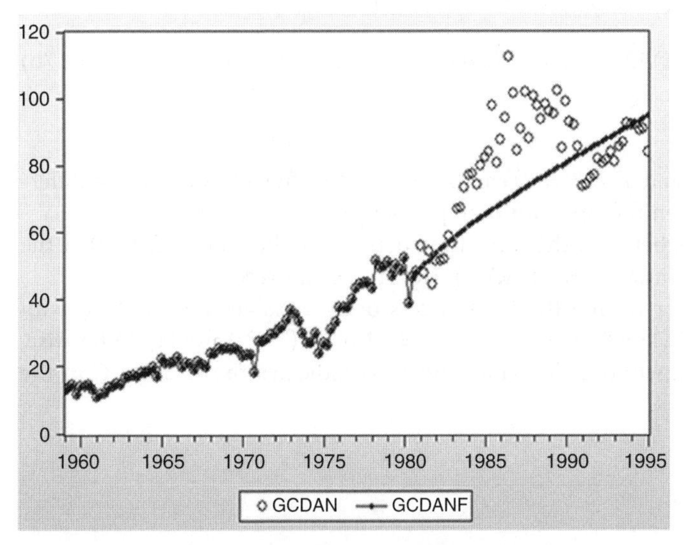

RMSE=14782, TIC=0.093, BP=0.268

Figure 6.4 The LS estimate of the LV(2) model in (6.6), with its forecast evaluation.

illustration, Figure 6.4 presents the statistical results of a model with the following ES, which is an extension of the model (6.1).

$$GCDAN\ GCDAN(-1)\ GCDAN(-2)\ C\ @Trend \tag{6.6}$$

1) Since the growth pattern of *GCDAN* is non-linear, then we can directly say that a model with a linear trend of *GCDAN* can never be a good forecast model, even without doing a regression.
2) The graph in Figure 6.4 clearly shows that the model (6.6) is not a good forecast model of *GCDAN*, even though each of the dependent variables has a significant adjusted effect on *GCDAN*. For a better forecast model, see Example 6.4.

Example 6.4 A LV(2) Model of GCDAN with Trend by Time Period
By observing the nonlinear growth pattern of *GCDAN* over time, I present a model with *@Trend* by a *Time Period* (TP). It is found that the scores reach a maximum at the time $T = 111$, and a relative minimum at $T = 129$ for the *@Year* > 1980. For this reason, I generate a variable *TP* of three levels using the following equation:

$$TP = 1 + 1(T > 111) + 1*(T> = 129).$$

Therefore, I apply either one of the following equations, which are in fact the same models. The first model has the intercept variable *C* and the function *@Droplast* can be replaced by *@Drop(3)*, *@Dropfirst* = *@Drop(2)*, or *@Drop(1)*. Its LS estimate has advantages for testing hypotheses using the *t*-statistic and the *F*-statistic, as presented in Figure 6.5. Based on these results, findings and notes presented are as follows:

$$\begin{array}{c} GCDAN\ GCDAN(-1)\ GCDAN(-2)\ T\ C \\ T*@Expand(TP,@droplast)\ @Expand(TP,@DropLast) \end{array} \tag{6.7a}$$

$$GCDAN\ GCDAN(-1)\ GCDAN(-2)\ T*@Expand(TP)\ @Expand(TP) \tag{6.7b}$$

1) The *t*-statistic of *T*(TP = 1)* and *T*(TP = 2)* can be used to test the adjusted effects differences of *T* on *GCDAN* between the first two time periods and the third *TP*.
2) Testing the effects differences between the three time periods can be done using either the Wald test or the Redundant Variables Test (RVT). Do this as an exercise.
3) The *F*-statistic can be used for testing the joint effects of all variables on *GCDAN*, with the statistical hypothesis H_0: C(1) = C(2) = C(3) = C(5) = C(6) = C(7) = 0 and H_1: Otherwise. Note that the equation of H_0 without the parameter C(4) indicates the variable *C* in the output.

Dependent Variable: GCDAN
Method: Least Squares
Date: 05/09/17 Time: 06:33
Sample (adjusted): 1959Q3 1995Q1
Included observations: 143 after adjustments

Variable	Coefficient	Std. Error	t-Statistic	Prob.
GCDAN(-1)	0.527327	0.083376	6.324688	0.0000
GCDAN(-2)	0.338170	0.083460	4.051878	0.0001
T	0.281756	0.248648	1.133153	0.2592
C	-27.44552	33.38145	-0.822179	0.4124
T*(TP=1)	-0.178062	0.243971	-0.729848	0.4667
T*(TP=2)	-0.609599	0.336513	-1.811518	0.0723
TP=1	27.31358	33.40221	0.817718	0.4150
TP=2	79.13269	43.74742	1.808854	0.0727

R-squared	0.974400	Mean dependent var	49.19538
Adjusted R-squared	0.973073	S.D. dependent var	29.74958
S.E. of regression	4.881782	Akaike info criterion	6.063216
Sum squared resid	3217.292	Schwarz criterion	6.228969
Log likelihood	-425.5199	Hannan-Quinn criter.	6.130570
F-statistic	734.0619	Durbin-Watson stat	2.006736
Prob(F-statistic)	0.000000		

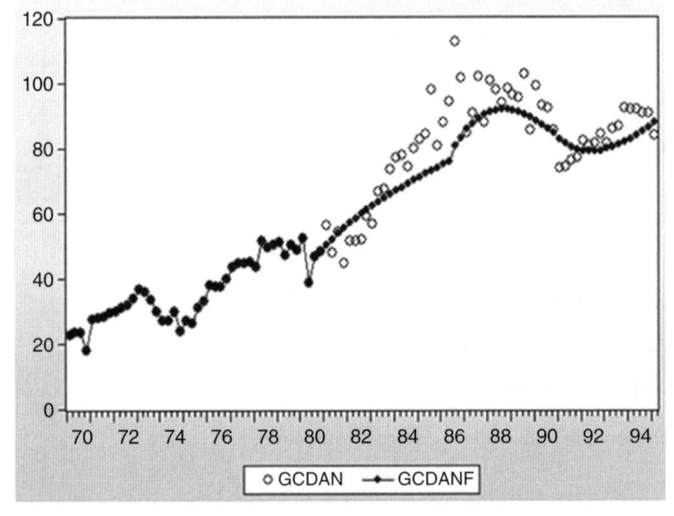

RMSE=9.263, TIC=0.057, BP=0.250

Figure 6.5 The LS estimate of the LV(2) model in (6.7a), with its forecast evaluation.

Example 6.5 LV(2) Model of *GCDAN* with Cubic Trend

As a good comparison with the model (6.7a) and (6.7b) with three-piece linear trends, a LV(2) model of *GCDAN* with the following ES is a model with a third degree polynomial trend and its statistical results are presented in Figure 6.6. Based on these results, the following findings and notes are presented.

$$GCDAN\ \ GCDAN(-1)\ GCDAN(-2)\ T\ \ T\hat{}2\ \ T\hat{}3\ \ C \qquad (6.8)$$

Dependent Variable: GCDAN
Method: Least Squares
Date: 05/09/17 Time: 07:01
Sample (adjusted): 1959Q3 1995Q1
Included observations: 143 after adjustments

Variable	Coefficient	Std. Error	t-Statistic	Prob.
GCDAN(-1)	0.542939	0.081546	6.658041	0.0000
GCDAN(-2)	0.314802	0.081263	3.873867	0.0002
T	-0.154509	0.115625	-1.336299	0.1837
T^2	0.004170	0.002011	2.073442	0.0400
T^3	-1.88E-05	8.68E-06	-2.165418	0.0321
C	3.807209	2.216085	1.717989	0.0881

R-squared	0.974000	Mean dependent var	49.19538
Adjusted R-squared	0.973052	S.D. dependent var	29.74958
S.E. of regression	4.883687	Akaike info criterion	6.050731
Sum squared resid	3267.505	Schwarz criterion	6.175046
Log likelihood	-426.6272	Hannan-Quinn criter.	6.101246
F-statistic	1026.464	Durbin-Watson stat	2.037114
Prob(F-statistic)	0.000000		

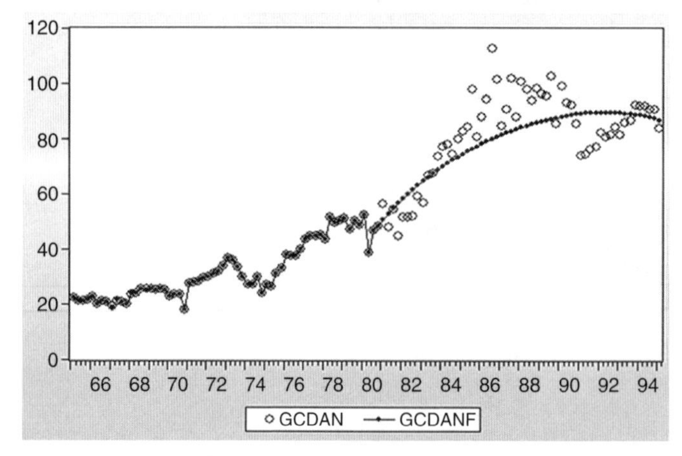

RMSE=10.267, TIC=0.063, BP=0.065

Figure 6.6 The LS estimate of the LV(2) model in (6.8), with its forecast evaluation.

	GCDAN	GCDANF_67	GCDANF_68
1994Q1	91.977	82.784	88.422
1994Q2	91.942	83.879	88.083
1994Q3	90.490	85.072	87.696
1994Q4	90.747	86.354	87.261
1995Q1	83.812	87.715	86.777

1) Compared to the model (6.7), this model has fewer greater values of RMSE, TIC, and Bias Proportion (BP). So, in the statistical sense, this forecast model is worse than (6.7).
2) On the other hand, by comparing the last five observed scores of GCDAN and its two forecast values based on the models (6.7) and (6.8), namely GCDANF_67 and GCDAN_68 as presented on the right, it is found that the first four forecast values of the model (6.8) are closer to the observed scores than (6.7), but the fifth is not. So, which one would you choose as a better forecast model?

Example 6.6 LV(2) Model of *GCDAN* with Heterogeneous Trends by @Year
By using the following ES, the unexpected error message "*Near singular matrix error. Regressors may be perfectly collinear*" is obtained. This is similar for each of the other variables in CONS1.wf1.

$$GCDAN\ C\ GCDAN(-1)\ GCDAN(-2)\ @Quarter^{*}@Expand(@Year)\\ @Expand(@Year, @Droplast) \tag{6.9}$$

Since I am very confident the model is valid, then I try to use other data set, namely CAUSE.wf1, and I obtain the LS estimates of the following two models where S = Personal Consumption Expenditures and W = Wages and Salaries.

$$S\ C\ S(-1)\ S(-2)\ @Quarter^{*}@Expand(@Year)\\ @Expand(@Year, @Droplast) \tag{6.10}$$

$$W\ C\ W(-1)\ W(-2)\ @Quarter^{*}@Expand(@Year)\\ @Expand(@Year, @Droplast) \tag{6.11}$$

As additional unexpected findings, by using the time series G and M in MACRO.wf1, the LS estimates also can be obtained but by using the other four variables; namely, *CN, GNP, I,* and *R* in MACRO.wf1.

As an illustration, Figure 6.7 presents only a part of the results of the model of S in (6.10), with the following ES

$$S\ C\ S(-1)\ S(-2)\ @Quarter\ @Quarter*@Expand(@Year, @Droplast)\\ @Expand(@Year, @Droplast) \tag{6.12}$$

1) In fact, both models are the same models. The model (6.12) has advantages for testing the adjusted effects differences of *@Quarter* on S between the levels of *@Year* as a categorical variable.
2) Figure 6.8 presents a result of RVT of the model (6.12), which shows that *@Quarter* as a numerical variable has significant linear effects differences on S between the 37 levels of *@Year* based on the F and LR (Likelihood Ratio) statistics.
3) The output represents a set of 37 multiple regressions with significant negative slopes of *S(−1)* and *S(−2)*. However, by using the simplest model with the following ES, it is found that each of *S(−1)* and *S(−2)* has a significant positive slope. Why the conclusions are contradicted?

   $$S\ S(-1)\ S(-2)\ C \tag{6.13}$$

   Note that the effects of *S(−1)* and *S(−2)* in (6.12) in fact are their effects, adjusted for the time variables, namely the numerical variable *@Quarter* and categorical variable *@Year*, which should be called "adjusted effects of *S(−1)* and *S(−2)*," in short. On the other hand, for (6.13) we have the "unadjusted effects of *S(−1)* and *S(−2)*," which can be stated or presented as the "effects of *S(-1)* and *S(−2)*," in short.
4) These results should be considered as additional illustrative time series models based on the data in CAUSE.wf1. Furthermore, (6.12) and (6.13) can easily be extended to various alternative models, as presented based on the data in CONS1.wf1. Do these as exercises.

Example 6.7 LV(2) Models with Special Heterogeneous Linear Trends

Referring to model (6.9), since its regressors are stated as being collinear, then I try to delete one of the regressors, because this make the regressors no longer collinear. I found two LS estimates of LV(2) models with special heterogeneous linear trends, as presented in Figure 6.9 with the following ESs, respectively.

Dependent Variable: S
Method: Least Squares
Date: 03/06/17 Time: 10:45
Sample (adjusted): 1959Q3 1995Q2
Included observations: 144 after adjustments

Variable	Coefficient	Std. Error	t-Statistic	Prob.
C	193.2490	17.11737	11.28964	0.0000
S(-1)	-0.703625	0.104392	-6.740189	0.0000
S(-2)	-0.269804	0.095974	-2.811201	0.0064
@QUARTER*(@YEAR=1959)	-1.977875	4.134749	-0.478354	0.6339
@QUARTER*(@YEAR=1960)	0.390192	1.310240	0.297802	0.7668
...	...	...	...	...
@QUARTER*(@YEAR=1994)	-1.130650	1.331710	-0.849021	0.3988
@QUARTER*(@YEAR=1995)	-4.692072	4.263830	-1.100436	0.2750
@YEAR=1959	-159.7656	21.34156	-7.486126	0.0000
@YEAR=1960	-166.9189	15.76691	-10.58666	0.0000
...	...	...	...	...
@YEAR=1993	-24.10528	7.802326	-3.089500	0.0029
@YEAR=1994	4.213817	7.491364	0.562490	0.5756

R-squared	0.995910	Mean dependent var	50.70967
Adjusted R-squared	0.991400	S.D. dependent var	31.51137
S.E. of regression	2.922311	Akaike info criterion	5.287876
Sum squared resid	580.7132	Schwarz criterion	6.855278
Log likelihood	-304.7271	Hannan-Quinn criter.	5.924780
F-statistic	220.7886	Durbin-Watson stat	3.060049
Prob(F-statistic)	0.000000		

RMSE=4.425, TIC=0.026, BP=0.000

Dependent Variable: S
Method: Least Squares
Date: 05/15/17 Time: 11:05
Sample (adjusted): 1959Q3 1995Q2
Included observations: 144 after adjustments

Variable	Coefficient	Std. Error	t-Statistic	Prob.
C	193.2490	17.11737	11.28964	0.0000
S(-1)	-0.703625	0.104392	-6.740189	0.0000
S(-2)	-0.269804	0.095974	-2.811201	0.0064
@QUARTER	-4.692072	4.263830	-1.100436	0.2750
@QUARTER*(@YEAR=1959)	2.714197	5.947631	0.456349	0.6496
@QUARTER*(@YEAR=1960)	5.082264	4.476034	1.135439	0.2602
...				
@QUARTER*(@YEAR=1994)	3.561422	4.406807	0.808164	0.4218
@YEAR=1959	-159.7656	21.34156	-7.486126	0.0000
@YEAR=1960	-166.9189	15.76691	-10.58666	0.0000
...				
@YEAR=1994	4.213817	7.491364	0.562490	0.5756

R-squared	0.995910	Mean dependent var	50.70967
Adjusted R-squared	0.991400	S.D. dependent var	31.51137
S.E. of regression	2.922311	Akaike info criterion	5.287876
Sum squared resid	580.7132	Schwarz criterion	6.855278
Log likelihood	-304.7271	Hannan-Quinn criter.	5.924780
F-statistic	220.7886	Durbin-Watson stat	3.060049
Prob(F-statistic)	0.000000		

Figure 6.7 The LS estimate of the models (6.10) and (6.12) with their forecast evaluations for the forecast sample *{@Year > 1980}*.

Redundant Variables Test
Equation: EQ14
Specification: S C S(-1) S(-2) @QUARTER @QUARTER
*@EXPAND(@YEAR,@DROPLAST) @EXPAND(@YEAR,@DR
OPLAST)
Redundant Variables: @QUARTER*@EXPAND(@YEAR,@DROPLAS
T)

	Value	df	Probability
F-statistic	3.550888	(36, 68)	0.0000
Likelihood ratio	152.3159	36	0.0000

Figure 6.8 RVT of the model (6.12).

$$gcdan\ gcdan(-1)\ gcdan(-2)\ @quarter{*}@expand(@year)\ @expand(@year,@droplast) \tag{6.14}$$

$$gcdan\ c\ gcdan(-1)\ gcdan(-2)\ @quarter{*}@expand(@year,@droplast)\ @expand(@year,@droplast) \tag{6.15}$$

Based on these results, the following findings and notes are presented.

1) Note that *@Expand(Year,@Droplast)* in ES (6.14) without the variable *C* indicates that the dummy variable *(@Year = 1995)* is not used as an independent variable. So, the multiple regression for 1995 goes to the origin, with the following equation, which can be easily copied from the output of the *Representations* of its LS estimate.

$$GCDAN = -0.688728^{*}GCDAN(-1) - 0.264555^{*}GCDAN(-2) + 170.251591^{*}\ @\ QUARTER$$

2) And *@Quarter∗@Expand(Year,@Droplast)* in ES (6.15) with the variable *C* indicates that the numerical variable *@Quarter∗(@Year=1995)* is not used as an independent variable. So, the multiple regression for 1995 without trend but with an intercept has the equation as follows:

$$GCDAN = 170.251591812 - 0.688728^{*}GCDAN(-1) - 0.264555^{*}GCDAN(-2)$$

3) These two models should be considered uncommon or special LV(2) models of *GCDAN* with heterogeneous trends. However, the model (6.14) is better than (6.15), since the multiple regressions for all levels of the *@Year* have trends.

6.2.4 Special Notes on Uncommon Models

Referring to the contradictory effects of *GCDAN(−1)* and *GCDAN(−2)* based on model (6.1) and uncommon models (6.14) and (6.15), Figure 6.10 presents a summary of the LS estimates of the models (6.14) and an additional uncommon LV(2) model of *GCDAN* with heterogeneous trends by *@Year* with the following ESs. Based on this summary, the following findings and notes are presented.

$$gcdan\ gcdan(-1)\ gcdan(-2)\ @quarter@quarter{*}@expand(@year,@droplast)\ @expand(@year,@droplast) \tag{6.16}$$

Dependent Variable: GCDAN
Method: Least Squares
Date: 05/07/17 Time: 08:12
Sample (adjusted): 1959Q3 1995Q1
Included observations: 143 after adjustments

Variable	Coefficient	Std. Error	t-Statistic	Prob.
GCDAN(-1)	-0.688728	0.099278	-6.937358	0.0000
GCDAN(-2)	-0.264555	0.092980	-2.845291	0.0059
@QUARTER*@EXPAND(@YEAR)	Yes			
@EXPAND(YEAR,@DROPLAST)	Yes			

R-squared	0.995822	Mean dependent var	49.19538
Adjusted R-squared	0.991275	S.D. dependent var	29.74958
S.E. of regression	2.778775	Akaike info criterion	5.187512
Sum squared resid	525.0682	Schwarz criterion	6.741451
Log likelihood	-295.9071	Hannan-Quinn criter.	5.818958
Durbin-Watson stat	3.090885		

Dependent Variable: GCDAN
Method: Least Squares
Date: 05/07/17 Time: 08:24
Sample (adjusted): 1959Q3 1995Q1
Included observations: 143 after adjustments

Variable	Coefficient	Std. Error	t-Statistic	Prob.
C	170.2516	14.03208	12.13303	0.0000
GCDAN(-1)	-0.688728	0.099278	-6.937358	0.0000
GCDAN(-2)	-0.264555	0.092980	-2.845291	0.0059
Q*@EXPAND(@Year,@Droplast)	Yes			
@Expand(@Year,@Droplast)	Yes			

R-squared	0.995822	Mean dependent var	49.19538
Adjusted R-squared	0.991275	S.D. dependent var	29.74958
S.E. of regression	2.778775	Akaike info criterion	5.187512
Sum squared resid	525.0682	Schwarz criterion	6.741451
Log likelihood	-295.9071	Hannan-Quinn criter.	5.818958
F-statistic	219.0248	Durbin-Watson stat	3.090885
Prob(F-statistic)	0.000000		

Figure 6.9 Incomplete LS estimates of the models (6.14) and (6.15).

$$gcdan\ gcdan(-1)\ gcdan(-2)\ C\ @quarter\ @quarter*@expand(@year,@droplast)\\ @expand(@year,@drop(1994),@drop(1995)) \tag{6.17}$$

1) Referring to the results of the model (6.1), in general, one would say that *GCDAN(−1)* has a positive significant effect on *GCDAN*, at the 1% level. However, the statement *"GCDAN(−1) has a positive significant effect on GCDAN"* is not a perfect statement, because the model has two numerical independent variables, namely *GCDAN(−1)* and *GCDAN(−2)*. The correct statement is *"GCDAN(−1) has a positive significant effect on GCDAN, adjusted for GCDAN(−2),"* or *"GCDAN(−1) has a positive significant adjusted effect on GCDAN,"* in short. This is similar for the effect of *GCDAN(−2)*. As a comparison, it is found that the pairs of

Variable	Model (6.14)			Model (6.16)			Model (6.17)		
	Coef.	t-Stat.	Prob.	Coef.	t-Stat.	Prob.	Coef.	t-Stat.	Prob.
GCDAN(-1)	-0.6887	-6.9374	0.0000	-0.6887	-6.9374	0.0000	-0.6887	-6.9374	0.0000
GCDAN(-2)	-0.2646	-2.8453	0.0059	-0.2646	-2.8453	0.0059	-0.2646	-2.8453	0.0059
C							179.7042	12.7346	0.0000
@QUARTER				170.2516	12.1330	0.0000	-9.4526	-2.1446	0.0356
		F-Stat			*F-Stat*			*F-Stat*	
Q*@EXPAND(@YEAR,*)		NA		Yes	3.8330	0.0000	Yes	3.8330	0.0000
:@EXPAND(@YEAR,*)		NA		Yes	NA		Yes	NA	
R-squared	0.9958			0.9958			0.9957		
Adj. R-squared	0.9913			0.9913			0.9914		
F-statistic	NA			NA			230.99		
Prob(F-statistic)	NA			NA			0.0000		
DW-Statistic	3.0909			3.0909			2.9329		

Figure 6.10 Summary of the LS estimates of the models (6.14)–(6.17).

variables *GCDAN & GCDAN(-1)*, and *GCDAN* and *GCDAN(-2))* also have significant positive correlations, which are consistent with the linear adjusted effect on each of *GCDAN(−1)* and *GCDA(−2)*.

2) On the other hand, the LS estimates of the three models (6.14)–(6.17) show that each of *GCDAN(−1)* and *GCDAN(−2)* has a negative significant adjusted effect on *GCDAN*, which are the impacts of the multicollinearity between the IVs. Even though the models are uncommon, they are acceptable models in the statistical sense because they present the statistical results. Refer to the Bezzecri's statement "The model must follow the data, and not the other way around" (1973, in Gifi 1991, p. 25).
3) Note that the three models (6.14)–(6.17) present exactly the same forecast model, even though they have different forms of the ESs, with their forecast evaluation presented in Figure 6.11. Compared to the other forecast LV(2) models of *GCDAN*, I would say these models are the best, because they have the *BP* of 0.000000.
4) In addition, Figure 6.10 also present the *F*-statistics of RVTs for testing the effects differences of the numerical variables *@Quarter* on *GCDAN* between all levels of the categorical variable *@Year*, specific for the last two models, with the complete results presented in Figure 6.12. Note that the RVTs are the same for both models, with exactly the same values of *F* and *LR* statistics.

6.3 Complete Heterogeneous LV(2) Models of *GCDAN* By *@Quarter*

6.3.1 Using the Simplest Equation Specification

To do analysis based on a heterogeneous LV(2) model by *@Quarter*, it is recommended to conduct analysis on each of the four subsamples of the *@Quarter* by using the following ES, because we can applied the object *"Forecast"* to each one. Note that this model is in fact exactly the same as (6.1). So readers can easily conduct the analysis based on each sample {*@Quarter* = *q*} for *q* = 1, 2, 3, and 4.

$$gcdan\ gcdan(-1)\ gcdan(-2)\ C \tag{6.18}$$

However, in order to conduct the forecasting, at first we should sort the data by *@Quarter* and *@Year*, similar to the data analysis presented in the Example 5.32. Then, based on each sample {*@Quarter* = *q*}, all models presented in Sections 6.2.2 and 6.2.3, as well as all more advanced or extended models presented in the following sections, can also be applied. Do these as exercises.

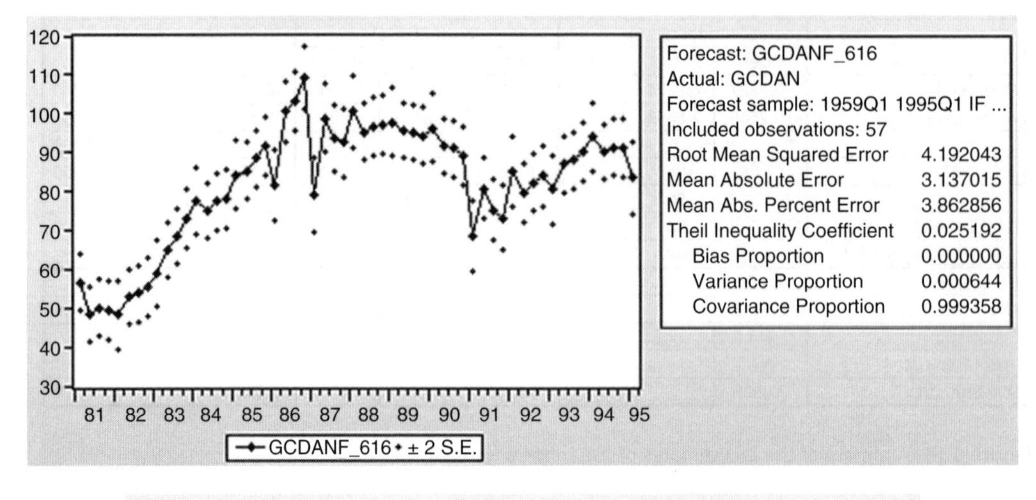

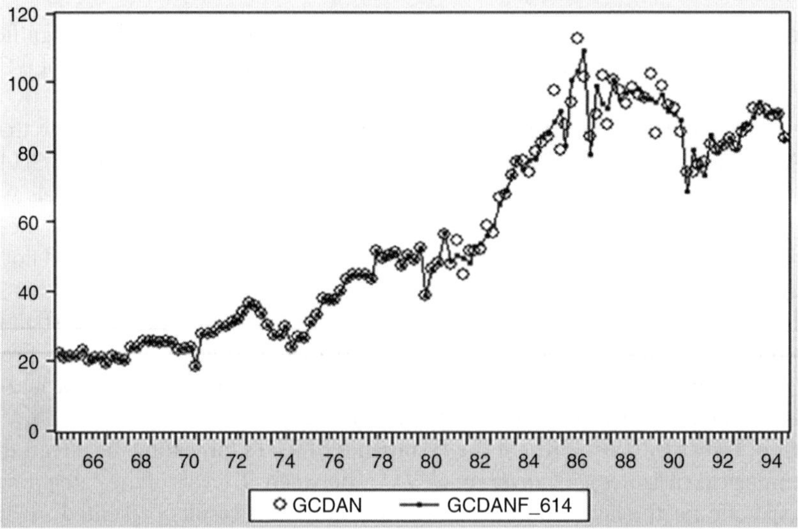

Figure 6.11 Forecast evaluation based on the models (6.14), (6.16), and (6.17).

Example 6.8 The Development of SORT_CONS1.wf1

Figure 6.13 presents the SORT_CONS1.wf1, which can be developed using the steps as follows:

1) At the first stage, we present all variables in the original CONS1.wf1, namely *GC, GCDAN, GWY, GYD,* and *R,* with additional variables that have been generated in previous data analysis, such as the time variables *q* = *@Quarter*, *t* = *@Trend+1* and *TP* = *1* + *1∗(T* > = *111)* + *1∗ (T* > *129),* on the screen.
2) At the second stage, we copy the data set to an Excel data file SORT_CONS1.xls.
3) At the third stage, the data file: SORT_CONS1.xls, can easily be transformed to SORT_-CONS1.wf1. However, note that Figure 6.13 presents a new variable SERIES01, which is a new ordering of the time variable, starting with 1959Q1, 1960Q1, …, 1994Q4.
4) Finally, by clicking *Proc/Sort Current Page.../Yes,* then insert *@Quarter @Year* as the *Sort Key(s)* … OK. Then we have Unstructured Panel Data in SORT_CONS1.wf1.

Redundant Variables Test
Equation: EQ_(6.16)
Specification: GCDAN GCDAN(-1) GCDAN(-2) @QUARTER
@QUARTER*@EXPAND(@YEAR,@DROPLAST)
@EXPAND(@YEAR,@DROPLAST)
Redundant Variables: @QUARTER*@EXPAND(@YEAR,@DROPLAS
T)

	Value	df	Probability
F-statistic	3.883349	(35, 68)	0.0000
Likelihood ratio	157.0435	35	0.0000

Redundant Variables Test
Equation: EQ_(6.17)
Specification: GCDAN GCDAN(-1) GCDAN(-2) @QUARTER C
@QUARTER*@EXPAND(@YEAR,@DROPLAST)
@EXPAND(@YEAR,@DROPLAST,@DROP(1994))
Redundant Variables: @QUARTER*@EXPAND(@YEAR,@DROPLAS
T)

	Value	df	Probability
F-statistic	3.883349	(35, 68)	0.0000
Likelihood ratio	157.0435	35	0.0000

Figure 6.12 Redundant variable tests based on the models (6.16) and (6.17).

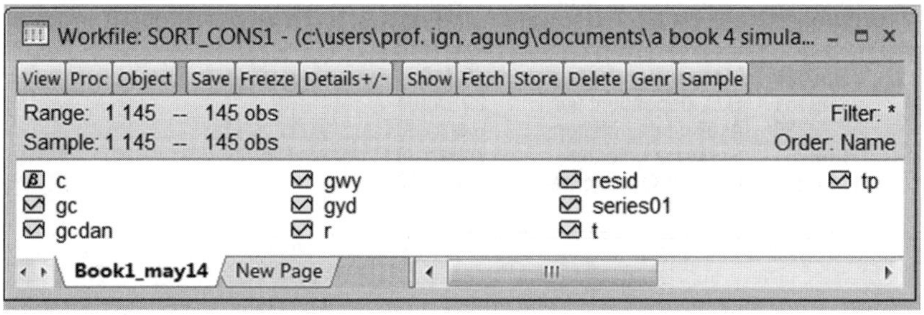

Workfile: SORT_CONS1 - (c:\users\prof. ign. agung\documents\a book 4 simula...
View Proc Object Save Freeze Details+/- Show Fetch Store Delete Genr Sample
Range: 1 145 -- 145 obs Filter: *
Sample: 1 145 -- 145 obs Order: Name

c	gwy	resid	tp
gc	gyd	series01	
gcdan	r	t	

Book1_may14 New Page

Figure 6.13 The variables in SORT_CONS1.wf1.

Example 6.9 An Application of the LV(2) Model of *GCDAN* for the Sample {*@Quarter* = 1}

As an illustration, the statistical results of an LV(2) model of *GCDAN* with heterogeneous trends by time period *TP* can be obtained by using the following two ESs, since both give the same LS regressions as well as forecast evaluation. The first model is considered the simplest. However, all required hypotheses should be tested using the Wald Test. By using the second ES, the output presents the *t*- and *F*-statistics that can be used to test several hypotheses, and the RVT (Redundant Variables Test) also can be applied. Do these as exercises – refer to previous examples.

$$GCDAN\ GCDAN(-1)\ GCDAN(-2)\ T*@Expand(TP)@Expand(TP) \quad (6.19a)$$

$$GCDAN\ GCDAN(-1)\ GCDAN(-2)\ T\ C$$
$$T^{*}@Expand(TP,@Drop(2))\ @Expand(TP,@Drop(2)) \quad (6.19b)$$

As an illustration, Figure 6.14 presents the LS estimate of the model (6.19a), which is a LV(2) model of *GCDAN* with heterogeneous trends by *TP*. Based on these results, findings and notes presented are as follows:

1) At the 10% level of significant, *GCDAN(−2)* has an insignificant adjusted effect on *GCDAN*. Should it be deleted from the model? The answer is "it should not," because at the level 15% of significance (Lapin 1973), GCDAN(−2) has a significant positive adjusted effect on *GCDAN* with a p-value $= 0.2053/2 = 0.10265 < 0.15$.

Dependent Variable: GCDAN
Method: Least Squares
Date: 05/14/17 Time: 21:59
Sample: 1 145 IF Q=1
Included observations: 35

Variable	Coefficient	Std. Error	t-Statistic	Prob.
GCDAN(-1)	0.396831	0.178641	2.221385	0.0349
GCDAN(-2)	0.277660	0.213944	1.297818	0.2053
T*(TP=1)	0.271303	0.121812	2.227237	0.0345
T*(TP=2)	0.196908	0.777694	0.253195	0.8020
T*(TP=3)	1.131443	0.571441	1.979982	0.0580
TP=1	-1.064120	2.861953	-0.371816	0.7129
TP=2	10.35072	88.35352	0.117151	0.9076
TP=3	-130.3862	85.59796	-1.523239	0.1393

R-squared	0.959620	Mean dependent var	50.71974
Adjusted R-squared	0.949151	S.D. dependent var	29.14564
S.E. of regression	6.572280	Akaike info criterion	6.801230
Sum squared resid	1166.261	Schwarz criterion	7.156738
Log likelihood	-111.0215	Hannan-Quinn criter.	6.923952
Durbin-Watson stat	1.710928		

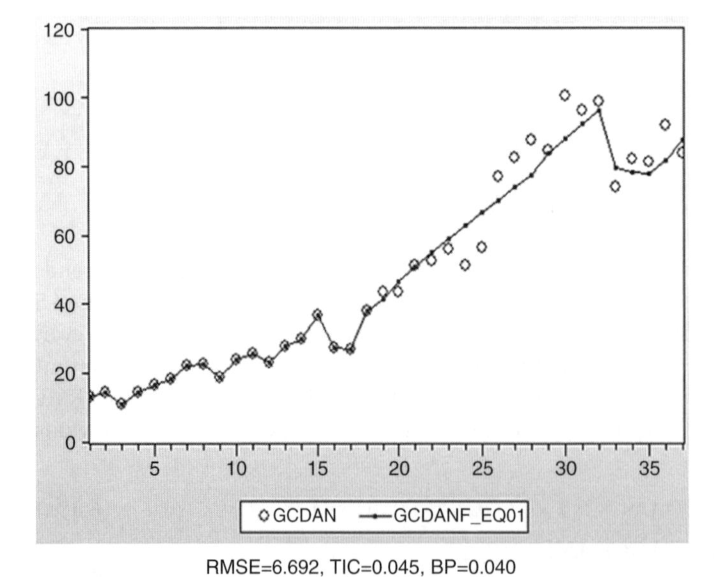

Figure 6.14 The LS estimate of the model (6.19a), and its forecast evaluation, for the forecast sample $\{t > 70\}$.

2) The in-sample forecast values can be easily obtained. However, for the forecast values beyond the sample period should be computed using Excel based on the following regression function. Do this as an exercise – refer to Example 6.8.

$$GCDAN = 0.527327350775*GCDAN(-1) + 0.338170417864*GCDAN(-2)$$
$$+ 0.281755932532*T*(TP = 3) - 27.4455182456*(TP = 3)$$

6.3.2 Using a Complete Equation Specification

In order to present the complete heterogeneous LV(2) model of *GCDAN* by *@Quarter*, the following two alternative ESs can be applied and the data analysis can be done using the original data set in CONS1.wf1 as well as SORT_CONS1.wf1.

$$\begin{gathered} gcdan\ gcdan(-1)*@expand(@quarter)\ gcdan(-2)*@expand(@quarter) \\ @expand(@quarter) \end{gathered} \tag{6.20a}$$

$$\begin{gathered} gcdan\ gcdan(-1)\ gcdan(-2)\ c\ gcdan(-1)*@expand(@quarter,@drop(1)) \\ gcdan(-2)*@expand(@quarter,@drop(1)@expand(@quarter,@drop(1)) \end{gathered} \tag{6.20b}$$

Example 6.10 The LS Estimates of the Model (6.20b)

Figure 6.15 presents the LS estimate of the model (6.20b) by using the original data and the SORT_CONS1.wf1. Based on these results, the findings and notes presented are as follows:

1) Both LS estimates present exactly the same regression function. So, we can use the original data to conduct the analysis based on any complete models by *@Quarter.*
2) The LS estimates can present four regression functions and the object *"Forecast"* is valid for the four quarters. However, the forecast values of *GCDAN* beyond the sample period can be computed using each of the regression functions.
3) In order to write the regression for each quarter, it is recommended to conduct analysis using the model (6.20a), so the four regressions can easily be written based on the output. Do this as an exercise.

6.4 LV(2) Models of *GCDAN* with Exogenous Variables

6.4.1 LV(2) Models with an Exogenous Variable

To be general, alternative LV(2) models are presented based on bivariate (X_t, Y_t). By using an additional independent variable (IV), for the simplest LV(2) model of *Y*, such as the LV(2) model in (6.1); in the theoretical sense, the effect of *X* should be dependent on *Y(−1)* and *Y(−2)*. Therefore, in the statistical sense, the model should have both interactions *Y(−1)*X* and *Y(−2)*X* as the IVs. For this reason, I propose the simplest alternative LV(2) models with a single exogenous or environmental variable *X*, *X(−1)*, or *X(−2)*, respectively, with a specific equation as follows:

$$Y\ Y(-1)*X\ Y(-2)*X\ X\ Y(-1)\ Y(-2)\ C \tag{6.21}$$

$$Y\ Y(-1)*X(-1)\ Y(-2)*X(-1)\ X(-1)\ Y(-1)\ Y(-2)\ C \tag{6.22}$$

$$Y\ Y(-1)*X(-2)\ Y(-2)*X(-2)\ X(-2)\ Y(-1)\ Y(-2)\ C \tag{6.23}$$

Dependent Variable: GCDAN
Method: Least Squares
Date: 06/10/17 Time: 09:12
Sample (adjusted): 1959Q3 1995Q1
Included observations: 143 after adjustments

Variable	Coefficient	Std. Error	t-Statistic	Prob.
GCDAN(-1)	0.675704	0.141641	4.770550	0.0000
GCDAN(-2)	0.279571	0.134212	2.083051	0.0392
C	3.079108	1.504713	2.046309	0.0427
GCDAN(-1)*(@QUARTER=2)	0.007499	0.199499	0.037591	0.9701
GCDAN(-1)*(@QUARTER=3)	0.368103	0.236515	1.556362	0.1220
GCDAN(-1)*(@QUARTER=4)	-0.557151	0.229824	-2.424247	0.0167
GCDAN(-2)*(@QUARTER=2)	0.042887	0.193782	0.221313	0.8252
GCDAN(-2)*(@QUARTER=3)	-0.273706	0.233747	-1.170952	0.2437
GCDAN(-2)*(@QUARTER=4)	0.589676	0.234113	2.518763	0.0130
@QUARTER=2	-2.748675	2.169629	-1.266887	0.2074
@QUARTER=3	-3.448752	2.128726	-1.620101	0.1076
@QUARTER=4	-2.212041	2.123594	-1.041650	0.2995

R-squared	0.977093	Mean dependent var	49.19538
Adjusted R-squared	0.975170	S.D. dependent var	29.74958
S.E. of regression	4.687841	Akaike info criterion	6.008006
Sum squared resid	2878.837	Schwarz criterion	6.256637
Log likelihood	-417.5725	Hannan-Quinn criter.	6.109038
F-statistic	507.9812	Durbin-Watson stat	2.029507
Prob(F-statistic)	0.000000		

Equation: EQ02 Workfile: SORT_CONS1_MAY20::Sort_cons1\

View | Proc | Object | Print | Name | Freeze | Estimate | Forecast | Stats | Resids

Dependent Variable: GCDAN
Method: Least Squares
Date: 05/19/17 Time: 15:04
Sample (adjusted): 1959Q3 1995Q1
Included observations: 143 after adjustments

Variable	Coefficient	Std. Error	t-Statistic	Prob.
GCDAN(-1)	0.675704	0.141641	4.770550	0.0000
GCDAN(-2)	0.279571	0.134212	2.083051	0.0392
C	3.079108	1.504713	2.046309	0.0427
GCDAN(-1)*(@QUARTER=2)	0.007499	0.199499	0.037591	0.9701
GCDAN(-1)*(@QUARTER=3)	0.368103	0.236515	1.556362	0.1220
GCDAN(-1)*(@QUARTER=4)	-0.557151	0.229824	-2.424247	0.0167
GCDAN(-2)*(@QUARTER=2)	0.042887	0.193782	0.221313	0.8252
GCDAN(-2)*(@QUARTER=3)	-0.273706	0.233747	-1.170952	0.2437
GCDAN(-2)*(@QUARTER=4)	0.589676	0.234113	2.518763	0.0130
@QUARTER=2	-2.748675	2.169629	-1.266887	0.2074
@QUARTER=3	-3.448752	2.128726	-1.620101	0.1076
@QUARTER=4	-2.212041	2.123594	-1.041650	0.2995

R-squared	0.977093	Mean dependent var	49.19538
Adjusted R-squared	0.975170	S.D. dependent var	29.74958
S.E. of regression	4.687841	Akaike info criterion	6.008006
Sum squared resid	2878.837	Schwarz criterion	6.256637
Log likelihood	-417.5725	Hannan-Quinn criter.	6.109038
F-statistic	507.9812	Durbin-Watson stat	2.029507
Prob(F-statistic)	0.000000		

Figure 6.15 The LS estimates of the models (6.20a) and (6.20b), using the original and sorted data.

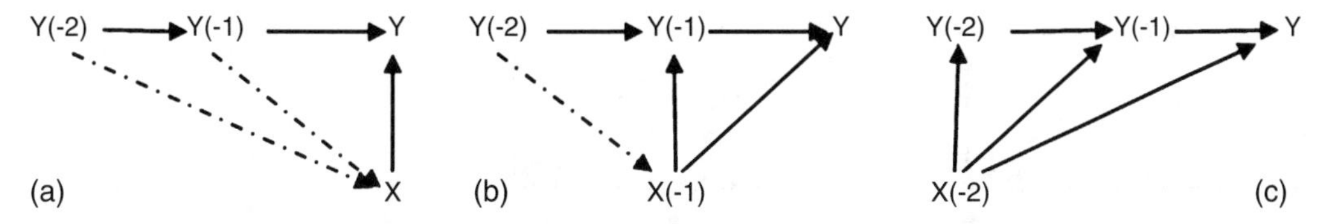

Figure 6.16 Three alternative path diagrams based on six time series *X, X(−1), X(−2), Y, Y(−1),* and *Y(−2)* out of many possible path diagrams, or up-and-downward relationships.

The up-and-downward or up-and-down-stream relationships between the six time series *X, X (−1), X(−2), Y, Y(−1),* and *Y(−2)* in the previous three models can be presented in the form of path diagrams, as presented in Figure 6.16.

Note that each of the path diagrams only present five pairs of up-and-downward or causal relationships out of the $C_6^2 = 15$ pairs. So I consider that these path diagrams represent the simplest LV(2) forecast two-way interaction models of *Y*. The dotted lines from each of *Y(−1)* and *Y (−2)* to *X* in Figure 6.16a indicate that *Y(−1)* and *Y(−2)* might not have causal relationships with *X*, but the effect of *X* on *Y* depends on both *Y(−1)* and *Y(−2)* in the theoretical sense. And the solid arrow from *X* to *Y* indicates that *X* is an upper variable or causal factor of *Y*. This is similar for the other two path diagrams.

Based on each of these models, the in-sample forecast values can directly be obtained by using the object "*Forecast*." However, for the forecast values beyond the sample period, a forecast model of *X* should be applied with the following ES where *p* should be selected using the trial-and-error method as one out of many possible models – refer to all possible forecast models based on a single time series.

$$X\,X(-1)...X(-p)\,C \tag{6.24}$$

Suppose we are using $p = 2$, the pairs of the models (6.21) and (6.24) would give the following pairs of LS regressions, which can be applied to compute the forecast values beyond the sample period by using Excel.

$$\begin{aligned} Y_t &= (\beta_1 Y_{t-1} + \beta_2 Y_{t-2} + \beta_3) * X_t + \beta_5 Y_{t-1} + \beta_6 Y_{t-2} + \beta_0 \\ X_t &= \delta_1 X_{t-1} + \delta_2 X_{t-2} + \delta_0 \end{aligned} \tag{6.25a}$$

or

$$\begin{aligned} Y_t &= (\beta_1 Y_{t-1} + \beta_2 Y_{t-2} + \beta_3) * (\delta_1 X_{t-1} + \delta_2 X_{t-2} + \delta_0) + \beta_5 Y_{t-1} + \beta_6 Y_{t-2} + \beta_0 \\ X_t &= \delta_1 X_{t-1} + \delta_2 X_{t-2} + \delta_0 \end{aligned} \tag{6.25b}$$

which shows the multiple regression of *Y* has four two-way interaction independent variables, namely *Y(−1)*X(−1), Y(−1)*X(−2), Y(−2)*X(−1),* and *Y(−2)*X(−2).*

For models (6.22) and (6.23) we have the following pairs of regressions, respectively.

$$\begin{aligned} Y_t &= (\beta_1 Y_{t-1} + \beta_2 Y_{t-2} + \beta_3) * X_{t-1} + \beta_5 Y_{t-1} + \beta_6 Y_{t-2} + \beta_0 \\ X_{t-1} &= \delta_1 X_{t-2} + \delta_2 X_{t-3} + \delta_0 \end{aligned} \tag{6.26}$$

$$\begin{aligned} Y_t &= (\beta_1 Y_{t-1} + \beta_2 Y_{t-2} + \beta_3) * X_{t-2} + \beta_5 Y_{t-1} + \beta_6 Y_{t-2} + \beta_0 \\ X_{t-2} &= \delta_1 X_{t-3} + \delta_2 X_{t-4} + \delta_0 \end{aligned} \tag{6.27}$$

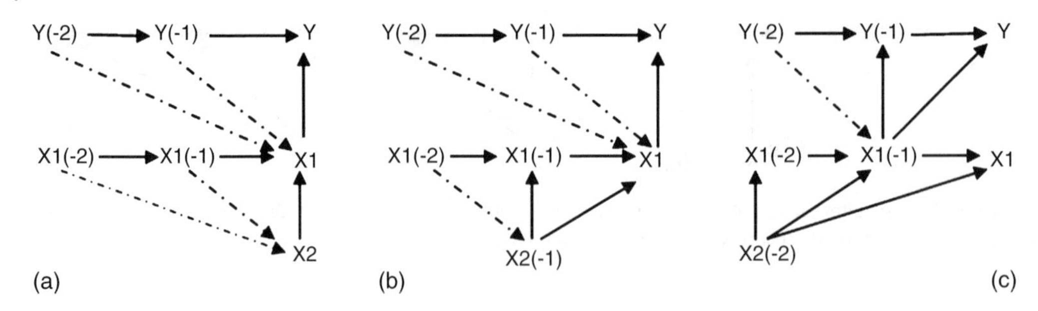

Figure 6.17 The path diagrams for three selected sets of the models (6.28a)–(6.28c).

6.4.2 LV(2) Models with Two Exogenous Variables

To be general, alternative LV(2) models are presented based on triple time series or tri-variate $(X1_t, X2_t, Y_t)$. Without loss of generality, we can assume that *X1* is the most important cause factor of *Y* and *X2* is an upper variable or cause factor of *X1*. As an extension of the previous models based on *(X,Y)*, I propose the following set of three ESs to forecast the endogenous variable Y_t. where *IV1* represents an independent variable, either one of the variables *X1*, *X1(−1)*, and *X1 (−2)*, and *IV2* represents an independent variable, either one of the variables *X2*, *X2(−1)*, and *X2(−2)*, so we would have nine possible sets of models to forecast Y_t.

$$Y\ Y(-1)*IV1\ Y1(-2)*IV1\ IV1\ Y(-1)\ Y(-2)\ C \quad (6.28a)$$

$$X1\ X1(-1)*IV2\ X1(-2)*IV2\ IV2\ X1(-1)\ X1(-2)\ C \quad (6.28b)$$

$$X2\ X2(-1)...X2(-p)\ C \quad (6.28c)$$

As an extension of the path diagrams in Figures 6.16, Figure 6.17 presents three selected sets of the first two models (6.28a) and (6.28b).

Similar to the path diagram in Figure 6.16, the arrows with solid lines represent the true causal effects between pairs of the variables, just as Figure 6.17(a) presents the true causal effects *Y (−2)→Y(−1)→Y, X1(−2)→X1(−1)→X1*, and *X2→X1→Y*; Figure 6.17(b) presents the true causal effects *Y(−2)→Y(−1)→Y, X1(−2)→X1(−1)→X1*, and *X2(−1)* has direct effects on both *X1* and *X1 (−1)*; and Figure 6.17(c) presents the true causal effects *Y(−2)→Y(−1)→Y, X1(−2)→X1(−1)→X1*, and *X2(−2)* has direct effects on *X1*, *X1(−1)*, and *X1(−2)*. The arrows with dotted lines represent the up-and-downward relationships between pairs of the variables. The other sets could easily be developed.

The in-sample forecast values based on each of these models can be directly obtained using the object "*Forecast*." However, for the forecast values beyond the sample period, a forecast model of *X1* and *X2* should be applied. See the following example.

Example 6.11 An Illustration of the Set of Models in (6.28a)

Figure 6.18 presents the LS estimate of the model (6.28a), with one of its acceptable reduced models, for *Y* = *GCDAN*, *X1* = *GC*, and *X2* = *GWY* in CONS1.wf1. In fact, the full model could be used to forecast the dependent variable *Y*, even though each of the interactions *Y(−1)∗X1* and *Y(−2)∗X1* has an insignificant adjusted effect with a very large *p*-value.

However, in general, one might want to see that the interactions are important predictors of *Y*. Hence the reduced model would be explored. The reduced model can be obtained using the trial-and-error method or the manual multistage selection method. This is similar for the LS regressions of the models (6.28b) and (6.28c), as presented in Figure 6.19.

Dependent Variable: Y
Method: Least Squares
Date: 05/21/17 Time: 08:59
Sample (adjusted): 1959Q3 1995Q1
Included observations: 143 after adjustments

Variable	Coefficient	Std. Error	t-Statistic	Prob.
Y(-1)*X1	6.19E-06	8.21E-05	0.075374	0.9400
Y(-2)*X1	-5.71E-05	8.27E-05	-0.690665	0.4909
X1	0.004822	0.001564	3.081923	0.0025
Y(-1)	0.524139	0.238346	2.199069	0.0296
Y(-2)	0.480989	0.241506	1.991620	0.0484
C	-1.748680	1.382468	-1.264897	0.2081

R-squared	0.974153	Mean dependent var	49.19538
Adjusted R-squared	0.973210	S.D. dependent var	29.74958
S.E. of regression	4.869294	Akaike info criterion	6.044827
Sum squared resid	3248.273	Schwarz criterion	6.169142
Log likelihood	-426.2052	Hannan-Quinn criter.	6.095343
F-statistic	1032.703	Durbin-Watson stat	2.037011
Prob(F-statistic)	0.000000		

Dependent Variable: Y
Method: Least Squares
Date: 05/21/17 Time: 09:00
Sample (adjusted): 1959Q3 1995Q1
Included observations: 143 after adjustments

Variable	Coefficient	Std. Error	t-Statistic	Prob.
Y(-1)*X1	-0.000146	3.07E-05	-4.756065	0.0000
Y(-2)*X1	9.81E-05	2.80E-05	3.507437	0.0006
X1	0.004842	0.001581	3.062446	0.0026
Y(-1)	0.990359	0.045315	21.85521	0.0000
C	-1.578474	1.394576	-1.131867	0.2597

R-squared	0.973405	Mean dependent var	49.19538
Adjusted R-squared	0.972634	S.D. dependent var	29.74958
S.E. of regression	4.921353	Akaike info criterion	6.059383
Sum squared resid	3342.320	Schwarz criterion	6.162979
Log likelihood	-428.2459	Hannan-Quinn criter.	6.101479
F-statistic	1262.743	Durbin-Watson stat	2.145354
Prob(F-statistic)	0.000000		

Figure 6.18 The LS estimate of the model (6.28a) and one of its reduced models.

Finally, Figure 6.20 presents the forecast values beyond the sample period for several time points. The steps of computation are as follows:

1) Copy the last five scores of the time T = *@Trend+1, X2, X1,* and Y to a spreadsheet, as presented in Figure 6.20.
2) The first forecast value of $X2$ for $T = 146$, can be computed using the following regression function, which can be copied from the output of "*Representations.*"

$$\begin{aligned} X2 &= 1.6085694806*X2(-1) - 0.603906692798*X2(-2) + 2.66921525192 \\ &= 1.6085694806*C6 - 0.603906692798*C5 + 2.66921525192 \end{aligned} \tag{6.29a}$$

Dependent Variable: X1
Method: Least Squares
Date: 05/21/17 Time: 11:38
Sample (adjusted): 1959Q3 1995Q1
Included observations: 143 after adjustments

Variable	Coefficient	Std. Error	t-Statistic	Prob.
X1(-1)*X2(-1)	0.000380	4.14E-05	9.170753	0.0000
X1(-2)*X2(-1)	-0.000394	4.25E-05	-9.276605	0.0000
X2(-1)	0.043514	0.062899	0.691806	0.4902
X1(-2)	1.024479	0.052991	19.33296	0.0000
C	-8.674297	3.141859	-2.760881	0.0065

R-squared	0.999911	Mean dependent var	1777.961
Adjusted R-squared	0.999908	S.D. dependent var	1393.778
S.E. of regression	13.36762	Akaike info criterion	8.057887
Sum squared resid	24659.67	Schwarz criterion	8.161483
Log likelihood	-571.1389	Hannan-Quinn criter.	8.099983
F-statistic	385894.5	Durbin-Watson stat	1.639690
Prob(F-statistic)	0.000000		

Dependent Variable: X2
Method: Least Squares
Date: 05/21/17 Time: 11:43
Sample (adjusted): 1959Q3 1995Q1
Included observations: 143 after adjustments

Variable	Coefficient	Std. Error	t-Statistic	Prob.
X2(-1)	1.608569	0.067819	23.71874	0.0000
X2(-2)	-0.603907	0.068630	-8.799451	0.0000
C	2.669215	1.256621	2.124121	0.0354

R-squared	0.999926	Mean dependent var	1325.132
Adjusted R-squared	0.999925	S.D. dependent var	964.3053
S.E. of regression	8.367947	Akaike info criterion	7.107450
Sum squared resid	9803.155	Schwarz criterion	7.169608
Log likelihood	-505.1827	Hannan-Quinn criter.	7.132708
F-statistic	942795.2	Durbin-Watson stat	2.022571
Prob(F-statistic)	0.000000		

Figure 6.19 The acceptable LS estimates of the models (6.28b) and (6.28c).

3) The second forecast value for $T = 146$, the forecast of $X1$, is computed using the function as follows:

$$\begin{aligned} X1 &= 0.000379898030562*X1(-1)*X2(-1) \\ &\quad -0.000393978804071*X1(-2)*X2(-1) \\ &\quad +0.043513626631*X2(-1) + 1.02447890306*X1(-2) - 8.67429744869 \qquad (6.29b) \\ &= 0.000379898030562*D6*C6 - 0.000393978804071*D5*C6 \\ &\quad +0.043513626631*D6 + 1.02447890306*D5 - 8.67429744869 \end{aligned}$$

	A	B	C	D	E
1		T	X2	X1	Y
2	1994Q1	141	3195.200	4599.200	91.977
3	1994Q2	142	3242.800	4665.100	91.942
4	1994Q3	143	3265.500	4734.400	90.490
5	1994Q4	144	3320.200	4796.027	90.747
6	1995Q1	145	3363.000	4836.326	83.812
7		**146**	**3407.197**	**4939.577**	**88.942**
8			3452.445	5062.598	86.979
9			3498.537	5193.333	89.147
10			3545.355	5328.196	88.703

Figure 6.20 Several forecast values beyond the sample period.

4) Then, the first forecast value of Y is computed using the following function.

$$Y = -0.000145815980686*Y(-1)*X1 + 9.80800954808e-05*Y(-2)*X1$$
$$+0.00484230752664*X1 + 0.990358994248*Y(-1) - 1.5784744557$$
$$= -0.000145815980686*E6*D7 + 0.0000980800954808*E5*D7 \tag{6.29c}$$
$$= -0.000145815980686*E6*D7 + 0.0000980800954808*E5*D7$$
$$+0.00484230752664*D7 + 0.990358994248*E6 - 1.5784744557$$

5) Finally, the forecast values for $T > 146$ can be obtained by using the block-copy-paste method of the forecast values of *X2, X1,* and *Y*.

6.5 Alternative Forecast Models Based on (*Y1,Y2*)

6.5.1 LV(2) Basic Interaction Models

As the extension or modification of the simplest interaction model based on (X,Y) presented in (6.22), goes under the assumption that the time series *Y1* and *Y2* have reciprocal/simultaneous causal effects. I propose similar pairs of models with the following ESs.

$$Y1\ Y1(-1)*Y2(-1)\ Y1(-2)*Y2(-1)\ Y2(-1)\ Y1(-1)\ Y1(-2)\ C \tag{6.30a}$$
$$Y2\ Y2(-1)*Y1(-1)\ Y2(-2)*Y1(-1)\ Y1(-1)\ Y2(-1)\ Y2(-2)\ C \tag{6.30b}$$

Referring to the model (6.22), which presents an LV(2) of *Y* with the effect of *X* on *Y* depends on both *Y(−1)* and *Y(−2)*, model (6.30a) presents an LV(2) model of *Y1* where the effect of *Y2(−1)* on *Y1* depends on both *Y1(−1)* and *Y1(−2)*. And model (6.30b) presents the LV(2) model of *Y2* where the effect of *Y1(−1)* on *Y2* depends on both *Y2(−1)* and *Y2(−2)*.

6.5.2 LV(2) Models of (Y1,Y2) with an Exogenous Variable and @*Trend*

6.5.2.1 Simple LV(2) Models *(Y1,Y2)* with an Exogenous Variable and Linear Trend

As an extension of the models in Eq. (6.30a) and (6.30b), the models based on *(Y1,Y2)* with an exogenous variable *X1* are presented as a set of the following three ESs where *IV* can be any one of *X1*, *X1(−1)*, and *X1(−2)*, similar to the model (6.28a), and T = @*Trend* or T = @*Trend* + 1. Referring to the models with alternative trends, this subsection presents the simplest LV(2) model with a linear trend as follows:

$$\begin{aligned} &Y1\ Y1(-1){*}Y2(-1)\ Y1(-2){*}Y2(-1)\ Y2(-1)\\ &Y1(-1){*}IV\ Y1(-2){*}IV\ IV\ Y1(-1)\ Y1(-2)\ T\ C \end{aligned} \tag{6.31a}$$

$$\begin{aligned} &Y2\ Y2(-1){*}Y1(-1)\ Y2(-2){*}Y1(-1)\ Y1(-1)\\ &Y2(-1){*}IV\ Y2(-2){*}IV\ IV\ Y2(-1)\ Y2(-2)\ T\ C \end{aligned} \tag{6.31b}$$

$$IV\ IV(-1)\ \ldots\ IV(-p)\ C \tag{6.31c}$$

Note that the specific models (6.31a) and (6.31b) present the forecast models of each *Y1* and *Y2* with a linear trend, so these are considered the simplest LV(2) models with *Trend* c ompared to the following one of the advanced LV(2) models. In general, or most cases, all independent variables of models (6.31a) and (6.31b) are highly correlated, then unexpected estimates of the parameters can be obtained. So, in order to have an acceptable reduced model so that most IVs have Prob. < 0.30, or a p-value = Prob./2 < 0.15; it is recommended to apply the manual multistage selection method, which is a combination of the trial-and-error and STEPLS-Combinatorial selection methods. For model (6.31c), the trial-and-error method is applied to obtain the best possible integer $p > 0$.

6.5.2.2 Models of *(Y1,Y2)* with an Exogenous Variable and Time-Related Effects

It is recognized that a lot of possible advanced LV(2) models with exogenous variables and/or trends could be presented. However, this section presents one of the simple LV(2) models. The following three ESs look simpler than the ESs (6.31a), (6.31b), and (6.31c), since only the ES (6.32c) has @*Trend*. However, by combining the first two with the third, we have LV(2) models of *Y1* and *Y2* with the time-related effects (TRE), or models with time interactions as independent variables. See the following example.

$$\begin{aligned} &Y1\ Y1(-1){*}Y2(-1)\ Y1(-2){*}Y2(-1)\ Y2(-1)\\ &Y1(-1){*}IV\ Y1(-2){*}IV\ IV\ Y1(-1)\ Y1(-2)\ C \end{aligned} \tag{6.32a}$$

$$\begin{aligned} &Y2\ Y2(-1){*}Y1(-1)\ Y2(-2){*}Y1(-1)\ Y1(-1)\\ &Y2(-1){*}IV\ Y2(-2){*}IV\ IV\ Y2(-1)\ Y2(-2)\ C \end{aligned} \tag{6.32b}$$

$$IV\ IV(-1)\ldots IV(-p)\ T\ C \tag{6.32c}$$

Example 6.12 Specific Models of (6.32a), (6.32b), and (6.32c)

For *IV* = *X1*, we have the following three multiple regressions

$$\begin{aligned} Y1 = {} & (\beta_1{}^*Y1(-1) + \beta_2{}^*Y1(-2) + \beta_3){*}Y2(-1)\\ & + (\beta_4{}^*Y1(-1) + \beta_5{}^*Y2(-1) + \beta_6){*}\boldsymbol{X1} + \beta_7{}^*Y1(-1) + \beta_8{}^*Y1(-2) + \beta_0 \end{aligned} \tag{6.33a}$$

$$\begin{aligned} Y2 = {} & (\delta_1{}^*Y2(-1) + \delta_2{}^*Y2(-2) + \delta_3){*}Y1(-1)\\ & + (\delta_4{}^*Y2(-1) + \delta_5{}^*Y2(-2) + \delta_6){*}\boldsymbol{X1} + \delta_7{}^*Y2(-1) + \delta_8{}^*Y2(-2) + \delta_0 \end{aligned} \tag{6.33b}$$

$$X1 = \gamma_1{}^*X1(-1) + \ldots + \gamma_2{}^*X1(-p) + \gamma_3{}^*T + \gamma_0 \quad (6.33c)$$

Then by inserting *X1* in the Eq. (6.33c) to the first two models, the model of *Y1* will have the interactions *Y1(−1)∗T* and *Y1(−2)∗T* as independent variables, and the model of *Y2* will have the interactions *Y2(−1)∗T* and *Y2(−2)∗T* as independent variables, in addition to the linear trend (*T*). So those models are considered to have time related-effects. In addition, each of the models would have $2(p + 1)$ two-way interaction independent variables.

Furthermore, note that the model (6.33c) is one out of many possible models of a single time series *X1*, which have been presented in the first three chapters and with various possible trends shown in Table 6.1.

Example 6.13 Unexpected Findings Based on the Set of Models (6.33a–c)

As an illustration of the models (6.33a–c), Figure 6.21 presents the LS estimates of the following two models for *Y1 = GCDAN*, *Y2 = GC*, and *X1 = GWY* in CONS1.wf1. Based on these estimates, the following findings and notes are presented.

$$\begin{array}{c} Y1\ Y1(-1)*Y2(-1)\ Y1(-2)*Y2(-1)\ Y1(-1)*X1\ Y1(-2)*X1 \\ Y2(-1)\ X1\ Y1(-1)\ Y1(-2)\ C \end{array} \quad (6.34a)$$

$$\begin{array}{c} Y2\ Y2(-1)*Y1(-1)\ Y2(-2)*Y1(-1)\ Y2(-1)*X1\ Y2(-2)*X1 \\ Y1(-1)\ X1\ Y2(-1)\ Y2(-2)\ C \end{array} \quad (6.34b)$$

1) Both models present unexpected estimates; (6.34a) shows each of the two-way interactions has a significant adjusted effect at the 10% level. But each of the interactions of (6.34b) has a very large p-value > 0.60. Despite this, both regressions are acceptable results, in the statistical sense, to forecast their dependent variables.
2) However, it is important to explore alternative reduced models for the model of *Y2*. Since its independent variables are highly or significantly correlated, then some of them can be omitted to present the best possible model of *Y2*, even though it will be very subjective. Refer to the special notes and comments presented in Chapter 5.
3) As additional analysis, Figure 6.22 presents two alternative reduced models.
   3.1 The first reduced model is obtained by using the variables *Y2* and *C* as the constant variables in the model and the interactions, *Y2(−1)∗Y1(−1)*, *Y2(−2)∗Y1(−1)*, *Y2(−1)∗X1*, and *Y2(−2)∗X1* as the search regressors, with the results presented in Figure 6.22a. It is found that not one of the main variables *Y1(−1)*, *X1*, *Y2(−1)*, and *Y2(−2)* can be inserted as an additional independent variable.
   3.2 The second reduced model is obtained by using the variables *Y2* and *C* as the constant variables in the model, and the main variables *Y2(−1)*, *Y2(−2)*, *Y1(−1)*, and *X1* as the search regressors, with the results presented in Figure 6.22b. It is also found that not one of the interactions can be inserted as an additional independent variable. It is surprising to see this model has an Adjusted R-squared = 0.999927, which is a little bit greater than the full model with R-squared = 0.999925. So it can be said that this model is a better fit than the full model. It is most unexpected that a reduced model has a greater Adjusted R-squared than the full model.
4) As a comparative study or additional reduced models, the following stages of analysis have been done.
   i) At the first stage, the variables *Y* and *C* are inserted as constant variables in the models, and the eight IVs as search regressors. By selecting 1, 2, 3, and 4 out of the search regressors,

Dependent Variable: Y1
Method: Least Squares
Date: 05/24/17 Time: 19:42
Sample (adjusted): 1959Q3 1995Q1
Included observations: 143 after adjustments

Variable	Coefficient	Std. Error	t-Statistic	Prob.
Y1(-1)*Y2(-1)	0.002943	0.001597	1.842546	0.0676
Y1(-2)*Y2(-1)	-0.003779	0.001603	-2.357868	0.0198
Y1(-1)*X1	-0.004257	0.002303	-1.848046	0.0668
Y1(-2)*X1	0.005324	0.002313	2.301553	0.0229
Y2(-1)	0.056278	0.024366	2.309730	0.0224
X1	-0.067846	0.032267	-2.102634	0.0374
Y1(-1)	1.177424	0.424544	2.773383	0.0063
Y1(-2)	-0.265796	0.439023	-0.605427	0.5459
C	1.853056	2.172214	0.853072	0.3951

R-squared	0.975950	Mean dependent var	49.19538
Adjusted R-squared	0.974514	S.D. dependent var	29.74958
S.E. of regression	4.749353	Akaike info criterion	6.014763
Sum squared resid	3022.552	Schwarz criterion	6.201236
Log likelihood	-421.0556	Hannan-Quinn criter.	6.090537
F-statistic	679.7021	Durbin-Watson stat	2.129092
Prob(F-statistic)	0.000000		

(a)

Dependent Variable: Y2
Method: Least Squares
Date: 05/25/17 Time: 06:29
Sample (adjusted): 1959Q3 1995Q1
Included observations: 143 after adjustments

Variable	Coefficient	Std. Error	t-Statistic	Prob.
Y2(-1)*Y1(-1)	-0.003188	0.006352	-0.501875	0.6166
Y2(-2)*Y1(-1)	0.003260	0.006455	0.505033	0.6144
Y2(-1)*X1	3.19E-05	0.000191	0.167195	0.8675
Y2(-2)*X1	-3.32E-05	0.000193	-0.172352	0.8634
Y1(-1)	-0.184360	0.491255	-0.375283	0.7080
X1	0.185645	0.054511	3.405630	0.0009
Y2(-1)	1.039137	0.309966	3.352421	0.0010
Y2(-2)	-0.153790	0.310052	-0.496015	0.6207
C	-9.283995	5.044742	-1.840331	0.0679

R-squared	0.999929	Mean dependent var	1777.961
Adjusted R-squared	0.999925	S.D. dependent var	1393.778
S.E. of regression	12.04916	Akaike info criterion	7.876736
Sum squared resid	19454.42	Schwarz criterion	8.063209
Log likelihood	-554.1866	Hannan-Quinn criter.	7.952510
F-statistic	237487.9	Durbin-Watson stat	1.849424
Prob(F-statistic)	0.000000		

(b)

Figure 6.21 The LS estimates of the models (a) (6.34a) and (b) (6.34b).

finally the results in Figure 6.23a are obtained, which is the best out of 70 sets of four regressors. Surprisingly, this model has an Adjusted R-squared = 0.999928, which is a little bit greater than the reduced model in Figure 6.22b with an Adjusted R-squared = 0.999927.

ii) Since each of $Y1(-1)$ and $Y2(-1)*Y1(-1)$ has very large p-values, we should select which one will be deleted from the eight search regressors in order to obtain the other reduced model.

Number of always included regressors: 1
Number of search regressors: 4
Selection method: Combinatorial
Number of search regressors: 4

Variable	Coefficient	Std. Error	t-Statistic	Prob.*
C	587.7658	27.46475	21.40074	0.0000
Y2(-2)*Y1(-1)	-0.214083	0.109183	-1.960771	0.0519
Y2(-1)*X1	-0.005489	0.003430	-1.600333	0.1118
Y2(-1)*Y1(-1)	0.215711	0.107279	2.010749	0.0463
Y2(-2)*X1	0.005731	0.003487	1.643798	0.1025

R-squared	0.971936	Mean dependent var	1777.961
Adjusted R-squared	0.971122	S.D. dependent var	1393.778
S.E. of regression	236.8513	Akaike info criterion	13.80708
Sum squared resid	7741595.	Schwarz criterion	13.91068
Log likelihood	-982.2063	Hannan-Quinn criter.	13.84918
F-statistic	1194.818	Durbin-Watson stat	0.383662
Prob(F-statistic)	0.000000		

Selection Summary

Number of combinations compared:	1

*Note: p-values and subsequent tests do not account for stepwise selection.

(a)

Dependent Variable: Y2
Method: Stepwise Regression
Date: 05/28/17 Time: 06:47
Sample (adjusted): 1959Q3 1995Q1
Included observations: 143 after adjustments
Number of always included regressors: 1
Number of search regressors: 4
Selection method: Combinatorial
Number of search regressors: 4

Variable	Coefficient	Std. Error	t-Statistic	Prob.*
C	-9.569624	2.749335	-3.480705	0.0007
Y2(-1)	0.883346	0.091161	9.689950	0.0000
X1	0.191681	0.026934	7.116567	0.0000
Y1(-1)	-0.153640	0.126014	-1.219228	0.2248
Y2(-2)	-0.000947	0.085737	-0.011040	0.9912

R-squared	0.999929	Mean dependent var	1777.961
Adjusted R-squared	0.999927	S.D. dependent var	1393.778
S.E. of regression	11.89032	Akaike info criterion	7.823666
Sum squared resid	19510.41	Schwarz criterion	7.927262
Log likelihood	-554.3921	Hannan-Quinn criter.	7.865762
F-statistic	487750.3	Durbin-Watson stat	1.841905
Prob(F-statistic)	0.000000		

Selection Summary

Number of combinations compared:	1

*Note: p-values and subsequent tests do not account for stepwise selection.

(b)

Figure 6.22 Two alternative reduced models of the model (6.34b).

Dependent Variable: Y2
Method: Stepwise Regression
Date: 05/25/17 Time: 16:00
Sample (adjusted): 1959Q2 1995Q1
Included observations: 144 after adjustments
Number of always included regressors: 1
Number of search regressors: 8
Selection method: Combinatorial
Number of search regressors: 4
Note: final equation sample is larger than stepwise sample (rejected regressors contain missing values)

Variable	Coefficient	Std. Error	t-Statistic	Prob.*
C	-9.681654	3.436387	-2.817394	0.0055
Y2(-1)	0.886523	0.036232	24.46796	0.0000
X1	0.186341	0.048217	3.864614	0.0002
Y1(-1)	-0.130830	0.226775	-0.576917	0.5649
Y2(-1)*Y1(-1)	-8.66E-06	7.63E-05	-0.113506	0.9098

R-squared	0.999930	Mean dependent var	1767.811
Adjusted R-squared	0.999928	S.D. dependent var	1394.226
S.E. of regression	11.85195	Akaike info criterion	7.816967
Sum squared resid	19525.15	Schwarz criterion	7.920085
Log likelihood	-557.8216	Hannan-Quinn criter.	7.858868
F-statistic	494689.3	Durbin-Watson stat	1.845424
Prob(F-statistic)	0.000000		

Selection Summary

Number of combinations compared: 70

*Note: p-values and subsequent tests do not account for stepwise selection.

(a)

Dependent Variable: Y2
Method: Stepwise Regression
Date: 05/25/17 Time: 15:41
Sample (adjusted): 1959Q3 1995Q1
Included observations: 143 after adjustments
Number of always included regressors: 1
Number of search regressors: 7
Selection method: Combinatorial
Number of search regressors: 4

Variable	Coefficient	Std. Error	t-Statistic	Prob.*
C	-10.85329	2.937161	-3.695163	0.0003
Y2(-1)	0.884289	0.045623	19.38263	0.0000
X1	0.185557	0.053419	3.473643	0.0007
Y2(-1)*Y1(-1)	-4.73E-05	4.00E-05	-1.182368	0.2391
Y2(-2)*X1	1.25E-06	2.62E-06	0.477814	0.6335

R-squared	0.999929	Mean dependent var	1777.961
Adjusted R-squared	0.999927	S.D. dependent var	1393.778
S.E. of regression	11.89328	Akaike info criterion	7.824163
Sum squared resid	19520.10	Schwarz criterion	7.927759
Log likelihood	-554.4276	Hannan-Quinn criter.	7.866259
F-statistic	487508.0	Durbin-Watson stat	1.853716
Prob(F-statistic)	0.000000		

Selection Summary

Number of combinations compar... 35

*Note: p-values and subsequent tests do not account for stepwise selection.

(b)

Figure 6.23 The first and second stages of the STEPLS-Combinatorial selection method.

In this case, I decide to delete the main variable *Y1(−1)* from the eight search regressors. So we have seven search regressors for the next analysis. Then, by doing the same STEPLS-Combinatorial selection method the best model out of 35 sets of regressors in Figure 6.23b is obtained, which shows that the interaction *Y2(−1)∗Y1(−1)* has a Prob. < 0.30, and the interaction *Y2(−2)∗X1* has a large p-value = 0.6335 with the Adjusted R-squared = 0.999927, which is greater the full model with the Adjusted R-squared = 0.999925.

5) So we have four reduced models where three of them have greater Adjusted R-squared values than the full model and one has a smaller value; that is, the model with the only two-way interaction in Figure 6.22a, however, it has a very large Adjusted R-squared = 0.97112. Which one would you choose as the best forecast model of *Y2*? Since it is defined that the effects of both *Y1(−1)* and *X1* depend on both *Y2(−1)* and *Y2(−2)*, then the LS regression in Figure 6.22a would be the best model, because at the 10% level of significance each of the interactions has either positive or negative significant effects on *Y2*, with a p-value = Prob./2 < 0.10.
6) To complete the analysis based on the three models in (6.33), Figure 6.24 presents the statistical results of the model (6.33c) for $p = 2$ as the best possible estimate.

Dependent Variable: X1
Method: Least Squares
Date: 05/27/17 Time: 12:43
Sample (adjusted): 1959Q3 1995Q1
Included observations: 143 after adjustments

Variable	Coefficient	Std. Error	t-Statistic	Prob.
X1(-1)	1.505771	0.072432	20.78873	0.0000
X1(-2)	-0.508407	0.072201	-7.041542	0.0000
T	0.194810	0.049207	3.958960	0.0001

R-squared	0.999931	Mean dependent var	1325.132
Adjusted R-squared	0.999930	S.D. dependent var	964.3053
S.E. of regression	8.062385	Akaike info criterion	7.033052
Sum squared resid	9100.288	Schwarz criterion	7.095209
Log likelihood	-499.8632	Hannan-Quinn criter.	7.058310
Durbin-Watson stat	1.966449		

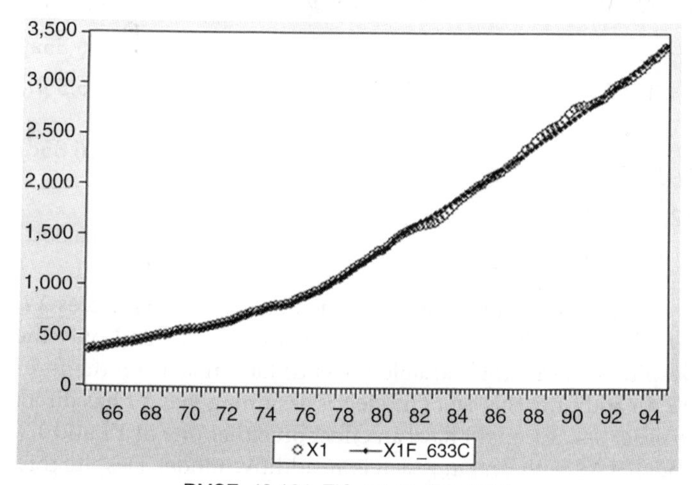

RMSE=42.181, TIC=0.009, BP=0.015

Figure 6.24 Statistical results of the LV(2) model of *X1* with a linear trend.

The in-sample forecast values of *X1* for {@*Year*>1990} are presented as the scores of the variable *X1F_633C* and the forecast values of *X1* beyond the sample period should be computed using the following function, starting with t = @*Trend* + 1 = 146 using Excel.

$$X1(t) = 1.50577126456*X1(t-1) - 0.508406830576*X1(t-2) + 0.194809700321*t$$

7) The forecast values of *Y1* and *Y2* beyond the sample period can be computed using the following regressions, which are obtained based on the LS estimates in Figures 6.21a and 6.22a respectively. Do these as exercises.

$$\begin{aligned}
Y1(t) = {} & 0.00294293801858*Y1(t-1)*Y2(t-1) - 0.00377912821443^{*}Y1(t-2)*Y2(t-1) \\
& - 0.00425691610612*Y1(t-1)*X1(t) + 0.00532382934267*Y1(t-2)*X1(t) \\
& + 0.0562783161219*Y2(t-1) - 0.0678459474891*X1(t) \\
& + 1.17742427949*Y1(t-1) - 0.265796469823*Y1(t-2) \\
& + 1.85305573565
\end{aligned}$$

$$\begin{aligned}
Y2(t) = {} & 0.21571107226*Y2(t-1)*Y1(t-1) - 0.214082821982*Y2(t-2)*Y1(t-1) \\
& - 0.00548918097569*Y2(t-1)*X1(t) + 0.00573126430999*Y2(t-2)*X1(t) \\
& + 587.765821662
\end{aligned}$$

6.5.3 LV(2) Models of *(Y1,Y2)* with two Exogenous Variables and Trend

6.5.3.1 Advanced Model of *(Y1,Y2)* with Exogenous *(X1,X2)* and a Linear Trend

As the extension of the models (6.32a), (6.31b), and (6.31c), the models based on *(Y1,Y2)* with two exogenous variables *X1* and *X2* are presented as a set of the following four ESs, where *IV1* can be either one of *X1*, *X1(−1)*, and *X1(−2)* similar to the ESs (6.32a) and (6.32b), and *IV2* can be one of either *X2*, *X2(−1)*, and *X2(−2)* with the simplest model for $p = q = 1$.

$$X2\ X2(-1) \ldots X2(-p)\ t\ C \tag{6.35a}$$

$$X1\ X1(-1)*IV2 \ldots X1(-p)*IV2\ X1(-1) \ldots X1(-q)\ IV2\ C \tag{6.35b}$$

$$\begin{gathered} Y1\ Y1(-1)*Y2(-1)\ Y1(-2)*Y2(-1)\ Y2(-1) \\ Y1(-1)*IV1\ Y1(-2)*IV1\ IV1\ Y1(-1)\ Y1(-2)\ C \end{gathered} \tag{6.35c}$$

$$\begin{gathered} Y2\ Y2(-1)*Y1(-2)\ Y2(-2)*Y1(-1)\ Y1(-1) \\ Y2(-1)*IV1\ Y2(-2)*IV1\ IV1\ Y2(-1)\ Y2(-2)\ C \end{gathered} \tag{6.35d}$$

This set of ESs is one out of many possible sets, which can be defined based on 12 variables *X1*, *X2*, *Y1*, *Y2*, *X1(−1)*, *X2(−1)*, *Y1(−1)*, *Y2(−1)*, *X1(−2)*, *X2(−2)*, *Y1(−2)*, and *Y2(−2)*, and only the model of *X2* has a linear trend. Despite this, the time variable t (trend) has impacts on the forecasting of both variables *Y1* and *Y2*. In addition, this set of models is assuming that *X1* has direct effects on both *Y1* and *Y2*. For the other set, *X1* might have an effect on either one of *Y1* and *Y2*. And this set of ESs only has the model of *X2* with *Trend*, which is an additive model. Despite this, the forecast model of *Y1* would have several three-way interactions of the time t as its independent variables. This is similar for the model of *Y2*. See Example 6.14.

Example 6.14 An Application of the Models (6.35a)–(6.35d)
As an illustration, the following four ESs are applied.

$$X2\,X2(-1)...X(2(-p))\,t\,C \tag{6.36a}$$

$$X1\,X1(-1)*X2...\,X1(-q)*X2\,X2\,X1(-1)...\,X1(-q)\,C \tag{6.36b}$$

$$\begin{gathered} Y1\,Y1(-1)*Y2(-1)\,Y1(-2)*Y2(-1)\,Y2(-1)\,Y1(-1)*X1\,Y1(-2)*X1\,X1 \\ Y1(-1)Y1(-2)\,C \end{gathered} \tag{6.36c}$$

$$\begin{gathered} Y2\,Y2(-1)*Y1(-1)\,Y2(-2)*Y1(-1)\,Y1(-1)\,Y2(-1)*X1\,Y2(-2)*X1\,X1 \\ Y2(-1)Y2(-2)\,C \end{gathered} \tag{6.36d}$$

where *X1* = *GYD*, *X2* = *GWY*, *Y1* = *GCDAN*, and *Y2* = *GC* in CONS1.wf1. In this case, it is defined that *X1* has a direct effect on both *Y1* and *Y2*, and *X2* has a direct effect on *X1*. Note that the ordering of the four ESs in (6.35a–d) and (6.36a–d) is the same as the process in computing the forecast values of *Y1* and *Y2* beyond the sample period. Refer to the previous examples. The LS estimates of the four ESs or their reduced regressions are as follows:

1) Figure 6.25 presents the LS estimate of the ES (6.36a) and its reduced model. So the forecast values of *X2* beyond the sample period can be computed using the following function.

$$\begin{aligned} X2(t) &= 1.50188036479^{*}X2(t-1) - 0.505515994255^{*}X2(t-2) \\ &\quad + 0.22590904134^{*}t - 1.12190577435 \end{aligned} \tag{6.37a}$$

2) Based on the model (6.36b), the following acceptable reduced regression is obtained, for q = 2, without *X1(t–2)* as the main independent variable.

$$\begin{aligned} X1(t) &= -0.000121441454913^{*}X1(t-1)*X2(t) \\ &\quad + 0.000125201074332^{*}X1(t-2)*X2(t) \\ &\quad + 0.431392871038^{*}X2(t) + 0.726840183201^{*}X1(t-1) \\ &\quad - 20.2020892987 \end{aligned} \tag{6.37b}$$

By inserting the function of *X2(t)* in (6.37a), we have the regression of *X1(t)* with the two-way interactions *X1(t–1)*t* and *X1(t–2)*t* as independent variables in addition to the other interactions.

3) The process of the analysis based on the models (6.36c) and (6.36d) are similar to the models (6.34a) and (6.34b), as follows:

3.1 Figure 6.26 presents the LS estimates of the models (6.36c and 6.36d), which show that each of the two-way interactions has an insignificant effect at the 10% level because of the impact of multicollinearity, which should be accepted.

However, I am very confident that by deleting one or two of the main variables, we would obtain alternative reduced models with at least one significant interaction. Do this as exercise using similar methods as presented in Example 6.13. In this example I present alternative selection methods, as follows:

By inserting the function of *X1(t)* in (6.37b) as the function of the time, *t*, we would have the regression of *Y1(t)* with the interactions *Y1(t–1)*X1(t)*t*, *Y1(t–2)*X1(t)*t*, *Y1(t–1)*X1(t–2)*t*, and. *Y1(t–2)*X1(t–2)*t* as independent variables in addition to the two-way interactions. So we have an advanced or a complex forecast model of *Y1*.

Dependent Variable: X2
Method: Least Squares
Date: 05/29/17 Time: 05:46
Sample (adjusted): 1959Q4 1995Q1
Included observations: 142 after adjustments

Variable	Coefficient	Std. Error	t-Statistic	Prob.
X2(-1)	1.518215	0.085485	17.76004	0.0000
X2(-2)	-0.551707	0.148608	-3.712502	0.0003
X2(-3)	0.029903	0.085587	0.349383	0.7273
T	0.228242	0.071177	3.206658	0.0017
C	-1.062308	1.721080	-0.617234	0.5381

R-squared	0.999931	Mean dependent var	1332.627
Adjusted R-squared	0.999929	S.D. dependent var	963.5302
S.E. of regression	8.132112	Akaike info criterion	7.064095
Sum squared resid	9059.980	Schwarz criterion	7.168173
Log likelihood	-496.5507	Hannan-Quinn criter.	7.106388
F-statistic	494826.6	Durbin-Watson stat	1.993578
Prob(F-statistic)	0.000000		

Dependent Variable: X2
Method: Least Squares
Date: 05/29/17 Time: 05:46
Sample (adjusted): 1959Q3 1995Q1
Included observations: 143 after adjustments

Variable	Coefficient	Std. Error	t-Statistic	Prob.
X2(-1)	1.501880	0.072801	20.63008	0.0000
X2(-2)	-0.505516	0.072468	-6.975757	0.0000
T	0.225909	0.067417	3.350900	0.0010
C	-1.121906	1.658791	-0.676339	0.4999

R-squared	0.999931	Mean dependent var	1325.132
Adjusted R-squared	0.999930	S.D. dependent var	964.3053
S.E. of regression	8.078054	Akaike info criterion	7.043752
Sum squared resid	9070.438	Schwarz criterion	7.126629
Log likelihood	-499.6283	Hannan-Quinn criter.	7.077429
F-statistic	674454.9	Durbin-Watson stat	1.965308
Prob(F-statistic)	0.000000		

Figure 6.25 The LS estimate of a model in (6.36a) with its reduced model.

3.2 Figure 6.27a presents the LS estimate of a reduced model of (6.36c) by using the STEPLS-Combinatorial method with *Y1, C,* and the four interactions as the variables always in the model, and the four main variables, namely *Y2(−1), X1, Y1(−1),* and *Y1(−2),* as the search regressors. The results show that only two of the main variables should be used in order to obtain an acceptable two-way interaction reduced model with each of the interactions having a significant adjusted effect at the 10% level. In fact, at the 5% level, each of them has either a positive or negative significant adjusted effect with a p-value = Prob./2 < 0.05. For instance, *Y1(−1)∗X1* has a negative adjusted effect with a p-value = 0.0761/2 < 0.05.

As a comparison, under the assumption that the eight independent variables are equally important, then the STEPLS: Stepwise selection method with *Y* and *C* as the variables always in the model and the eight independent variables as the search regressors can be applied. The output presents a regression with only the four main variables.

Dependent Variable: Y1
Method: Least Squares
Date: 05/31/17 Time: 19:10
Sample (adjusted): 1959Q3 1995Q1
Included observations: 143 after adjustments

Variable	Coefficient	Std. Error	t-Statistic	Prob.
Y1(-1)*Y2(-1)	0.001631	0.002309	0.706344	0.4812
Y1(-2)*Y2(-1)	-0.002695	0.002342	-1.150999	0.2518
Y2(-1)	0.046499	0.053902	0.862645	0.3899
Y1(-1)*X1	-0.001486	0.002135	-0.695808	0.4878
Y1(-2)*X1	0.002422	0.002164	1.119159	0.2651
X1	-0.037761	0.047406	-0.796541	0.4271
Y1(-1)	0.628826	0.311715	2.017308	0.0457
Y1(-2)	0.280370	0.320512	0.874758	0.3833
C	0.350466	1.732984	0.202232	0.8400

R-squared	0.975436	Mean dependent var	49.19538
Adjusted R-squared	0.973969	S.D. dependent var	29.74958
S.E. of regression	4.799825	Akaike info criterion	6.035905
Sum squared resid	3087.134	Schwarz criterion	6.222378
Log likelihood	-422.5672	Hannan-Quinn criter.	6.111679
F-statistic	665.1323	Durbin-Watson stat	2.073491
Prob(F-statistic)	0.000000		

(a)

Dependent Variable: Y2
Method: Least Squares
Date: 05/31/17 Time: 19:13
Sample (adjusted): 1959Q3 1995Q1
Included observations: 143 after adjustments

Variable	Coefficient	Std. Error	t-Statistic	Prob.
Y2(-1)*Y1(-1)	-0.006001	0.006149	-0.975928	0.3309
Y2(-2)*Y1(-1)	0.006219	0.006255	0.994304	0.3219
Y1(-1)	-0.299825	0.451326	-0.664321	0.5076
Y2(-1)*X1	8.42E-06	0.000114	0.073691	0.9414
Y2(-2)*X1	-1.15E-05	0.000115	-0.099567	0.9208
X1	0.163403	0.044186	3.698043	0.0003
Y2(-1)	1.272901	0.297398	4.280128	0.0000
Y2(-2)	-0.436481	0.310330	-1.406509	0.1619
C	1.056495	5.514829	0.191574	0.8484

R-squared	0.999931	Mean dependent var	1777.961
Adjusted R-squared	0.999926	S.D. dependent var	1393.778
S.E. of regression	11.95722	Akaike info criterion	7.861417
Sum squared resid	19158.66	Schwarz criterion	8.047889
Log likelihood	-553.0913	Hannan-Quinn criter.	7.937190
F-statistic	241154.3	Durbin-Watson stat	1.862237
Prob(F-statistic)	0.000000		

(b)

Figure 6.26 The LS estimates of the models (6.36c) and (6.36d).

However, to compute the forecast of *Y1* beyond the sample period, I prefer to apply the regression in Figure 6.36a with the following function, since an interaction model is acceptable in the theoretical sense whenever the model has at least two independent variables, and the assumption that all independent variables are equally important cannot be accepted.

Dependent Variable: Y1
Method: Stepwise Regression
Date: 05/31/17 Time: 21:10
Sample (adjusted): 1959Q3 1995Q1
Included observations: 143 after adjustments
Number of always included regressors: 5
Number of search regressors: 4
Selection method: Combinatorial
Number of search regressors: 2

Variable	Coefficient	Std. Error	t-Statistic	Prob.*
C	0.782083	1.678498	0.465942	0.6420
Y1(-1)*X1	-0.002824	0.001580	-1.787267	0.0761
Y1(-1)*Y2(-1)	0.003041	0.001756	1.731842	0.0856
Y1(-2)*X1	0.003373	0.001601	2.106413	0.0370
Y1(-2)*Y2(-1)	-0.003658	0.001782	-2.053292	0.0420
Y2(-1)	0.003571	0.001650	2.164338	0.0322
Y1(-1)	0.882776	0.063289	13.94824	0.0000

R-squared	0.975208	Mean dependent var	49.19538
Adjusted R-squared	0.974115	S.D. dependent var	29.74958
S.E. of regression	4.786387	Akaike info criterion	6.017141
Sum squared resid	3115.691	Schwarz criterion	6.162175
Log likelihood	-423.2256	Hannan-Quinn criter.	6.076076
F-statistic	891.6221	Durbin-Watson stat	2.090831
Prob(F-statistic)	0.000000		

Selection Summary

Number of combinations compar... 6

*Note: p-values and subsequent tests do not account for stepwise selection.

(a)

Dependent Variable: Y1
Method: Stepwise Regression
Date: 06/01/17 Time: 19:59
Sample (adjusted): 1959Q3 1995Q1
Included observations: 143 after adjustments
Number of always included regressors: 1
Number of search regressors: 8
Selection method: Stepwise forwards
Stopping criterion: p-value forwards/backwards = 0.5/0.5

Variable	Coefficient	Std. Error	t-Statistic	Prob.*
C	0.715789	0.859910	0.832400	0.4066
Y1(-1)	0.582639	0.077853	7.483825	0.0000
Y1(-2)	0.314435	0.079768	3.941859	0.0001
X1	0.036686	0.010139	3.618206	0.0004
Y2(-1)	-0.038201	0.010801	-3.536924	0.0006

R-squared	0.974744	Mean dependent var	49.19538
Adjusted R-squared	0.974012	S.D. dependent var	29.74958
S.E. of regression	4.795830	Akaike info criterion	6.007710
Sum squared resid	3173.998	Schwarz criterion	6.111306
Log likelihood	-424.5513	Hannan-Quinn criter.	6.049806
F-statistic	1331.537	Durbin-Watson stat	2.049060
Prob(F-statistic)	0.000000		

Selection Summary

Added Y1(-1)
Added Y1(-2)
Added X1
Added Y2(-1)

*Note: p-values and subsequent tests do not account for stepwise selection.

(b)

Figure 6.27 Two alternative reduced models of the model (6.36c).

$$
\begin{aligned}
Y1(t) = &(0.00304105669283*Y1(t-1) - 0.00365811821926*Y1(t-2) \\
&+ 0.00357101251661)*Y2(t-1) \\
&+ (-0.00282389378024^{*}Y1(t-1) \\
&+ 0.00337319539428^{*}Y1(t-2))*X1(t) \\
&+ 0.882776080587*Y1(t-1) + 0.78208349187
\end{aligned} \tag{6.37c}
$$

3.3 Similar to the results in Figure 6.27a, the LS estimate of a reduced model of the model (6.36d) by using the STEPLS-Combinatorial method with *Y2*, *C*, and the four interactions as the variables always in the model, along with the four main variables, namely *Y1(−1)*, *X1*, *Y2(−1)*, and *Y2(−2)* as the search regressors, shows each of the interactions has a very large *p*-value by inserting one of the main variables. For this reason, the LS estimate is presented in Figure 6.28a, which shows that each of the interactions has either positive or negative significant effects with a p-value = Prob./2 < 0.05.

3.4 Figure 6.28b presents the results by using the default Stepwise Selection method, with *Y* and *C* as the variables always in the model and the eight independent variables as the search regressors. Even though this regression has only three independent variables, it is better than the regression in Figure 6.28a, since it has a greater Adjusted R-squared value.

3.5 Then, to compute the forecast values of *Y2* beyond the sample period, the following regression should be applied together with the three regressions (6.37a)–(6.37c) by using Excel.

$$
\begin{aligned}
Y2(t) = &(0.221130839645^{*}Y2(t-1) - 0.219008010802^{*}Y2(t-2))*Y1(t-1) \\
&+ (-0.00384410243762^{*}Y2(t-1) + 0.00399398122594^{*}Y2(t-2))*X1(t) \\
&+ 590.371180834
\end{aligned} \tag{6.37d}
$$

4) Figure 6.29 presents the first three forecast values of *X2*, *X1*, *Y1*, and *Y2* beyond the sample period. The steps of the computation are as follows:

4.1 First, copy the last five observed scores of the variables *T*, *X2*, *X1*, *Y1*, and *Y2*, which are for the @*Year*>1993, to the spreadsheet.

4.2 The first forecast values of *X2*, *X1*, *Y1*, and *Y2*, respectively, can be computed using the following equations for the time t = 146.

$$
\begin{aligned}
X2(t) &= 1.50188036479^{*}X2(t-1) - 0.505515994255^{*}X2(t-2) \\
&\quad + 0.22590904134^{*}t - 1.12190577435 \\
&= 1.50188036479^{*}C6 - 0.505515994255^{*}C5 \\
&\quad + 0.22590904134^{*}B7 - 1.12190577435
\end{aligned} \tag{6.38a}
$$

$$
\begin{aligned}
X1(t) &= -0.000121441454913^{*}X1(t-1)*X2(t) + 0.000125201074332^{*}X1(t-2)*X2(t) \\
&\quad + 0.431392871038^{*}X2(t) + 0.726840183201^{*}X1(t-1) - 20.2020892987 \\
&= -0.000121441454913^{*}D6^{*}C7 + 0.000125201074332^{*}D5^{*}C7 \\
&\quad + 0.431392871038^{*}C7 + 0.726840183201^{*}D6 - 20.2020892987
\end{aligned} \tag{6.38b}
$$

Dependent Variable: Y2
Method: Least Squares
Date: 05/31/17 Time: 20:45
Sample (adjusted): 1959Q3 1995Q1
Included observations: 143 after adjustments

Variable	Coefficient	Std. Error	t-Statistic	Prob.
C	590.3712	28.05481	21.04350	0.0000
Y2(-1)*Y1(-1)	0.221131	0.100404	2.202411	0.0293
Y2(-2)*Y1(-1)	-0.219008	0.102182	-2.143313	0.0338
Y2(-1)*X1	-0.003844	0.002115	-1.817645	0.0713
Y2(-2)*X1	0.003994	0.002149	1.858129	0.0653

R-squared	0.971151	Mean dependent var	1777.961
Adjusted R-squared	0.970315	S.D. dependent var	1393.778
S.E. of regression	240.1400	Akaike info criterion	13.83466
Sum squared resid	7958078.	Schwarz criterion	13.93826
Log likelihood	-984.1782	Hannan-Quinn criter.	13.87676
F-statistic	1161.377	Durbin-Watson stat	0.399997
Prob(F-statistic)	0.000000		

(a)

Dependent Variable: Y2
Method: Stepwise Regression
Date: 06/04/17 Time: 07:48
Sample (adjusted): 1959Q3 1995Q1
Included observations: 143 after adjustments
Number of always included regressors: 1
Number of search regressors: 8
Selection method: Stepwise forwards
Stopping criterion: p-value forwards/backwards = 0.5/0.5

Variable	Coefficient	Std. Error	t-Statistic	Prob.*
C	-2.553224	2.716158	-0.940013	0.3488
Y2(-1)	0.858445	0.044767	19.17584	0.0000
X1	0.147912	0.037227	3.973273	0.0001
Y2(-2)*X1	-1.41E-06	1.02E-06	-1.385909	0.1680

R-squared	0.999929	Mean dependent var	1777.961
Adjusted R-squared	0.999928	S.D. dependent var	1393.778
S.E. of regression	11.83456	Akaike info criterion	7.807498
Sum squared resid	19467.88	Schwarz criterion	7.890375
Log likelihood	-554.2361	Hannan-Quinn criter.	7.841175
F-statistic	656477.3	Durbin-Watson stat	1.790856
Prob(F-statistic)	0.000000		

Selection Summary

Added Y2(-1)
Added X1
Added Y2(-2)*X1

*Note: p-values and subsequent tests do not account for stepwise selection.

(b)

Figure 6.28 Two alternative reduced models of the model (6.36d).

	A	B	C	D	E	F
1		*T*	*X2*	*X1*	*Y1*	*Y2*
2	1994Q1	141	3195.20	4856.90	91.98	4599.20
3	1994Q2	142	3242.80	5002.20	91.94	4665.10
4	1994Q3	143	3265.50	5070.40	90.49	4734.40
5	1994Q4	144	3320.20	5145.80	90.75	4796.03
6	1995Q1	145	3363.00	5225.50	83.81	4836.33
7	1995Q2	146	***3404.27***	***5279.39***	***85.78***	***5167.65***
8	1995Q3	147	*3444.84*	*5348.28*	*75.13*	*4820.78*
9	1995Q4	148	*3485.14*	*5410.63*	*90.68*	*7056.76*

Figure 6.29 The first three forecast values of *X2*, *X1*, *Y1*, and *Y2* beyond the sample period.

$$\begin{aligned} Y1(t) &= (0.00304105669283^{*}Y1(t-1) - 0.00365811821926^{*}Y1(t-2) \\ &\quad + 0.00357101251661) * Y2(t-1) \\ &\quad + (-0.00282389378024^{*}Y1(t-1) + 0.00337319539428^{*}Y1(t-2)) * X1(t) \\ &\quad + 0.882776080587^{*}Y1(t-1) + 0.78208349187 \\ &= (0.00304105669283^{*}\mathrm{E6} - 0.00365811821926^{*}\mathrm{E5} + 0.00357101251661) * \mathrm{F6} \\ &\quad + (-0.00282389378024^{*}\mathrm{E6} + 0.00337319539428^{*}\mathrm{E5}) * \mathrm{D7} \\ &\quad + 0.882776080587^{*}\mathrm{E6} + 0.78208349187 \end{aligned} \tag{6.38c}$$

$$\begin{aligned} Y2(t) &= (0.221130839645^{*}Y2(t-1) - 0.219008010802^{*}Y2(t-2)) * Y1(t-1) \\ &\quad + (-0.00384410243762 * Y2(t-1) + 0.00399398122594^{*}Y2(t-2)) * X1 \\ &\quad + 590.371180834 \\ &= (0.221130839645^{*}\mathrm{F6} - 0.219008010802^{*}\mathrm{F5}) * \mathrm{E6} \\ &\quad + (-0.00384410243762^{*}\mathrm{F6} + 0.00399398122594^{*}\mathrm{F5}) * \mathrm{D7} + 590.371180834 \end{aligned} \tag{6.38d}$$

4.3 Then their forecast values for $t = 147$ and $t = 148$ can be obtained using the block-copy-paste method for the values for $t = 146$. Based on these results, the following findings and notes are presented.

a) The forecast values of *X2* and *X1* increase for the three time points. Their values are reasonable or acceptable.
b) The first forecast value of *Y1* is reasonable, but the second forecast value for the time $t = 147$ is very low compared to the first and third forecast values and the observed scores. So, it is an unexpected forecast value.
c) On the other hand, an unbelievably large forecast value of *Y2* is obtained for the time $t = 148$, even though its first two forecast values could be considered as reasonable or acceptable values.

5) As an additional illustration, Figure 6.30 presents the forecast of *Y1* and *Y2* based on the full models with their LS estimates presented in Figure 6.26, and their simplest reduced model presented in Figures 6.27b and 6.28b, respectively, which are obtained using the default Stepwise Selection method. Based on these results, the following findings and notes are presented.
   5.1 By using the same type of models or the same estimation method, their first forecast values for the time $t = 146$ can be considered acceptable values, which are very close to the last observed scores for $t = 145$. This is similar for their first forecast values presented in Figure 6.29.
   5.2 Compared to the forecast values of *Y1* in Figure 6.29, the forecast values of *Y1* for $t = 147$ and $t = 148$ are far too far from the first forecast value; one of them is even negative. So it can be said that both regressions in Figures 6.27b and 6.28b are not acceptable forecast models of *Y1* and the forecast model in Figure 6.29 is the best among the three models of *Y1*.
   5.3 On the other hand, the forecast values of *Y2* based on both its LS estimates are acceptable and they are better than its forecast in Figure 6.29.
6) Based on these findings, the unexpected or unbelievable forecast values beyond the sample period presented previously, and notes and comments presented in Sections 1.6 and 1.7, I have the following notes and statements.
   - *We can easily evaluate the in-sample forecast values obtained based on any multiple regression. However, we never know what type of model would give the best possible consecutive forecast values beyond the sample period. So, a researcher should try to develop alternative good fit models for the data with acceptable in-sample forecast values, supported by his or her best expert judgment. However, a good fit model may present worse in-sample forecast values – refer to the Example 5.24.*
   - *Since the OLS estimates of the model parameters are obtained without using the basic assumptions of the error terms, then the forecast values are also computed without using the assumptions of the error terms of the model. Refer to special notes and comments presented in Sections 1.6 and 1.7.*
   - *In the theoretical sense, the effect of an exogenous variable always depends on at least one of the other exogenous variables. Hence, I would say that an interaction model is most likely to be a better model than its corresponding additive model. However, in almost all cases a*

	A	B	C	D	E	F	G	H
1		T	X2	X1	Y1		Y2	
2	1994Q1	141	3195.20	4856.90	91.98		4599.20	
3	1994Q2	142	3242.80	5002.20	91.94		4665.10	
4	1994Q3	143	3265.50	5070.40	90.49		4734.40	
5	1994Q4	144	3320.20	5145.80	90.75		4796.03	
6	1995Q1	145	3363.00	5225.50	83.81		4836.33	
7	1995Q2	146	***3402.73***	***5278.71***	***85.33***	***86.99***	**4892.93**	**4894.178**
8	1995Q3	147	***3440.47***	***5346.13***	*279.80*	*269.59*	**4950.69**	**4951.984**
9	1995Q4	148	***3476.87***	***5406.01***	*-1060.0*	*378.59*	**5089.88**	**5009.579**

Figure 6.30 Forecast values of *Y1* and *Y2* based on full models, and their simplest reduced models.

nonhierarchical interaction model would be acceptable, in the statistical sense, because of the impacts of multicollinearity. Refer to notes by Hamel, Bezzecri, and Tukey presented in Section 5.11.

6.5.3.2 Advanced Model of *(Y1,Y2)* with Exogenous *(X1,X2)* and Interaction Trend

As more advanced models, for the model of *X2* in (6.35a) we can apply various alternative *Trends*, shown in Table 6.1. Furthermore, the alternative trends also could be used for the other models in the set (6.35a)–(6.35d). So we can have lots of possible forecast models that are more advanced for *Y1* and *Y2*. However, this section presents a modification of the set of models in (6.35a)–(6.35d), specifically to only replace the model of *X2* in (6.35a) with the following ES, where the integer p is selected using the trial-and-error method.

$$X2\,X2(-1)*T...X2(-p)*T\,X2(-1)...X2(-p)\;T\;C \tag{6.39a}$$

with the other three ESs, namely (6.39b, (6.39c), and (6.39d), which are exactly the same as the ESs (6.35b–6.35d).

Example 6.15 An Application of the Model (6.39a)

Figure 6.31 presents an acceptable LS estimate of the model (6.39a) for $p = 2$ with its reduced model. Based on these outputs, the findings and notes presented are as follows:

1) In fact, the regression in Figure 6.31a is an acceptable forecast model of *X2*, even though both interactions have insignificant effects. Refer to the regression of *Y2* in (6.36d) in which most of its independent variables have large *p*-values, but it presents acceptable forecast values as presented in Figure 6.30.
2) As a comparison or additional illustration, Figure 6.31b presents its reduced model. So we could present two alternative forecast values of *X2*, based on the following functions. Do these as exercises.

$$\begin{aligned} X2(t) = & -0.00198846154209^{*}X2(t-1)*t + 0.00195944450807^{*}X2(t-2)*t \\ & + 1.70555981664^{*}X2(t-1) - 0.700323531209^{*}X2(t-2) \\ & + 0.146266276174^{*}t - 2.99612515751 \end{aligned} \tag{6.40a}$$

$$\begin{aligned} X2(t) = & \,0.00410405857532^{*}X2(t-1)*t - 0.00418817714535^{*}X2(t-2)*t \\ & + 1.00642356851^{*}X2(t-1) + 0.241082080018^{*}t - 1.94023486881 \end{aligned} \tag{6.40b}$$

6.5.4 LV(2) Models of (*Y1,Y2*) with Three Exogenous Variables and Trend

As an extension of the models (6.35a)–(6.35d), this section presents alternative models with an additional exogenous or environmental variable, say *Z1*, which has a direct effect on at least one of the four dependent variables in (6.35a)–(6.35d), namely *X1*, *X2*, *Y1*, and *Y2*. So we would have a lot of possible models since we can insert at least one of *Z*, *Z(−1)*, or *Z(−2)* as an additional main independent variable, besides its interaction with specific independent variables of the models in (6.35a)–(6.35d). However, this section only presents a special set of five models and one of its extensions in the following illustrative examples.

Dependent Variable: X2
Method: Least Squares
Date: 06/02/17 Time: 20:08
Sample (adjusted): 1959Q3 1995Q1
Included observations: 143 after adjustments

Variable	Coefficient	Std. Error	t-Statistic	Prob.
X2(-1)*T	-0.001988	0.002817	-0.705989	0.4814
X2(-2)*T	0.001959	0.002837	0.690711	0.4909
X2(-1)	1.705560	0.313270	5.444384	0.0000
X2(-2)	-0.700324	0.313545	-2.233567	0.0271
T	0.146266	0.107407	1.361798	0.1755
C	-2.996125	2.703520	-1.108231	0.2697

R-squared	0.999932	Mean dependent var	1325.132
Adjusted R-squared	0.999929	S.D. dependent var	964.3053
S.E. of regression	8.108185	Akaike info criterion	7.064678
Sum squared resid	9006.746	Schwarz criterion	7.188993
Log likelihood	-499.1244	Hannan-Quinn criter.	7.115193
F-statistic	401671.0	Durbin-Watson stat	1.941910
Prob(F-statistic)	0.000000		

(a)

Dependent Variable: X2
Method: Least Squares
Date: 06/02/17 Time: 20:04
Sample (adjusted): 1959Q3 1995Q1
Included observations: 143 after adjustments

Variable	Coefficient	Std. Error	t-Statistic	Prob.
X2(-1)*T	0.004104	0.000712	5.764594	0.0000
X2(-2)*T	-0.004188	0.000697	-6.008997	0.0000
X2(-1)	1.006424	0.012848	78.33080	0.0000
T	0.241082	0.100078	2.408950	0.0173
C	-1.940235	2.700064	-0.718589	0.4736

R-squared	0.999929	Mean dependent var	1325.132
Adjusted R-squared	0.999927	S.D. dependent var	964.3053
S.E. of regression	8.224532	Akaike info criterion	7.086459
Sum squared resid	9334.724	Schwarz criterion	7.190055
Log likelihood	-501.6818	Hannan-Quinn criter.	7.128555
F-statistic	487982.6	Durbin-Watson stat	1.874190
Prob(F-statistic)	0.000000		

(b)

Figure 6.31 An LS acceptable estimate of the model (6.39a), and its reduced model.

Example 6.16 A Special Set of Five Models

As an extension of the set of models (6.35a)–(6.35d), a set of five models proposed has the set of ESs as follows:

$$\begin{gathered} Y1\ Y1(-1)*Y2(-1)\ Y1(-2)*Y2(-1)\ Y2(-1) \\ Y1(-1)*IV\ Y1(-2)*IV1\ IV1\ Y1(-1)\ Y1(-2)\ Z1(-1)\ C \end{gathered} \tag{6.41a}$$

$$\begin{gathered} Y2\ Y2(-1)*Y1(-1)\ Y2(-2)*Y1(-1)\ Y1(-1) \\ Y2(-1)*IV\ Y2(-2)*IV1\ IV1\ Y2(-1)\ Y2(-2)\ C \end{gathered} \tag{6.41b}$$

Y1	C	Z1(-1)	Y1(-1)	Y1(-2)	Y1(-2)*X1(-1)	Y2(-1)	Y1(-1)*Y2(-1)	Y1(-1)*X1(-1)
Coef.	-0.7212	-0.0074	0.4800	0.5934	-0.0001	0.0071	-0.0005	0.0005
t-Stat.	-0.4601	-3.3823	2.0582	2.4612	-1.1106	3.8138	-2.4276	2.2700
Prob.	0.6462	0.0009	0.0415	0.0151	0.2687	0.0002	0.0165	0.0248
Y2	C		Y2(-1)	Y2(-1)*X1(-1)	X1(-1)	Y2(-2)*X1(-1)	Y2(-1)*Y1(-1)	Y2(-2)*Y1(-1)
Coef.	-11.9804		1.1323	0.0001	-0.0752	-0.0001	-0.0060	0.0060
t-Stat.	-3.5619		20.7114	1.0123	-1.6524	-1.0600	-1.1132	1.0961
Prob.	0.0005		0.0000	0.3132	0.1008	0.2910	0.2676	0.2750
X1	C		X1(-1)*X2(-1)	X1(-2)*X2(-1)	X1(-1)	X1(-2)	X2(-1)	
Coef.	-18.5449		-0.0003	0.0003	1.1602	-0.3646	0.3367	
t-Stat.	-3.5970		-2.7993	2.7785	4.6315	-1.4814	4.2301	
Prob.	0.0004		0.0059	0.0062	0.0000	0.1408	0.0000	
X2	C		X2(-1)	X2(-2)	T			
Coef.	-1.1219		1.5019	-0.5055	0.2259			
t-Stat.	-0.6763		20.6301	-6.9758	3.3509			
Prob.	0.4999		0.0000	0.0000	0.0010			
Z1	C	Z1(-1)	Z1(-2)	Z1(-3)	Z1(-4)			
Coef.	34.6910	1.2834	-0.6487	0.5824	-0.2725			
t-Stat.	2.1218	15.5741	-5.0657	4.5462	-3.3100			
Prob.	0.0357	0.0000	0.0000	0.0000	0.0012			

Figure 6.32 Summary of the LS estimates of the five models in (6.41a) and (6.41b).

$$X1\ X1(-1)*IV2\ \ldots\ X1(-p1)*IV2\ X1(-1)\ \ldots\ X1(-p1)\ IV2\ C \tag{6.41c}$$

$$X2\ X2(-1)\ \ldots\ X2(-p2)\ T\ C \tag{6.41d}$$

$$Z1\ Z1(-1)\ \ldots\ Z(-p3)\ C \tag{6.41e}$$

where *IV1* is one of *X1*, *X1(−1)*, and *X1(−2)*, *IV2* is one of *X2*, *X2(−1)*, and *X2(−2)*, and *Z1(−1)* is defined to have a direct effect only on *Y1* so is used as a main independent variable of the model.

As an illustration, Figure 6.32 presents the summary of the LS estimates of the five models (6.41a)–(6.41e) with the variables *Y1* = *GCDAN* (Expenditures New Autos), *Y2* = *GC* (Consumption Expenditures), *X1* = *GYD* (Disposable Income), *X2* = *GW* (Wages and salaries), and $Z1 = 100*R$ (Interest Rates), in CONS1.wf1. In this case, it is important to note that the interest rate *Z1(−1)* has a direct effect on *GCDAN* as the "expenditures of new autos." The steps of the analysis are as follows:

1) First, the analysis of the model (6.41a) is done using the combinatorial selection method, with *Y1*, *C*, and *Z1(−1)* as the variables always in the model, and the other eight independent variables of the model as the search regressors. Then by using the trial-and-error method to select the subset of the eight search regressors, the regression in Figure 6.32 is obtained, which shows each of the independent variables has a significant adjusted effect at the 1 or 5% levels, except $Y2(-1)*X1(-1)$ with a Prob. = 0.2867 < 0.30. So, at the 15% level, it has a negative significant adjusted effect with a p-value = Prob./2 = 0.14335 < 0.15.
2) Second, the analysis of the model (6.41b) is done using the combinatorial selection and the trial-and-error methods, with *Y1* and *C* as the variables always in the model and the eight independent variables of the model as the search regressors. Based on the results, the following findings and notes are presented.

   2.1 Even though, one of the independent variables $Y2(-1)*X1(-1)$ has a Prob. = 0.3132 > 0.30, I consider this model acceptable, since the null hypothesis H_0: C(3) = C(4) = C(5) = 0 is rejected based on the Wald (F-stat) of F_0 = 11.19254, with df = (3136), and

p-value = 0.0000, which gives the conclusion that effect of *X1(−1)* on *Y2* is significantly dependent on both *Y2(−1)* and *Y2(−2)*.

2.2 At the 10% level, the null hypothesis H_0: C(6) = C(7) = 0 is rejected based on the Wald (*F*-stat) of F_0 = 2.559715, with *df* = (2136), and *p*-value = 0.0810, which gives the conclusion that the effect of *Y1(−1)* on *Y2* is significantly dependent on both *Y2(−1)* and *Y2(−2)*.

2.3 In this case, we have the following regression function, which shows the effect of *X1(−1)* on *Y2* depends on the function (0.00012∗*Y2(−1)* – 0.07524 – 0.00012∗*Y2(−2))*∗*X1(−1)* and the effect of *Y1(−1)* on *Y2* depends on the function (– 0.00598∗*Y2(−1)* + 0.00599∗*Y2(−2)*).

$$\begin{aligned} Y2 &= -11.98035 + 1.13233^{*}Y2(-1) \\ &\quad + [0.00012^{*}Y2(-1) - 0.07524 - 0.00012^{*}Y2(-2)]*X1(-1) \\ &\quad + [-0.00598^{*}Y2(-1) + 0.00599^{*}Y2(-2)]*Y1(-1) \end{aligned}$$

3) The LS estimate of each of the three other models is obtained using the trial-and-error method and they are acceptable results. By inserting these three functions in the regressions of *Y1* and *Y2*, we will have complex forecast models of *Y1* and *Y2*, with three-way interactions and two-way time-interactions, or the TRE.

Example 6.17 An Extension of the Set of Five Models in (6.41a)–(6.41e)

Under the assumption that the effect *X1* = *GYD* on *Y1* = *GCDAN* and *Y2* = *GC* depends on *Z1* = *100*∗*R*, then at least one of the interactions of *IV1*∗*Z1*, where *IV1* can be any one of *X1*, *X1(−1)*, and *X1(−2)* should be used as the independent variable of the models (6.41a) and (6.41b). From all possible models, I propose the following two ESs to replace (6.41a) and (6.41b).

$$\begin{gathered} Y1\ Y1(-1)*Y2(-1)\ Y1(-2)*Y2(-1)\ Y2(-1) \\ Y1(-1)*IV\ Y1(-2)*IV1\ IV1\ Y1(-1)\ Y1(-2)\ IV1^{*}Z1(-1)\ Z1(-1)\ C \end{gathered} \tag{6.42a}$$

$$\begin{gathered} Y2\ Y2(-1)*Y1(-1)\ Y2(-2)*Y1(-1)\ Y1(-1) \\ Y2(-1)*IV\ Y2(-2)*IV1\ IV1\ Y2(-1)\ Y2(-2)\ IV1^{*}Z1(-1)\ Z1(-1)\ C \end{gathered} \tag{6.42b}$$

As an illustration, for *IV* = *X1(−1)*, by using the STEPLS-Combinatorial Selection and trial-and-error methods with *Y1/Y2*, *C*, *X1(−1)*∗*Z1(−1)*, and *Z1(−1)* as the variables always in the model, and the other eight independent variables as the search regressors, the LS estimates in Figure 6.33 are obtained, which are acceptable reduced models. Based on these results, the following findings and notes are presented.

1) Even though *X1(−1)*∗*Z1(−1)* of the first LS estimate of *Y1* has a Prob. = 0.2275, but at the 15% level (Lapin 1973), it has a positive significant adjusted effect with a *p*-value = Prob./2 = 0.2275/2 < 0.15. In addition, both *X1(−1)*∗*Z1(−1)* and *Z1(−1)* has a significant partial joint effects on *Y1*, based on the Wald (*F*-stat) of F_0 = 8.474828, with *df* = (2134), and *p*-value = 0.0021.
2) As an additional illustration, Figure 6.33 also presents an alternative LS estimate of *Y1-alt*, which is the best combination of four out of 35 possible combinations. Compared to the first regression of *Y1*, we never know which one gives better forecast values beyond the sample period before we do the computation – refer the unbelievable results in Figures 6.30 and comparative models presented in the Example 5.24. Do these as exercises.

Y1	C	X1(-1)*Z1(-1)	Z1(-1)	Y1(-1)	Y1(-2)	Y1(-2)*X1(-1)	Y2(-1)	Y1(-1)*X1(-1)	Y1(-1)*Y2(-1)
Coef.	0.7572	0.0000	-0.0128	0.4215	0.6686	-0.0001	0.0074	0.0004	-0.0004
t-stat.	0.3817	1.2123	-2.5741	1.7730	2.6901	-1.4366	3.9471	1.7667	-1.7659
Prob.	0.7033	0.2275	0.0111	0.0785	0.0081	0.1532	0.0001	0.0796	0.0797
Y1-alt	C	X1(-1)*Z1(-1)	Z1(-1)	Y1(-1)	Y1(-2)	Y1(-2)*Y2(-1)	X1(-1)		
Coef.	0.4749	0.0000	-0.0145	0.4812	0.6414	-0.0001	0.0072		
t-stat.	0.2467	1.7448	-3.1485	6.0636	6.5650	-4.9587	4.5406		
Prob.	0.8055	0.0833	0.0020	0.0000	0.0000	0.0000	0.0000		
Y2	C	X1(-1)*Z1(-1)	Z1(-1)	Y2(-1)	Y2(-1)*X1(-1)	Y1(-1)	Y2(-1)*Y1(-1)		
Coef.	-8.1479	0.0000	-0.0018	1.0852	0.0000	-0.9816	0.0002		
t-stat.	-1.6503	-2.0949	-0.1591	85.2170	-4.5537	-2.3301	1.4744		
Prob.	0.1012	0.0380	0.8739	0.0000	0.0000	0.0213	0.1427		

Figure 6.33 Summary of the LS estimates of a set of models in (6.42a) and (6.42b).

3) Even though *Z1(−1)* of the *Y2*-regression has a large Prob. = 0.8739, it can be kept in the model because its interaction *X1(−1)∗Z1(−1)* has a significant adjusted effect on *Y2*. In addition, both *X1(−1)∗Z1(−1)* and *Z1(−1)* has significant partial joint effects on *Y2* based on the Wald (F-stat) of $F_0 = 8.791009$, with $df = (2137)$, and p-value = 0.0003. So there is no problem with this regression.

6.6 Triangular Effects Models Based on (*X1,X2,Y1*)

It should be recognized that *X1* = *GYD*, *X2* = *GY*, and *Y1* = *GCDAN* have triangular causal effects, because both *X1* and *X2* have direct effects on *Y1* and *X2* has a direct effect on *X1*. So all triangular effects models (TE_Models) presented in Chapter 5 are acceptable models for the triple time series *(X1,X2,Y1)* with adjustment or modification, since in that chapter the models are presented using the triple time series *(Y1,Y2,Y3)*.

6.6.1 Partial Two-Way Interaction LV(p) TE_Models

Since *Y1(−1)* has been known as one of the best predictors of *Y1*, then the effects of *X1* and *X2* on *Y1* should depend on *Y1(−1)*. Hence, in the statistical sense, the interaction *X1∗Y1(−1)* and *X2∗Y1(−1)* should be used as independent variables of the *Y1*-regression. Similarly, the effect of *X2* on *X1* depends on *X1(−1)*. In this section the defined models are presented as the following set of three ESs, called LV(p) partial two-way interaction triangular effects models, where the integers $p1$, $p2$, and $p3$ can easily be selected using the trial-and-error method with the simplest models for $p1 = p2 = p3 = 1$.

$$Y1\, \boldsymbol{X1^{*}Y1(-1)}\, \boldsymbol{X2^{*}Y1(-1)}\, \boldsymbol{X1}\, \boldsymbol{X2}\, Y1(-1) ... Y1(-p1)\, C \tag{6.43a}$$

$$X1\, \boldsymbol{X2^{*}X1(-1)}\, \boldsymbol{X2}\, X1(-1) ... X1(-p2)\, C \tag{6.43b}$$

$$X2\, X2(-1) ... X2(-p3)\, C \tag{6.43c}$$

DV=Y1	C	X1*Y1(-1)	X2*Y1(-1)	X1	X2	Y1(-1)
Coef.	1.7641	-0.0007	0.0010	0.0591	-0.0804	0.9295
t-Stat.	0.7222	-1.6472	1.5175	1.9418	-1.7678	10.9741
Prob.	0.4714	0.1018	0.1314	0.0542	0.0793	0.0000
DV=X1	**C**	**X2*X1(-1)**	**X2**	**X1(-1)**	**X2(-2)**	**X2(-3)**
Coef.	-16.2697	3.71E-06	0.8098	0.7504	-0.9657	0.5391
t-Stat.	-3.0743	1.6498	7.0589	13.7261	-3.5752	2.6276
Prob.	0.0026	0.1013	0.0000	0.0000	0.0005	0.0096
DV=X2	**C**	**X2(-1)**	**X2(-2)**			
Coef.	2.6692	1.6086	-0.6039			
t-Stat.	2.1241	23.7187	-8.7995			
Prob.	0.0354	0.0000	0.0000			

Figure 6.34 The statistical results of a set of models in (6.43a)–(6.43c).

Example 6.18 Statistical Results of a Set of Models in (6.43a)–(6.43c)
Figure 6.34 presents a summary of acceptable LS regressions of the models (6.43a)–(6.43c), which are obtained by using the trial-and-error method. Based on these results, the following findings and notes are presented.

1) Since, at the 10% level of significance, the interaction $X1*Y1(-1)$ has a negative adjusted effect on $Y1$ with a p-value = 0.1018/2 = 0.0509, and $X2*Y1(-1)$ has a positive significant effect on $Y1$ with a p-value = 0.1314/2 = 0.0657, the model is an acceptable model, in the statistical sense.
2) As a comparison, if $Y1(-2)$ is used as an additional independent variable, the $X1*Y(-1)$ has a negative effect with Prob. = 0.2062 and $X2*Y1(-1)$ has a positive effect with Prob. = 0.2621.
3) Note that the very small coefficient of $X1*X2$ in the model (6.43b) could be modified by doing analysis using the variables $X1/10$ and $X2/10$. Do this as an exercise.

6.6.2 A Complete Two-Way Interaction LV(p) TE_Models

As an extension of the set of models in (6.43a)–(6.43c), the following set of a complete two-way interaction LV(p) TE_Models is defined to present the effects of each of $X1$ and $X2$ on $Y1$ depending on $Y1(-1)$, since $Y1(-1)$ is one of the best predictors for $Y1$.

$$Y1\ \boldsymbol{X1^{*}X2}\ \boldsymbol{X1^{*}Y1(-1)}\ \boldsymbol{X2^{*}Y1(-1)}\ \boldsymbol{X1}\ \boldsymbol{X2}\ Y1(-1) \ldots Y1(-p1)\ C \quad (6.44a)$$

$$X1\ \boldsymbol{X2^{*}X1(-1)}\ \boldsymbol{X2}\ X1(-1) \ldots X1(-p2)\ C \quad (6.44b)$$

$$X2\ X2(-1) \ldots X2(-p3)\ C \quad (6.44c)$$

Example 6.19 Statistical Results of the Models (6.44a) and (6.44b)
Figure 6.35 presents a summary of acceptable LS regressions of the models (6.44a) and (6.44b), which are obtained by using the manual multistage selection and the STEPLS-Combinatorial selection methods because the LS estimate of the model (6.44c) is the same as (6.43c), which has been presented in Figure 6.33. Based on these results, the following findings and notes are presented.

DV=Y1	C	X1*X2	X1*Y1(-1)	X2*Y1(-1)	Y1(-2)	Y1(-1)	X1
Coef.	0.1900	-3.29E-06	0.0004	-0.0005	0.3252	0.4327	0.0104
t-Stat.	0.1002	-2.0611	1.7752	-1.6552	3.9291	2.4933	2.8025
Prob.	0.9203	0.0412	0.0781	0.1002	0.0001	0.0139	0.0058
DV=X1	C	X2*X1(-1)	X2	X1(-1)			
Coef.	-18.7012	4.43E-06	0.4644	0.6978			
t-Stat.	-3.8357	1.9379	6.7614	13.9486			
Prob.	0.0002	0.0546	0.0000	0.0000			

Figure 6.35 The LS estimates of the models (6.44a) and (6.44b).

1) The results of the model (6.44a) are obtained using the steps as follows:

   1.1 At the first stage, the STEPLS-Combinatorial selection method is applied with *Y1* and *C* as the variables always in the model, and *X1*X2, X1*Y1(−1),* and *X2*Y1(−1)* as the search regressors. An acceptable regression of *Y1* is obtained with the three two-way interactions as independent variables.

   1.2 At the second stage, the STEPLS-Combinatorial selection method is also applied with *Y1, C, X1*X2, X1*Y1(−1),* and *X2*Y1(−1)* as the variables always in the model, and the main variables *X1, X2, Y1(−1), Y1(−2),* and *Y1(−3)* as the search regressors. An acceptable regression of *Y1* is obtained, as presented in Figure 6.35.

6.6.3 Three-Way Interaction LV(p) TE_ Models

Referring to *Y1(−1)* as one of the best predictors of *Y1*, then the effects of the exogenous variables *X1*X2, X1,* and *X2* on *Y1* should depend on *Y1(−1)*. Similarly, the effect of *X2* on *X1* depends on *X1(−1)*. For these reasons, as an extension of the models (6.44a)–(6.44c), the following set of general ESs is acceptable.

$$Y1\ \mathbf{X1^{*}X2^{*}Y1(-1)}\ \mathbf{X1^{*}Y1(-1)}\ \mathbf{X2^{*}Y1(-1)}\ \mathbf{X1^{*}X2}\ \mathbf{X1}\ \mathbf{X2}\ Y1(-1)...Y1(p1)\ C \quad (6.45a)$$

$$X1\ \mathbf{X2^{*}X1(-1)}\ \mathbf{X2}\ X1(-1)...X1(-p2)\ C \quad (6.45b)$$

$$X2\ X2(-1)...X2(-p3)\ C \quad (6.45c)$$

Note that (6.45a) is a hierarchical three-way interaction model. However, its acceptable LS regression would be a nonhierarchical reduced regression in almost all cases because, in general, the independent variables of the model are highly or significantly correlated. See Example 6.20. And note that (6.45b) is a hierarchical two-way interaction model.

Example 6.20 Two Alternative LS Regressions, Based on the Model (6.45a)
Figure 6.36 presents the LS estimates of two alternative reduced models, namely RM1 and RM2, of the full model (6.45a). Based on these results, the findings and notes presented are as follows:

1) The very small (−4.09E-8) coefficients of the interactions can easily be adjusted or modified by using the rescaled of the independent variables, such as *X1S* = *X1*/100, and *X2S* = *X2/100*. Do this as an exercise.

Dependent Variable Y1							
RM1	C	X1*X2*Y1(-1)	X2*Y1(-1)	X1*X2	Y1(-1)	Y1(-2)	
Coef.	2.5592	-4.09E-08	0.0001	6.16E-07	0.5231	0.2932	
t-stat	1.0582	-2.7664	1.3193	3.4153	3.5196	3.5626	
Prob.	0.2918	0.0065	0.1893	0.0008	0.0006	0.0005	
Y1f	RMSE=8.515		TIC=0.052		BP=0.005		
RM2	C	X1*X2*Y1(-1)	Y1(-1)	Y1(-2)	X1	X2	Y1(-3)
Coef.	0.7641	-1.37E-08	0.5915	0.2769	0.0290	-0.0389	0.1200
t-stat	0.6076	-3.8799	7.1487	3.0026	2.3497	-2.0032	1.3883
Prob.	0.5445	0.0002	0.0000	0.0032	0.0202	0.0472	0.1673
Y1f	RMSE=8.070		TIC=0.049		BP=0.012		

Figure 6.36 The LS estimates of two reduced models RM1 and RM2 of the model (6.45a).

2) Based on the results of RM1, we have the findings as follows:
   2.1 By looking at the three independent variables *X1∗X2∗Y1(−1), X2∗Y1(−1),* and *Y1(−1)* we have the term/function *{(−4.09E-08∗X1+ 0.0001)∗X2 + 5.321}∗Y1(−1),* with *(−4.09E-08∗X1 + 0.0001) ≤ 0* for *X1* ≤ 2444.998, and, otherwise, it is greater than zero. Hence, the effect of *Y1(−1)* on *Y1* would decrease with increasing scores of *X2* for *X1* ≤ 2444.998; and it would increase with increasing scores of *X2* for *X1* > 2444.998. However, the effects of *Y1(−1)* on *Y1* are positive for all scores of *X1* and *X2* because *−4.09E-08∗X1∗X2* has a minimum value of −0.7875.
   2.2 Even though *X2∗Y(−1)* has a Prob. = 0.1890, it does not have to be deleted from the model, because the interaction *X1∗X2∗Y1(−1)* has significant effects on *Y1*. In addition, the joint effects of *X1∗X2* and *X2* are significantly dependent on *Y1(−1),* since the H_0: C (2) = C(3) = 0 is rejected based on the Wald (*F*-stat) of F_0 = 6.897113, with *df* = (2137), and *p*-value = 0.0014.
   2.3 By looking at the two independent variables *X1∗X2∗Y1(−1)* and *X1∗X2,* we have the term/function (−4.09E-08∗*Y1(−1)* + 6.16E-07)∗*X1∗X2.* Then, we can conclude that the effect of *X1∗X2* on *Y1* would decrease with increasing values of *Y1(−1)* because *Y1(−1)* has a negative coefficient.
3) Based on the results of RM2, we have the findings as follows:
   3.1 By looking at the two independent variables *X1∗X2∗Y1(−1)* and *Y1(−1),* we have the term/function (−1.37E-08∗*X1∗X2* + 0.5915)∗*Y1(−1).* Then we find that the effect of *Y1 (−1)* would decrease with increasing scores of *X1∗X2,* but the effects of *Y(−1)* on *Y1* are positive for all values of *X1∗X2,* with a minimum of (−0.24075 + 0.5915) = 0.350745. In addition, the partial joint effects of *X1∗X2∗Y1(−1)* and *Y1(−1)* is significant, because each has a significant effect at the 1% level.
   3.2 Since *X1∗X2∗Y1(−1)* has a significant effect on *Y1,* then it can be said that the effect of *X1∗X2* on *Y1* is significantly dependent on *Y1(−1).*

6.7 Bivariate Triangular Effects Models Based on (*X1,X2,Y1,Y2*)

Similar to the triangular effects models (TE_Models) based on *(X1,X2,Y1)* presented previously, all triangular effects models are also valid for the triple time series *(X1,X2,Y2)* with *Y2* = *GC*. So we would have a set of bivariate triangular effects models based on *(X1,X2,Y1,Y2)*. However, in addition, we also have to take into account the possible causal effects between *Y1* = *GCDAN* and *Y2* = *GC*, which have been presented previously. Therefore, under the criterions or assumptions:

A1. *Y1* and *Y2* have reciprocal causal effects. However, for their forecasting they should be represented by *Y1(−1)* and *Y2(−1)*.
A2. *X1* and *X2* have direct effects on each of *Y1* and *Y2*, and
A3. *X2* has a direct effects on *X1*.

I propose the following alternative sets of LV(p) models.

6.7.1 Partial Two-Way Interaction Models

As an extension of the models (6.43a)–(6.43c), based on the variables *X1*, *X2*, *Y1*, and *Y2*, I propose LV(p) bivariate triangular effects models with the following set of specific equations.

$$\begin{aligned} &Y1\ \boldsymbol{Y1(-1)*Y2(-1)}\ \boldsymbol{Y1(-1)*X1}\ \boldsymbol{Y1(-1)*X2}\ \boldsymbol{Y2(-1)} \\ &\quad X1\ X2\ Y1(-1)...Y1(-p1)\ C \end{aligned} \tag{6.46a}$$

$$\begin{aligned} &Y2\ \boldsymbol{Y1(-1)*Y2(-1)}\ \boldsymbol{Y2(-1)*X1}\ \boldsymbol{Y2(-1)*X2}\ \boldsymbol{Y1(-1)} \\ &\quad X1\ X2\ Y2(-1)...Y2(-p2)\ C \end{aligned} \tag{6.46b}$$

$$X1\ \boldsymbol{X1(-1)*X2}\ \boldsymbol{X2}\ X1(-1)...X1(-p3)\ C \tag{6.46c}$$

$$X2\ X2(-1)...X2(-p4)\ C \tag{6.46d}$$

Referring to the assumption A1, *Y2(−1)* should be used as the independent variable of the forecast model of *Y1* in (6.46a) and *Y1(−1)* should be used as the independent variable of the forecast model of *Y2* in (6.46b). If *Y2* is used as an independent variable of the model (6.46a) and *Y1* is used as an independent variable of the model (6.46b), then the forecast values beyond the sample period for *Y1* and *Y2* cannot be computed. Refer to the models presented in Chapter 5, specifically (5.54).

6.7.1.1 A Complete Two-Way Interaction Model

As an extension of the models (6.46a)–(6.46d), based on the variables *X1*, *X2*, *Y1*, and *Y2*, I propose LV(p) bivariate triangular effects models, with the following set of specific equations.

$$\begin{aligned} &Y1\ \boldsymbol{Y1(-1)*Y2(-1)}\ \boldsymbol{Y1(-1)*X1}\ \boldsymbol{Y1(-1)*X2}\ \boldsymbol{X1^{*}X2}\ \boldsymbol{Y2(-1)} \\ &\quad X1\ X2\ Y1(-1)...Y1(-p1)\ C \end{aligned} \tag{6.47a}$$

$$\begin{aligned} &Y2\ \boldsymbol{Y1(-1)*Y2(-1)}\ \boldsymbol{Y2(-1)*X1}\ \boldsymbol{Y2(-1)*X2}\ \boldsymbol{X1^{*}X2}\ \boldsymbol{Y1(-1)} \\ &\quad X1\ X2\ Y2(-1)...Y2(-p2)\ C \end{aligned} \tag{6.47b}$$

$$X1\ \boldsymbol{X1(-1)*X2}\ X2\ X1(-1)...X1(-p3)\ C \tag{6.47c}$$

$$X2\ X2(-1)...X2(-p4)\ C \tag{6.47d}$$

6.7.2 Three-Way Interaction TE_Models

As an extension of the models (6.47a)–(6.47d), I propose a LV(p) bivariate triangular effect model, with the following set of specific equations.

$$Y1\ \boldsymbol{Y1(-1)*Y2(-1)*X1}\ \boldsymbol{Y1(-2)*Y2(-1)*X2}\ \boldsymbol{Y1(-1)*X1^{*}X2}\ Y1(-1)*Y2(-1)$$
$$Y1(-1)*X1\ Y1(-1)*X2\ X1^{*}X2\ Y2(-1)\ X1\ X2\ Y1(-1)...Y1(-p1)\ C \tag{6.48a}$$

$$Y2\ \boldsymbol{Y1(-1)*Y2(-1)*X1}\ \boldsymbol{Y1(-1)*Y2(-1)*X2}\ \boldsymbol{Y2(-1)*X1^{*}X2}\ Y1(-1)*Y2(-1)$$
$$Y2(-1)*X1\ Y2(-1)*X2\ X1^{*}X2\ Y1(-1)\ X1\ X2\ Y2(-1)...Y1(-p2)\ C \tag{6.48b}$$

$$X1\ \boldsymbol{X1(-1)*X2}\ X2\ X1(-1)...X1(-p3)\ C \tag{6.48c}$$

$$X2\ X2(-1)...X2(-p4)\ C \tag{6.48d}$$

Note that the three-way interactions of the model (6.48a) are developed or defined based on the two-way interactions *Y1(−1)∗Y2(−1)*, *Y1(−1)∗X1*, and *Y1(−1)∗X2*. Similarly, the three-way interactions of the model (6.48b) are defined based on the two-way interactions *Y1(−1)∗Y2(−1)*, *Y2 (−1)∗X1*, and *Y2(−1)∗X2*.

Example 6.21 Analysis of the Models (6.48a)–(6.48d)

In general, the analysis of a three-way interaction model should be done using the stages as follows;

1) First, to do the analysis based on a model with only three-way interactions as independent variables, such as that presented in Figure 6.37 for the model of *Y1*, and in Figure 6.38 for the model of *Y2*, with the reduced models RM-1. In these cases, both regressions are acceptable for use in the second stage of analysis.
2) At the second stage, the two-way interactions are inserted as the search regressors. It is recommended to apply the STEPLS-Combinatorial and the trial-and-error method in order to obtain acceptable subsets for the models of *Y1* and *Y2*, namely RM-2.
3) At the third or final stage, the main variables should be inserted as additional independent variables of RM-2, the regressions RM-3 for *Y1* and *Y2* are obtained by selecting *Y1(−p1)* and *Y2(−p2)* for alternative *p1* and *p2*, respectively, using the trial-and error method.
4) Specific for the *Y1*-regression RM-3, RM-4 is presented as a comparison. In fact, in the statistical sense, RM-3 is an acceptable regression even though the interaction *X1∗X2* has a large Prob. = 0.5666, because its interaction *Y1(−1)∗X1∗X2* has a significant adjusted effect.
5) The OLS estimates of the models (6.48c) and (6.48d) can easily be obtained, as presented in Figure 6.39.

Example 6.22 Computing the Forecast Values of *Y1* and *Y2*

First of all, note that the models of *Y1* and *Y2* have *X1* and *X2* as independent variables. Hence, in order to compute the forecast values *Y1(T + 1)* and *Y2(T + 1)*, first we have to compute the forecast values *X2(T + 1)* and *X1(T + 1)* using the model (6.48d) and (6.48c), respectively, with the results presented in Figure 6.40. The computation should be done using a spreadsheet containing at least the last three observed scores of the variables *X2, X1, Y1, Y2,* and *t* since the *Y1*-regression in RM-4 has *Y1(−3)* as an independent variable. In this case, Figure 6.40 presents the last five of their observed scores. Then, the steps of the computation are as follows, with a specific ordering:

DV = Y1	RM-1		RM-2		RM-3		RM-4	
Variable	Coef.	Prob.	Coef.	Prob.	Coef.	Prob.	Coef.	Prob.
C	23.94718	0.0000	13.14932	0.0000	10.18829	0.0000	10.03228	0.0000
Y1(-1)*Y2(-1)*X1	-5.72E-07	0.0000	3.59E-07	0.0008	3.45E-07	0.0011	3.78E-07	0.0000
Y1(-1)*Y2(-1)*X2	-3.08E-07	0.0654	-1.90E-07	0.1099	-2.58E-07	0.0313	-2.98E-07	0.0021
Y1(-1)*X1*X2	1.14E-06	0.0000	-4.22E-07	0.0011	-3.25E-07	0.0133	-3.27E-07	0.0123
Y1(-1)*X2			0.00217	0.0002	0.001701	0.0036	0.001781	0.0017
Y1(-1)*Y2(-1)			-0.00109	0.0120	-0.00083	0.0561	-0.00089	0.0344
X1*X2			1.19E-06	0.1636	4.98E-07	**0.5666**		
Y1(-3)					0.232782	0.0063	0.246369	0.0026
R-squared	0.9134		0.9711		0.9722		0.9722	
Adj.	0.9115		0.9699		0.9708		0.9709	
F-statistic	492.15		768.29		670.26		785.82	
Prob(F-statistic)	0.0000		0.0000		0.0000		0.0000	
DW-Stat	0.6911		2.1719		2.1009		2.1538	

Figure 6.37 OLS estimates based on alternative reduced models (RMs) of the model (6.48a).

DV=Y2	RM-1			RM-2			RM-3		
Variable	Coef.	t-Stat.	Prob.	Coef.	t-Stat.	Prob.	Coef.	t-Stat.	Prob.
C	761.793	18.44	0.0000	262.0496	31.318	0.0000	-17.0068	-1.499	0.1363
Y1(-1)*Y2(-1)*X1	-1.84E-05	-3.416	0.0008	3.15E-05	14.472	0.0000	-2.83E-06	-2.185	0.0307
Y1(-1)*Y2(-1)*X2	2.98E-05	3.962	0.0001	-5.12E-05	-16.657	0.0000	3.92E-06	1.977	0.0501
Y2(-1)*X1*X2	3.11E-08	1.943	0.0540	-1.55E-08	-1.568	0.1192	6.66E-09	1.316	0.1904
Y1(-1)*Y2(-1)				0.008275	8.874	0.0000	0.00137	1.663	0.0986
Y2(-1)*X2				0.004487	18.574	0.0000	-0.00040	-2.355	0.0200
Y2(-1)*X1				-0.00275	-18.513	0.0000	0.00031	2.672	0.0085
X1*X2				0.00011	1.311	0.1921	-8.51E-05	-2.037	0.0437
X2							0.78659	6.812	0.0000
Y2(-2)							0.46330	4.889	0.0000
Y1(-1)							-1.44269	-1.546	0.1244
R-squared	0.9282			0.9991			0.9999		
Adj.	0.9266			0.9991			0.9999		
F-statistic	603.03			21488.5			135064.6		
Prob(F-statistic)	0.0000			0.0000			0.0000		
DW-stat.	0.2101			0.6085			0.9807		

Figure 6.38 OLS estimates based on alternative reduced models (RMs) of the model (6.48b).

1) The first forecast value of $X2$ in cell B7, for the time 1995Q2 or $t = 146$, is computed using the following function, which can be done directly by using the block-copy-paste method of the equation in the output of "*View/Representations.*"

$$X2 = 2.66921525194 + 1.6085694806^{*}X2(-1) - 0.603906692796^{*}X2(-2)$$

$$\mathbf{B7} = 2.66921525194 + 1.6085694806^{*}\mathrm{B6} - 0.603906692796^{*}\mathrm{B5}$$

2) Similarly, the first forecast value of $X1$ in C7, is computed as follows:

Dependent Variable: X1
Method: Least Squares
Date: 07/10/17 Time: 08:56
Sample (adjusted): 1959Q2 1995Q1
Included observations: 144 after adjustments

Variable	Coefficient	Std. Error	t-Statistic	Prob.
C	-18.70116	4.875585	-3.835675	0.0002
X1(-1)*X2	4.43E-06	2.29E-06	1.937947	0.0546
X2	0.464389	0.068683	6.761367	0.0000
X1(-1)	0.697789	0.050026	13.94856	0.0000

R-squared	0.999820	Mean dependent var	1942.487
Adjusted R-squared	0.999816	S.D. dependent var	1507.184
S.E. of regression	20.46154	Akaike info criterion	8.902356
Sum squared resid	58614.47	Schwarz criterion	8.984851
Log likelihood	-636.9696	Hannan-Quinn criter.	8.935877
F-statistic	258578.4	Durbin-Watson stat	2.441359
Prob(F-statistic)	0.000000		

(a)

Dependent Variable: X2
Method: Least Squares
Date: 07/10/17 Time: 09:01
Sample (adjusted): 1959Q3 1995Q1
Included observations: 143 after adjustments

Variable	Coefficient	Std. Error	t-Statistic	Prob.
C	2.669215	1.256621	2.124121	0.0354
X2(-1)	1.608569	0.067819	23.71874	0.0000
X2(-2)	-0.603907	0.068630	-8.799451	0.0000

R-squared	0.999926	Mean dependent var	1325.132
Adjusted R-squared	0.999925	S.D. dependent var	964.3053
S.E. of regression	8.367947	Akaike info criterion	7.107450
Sum squared resid	9803.155	Schwarz criterion	7.169608
Log likelihood	-505.1827	Hannan-Quinn criter.	7.132708
F-statistic	942795.2	Durbin-Watson stat	2.022571
Prob(F-statistic)	0.000000		

(b)

Figure 6.39 OLS Estimates of the models of (a) (6.48c) and (b) (6.48d).

	A	B	C	D	E	F
1		X2	X1	Y1	Y2	t
2	1994Q1	3195.2	4856.9	91.977	4599.2	141
3	1994Q2	3242.8	5002.2	91.942	4665.1	142
4	1994Q3	3265.5	5070.4	90.490	4734.4	143
5	1994Q4	3320.2	5145.8	90.747	4796.027	144
6	1995Q1	3363	5225.5	83.812	4836.326	145
7	**1995Q2**	3407.197	5260.653	81.945	4931.616	**146**
8	**1995Q3**	3452.445	5307.546	80.101	5001.427	**147**
9	**1995Q4**	3498.537	5363.070	78.160	5089.042	**148**

Figure 6.40 Spreadsheet for computing forecast values *X2*, *X1*, *Y1*, and *Y2*.

$$X1 = -18.7011585418 + 4.43308384377e-06^{*}X1(-1)*X2$$
$$+0.464388628652^{*}X2 + 0.697788639974^{*}X1(-1)$$
$$\mathbf{C7} = -18.7011585418 + 0.00000443308384377^{*}B6^{*}B7$$
$$+0.464388628652^{*}B7 + 0.697788639974^{*}C6$$

3) Then the first forecast values of both *Y1* and *Y2*, respectively, for 1995Q2 or $t = 146$ in cells D7 and E7 are computed using the following functions,

$$Y1(RM-2) = 13.1493207618 + 3.59039954674e-07^{*}Y1(-1)*Y2(-1)*X1$$
$$-1.89825478168e-07^{*}Y1(-1)*Y2(-1)*X2 - 4.22116802406e$$
$$-07^{*}Y1(-1)*X1^{*}X2$$
$$+0.00216966164614^{*}Y1(-1)*X2 - 0.00108888558134^{*}Y1(-1)*Y2(-1)$$
$$+1.18703491578e-06^{*}X1^{*}X2$$
$$\mathbf{D7} = 13.1493207618 + 0.000000359039954674^{*}D6^{*}E6^{*}C7$$
$$-0.000000189825478168^{*}D6^{*}E6^{*}B7 - 0.000000422116802406^{*}D6^{*}B7^{*}C7$$
$$+0.00216966164614^{*}D6^{*}B7 - 0.00108888558134^{*}D6^{*}E6$$
$$+0.00000118703491578^{*}C7^{*}B7$$
$$Y2^{*}(RM-3) = -17.0068200786 - 2.82516695963e-06^{*}Y1(-1)*Y2(-1)*X1$$
$$+3.92317317025e-06^{*}Y1(-1)*Y2(-1)*X2$$
$$+6.65564014928e-09^{*}Y2(-1)*X1^{*}X2$$
$$+0.00137092101058^{*}Y1(-1)*Y2(-1)$$
$$-0.000402514485887^{*}Y2(-1)*X2$$
$$+0.000306510838787^{*}Y2(-1)*X1 - 8.50620438662e-05^{*}X1^{*}X2$$
$$+0.78658941364^{*}X2 + 0.463298233325^{*}Y2(-2)$$
$$-1.44268452504^{*}Y1(-1)$$
$$\mathbf{E7} = -17.0068200786 - 0.00000282516695963^{*}D6^{*}E6^{*}C7$$
$$+0.00000392317317025^{*}D6^{*}E6^{*}B7 + 6.65564014928E-09^{*}E6^{*}C7^{*}B7$$
$$+0.00137092101058^{*}D6^{*}E6 - 0.000402514485887^{*}E6^{*}B7$$
$$+0.000306510838787^{*}E6^{*}C7 - 0.0000850620438662^{*}C7^{*}B7$$
$$+0.78658941364^{*}B7 + 0.463298233325^{*}E5 - 1.44268452504^{*}D6$$

4) In addition, it is important to note that if *Y2* is used as an independent variable of the model (6.48a) and *Y1* is used as an independent variable of the model (6.48b), the forecast values of *Y1(T + 1)* and *Y2(T + 1)* cannot be computed. If the independent and *X2* of both models of *Y1* and *Y2* are replaced by *X1(−1)* and *X2(−1)*, both forecast values *Y1(T + 1)* and *Y2(T + 1)* could be computed directly, based on new functions.

DV=Y1	C	Y1(-1)*X2	X1*X2	Y1(-1)*Y2(-1)*X1	Y1(-1)*X1	Y1(-1)*X1*X2	Y1(-1)*Y2(-1)	
Coef.	13.334	0.001961	1.14E-06	3.13E-07	0.000446	-5.28E-07	-0.001438	
t-Stat.	13.563	3.586	1.325	3.376	1.646	-4.205	-2.817	
Prob.	0.0000	0.0005	0.1874	0.0010	0.1021	0.0000	0.0056	
DV=Y2	C	Y1(-1)*Y2(-1)	Y2(-1)*X1	Y1(-1)*Y2(-1)*X1	Y2(-1)*X2	Y1(-1)*Y2(-1)*X2	Y2(-1)*X1*X2	X1*X2
Coef.	262.0496	0.008275	-0.002746	3.15E-05	0.004487	-5.12E-05	-1.55E-08	0.00011
t-Stat.	31.3181	8.87403	-18.51347	14.47205	18.57387	-16.6569	-1.567909	1.3108
Prob.	0.0000	0.0000	0.0000	0.0000	0.0000	0.0000	0.1192	0.1921

Figure 6.41 OLS estimates of the interactions reduced models of (6.48a) and (6.48b).

Example 6.23 Alternative Analysis of the Models (6.48a)–(6.48d)
Under the assumption that all interaction regressors are equally important, then we can apply the STEPLS-Combinatorial and the trial-and-error selection methods specific for the models of *Y1* in (6.48a), and *Y2* in (6.48b). The steps of the analysis are as follows:

1) The variables *Y∗* and *C* are the dependent and independent variables, which are always in the model with all interactions as the search regressors.
2) The trial-and-error method is used to select the maximum number of the search regressors that give the good fit models of *Y1* and *Y2*, with the results presented in Figure 6.41.
3) Compared to the *Y1*-regression RM-2 in Figure 6.37, this method of analysis presents different regression functions, with two three-way interactions and four two-way interactions. Both models are acceptable in the statistical sense.
4) However, compared to the *Y2*-regression RM-2 in Figure 6.39, this method of analysis presents exactly the same regression function.

6.8 Models with Exogenous Variables and Alternative Trends

Referring to all alternative models of *Y1* = *GCDAN* with the time variables presented in Section 6.2.3, and the summary of alterative trends presented in Table 6.1, we can have a lot of possible models of *Y1* with exogenous variables and alternative trends. This is similar for the of *Y2* = *GC*.

6.8.1 Models Based on (X1,X2,Y1)

It should be understood that all models based on *(X1,X2,Y1)* can easily be extended to the models with alternative trends as presented in Table 6.1. However, the following examples present only two specific models

Example 6.24 A Model Based on *(X1,X2,Y1)* with Heterogeneous Trends by *@Quarter*
As an extension or a modification of the partial two-way interaction model (6.43a), I defined a model of *Y1* with exogenous variables *(X1,X2)* and heterogeneous trends by *@Quarter*, with the following ES. Based on its OLS regression in Figure 6.42, the following findings and notes are presented.

$$\begin{gathered} Y1\,Y1(-1)*X1(-1)\,Y1(-1)*X2(-1)\,Y1(-1)\,X1(-1)\,X2(-1)\,T\,C \\ T^{*}@Expand(@Quarter,@Dropfirst)\ @Expand(@Quarter,@Dropfirst) \end{gathered} \tag{6.49}$$

Dependent Variable: Y1
Method: Least Squares
Date: 07/18/17 Time: 13:04
Sample (adjusted): 1959Q2 1995Q1
Included observations: 144 after adjustments

Variable	Coefficient	Std. Error	t-Statistic	Prob.
Y1(-1)*X1(-1)	-0.001071	0.000572	-1.872625	0.0633
Y1(-1)*X2(-1)	0.001574	0.000900	1.748821	0.0827
Y1(-1)	0.844871	0.138049	6.120076	0.0000
X1(-1)	0.084518	0.043895	1.925472	0.0563
X2(-1)	-0.124144	0.071531	-1.735519	0.0850
T	0.071611	0.117417	0.609890	0.5430
C	5.595211	4.056630	1.379276	0.1702
T*(@QUARTER=2)	0.007865	0.028623	0.274796	0.7839
T*(@QUARTER=3)	0.036861	0.028462	1.295081	0.1976
T*(@QUARTER=4)	-0.006644	0.028468	-0.233379	0.8158
@QUARTER=2	-1.696742	2.413542	-0.703009	0.4833
@QUARTER=3	-1.959156	2.415685	-0.811015	0.4188
@QUARTER=4	-1.991227	2.425598	-0.820922	0.4132

R-squared	0.974194	Mean dependent var	48.94819
Adjusted R-squared	0.971830	S.D. dependent var	29.79341
S.E. of regression	5.000510	Akaike info criterion	6.142897
Sum squared resid	3275.669	Schwarz criterion	6.411005
Log likelihood	-429.2886	Hannan-Quinn criter.	6.251841
F-statistic	412.1090	Durbin-Watson stat	2.535592
Prob(F-statistic)	0.000000		

Figure 6.42 OLS estimate of the model (6.49).

1) All independent variables have significant joint effects on *Y1*, based on the *F*-statistic of $F_0 = 421.1090$, with a *p*-value = 0.0000.
2) And aside from the time *t*, each of the independent variables has a significant effect on *Y1*, at the 1, 5, or 10% levels. So it can be concluded that the joint effects of *X1(−1)* and *X2(−1)* on *Y1* are significantly dependent on *Y1(−1)*. Also, this can be tested using the Wald test with the null hypothesis H_0: C(1) = C(2) = 0 or the RVT of *Y1(−1)∗X1(−1) Y1(−1)∗X2(−1)*. Do this as an exercise.
3) Figure 6.43 presents the forecast evaluation based on the model (6.49), with the forecast sample {@*YEAR*>1980}, and the graphs of the scores for *Y1* and *Y1F*.

Example 6.25 Another Model Based on *(X1,X2,Y1)*

As an extension of the model of *Y1* = *GCDAN* by *TP* in (6.7a), this example presents a partial two-way interaction model based on *(X1,X2,Y1)* with heterogeneous trends by *TP*. Figure 6.44 presents the statistical results based on the model with the ES as follows:

$$\begin{aligned} &Y1\;Y1(-1)*X1(-1)\;Y1(-1)*X2(-1)\;Y1(-1)\;X1(-1)\;X2(-1)\;T\;C \\ &T^{*}@Expand(TP,@Drop(2))@Expand(TP,@Drop(2)) \end{aligned} \tag{6.50}$$

Comparing both models (6.7a) and (6.50), the statistical results in Figure 6.44, specifically the graphs of *Y1* and *Y1F_Eq652*, clearly show that the model (6.50) is the best forecast model of *Y1*. In order to compute the forecast values of *Y1* beyond the sample period, the forecast models of *X1* and *X2* in Figure 6.44 can be applied directly.

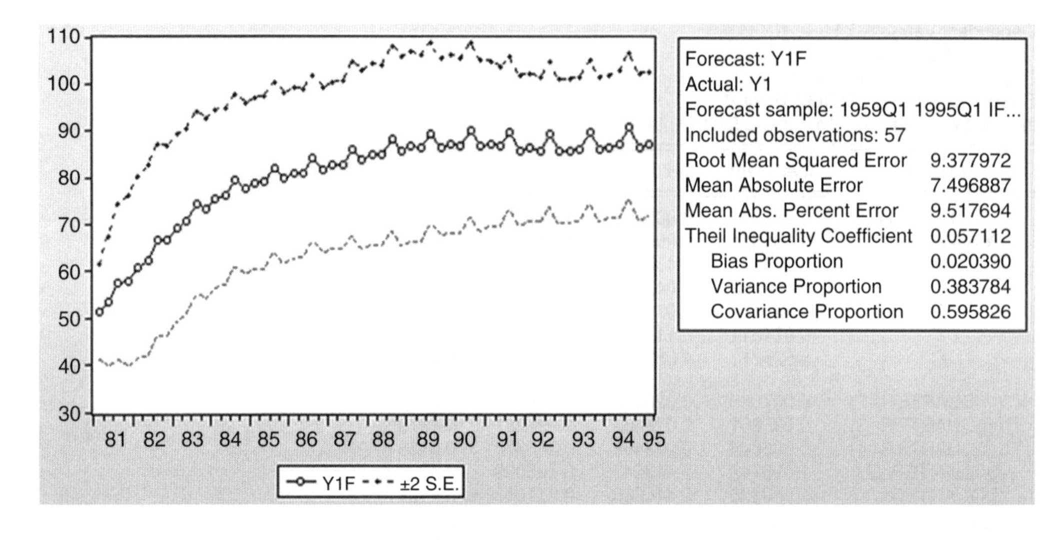

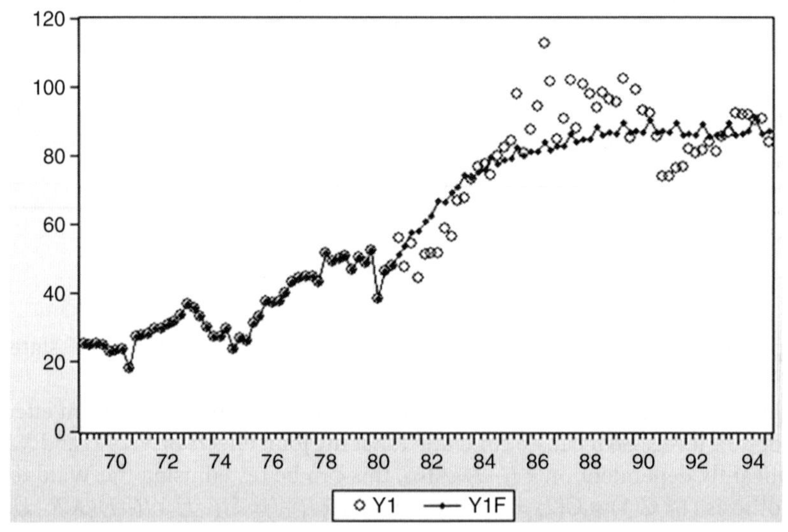

Figure 6.43 The forecast evaluation based on the model (6.49), and the graphs of *Y1* and *Y1F*.

6.8.2 Models Based on (*X1,X2,Y1,Y2*) with Trend

All models based on *(X1,X2,Y1,Y2)* presented previously can also easily be extended to the models with alternative trends, as presented in Table 6.1. However, this section presents only one example as an extension of the models (6.48a) and (6.48b).

Example 6.26 An Extension of the Models of *Y1* and *Y2* in (6.48a) and (6.48b)
Referring to the models with heterogenous trends by *TP* (time period) in (6.19a) and (6.19b), this example also presents the models with heterogenous trends by *TP* as an extension of the models of *Y1* and *Y2* in (6.48a and 6.48b), specifically the RM-4 of *Y1* in Figure 6.37 and the RM-3 of *Y2* in Figure 6.38 with the following equation specifications, since the models (6.48c) and (6.48d) can be used directly to compute the forecast values beyond the sample period of *Y1* and *Y2*.

Dependent Variable: Y1
Method: Least Squares
Date: 07/18/17 Time: 15:42
Sample (adjusted): 1959Q2 1995Q1
Included observations: 144 after adjustments

Variable	Coefficient	Std. Error	t-Statistic	Prob.
Y1(-1)*X1(-1)	-5.48E-05	0.000616	-0.088908	0.9293
Y1(-1)*X2(-1)	3.94E-05	0.000963	0.040977	0.9674
Y1(-1)	0.525379	0.202850	2.589996	0.0107
X1(-1)	0.052533	0.043262	1.214294	0.2268
X2(-1)	-0.062658	0.069983	-0.895324	0.3722
T	-0.950589	0.327088	-2.906221	0.0043
C	129.9468	39.21653	3.313572	0.0012
T*(TP=1)	1.028220	0.334352	3.075262	0.0026
T*(TP=3)	0.999660	0.345101	2.896715	0.0044
TP=1	-125.4058	37.77874	-3.319481	0.0012
TP=3	-141.1333	45.09527	-3.129670	0.0022

R-squared	0.977427	Mean dependent var	48.94819
Adjusted R-squared	0.975730	S.D. dependent var	29.79341
S.E. of regression	4.641464	Akaike info criterion	5.981250
Sum squared resid	2865.244	Schwarz criterion	6.208111
Log likelihood	-419.6500	Hannan-Quinn criter.	6.073434
F-statistic	575.9052	Durbin-Watson stat	2.146855
Prob(F-statistic)	0.000000		

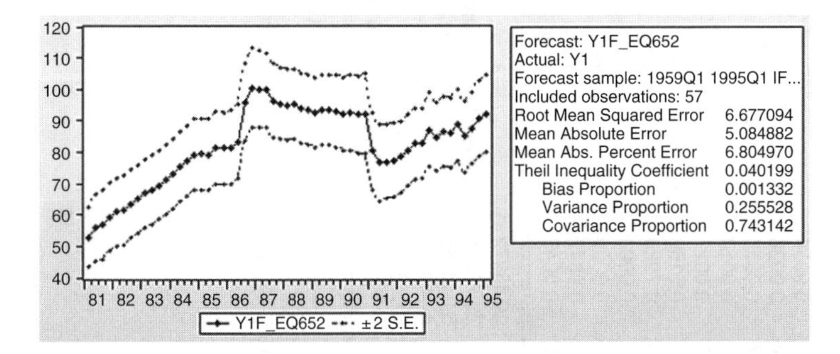

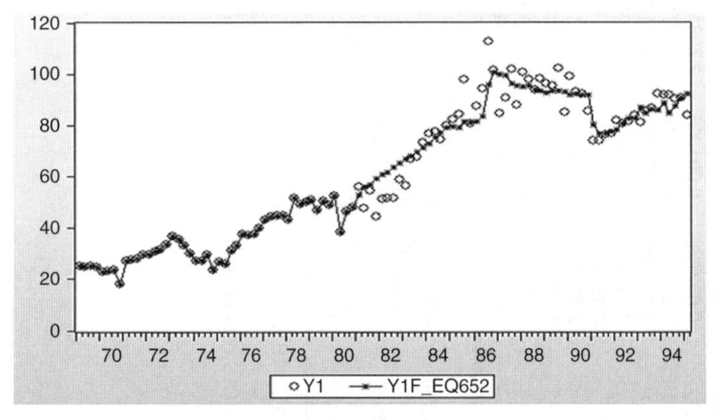

Figure 6.44 Statistical results based on the model (6.50).

Dependent Variable: Y1
Method: Least Squares
Date: 07/18/17 Time: 06:44
Sample (adjusted): 1959Q4 1995Q1
Included observations: 142 after adjustments

Variable	Coefficient	Std. Error	t-Statistic	Prob.
Y1(-1)*Y2(-1)*X1	-4.38E-07	2.19E-07	-1.996695	0.0479
Y1(-1)*Y2(-1)*X2	-3.36E-07	1.19E-07	-2.828065	0.0054
Y1(-1)*X1*X2	7.96E-07	2.61E-07	3.047494	0.0028
Y1(-1)*X2	-0.002722	0.001006	-2.706137	0.0077
Y1(-1)*Y2(-1)	0.002376	0.000749	3.170714	0.0019
Y1(-3)	0.243940	0.078942	3.090129	0.0024
T	0.380449	0.585311	0.649995	0.5168
T*(TP=1)	-0.157410	0.613926	-0.256399	0.7980
T*(TP=3)	2.698234	0.611125	4.415192	0.0000
TP=1	8.325989	1.362200	6.112163	0.0000
TP=2	-1.306875	68.10186	-0.019190	0.9847
TP=3	-363.6354	125.7618	-2.891461	0.0045

R-squared	0.978555	Mean dependent var	49.43972
Adjusted R-squared	0.976741	S.D. dependent var	29.71056
S.E. of regression	4.531139	Akaike info criterion	5.940545
Sum squared resid	2669.059	Schwarz criterion	6.190334
Log likelihood	-409.7787	Hannan-Quinn criter.	6.042049
Durbin-Watson stat	1.976145		

(a). Y1F with RMSE=5.686 & TIC=0.0334

Dependent Variable: Y2
Method: Least Squares
Date: 07/18/17 Time: 06:38
Sample (adjusted): 1959Q3 1995Q1
Included observations: 143 after adjustments

Variable	Coefficient	Std. Error	t-Statistic	Prob.
Y1(-1)*Y2(-1)*X1	-3.78E-06	1.51E-06	-2.498688	0.0137
Y1(-1)*Y2(-1)*X2	5.41E-06	2.44E-06	2.218642	0.0283
Y2(-1)*X1*X2	-8.27E-09	6.72E-09	-1.229385	0.2212
Y1(-1)*Y2(-1)	0.000825	0.001185	0.696105	0.4876
Y2(-1)*X2	-0.000541	0.000207	-2.607158	0.0102
Y2(-1)*X1	0.000440	0.000149	2.951224	0.0038
X1*X2	-8.09E-05	5.44E-05	-1.487066	0.1395
X2	0.683529	0.161357	4.236138	0.0000
Y2(-2)	0.505507	0.086770	5.825800	0.0000
Y1(-1)	-0.412158	1.086675	-0.379283	0.7051
T	8.149568	2.804005	2.906403	0.0043
T*(TP=1)	-8.361844	2.523259	-3.313906	0.0012
T*(TP=3)	9.121881	2.180318	4.183739	0.0001
TP=1	-12.29878	24.88525	-0.494220	0.6220
TP=2	-930.8851	268.7106	-3.464266	0.0007
TP=3	-2181.177	485.9298	-4.488667	0.0000

R-squared	0.999932	Mean dependent var	1777.961
Adjusted R-squared	0.999924	S.D. dependent var	1393.778
S.E. of regression	12.15203	Akaike info criterion	7.937988
Sum squared resid	18754.32	Schwarz criterion	8.269495
Log likelihood	-551.5662	Hannan-Quinn criter.	8.072697
Durbin-Watson stat	1.340828		

(b). Y2F with RMSE=17.277 &TIC=0.002

Figure 6.45 The OLS estimates of the models (6.51a) and (6.51b). (a) *Y1F* with RMSE = 5.686 and TIC = 0.0334 and (b) *Y2F* with RMSE = 17.277 and TIC = 0.002.

$$Y1\,Y1(-1)*Y2(-1)*X1\,Y1(-1)*Y2(-1)*X2\,Y1(-1)*X1^{*}X2$$
$$Y1(-1)*X2\,Y1(-1)*Y2(-1)\,Y1(-3) \quad (6.51a)$$
$$T\,T^{*}@Expand(TP,@Drop(2))\ @Expand(TP)$$
$$Y2\,Y1(-1)*Y2(-1)*X1\,Y1(-1)*Y2(-1)*X2\,Y2(-1)*X1^{*}X2$$
$$Y1(-1)*Y2(-1)\,Y2(-1)*X2\,Y2(-1)*X1\,X1^{*}X2\,X2\,Y2(-2)\,Y1(-1) \quad (6.51b)$$
$$T\,T^{*}@Expand(TP,@Drop(2))\ @Expand(TP)$$

1) Based on the *Y1*-regression in Figure 6.45a, the following findings and notes are presented.
   1.1 Each of the interaction independent variables has a significant effect on *Y1* at the 1 or 5% levels of significance.
   1.2 It is found that the time variable *T* has significant difference adjusted effects on *Y1* between the three time periods, based on the F-statistic of $F_0 = 11.15545$, with $df = (2130)$, and a *p*-value = 0.0000. So the model has significant heterogeneous trends by *TP*.
   1.3 Compared to (6.50) this model has smaller values of RMSE and TIC. Hence, this model is a better forecast model of *Y1* than (6.50).
2) Based on the *Y2*-regression in Figure 6.45b, the following findings and notes are presented.
   2.1 Two of the three-way interactions, namely *Y1(−1)∗Y2(−1)∗X1*, and *Y1(−1)∗Y2(−1)∗X2* have significant effects on *Y2* at the 5% level. However, at the 15% level, *Y2(−1)∗X1∗X2* has a negative significant effect on *Y2*, with a *p*-value = 0.2212/2 = 0.1106 < 0.15. For these reasons, this model is an acceptable three-way interaction forecast model of *Y2*, even though some of the other independent variables have large *p*-values.
   2.2 Compared to the four reduced models of *Y2* in Figure 6.38, this model is the best forecast model of *Y1* since it has the smallest values of RMSE and TIC.
   2.3 It is found that the time variable *T* has significant difference adjusted effects on *Y2* between the three time periods, based on the *F*-statistic of $F_0 = 9.415469$, with $df = (2.127)$, and a *p*-value = 0.0002. So the model has significant heterogeneous trends by *TP*.

6.9 Special LV(4) Models with Exogenous Variables

It is recognized that all models presented previously could be modified to special LV(4) models in order to match the data by quarters in two consecutive years. However, as an illustration, I only present special three-way interaction LV(4) TE_Models based on the three-way interaction LV(*p*) TE_Models in (6.48a)–(6.48d), with the following ESs.

$$Y1\ \mathbf{Y1(-4)*Y2(-4)*X1\ Y1(-4)*Y2(-4)*X2\ Y1(-4)*X1^{*}X2}$$
$$Y1(-4)*Y2(-4)\ Y1(-4)*X1\ Y1(-4)*X2\ X1^{*}X2\ Y2(-4)\ X1\ X2\ Y1(-4)\ C \quad (6.52a)$$
$$Y2\ \mathbf{Y1(-4)*Y2(-4)*X1\ Y1(-4)*Y2(-4)*X2\ Y2(-4)*X1^{*}X2}$$
$$Y1(-4)*Y2(-4)\ Y2(-4)*X1\ Y2(-4)*X2\ X1^{*}X2\ Y1(-4)\ X1\ X2\ Y2(-4)\ C \quad (6.52b)$$
$$X1\,\mathbf{X1(-4)*X2}\,\mathbf{X2}\,X1(-4)\,C \quad (6.52c)$$
$$X2\,X2(-4)...X2(-4p)\,C \quad (6.52d)$$

DV=Y1	RM-1			RM-2			RM-3			RM-4		
Variable	Coef.	t-Stat.	Prob.*	Coef.	t-Stat.	Prob.*	Coef.	t-Stat.	Prob.*	Coef.	t-Stat.	Prob.*
C	15.32564	16.495	0.0000	10.959	2.562	0.0115	12.15832	8.569	0.0000	4.924582	1.332	0.1851
Y1(-4)*Y2(-4)*X2	6.18E-07	2.106	0.0370	8.37E-07	2.182	0.0309	8.42E-07	2.204	0.0292	5.38E-07	1.333	0.1850
X1*X2	3.77E-06	3.892	0.0002	-1.37E-06	-0.297	0.7666						
Y1(-4)*Y2(-4)	-0.001681	-2.314	0.0222	-0.00382	-2.372	0.0191	-0.00357	-2.613	0.0100	-0.0032	-2.355	0.0200
Y1(-4)*Y2(-4)*X1	-1.22E-07	-1.569	0.1190	-1.20E-07	-1.538	0.1265	-1.25E-07	-1.623	0.1069	-1.09E-07	-1.433	0.1541
Y1(-4)*X1*X2	-4.78E-07	-2.208	0.0290	-6.16E-07	-2.016	0.0458	-6.29E-07	-2.089	0.0386	-3.47E-07	-1.066	0.2884
Y1(-4)*X1	0.001829	2.992	0.0033	0.00353	2.667	0.0086	0.00335	2.862	0.0049	0.002811	2.377	0.0189
Y2(-4)				0.12888	1.527	0.1293	0.10874	2.164	0.0322	0.134167	2.629	0.0096
X1				-0.09643	-1.538	0.1263	-0.08273	-1.952	0.0530	-0.10275	-2.395	0.0180
Y1(-4)										0.454469	2.115	0.0363
R-squared	0.9618			0.9625			0.9625			0.9637		
Adj. R-squared	0.9601			0.9602			0.9605			0.9615		
F-statistic	562.79			423.55			487.38			438.15		
Prob(F-statistic)	0.0000			0.0000			0.0000			0.0000		
DW-Stat.	0.9375			1.0101			0.9981			1.0870		
Y1f	RMSE=8.193 & TIC=0.049			RMSE=7.916 & TIC=0.048			RMSE=7.898 & TIC=0.012			RMSE=7.714 & TIC=0.003		

Figure 6.46 OLS estimates of four special LV(4) TE_Models of *Y1* in (6.52a)–(6.52d), with their forecast evaluations; RMSE, and TIC.

Example 6.27 Application of the Models (6.52a)–(6.52d)
By using the same methods as presented in Example 6.26, the steps of analysis based on the four models in (6.52a)–(6.52d) are as follows:

1) OLS estimates of the four reduced models of *Y1* in (6.52a) are presented in Figure 6.46. Based on these results, the following findings and notes are presented
   1.1 At the 10% level of significance, the interaction *Y1(−4)*∗*Y2(−4)*∗*X1* has a negative adjusted effect on *Y1* with a p-value = Prob./2 = 0.1190/2 = 0.0595. Hence, it can be concluded that each of the independent variables of RM-1 has either s positive or negative significant adjusted effect and all independent variables have a joint significant effect.
   1.2 Note that, even though the interaction *X1*∗*X2* in RM-2 has a large p-value, it does not have to be deleted from the model because its interaction, namely *Y1(−4)*∗*X1*∗*X2*, has a significant adjusted effect with a p-value = 0.0458. Furthermore, at the 10% level, each of the interactions *Y1(−4)*∗*Y2(−4)*∗*X1* and the main variable *X1* has a negative significant adjusted effect on *Y1*, with a p-value of 0.1265/2 and 01263/2, respectively, and *Y2 (−4)* has a positive significant adjusted effect with a p-value = 0.1293/2. So, RM-2 can be considered an acceptable forecast model of *Y1*.
   1.3 However, as a comparison, by deleting *X1*∗*X2* from RM-2, RM-3 is obtained. Note that each of the independent variables of this model has either a positive or negative adjusted effect on *Y1* at 1, 5, or 10% levels of significance, with a p-value = Prob./2. For instance, at the 1% level, *Y1(−4)*∗*Y2(−4)* has a negative significant effect with a p-value = 0.0050, and at the 10% level, *Y1(−4)*∗*Y2(−4)*∗*X1* a negative significant effect with a p-value = 0.05345. Compared to RM-2, RM-3 has smaller values of RMSE and TIC, so it can be concluded that RM-3 is a better forecast model for *Y1*. Would you agree?
   1.4 As an additional model, I try to use *Y1(−4)* and *X2* as the search regressors for RM-3, and RM-4 is obtained, with the three-way interactions *Y1(−4)*∗*Y2(−4)*∗*X1*, *Y1(−4)*∗*y2(−4)*∗*X2*, and *Y1(−4)*∗*X1*∗*X2*, which have p-values of 0.1850, 0.1541, and 0.2884, respectively. Hence, at the 10% level, the interactions *Y1(−4)*∗*Y2(−4)*∗*X1* and *Y1(−4)*∗*y2(−4)*∗*X2*, respectively, have positive and negative significant effects with a p-value of 0.0925 and

DV=Y2	RM-1			RM-2			RM-3			RM-4		
Variable	Coef.	t-Stat.	Prob.*	Coef.	t-Stat.	Prob.*	Coef.	t-Stat.	Prob.*	Coef.	t-Stat.	Prob.*
C	276.2552	26.700	0.0000	274.2064	23.418	0.0000	-1.02E+01	-0.810	0.4197	-36.59351	-2.832	0.0054
X1*X2	0.000375	6.229	0.0000							0.000197	4.363	0.0000
Y2(-4)*X1*X2	-6.84E-09	-0.796	0.4277	-1.96E-08	-2.067	0.0407	1.87E-08	4.318	0.0000	2.40E-08	5.093	0.0000
Y1(-4)*Y2(-4)*X1	3.00E-05	14.604	0.0000	2.91E-05	12.541	0.0000	-3.80E-06	-3.567	0.0005	-3.75E-06	-3.201	0.0017
Y1(-4)*Y2(-4)	0.008868	8.283	0.0000	9.10E-03	7.509	0.0000	1.15E-03	1.298	0.1964	0.001409	1.435	0.1536
Y2(-4)*X2	0.004136	16.612	0.0000	4.38E-03	15.706	0.0000	-0.000262	-1.794	0.0752	-0.000585	-3.780	0.0002
Y2(-4)*X1	-0.00272	-16.469	0.0000	-0.00255	-13.821	0.0000	0.000106	1.162	0.2475	0.000135	1.147	0.2536
Y1(-4)*Y2(-4)*X2	-4.93E-05	-16.013	0.0000	-4.82E-05	-13.840	0.0000	5.42E-06	3.119	0.0022	5.31E-06	2.836	0.0053
X1							0.668313	14.457	0.0000			
Y2(-4)							0.367444	5.189	0.0000	1.00938	6.807	0.0000
Y1(-4)							-1.14826	-1.213	0.2272	-1.938433	-1.732	0.0857
X2										0.292121	1.735	0.0851
R-squared	0.9987						0.9999			0.9999		
Adj. R-squared	0.9987						0.9999			0.9999		
F-statistic	14859.2						126237.6			118900.8		
Prob(F-statistic)	0.0000						0.0000			0.0000		
DW-Stat.	0.4791						0.8157			0.7225		
Y2F	RMSE= 53.638 & TIC=0.007			RMSE= 117.18 & TIC=0.017			RMSE= 17.955 & TIC=0.003			RMSE=17.744 & TIC=0.003		

Figure 6.47 OLS estimates of three special LV(4) TE_Models of *Y2* in (6.52b).

0.07705, and at the 15% level (Lapin 1973), *Y1(−4)∗X1∗X2* has a negative adjusted effect with a p-value = 0.1442 < 0.15. I would consider RM-4 an acceptable forecast model of *Y1*, out of many possible options.

2) OLS estimates of the three reduced models of *Y2* in (6.52b) are presented in Figure 6.47. Based on these results, the following findings and notes are presented.
   2.1 Since the interaction *Y2(−4)∗X1∗X2* in RM-1 has a large p-value = 0.4277, then a reduced model should be explored. For this reason, I try to delete one of its two-way interactions, namely *X1∗X2*, because *Y2(−4)* is one of the best predictors for *Y2*, and I consider this interaction as the least important. RM-2 is obtained with all independent variables having a Prob. < 0.05. However, its in-sample forecast values have a very large RMSE = 117.18, then this model is a worse forecast model than REM-1. Refer to a similar case presented in Example 5.24.
   2.2 Then I try to insert the main variables as additional independent variables.
   a) By inserting the main variables as the additional independent variables of RM-2, I obtain RM-3 as an acceptable model, in the statistical sense, since each of the three-way interactions has either a positive or negative significant adjusted effect on *Y2* at the 1% level. In addition, note that at the 15% level (Lapin 1973), *Y1(−4)∗X1* and *Y1(−4)*, respectively, have positive and negative significant adjusted effects on *Y2*, with p-values of 0.2475/2 = 0.12375 and 0.2272/2 = 0.1136, which are less than 0.15.
   b) As an additional model, by inserting the main variables as the additional independent variables of RM-1, I obtain RM-4 as an acceptable model, in the statistical sense, since each of the three-way interactions has either a positive or negative significant adjusted effect on *Y2* at the 1% level.
   c) Note that the values of RMSE and TIC of both RM-3 and RM-4 do not have large differences. Therefore, both models can be considered as acceptable forecast models of *Y2*.

Dependent Variable: X1
Method: Least Squares
Date: 07/11/17 Time: 15:35
Sample (adjusted): 1961Q1 1995Q1
Included observations: 137 after adjustments

Variable	Coefficient	Std. Error	t-Statistic	Prob.
X1(-4)*X2	-6.00E-07	3.90E-06	-0.153722	0.8781
X2	0.876607	0.057890	15.14252	0.0000
X1(-4)	0.327498	0.067389	4.859806	0.0000
X1(-8)	0.145290	0.063368	2.292805	0.0234
C	-48.44696	6.436286	-7.527160	0.0000

R-squared	0.999741	Mean dependent var	2023.447
Adjusted R-squared	0.999734	S.D. dependent var	1500.899
S.E. of regression	24.49514	Akaike info criterion	9.270640
Sum squared resid	79201.55	Schwarz criterion	9.377209
Log likelihood	-630.0388	Hannan-Quinn criter.	9.313947
F-statistic	127617.4	Durbin-Watson stat	0.816080
Prob(F-statistic)	0.000000		

(a)

Dependent Variable: X2
Method: Least Squares
Date: 07/11/17 Time: 15:34
Sample (adjusted): 1961Q1 1995Q1
Included observations: 137 after adjustments

Variable	Coefficient	Std. Error	t-Statistic	Prob.
X2(-4)	1.420641	0.079501	17.86942	0.0000
X2(-8)	-0.392407	0.083424	-4.703769	0.0000
C	19.18449	4.918164	3.900742	0.0002

R-squared	0.998985	Mean dependent var	1371.371
Adjusted R-squared	0.998970	S.D. dependent var	958.9459
S.E. of regression	30.77187	Akaike info criterion	9.712734
Sum squared resid	126885.7	Schwarz criterion	9.776675
Log likelihood	-662.3222	Hannan-Quinn criter.	9.738718
F-statistic	65970.30	Durbin-Watson stat	0.318364
Prob(F-statistic)	0.000000		

(b)

Figure 6.48 The statistical results of special models of the models of (a) (6.52c) and (b) (6.52d).

3) Finally, Figure 6.48 presents the statistical results of special models of (6.52c) and (6.52d), which are obtained using the trial-and-error method.

Example 6.28 Forecast Values Beyond the Sample Period Based on the Previous Example
Figure 6.49 presents the results on the forecast values beyond the sample period based on selected reduced models (RM-3 and RM-4) of *Y1* and *Y2* in Figures 6.46 and 6.47, using Excel. The steps of the computation are as follows:

1) Make an excel file of the variables *t, X2, X1, Y1-RM3, Y1-RM4, Y2-RM3,* and *Y2-RM4* for the last 10 of their observed scores, namely the observed scores from $t = 136$ up to $t = 145$.

	A	B	C	D	E	F	G	H
1		t	X2	X1	Y1-RM3	Y1-RM4	Y2-RM3	Y2-RM4
2	1992Q4	136	3021.7	4740.5	84.11	84.11	4329.6	4329.6
3	1993Q1	137	3045.9	4686.3	81.233	81.233	4367.8	4367.8
4	**1993Q2**	**138**	3075.1	4771.6	85.68	85.68	4424.7	4424.7
5	1993Q3	139	3114.9	4804.1	86.795	86.795	4481	4481
6	1993Q4	140	3144.9	4895.3	92.347	92.347	4543	4543
7	1994Q1	141	3195.2	4856.9	91.977	91.977	4599.2	4599.2
8	**1994Q2**	**142**	3242.8	5002.2	91.942	91.942	4665.1	4665.1
9	1994Q3	143	3265.5	5070.4	90.49	90.49	4734.4	4734.4
10	1994Q4	144	3320.2	5145.8	90.747	90.747	4796.03	4796.03
11	1995Q1	145	3363	5225.5	83.812	83.812	4836.33	4836.33
12	1995Q2	**146**	3419.347	5270.197	85.363	85.697	4964.578	4935.001
13	1995Q3	**147**	3435.978	5311.642	84.233	82.338	5011.244	4988.563
14	1995Q4	**148**	3501.915	5407.028	84.438	83.115	5119.321	5085.895

Figure 6.49 Spreadsheet for computing the forecast values based on the regression functions in the previous example.

2) Then the forecast values for $t = 146$ are computed as follows:

2.1 The first step is to compute *X2(146)* using the equation as follows:

$$X2 = 1.42064053148^{*}X2(-4) - 0.392406802242^{*}X2(-8) + 19.1844877224$$

$$\boldsymbol{X2}(146) = 1.42064053148^{*}\boldsymbol{X2}(142) - 0.392406802242^{*}\boldsymbol{X2}(138) + 19.1844877224$$

$$\mathbf{C12} = 1.42064053148^{*}\mathbf{C8} - 0.392406802242^{*}\mathbf{C4} + 19.1844877224$$

2.2 The second step is to compute *X1(146)* using the equation as follows:

$$X1(146) = -5.99893586947\text{e}-07^{*}\boldsymbol{X1}(142)*\boldsymbol{X2}(146) + 0.876607427662^{*}\boldsymbol{X2}(146) + 0.327498256312^{*}\boldsymbol{X1}(142) + 0.145290308085^{*}\boldsymbol{X1}(138) - 48.4469567636$$

$$\mathbf{D12} = -0.000000599893586947^{*}\mathbf{D8}^{*}\mathbf{C12} + 0.876607427662^{*}\mathbf{C12} + 0.327498256312^{*}\mathbf{D8} + 0.145290308085^{*}\mathbf{D4} - 48.4469567636$$

2.3 Then we can compute the forecast values of any one of the variables *Y1-RM1, Y1-RM4, Y2-RM3,* and *Y2-RM4* using the following equations.

$$Y1 = 12.1583180778 + 8.41869959086\text{e}-07^{*}Y1(-4)*Y2(-4)*X2 - 0.00356869149389^{*}Y1(-4)*Y2(-4) - 1.24673991939\text{e}-07^{*}Y1(-4)*Y2(-4)*X1 - 6.29320810593\text{e}-07^{*}Y1(-4)*X1^{*}X2 + 0.00334827414828^{*}Y1(-4)*X1 + 0.108742419667^{*}Y2(-4) - 0.0827321729764^{*}X1$$

$$\mathbf{E12} = 12.1583180778 + 0.000000841869959086^{*}\text{E8}^{*}\text{G8}^{*}\text{C12} - 0.00356869149389^{*}\mathbf{E8}^{*}\mathbf{G8} - 0.000000124673991939^{*}\mathbf{E8}^{*}\mathbf{G8}^{*}\mathbf{D12} - 0.000000629320810593^{*}\mathbf{E8}^{*}\mathbf{D12}^{*}\mathbf{C12} + 0.00334827414828^{*}\mathbf{E8}^{*}\mathbf{D12} + 0.108742419667^{*}\mathbf{G8} - 0.0827321729764^{*}\text{D12}$$

$$Y1 = 6.65431331983 + 2.58299624886e-07^{*}Y1(-4)*Y2(-4)*X2 - 0.00196849376788^{*}Y1(-4)*Y2(-4) - 1.62672259201e-07^{*}Y1(-4)*Y2(-4)*X1 - 3.96648590535e-08^{*}Y1(-4)*X1^{*}X2 + 0.00176788484672^{*}Y1(-4)*X1 + 0.108943337412^{*}Y2(-4) - 0.0785693049996^{*}X1 + 0.471266459753^{*}Y1(-4) - 0.188407157095^{*}Y1(-8)$$

$$F\mathbf{12} = 6.65431331983 + 0.000000258299624886^{*}\mathbf{F8}^{*}\mathbf{H8}^{*}\mathbf{C12} - 0.00196849376788^{*}\mathbf{F8}^{*}\mathbf{H8} - 0.000000162672259201^{*}\mathbf{F8}^{*}\mathbf{H8}^{*}\mathbf{D12} - 0.0000000396648590535^{*}F8^{*}\mathbf{D12}^{*}\mathbf{C12} + 0.00176788484672^{*}\mathbf{F8}^{*}\mathbf{D12} + 0.108943337412^{*}\mathbf{H8} - 0.0785693049996^{*}\mathbf{D12} + 0.471266459753^{*}\mathbf{F8} - 0.188407157095^{*}\mathbf{F4}$$

$$Y2 = -10.1707237054 - 3.803599364e-06^{*}Y1(-4)*Y2(-4)*X1 + 5.41857477751e-06^{*}Y1(-4)*Y2(-4)*X2 + 1.8705155357e-08^{*}Y2(-4)*X1^{*}X2 + 0.00115132792669^{*}Y1(-4)*Y2(-4) + 0.000106151707072^{*}Y2(-4)*X1 - 0.000262116606287^{*}Y2(-4)*X2 + 0.668313498961^{*}X1 + 0.36744364889^{*}Y2(-4) - 1.14826008263^{*}Y1(-4)$$

$$G1\mathbf{2} = -10.1707237054 - 0.000003803599364^{*}\mathbf{F8}^{*}\mathbf{G8}^{*}\mathbf{D12} + 0.00000541857477751^{*}\mathbf{F8}^{*}\mathbf{G8}^{*}\mathbf{C12} + 0.000000018705155357^{*}\mathbf{G8}^{*}\mathbf{D12}^{*}\mathbf{C12} + 0.00115132792669^{*}\mathbf{F8}^{*}\mathbf{G8} + 0.000106151707072^{*}\mathbf{G8}^{*}\mathbf{D12} - 0.000262116606287^{*}\mathbf{G8}^{*}\mathbf{C12} + 0.668313498961^{*}\mathbf{D12} + 0.36744364889^{*}\mathbf{G8} - 1.14826008263^{*}\mathbf{F8}$$

$$Y2 = -24.1013461394 - 5.76431467283e-06^{*}Y1(-4)*Y2(-4)*X1 + 8.18235411904e-06^{*}Y1(-4)*Y2(-4)*X2 + 1.71252464027e-08^{*}Y2(-4)*X1^{*}X2 + 0.00232196989492^{*}Y1(-4)*Y2(-4) + 0.000465229272534^{*}Y2(-4)*X1 - 0.000808521661374^{*}Y2(-4)*X2 + 0.4425844169^{*}Y2(-4) - 3.46277575357^{*}Y1(-4) + 0.955144998472^{*}X2$$

$$\mathbf{H12} = -24.1013461394 - 0.00000576431467283^{*}\mathbf{F8}^{*}\mathbf{H8}^{*}\mathbf{D12} + 0.00000818235411904^{*}\mathbf{F8}^{*}\mathbf{H8}^{*}\mathbf{C12} + 0.0000000171252464027^{*}\mathbf{H8}^{*}\mathbf{D12}^{*}\mathbf{C12} + 0.00232196989492^{*}\mathbf{F8}^{*}\mathbf{H8} + 0.000465229272534^{*}\mathbf{H8}^{*}\mathbf{D12} - 0.000808521661374^{*}\mathbf{H8}^{*}\mathbf{C12} + 0.4425844169^{*}\mathbf{H8} - 3.46277575357^{*}\mathbf{F8} + 0.955144998472^{*}\mathbf{C12}$$

2.4 Finally, the forecast values for $t > 146$ of the variables *X2, X1, Y1-RM3, Y1_RM4, Y2_RM3,* and *Y2-RM4* can be obtained by using the block-copy-paste method of their values for $t = 146$.

6.10 Models with Exogenous Variables by @*Quarter*

Referring to the LV(2) models of *Y1* = *GCDAN* by @*Quarter* in (6.20), it is recognized that all ESs of the models with exogenous variables and trends presented previously can easily be applied for each subsample @*Quarter* = 1, 2, 3, and 4, in order to obtain forecast values for each of @*Quarters*. In this case, I propose two alternative data analyses. The first one conducts the analysis using the whole sample, and the second uses each of the four work files, such as CONS1_Q1.wf1 for the sample @*Quarter* = 1. Then, for each new work file, we have annual time series.

6.10.1 Alternative Models Based on the Whole Sample

All models based on *(X1,Y1)*, *(X1,X2,Y1)*, and *(X1,X2,Y1,Y2)* discussed previously can easily be presented as the models by @*Quarter*. However, this section presents only the simplest two-way interaction model of *Y1* by @*Quarter* based on *(X1,Y1)*, because all models based on *(X1,X2,Y1)* and *(X1,X2,Y1,Y2)*, can easily be modified to similar equations by @*Quarter*.

6.10.1.1 Explicit Estimation Function

Based on *(X1,Y1)*, I propose the following explicit estimation function since, by using this, the equations of the regressions for the four quarters can easily be presented based on the output of its "*Representations*."

$$\begin{aligned} Y1 = {} & (C(11)*Y1(-1)*X1(-1) + C(12)*Y1(-1) + C(13)*X1(-1) + C(10)) \\ & *(@QUARTER = 1) \\ & + (C(21)*Y1(-1)*X1(-1) + C(22)*Y1(-1) + C(23)*X1(-1) + C(20)) \\ & *(@QUARTER = 2) \\ & + (C(31)*Y1(-1)*X1(-1) + C(32)*Y1(-1) + C(33)*X1(-1) + C(30)) \\ & *(@QUARTER = 3) \\ & + (C(41)*Y1(-1)*X1(-1) + C(42)*Y1(-1) + C(43)*X1(-1) + C(40)) \\ & *(@QUARTER = 4) \end{aligned} \tag{6.53}$$

Example 6.29 An Application of the Model (6.53)

Figure 6.50 presents the OLS estimate of the model (6.53) for *Y1* = *GCDAN* and *X1* = *GYD*. Its in-sample forecast values are presented as a single variable, namely *Y1F*. Based on the output of its "*Representations*," we have the following four regression functions, each of which can be used to compute the forecast values beyond the sample period of *Y1* for each quarter, together with a dynamic model of *X1*. Either one of the regression function of *X1* can be applied directly or use a new alternative model. Do this as an exercise.

Dependent Variable: Y1
Method: Least Squares
Date: 07/22/17 Time: 06:20
Sample (adjusted): 1959Q2 1995Q1
Included observations: 144 after adjustments
Y1=(C(11)*Y1(-1)*X1(-1)+C(12)*Y1(-1)+C(13)*X1(-1)+C(10))
*(@QUARTER=1)+(C(21)*Y1(-1)*X1(-1)+C(22)*Y1(-1)+C(23)*X1(
-1)+C(20))*(@QUARTER=2)+(C(31)*Y1(-1)*X1(-1)+C(32)*Y1(-1)
+C(33)*X1(-1)+C(30))*(@QUARTER=3)+(C(41)*Y1(-1)*X1(-1)
+C(42)*Y1(-1)+C(43)*X1(-1)+C(40))*(@QUARTER=4)

	Coefficient	Std. Error	t-Statistic	Prob.
C(11)	-8.28E-05	3.10E-05	-2.672675	0.0085
C(12)	0.988534	0.088292	11.19623	0.0000
C(13)	0.007834	0.002977	2.631026	0.0096
C(10)	-2.141884	2.668668	-0.802604	0.4237
C(21)	-4.46E-06	3.07E-05	-0.145439	0.8846
C(22)	0.964115	0.093846	10.27335	0.0000
C(23)	0.001208	0.002891	0.417673	0.6769
C(20)	0.161824	2.657652	0.060890	0.9515
C(31)	-2.15E-05	3.07E-05	-0.699347	0.4856
C(32)	1.177525	0.090615	12.99478	0.0000
C(33)	-0.000526	0.002786	-0.188842	0.8505
C(30)	-2.671922	2.705983	-0.987413	0.3253
C(41)	-6.15E-05	2.94E-05	-2.091872	0.0384
C(42)	0.807591	0.083908	9.624684	0.0000
C(43)	0.009307	0.002426	3.836534	0.0002
C(40)	-1.277019	2.745360	-0.465156	0.6426

R-squared	0.977120	Mean dependent var	48.94819
Adjusted R-squared	0.974439	S.D. dependent var	29.79341
S.E. of regression	4.763354	Akaike info criterion	6.064220
Sum squared resid	2904.262	Schwarz criterion	6.394200
Log likelihood	-420.6239	Hannan-Quinn criter.	6.198305
Durbin-Watson stat	2.551137		

Figure 6.50 The OLS estimate of the model in (6.53).

1) $$Y1(Q1) = (-8.2772151119e-05^{*}Y1(-1)*X1(-1) + 0.988533636122^{*}Y1(-1) + 0.00783350716581^{*}X1(-1) - 2.14188383877)$$

2) $$Y1(Q2) = (-4.46106096321e-06^{*}Y1(-1)*X1(-1) + 0.964115022406^{*}Y1(-1) + 0.0012075039487^{*}X1(-1) + 0.161824369353)$$

3) $$Y1(Q3) = (-2.15007937388e-05^{*}Y1(-1)*X1(-1) + 1.17752465811^{*}Y1(-1) - 0.00052602039198^{*}X1(-1) - 2.67192159831)$$

4) $$Y1(Q4) = (-6.14658403519e-05^{*}Y1(-1)*X1(-1) + 0.807591132657^{*}Y1(-1) + 0.00930691245029^{*}X1(-1) - 1.27701946265)$$

Based on the results previously, the following findings and notes are presented.

1) The interaction *Y1(−1)∗X1(−1)* has a significant effect on *Y1* in two regressions for the first and fourth quarters at the 1 and 5% levels.
2) Despite the fact it has large *p*-values than the other two regressions, the models do not have to be reduced in order to present four regression functions with exactly the same set of numerical independent variables, which are required for testing differential partial effects of the independent variables between the quarters. Whenever those testing are not considered, then the reduced models can be explored by deleting one or both of *Y1(−1)* and *X1(−1)* from the two models. Do these as exercises.
3) A single forecast evaluation can be presented for the whole model, but it cannot be computed for each quarter.

6.10.1.2 Implicit Estimation Function

As modified models of the model in (6.53), the following ESs are two alternative implicit estimation functions of the model.

$$Y1\ Y1(-1)*X1(-1)*@Expand(@Quarter)\ Y1(-1)*@Expand(@Quarter)\ X1(-1)*@Expand(@Quarter)\ @Expand(@Quarter) \tag{6.54a}$$

$$Y1\ C\ Y1(-1)*X1(-1)\ Y1(-1)\ X1(-1)\ Y1(-1)*X1(-1)*@Expand(@Quarter,@Dropfirst)\ Y1(-1)*@Expand(@Quarter,@Dropfirst)\ X1(-1)*@Expand(@Quarter,@Dropfirst)\ @Expand(@Quarter,@Dropfirst) \tag{6.54b}$$

Note that the OLS estimates of these models would be exactly the same as the model (6.53). However, model (6.54b) uses *@Quarter* = 1 as the reference group, indicated by the function *@Dropfirst* in its ES.

6.10.2 Forecasting Based on each Quarter's Level

I have found that in order to forecasting based on each quarter's level, we have to develop four new work files for the four quarters' levels, namely DATA_Q1.wf1, DATA_Q2.wf1, DATA_Q3.wf1, and DATA_Q4.wf1 so that we would have four annual time series data sets. Therefore, all annual time-series models presented in Agung (2009a), as well as in other books, can be applied for each of the data sets. Specific for the data in CONS1.wf1, the following examples present illustrative data analysis based on only CONS1_Q1,wf1, using selected ESs previously presented.

Example 6.30 Application of the Models (6.53) or (6.54b)

Based on CONS1_Q1.wf1, we would have the following ES with its statistical results presented in Figure 6.51 and a forecast sample {*@Year*>1980}. Based on these results, findings and notes presented are as follows:

$$Y1\ Y1(-1)*X1(-1)\ Y1(-1)\ X1(-1)\ C \tag{6.55}$$

1) The regression can be considered an acceptable forecast model of *Y1*, since it has a small value of TIC = 0.060.
2) However, I would say this regression is not the best forecast model, because the graph of the observed scores of *Y1* have three extreme values for $t > 1980$, namely a maximum of 100.595

Dependent Variable: Y1
Method: Least Squares
Date: 07/23/17 Time: 12:28
Sample (adjusted): 1960 1995
Included observations: 36 after adjustments

Variable	Coefficient	Std. Error	t-Statistic	Prob.
Y1(-1)*X1(-1)	-0.000139	4.35E-05	-3.203055	0.0031
Y1(-1)	0.984211	0.133214	7.388199	0.0000
X1(-1)	0.013533	0.004104	3.297691	0.0024
C	-4.484292	3.772517	-1.188674	0.2433

R-squared	0.951522	Mean dependent var	49.70531
Adjusted R-squared	0.946977	S.D. dependent var	29.36400
S.E. of regression	6.761545	Akaike info criterion	6.764819
Sum squared resid	1462.992	Schwarz criterion	6.940766
Log likelihood	-117.7667	Hannan-Quinn criter.	6.826229
F-statistic	209.3651	Durbin-Watson stat	2.111323
Prob(F-statistic)	0.000000		

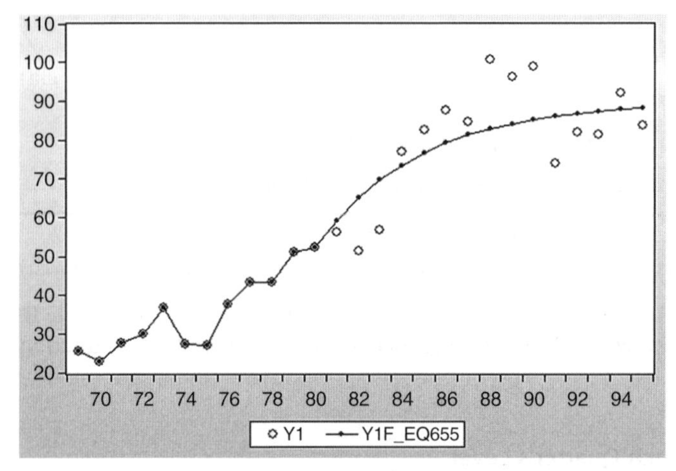

RMSE=9.615 & TIC=0.060

Figure 6.51 Statistical results of the model (6.55), with the forecast sample *{@Year>1980}*.

at $t = 29$, a relative minimum of 73.899 at $t = 32$, and a relative maximum of 91.877 at $t = 35$. For this reason, models with polynomial trend or heterogeneous trends should be explored, which are presented in the following examples.

3) In order to compute the forecast values of *Y1* beyond the sample period, we can apply any one of the previous forecast models of *X1*, or define another alternative model.

Example 6.31 Application of the Model (6.55) with Fourth Degree Polynomial Trend
Since the graph of the observed scores of *Y1* has three extreme values for $t > 1980$, then I try to apply the model in (6.55) with the fourth degree polynomial trend using the combinatorial and the trial-and-error selection methods with the regression in Figure 6.51 as the estimation function and "*t t^2 T^3 T^4*" as the search regressors. In this case, it so happens that the four time variables can be used as additional independent variables with the statistical results presented in Figure 6.52 using the forecast sample {@*Year*>1980}. Based on these results, the following findings and notes are presented.

1) At the 10% level, the interaction $Y1(-1)*X1(-1)$ has a negative significant adjusted effect on $Y1$ with a p-value = 9.1775/2 = 0.08875 < 0.10.
2) Each of the time variables has either a positive of negative significant adjusted effect on $Y1$ at the 1% level of significance.
3) Compared to the regression in Figure 6.51, this regression is a better forecast model than (6.55), since it has smaller values of RMSE and TIC.

Example 6.32 Application of the Model (6.55) with Heterogeneous Trends

As an additional illustration, referring to the three extreme scores of the variables $Y1$ for t = 29, 32, and 35, presented previously, I generate an ordinal TP variable with four levels using the following equation in order to present a model of (6.55) with heterogeneous trends.

$$TP = 1 + 1^*(t >= 29) + 1^*(t >= 32) + 1^*(t >= 35)$$

Dependent Variable: Y1
Method: Stepwise Regression
Date: 08/06/17 Time: 12:58
Sample (adjusted): 1960 1995
Included observations: 36 after adjustments
Number of always included regressors: 4
Number of search regressors: 4
Selection method: Stepwise forwards
Stopping criterion: p-value forwards/backwards = 0.5/0.5

Variable	Coefficient	Std. Error	t-Statistic	Prob.*
Y1(-1)*X1(-1)	-0.000157	0.000113	-1.383354	0.1775
Y1(-1)	0.823334	0.399347	2.061700	0.0486
X1(-1)	-0.104420	0.034864	-2.995085	0.0057
C	25.78545	12.05653	2.138712	0.0413
T^4	-0.001155	0.000285	-4.048158	0.0004
T^3	0.080189	0.020323	3.945724	0.0005
T^2	-1.222479	0.344990	-3.543518	0.0014
T	9.537894	2.944250	3.239498	0.0031

R-squared	0.970639	Mean dependent var	49.70531
Adjusted R-squared	0.963299	S.D. dependent var	29.36400
S.E. of regression	5.625405	Akaike info criterion	6.485593
Sum squared resid	886.0651	Schwarz criterion	6.837486
Log likelihood	-108.7407	Hannan-Quinn criter.	6.608413
F-statistic	132.2363	Durbin-Watson stat	1.980561
Prob(F-statistic)	0.000000		

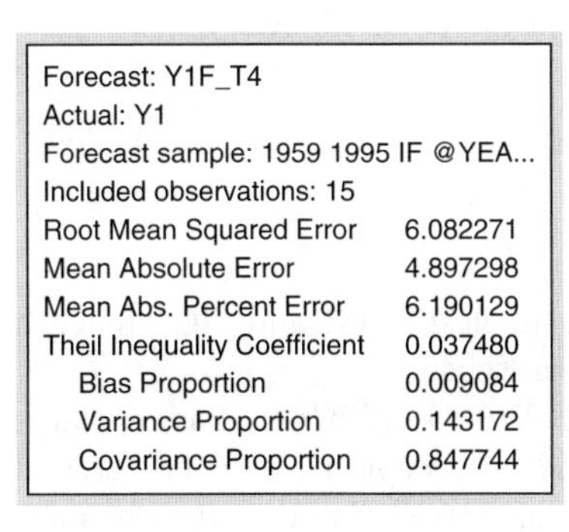
Forecast: Y1F_T4
Actual: Y1
Forecast sample: 1959 1995 IF @YEA...
Included observations: 15

Root Mean Squared Error	6.082271
Mean Absolute Error	4.897298
Mean Abs. Percent Error	6.190129
Theil Inequality Coefficient	0.037480
Bias Proportion	0.009084
Variance Proportion	0.143172
Covariance Proportion	0.847744

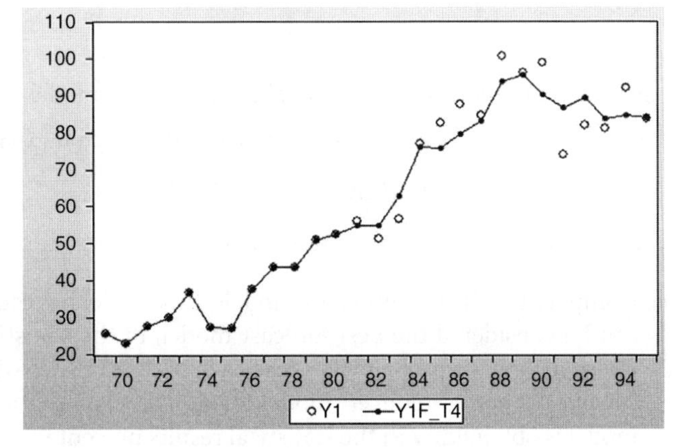

Figure 6.52 Statistical result of the model (6.55) with fourth degree polynomial trend.

Dependent Variable: Y1
Method: Least Squares
Date: 08/06/17 Time: 16:02
Sample (adjusted): 1960 1995
Included observations: 36 after adjustments

Variable	Coefficient	Std. Error	t-Statistic	Prob.
C	6.275456	4.944616	1.269149	0.2161
Y1(-1)*X1(-1)	1.70E-05	0.000106	0.161421	0.8731
Y1(-1)	0.047569	0.371067	0.128194	0.8990
X1(-1)	0.019216	0.008025	2.394491	0.0245
T*(TP=1)	0.441136	0.665300	0.663063	0.5134
T*(TP=2)	-6.960154	3.748849	-1.856611	0.0752
T*(TP=3)	0.112775	4.271732	0.026400	0.9791
T*(TP=4)	-13.08019	7.433714	-1.759577	0.0907
TP=2	222.5057	115.1240	1.932748	0.0647
TP=3	-23.48783	136.9761	-0.171474	0.8652
TP=4	443.1032	267.3165	1.657597	0.1099

R-squared	0.980304	Mean dependent var	49.70531
Adjusted R-squared	0.972425	S.D. dependent var	29.36400
S.E. of regression	4.876096	Akaike info criterion	6.253035
Sum squared resid	594.4077	Schwarz criterion	6.736888
Log likelihood	-101.5546	Hannan-Quinn criter.	6.421912
F-statistic	124.4270	Durbin-Watson stat	1.777639
Prob(F-statistic)	0.000000		

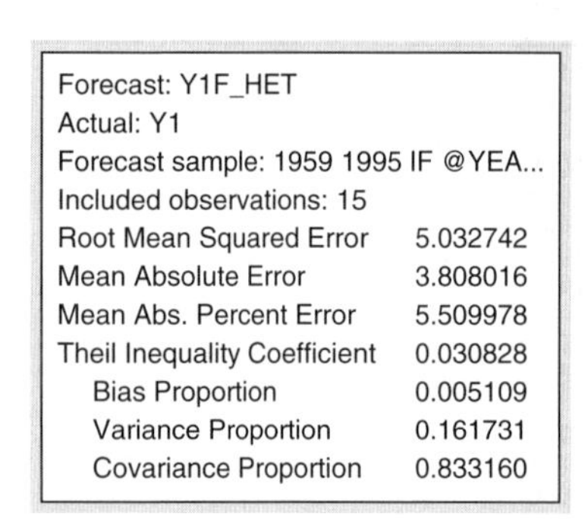

Forecast: Y1F_HET
Actual: Y1
Forecast sample: 1959 1995 IF @YEA...
Included observations: 15

Root Mean Squared Error	5.032742
Mean Absolute Error	3.808016
Mean Abs. Percent Error	5.509978
Theil Inequality Coefficient	0.030828
Bias Proportion	0.005109
Variance Proportion	0.161731
Covariance Proportion	0.833160

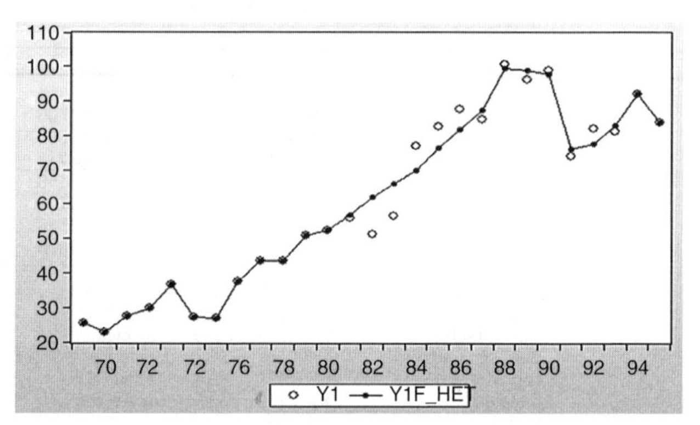

Figure 6.53 Statistical result of the model (6.55) with heterogeneous trends.

Figure 6.53 presents the statistical results of the model with the ES as follows:

$$YI\,C\,YI(-1)*XI(-1)\,YI(-1)\,XI(-1)\,T^{*}@EXPAND(TP)$$
$$@EXPAND(TP, @DROPFIRST) \tag{6.56}$$

Based on these results, the findings and notes presented are as follows;

1) Compared to the two previous models, this model has the smallest RMSE and TIC. Hence, it can be considered the best forecast model, in the statistical sense.
2) However, it has a problem, because each of the IVs, $Y1(-1)*X1(-1)$ and $Y1(-1)$, has very large p-values. Then, a reduced model should be explored and an acceptable reduced forecast model is obtained, with the statistical results presented in Figure 6.54. Based on these results, the following findings and notes are presented.

Dependent Variable: Y1
Method: Least Squares
Date: 08/07/17 Time: 08:16
Sample (adjusted): 1960 1995
Included observations: 36 after adjustments

Variable	Coefficient	Std. Error	t-Statistic	Prob.
C	12.17261	4.661633	2.611233	0.0148
Y1(-1)*X1(-1)	0.000210	7.44E-05	2.816945	0.0091
Y1(-1)	-0.328846	0.365442	-0.899859	0.3765
T*(TP=1)	1.759180	0.406266	4.330113	0.0002
T*(TP=2)	-8.271677	4.032110	-2.051451	0.0504
T*(TP=3)	4.301210	4.237116	1.015127	0.3194
T*(TP=4)	-18.47668	7.701745	-2.399025	0.0239
TP=2	295.1833	120.7374	2.444837	0.0216
TP=3	-123.1805	141.8780	-0.868214	0.3932
TP=4	673.3952	271.1717	2.483280	0.0198

R-squared	0.975786	Mean dependent var	49.70531
Adjusted R-squared	0.967405	S.D. dependent var	29.36400
S.E. of regression	5.301419	Akaike info criterion	6.403959
Sum squared resid	730.7312	Schwarz criterion	6.843826
Log likelihood	-105.2713	Hannan-Quinn criter.	6.557485
F-statistic	116.4197	Durbin-Watson stat	2.056379
Prob(F-statistic)	0.000000		

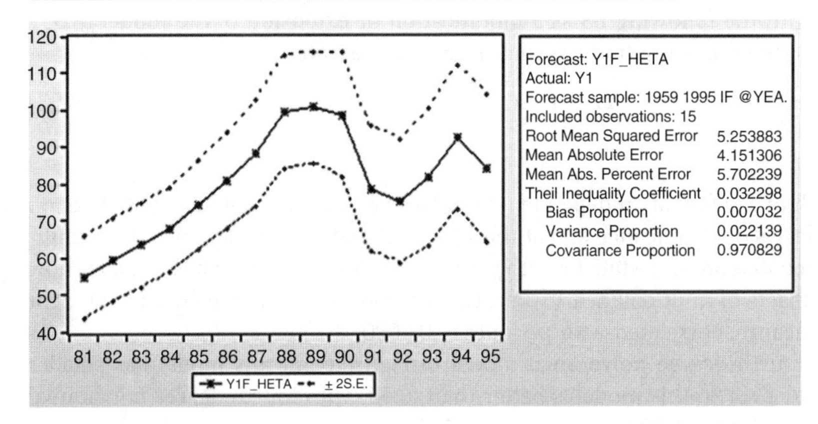

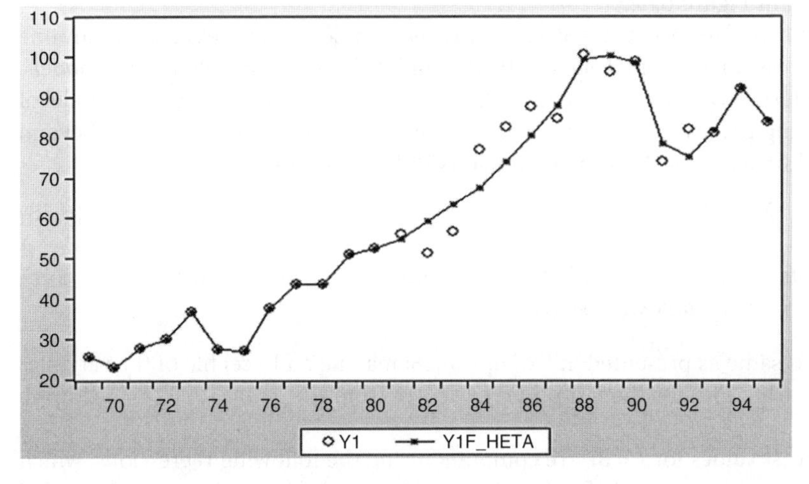

Figure 6.54 Statistical results of an acceptable reduced model of the model (6.56).

2.1 The reduced model is obtained by using the trial-and-error method. Unexpectedly, the acceptable reduced model with the interaction *Y1(−1)*X1(−1)* has a significant effect at the 1% level, and is obtained by deleting the main variable *X1(−1)*, which has a significant adjusted effect at the 5% level. If *Y1(−1)* is deleted instead of the *X1(−1)* a reduced model with *Y1(−1)*X1(−1)* is obtained that has a *p*-value = 0.5657.

2.2 Even though *Y1(−1)* has an insignificant effect, it does not have to be deleted from the model because its interaction has a significant adjusted effect.

2.3 This model has a greater RMSE and TIC than (6.56), which are the characteristics of any reduced model. If *Y1(−1)* were to be deleted then the reduced model would have even greater values of RMSE and TIC, but it is an acceptable forecast model of *Y1*.

2.4 As an additional analysis, the null hypothesis H_0: C(4) = C(5) = C(6) = C(7) = 0 is rejected, based on the Wald test of $F_0 = 9.546727$, with $df = (4,26)$, and *p*-value = 0.0001. Hence, it can be concluded that the model has significant heterogeneous trends.

Example 6.33 An Application of the Time-Related Effects Model

The TRE models have been presented for the monthly time series in Chapter 2, and for annual time series in Agung (2009a). The same models also can be presented for a quarterly time series based on the whole data set, as well as based on each sub-data set by *@Quarter*.

As an additional illustration, Figure 6.55 presents the statistical results of a TRE-Model based on CONS1_Q1.wf1, with the following ES as a modification or extension of the model (6.55). Based on these results, the following findings and notes are presented.

$$Y1\,C\,Y1(-1)*X1(-1)*T\;Y1(-1)*T\;X1(-1)*T\;Y1(-1)*X1(-1)\;Y1(-1)\;X1(-1)\;T \qquad (6.57)$$

1) Unexpectedly, each of the first six independent variables has a significant adjusted effect on *Y1* at the 1 or 5% levels of significance and at the 10% level, the time *T* has a positive significant adjusted effect based on the *t*-statistic with *p*-value = 0.1597/2 < 0.10. These results show the unexpected impacts of multicollinearity because it is found that all pairs of independent variables are significantly correlated with *p*-values = 0.0000.
2) Compared to the fourth degree polynomial model in Figure 6.52, this model has smaller values of RMSE and TIC. So this model is better than (6.55) with fourth degree polynomial trend, as presented in Figure 6.52.
3) In order to compute the forecast values of *Y1* beyond the sample period, we can apply any one of the previous forecast models of *X1*, such as (6.48c) and (6.48d), or other alternative models.
4) As an additional illustration, Figure 6.56 presents an OLS estimate of a simple TRE-model of *X1* with the following ES, which can be applied to compute the forecast values of *Y1* beyond the sample period based on each of the regressions in Figures 5.1–5.5.

$$X1\;C\;X1(-1)*t\;X1(-1)\,t \qquad (6.58)$$

5) Figure 6.57 presents the forecast values based on the model (6.57) and (6.58), which are computed using Excel with the following steps.

5.1 The first step is same as presented in Example 6.28; making an Excel file of the variables *Year*, *t*, *X1*, and *Y1_TRE* for the last observed scores from the *Year*=1990 (t = 31) up to t = 36.

5.2 Then the forecast values for t = 37 are computed using the following regressions, which are obtained from the output of "*Representations*," and each of them can directly copied

Dependent Variable: Y1
Method: Least Squares
Date: 08/07/17 Time: 13:52
Sample (adjusted): 1960 1995
Included observations: 36 after adjustments

Variable	Coefficient	Std. Error	t-Statistic	Prob.
C	92.61612	21.12962	4.383236	0.0001
Y1(-1)*X1(-1)*T	-5.85E-05	1.41E-05	-4.134422	0.0003
Y1(-1)*T	0.165676	0.050548	3.277576	0.0028
X1(-1)*T	0.005102	0.001316	3.876313	0.0006
Y1(-1)*X1(-1)	0.001143	0.000516	2.216496	0.0350
Y1(-1)	-2.367107	0.820418	-2.885243	0.0074
X1(-1)	-0.175927	0.049087	-3.584001	0.0013
T	1.407428	0.974316	1.444530	0.1597

R-squared	0.974252	Mean dependent var	49.70531
Adjusted R-squared	0.967815	S.D. dependent var	29.36400
S.E. of regression	5.267971	Akaike info criterion	6.354298
Sum squared resid	777.0424	Schwarz criterion	6.706191
Log likelihood	-106.3774	Hannan-Quinn criter.	6.477118
F-statistic	151.3509	Durbin-Watson stat	2.206053
Prob(F-statistic)	0.000000		

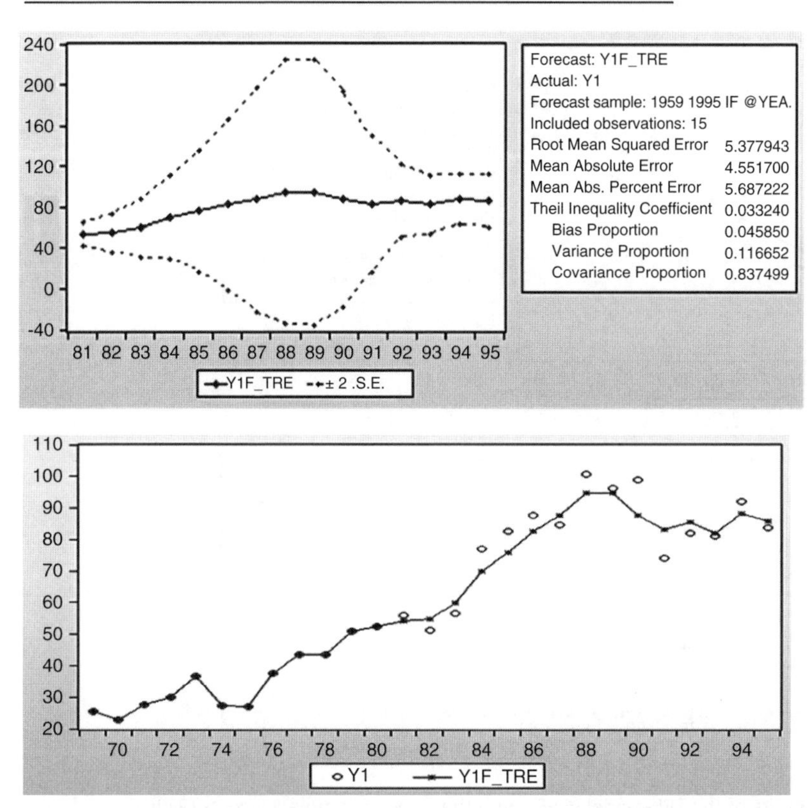

Figure 6.55 Statistical results of the TRE model (6.57), with the forecast sample *{@Year>1980}*.

Dependent Variable: X1
Method: Least Squares
Date: 08/08/17 Time: 16:32
Sample (adjusted): 1960 1995
Included observations: 36 after adjustments

Variable	Coefficient	Std. Error	t-Statistic	Prob.
C	-35.86561	23.80552	-1.506609	0.1417
X1(-1)*T	-0.001776	0.001675	-1.060308	0.2969
X1(-1)	1.071481	0.081238	13.18943	0.0000
T	6.775928	3.607031	1.878533	0.0694

R-squared	0.999282	Mean dependent var	1992.417
Adjusted R-squared	0.999214	S.D. dependent var	1549.546
S.E. of regression	43.43665	Akaike info criterion	10.48492
Sum squared resid	60375.77	Schwarz criterion	10.66087
Log likelihood	-184.7286	Hannan-Quinn criter.	10.54633
F-statistic	14836.47	Durbin-Watson stat	1.913800
Prob(F-statistic)	0.000000		

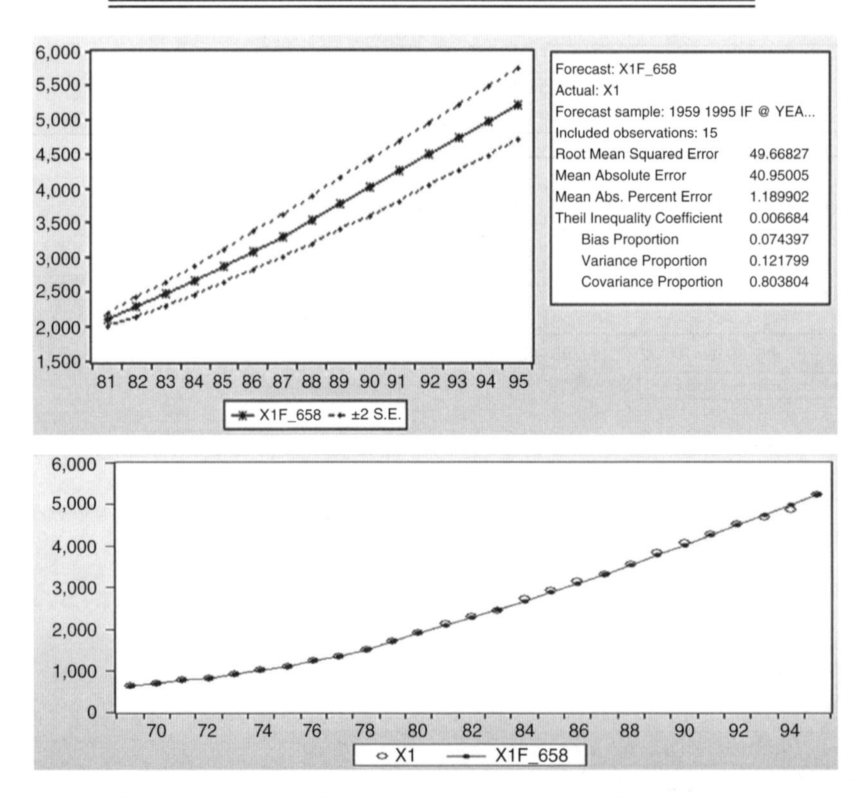

Figure 6.56 Statistical results of the TRE-model (6.58), with the forecast sample *{@Year > 1980}*.

to the Excel file for computing the forecast values of $X1(t = 37)$ and $Y1(t = 37)$, respectively, with adjustment as presented on the right-hand side of Figure 6.57.

$$X1(t) = -35.8656124802 - 0.00177583821515^{*}X1(t-1)*T + 1.0714808591^{*}X1(t-1) + 6.77592789634^{*}T \tag{6.59a}$$

$$Y1(t) = 92.6161168441 - 5.84710399973e-05^{*}Y1(36)*X1(t-1)*T + 0.165675624932^{*}Y1(t-1)*T + 0.00510182092697^{*}X1(t-1)*T + 0.00114268843041^{*}Y1(t-1)*X1(t-1) - 2.36710660718^{*}Y1(t-1) - 0.175927022859^{*}X1(t-1) + 1.4074282965^{*}T \tag{6.59b}$$

$$Y1(t) = -0.000139461592707^{*}Y1(t-1)*X1(t-1) + 0.984211052756^{*}Y1(t-1) + 0.0135330359235^{*}X1(t-1) - 4.48429242534 \tag{6.59c}$$

5.3 Finally, the forecast values for $t > 37$ of the variables *X1* and *Y1-TRE* can easily be obtained by using the block-copy-paste method of their values for $t = 37$.

5.4 Based on the results in Figure 6.57, the findings and notes presented are as follows:

a) The forecast values of *X1* increase, corresponding to the increase of observed scores from the *Year*=1990 ($t = 31$).

b) The forecast values of *Y1* fluctuate, corresponding to the fluctuation of observed scores from the *Year*=1990 ($t = 31$). Note that *Y1* has an unacceptable negative value for the *Year*=2000 ($t = 41$). However, the forecast values of 80.121 and 85.991 are for $t = 37$ and $t = 38$, respectively, then I would consider they are reasonable values compared to the observed score of 83.812 for $t = 36$. For this reason, I would recommend re-estimating the model after having a new true observed score, since doing a forecast based on an annual time series for sure would have greater risk errors compared to the forecast based on quarterly and monthly time series.

	A	B	C	D	E
1	Year	t	X1	Y1_TRE	Y1_655
2	1990	31	4074.8	98.899	
3	1991	32	4263.3	73.899	
4	1992	33	4515.2	82.064	
5	1993	34	4686.3	81.233	
6	1994	35	4856.9	91.977	
7	1995	36	5225.5	83.812	
8	**1996**	**37**	**5470.5**	**80.121**	**87.643**
9	1997	38	5714.0	85.991	87.278
10	1998	39	5955.1	72.244	88.952
11	1999	40	6193.0	109.667	87.211
12	2000	41	6426.7	-10.532	92.544

C8 = -35.8656124802
- 0.00177583821515*C7*B8
+ 1.0714808591*C7
+ 6.77592789634*B8

D8 = 92.6161168441
- 0.0000584710399973*D7*C7*B8
+ 0.165675624932*D7*B8
+ 0.00510182092697*C7*B8
+ 0.00114268843041*D7*C7
- 2.36710660718*D7
- 0.175927022859*C7
+ 1.4074282965*B8

E8 = -0.000139461592707*D7*C7
+ 0.984211052756*D7
+ 0.0135330359235*C7 -4.48429242534

Figure 6.57 Forecast values based on the regressions (6.59a)–(6.59c).

c) However, as a comparison, Figure 6.57 presents the forecast values of *Y1* based on the model (6.55), namely the regression (6.59c), which shows a small fluctuation with a minimum of 87.211 and a maximum of 92.544 for $t = 37$ to $t = 41$. Note that this model has a TIC = 0.060 and the TRE-model (6.57) has a smaller TIC = 0.033. So, in the statistical sense, the TRE model should be a better forecast model than the model (6.55), but the results show otherwise – refer to the previous note in point 3.2.

5.5 Finally, it should be noted that doing a forecast based on annual time series would have greater risk errors compared to the forecast based on quarterly and monthly time series.

Example 6.34 Two Alternative TRE-Models of *Y1*

As special additional forecast models, I present two alternative TRE-Models of *Y1* with the following set of four ESs based on the triple time series *(X1,X2,Y1)* in which only the forecast model of *X2* is an TRE-model. In fact, the other three models of *X1* and *Y1* are also TRE-models, since the interaction *X2(−1)*t* is inserted as the independent variable of the three models.

$$X2\ C\ X2(-1)*t\ X2(-1)\ t \tag{6.60a}$$

$$X1\ C\ X1(-1)*X2(-1)\ X1(-1)\ X2(-1) \tag{6.60b}$$

$$Y1\ C\ Y1(-1)*X1(-1)\ Y1(-1)\ X1(-1) \tag{6.60c}$$

$$Y1\ C\ Y1(-1)*X1(-1)\ Y1(-1)*X2(-1)\ Y1(-1)\ X1(-1)\ X2(-1) \tag{6.60d}$$

Figure 6.58 presents the forecast values based on the three OLS regressions of (6.60a)–(6.60d) using the following equations.

$$\begin{aligned} X2(t) = &-25.3531801276 - 0.00232192393323^{*}X2(t-1)*T \\ &+ 1.09246701486^{*}X2(t-1) + 3.82437814467^{*}T \end{aligned} \tag{6.61a}$$

$$\begin{aligned} X1(t) = &-56.6707072277 - 1.30402954144e-05^{*}X1(t-1)*X2(t-1) \\ &+ 0.480365323884^{*}X1(t-1) + 0.953273741051^{*}X2(t-1) \end{aligned} \tag{6.61b}$$

	B	C	D	E	F	G
1	t	X2	X1	Y1	Y1f_6.61c	Y1f_6.61d
2	31	2704.0	4074.8	98.899	86.963	86.328
3	32	2789.5	4263.3	73.899	87.763	86.227
4	33	2916.5	4515.2	82.064	87.579	85.984
5	34	3045.9	4686.3	81.233	87.640	85.758
6	35	3195.2	4856.9	91.977	86.780	85.806
7	36	3363.0	5225.5	83.812	85.596	85.930
8	**37**	**3501.2**	**5430.2**		**87.643**	**85.986**
9	38	3636.0	5673.5		69.002	103.410
10	39	3766.7	5914.5		72.296	112.891
11	40	3892.8	6152.3		75.557	122.910
12	41	4013.6	6386.1		78.776	133.465
13	RMSE	33.474	49.390		9.615	6.730
14	TIC	0.007	0.007		0.060	0.053

C8 = -25.3531801276
- 0.00232192393323*C7*B8
+ 1.09246701486*C7 + 3.82437814467*B8

D8 = -56.6707072277
- 0.0000130402954144*D7*C7
+ 0.480365323884*D7 + 0.953273741051*C7

F8 = -0.000139461592707*E7*D7
+ 0.984211052756*E7 + 0.0135330359235*D7
- 4.48429242534

G8 = 3.99274273148 -0.00195178602454*E7*D7
+ 0.00269394408642*E7*C7
+ 0.959238353647*E7 + 0.164976163746*D7
- 0.22747445302*C7

Figure 6.58 Forecast values of *X1*, *X1*, and *Y1* using the four regressions (6.61a)–(6.61d).

$$Y1(t) = -0.000139461592707^{*}Y1(t-1)*X1(t-1) + 0.984211052756^{*}Y1(t-1)$$
$$+ 0.0135330359235^{*}X1(-1) - 4.48429242534 \quad (6.61c)$$

$$Y1(t) = 3.99274273148 - 0.00195178602454^{*}Y1(t-1)*X1(t-1)$$
$$+ 0.00269394408642^{*}Y1(t-1)*X2(t-1) + 0.959238353647^{*}Y1(t-1) \quad (6.61d)$$
$$+ 0.164976163746^{*}X1(t-1) - 0.22747445302^{*}X2(t-1)$$

Based on these results, the findings and notes presented are as follows:

1) The two forecast values of *Y1*, namely *Y1f_6.61c* and *Y1f_6.61d* based on the models (6.60c) and (6.60d), at the last sample point are 85.596 and 85.930, respectively, compared to the observed score of 83.812, which shows that the first model has a smaller deviation from the true observed value. Hence, one might say the first model is a better forecast model. In fact, there is a contradiction, since the first model has greater values of RMSE and TIC, as presented previously. So, in the statistical sense, the second model is a better model, specifically for the in-sample forecast values.
2) However, for the first forecast values beyond the sample period, for $t = T + 1 = 37$, the two models present the values of 87.643 and 85.986, but we never know their deviation from the true value before we have the data for $t = 37$. So it is very difficult to judge which is a better forecast model. It would be even more difficult for $t > 37$. For these reasons, I propose to re-estimate the models after having the first additional true observed scores, or develop alternative models because there might be an additional variable(s) that need(s) to be taken into account after 1 year. Refer also to the unexpected forecast values of the models of *Y1* presented in Figure 6.57, as well as the other models.

7

Forecasting Based on Time Series by States

7.1 Introduction

In this chapter, the panel data considered is unstacked panel data, namely the annual time series by states, where the units of the analysis are the time observations. So that the sets or multidimensional of endogenous, exogenous, and environmental variables, respectively, for the state i can be presented using the symbols $Y\_i_t = (Y1\_i,\ldots,Yg\_i,\ldots)_t$, $X\_i_t = (X1\_i,\ldots,Xk\_i,\ldots)_t$, and $Z_t = (Z1,\ldots,Zm,\ldots)_t$ for $i = 1,\ldots,N$; and $t = 1,\ldots,T$. Note that the scores of the environmental variables are constant for all states or individuals. Using these symbols, the panel data is considered the data of multivariate time series by states (countries, regions, agencies, firms, industries, households, or individuals).

Agung (2014) has presented various models, such as the VAR and System Equation Models (or SCM = Seemingly Causal Models) based on unstacked panel data with a small number of N. Referring to the multiple OLS regression analyses presented in previous chapters, then all data analyses using the VAR and SCM models presented in Agung (2014), can be reanalyzed using sets of equation specifications in order to compute the in-sample forecast values of each of their dependent variables. Do these as exercises. However, this chapter presents alternative models as their modifications or extended time series models.

However, for panel data with a large number of firms, the time series models can be applied by a few selected firms or by means of groups or sectors of firms, with the data presented as unstacked panel data. In addition, note that Tables 4.1 and 6.1 present alternative specific trends for all forecasting models of monthly and quarterly time series, respectively, which could also be applied for the unstacked panel data, and Table 7.1 presents alternative specific trends for the models of annual panel data. However, this chapter presents only illustrative examples based on annual panel data.

Illustrative examples are presented using the unstacked panel data in POOL7.wf1, as presented in Figure 7.1, which shows a time series of a single variable *GDP* for seven states. For illustrative examples, some of the seven time series could be selected either as *Yg_i, Xk_i*, or *Zm_i*.

7.2 Models Based on a Bivariate *(Y1_1,Y1_2)*

For a pair of correlated states, Figure 7.2 presents three alternative up-and-down-stream, or causal relationships based on a bivariate *(Y1_1,Y2_2)* with the following alternative models.

Advanced Time Series Data Analysis: Forecasting Using EViews, First Edition. I Gusti Ngurah Agung.

Table 7.1 Alternative specific trends for all forecasting models based on a single annual time series *Y*, bivariate, and triple time series.

Continuous Regressions with Trend	
Additional Time IV to insert	**Forecast Model (FM) with**
1) *@Trend*	Linear Trend
2) *@Trend^2 @trend*	Quadratic Trend
3) *@Trend^3 @Trend^2 @Trend*	Cubic Trend
4) *log(@Trend + 1) = log(t)*	Logarithmic Trend
Regressions with Heterogeneous Trends by TP([a])	
5) t*@Expand(TP[a]) @Expand(TP[a],@Droplast)	
6) t*@Expand(TP[a]) t^2*@Expand(TP[a]) @Expand(TP[a],@Droplast)	
7) t*@Expand(TP[a]) t^2*@Expand(TP[a]) t^3*@Expand(TP[a]) @Expand(TP[a],@Droplast)	

[a] TP = Time Period.

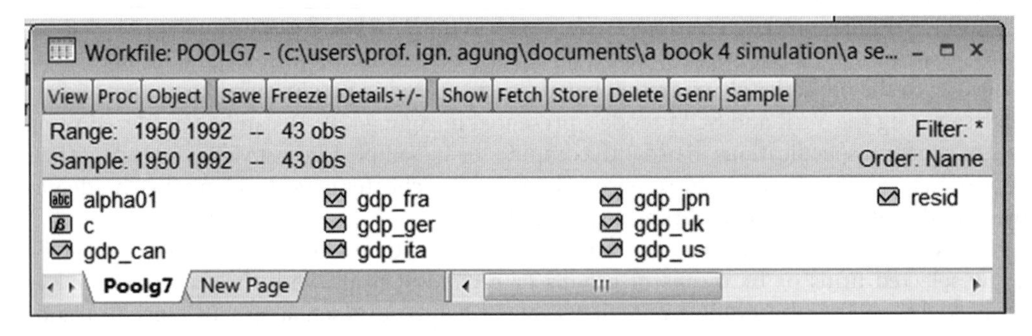

Figure 7.1 Variables in the work file FOOLG7.wf1.

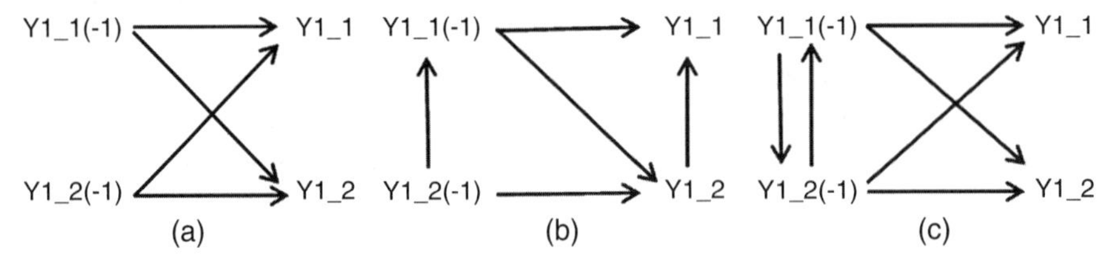

Figure 7.2 Alternative path diagrams based on bivariate *(Y1_1,Y1_2)*.

7.2.1 Alternative LV(p) Models Based on Figure 7.1a

7.2.1.1 Additive LV(p) Models

Whenever both states considered are correlated, then in the theoretical sense, both *Y1_1(−1)* and *Y1_2(−1)* should have direct effects on each of *Y1_1* and *Y1_2* without any assumption. For this reason, the path diagram presented in Figure 7.1a is the basic up-and-downstream or pairwise causal relation between a correlated *Y1_1* and *Y1_2*. Then I propose two additive LV(p) models with the following general equation specification (ES) each. So, based on each regression analysis, we can easily obtain its forecast evaluation. However, I would recommend applying the following pairs of interaction models.

$$Y1\_1 \;\; C \;\; Y2\_(-1) \;\; Y1\_1(-1) \ldots Y1\_1(-p1) \tag{7.1a}$$

$$Y1\_2 \;\; C \;\; Y1\_1(-1) \;\; Y1\_2(-1) \ldots Y1\_2(-p2) \tag{7.1b}$$

	AM-1			AM-2			AM-3			AM-4		
Variable	Coef.	t-Stat.	Prob.	Coef.	t-Stat.	Prob.	Coef.	t-Stat.	Prob.	Coef.	t-Stat.	Prob.
C	4892.5	15.805	0.000	273.3	1.109	0.274	36.896	0.065	0.949	1424.9	1.901	0.065
GDP_CAN(-1)	0.769	28.778	0.000				-0.033	-0.364	0.718	0.165	1.357	0.183
GDP_US(-1)				0.996	54.293	0.000	1.097	6.312	0.000			
GDP_US(-2)							-0.057	-0.328	0.745	0.783	5.005	0.000
R-squared	0.954			0.987			0.986			0.971		
Adj. R-sq.	0.953			0.986			0.985			0.970		
F-statistic	828.2			2947.8			879.6			642.7		
Prob(F-stat.)	0.000			0.000			0.000			0.000		
DW-stat	0.553			1.829			1.940			0.971		
RMSE	516.12			573.41			758.86			481.10		
TIC	0.015			0.017			0.023			0.014		

Figure 7.3 Statistical summary of four alternative models based on the ES (7.1a).

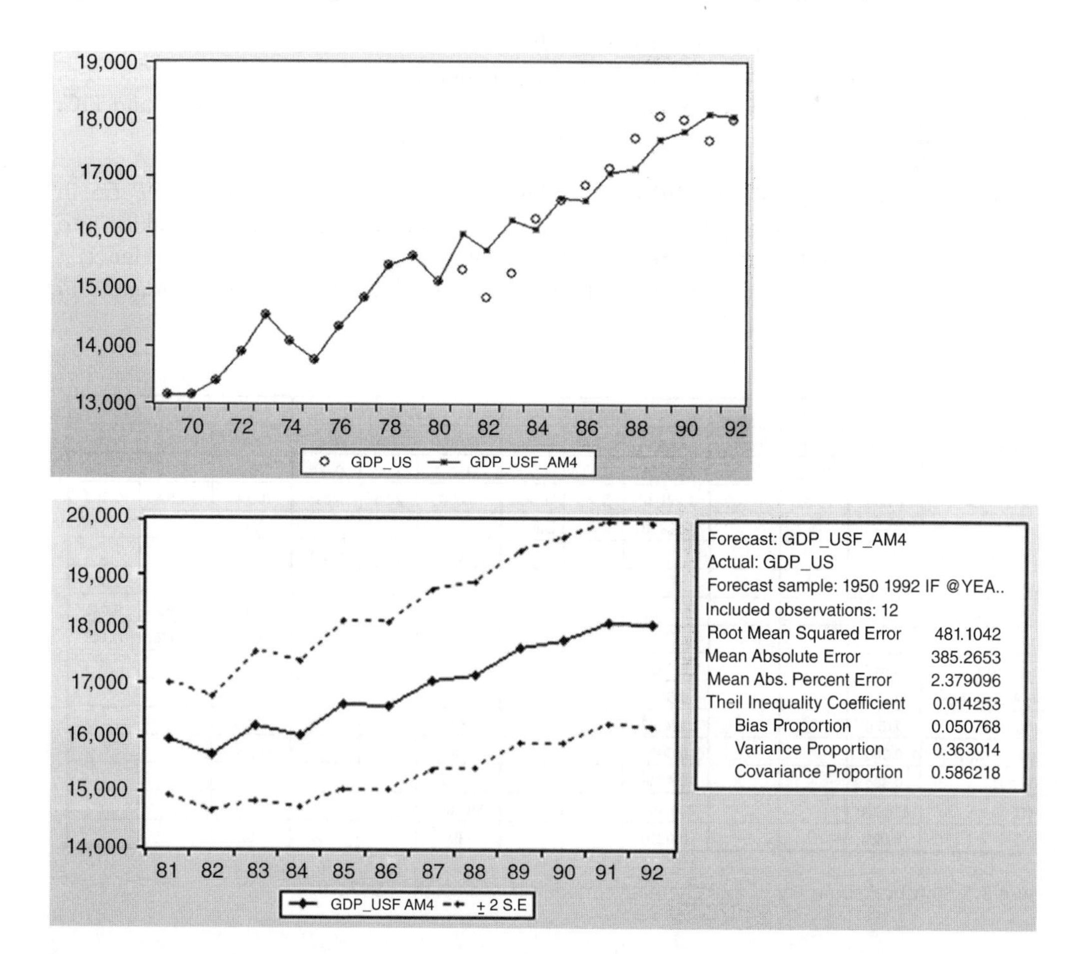

Figure 7.4 The graphs of *GDP_US* and *GDP _USF_ AM-4*, with its forecast evaluation.

Example 7.1 Applications of the Models (7.1a)
Figure 7.3 presents the summary of four *alternative models* (*AMs*) of *GDP_US*, namely *AM-1*, *AM-2*, *AM-3*, and *AM-4* based on the general ES (7.1a) with their forecast evaluations RMSE and TIC, which show that the *AM-4* is the best model to forecast *GDP_US*.

As an additional analysis based on the *AM-4*, Figure 7.4 presents the graphs of *GDP_US* and *GDP_USF_AM-4* with its forecast evaluation. The graphs show that scores of *GDP_US* are very close to its forecast values. So it can be concluded that the *AM-4* is an acceptable forecast model for *GDP_US*, which will be compared with the results of selected models presented later.

7.2.1.2 Two-Way Interaction LV(*p*) Models

In the theoretical sense and empirical findings, *Y1_1(−1)* is one of the best predictors of *Y1_1*. For this reason, then the effect of *Y1_2(−1)* on *Y1_1* should depend on *Y1_1(−1)*. So, in the statistical sense, the model of *Y1_1* should have the interaction *Y1_1(−1)∗Y1_2(−1)* as one of its independent variables. Similarly, the model of *Y1_2* also should have the interaction *Y1_1 (−1)∗Y1_2(−1)* as an independent variable. Therefore, as an extension of the models (7.1a) and (7.1b), we have a two-way interaction LV(*p*) model with the following ESs.

$$Y1\_1\ \ C\ \ Y1\_1(-1)*Y1\_2(-1)\ \ Y1\_2(-1)\ \ Y1\_1(-1) \ldots Y1\_1(-p1) \tag{7.2a}$$

$$Y1\_2\ \ C\ \ Y1\_1(-1)*Y1\_2(-1)\ \ Y1\_1(-1)\ \ Y1\_2(-1) \ldots Y1\_2(-p2) \tag{7.2b}$$

Example 7.2 Applications of the Models (7.2a)
Figure 7.5 presents the summary of four alternative models of *GDP_US*, namely *AM-5*, *AM-6*, *AM-7*, and *AM-8*, based on the general ES (7.2a) with their forecast evaluations RMSE and TIC. Based on these results, the findings and notes presented are as follows:

1) The interaction *GDP_US(−1)∗GDP_Can(−1)* has a positive significant effect on *GDP_US*, in *AM-5*, *AM-6*, and *AM-7*, but in *AM-8* it has a negative effect, adjusted for or by taking into account the other independent variables with a *p*-value = 0.063/2 = 0.0365 < 0.05. Hence, it can be concluded that the data support the hypothesis "the effect of *GDP_Can(−1)* on *GDP_US* depends on *GDP_US(−1)*."

Variable	AM-5			AM-6			AM-7			AM-8		
	Coef.	t-Stat.	Prob.	Coef.	t-Stat.	Prob.	Coef.	t-Stat.	Prob.	Coef.	t-Stat.	Prob.
C	8042.2	35.91	0.000	7835.391	10.54	0.000	3394.7	2.99	0.005	-1252.8	-1.31	0.200
_US(-1)*_CAN(-1)	0.000	26.81	0.000	2.98E-05	3.79	0.001	0.000	2.55	0.015	0.000	-1.92	0.063
GDP_CAN(-3)				0.088707	0.49	0.628						
GDP_US(-3)							0.630	4.27	0.000			
GDP_US(-1)										1.372	6.212	0.000
GDP_CAN(-2)										0.303	1.763	0.086
GDP_US(-2)										-0.344	-1.476	0.149
R-squared	0.947			0.9460			0.964			0.987		
Adj.R-squared	0.946			0.9431			0.962			0.986		
F-statistic	718.6			324.37			490.2			706.2		
Prob(F-stat.)	0.000			0.0000			0.000			0.000		
DW-stat.	0.406			0.4646			0.692			2.054		
RMSE	850.770			723.5500			594.800			631.154		
TIC	0.025			0.0210			0.017			0.019		

Figure 7.5 Statistical summary of four alternative models based on the ES (7.2a).

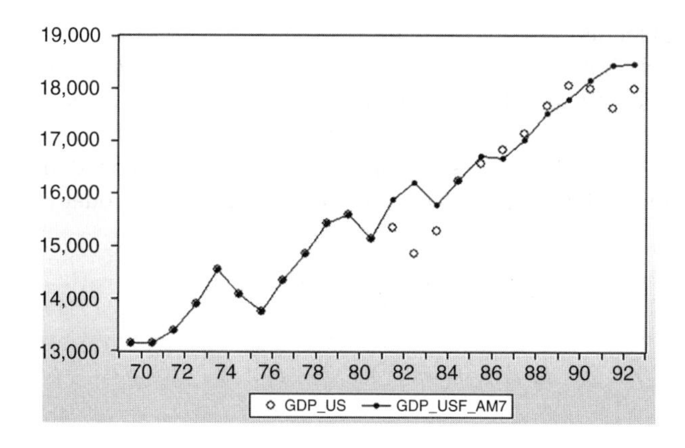

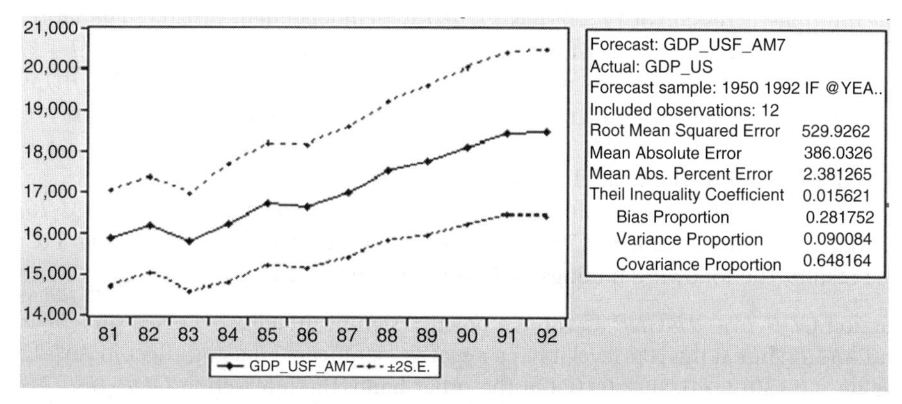

Figure 7.6 The graphs of *GDP_US* and *_USF_ AM-7*, and its forecast evaluations.

2) *AM-7* is the best forecast interaction model of *GDP_US*, since it has the smallest values of RMSE and TIC. However, compared to the additive *AM-4*, it has greater values, which will be discussed in the following subsection, because *AM-7* and *AM-4* present different pattern of forecast values. By looking at the graphs in Figure 7.6, I would consider *AM-7* to be an acceptable forecast model of *GDP_US*.

7.2.2 Alternative LV(*p*) Models Based on Figure 7.1b

As an extension of the path diagram in Figure 7.1a, Figure 7.2b presents the path diagram defining *Y1_2* as a cause factor of *Y1_1*. For this reason, then *Y1_2(−1)* also is a cause factor of *Y1_1 (−1)*. Similarly, the effect of *Y1_2(−1)* on *Y1_1* should be valid or a true effect, in a theoretical sense, because *Y1_2* has a direct effect on *Y1_1*. For this reason, the arrow can be omitted. In addition, *Y1(−1)* should be an upper variable of *Y2*.

7.2.2.1 A Partial Two-Way Interaction LV(*p*) Model

By looking at the indirect effect of *Y1_1(−1)* on *Y1_1* through *Y1_2*, the interaction *Y1_1(−1)* Y1_2* should be used in the model of *Y1_1*. In other words, an additive model is not a suitable model. Then, I propose a partial two-way interaction LV(*p*) model as follows:

$$Y1\_1\ \ C\ \ Y1\_1(-1)*Y1\_2\ \ Y1\_2\ \ Y1\_2(-1)\ \ Y1\_1(-1) \ldots Y1\_1(-p1) \tag{7.3a}$$

$$Y1\_2\ \ C\ \ Y1\_2(-1)\ \ Y1\_1(-1)\ \ Y1\_2(-1) \ldots Y1\_2(-p2) \tag{7.3b}$$

7.2.2.2 A Full Two-Way Interaction LV(*p*) Model

As an extension of the model (7.3a) and (7.3b), referring to the indirect effect of *Y1_2(−1)* on *Y1_1* through *Y_1(−1)*, then the interaction *Y1_1(−1)∗Y1_2(−1)* should be used as an independent variable of the model of *Y1_1*. Similarly, the indirect effect of *Y1_2(−1)* on *Y1_2* through *Y1_1(−1)* indicates that the interaction *Y1_1(−1)∗Y1_2(−1)* should be used as an independent variable of the model of *Y1_2*. So, we would have a full two-way interaction LV(*p*) model as follows:

$$Y1\_1 \;\; C \;\; Y1\_1(-1)*\mathbf{Y1\_2} \;\; Y1\_1(-1)*Y1\_2(-1) \;\; Y1\_2 \;\; Y1\_2(-1) \;\; Y1\_1(-1) \ldots Y1\_1(-p1) \qquad (7.4a)$$

$$Y1\_2 \;\; C \;\; Y1\_1(-1)*Y1\_2(-1) \;\; Y1\_2(-1) \;\; Y1\_1(-1) \;\; Y1\_2(-1) \ldots Y1\_2(-p2) \qquad (7.4b)$$

Note that the multiple regression of *Y1_1(t)* has *Y1_2(t)* as an independent variable. Therefore, to compute the forecast values *Y1_1(T + k)*, for $k = 1, 2, \ldots$, we should have the values of *Y1_2(T + k)* that can be predicted using the ES (7.4b) because this multiple regression does not have *Y1_1(t)* as an independent variable, compared to the triangular-effects model presented in Chapter 6.

Example 7.3 Application of the Model (7.4a)

Figure 7.7 presents the summary of four alternative models of *GDP_US*, namely *AM-9, AM-10, AM-11*, and *AM-12* with the general ES (7.4a) and their forecast evaluations RMSE and TIC. Based on these results, the following findings and notes are presented.

1) The interaction *GDP_US(−1)∗GDP_Can* has a positive significant adjusted effect on *AM-9, AM-10*, and *AM-11*, but at the 10% level it has a negative significant adjusted effect in *AM-12* with a *p*-value = 0.130/2 = 0.065 < 0.10. On the other hand, the interaction *GDP_US(−1)∗GDP_Can(−1)* has a negative significant adjusted effect in *AM-9, AM-10*, and *AM-11*, but at the 10% level it has a significant positive adjusted effect in *AM-12*, with a *p*-value = 0.114/2 = 0.057 < 0.10. Hence, the four models show that the effect of each of *GDP_Can* and *GDP_Can(−1)* on *GDP_US* is significantly dependent on *GDP_US(−1)*.

Variable	AM-9			AM-10			AM-11			AM-12		
	Coef.	t-Stat.	Prob.	Coef.	t-Stat.	Prob.	Coef.	t-Stat.	Prob.	Coef.	t-Stat.	Prob.
C	7833	42.69	0.000	2691.458	3.177	0.003	3017.874	2.976	0.005	764.7407	1.054	0.299
_US(-1)*_Can	6.98E-05	4.897	0.000	5.53E-05	5.371	0.000	5.66E-05	5.331	0.000	-6.61E-05	-1.551	0.130
_US(-1)*_Can(-1)	-3.56E-05	-2.491	0.017	-4.46E-05	-4.374	0.000	-4.10E-05	-3.435	0.002	6.76E-05	1.622	0.114
GDP_US(-1)										0.862666	10.239	0.000
GDP_US(-2)				0.684	6.212	0.000	0.712	5.893	0.000			
GDP_Can(-1)							-0.133	-0.598	**0.554**	-1.571	-2.489	0.018
GDP_Can										1.648	2.504	0.017
R-squared	0.9673			0.9838			0.9840			0.9933		
Adj. R-squared	0.9657			0.9825			0.9822			0.9923		
F-statistic	577.7			749.3			552.3			1064.6		
Prob(F-stat)	0.0000			0.0000			0.0000			0.0000		
DW-stat.	0.3074			1.0171			1.0763			1.5367		
RMSE	739.73			405.45			394.26			188.19		
TIC	0.022			0.012			0.011			0.006		

Figure 7.7 Statistical summary of four alternative models based on the ES (7.4a).

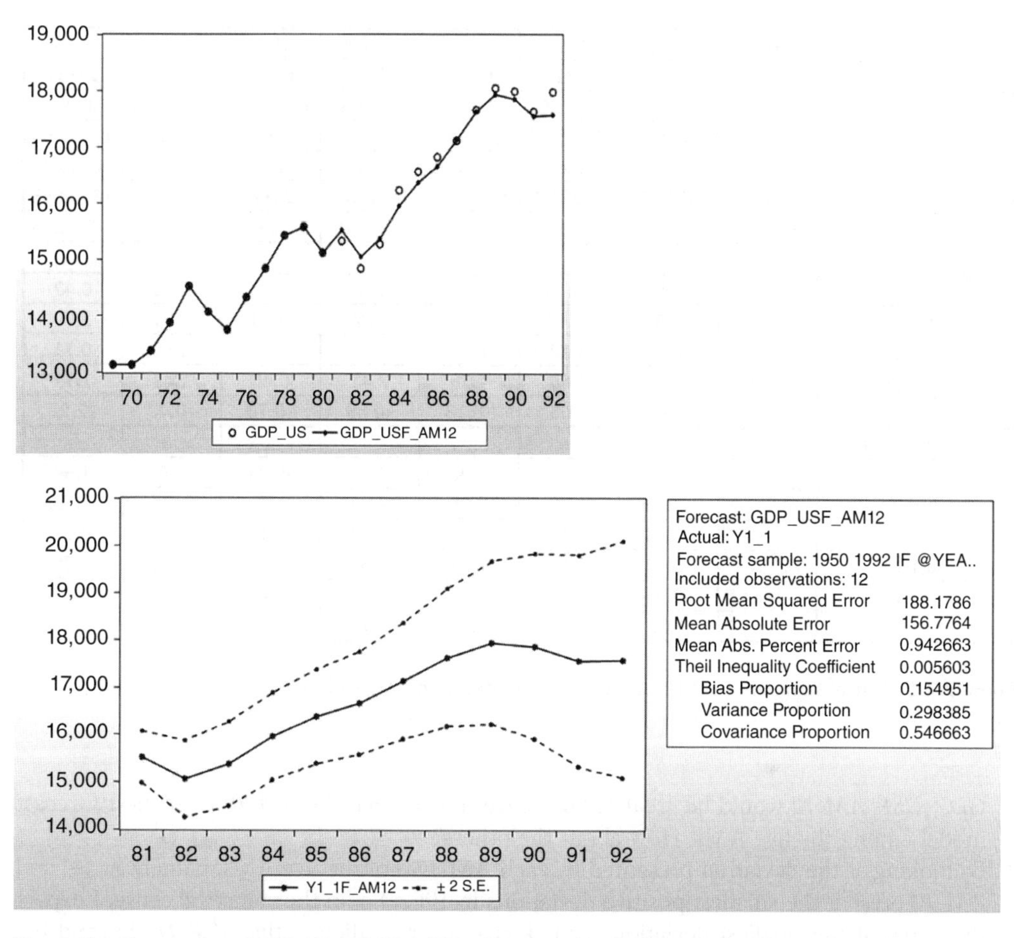

Figure 7.8 The graphs of *GDP_US* and *GDP_USF_ AM-12*, and its forecast evaluations.

2) By looking at their forecast evaluations, RMSE and TIC, it can be concluded that *AM-12* is the best forecasting model among the four models, with its graphs presented in Figure 7.8. However, compared to the nine *AM-1* to *AM-9*, *AM-10*, *AM-11*, and *AM-12* are the best three forecast models of *GDP_US*. Then there is a question, which one would be the best to predict the future (beyond the sample period)? See the comparative analysis presented in Section 7.2.2.3.

7.2.2.3 Comparison Between Selected Forecasts

As an illustration, Figure 7.9 presents a statistical summary of *GDP_US* and its forecast values *GDP_USF* for selected AMs (Alternative Models) and *YEAR* for the last eight sample time points with their Deviations from US and their Growth Rates. Based on these statistics, we have to use our judgment to select which one would be the best possible model for predicting the observed *GDP_US* beyond the sample period.

1) Since *AM-12* has the smallest RMSE, then it is expected to be the best forecast model. However, its forecast value in 1962 has a large negative deviation of 2.22% from the *GDP_US* because *GDP_US* had such a sudden high growth rate of 1.94 in 1962. Under the assumption that the *GDP_US* would be greater in 1963, then I predict that its deviation with

Year	1985	1986	1987	1988	1989	1990	1991	1992
US	16570	16835	17123	17664	18045	17990	17632	17981
AM-4	16592.84	16568.19	17045.63	17141.65	17641.37	17776.61	18093.64	18056.99
AM-7	16703	16655.15	17015.04	17528.8	17778.53	18143.31	18445.38	18465.94
AM-10	16791.73	16704.12	17656.64	17989.72	18463.86	18030.2	17783.76	18108.28
AM-11	16693.04	16640.78	17583.4	17980.74	18479.02	18112.8	17883.51	18288.93
AM-12	16377.74	16659.56	17131.95	17632.9	17938.09	17863.52	17555.48	17581.35
Dev. from US (%)								
AM-4	0.14	-1.58	-0.45	-2.96	-2.24	-1.19	2.62	0.42
AM-7	0.80	-1.07	-0.63	-0.77	-1.48	0.85	4.61	2.70
AM-10	1.34	-0.78	3.12	1.84	2.32	0.22	0.86	0.71
AM-11	0.74	-1.15	2.69	1.79	2.41	0.68	1.43	1.71
AM-12	-1.16	-1.04	0.05	-0.18	-0.59	-0.70	-0.43	-2.22
Growth Rate (%)								
US		1.57	1.68	3.06	2.11	-0.31	-2.03	1.94
AM-4		-0.15	2.80	0.56	2.83	0.76	1.75	-0.20
AM-7		-0.29	2.12	2.93	1.40	2.01	1.64	0.11
AM-10		-0.52	5.39	1.85	2.57	-2.41	-1.39	1.79
AM-11		-0.31	5.36	2.21	2.70	-2.02	-1.28	2.22
AM-12		1.69	2.76	2.84	1.70	-0.42	-1.75	0.15

Figure 7.9 Statistical summary *GDP_US* and *GDP_USF* by selected AMs and YEAR.

GDP_USF_AM-12 would be greater. So, we cannot say that the *AM-12* is the best forecast model among the five AMs. How about the others?

2) By looking at the deviation presented by each AM, two of the five AMs, namely *AM-4* and *AM-10* present the smallest positive deviations in 1962. So, in the statistical sense, I expect they present the smallest deviations or risk errors for predicting the *GDP_US* beyond the sample period, at least in 1963. However, we never know before we compute their forecast values beyond the sample period and the observed *GDP_US* is available. See the following note.
3) Under the assumption that the growth rate of the *GDP_US* is equal to 1.8282%, which is its growth rate over the whole sample period based on the classical exponential growth model, then it is predicted that the *GDP_US* in 1963 would be $(1 + 0.018282) * 17981 = 18309.73$. This can be compared to the forecast values in 1963 based on the AMs. Do these as exercises and compare to the unexpected and unbelievable forecast value presented in Chapter 6.

7.2.3 Alternative LV(*p*) Models Based on Figure 7.1c

By using the assumption that the variables *Y1_1(−1)* and *Y1_2(−1)* have reciprocal (or two-way) causal effects, then *Y1_1* and *Y1_2* also have reciprocal causal effects, but each of *Y1_1* and *Y1_2* will not be used as independent variables of the models. For this reason, no arrow is presented between *Y1_1* and *Y1_2*. Therefore, based on the path diagram in Figure 7.1c, we have the models with the following pairs of ESs.

$$Y1\_1\ C\ Y1\_1(-1)*\mathbf{Y1\_2(-1)}\ \mathbf{Y1\_2(-1)}\ Y1\_1(-1)\ldots Y1\_1(-p1) \tag{7.5a}$$

$$Y1\_2\ C\ \mathbf{Y1\_1(-1)}*Y1\_2(-1)\ \mathbf{Y1\_1(-1)}\ Y1\_2(-1)\ldots Y1\_2(-p2) \tag{7.5b}$$

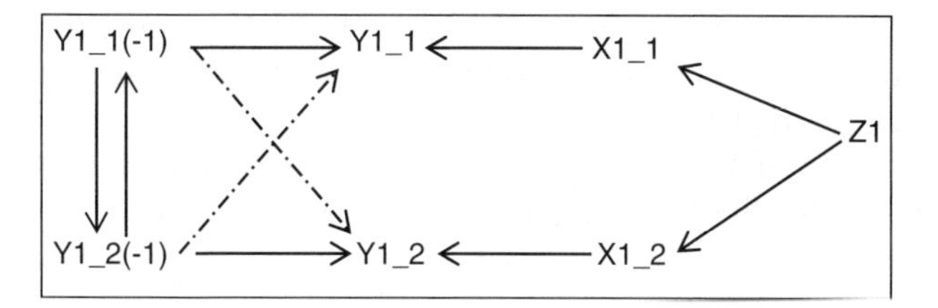

Figure 7.10 An extension of the path diagram in Figure 7.2a.

Note that the model (7.5a) represents the triangular-effects model between the time series *Y1_1*, *Y1_1(−1)*, and *Y1_2(−1)*, and the model (7.5b) represents the triangular-effects model between the time series *Y1_2*, *Y2_2(−1)*, and *Y1_1(−1)*.

7.3 Advanced LP(*p*) Models of *(Y1_1,Y1_2)*

As an extension of the path diagram in Figure 7.2, Figure 7.10 presents a path diagram of endogenous time series *(Y1_1,Y1_2)*, exogenous time series *(X1_1,X1_2)*, and an environmental time series *Z1*. I propose the following alternative models. First of all, note that the dotted line arrows indicate that the causal effects between the corresponding pairs of variables are the cause of defined causal effects between the other pairs of variables, which are represented by the solid line arrows. This is specific for the reciprocal causal effects of *Y1_1(−1)* and *Y1_2(−1)* also indicates that *Y1_1* and *Y1_2* have reciprocal causal effects, but they are not used as independent variables in the models. In addition, the important assumptions of the models are that the exogenous variables only have effects on the endogenous variables within the same states, and the environmental variable *Z1* only has an effect on all exogenous variables. This path diagram could easily be extended to alternative models by modifying the arrows or inserting additional arrows between the variables, such as the environmental variable *Z1* that has a direct effect on both *Y1_1* and *Y1_2*.

7.3.1 Two-Way Interaction LV(*p*) Models

Based on the path diagram in Figure 7.10, I propose a set of five multiple two-way interaction regression models using a set of five general equation specifications (ESs), with the following explanation:

$$Y1\_1\ \ C\ \ \boldsymbol{Y1\_1(-1){*}Y1\_2(-1)}\ \ \boldsymbol{Y1\_1(-1){*}X1\_1}\\ Y1\_1(-1) \ldots Y1\_1(-p1)\ \ Y1\_2(-1)\ \ X1\_1 \tag{7.6a}$$

$$Y1\_2\ \ C\ \ \boldsymbol{Y1\_1(-1){*}Y1\_2(-1)}\ \ \boldsymbol{X1\_2^{*}Y1\_2(-1)}\\ Y1\_2(-1) \ldots Y1\_2(-p2)\ \ Y1\_1(-1)\ \ X1\_2\ \ X2\_2 \tag{7.6b}$$

$$X1\_1\ \ C\ \ \boldsymbol{X1\_1(-1){*}Z1}\ \ X1\_1(-1) \ldots X1\_1(-p3)\ \ Z1 \tag{7.6c}$$

$$X1\_2\ \ C\ \ \boldsymbol{X1\_2(-1){*}Z1}\ \ X1\_2(-1) \ldots X1\_2(-p4)\ \ Z1 \tag{7.6d}$$

$$Z1\ \ C\ \ Z1(-1) \ldots Z1(-p5) \tag{7.6e}$$

This is specific for the model of *Y1_1* since *Y1_1(−1)* is known as the best predictor of *Y1_1*, then the effects of *Y1_2(−1)* and *X1_1* on *Y1_1* should depend on *Y1_1(−1)* in the theoretical sense. For this reason, in the statistical sense, both interactions *Y1_1(−1)∗Y1_2(−1)* and *Y1_1(−1)∗X1_1* should be used as independent variables of the model (7.6a).

This is similar for the model of *Y1_2* in (7.6b); the effects of *Y1_1(−1)* and *X1_2* on *Y1_2* depend on *Y1_2(−1)*. Then, the interactions *Y1_1(−1)*Y1_2(−1)* and *X2_1(−1)*Y1_2(−1)* should be used as the independent variables of the model.

Likewise, for each of the exogenous variables *X1_1* and *X1_2* within the two states, we have the models (7.6c)–(7.6d), respectively. Finally, the ES (7.6e) presents the LV(*p5*) model of the environmental variable *Z1*.

7.3.2 Three-Way Interaction LV(*p*) Models

As an extension of the set of models in (7.6), specifically as the extension of the models (7.6a) and (7.6b), we have the following three-way interaction LV(*p*) models with the following explanation. Note that the other three equation specifications would be exactly the same as (7.6c)–(7.6e).

$$\begin{aligned} &Y1\_1\ \ C\ \ \mathbf{Y1\_1(-1)*X1\_1*Z1}\ \ \ Y1\_1(-1)*Y1\_2(-1)\ \ Y1\_1(-1)*X1\_1 \\ &\quad Y1\_1(-1)\ldots Y1\_1(-p1))\ \ Y1\_2(-1)\ \ X1\_1\ \ Z1 \end{aligned} \tag{7.7a}$$

$$\begin{aligned} &Y1\_2\ \ C\ \ \mathbf{X1\_2*Y1\_2(-1)*Z1}\ \ \ \mathbf{X1\_2(-1)*Z1}\ \ \ Y1\_1(-1)*Y1\_2(-1) \\ &\quad X1\_2*Y1\_2(-1)\ \ Y1\_2(-p2)\ldots Y1\_1(-1)\ \ X1\_2 \end{aligned} \tag{7.7b}$$

The interaction *Y1_1(−1)*X1_1*Z1* is used as an independent variable of the model (7.7a) because *Z1* has a direct effect on *X1_1* and the effect of *X1_1* on *Y1_1* depends on *Y1(−1)*. In other words, *Z1* has an indirect effect on *Y1_1* through *X1_1* and *Y1_1(−1)*. This is similar for the interactions *X1_2*Y1_2(−1)*Z1* because *Z1* has an indirect effect on *Y1_2* through *X1_2*, which depends on *Y1_2(−1)*.

7.3.3 Alternative Additive Models

Since I found many students as well as researchers had been using additive multivariate models based on the path diagram in Figure 7.10, I now present two alternative additive set of five additive models, as follows.

7.3.3.1 Additive Translog-Linear Models

Additive translog-linear models, in fact, are derived from the multiplicative Cobb–Douglas production functions, which are nonlinear models. Based on the path diagram in Figure 7.10, we have the set of five equation specifications of the models.

$$log(Y1\_1)C\,log(Y1\_1(-1))\ldots log(Y1\_1(-p1))\,log(Y1\_2(-1))\,log(X1\_1) \tag{7.8a}$$

$$log(Y1\_2)C\,log(Y1\_2(-1))\ldots log(Y1\_2(-p1))\,log(Y1\_1(-1))\,log(X1\_1) \tag{7.8b}$$

$$log(X1\_1)C\,log(X1\_1(-1))\ldots log(X1\_(-p3))\,log(Z1) \tag{7.8c}$$

$$log(X1\_2)C\,log(X1\_2(-1))\ldots log(X1\_2(-p4))\,log(X2\_2)\,log(Z1) \tag{7.8d}$$

$$log(Z1)C\,log(Z1(-1))\ldots log(Z1(-p5)) \tag{7.8e}$$

The OLS estimate of each the models can be easily obtained using various alternative options – refer to the notes and comments on the options presented in Sections 1.6 and 1.7. Note that the models should be applicable for a positive time series and they can easily be extended to lower and upper bounds models by the using the transformed dependent variables, such as by using *DVul = log((V−L)/(U−V))* for each dependent variable *DV*, where *U* and *L* are fixed upper and lower bounds of *DV*, which should be subjectively selected and supported by theoretical and substantial perspectives. Refer to various alternative translog-linear models presented in Chapters 4 and 5.

7.3.3.2 Additive Models as the Worst Models

For illustrative purposes, based on Figure 7.10, a set of five additive models can be presented using the following equation specification. However, I would consider these set of models to be the worst, since the effect of an independent variable on the dependent variable should depend on at least one of the other independent variables.

$$Y1\_1\ \ C\ \ Y1\_1(-1)\ \ Y1\_1(-1)\ldots Y1\_1(-p1)\ \ Y1\_2(-1)\ \ X1\_1 \tag{7.9a}$$

$$Y1\_2\ \ C\ \ Y1\_2(-1)\ \ Y1\_2(-1)\ \ \ldots Y1\_2(p2)\ \ Y1\_1(-1)\ \ X1\_2\ \ X2\_2 \tag{7.9b}$$

$$X1\_1\ \ C\ \ X1\_1(-1)\ldots X1\_(-p3)\ \ Z1 \tag{7.9c}$$

$$X1\_2\ \ C\ \ X1\_2(-1)\ldots X1\_2(-p4)\ \ X2\_2\ \ Z1 \tag{7.9d}$$

$$Z1\ \ C\ \ Z1(-1)\ldots Z1(-p5) \tag{7.9e}$$

7.4 Advanced LP(*p*) Models of *(Y1_1,Y1_2,Y1_3)*

As an extension of the LV(*p*) models based on bivariate *(Y1_1,Y1_2)*, this section presents advanced models based on three endogenous time series *(Y1_1,Y1_2,Y1_3)*, with exogenous time series *(X1_1,X1_2,X1_3)*, and an environmental time series *Z1* for three correlated states, First of all, it is assumed that *Y1_1*, *Y1_2*, and *Y1_3* have triangular up-and-downstream relationships or causal effects, so that their lag variables *Y1_1(−1)*, *Y1_2(−1)*, and *Y1_3(−1)* also have triangular causal effects. For this reason, I present two alternative models as in the following subsections.

7.4.1 Triangular Effects Model of *(Y1_1,Y1_2,Y1_3)*

As an extension of the path diagram in Figure 7.1, Figure 7.11 presents a unidirectional (one-way) up-and-downstream or causal relationship between the seven selected variables and their first-order lags of the three endogenous variables, based on three correlations:

i) *Y1_1*, *Y1_2*, and *Y1_3* have triangular causal relationships, specifically both *Y1_2* and *Y1_3* have direct effects on *Y1_1*, and *Y1_3* has a direct effect on *Y1_2*,
ii) each of the exogenous variables has a direct effect on the endogenous variable within the same state, and
iii) *Z1* has a direct effect on all of the exogenous variables.

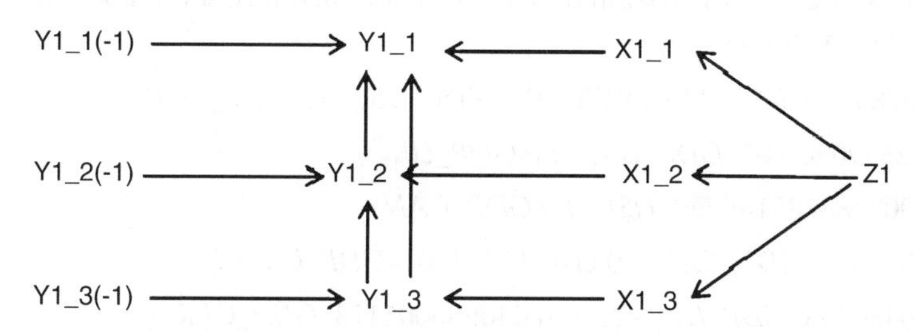

Figure 7.11 Unidirectional causal relationships between selected pairs of variables.

However, note that the direct effects from *Y1_i(−1)* on *Y1_i* for $i = 1$, 2, and 3 are not the assumptions but the "true" causal effects, in the theoretical sense. Based on this path diagram, I propose a set of seven two-way interaction multiple regression models, presented by the following set of ESs, with the explanations or backgrounds as follows:

$$Y1\_1\ C\ \boldsymbol{Y1\_1(-1)*Y1\_2\ Y1\_(-1)*Y1\_3\ Y1\_(-1)*X1\_1}\ Y1\_1(-1)\ldots Y1\_1(-p1))\ Y1\_2\ Y1\_3\ X1\_1 \quad (7.10a)$$

$$Y1\_2\ C\ \boldsymbol{Y1\_2(-1)*Y1\_3\ X1\_2^{*}Y1\_2(-1)}\ Y1\_2(-1)\ldots Y1\_2(-p2)\ Y1\_3(-1)\ X1\_2 \quad (7.10b)$$

$$Y1\_3\ C\ \boldsymbol{Y1\_3(-1)*X1\_3}\ Y1\_3(-1)\ldots Y1\_3(-p3)\ X1\_3 \quad (7.10c)$$

$$X1\_1\ C\ \boldsymbol{X1\_1(-1)*Z1}\ X1\_1(-1)\ldots X1\_1(-p3)\ Z1 \quad (7.10d)$$

$$X1\_2\ C\ \boldsymbol{X1\_2(-1)*Z1}\ X1\_2(-1)\ldots X1\_2(-p4)\ Z1 \quad (7.10e)$$

$$Z1\ C\ Z1(-1)\ldots Z1(-p5) \quad (7.10f)$$

1) Since *Y1_1(−1)* is the best predictor of *Y1_1*, and variables *Y1_2, Y1_3*, and *X1_1* have direct effects on *Y1_1*, then the effects of each of the variables *Y1_2, Y1_3*, and *X1_1* on *Y1_1* should depend on *Y1_1(−1)*, in the theoretical sense. For this reason, the two-way interactions *Y1_1(−1)*Y1_2, Y1_1(−1)*Y1_3*, and *Y1_1(−1)*X1_1* should be used as the independent variables of the model of *Y1_1*.
2) Similarly, the interactions *Y1_2(−1)*Y1_3* and *X1_2*Y1_2(−1)* should be used as the independent variables of the model of *Y1_2*, because the effects of *Y1_3 and X1_2* on *Y1_2* depend on *Y1_2(−1)*.
3) In addition, the interaction *Y1_3(−1)*X1_3* should be used as the independent variable of the model of *Y1_3*; the interaction *X1_1(−1)*Z1* should be used as an independent variable of the dynamic model of *X1_1*; and the interactions *X1_2(−1)*Z1* should be used as independent variables of the dynamic model of *X1_2*.
4) It is important to note that the models (7.10a)–(7.10c) also indirectly represent the triangular relationships between *Y1_1(−1), Y1_2(−1)*, and *Y1_3(−1)*; as well as between *X1_1, X1_2*, and *X1_3*, because of assumption A1.
5) Note that the assumption A2 and the ES (7.10b) do not indicate that *X1_2* not only has an effect on *Y1_2*, but it also has an indirect effect on *Y1_1* because of assumption A1. Similarly, *X1_3* has an indirect effect on both *Y1_1* and *Y1_2*.

Example 7.4 Application of the Models (7.10)

As an illustration, the data in POOLG7.wf1 is used by selecting *Y1_1* = *GDP_US*, *Y1_2* = *GDP_GER*, and *Y1_3* = *GDP_UK*, *X1_1* = *GDP_CAN*, *X1_2* = *GDP_FRA*, and *X1_3* = *GDP_ITA*, and *Z1* = *GDP_JPN*. By using the manual stepwise selection method, which is a combination of the combinatorial and the trial-and-error selection methods, seven good fit regression functions are obtained as follows:

$$\begin{aligned} GDP\_US = & -2267.00678579 + 3.51048028817\text{e}-05*GDP\_US(-1)*GDP\_GER \\ & + 6.77865255908\text{e}-05*GDP\_US(-1)*GDP\_UK \\ & - 0.000120034837851*GDP\_US(-1)*GDP\_CAN \\ & + 1.86282522384*GDP\_CAN + 0.661313314501*GDP\_US(-1) \\ & - 0.306386480455*GDP\_US(-2) - 0.440586001113*GDP\_GER \end{aligned} \quad (7.11a)$$

$$GDP\_GER = -2142.83545237 - 3.1573109518e-05*GDP\_GER(-1)*GDP\_UK + 1.48748409356e-05*GDP\_GER(-1)*GDP\_FRA + 1.03561768235*GDP\_GER(-1) + 0.686701967262*GDP\_UK - 0.291306479074*GDP\_GER(-2) \quad (7.11b)$$

$$GDP\_UK = 2518.72626088 + 1.28935558463e-05*GDP\_UK(-1)*GDP\_ITA + 0.895231246174*GDP\_UK(-1) - 0.511444723089*GDP\_UK(-2) + 0.256934305397*GDP\_ITA \quad (7.11c)$$

$$GDP\_CAN = 10.7117971616 - 8.49918724309e-06*GDP\_CAN(-1)*GDP\_JPN + 1.18415047846*GDP\_CAN(-1) - 0.225635421662*GDP\_CAN(-2) + 0.193432676266*GDP\_JPN \quad (7.11d)$$

$$GDP\_FRA = 899.061015057 - 1.65245590816e-05*GDP\_FRA(-1)*GDP\_JPN + 0.810791925362*GDP\_FRA(-1) - 0.156971059072*GDP\_FRA(-2) + 0.503334092064*GDP\_JPN \quad (7.11e)$$

$$GDP\_ITA = 634.053444519 - 3.37540800964e-06*GDP\_ITA(-1)*GDP\_JPN + 0.736883877923*GDP\_ITA(-1) + 0.241343575858*GDP\_JPN \quad (7.11f)$$

$$GDP\_JPN = 135.462371389 + 1.40841116379*GDP\_JPN(-1) - 0.399189743001*GDP\_JPN(-2) \quad (7.11g)$$

In general, at the first stage, the analysis is done for the full model. If the results show one or more two-way interactions have large *p*-values, say greater or than 0.30, then redo the analysis using the manual multistage selection method. As an illustration, the results of the analysis based on the full model of *GDP_US* (7.10a) show the interaction *GDP_US (−1)*GDP_UK* has a very large Prob. = 0.8455. Hence, the second analysis is done using the stages as follows:

1) At the first stage, an acceptable two-way interaction regression function is obtained by using the following ES, with the results presented in Figure 7.12.

   *GDP_US C GDP_US(−1)*GDP_GER*
   *GDP_US(−1)*GDP_UK GDP_US(−1)*GDP_CAN*

   Since, at the 10% level of significance, *GDP_US(−1)*GDP_UK* has a negative significant adjusted effect with a *p*-value = 0.1610/2 = 0.0805, and *GDP_US(_1)*GDP_CAN* has a positive significant adjusted effect with a *p*-value = 0.1104/2 = 0.0552, the model is a good fit two-way interaction model.
2) At the second stage, the combination of STEPLS- Combinatorial and the trial-and-error methods are applied, with the final regression presented in Figure 7.12.
3) Do these as exercises for the other models.

Variable	The first stage			The second stgae		
	Coef.	t-Stat.	Prob.	Coef.	t-Stat.	Prob.
C	8231.667	28.417	0.0000	-2267.01	-2.2092	0.0342
GDP_US(-1)*GDP_GER	4.43E-05	4.3810	0.0001	3.51E-05	2.2208	0.0333
GDP_US(-1)*GDP_UK	-2.32E-05	-1.4296	0.1610	6.78E-05	5.7114	0.0000
GDP_US(-1)*GDP_CAN	1.26E-05	1.6345	0.1104	-0.0001	-5.4276	0.0000
GDP_CAN				1.8628	5.9358	0.0000
GDP_US(-1)				0.6613	5.6791	0.0000
GDP_US(-2)				-0.3064	-2.8268	0.0079
GDP_GER				-0.4406	-2.1213	0.0415
R-squared	0.9774			0.9961		
Adjusted R-squared	0.9756			0.9953		
F-statistic	547.73			1210.7		
Prob(F-statistic)	0.0000			0.0000		
DW-stat.	0.5536			2.0020		

Figure 7.12 Statistical results summary of the model (7.10a).

Example 7.5 Computing Forecast Values Based on the Regressions (7.11)

Figure 7.13 presents the spreadsheet for computing the forecast values beyond the sample period using the regression functions (7.11a) to (7.11g). However, the computation should be done using reversed ordering. The first stage is to compute the forecast value of *GDP_** in 1993, starting with the *GDP_JPN* in cell **H7** using the regression (7.11g), and the last stage is to compute the forecast value of *GDP_US* in cell **B7** using the regression (7.11a). The computations use the following set of equations.

$$GDP\_JPN = 135.462371389 + 1.40841116379*GDP\_JPN(-1)$$
$$-0.399189743001*GDP\_JPN(-2)$$

$$\mathbf{H7} = 135.462371389 + 1.40841116379*\mathrm{H6} - 0.399189743001*\mathrm{H5}$$

$$GDP\_ITA = 634.053444519 - 3.37540800964e-06*GDP\_ITA(-1)*GDP\_JPN$$
$$+ 0.736883877923*GDP\_ITA(-1) + 0.241343575858*GDP\_JPN$$

$$\mathbf{G7} = 634.053444519 - 0.00000337540800964*\mathrm{G6}*\mathrm{H7}$$
$$+ 0.736883877923*\mathrm{G6} + 0.241343575858*\mathrm{H7}$$

	A	B	C	D	E	F	G	H
1		GDP_US	GDP_GER	GDP_UK	GDP_CAN	GDP_FRA	GDP_ITA	GDP_JPN
2	1988	17664	13945	12900	17394	13516	12153	13537
3	1989	18045	14278	13216	17758	13866	12464	14066
4	1990	17990	14840	13222	17308	14141	12782	14597
5	1991	17632	15263	12866	16444	14141	13066	15258
6	1992	17981	15357	12813	16413	14237	13113	15496
7	**1993**	18165.90	15261.08	13076.03	16591.75	14476.74	13424.38	15869.36
8	**1994**	18548.73	15294.61	13510.63	16808.88	14706.97	13721.60	16300.21
9	**1995**	19068.88	15490.41	13967.91	17018.73	14913.16	14013.55	16757.97

Figure 7.13 Spreadsheet for computing the forecast values beyond the sample period based on the regressions (7.11a)–(7.11g).

$$
\begin{aligned}
GDP\_FRA = {} & 899.061015057 - 1.65245590816e-05*GDP\_FRA(-1)*GDP\_JPN \\
& + 0.810791925362*GDP\_FRA(-1) - 0.156971059072*GDP\_FRA(-2) \\
& + 0.503334092064*GDP\_JPN \\
\mathbf{F7} = {} & 899.061015057 - 0.0000165245590816*F6*H7 \\
& + 0.810791925362*F6 - 0.156971059072*F5 + 0.503334092064*H7 \\
GDP\_CAN = {} & 10.7117971616 - 8.49918724309e-06*GDP\_CAN(-1)*GDP\_JPN \\
& + 1.18415047846*GDP\_CAN(-1) - 0.225635421662*GDP\_CAN(-2) \\
& + 0.193432676266*GDP\_JPN \\
\mathbf{E7} = {} & 10.7117971616 - 0.00000849918724309*E6*H7 \\
& + 1.18415047846*E6 - 0.225635421662*E5 + 0.193432676266*H7 \\
GDP\_UK = {} & 2518.72626088 + 1.28935558463e-05*GDP\_UK(-1)*GDP\_ITA \\
& + 0.895231246174*GDP\_UK(-1) - 0.511444723089*GDP\_UK(-2) \\
& + 0.256934305397*GDP\_ITA \\
\mathbf{D7} = {} & 2518.72626088 + 0.0000128935558463*D6*G7 \\
& + 0.895231246174*D6 - 0.511444723089*D5 + 0.256934305397*G7 \\
GDP\_GER = {} & -2142.83545237 - 3.1573109518e-05*GDP\_GER(-1)*GDP\_UK \\
& + 1.48748409356e-05*GDP\_GER(-1)*GDP\_FRA \\
& + 1.03561768235*GDP\_GER(-1) + 0.686701967262*GDP\_UK \\
& - 0.291306479074*GDP\_GER(-2) \\
\mathbf{C7} = {} & -2142.83545237 - 0.000031573109518*C6*D7 \\
& + 0.0000148748409356*C6*F7 + 1.03561768235*C6 \\
& + 0.686701967262*D7 - 0.291306479074*C5 \\
GDP\_US = {} & -2267.00678579 + 3.51048028817e-05*GDP\_US(-1)*GDP\_GER \\
& + 6.77865255908e-05*GDP\_US(-1)*GDP\_UK \\
& - 0.000120034837851*GDP\_US(-1)*GDP\_CAN \\
& + 1.86282522384*GDP\_CAN + 0.661313314501*GDP\_US(-1) \\
& - 0.306386480455*GDP\_US(-2) - 0.440586001113*GDP\_GER \\
\mathbf{B7} = {} & -2267.00678579 + 0.0000351048028817*B6*C7 \\
& + 0.0000677865255908*B6*D7 - 0.000120034837851*B6*E7 \\
& + 1.86282522384*E7 + 0.661313314501*B6 \\
& - 0.306386480455*B5 - 0.440586001113*C7
\end{aligned}
$$

Then the forecast values for @*Year*>1993 can easily be obtained using the block-copy-paste method of the forecast values in 1993, as presented in Figure 7.13. For an additional analysis, Figure 7.14 presents the annual growth rates of the seven *GDP_**. Note that the growth rates of *GDP_GER*, *GDP_UK*, and *GDP_CAN* are unexpected, which show the growth rates in 1962 and 1963 have different signs.

YEAR	GDP_US	GDP_GER	GDP_UK	GDP_CAN	GDP_FRA	GDP_ITA	GDP_JPN
1988							
1989	2.157	2.388	2.450	2.093	2.590	2.559	3.908
1990	-0.305	3.936	0.045	-2.534	1.983	2.551	3.775
1991	-1.990	2.850	-2.692	-4.992	0.000	2.222	4.528
1992	1.979	**0.616**	**-0.412**	**-0.189**	0.679	0.360	1.560
1993	1.028	**-0.625**	**2.053**	**1.089**	1.684	2.375	2.409
1994	2.107	0.220	3.324	1.309	1.590	2. 214	2.715
1995	2.804	1.280	3.385	1.248	1.402	2.128	2.808

Figure 7.14 Annual growth rates of each of the seven *GDP_∗*.

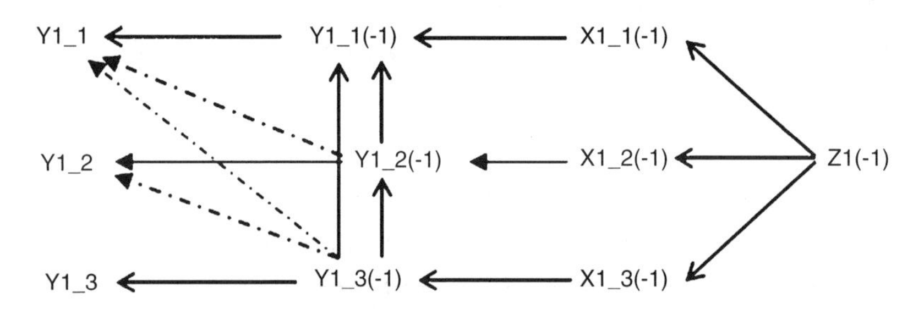

Figure 7.15 One-way causal relationships between 10 selected variables.

7.4.2 Full-Lag Variables Triangular Effects Model

As a modification of the path diagram in Figure 7.11, Figure 7.15 presents a full-lag-variables model of the endogenous variables *Y1_1, Y1_2,* and *Y1_3*. In addition, the models of each exogenous variables *X1_1, X1_2,* and *X1_3* and the environmental variable *Z1* also use lag independent variables.

Based on this figure, I propose a set of seven multiple regressions, which are the modification of the seven multiple regressions in (7.10). Since each of the dependent variables is predicted using the lag variables, the models are called "Full-Lag Variables Triangular Effects Models." The advantage of these models is that the first forecast values beyond the sample period of any one of the seven time series can be computed directly.

7.4.2.1 Two-Way Interaction Triangular Effects Models

As a modification of the set of models in (7.10), we have the set of full-lag variables models with the equation specifications as follows:

$$\begin{aligned}&Y1\_1\ C\ \boldsymbol{Y1\_1(-1)*Y1\_2(-1)}\ \boldsymbol{Y1\_1(-1)*Y1\_3(-1)}\ \boldsymbol{Y1\_1(-1)*X1\_1(-1)}\\&\quad Y1\_1(-1)\ldots Y1\_1(-p1))\ Y1\_2(-1)\ Y1\_3(-1)\ X1\_1(-1)\end{aligned}\tag{7.12a}$$

$$\begin{aligned}&Y1\_2\ C\ \boldsymbol{Y1\_2(-1)*Y1\_3(-1)}\ \boldsymbol{Y1\_2(-1)*X1\_2(-1)}\\&\quad Y1\_2(-1)\ldots Y1\_2(-p2)\ Y1\_3(-1)\ X1\_2(-1)\end{aligned}\tag{7.12b}$$

$$Y1\_3\ C\ \boldsymbol{Y1\_3(-1)*X1\_3(-1)}\ Y1\_3(-1)\ldots Y1\_3(-p3)\ X1\_3(-1)\tag{7.12c}$$

$$X1\_1\ C\ \boldsymbol{X1\_1(-1)*Z1(-1)}\ X1\_1(-1)\ldots X1\_1(-p3)\ Z1(-1)\tag{7.12d}$$

$$X1\_2\ \ C\ \ \boldsymbol{X1\_2(-1)*Z1(-1)}\ \ X1\_2(-1)\ldots X1\_2(-p4)\ \ Z1(-1) \tag{7.12e}$$

$$X1\_3\ \ C\ \ \boldsymbol{X1\_3(-1)*Z1\ \ (-1)}\ \ X1\_3(-1)\ldots X1\_3(-p4)\ \ Z1(-1) \tag{7.12f}$$

$$Z1\ \ C\ \ Z1(-1)\ldots Z1(-p5) \tag{7.12g}$$

Note that the interactions *Y1_1(−1)∗Y1_2(−1), Y1_1(−1)∗Y1_3(−1),* and *Y1_1(−1)∗X1_1(−1)* in the model (7.12a), represent the indirect effect of *Y1_2(−1), Y1_3(−1),* and *X1_1(−1)* on *Y1_1* through *Y1_1(−1),* since *Y1_1(−1)* is known as the best predictor of *Y1_1*. The interactions of independent variables in the other models have similar characteristics to (7.12a).

7.4.2.2 Three-Way Interaction Models

As an extension of the set of two-way interaction models in (7.12), the following set of equation specifications present a set of three-way interaction models. The equation specifications (7.13c) up to (7.13g) will be same as the ESs (7.12c) to (7.12g)

$$\begin{aligned}&Y1\_1\ \ C\ \ \boldsymbol{Y1\_1(-1)*Y1\_2(-1)*Y1\_3(-1)}\ \ \boldsymbol{Y1\_1(-1)*Y1\_2(-1)*X1\_1(-1)}\\&\boldsymbol{Y1\_1(-1)*Y1\_3(-1)*X1\_1(-1)}\\&Y1\_1(-1)*Y1\_2(-1)\ \ Y1\_1(-1)*Y1\_3(-1)\ \ Y1\_1(-1)*X1\_1(-1)\\&Y1\_1(-1)\ldots Y1\_1(-1))\ \ Y1\_2(-1)\ \ Y1\_3(-1)\ \ X1\_1(-1)\end{aligned} \tag{7.13b}$$

$$\begin{aligned}&Y1\_2\ \ C\ \ \boldsymbol{Y1\_2(-1)*Y1\_3(-1)*X1\_2(-1)}\\&Y1\_2(-1)*Y1\_3(-1)\ \ Y1\_2(-1)*X1\_2(-1)\\&Y1\_2(-1)\ldots Y1\_2(-p2)\ \ Y1\_3(-1)\ \ X1\_2(-1)\end{aligned} \tag{7.13b}$$

Example 7.6 Application of the Models (7.13a)

Referring to the path diagram in Figure 7.11, I consider *Y1_1(−1)∗Y1_2(−1)∗Y1_3(−1)* as the most important interaction of the model (7.13a). For this reason, the stages of analysis are done as follows:

1) At the first stage is the OLS estimation of the ES: "*Y1_1 C Y1_1(−1)∗Y1_2(−1)∗Y1_3(−1)*" with the results showing that the interaction *Y1_1(−1)∗Y1_2(−1)∗Y1_3(−1)* has a significant effect on *Y1_1* based on the *t*-statistic of $t_0 = 20.71$, with $df = 40$, and $p\text{-value} = 0.0000$.
2) At the second stage, the STEPLS-Combinatorial and the trial-and-error methods are applied using the ES "*Y1_1 C Y1_1(−1)∗Y1_2(−1)∗Y1_3(−1)*" as the variables always in the model and all the other interactions in (7.13a) as the search regressors, so the results in Figure 7.16 are obtained where each of the three-way interactions has a significant effect at the 1% level.
3) At the third stage, by using the STEPLS-Combinatorial selection method, then selecting 1, 2, 3, or 4 of the main variables *Y1_1(−1), Y1_2(−1), Y1_3(_1),* and *X1_1(−1)* as additional independent variables, the results always show that some of the interactions have very large *p*-values. For these reasons, I present the results using the three main variables as an illustration.
4) At the final stage, the trial-and-error selection method should be applied in order to obtain a regression with the interactions having significant effects at the 1 or 5% levels. For instance, *Y1_1(−1)∗Y1_2(_1)∗X1_1(−1)* has a positive significant adjusted effect on *Y1* with a $p\text{-value} = 0.0444/2 = 0.0222$, and *Y1_1(−1)∗Y1_3(_1)∗X1_1(−1)* has a negative significant adjusted effect with a $p\text{-value} = 0.0242/2 = 0.0121$.

DV: Y1	The second stage			The third stage			The final stage		
Variables	Coef.	t-Stat.	Prob.	Coef.	t-Stat.	Prob.	Coef.	t-Stat.	Prob.
C	4004.351	8.765	0.0000	263.0348	0.085	0.9328	1251.455	0.417	0.6792
Y1_1(-1)*Y1_2(-1)*Y1_3(-1)	-3.79E-09	-4.504	0.0001	-9.94E-10	-0.332	**0.7418**	-4.35E-09	-4.325	0.0001
Y1_1(-1)*Y1_3(-1)	0.000126	13.609	0.0000	1.35E-09	0.430	**0.6699**	0.000125	5.058	0.0000
Y1_1(-1)*Y1_2(-1)*X1_1(-1)	2.26E-09	2.872	0.0067	-4.77E-09	-1.976	0.0563	4.22E-09	2.086	0.0444
Y1_1(-1)*Y1_3(-1)*X1_1(-1)	-3.17E-09	-4.547	0.0001	9.98E-05	3.052	0.0044	-5.52E-09	-2.356	0.0242
X1_1(-1)				0.39158	1.188	0.2431			
Y1_2(-1)				-0.434156	-1.149	0.2586	-0.399266	-1.054	0.2993
Y1_3(-1)				0.8039	0.863	0.3943	0.86722	0.927	0.3604
R-squared	0.9887			0.9895			0.9891		
Adjusted R-squared	0.9875			0.9873			0.9872		
F-statistic	810.22			457.56			527.38		
Prob(F-statistic)	0.0000			0.0000			0.0000		
DW-stat.	1.7381			1.7105			1.7715		

Figure 7.16 Statistical results base on the equation specification (7.13a).

Example 7.7 Application of the Models (7.13b)
Referring to the path diagram in Figure 7.11, I consider *Y1_2(−1)∗Y1_3(−1)∗X1_2(−1)* as the most important interaction of the model (7.13b). For this reason, the stages of analysis are done as follows:

1) At the first stage, the OLS estimates of the ES: "*Y1_1 C Y1_2(−1)∗Y1_3(−1)∗X1_2(−1)*" shows the interaction *Y1_2(−1)∗Y1_3(−1)∗X1_2(−1)* has a significant effect on *Y1_1*, based on the *t*-statistic of t_0 = 17.75050, with df = 40, and p-value = 0.0000.
2) At the second stage, by using the ES "*Y1_1 C Y1_2(−1)∗Y1_3(−1)∗X1_2(−1)*" as the variables always in the model and all the other interactions in (7.13b) as the search regressors, the results in Figure 7.17 are obtained, which show that each of the interactions has a significant adjusted effect at the 1 or 5% levels.
3) At the final stage, the three main variables *Y1_2(−1)*, *Y1_3(−1)*, and *X1_2(−1)* are used as the search regressors. By using the trial-and-error method, the results with *Y1_2(−1)* and *X1_2(−1)* as additional independent variables of the model found at the second stage are obtained. Note that, at the 10% level of significance, the interaction *Y1_2(−1)∗X1_2(−1)* has an insignificant effect on *Y1_2*, but it has a positive significant effect with a p-value = 0.164/2 < 0.10.

Example 7.8 Application of the Models (7.12c)–(7.12g)
By using the STEPLS-combinatorial and the trial-and-error selection methods, an acceptable regression function of each of the equation specification in (7.12c) up to (7.12g), which are representing the models (7.13c) and (7.13g), with their statistical results summary presented in Figure 7.18. Based on these results, the following findings and notes are presented.

1) At the first stage of analysis of the ES (7.12c), the OLS estimate of the ES "*Y1_3 C Y1_3(−1)∗X1_3(−1)*" shows that the interaction *Y1_3(−1)∗X1_3(−1)* has a significant effect on *Y1_3* based on the *F*-statistic of F_0 = 1728.808, with df = (1.40), and a p-value = 0.0000. Then at the second stage, the additional independent variables are selected from the regressors *Y1_3(−1)* ... *Y1_3(−p3)*, and *X1_3(−1)*, using the combinatorial and trial-and-error selection

DV: Y2	The second stage			The final stage		
Variable	Coef.	t-Stat.	Prob.*	Coef.	t-Stat.	Prob.*
C	6964.986	24.225	0.0000	-1138.332	-1.978	0.0556
Y1_2(-1)*Y1_3(-1)*X1_2(-1)	-6.15E-09	-9.110	0.0000	4.17E-09	2.271	0.0292
Y1_2(-1)*X1_2(-1)	3.47E-05	2.355	0.0238	-4.69E-05	-1.420	0.1642
Y1_3(-1)*X1_2(-1)	0.000116	4.950	0.0000	-6.05E-05	-3.172	0.0031
Y1_2(-1)				1.128158	4.690	0.0000
X1_2(-1)				0.751945	4.114	0.0002
R-squared	0.9869			0.9971		
Adjusted R-squared	0.9858			0.9967		
F-statistic	950.78			2458.5		
Prob(F-statistic)	0.0000			0.0000		
DW-statistic	0.9979			1.3447		

Figure 7.17 Statistical results base on the equation specification (7.13b).

DV: Y1_3	C	Y1_3(-1)*X1_3(-1)	Y1_3(-2)	X1_3(-1)	R^2	Adj. R^2	DW-Stat
Coef.	5804.544	3.67E-05	-0.3250	0.4326	0.9851	0.9839	0.8098
t-Stat	5.0816	3.951037	-1.2242	3.7409			
Prob.	0.0000	0.0003	0.2286	0.0006			
DS: X1_1	C	X1_1(-1)*Z1(-1)	X1_1(-1)	Z1(-1)	R^2	Adj. R^2	DW-Stat
Coef.	-384.153	-1.20E-05	1.0368	0.1874	0.9913	0.9907	1.6569
t-Stat	-0.5909	-2.44914	9.3720	2.3503			
Prob.	0.5581	0.019	0.0000	0.0241			
DV: X1_2	C	X1_2(-1)*Z1(-1)	X1_2(-4)	Z1(-4)	R^2	Adj. R^2	DW-Stat
Coef.	859.751	1.13E-05	1.1124	-0.2244	0.9876	0.9865	0.7088
t-Stat	1.6244	2.34061	5.4384	-1.4755			
Prob.	0.1135	0.0253	0.000	0.1493			
DV: X1_3	C	X1_3(-1)*Z1(-1)	X1_3(-4)	Z1(-4)	R^2	Adj. R^2	DW-Stat
Coef.	957.262	1.16E-05	1.0738	-0.1971	0.9884	0.9874	0.6998
t-Stat	1.9488	2.63694	5.0630	-1.1957			
Prob.	0.0594	0.0124	0.0000	0.2398			
DV: Z1	C	Z1(-1)	Z1(-2)		R^2	Adj. R^2	DW-Stat
Coef.	135.462	1.40841	-0.3992		0.9977	0.9976	1.9554
t-Stat	1.9060	9.22442	-2.5529				
Prob.	0.0642	0.0000	0.0148				

Figure 7.18 The statistical results summary based on the equation specifications (7.12c)–(7.12g).

methods in order to obtain a regression with a significant interaction. Similar is done based on the other models.

2) However, it is very important to mention that the two regressions of *X1_2* and *X1_3* are really unexpected regressions, because both of them use fourth lags of the main variables. It is found that if the lower lag main variables are inserted as additional independent variables, the interactions have insignificant effects. Do the analysis using other lag main variables as an exercise.

7.4.3 Translog-Linear Triangular Effects Model

The translog-linear model in fact is a derivation of the Cobb–Douglas product function, which is nonlinear and also known as a multiplicative model. So a translog-linear model can be considered a complex model, in the theoretical sense. The set of seven ESs is as follows:

$$log(Y1\_1)\ \ C\ \ log(Y1\_1(-1))\ldots log(Y1\_1(-p1))$$
$$\log(\boldsymbol{Y1\_2(-1)})\ \log(\boldsymbol{Y1\_3(-1)})\ \ log(X1\_1(-1)) \tag{7.14a}$$

$$log(Y1\_2)\ \ C\ \ log(Y1\_2(-1))\ldots log(Y1\_2(-p2))\ \log(\boldsymbol{Y1\_3(-1)})\ \ log(X1\_2(-1)) \tag{7.14b}$$

$$log(Y1\_3)\ \ C\ \ log(Y1\_3(-1))\ldots log(Y1\_3(-p3))\ \ log(X1\_3(-1)) \tag{7.14c}$$

$$log(X1\_1)\ \ C\ \ log(X1\_1(-1))\ldots log(X1\_1(-p4))\ \ log(Z1(-1)) \tag{7.14d}$$

$$log(X1\_2)\ \ C\ \ log(X1\_2(-1))\ldots log(X1\_2(-p5))\ \ log(Z1(-1)) \tag{7.14e}$$

$$log(X1\_3)\ \ C\ \ log(X1\_3)(-1))\ldots log(X1\_3(-p6))\ \ log(Z1(-1)) \tag{7.14f}$$

$$log(Z1)\ \ C\ \ log(Z1(-1))\ldots log(Z1(-p7)) \tag{7.14g}$$

Note that the first three equations represent a triangular effects model based on the three variables *log(Y1_1)*, *log(Y1_2)*, and *log (Y1_3)*.

7.5 Full-Lag Variables Circular Effects Model

As a modification of the path diagram in Figure 7.15, Figure 7.19 presents a full-lag-variables circular effects model of *Y1_1*, *Y1_2*, and *Y1_3*, which is represented by the circular relationships between their first-lag of the endogenous variables. So, in the theoretical sense, the variables *X1_1(−1)*, *X1_2(−1)*, and *X1_3(−1)* also have circular relationships. However, their relationships are not taken into account in the model.

7.5.1 Two-Way Interaction Circular Effects Models

As a modification of the set of models in (7.10), we have the set of full-lag variables models with the ESs as follows:

$$Y1\_1\ \ C\ \ \boldsymbol{Y1\_1(-1)*Y1\_2(-1)}\ \ \boldsymbol{Y1\_1(-1)*X1\_1(-1)}$$
$$Y1\_1(-1)\ldots Y1\_1(-p1))\ \ Y1\_2(-1)\ \ X1\_1(-1) \tag{7.15a}$$

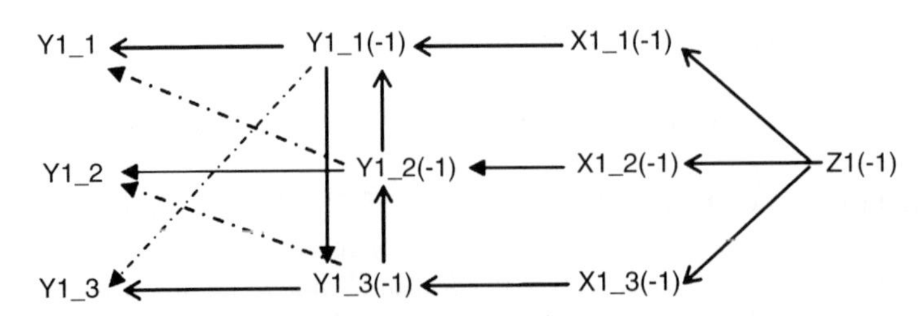

Figure 7.19 A circular effects model between *Y1_1*, *Y1_2*, and *Y1_3*.

$$Y1\_2\ \ C\ \ \boldsymbol{Y1\_2(-1)*Y1\_3(-1)Y1\_2(-1)*X1\_2(-1)}$$
$$Y1\_2(-1)\ldots Y1\_2(-p2)\ \ Y1\_3(-1)\ \ X1\_2(-1) \tag{7.15b}$$
$$Y1\_3\ \ C\ \ \boldsymbol{Y1\_3(-1)*Y1\_1(-1)\ \ Y1\_3(-1)*X1\_3(-1)}$$
$$Y1\_3(-1)\ldots Y1\_3(-p3)\ \ Y1\_1(-1)\ \ X1\_3(-1) \tag{7.15c}$$

7.5.2 Three-Way Interaction Circular Effects Models

As a modification of the set of models in (7.15), we have the set of full-lag variables models with the ESs as follows:

$$Y1\_1\ \ C\ \ \boldsymbol{Y1\_1(-1)*Y1\_2(-1)*X1\_1(-1)}\ \ Y1\_1(-1)*Y1\_2(-1)\ \ Y1\_1(-1)*X1\_1(-1)$$
$$Y1\_1(-1)\ldots Y1\_1(-p1))\ \ Y1\_2(-1)\ \ X1\_1(-1) \tag{7.16a}$$
$$Y1\_2\ \ C\ \ \boldsymbol{Y1\_2(-1)*Y1\_3(-1)*X1\_2(-1)}\ \ Y1\_2(-1)*Y1\_3(-1)\ \ Y1\_2(-1)*X1\_2(-1)$$
$$Y1\_2(-1)\ldots Y1\_2(-p2)\ \ Y1\_3(-1)\ \ X1\_2(-1) \tag{7.16b}$$
$$Y1\_3\ \ C\ \ \boldsymbol{Y1\_3(-1)*Y1\_1(-1)*X1\_3(-1)}\ \ \boldsymbol{Y1\_3(-1)}*Y1\_1(-1)\ \ Y1\_3(-1)*X1\_3(-1)$$
$$Y1\_3(-1)\ldots Y1\_3(-p3)\ \ Y'1\_1(-1)\ \ X1\_3(-1) \tag{7.16c}$$

7.6 Full-Lag Variables Reciprocal-Effects Model

As a modification of the path diagram in Figure 7.15, Figure 7.20 presents full-lag-variables reciprocal relationships between the first-lag of the endogenous variables *Y1_1*, *Y1_2*, and *Y1_3*.

7.6.1 Two-Way Interaction Reciprocal-Effects Models

As a modification of the set of models in Eq. (7.15), we have the set of full-lag variables two-way interaction reciprocal-effects models with the ESs as follows:

$$Y\ 1\_1\ \ C\ \ \boldsymbol{Y1\_1(-1)*Y1\_2(-1)\ \ Y1\_(1)*Y1\_3(-1)\ \ Y1\_1(-1)*X1\_1(-1)}$$
$$Y1\_1(-1)\ldots Y1\_1(-p1))\ \ Y1\_2(-1)\ \ Y1\_3(-1)\ \ X1\_1(-1) \tag{7.17a}$$
$$Y1\_2\ \ C\ \ \boldsymbol{Y1\_2(-1)*Y1\_1(-1)\ \ Y1\_2(-1)*Y1\_3(-1)\ \ Y1\_2(-1)*X1\_2(-1)}$$
$$Y1\_2(-1)\ldots Y1\_2(-p2)\ \ Y1-1(-1)\ \ Y1\_3(-1)\ \ X1\_2(-1) \tag{7.17b}$$

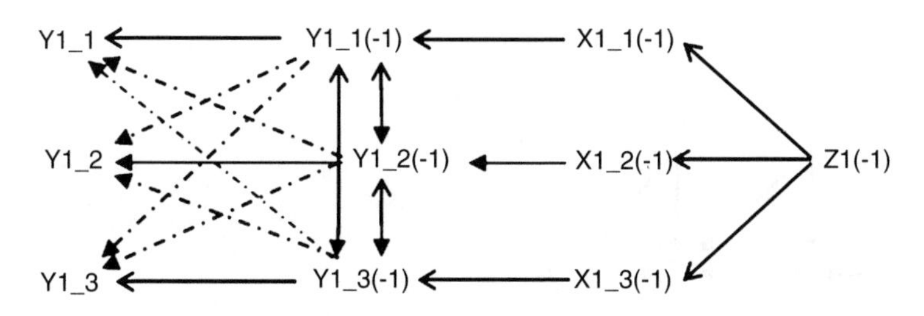

Figure 7.20 A reciprocal-effects model between *Y1_1(−1)*, *Y1_2(−1)*, and *Y1_3(−1)*.

$$Y1\_3\ \ C\ \ \boldsymbol{Y1\_3(-1)*Y1\_1(-1)}\ \ \boldsymbol{Y1\_3(-1)*Y1\_2(-1)}\ \ \boldsymbol{Y1\_3(-1)*X1\_3(-1)}$$
$$Y1\_3(-1) \ldots Y1\_3(-p3)\ \ Y1\_1(-1)\ \ Y1\_2(-1)\ \ X1\_3(-1) \tag{7.17c}$$

7.6.2 Three-Way Interaction Reciprocal-Effects Models

As an extension of the models (7.17), we have the set of full-lag variables three-way interaction reciprocal-effects models with the ESs as follows:

$$Y1\_1\ \ C\ \ \boldsymbol{Y1\_1(-1)*Y1\_2(-1)*Y1\_3(-1)}$$
$$Y1\_1(-1)*Y1\_2(-1)\ \ Y1\_(1)*Y1\_3(-1)\ \ Y1\_1(-1)*X1\_1(-1)$$
$$Y1\_1(-1) \ldots Y1\_1(-p1))\ \ Y1\_2(-1)\ \ Y1\_3(-1)\ \ X1\_1(-1) \tag{7.18a}$$
$$Y1\_2\ \ C\ \ \boldsymbol{Y1\_1(-1)*Y1\_2(-1)*Y1\_3(-1)}$$
$$Y1\_2(-1)*Y1\_1(-1)\ \ Y1\_2(-1)*Y1\_3(-1)\ \ Y1\_2(-1)*X1\_2(-1)$$
$$Y1\_2(-1) \ldots Y1\_2(-p2)\ \ Y1\_1(-1)\ \ Y1\_3(-1)\ \ X1\_2(-1) \tag{7.18b}$$
$$Y1\_3\ \ C\ \ \boldsymbol{Y1\_1(-1)*Y1\_2(-1)*Y1\_3(-1)}$$
$$Y1\_3(-1)*Y1\_1(-1)\ \ Y1\_3(-1)*Y1\_2(-1)\ \ Y1\_3(-1)*X1\_3(-1)$$
$$Y1\_3(-1) \ldots Y1\_3(-p3)\ \ Y1\_1(-1)\ \ Y1\_2(-1)\ \ X1\_3(-1) \tag{7.18c}$$

7.7 Successive Up-and-Downstream Relationships

Suppose the seven variables *X1, X2, X3, Y1_1, Y1_2, Y1_3,* and *Z1* have successive up-and-downstream or causal relationships as presented in Figure 7.21. In addition, note that the first-lag variables *Y1_1(−1), Y1_2(−1), Y1_3(−1), X1_1(−1), X1_2(−1), X1_3(−1),* and *Z1(−1)* are the upper (source, predictor, or cause) variables for each of the seven variables, in the theoretical sense.

Then many sets of dynamic or lag variables models can be assumed to be the true population models, starting with the set of seven simple regression models. So we could have a lot of possible sets of seven multiple regression models. However, in this section I present only the following three successive lag variables interaction effects models, starting with a set of the simplest two-way interaction models.

7.7.1 A Set of the Simplest Two-Way Interaction Models

Based on the path diagram in Figure 7.21, as an extension of the model (7.2a), the following set of ESs present a set (or series) of the simplest two-way interaction LV(1) models, where the first one is exactly the same as the two-way interaction model (7.2a), which indicates that the effect of *Y1_2* on *Y1_1* depends on *Y1_1(−1).*

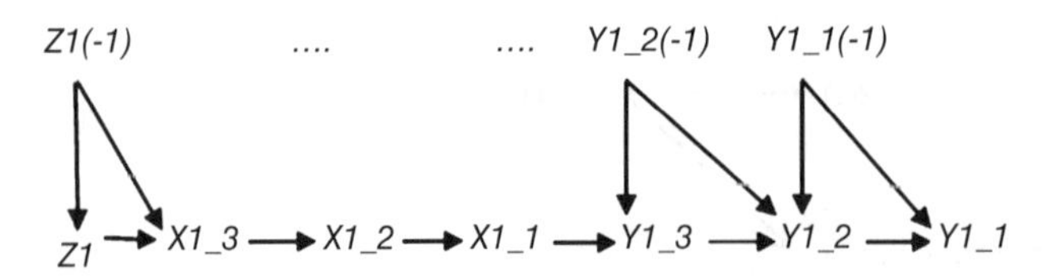

Figure 7.21 Successive up-and-downstream relationships.

$$Y1\_1 \ \ C \ \ \mathbf{Y1\_1(-1)*Y1\_2} \ \ Y1\_1(-1) \ldots Y1\_1(-p1) \ \ Y1\_2 \tag{7.19a}$$

$$Y1\_2 \ \ C \ \ \mathbf{Y1\_2(-1)*Y1\_3} \ \ Y1\_2(-1) \ldots Y1\_2(-p2) \ \ Y1\_3 \tag{7.19b}$$

$$Y1\_3 \ \ C \ \ \mathbf{Y1\_3(-1)*X1\_1} \ \ Y1\_3(-1) \ldots Y1\_3(-p3) \ \ X1\_1 \tag{7.19c}$$

$$X1\_1 \ \ C \ \ \mathbf{X1\_1(-1)*X1\_2} \ \ X1\_1(-1) \ldots X1\_1(-p4) \ \ X1\_2 \tag{7.19d}$$

$$X1\_2 \ \ C \ \ \mathbf{X1\_2(-1)*X1\_3} \ \ X1\_2(-1) \ldots X1\_2(-p5) \ \ X1\_3 \tag{7.19e}$$

$$X1\_3 \ \ C \ \ \mathbf{X1\_3(-1)*Z1} \ \ X1\_3(-1) \ldots X1\_3(-p6) \ \ Z1 \tag{7.19f}$$

$$Z1 \ \ C \ \ Z1(-1) \ldots Z1(-p7) \tag{7.19g}$$

It is important to mention that the in-sample forecast values of each of the dependent variables can be computed directly. However, its forecast values beyond the sample period should be computed using a special ordering, starting with the forecast values of *Z1*, then the next are forecast values of *X1_3* depending on the fitted values $\hat{Z}1$, the forecast values of *X1_2* depending on the interaction $X1\_2(-1)*\hat{X}1\_3$, and at the final stage the forecast values of *Y1_1* in (7.19A) depending on *Y1_1(−1)* and the fitted values of all other variables. Hence, this process clearly shows that the forecast values beyond the sample period of *Y1_1* are computed based on a complex regression function, where one of its independent variable is the following interaction

$$Y1\_1(-1)*\hat{Y}1\_2*\hat{Y}1\_3*\hat{X}1\_1*\hat{X}1\_2*\hat{X}1\_3*\hat{Z}1.$$

Example 7.9 An Application of the Models (7.19)

Figure 7.22 presents the summary of acceptable OLS regressions of the seven models in (7.19). Based on these results, the findings and notes presented are as follows:

1) By using only the two-way interaction as an independent variable of each model (7.19A)–(7.19f), it is found that the interaction has significant effect on the corresponding dependent variable with a p-value 0.0000.
2) Then additional independent variables are inserted by using the trial-and-error and combinatorial selected methods in order to obtain acceptable interaction regressions. Note that at the 10% level of significance, *Y1_3(−1)*X1_1* has a positive significant adjusted effect on *Y1_3*, with a p-value = 0.1635/2 = 0.08175.
3) The in-sample forecast values of each of the dependent variables can easily be obtained. However, for forecast values beyond the sample, these should be computed starting with the regression of *Z1*, then the regression of *X1_3*, and the last being the regression of *Y1_1*. Do this as an exercise – refer to Example 6.34.

7.7.2 Successive Two-Way Interaction Triangular Effects Models

Note that the up-and-downstream relationships between the seven variables presented in Figure 7.21, also presenti *partial* direct effects of each of the variables *Y1_3*, *X1_1*, *X1_2*, *X1_3*, and *Z1* on *Y1_1* in addition to their direct effects on a downward variable.

For this reason, an alternative path diagram is presented in Figure 7.23. Note that the arrows with dotted lines represent the partial direct effects of a variable on a downward variable in addition to its indirect effect, such as the triangular relationships between the three variables *Y1_3 (−1)*, *Y1_2(−1)*, and *Y1_1(−1)*; and the three variables *Z1(−1)*, *X1_3(−1)*, and *X1_2(−1)*. However, for the successive triangular effects models, the seven up-and-down stream relationships are presented using their first-lag variables, which are the extension of the path diagram in

DV: Y1_1	C	Y1_1(-1)*Y1_2	Y1_1(-2)	Y1_2	
Coef.	4622.686	1.13E-05	0.231	0.438	
t-Stat.	3.948	1.47866	1.417	3.062	
Prob.	0.000	0.14770	0.165	0.004	
DV: Y1_2	C	Y1_2(-1)*Y1_3	Y1_2(-1)	Y1_2(-2)	Y1_3
Coef.	-2034.802	-1.57E-05	1.097	-0.248	0.567
t-Stat.	-3.112	-2.65312	7.539	-1.912	4.242
Prob.	0.004	0.01180	0.000	0.064	0.000
DV: Y1_3	C	Y1_3(-1)*X1_1	Y1_3(-1)	Y1_3(-2)	
Coef.	846.133	5.53E-06	1.167	-0.317	
t-Stat.	1.639	1.41465	5.767	-2.010	
Prob.	0.110	0.16550	0.000	0.052	
DV: X1_1	C	X1_1(-1)*X1_2	X1_1(-1)	X1_2(-2)	X1_2
Coef.	-1061.512	-1.39E-05	1.081	-0.421	0.619
t-Stat.	-1.201	-1.61524	6.139	-1.534	2.502
Prob.	0.238	0.11520	0.000	0.134	0.017

DV: X1_2	C	X1_2(-1)*X1_3	X1_2(-5)	X1_3(-1)
Coef.	978.225	8.62E-06	0.147	0.727
t-Stat.	2.865	1.56263	1.013	4.270
Prob.	0.007	0.12770	0.318	0.000
DV: X1_3	C	X1_3(-1)*Z1	X1_3(-1)	Z1
Coef.	634.053	-3.38E-06	0.737	0.241
t-Stat.	3.082	-1.50655	7.825	3.115
Prob.	0.004	0.14020	0.000	0.004
DV: Z1	C	Z1(-1)	Z1(-2)	
Coef.	135.462	1.40841	-0.399	
t-Stat.	1.906	9.22442	-2.553	
Prob.	0.064	0.00000	0.015	

Figure 7.22 The summary of acceptable OLS regressions of the models in (7.19).

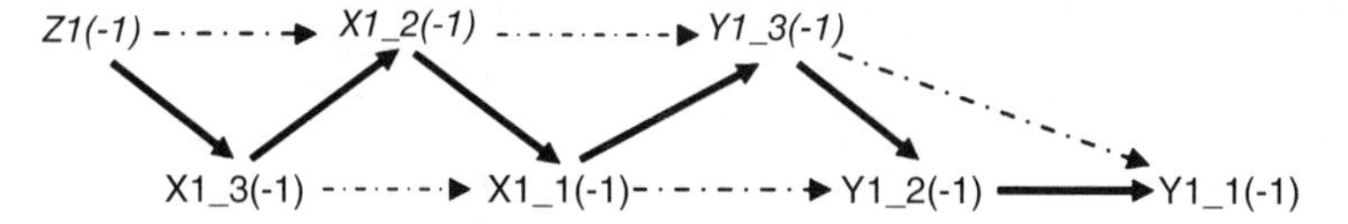

Figure 7.23 Alternative triangular relationships representing the seven variables.

Figure 7.15. Hence, I propose the following successive first-lag variables relationships, which represent the successive relationships of the original variables.

$$Y1\_1\ C\ \boldsymbol{Y1\_1(-1)*Y1\_2(-1)}\ \boldsymbol{Y1\_1(-1)*Y1\_3(-1)}\ Y1\_1(-1)\ldots Y1\_1(-p1))\ Y1\_2(-1)\ Y1\_3(-1) \tag{7.20a}$$

$$Y1\_2\ C\ \boldsymbol{Y1\_2(-1)*Y1\_3(-1)}\ \boldsymbol{Y1\_2(-1)*X1\_1(-1)}\ Y1\_2(-1)\ldots Y1\_2(-p2)\ Y1\_3\ X1\_1(-1) \tag{7.20b}$$

$$Y1\_3\ C\ \boldsymbol{Y1\_3(-1)*X1\_1(-1)}\ \boldsymbol{Y1\_3(-1)*X1\_2(-1)}\ Y1\_3(-1)\ldots Y1\_3(-p3)\ X1\_1(-1)\ X1\_2(-1) \tag{7.20c}$$

$$X1\_1\ C\ \boldsymbol{X1\_1(-1)*X1\_2(-1)}\ \boldsymbol{X1\_1(-1)*X1\_3(-1)}\ X1\_1(-1)\ldots X1\_1(-p4)\ X1\_2(-1)\ X1\_3(-1) \tag{7.20d}$$

$$X1\_2\ C\ \boldsymbol{X1\_2(-1)*X1\_3(-1)}\ \boldsymbol{X1\_2(-1)*Z1(-1)}\ X1\_2(-1)\ldots X1\_2(-p5)\ X1\_3(-1)\ Z1(-1) \tag{7.20e}$$

$$X1\_3\ C\ \boldsymbol{X1\_3(-1)*Z1(-1)}\ X1\_3(-1)\ldots X1\_3(-p6)\ Z1(-1) \tag{7.20f}$$

$$Z1\,C\,Z1(-1)\ldots Z1(-p7) \tag{7.20g}$$

It is well-known that each of the first-lag variables has a direct effect on each of the seven original variables. However, the original variables are not presented in Figure 7.23, so the two-way interaction *Y1_1(−1)∗Y1_2(−1)* and *Y1_1(−1)∗Y1_3(−1)* in ES (7.20a) represent the indirect effects of both *Y1_2(−1)* and *Y1_3(−1)* on *Y1_1* through *Y1_1(−1)*. This is similar for the other models. Hence, we have successive two-way interaction triangular effects models.

7.7.3 Successive Three-Way Interaction Triangular Effects Models

As an extension of the two-way interaction triangular effects models in (7.20), we have a set of three-way interaction triangular effects models, with the following set of equations.

$$Y1\_1\ C\ \boldsymbol{Y1\_1(-1)*Y1\_2(-1)*Y1\_3(-1)}\ Y1\_1(-1)*Y1\_2(-1)\ Y1\_1(-1)*Y1\_3(-1)Y1\_1(-1)\ldots Y1\_1(-p1)\ Y1\_2(-1)\ Y1\_3(-1) \tag{7.21a}$$

$$Y1\_2\ C\ \boldsymbol{Y1\_2(-1)*Y1\_3(-1)*X1\_1(-1)}\ Y1\_2(-1)*Y1\_3(-1)\ Y1\_2(-1)*X1\_1(-1)Y1\_2(-1)\ldots Y1\_2(-p2)\ Y1\_3(-1)\ X1(-1) \tag{7.21b}$$

$$Y1\_3\ C\ \boldsymbol{Y1\_3(-1)*X1\_1(-1)*X1\_2(-1)}\ Y1\_3(-1)*X1\_1(-1)\ Y1\_3(-1)*X1\_2(-1)Y1\_3(-1)\ldots Y1\_3(-p3)\ x1\_1(-1)\ x1\_2(-1) \tag{7.21c}$$

$$X1\_1\ C\ \boldsymbol{X1\_1(-1)*X1\_2(-1)*X1\_3(-1)}\ X1\_1(-1)*X1\_2(-1)\ X1\_1(-1)*X1\_3(-1)X1\_1(-1)\ldots X1\_1(-p4)\ X1\_2(-1)\ X1\_3(-1) \tag{7.21d}$$

RM-1	C	Y1_1(-1)*Y1_2(-1)*Y1_3(-1)	Y1_1(-1)*Y1_2(-1)	Y1_1(-1)*Y1_3(-1)	Y1_1(-1)			RMSE	TIC
Coef.	3727.156	-3.10E-09	2.59E-05	5.98E-05	0.229431			363.84	0.0108
t-Stat	1.930062	-2.144075	1.747144	1.929978	0.607313				
Prob.	0.0613	0.0387	0.0889	0.0613	0.5474				
RM-2	C	Y1_1(-1)*Y1_2(-1)*Y1_3(-1)	Y1_1(-1)*Y1_2(-1)	Y1_1(-1)*Y1_3(-1)		Y1_2(-1)		RMSE	TIC
Coef.	5157.949	-4.13E-09	4.42E-05	7.32E-05		-0.117882		339.16	0.0101
t-Stat	6.85738	-5.82934	2.207439	4.295562		-0.581339			
Prob.	0	0	0.0336	0.0001		0.5645			
RM-3	C	Y1_1(-1)*Y1_2(-1)*Y1_3(-1)	Y1_1(-1)*Y1_2(-1)	Y1_1(-1)*Y1_3(-1)			Y1_3(-1)	RMSE	TIC
Coef.	5016.205	-3.93E-09	3.40E-05	7.74E-05			-0.044956	343..23	0.0102
t-Stat	3.406289	-6.204512	3.796118	3.561748			-0.118986		
Prob.	0.0016	0	0.0005	0.001			0.9059		
RM-4	C	Y1_1(-1)*Y1_2(-1)*Y1_3(-1)	Y1_1(-1)*Y1_2(-1)	Y1_1(-1)*Y1_3(-1)		Y1_2(-1)	Y1_3(-1)	RMSE	TIC
Coef.	-255.4437	-5.07E-09	0.00013	-4.15E-05		-1.351846	2.373685	285.66	0.008472
t-Stat	-0.078667	-5.748876	2.417072	-0.600942		-1.808099	1.711338		
Prob.	0.9377	0	0.0208	**0.5516**		0.079	0.0956		
RM-5	C		Y1_1(-1)*Y1_2(-1)	Y1_1(-1)*Y1_3(-1)	Y1_1(-1)	Y1_2(-1)	Y1_3(-1)	RMSE	TIC
Coef.	-10910.8		8.62E-05	-0.000184	1.267492	-1.49538	3.714009	285.36	0.008473
t-Stat.	-2.430111		1.816779	-2.247658	5.986522	-2.013895	2.481058		
Prob.	0.0202		0.0776	0.0308	0	0.0515	0.0179		

Figure 7.24 Statistical results of five acceptable reduced models of the model (7.21a), with their forecast evaluation, RMSE and TIC, using a forecast sample {@*Year*>*1980*}.

$$X1\_2\ \ C\ \ \boldsymbol{x1\_2(-1)*X1\_3(-1)*Z1(-1)}\ \ X1\_2(-1)*X1\_3(-1)\ \ X1\_2(-1)*Z1(-1)$$
$$X1\_2(-1)\ldots X1\_2(-p5)\ \ X1\_3(-1)\ \ Z1(-1) \tag{7.21e}$$
$$X1\_3\ \ C\ \ \boldsymbol{X1\_3(-1)*Z1(-1)}\ \ X1\_3(-1)\ldots X1\_3(-p6)\ \ Z1(-1) \tag{7.21f}$$
$$Z1\ \ C\ \ Z1(-1)\ldots Z1(-p7) \tag{7.21g}$$

Example 7.10 Alternative OLS Regressions Based on the Model (7.21a), for $p1 = 1$
In general, the OLS estimate of a full three-way interaction model presents unexpected insignificant independent variables because of their multicollinearity or if they are highly correlated. For this reason, Figure 7.24 presents the statistical results of five acceptable reduced models, namely *RM-1* to *RM-5* of the model (7.21a), which are obtained by using the trial-and-error (manual) and STEPLS-Combinatorial selection methods. The steps of regression analysis are as follows:

1) The first four reduced models were intentionally developed in order to have a significant three-way interaction independent variable. The steps of the analysis are as follows:
   1.1 The first OLS regression analysis uses the ES: "*Y 1_1 C Y1_1(−1)*∗*Y1_2(−1)*∗*Y1_3(−1) Y1_1(−1)*∗*Y1_2(−1) Y1_2(−1)*∗*Y1_3(−1)*". It happens that each of the independent variables has a significant adjusted effect. So the model is an acceptable model, to be used at the second stage of analysis.
   1.2 The trial-and-error and combinatorial selection methods are applied to select one of the three main variables *Y1_1(−1)*, *Y1_2(−1)*, and *Y1_3(−1)*. The first three reduced models are obtained.
   1.3 At the third stage, two of the three main variables are selected, then the *RM-4* is obtained.

2) As an alternative model, I try to do the following steps of regression analysis.
   2.1 At the first stage, the OLS regression analysis is applied by using the ES "*Y1_1 C'Y1_1(-1), Y1_2(-1) Y1_3(-1)*". It happens each of the main variables has a significant adjusted effect on *Y1_1*. So the full additive regression can be used as the estimation function at the second stage of analysis.
   2.2 At the second stage, the three interactions are used as the search regressors. Then by using the STEPLS-Combinatorial selection method, *RM-5* is obtained as an acceptable regression.

3) Finally, based on the results in Figure 7.23, the findings and notes presented are as follows:
   3.1 The first four reduced models are good three-way interaction models, since the three-way interaction has a significant effect on *Y1_1*, even though *Y1_1(−1)* *Y1_3(−1)* in *RM-4* has a large p-value = 0.5516.
   3.2 Among the four three-way interaction reduced models, *RM-4* is the best forecast model since it has the smallest RMSE and TIC.
   3.3 The reduced model *RM-5* of (7.21a) is in fact the estimate of the full two-way interaction model in (7.20a), for $p1 = 1$.
   3.4 Note the two reduced models *RM-4* and *RM-5* have very small and different values of RMSE and TIC. Since *RM-4* has smaller RMSE = 285.16 and TIC = 0.008 72 compared to *RM-5* with RMSE = 285.36 and TIC = 0.008 73, then *RM-4* could be considered as a better forecast model, in the statistical sense. Despite this, we never know which one gives the best forecast values beyond the sample period. Refer to the Examples 6.33 and 6.34.

4) The first forecast value beyond the sample period, namely for $t = T + 1$, can be computed directly.

Example 7.11 Acceptable OLS Regressions Based on the Models (7.21b)–(7.21g)
By using the same process as the analysis based on the ES (7.21a) presented in Example 7.10, I obtain acceptable forecast models for *Y1_2, Y1_3, X1_1, X1_2, X1_3,* and *Z1* with the summary of their OLS estimates presented in Figure 7.25. Based on this summary, it is important to identify the problems of the following two regressions.

1) The first problem is with the regression of *Y1_2*, as follows:
   1.1 Even though *Y1_2(−1)*Y1_3(−1)* has a large p-value, it does not have to be deleted from the model, because its interaction *Y1_2(−1)*Y1_3(−1)*X1_1(−1)* has a significant adjusted effect on *Y1_2* at the 1% level.
   1.2 At the 10% level, the main independent variable (IV) *X1_1(−1)* has insignificant effect on *Y1_2*. However, it has a negative significant effect with a p-value = 0.1411/2 = 0.070 55 < 0.10.
   1.3 Looking at the special function (− 4.39e-09**X1_1(−1)* + 2.91e-05)**Y1_2(−1)*Y1_3(−1)* it can be concluded that the effect *Y1_2(−1)*Y1_3(−1)* on *Y1_2* would decrease with increasing values of *X1_1(−1)* because *X1_1(−1)* has a negative coefficient.

2) The second problem is with the regression of *X1_3* as follows:
   2.1 The interaction *X1_3(−1)*Z1(−1)* has an insignificant adjusted effect at the 10% level of significance. However, it has a significant negative adjusted effect on *X1_3* with a p-value = 0.1408/2 = 0.0704 < 0.10.
   2.2 It can be concluded that the effect of *Z1(−1)* decreases with increasing scores of *X1_3(−1)* because *(−6.31E-06*X1_3(−1)* + 0.826393*)* would decrease with increasing scores *X1_3(−1)*.
   2.3 An alternative acceptable forecast model of *X1_3(−1)* is presented in Figure 7.22.

DV: Y1_2	C	Y1_2(-1)*Y1_3(-1)*X1_1(-1)	Y1_2(-1)*Y1_3(-1)	Y1_2(-1)*X1_1(-1)	Y1_3(-1)	X1_1(-1)	
Coef.	-770.593	-4.39E-09	2.91E-05	7.09E-05	1.143989	-0.476006	
t-Stat.	-0.676577	-10.86138	0.825839	2.16142	2.867825	-1.504654	
Prob.	0.503	0	0.4143	0.0374	0.0069	0.1411	
DV: Y1_3	C	Y1_3(-1)*X1_1(-1)*X1_2(-1)	Y1_3(-1)*X1_1(-1)	Y1_3(-1)*X1_2(-1)	X1_1(-1)		
Coef.	4953.958	-4.30E-09	7.88E-05	5.74E-05	-0.390238		
t-Stat.	13.70314	-4.318172	2.95762	6.716923	-2.33359		
Prob.	0	0.0001	0.0054	0	0.0252		
DV: X1_1	C	X1_1(-1)*X1_2(-1)*X1_3(-1)	X1_1(-1)*X1_2(-1)	X1_1(-1)*X1_3(-1)	X1_2(-1)	X1_3(-1)	X1_1(-1)
Coef.	5198.835	-5.77E-09	0.000337	-0.000195	-2.931053	2.712668	-0.054161
t-Stat.	1.731141	-1.861194	2.546094	-1.364133	-1.818705	1.606008	-0.095924
Prob.	0.0922	0.0711	0.0155	0.1812	0.0775	0.1173	0.9241
DV: X1_2	C	X1_2(-1)*X1_3(-1)*Z1(-1)	X1_2(-1)*X1_3(-1)	X1_2(-1)*Z1(-1)	X1_3(-1)	Z1(-1)	
Coef.	-461.2518	-5.17E-09	-0.000109	0.000194	2.010727	-1.257465	
t-Stat.	-0.411494	-2.376817	-1.245254	1.862598	2.43381	-1.743095	
Prob.	0.6832	0.0231	0.2213	0.0709	0.0202	0.0901	
DV: X1_3	C	X1_3(-1)*Z1(-1)	Z1(-1)				
Coef.	2214.98	-6.31E-06	0.826393				
t-Stat.	14.58807	-1.503219	14.53831				
Prob.	0	0.1408	0				
DV: Z1	C	Z1(-1)	Z1(-2)				
Coef.	135.4624	1.408411	-0.39919				
t-Stat.	1.906018	9.224421	-2.552863				
Prob.	0.0642	0	0.0148				

Figure 7.25 Acceptable forecasting models obtained based on the ESs (7.21b)–(7.21g).

3) For the reasons before, I would conclude that the forecast models RM-4 of *Y1_1* and the six models presented in Figure 7.24 can be used directly to compute the first forecast values beyond the sample period, namely for $t = T + 1$, of the seven time series, by using Excel. Then by using the block-copy-paste method, the forecast values for $t > T + 1$ could easily be obtained. Do this as an exercise. Refer to the method presented in Example 7.15.

7.8 Forecast Models with the Time Independent Variable

It should be well accepted that the forecasting models of any time series Y_t are a function of the time t = *@trend*, because the time t can represent all unobserved exogenous variables, mainly for the exogenous and environmental variables that are highly or significantly correlated with it. For this reason, all models presented in (7.1) up to (7.21), can be modified or extended to more advanced models, where the time t should be used as an additional independent variable. So we can have models with alternative trends as presented in Table 7.1, or time-related effects (TRE) models as presented in Agung (2014, p. 97).

7.8.1 Forecast Models with Alternative Trends

As the simplest extension model is to transform selected models in each of the sets of models to the models with linear trends. Here, I present only the following two alternative extensions.

7.8.1.1 Forecast Models with Linear Trends

As an illustration, based on the set of models (7.6), we have the following set of five models (7.21a)–(7.21e), where only (7.22a) and (7.22b) are the models with trends. Hence, (7.22c)–(7.22e) are exactly the same as the models (7.6c)–(7.6e).

$$\begin{aligned} &Y1\_1\ \ C\ \ \boldsymbol{Y1\_1(-1)*Y1\_2(-1)}\ \ \boldsymbol{Y1\_1(-1)*X1\_1} \\ &\qquad Y1\_1(-1)\ldots Y1\_1(-p1))\ \ Y1\_2(-1)\ \ X1\_1\ \ t \end{aligned} \tag{7.22a}$$

$$\begin{aligned} &Y1\_2\ \ C\ \ \boldsymbol{Y1\_1(-1)*Y1\_2(-1)}\ \ \boldsymbol{X1\_2^{*}Y1\_2(-1)} \\ &\qquad Y1\_2(-1)\ldots Y1\_2(-p2)\ \ Y1\_1(-1)\ \ X1\_2\ \ t \end{aligned} \tag{7.22b}$$

$$X1\_1\ \ C\ \ \boldsymbol{X1\_1(-1)*Z1}\ \ X1\_1(-1)\ldots X1\_1(-p3)\ \ Z1 \tag{7.22c}$$

$$X1\_2\ \ C\ \ \boldsymbol{X1\_2(-1)*X2\_2}\ \ \boldsymbol{X1\_2(-1)*Z1}\ \ X1\_2(-1)\ldots X1\_2(-p4)\ \ X2\_2\ \ Z1 \tag{7.22d}$$

$$Z1\ \ C\ \ Z1(-1)\ldots Z1(-p5) \tag{7.22e}$$

Note that due to having the time t as an independent variable of the model (7.22b), then the model of *Y1_1* in (7.22a) would be a TRE-Model, specifically in computing its forecast values beyond the sample period.

Example 7.12 An Application of a Set of Models (7.22)

Figure 7.26 presents the summary of OLS estimates of a set of the simplest models in (7.22), which are LV(1) models for $p1 = p2 = p3 = p4 = p5 = 1$. Based on these results, the following findings and notes are presented.

1) The models of *Y1_1* and *Y1_2* are nonhierarchical reduced models, since they do not have all of the main variables; for instance, the model of *Y1_1* does not have *X1_1* as its independent variable. Their estimates are obtained by using the trial-and-error method in order to have acceptable two-way interaction models.

DV: Z1	C	Z1(-1)					Adj. R^2	F-Stat	Prob.	DW-Stat
Coef.	192.1386	1.019567					0.9974	15538.91	0.0000	1.2200
t-Stat.	2.8182	124.6552								
Prob.	0.0075	0								
DV: X1_2	C	X1_2(-1)*Z1	X1_2(-1)	Z1			Adj. R^2	F-Stat	Prob.	DW-Stat
Coef.	927.193	-1.76E-05	0.645203	0.526366			0.9985	9389.90	0.0000	1.4608
t-Stat.	5.671022	-6.719562	11.52917	7.275871						
Prob.	0.0000	0.0000	0.0000	0.0000						
DV: X1_1	C	X1_1(-1)*Z1	X1_1(-1)	Z1			Adj. R^2	F-Stat	Prob.	DW-Stat
Coef.	-339.005	-1.11E-05	1.026734	0.181997			0.9914	2293.85	0.0000	0.3803
t-Stat.	-0.54977	-2.363107	9.801401	2.441101						
Prob.	0.5857	0.0233	0.0000	0.0194						
DV: Y1_2	C	Y1_1(-1)*Y1_2(-1)	X1_2*Y1_2(-1)	Y1_1(-1)	X1_2	T	Adj. R^2	F-Stat	Prob.	DW-Stat
Coef.	6516.839	9.88E-05	-8.17E-05	-1.12669	1.303392	29.2596	0.9971	2799.88	0.0000	1.1839
t-Stat.	8.606303	6.685966	-5.461848	-7.43011	10.25272	1.235493				
Prob.	0.0000	0.0000	0.0000	0.0000	0.0000	0.2246				
DV: Y1_1	C	Y1_1(-1)*Y1_2(-1)	Y1_1(-1)*X1_1	Y1_2(-1)	Y1_1(-1)	T	Adj. R^2	F-Stat	Prob.	DW-Stat
Coef.	1805.658	-1.90E-05	1.00E-05	0.162558	0.731466	62.02404	0.9884	700.15	0.0000	1.5351
t-Stat.	1.442769	-1.760176	1.481967	0.674686	4.435078	1.189666				
Prob.	0.1577	0.0869	0.1471	0.5042	0.0001	0.242				

Figure 7.26 The summary of the OLS estimates of the set of models (7.22).

	A	B	C	D	E	F	G
1	Year	T	Z1	X1_2	X1_1	Y1_2	Y1_1
2	1990	40	14597	14141	17308	14840	17990
3	1991	41	15258	14141	16444	15263	17632
4	1992	42	15496	14237	16413	15357	17981
5		43	**15893.28**	**14499.22**	**16517.30**	**15506.91**	**17866.10**
6		44	16333.90	14714.69	16605.66	15232.41	17734.26
7		45	16795.87	14915.43	16679.55	15066.34	17685.52

Figure 7.27 The forecast values beyond the sample period based on the models (7.20).

2) Figure 7.27 presents the forecast values beyond the sample period of each of their dependent variables. Based on these forecast values, the findings and notes presented are as follows:

   2.1 The forecast value of *Y1_1* for $t = T+1 = 43$ is smaller than its observed score/value at $t = T = 42$, but the forecast value of *Y1_2* for $t = T+1 = 43$ is greater than its observed score at $t = T = 42$. We never know whether they are good predictors. Whenever a forecast value is considered to be an inappropriate prediction, then another model should be explored.

   2.2 The graphs of *Y1_1* and its forecast values *Y1_1F* in Figure 7.28 show that the forecast values decrease starting form $t = 39$ or @*Year*=1989 compared to the observed values. I would say that alternative models should be explored in order to have a better prediction at least for the time $t = T+1$. Do these as exercises.

   2.3 On the other hand, there is no problem with the forecast values of *Y1_2*, even though its forecast values beyond the sample period are decreasing from the time $t = T+1$, since after having the true observed score at $t = T+1$, the forecast model should be modified. In other words, a new regression function should developed.

7.8.1.2 Extension of the Models with Polynomial Trends

Referring to the graphs of *Y1_1* and its forecast values *Y1_1F* in Figure 7.28, I would say that the model of *Y1_1* with a polynomial trend is a better forecasting model. Similarly, I also apply the model of *Y1_2* with the following general ESs. In addition, note that (7.23c)–(7.23e) are the same as the previous models.

$$Y1\_1\ C\ \mathbf{Y1\_1(-1)*Y1\_2(-1)}\ \mathbf{Y1\_1(-1)*X1\_1}$$
$$Y1\_1(-1)\ldots Y1\_1(-p1))\ Y1\_2(-1)\ X1\_1\ t\ t\hat{\ }2\ldots t\hat{\ }k \tag{7.23a}$$

$$Y1\_2\ C\ \mathbf{Y1\_1(-1)*Y1\_2(-1)}\ \mathbf{X1\_2*Y1\_2(-1)}\ \mathbf{X2\_2*Y1\_2(-1)}$$
$$Y1\_2(-1)\ Y1\_2(-1)\ldots Y1\_2(-p2)\ Y1\_1(-1)\ X1\_2\ X2\_2\ t\ t\hat{\ }2\ldots t\hat{\ }m \tag{7.23b}$$

Example 7.13 An Application of Specific Models in (7.23a) and (7.23b)

As the extension of the regressions of *Y1_1* and *Y1_2* with linear trends in Figure 7.26, Figure 7.29 presents the statistical results based on specific models in (7.23a) and (7.23b).

Based on these results, the following findings and notes are presented.

1) The regression functions of *Y1_1* and *Y1_2* are obtained using the trial-and-error-method in order to have acceptable regressions with significant two-way interaction independent variables.

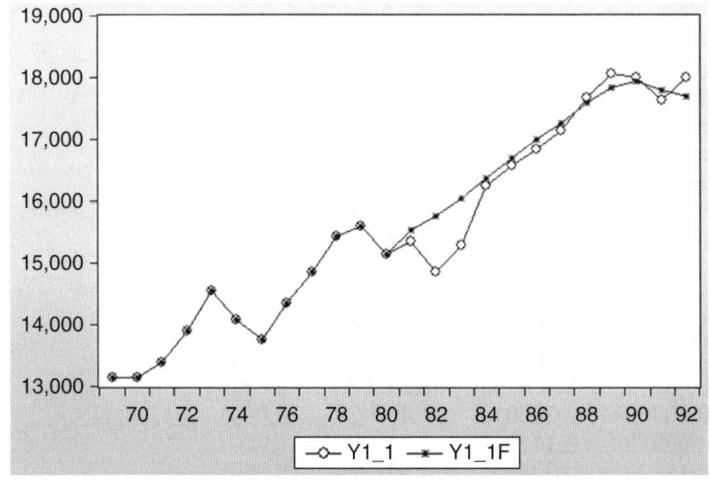

RSME=366.2547 & TIC=0.0108

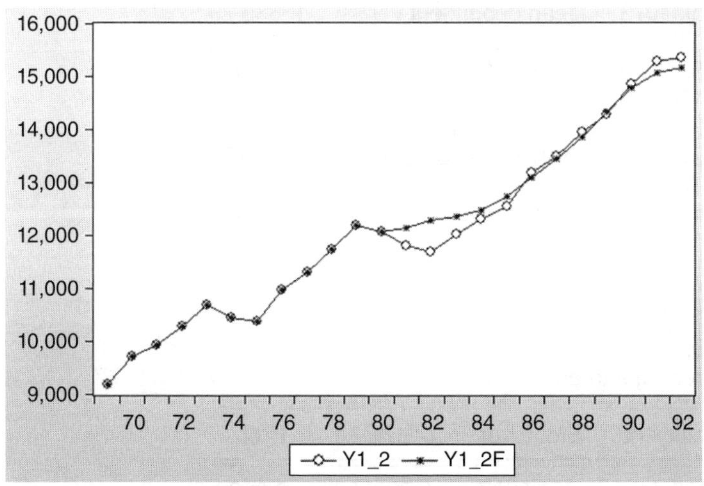

RSME=243.8000 & TIC=0.0090

Figure 7.28 The graphs of *Y1_1* and *Y1* with their forecast evaluations for the forecast sample {@*Year>1980*}.

2) These models are better forecast models since they have smaller RMSE and TIC, as presented in Figure 7.30, compared to the forecast evaluation of the models (7.22a and 7.22b) in Figure 7.28.
3) The forecast values of *Y1_2* and *Y1_1* beyond the sample period are computed using the following regression functions, respectively.

i) $Y1\_2 = 6176.41113612 + 0.000124019929787*Y1\_1(-1)*Y1\_2(-1)$
$-0.000129115181099*X1\_2*Y1\_2(-1) - 1.39578301085*Y1\_1(-1)$
$+1.94705845889*X1\_2 - 36.8038069073*T + 0.0296158296154*T\hat{}3$

$Y1\_1 = -69.8411770869 - 8.83118461716e-05^{*}Y1\_1(-1)*Y1\_2(-1)$

ii) $+3.45908694514e-05*Y1\_1(-1)*X1\_1 + 0.808812473666*Y1\_1(-1)$
$+0.931645698749*Y1\_2(-1) - 336.594884167*T + 40.0530788954*T\hat{}2$
$-1.50891333054*T\hat{}3 + 0.0188087789767*T\hat{}4$

Dependent Variable: Y1_1
Method: Least Squares
Date: 08/30/17 Time: 21:29
Sample (adjusted): 1951 1992
Included observations: 42 after adjustments

Variable	Coefficient	Std. Error	t-Statistic	Prob.
C	-69.84118	4352.861	-0.016045	0.9873
Y1_1(-1)*Y1_2(-1)	-8.83E-05	4.18E-05	-2.113611	0.0422
Y1_1(-1)*X1_1	3.46E-05	7.70E-06	4.492620	0.0001
Y1_1(-1)	0.808812	0.389456	2.076773	0.0457
Y1_2(-1)	0.931646	0.687022	1.356064	0.1843
T	-336.5949	149.8952	-2.245535	0.0316
T^2	40.05308	9.087610	4.407438	0.0001
T^3	-1.508913	0.341943	-4.412760	0.0001
T^4	0.018809	0.004091	4.597899	0.0001

R-squared	0.994073	Mean dependent var	13335.29
Adjusted R-squared	0.992636	S.D. dependent var	2944.502
S.E. of regression	252.6724	Akaike info criterion	14.08947
Sum squared resid	2106831.	Schwarz criterion	14.46183
Log likelihood	-286.8790	Hannan-Quinn criter.	14.22596
F-statistic	691.8631	Durbin-Watson stat	1.505778
Prob(F-statistic)	0.000000		

Dependent Variable: Y1_2
Method: Least Squares
Date: 09/01/17 Time: 11:07
Sample (adjusted): 1951 1992
Included observations: 42 after adjustments

Variable	Coefficient	Std. Error	t-Statistic	Prob.
C	6176.411	751.0627	8.223562	0.0000
Y1_1(-1)*Y1_2(-1)	0.000124	1.93E-05	6.413544	0.0000
X1_2*Y1_2(-1)	-0.000129	2.85E-05	-4.530585	0.0001
Y1_1(-1)	-1.395783	0.201999	-6.909850	0.0000
X1_2	1.947058	0.355223	5.481231	0.0000
T	-36.80381	41.13814	-0.894639	0.3771
T^3	0.029616	0.015340	1.930573	0.0617

R-squared	0.997682	Mean dependent var	9583.595
Adjusted R-squared	0.997285	S.D. dependent var	3335.595
S.E. of regression	173.8186	Akaike info criterion	13.30491
Sum squared resid	1057452.	Schwarz criterion	13.59452
Log likelihood	-272.4032	Hannan-Quinn criter.	13.41107
F-statistic	2510.608	Durbin-Watson stat	1.154915
Prob(F-statistic)	0.000000		

Figure 7.29 OLS estimates of specific models in (7.23a) and (7.23b).

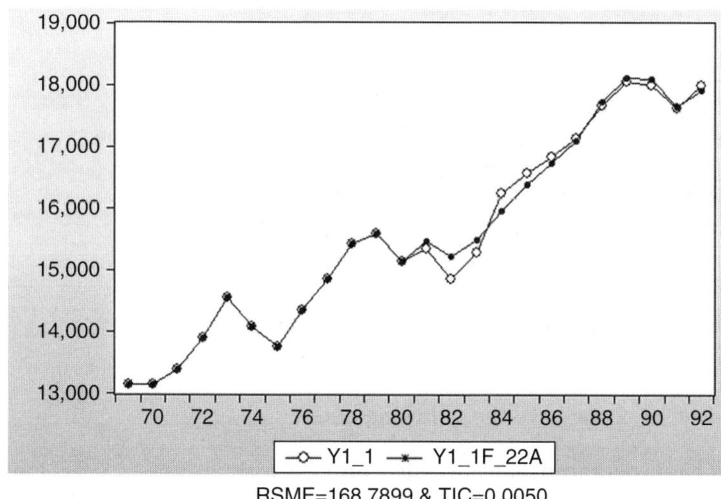

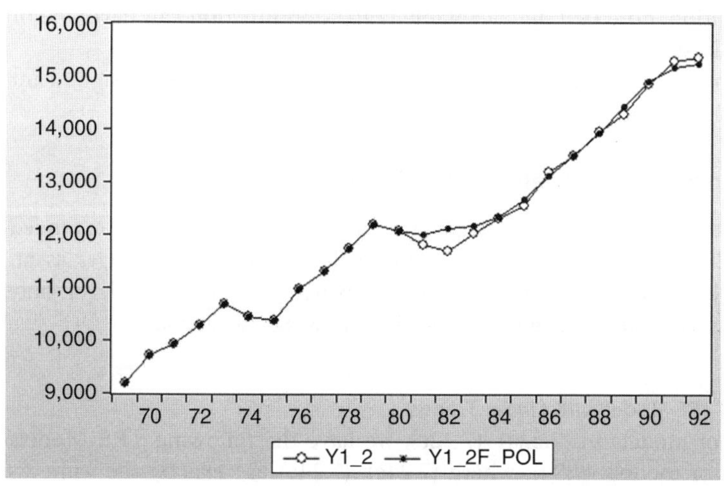

Figure 7.30 The graphs of *Y1_1* and *Y1* with their forecast evaluations for the forecast sample {@*Year*>*1980*}.

4) Specific to the regression of *Y1_1*, we have the following notes.

4.1 At the 1% level, *Y1_2(−1)* has a positive significant adjusted effect on *Y1_1* with a *p*-value = 0.1843/2 = 0.09215.

4.2 Each of the other independent variables has either a positive or negative significant adjusted effect on *Y1_1* at the 1% or 5% levels. So, in the statistical sense, the regression of *Y1_1* is an acceptable forecast model with a fourth degree polynomial trend.

4.3 Its forecast values beyond the sample period are presented in Figure 7.31, and show a value of 18,587.05 at the time $t = T+1 = 43$, which is greater than the observed score of 17,981 of *Y1_1* at the time $t = T = 42$, and the forecast values of *Y1_1* based on the model (7.22a). Then the value decreases to 18,150.97 at $t = T+2 = 44$, and to 17,913.93 at the time $t = T+2 = 45$.

	A	B	C	D	E	F	G
1	Year	T	Z1	X1_2	X1_1	Y1_2	Y1_1
2	1990	40	14597	14141	17308	14840	17990
3	1991	41	15258	14141	16444	15263	17632
4	1992	42	15496	14237	16413	15357	17981
5		43	15893.28	14499.22	16517.30	15578.54	18587.05
6		44	16333.90	14714.69	16605.66	15663.41	18150.97
7		45	16795.87	14915.43	16679.55	15440.63	17913.93

Figure 7.31 The forecast values beyond the sample period based on the models (7.23).

5) Specific for the regression of *Y1_2*, we have the following notes.

5.1 Even though the time *t* has insignificant adjusted effect on *Y1_2* with a p-value = 0.3771, it still can be used as an independent variable because *t^3* has significant adjusted effect at the 10% level. In addition, note that an acceptable regression function can have one or more insignificant adjusted effects.

5.2 Each of the other independent variables has either a positive or negative significant adjusted effect on *Y1_1* at the 1% level of significance.

7.8.2 Two-Way Interaction with Time-Related Effects Models

All sets of models in (7.1) up to (7.21), as well as all models with alternative trends, can be extended to two-way Time-Related Effects (TRE) Models by inserting the interaction *DV(−1)*t* as an additional independent variable of the model of selected *DV* (dependent variable). However, only two selected cases are presented as illustrative models in the following subsections.

7.8.2.1 Two-Way Interaction TRE-Model with Linear Trends

As an extension of the set of models in (7.22a)–(7.22e), we have the following TRE-Models for the first two in the set of models (7.24), where (7.24c)–(7.24e) are exactly the same as (7.22c)–(7.22e) and (7.6c)–(7.6e).

$$Y1\_1\ C\ \boldsymbol{Y1\_1(-1)*t}\ \boldsymbol{Y1\_1(-1)*Y1\_2(-1)}\ \boldsymbol{Y1\_1(-1)*X1\_1}$$
$$Y1\_1(-1)\ldots Y1\_1(-p1))\ Y1\_2(-1)\ X1\_1\ t \tag{7.24a}$$

$$Y1\_2\ C\ \boldsymbol{Y1\_2(-1)*t}\ \boldsymbol{Y1\_1(-1)*Y1\_2(-1)}\ \boldsymbol{X1\_2*Y1\_2(-1)}$$
$$Y1\_2(-1)\ldots Y1\_2(-p2)\ Y1\_1(-1)\ X1\_2\ t \tag{7.24b}$$

Example 7.14 An Application of Specific Models (7.24a) and (7.24b)

Figure 7.32 presents the OLS regressions of three alternative specific models in (7.24a), which shows that the Reduced Model 1 (*RM-1*) has the smallest RMSE and TIC. So it is the best forecast model of *Y1_1* among the three, even though *X1_1* has an insignificant adjusted effect with a large p-value = 0.5698. At the 15% level (applied in Lapin 1973), *Y1_1(−1)*X1_1* has a positive significant adjusted effect on *Y1_1* with a p-value = 0.2268/2 = 0.1134 < 0.15. In addition, note that a good fit model can have one or more insignificant independent variables.

Figure 7.33 presents the OLS regressions of four alternative specific models in (7.24b), of which two are LV(1) TRE_Models, and the last two are LV(2) TRE_Models. Specific for the fourth model, that is, the LV(2) Reduced Model, the following findings and notes are presented.

DV: Y1_1	Full-Model			Reduced Model 1			Reduced Model 2		
Variable	Coef.	t-Stat.	Prob.	Coef.	t-Stat.	Prob.	Coef.	t-Stat.	Prob.
C	2207.31	0.3151	0.7546	11343.83	6.3468	0.0000	10690.83	7.8315	0.0000
Y1_1(-1)*T	-0.003626	-0.0982	0.9223	-0.05037	-3.9314	0.0004	-0.0471	-4.1421	0.0002
Y1_1(-1)*Y1_2(-1)	4.36E-06	0.0397	0.9685	0.000142	3.5029	0.0013	0.000142	3.5442	0.0011
Y1_1(-1)*X1_1	-1.68E-06	-0.0541	0.9572	2.77E-05	1.2303	**0.2268**	1.53E-05	2.3400	0.0249
Y1_1(-1)	0.668475	1.3478	0.1866						
Y1_2(-1)	-0.151382	-0.0922	0.9271	-2.1600	-3.1039	0.0038	-2.12251	-3.0925	0.0038
X1_1	0.188648	0.4244	0.6739	-0.1973	-0.5737	**0.5698**			
T	115.964	0.2126	0.8329	800.818	3.9934	0.0003	752.4349	4.1748	0.0002
R-squared	0.9900			0.9894			0.9893		
Adj. R-squared	0.9879			0.9876			0.9879		
F-statistic	479.64			546.53			668.22		
Prob(F-statistic)	0.0000			0.0000			0.0000		
DW-Stat	1.4416			1.2658			1.1627		
	RMSE=384.49 & TIC=0.0114			RMSE=154.86 & TIC=0.0105			RMSE=384.58 & TIC=0.0108		

Figure 7.32 The statistical results summary of the model (7.24a) and two of its reduced models.

DV: Y1_2	LV(1) Full Model			LV(1) Reduced Model			LV(2) Full Model			LV(2) Reduced Model		
Variable	Coef.	t-Stat.	Prob.	Coef.	t-Stat.	Prob.	Coef.	t-Stat.	Prob.	Coef.	t-Stat.	Prob.
C	9162.93	4.178	0.0002	9311.89	4.159	0.0002	8517.12	6.024	0.0000	8701.12	11.403	0.0000
Y1_2(-1)*T	-0.013568	-1.263	0.2154	-0.02711	-4.024	0.0003	0.0050	2.734	0.0100	0.0049	2.805	0.0083
Y1_1(-1)*Y1_2(-1)	0.00011	3.253	0.0026	7.09E-05	2.957	0.0055	0.0001	5.020	0.0000	0.0001	9.830	0.0000
X1_2*Y1_2(-1)	-3.67E-05	-0.650	**0.5200**	5.07E-05	3.662	0.0008	-0.0001	-5.249	0.0000	-0.0001	-7.751	0.0000
Y1_2(-1)	-0.361125	-1.242	0.2229	-0.36666	-1.234	0.2255	0.0362	0.156	**0.8773**			
Y1_2(-2)							-0.4025	-3.669	0.0009	-0.3927	-4.426	0.0001
Y1_1(-1)	-1.231387	-3.583	0.0010	-0.83367	-3.451	0.0015	-1.5826	-5.880	0.0000	-1.6168	-10.566	0.0000
X1_2	0.92651	1.595	0.1200				2.0228	7.030	0.0000	2.0585	11.990	0.0000
T	174.520	1.509	0.1404	318.993	4.345	0.0001						
R-squared	0.99758			0.99739			0.99858			0.99857		
Adj. R-squared	0.99708			0.99695			0.99827			0.99832		
F-statistic	1998.19			2232.35			3304.09			3968.66		
Prob(F-statistic)	0.00000			0.00000			0.00000			0.00000		
DW-Statistic	1.14767			1.04296			1.80716			1.78749		
	RMSE=219.79 & TIC=0.0082			RMSE=251.55 & TIC=0.0093			RMSE=159.85 & TIC=0.0059			***RMSE=138.35 & TIC=0.0051***		

Figure 7.33 OLS regressions of four alternative specific models of *Y1_2* in (7.24b).

1) Each of the independent variables has either a positive or negative significant adjusted effect on *Y1_2* with a *p*-value = Prob. /2, at the 1% level of significance.
2) Since it has the smallest RMSE and TIC, then it can be considered the best forecast model among the four.
3) As special analyses, I present the following findings.
   3.1 The three independent variables *Y1_2(−1)∗t, Y1_1(−1)∗Y1_2(−1),* and *X1_2∗Y1_2(−1)* indicate that the effect of *Y1_2(−1)* on *Y1_2* depends on an additive function of the variables *t, Y1_1(−1)* and *X1_2.*
   3.2 The independent variable *Y1_2(−1)∗t* with a positive coefficient of 2.85 indicates that the effect of *Y1_2(−1)* on *Y1_2* increases as the time *t* increases.
   3.3 The independent variables {−7.752∗*Y1_2(−1)* + 11.990}∗*X1_2* indicate that the effect of the exogenous variable *X1_2* on *Y1_2* decreases as *Y1_2(−1)* increases. And *X1_2* has a zero effect on *Y1_2(−1)* whenever {−7.752∗*Y1_2(−1)* + 11.990} = 0.

4) Since the Reduced Model 3 (*RM-3*) of *Y1_1* and the LV(2) Reduced Model of *Y1_2* are the best forecast models, then the following regressions should be used to compute the forecast values of *Y1_1* and *Y1_2* together with the regressions of *X1_1*, *X1_2*, and *Z1*, Do this as an exercise and refer to the results in Figure 7.31.

   i) $Y1\_1 = 11343.8256568 - 0.0503678936248 * Y1\_1(-1) * T$
   $+ 0.000142016794778 * Y1\_1(-1) * Y1\_2(-1) + 2.76662318644e-05 * Y1\_1(-1) * X1\_1$
   $- 2.16004225683 * Y1\_2(-1) - 0.197301571501 * X1\_1 + 800.818015769 * T$

   $Y1\_2 = 8701.11860891 + 0.00490085865762 * Y1\_2(-1) * T$

   ii) $+ 0.000134926839444 * Y1\_1(-1) * Y1\_2(-1) - 0.000128448958291 * X1\_2 * Y1\_2(-1)$
   $- 0.392702045636 * Y1\_2(-2) - 1.6168350495 * Y1\_1(-1) + 2.05845230469 * X1\_2$

7.8.2.2 Two-Way Interaction TRE-Model with Polynomial Trends

As an extension of the set of models in (7.23a) and (7.23b), we have the following TRE-Models as the first two in the set (7.25), where (7.25c)–(7.25e) are exactly the same as (7.23c)–(7.23e), and (7.6c)–(7.6e).

$$Y1\_1\ \ C\ \ \mathbf{Y1\_1(-1)*t}\ \ \mathbf{Y1\_1(-1)*Y1\_2(-1)}\ \ \mathbf{Y1\_1(-1)*X1\_1}$$
$$Y1\_1(-1) \ldots Y1\_1(-p1))\ \ Y1\_2(-1)\ \ X1\_1\ \ t\ \ t\hat{}2 \ldots t\hat{}k \tag{7.25a}$$
$$Y1\_2\ \ C\ \ \mathbf{Y1\_2(-1)*t}\ \ \mathbf{Y1\_1(-1)*Y1\_2(-1)}\ \ \mathbf{X1\_2^{*}Y1\_2(-1)}$$
$$Y1\_2(-1) \ldots Y1\_2(-p2)\ \ Y1\_1(-1)X1\_2\ \ X2\_2\ \ t\ \ t\hat{}2 \ldots t\hat{}m \tag{7.25b}$$

Example 7.15 An Application of Specific Models (7.25a) and (7.25b)

Figure 7.34 presents the OLS regressions of four alternative specific models *Y1_1* in (7.25a), which shows that the *RM-3* has the smallest RMSE and TIC. So it is the best forecasting model of *Y1_1* among the four, since it has the smallest RMSE and TIC, even though two of its independent variables have large *p*-values. The steps of analysis are as follows:

1) Referring to alternative two-way interaction TRE-models of *Y1_1*, then the interaction *Y1_1 (−1)*t* should always be one of their independent variables. For this reason, at the first stage I try to estimate the *RM-1*, which shows that the interaction *Y1_1(−1)*t* has a large *p*-value = 0.836. For this reason, we have to explore a reduced model.
2) Even though the time variable *t* has significant effect with a small *p*-value = 0.037, at the second stage of analysis I try to delete the time *t* in order to have small *p*-value for the interaction *Y1_1(−1)*t*. It happens that at the 10% level, the interaction has a negative significant adjusted effect on *Y1_1* with a *p*-value = 0.142/2 = 0.071 < 0.10. So the *RM-2* is an acceptable two-way interaction TRE-model.
3) For the third stage of analysis, we have the main variables *Y1_1(−1)*, *Y1_2(−1)*, and *X1_1* as the search regressors. By using the STEPLS-Combinatorial selection method, *RM-3* is obtained, which shows that the interaction *Y1_1(−1)*t* has a negative significant effect on *Y1_1* with a *p*-value = 0.148/2 = 0.074. So this is also an acceptable two-way interaction TRE-model.
4) However, since *Y1_1(−1)*X1_1* has a large *p*-value = 0.452, at the fourth stage of analysis I try to delete either one of *Y1_1(−1)* and *X1_1*. *RM-4* is found to be a better model by deleting the *Y1_1(−1)*.

DV: Y1_1	RM-1			RM-2			RM-3			RM-4		
Variable	Coef	t-Stat.	Prob.	Coef	t-Stat.	Prob.	Coef	t-Stat.	Prob.	Coef	t-Stat.	Prob.
C	8216.88	12.400	0.000	7025.46	18.082	0.000	873.505	0.279	0.782	8858.98	5.169	0.000
Y1_1(-1)*T	0.00235	0.208	0.836	-0.01354	-1.504	0.142	-0.01685	-1.487	0.146	-0.02282	-1.851	0.073
Y1_1(-1)*Y1_2(-1)	-2.68E-05	-2.172	0.037	-1.26E-05	-1.146	0.260	-2.54E-05	-2.379	0.023	-1.45E-05	-1.312	0.199
Y1_1(-1)*X1_1	3.64E-05	4.189	0.000	4.40E-05	5.263	0.000	2.20E-05	0.762	**0.452**	7.10E-05	2.735	0.010
T	-219.476	-2.166	0.037									
T^2	45.2218	5.132	0.000	33.9191	4.543	0.000	31.0245	2.862	0.007	42.9806	3.867	0.001
T^3	-1.80681	-5.548	0.000	-1.3816	-5.057	0.000	-1.24183	-3.727	0.001	-1.6083	-4.705	0.000
T^4	0.02181	5.468	0.000	0.0174	4.828	0.000	0.01629	4.276	0.000	0.0194	4.800	0.000
Y1_1(-1)							0.71681	2.936	0.006			
X1_1							0.24324	0.524	**0.604**	-0.4788	-1.098	0.280
R-squared	0.99321			0.99228			0.99409			0.99254		
Adj. R-squared	0.99182			0.99095			0.99265			0.99101		
F-statistic	710.889			749.537			693.490			646.414		
Prob(F-statistic)	0.00000			0.00000			0.00000			0.00000		
DW-Statistic	1.09986			0.91419			1.43227			1.06903		
	RMSE=213.99 & TIC=0.0064			RMSE=197.71 & TIC=0.0059			***RMSE=163.66 & TIC=0.0049***			RMSE=271.62 & TIC=0.0100		

Figure 7.34 OLS regressions of four alternative specific models of *Y1_1* in (7.25a).

DV: Y1_2	RM-1			RM-2			RM-3			RM-4		
Variable	Coef.	t-Stat.	Prob.	Coef.	t-Stat.	Prob.	Coef.	t-Stat.	Prob.	Coef.	t-Stat.	Prob.
C	2589.72	12.033	0.000	4358.06	3.922	0.000	4310.32	3.858	0.001	3003.49	3.380	0.002
Y1_2(-1)*T	-0.016836	-2.419	0.021	-0.019987	-1.631	0.112	-0.028415	-2.827	0.008	-0.0219	-1.727	0.093
Y1_1(-1)*Y1_2(-1)	-1.05E-05	-1.282	0.208	3.89E-05	1.423	0.164	1.37E-05	0.788	0.436	-1.11E-05	-1.326	0.194
X1_2*Y1_2(-1)	7.52E-05	5.831	0.000	2.14E-05	0.448	**0.657**	7.63E-05	6.029	0.000	8.98E-05	2.719	0.010
T	350.088	12.711	0.000	251.1787	3.050	0.005	343.612	12.587	0.000	372.559	6.842	0.000
T^2	-8.425246	-6.545	0.000	-4.968849	-2.076	0.046	-5.226733	-2.180	0.036	-7.70011	-3.863	0.001
T^3	0.13875	6.489	0.000	0.11174	4.396	0.000	0.11673	4.628	0.000	0.13459	5.779	0.000
Y1_1(-1)				-0.535854	-1.912	0.065	-0.274648	-1.568	0.126			
X1_2				0.568101	1.189	0.243				-0.14797	-0.480	0.634
R-squared	0.99827			0.99845			0.99839			0.99828		
Adj. R-squared	0.99797			0.99808			0.99805			0.99793		
F-statistic	3364.55			2661.09			3004.53			2820.54		
Prob(F-statistic)	0.0000			0.0000			0.0000			0.0000		
DW-Statistic	1.2433			1.2377			1.2679			1.2656		
RMSE/TIC	161.98/0.0060			***150.38/0.0056***			164.48/0.0061			165.98/0.0062		

Figure 7.35 OLS regressions of four specific models of *Y1_2* in (7.25b).

5) By looking at the forecast evaluation of the four RMs, *RM-3* has the smallest RMSE and TIC. So, in the statistical sense, it should be used to compute the forecast values of *Y1_1* beyond the sample period, even though *Y1_1(−1)∗X1_1* and *X1_1* have large *p*-values.
6) In addition, Figure 7.35 also presents the OLS regressions of four alternative specific models of *Y1_2* in (7.25a), which are obtained using a similar method to the reduced models of *Y1_1*. The results show that *RM-2* has the smallest RMSE and TIC. So it can be considered the best forecast model of *Y1_2* among the four.

7.8.3 Three-Way Interaction Time-Related Effects Models

All sets of models in (7.1) up to (7.21), as well as all those with alternative trends, can be extended to the three-way interaction TRE Models by inserting the interaction *DV(−1)∗IV∗t* as an additional independent variable of the model of *DV* (dependent/endogenous variable) and *IV* is another independent variable in the model. However, only two selected cases are presented as the illustrative models in the following subsections.

7.8.3.1 Three-Way Interaction TRE-Model with Linear Trends

As an extension of the set of models in (7.10) we have the following Three-Way Interaction TRE-Models as the first three models in the set (7.26) where (7.26d)–(7.26f) are exactly the same as (7.10d)–(7.10f).

$$\begin{aligned} &Y1\_1\ \ C\ \ \mathbf{Y1\_1(-1)*Y1\_2^*t\ \ Y1\_1(-1)*Y1\_3^*t\ \ Y1\_1(-1)*X1\_1^*t\ \ Y1\_1(-1)*t} \\ &\quad Y1\_1(-1)*Y1\_2\ \ Y1\_1(-1)*Y1\_3\ \ Y1\_1(-1)*X1\_1 \\ &\quad Y1\_1(-1)\ldots Y1\_1(-p1))\ \ Y1\_2\ \ Y1\_3\ \ X1\_1\ \ t \end{aligned} \tag{7.26a}$$

$$\begin{aligned} &Y1\_2\ \ C\ \ \mathbf{Y1\_2(-1)*Y1\_3^*t\ \ Y1\_2(-1)*X1\_2^*t\ \ Y1\_2(-1)*t} \\ &\quad Y1\_2(-1)*Y1\_3\ \ Y1\_2(-1)*X1\_2\ \ Y1\_2(-1)\ldots Y1\_2(-p2)\ \ Y1\_3\ \ X1\_2\ \ t \end{aligned} \tag{7.26b}$$

$$\begin{aligned} &Y1\_3\ \ C\ \ \mathbf{Y1\_3(-1)*X1\_3*t\ \ Y1\_3(-1)*t} \\ &\quad Y1\_3(-1)*X1\_3Y1\_3(-1)\ldots Y1\_3(-p3)\ \ X1\_3\ \ t \end{aligned} \tag{7.26c}$$

$$X1\_1\ \ C\ \ \mathbf{X1\_1(-1)*Z1}\ \ X1\_1(-1)\ldots X1\_1(-p3)\ \ Z1 \tag{7.26d}$$

$$X1\_2\ \ C\ \ \mathbf{X1\_2(-1)*Z1}\ \ X1\_2(-1)\ldots X1\_2(-p4)\ \ Z1 \tag{7.26e}$$

$$Z1\ \ C\ \ Z1(-1)\ldots Z1(-p5) \tag{7.26f}$$

Note the model of *Y1_1* in (7.26a) is developed from the model (7.10a) by inserting four additional interaction independent variables, which are presented in bold. This model shows that the effect of *Y1_1(−1)* on *Y1_1* depends on *Y1_3∗t, X1_2∗t, X1_1∗t, t, Y1_2, Y1_3*, and *X1−1.*

The model of *Y1_2* in (7.26b) is developed from (7.10b) by inserting three additional interaction independent variables, which are presented in bold. This model shows that the effect of *Y1_2(−1)* on *Y1_2* depends on *Y1_3∗t, X1_2∗t, t, Y1_3*, and *X1_2.*

Finally, the model of *Y1_3* in (7.26c) is developed from (7.10c) by inserting two additional interaction independent variables, which are presented in bold. This model shows that the effect of *Y1_3(−1)* on *Y1_3* depends on *X1_3∗t* and *t.*

Example 7.16 Application of Specific Models of *Y1_1* in (7.26a)

Figure 7.36 presents the OLS regressions of four alternative specific models of *Y1_1* in (7.26a), which shows that the *RM-4* has the smallest RMSE and TIC. So it is the best forecast model among the four, even though *Y1_1(−1)∗t* has a large p-value = 0.448. The steps of analysis are as follows:

1) As the first stage of analysis, the *RM-1* is an acceptable TRE-model because each of the time interactions has either a positive or negative significant adjusted effect on *Y1_1* at the 1 or 10% levels of significance. For instance, the interactions *Y1_1(−1)∗Y1_2∗t* have a negative significant adjusted effect with a p-value = 0.015/2 = 0.0075, and *Y1_1(−1)∗Y1_3∗t* has a positive significant adjusted effect with a p-value = 0.167/2 = 0.0835 < 0.10.
2) At the second stage, the interactions *Y1_1(−1)∗Y1_2, Y1_1(−1)∗Y1_3*, and *Y1_1(−1)∗X1_1* are used as the search regressors to be selected for additional independent variables using the combinatorial selection method. *RM-2* is obtained, which shows that at the 10% level, *Y1_1(−1)∗Y1_2∗t* has a negative significant adjusted effect with a p-value = 0.140/2 = 0.071.
3) At the third stage, it is found that the four main variables *X1_1, Y1_1(−1), Y1_2*, and *t* can be inserted as additional independent variables of *RM-2* by using the STEPLS selection method.

DV: Y1_1	RM-1			RM-2			RM-3			RM-4		
Variable	Coef.	t-Stat.	Prob.	Coef.	t-Stat.	Prob.	Coef.	t-Stat.	Prob.	Coef.	t-Stat.	Prob.
C	8511.888	47.238	0.000	6233.15	16.022	0.000	-12636.5	-2.046	0.049	-14934.9	-2.638	0.014
Y1_1(-1)*Y1_2*T	-9.08E-07	-2.556	0.015	-3.98E-07	-1.511	0.140	-1.30E-06	-1.545	0.132	-1.38E-06	-1.823	0.079
Y1_1(-1)*Y1_3*T	7.84E-07	1.409	0.167	1.62E-06	3.907	0.000	1.54E-06	2.455	0.020	1.37E-06	2.240	0.033
Y1_1(-1)*X1_1*T	-4.19E-07	-1.547	0.130	-1.69E-06	-6.026	0.000	8.87E-07	0.728	0.472	1.38E-06	1.179	0.248
Y1_1(-1)*T	0.02353	12.270	0.000	0.01025	4.048	0.000	0.01505	0.609	0.547	0.01849	0.770	0.448
Y1_1(-1)*X1_1				4.64E-05	6.197	0.000	-0.00018	-2.265	0.030	-0.00022	-3.151	0.004
X1_1							2.37888	3.325	0.002	2.90207	4.251	0.000
Y1_1(-1)							1.55296	2.500	0.018	1.86544	3.215	0.003
Y1_2							0.72003	1.778	0.085	0.91280	2.457	0.021
T							-431.91	-1.369	0.180	-498.031	-1.706	0.099
Y1_1(-4)										-0.25128	-2.386	0.024
R-squared	0.98280			0.99168			0.99483			0.99576		
Adj. R-squared	0.98094			0.99052			0.99338			0.99424		
F-statistic	528.42360			857.821			684.076			656.841		
Prob(F-stat.)	0.00000			0.00000			0.00000			0.00000		
DW-stat.	0.98741			1.70527			1.29048			1.66534		
RMSE/TIC	506.76/0.0149			329.35/0.0098			209.07/0.0062			*151.94/0.0045*		

Figure 7.36 OLS regressions of four specific models of *Y1_1* in (7.26a).

So *RM-3* can be considered an acceptable regression, in the statistical sense. Since it has the smallest values of RMSE and TIC compared to *RM-1* and *RM-2*, then *RM-3* is the best forecast model of *Y1_1*. Based on *RM-3*, the following notes are presented.

3.1 Each of the main independent variables *X1_1*, *Y1_1(−1)*, and *Y1_2* has a positive significant adjusted effect at the 1 or 5% levels, and at the 10% level, the time t has a negative significant adjusted effect with a p-value = 0.180/2 = 0.090 < 0.10.

3.2 Because the interactions *Y1_1(−1)∗X1_1∗t* and *Y1_1(−1)∗t* have large p-values, then I have tried several alternative models besides *RM-4*, which could be considered an unexpected model with the smallest RMSE and TIC compared to *RM-1*, *RM-2*, *RM-3*, and the other alternatives. For alternative models and a comparison, Figure 7.37 presents three alternative reduced models of *RM-3* that have greater values of RMSE and TIC. Since their TICs are less than 0.01, then they can be considered perfect forecast models, referring to the classification presented in Chapter 1.

DV: Y1_1	RM-3a			RM-3b			RM-3c		
Variable	Coef.	t-Stat.	Prob.*	Coef.	t-Stat.	Prob.*	Coef.	t-Stat.	Prob.*
C	9537.942	4.7624	0	5845.491	10.5044	0.0000	5742.10	1.8253	0.0770
Y1_1(-1)*Y1_2*T	-3.36E-07	-1.2924	0.2047	7.55E-07	1.3768	0.1776	7.66E-07	1.1880	0.2433
Y1_1(-1)*Y1_3*T	2.26E-06	4.0693	0.0003	1.99E-06	4.0582	0.0003	1.97E-06	2.8117	0.0082
Y1_1(-1)*X1_1*T	-2.65E-06	-4.1669	0.0002	-2.22E-06	-4.4472	0.0001	-2.20E-06	-2.4406	0.0202
Y1_1(-1)*T	0.014725	4.0554	0.0003	-0.028664	-1.8273	0.0764	-0.029255	-1.2299	0.2274
Y1_1(-1)*X1_1	7.47E-05	4.0643	0.0003	7.42E-05	4.9087	0.0000	7.37E-05	3.3172	0.0022
Y1_1(-1)	-0.55940	-1.6806	0.1017				0.01584	0.0334	0.9735
Y1_2				-0.36271	-1.3768	0.1776	-0.364743	-1.3300	0.1926
T				399.931	2.3119	0.0270	404.218	1.8587	0.0720
R-squared	0.99230			0.99304			0.99304		
Adj. R-squared	0.99098			0.99161			0.99136		
F-statistic	751.548			693.223			588.750		
Prob(F-stat.)	0.00000			0.00000			0.00000		
DW-Stat.	1.41472			1.32492			1.32742		
RMSE/TIC	331.99/0.0099			286.64/0.0085			285.83/0.0086		

Figure 7.37 OLS regressions of three alternative reduced model of *RM-3*.

DV: Y1_2	RM-1			RM-2			RM-3			RM-4		
Variable	Coef.	t-Stat.	Prob.	Coef.	t-Stat.	Prob.	Coef.	t-Stat.	Prob.	Coef.	t-Stat.	Prob.
C	4195.19	23.065	0.000	1213.74	2.669	0.011	-4617.96	-2.181	0.036	-3080.49	-1.641	0.111
Y1_2(-1)*Y1_3*T	7.81E-07	1.321	0.194	-4.51E-06	-3.307	0.002	-4.74E-06	-2.740	0.010	-3.98E-06	-2.658	0.012
Y1_2(-1)*X1_2*T	-2.79E-06	-4.086	0.000	2.40E-06	2.030	0.050	5.62E-06	3.813	0.001	5.29E-06	4.160	0.000
Y1_2(-1)*T	0.046824	13.681	0.000	0.00882	2.258	0.030	-0.01836	-1.565	0.127	-0.0305	-2.686	0.012
Y1_2(-1)*Y1_3				0.00020	4.504	0.000	0.00016	2.650	0.012	0.00012	2.263	0.031
Y1_2(-1)*X1_2				-6.91E-05	-2.167	0.037	-0.00019	-2.604	0.014	-0.00014	-2.166	0.038
T							2.93051	0.025	0.980	147.334	1.312	0.199
Y1_2(-1)							0.83162	3.734	0.001	1.11848	5.429	0.000
X1_2							1.26150	2.652	0.012	0.93165	2.179	0.037
Y1_2(-2)										-0.41554	-3.728	0.001
R-squared	0.98420			0.99631			0.99845			0.99884		
Adj. R-squared	0.98295			0.99580			0.99808			0.99850		
F-statistic	788.79			1943.15			2659.25			2964.92		
Prob(F-stat.)	0.0000			0.0000			0.0000			0.0000		
DW-Stat.	0.5360			0.5360			1.3741			2.0457		

Figure 7.38 OLS regressions of four specific models of *Y1_2* in (7.26b).

Example 7.17 Application of Specific Models on *Y1_2* in (7.26b)
Figure 7.38 presents the OLS regressions of four alternative specific models of *Y1_2* in (7.26b), which show that the *RM-4* has the smallest RMSE and TIC. So it is the best forecast model of the four, even though $Y1\_1(-1)*t$ has a large p-value = 0.448. The results are obtained using the STEPLS-Combinatorial and the trial-and-error methods. The steps of the analysis are as follows:

1) At the first two stages of the regression analysis, I obtain the two good fit *RM-1* and *RM-2* models with only the interaction of independent variables.
2) At the third stage, we have four main variables: t, $Y1\_2(-2)$, $Y1\_3$, and $X1\_2$ as the search regressors for *RM-2*. By using the trial-and-error method, I obtain *RM-3* with each of the interaction independent variables having either a positive or negative significant adjusted effect at the 1 or 10% levels. For instance, $Y1\_2(-1)*Y1\_3*t$ has a negative significant effect with a p-value $= 0.010/2 = 0.005 < 0.01$, and $Y1\_2(-1)*t$ has a negative significant effect with a p-value $= 0.127/2 = 0.0635 < 0.10$.

   Even though the time t has a large p-value, it should be kept in the model since its interactions with the other variables have significant adjusted effects.
3) At the fourth stage, I try to insert a higher order lag-variable of *Y1_2* with the objective to have a model with a greater Durbin–Watson statistical value. Unexpectedly, the time variable t in RM-4 has a positive significant adjusted effect on *Y1_2* with a p-value $= 0.199/2 = 0.0995 < 0.10$.

Example 7.18 Application of Specific Models on *Y1_3* in (7.26c)
Figure 7.39 presents the OLS regressions of four alternative specific models *Y1_3* in (7.26c), which show that the *RM-4* has the smallest RMSE and TIC. The steps of analysis are as follows:

1) At the first stage of analysis, we have *RM-1* with the interaction independent variables only as a TRE-model, which shows that the effect of $Y1\_3(-1)$ on *Y1_3* significantly depends on $X1\_3*t$, t, and *X1_3*.
2) At the second stage of analysis, the *RM-2* shows an TRE-model with a negative insignificant linear trend based on the t-statistic of $t_0 = -1.026$ with a p-value $= 0.312/2 = 0.156$.

DV: Y1_3	RM-1			RM-2			RM-3			RM-4		
Variable	Coef.	t-Stat.	Prob.	Coef.	t-Stat.	Prob.	Coef.	t-Stat.	Prob.	Coef.	t-Stat.	Prob.
C	4956.66	21.466	0.000	5100.81	18.881	0.000	6996.12	8.991	0.000	9290.82	7.994	0.000
Y1_3(-1)*X1_3*T	-6.82E-07	-6.360	0.000	-1.27E-06	-2.172	0.036	-1.11E-06	-1.963	0.058	-6.60E-07	-1.194	0.241
Y1_3(-1)*T	0.01190	2.807	0.008	0.02734	1.749	0.089	0.02122	1.400	0.170	-0.019873	-0.923	0.363
Y1_3(-1)*X1_3	3.85E-05	2.529	0.016	3.26E-05	2.010	0.052	5.33E-05	3.072	0.004	0.00017	3.468	0.001
T				-81.5639	-1.026	0.312	-32.7165	-0.409	0.685	334.382	2.047	0.048
Y1_3(-2)							-0.41212	-2.5119	0.0168	-0.59319	-3.515	0.001
X1_3										-1.23025	-2.525	0.016
R-squared	0.99165			0.99189			0.99289			0.99401		
Adj. R-squared	0.99100			0.99101			0.99188			0.99296		
F-statistic	1505.11			1130.66			977.56			940.83		
Prob(F-stat.)	0.00000			0.00000			0.00000			0.00000		
DW-Stat.	0.97084			1.06332			1.27774			1.73429		
RMSE/TIC	450.79/0.0190			489.35/0.0209			421.51/0.0178			*364.21/0.0154*		

Figure 7.39 OLS regressions of four specific models of *Y1_3* in (7.26c).

3) At the third stage of analysis, I try to insert the lag variables of *Y1_3(−1)* and *Y1_3(−2)* as additional independent variables of *RM-2*. However, it is found that only *Y1_3(−2)* is an appropriate additional independent variable of *RM-2*. For comparison, do this as an exercise to insert *X1_3* in *RM-2* instead of *Y1_3(−2)*.
4) Unexpectedly, it is found that *RM-4* has a positive significant linear trend with a p-value = 0.048/2 = 0.024. Even though this model has two interactions, namely *Y1_3 (−1)∗X1_3∗t* and *Y1_3(−1)∗t* with large p-values of 0.248 and 0.363, respectively, this model is the best forecast model, in the statistical sense, since it has the smallest RMSE and TIC among the four.

7.8.3.2 Three-Way TRE-Model with Polynomial Trends

As an extension of the set of models in (7.26a)–(7.26c) we have the following Three-Way TRE-Models with Polynomial Trends. Note that the bolded independent variables are exactly the same as the independent variables in (7.26a)–(7.26c)

$$Y1\_1\ C\ \boldsymbol{Y1\_1(-1)*Y1\_2*t\ \ Y1\_1(-1)*Y1\_3*t\ \ Y1\_1(-1)*X1\_1*t}$$
$$\boldsymbol{Y1\_1(-1)*t}\ \ Y1\_1(-1)*Y1\_2\ \ Y1\_1(-1)*Y1\_3\ \ Y1\_1(-1)*X1\_1$$
$$t\ \ t\hat{\ }2\ldots t\hat{\ }k\ \ Y1\_1(-1)\ldots Y1\_1(-p1)\ \ Y1\_2\ \ Y1\_3\ \ X1\_1 \quad (7.27a)$$

$$Y1\_2\ C\ \boldsymbol{Y1\_2(-1)*Y1\_3*t\ \ Y1\_2(-1)*X1\_2*t\ \ Y1\_2(-1)*t}\ \ Y1\_2(-1)*Y1\_3$$
$$Y1\_2(-1)*X1\_2\ \ t\ \ t\hat{\ }2\ldots t\hat{\ }m\ \ Y1\_2(-1)\ldots Y1\_2(-p2)\ldots Y1\_3(-1)\ \ X1\_2 \quad (7.27b)$$

$$Y1\_3\ C\ \boldsymbol{Y1\_3(-1)*X1\_3*t\ \ Y1\_3(-1)*t}$$
$$Y1\_3(-1)*X1\_3\ \ t\ \ t\hat{\ }2\ldots t\hat{\ }n\ \ Y1\_3(-1)\ldots Y1\_3(-p3)\ \ X1\_3 \quad (7.27c)$$

Example 7.19 Application of Specific Models on *Y1_1* in (7.27a)

Figure 7.40 presents the OLS regressions of three alternative specific models of *Y1_1* in (7.27a), which show that the *RM-3* has the smallest RMSE and TIC. The steps of analysis are as follows:

1) The first two stage of analyses are not presented, because the results are exactly the same as those in Figure 7.36.

DV: Y1_1	RM-1			RM-2			RM_3		
Variable	Coef.	t-Stat.	Prob.	Coef.	t-Stat.	Prob.	Coef.	t-Stat.	Prob.
C	5131.44	4.5639	0.0001	5673.21	7.6255	0.0000	-2607.17	-1.1523	0.2580
Y1_1(-1)*Y1_2*T	-3.00E-07	-0.7558	0.4553	-4.23E-07	-1.2190	0.2315	1.07E-06	1.6336	0.1125
Y1_1(-1)*Y1_3*T	1.91E-06	3.4533	0.0016	1.80E-06	3.4511	0.0015	-1.84E-06	-1.5493	0.1315
Y1_1(-1)*X1_1*T	-1.95E-06	-3.2009	0.0031	-1.72E-06	-3.5360	0.0012	-1.04E-06	-1.9808	0.0565
Y1_1(-1)*T	-0.023019	-2.0598	0.0476	-0.018242	-2.1935	0.0354	-0.008739	-1.1368	0.2643
Y1_1(-1)*X1_1	7.25E-05	4.0670	0.0003	6.50E-05	4.8631	0.0000	5.64E-05	4.0493	0.0003
T	83.8318	0.6472	0.5221						
T^2	20.8969	1.9778	0.0566	25.9574	3.6858	0.0008	16.4099	2.2419	0.0323
T^3	-0.651088	-1.6477	0.1092	-0.841739	-3.2245	0.0028	-0.589015	-2.0406	0.0499
T^4	0.00827	1.7713	0.0860	0.01033	3.0455	0.0045	0.00856	2.5150	0.0173
Y1_3							1.93348	3.4009	0.0019
Y1_2							-0.523438	-1.7571	0.0888
R-squared	0.99453			0.99446			0.99623		
Adj. R-squared	0.99300			0.99312			0.99501		
F-statistic	646.764			740.599			819.128		
Prob(F-statistic)	0.00000			0.00000			0.00000		
DW-Stat.	1.311543			1.336706			1.539267		

Figure 7.40 OLS regressions of three alternative specific models *Y1_1* in (7.27a).

2) At the third stage of analysis, by using *t, t^2, t^3,* and *t^4* as additional IVs, the results are presented as *RM-1* in Figure 7.40. Since each of the time variables *t^2, t^3,* and *t^4* has either a positive or negative adjusted effect on *Y1_1,* we can keep the time *t* in the model, even though it has a large *p*-value. However, because the interaction *Y1_1(−1)*∗*Y1_2*∗*t* has a large *p*-value = 0.4553 then I delete the time *t* from *RM-1* and *RM-2* is obtained. Note that, at the 15% level, the interaction *Y1_1(−1)*∗*Y1_2)*∗*t* has a negative significant adjusted effect on *Y1_1* with a *p*-value = 0.2315/2 = 0.115 75 < 0.15, and the other two three-way interactions have a significant effect at the 1% level of significance.
3) By using *Y1_2, Y1_3,* and *X1_1* as the search regressors for *RM-2, RM-3* is obtained as the best possible model. At the 10% level, the two three-way interactions *Y1_1(−1)*∗*Y1_2)*∗*t* and *Y1_1(−1)*∗ *Y1_2)*∗*t,* respectively, have positive and negative significant adjusted effects, with *p*-values of 0.1125/2 and 0.1315/2; and the interaction *Y1_1(−1)*∗*X1_1*∗*t* has a negative significant adjusted effect with a *p*-value = 0.0565/2.

Example 7.20 Application of Specific Models on *Y1_2* in (7.27b)
Figure 7.41 presents the OLS regressions of four alternative specific models *Y1_2* in (7.27b), which shows that *RM-4* has the smallest RMSE and TIC. The steps of analysis are as follows:

1) The first two stage of analyses are not presented, because the results are exactly the same as those in Figure 7.38.
2) At the third stage of analysis, by using *t, t^2, t^3,* and *t^4* as the search regressors, an acceptable model is *RM-1* in Figure 7.41, with only the time variables *t, t^2,* and *t^3* as additional independent variables. Even though *Y1_2(−1)*∗*X1_2* has an insignificant adjusted effect, it can be kept in the model because at the 10% level the three-way interaction *Y1_2(−1)*∗ *X1_2*∗*t* has a significant positive adjusted effect with a *p*-value = 0.111/2 = 0.0555.
3) At the fourth stage, *Y1_2, Y1_3,* and *X1_2* are used as the search regressors for *RM-1,* and *RM-2* with *X1_2* and *Y1_3* as two additional independent variables for *RM-1* are obtained.

Variable	RM-1			RM-2			RM-3		
	Coef.	t-Stat.	Prob.	Coef.	t-Stat.	Prob.	Coef.	t-Stat.	Prob.
C	1391.104	2.461	0.019	-703.3243	-0.362	0.720	1449.11	0.875	0.389
Y1_2(-1)*Y1_3*T	-2.80E-06	-2.128	0.041	-4.94E-06	-1.865	0.072	-2.47E-06	-3.388	0.002
Y1_2(-1)*X1_2*T	1.90E-06	1.640	0.111	4.11E-06	1.590	0.122	2.76E-06	2.957	0.006
Y1_2(-1)*T	-0.02622	-2.752	0.010	-0.008466	-0.517	**0.609**	-0.02162	-1.577	0.126
Y1_2(-1)*Y1_3	8.97E-05	2.052	0.048	0.00018	1.221	0.231			
Y1_2(-1)*X1_2	2.28E-05	0.747	**0.461**	-0.000115	-0.809	0.425	6.74E-05	1.618	0.117
T	305.0871	5.122	0.000	180.0383	1.563	0.128	631.4681	6.897	0.000
T^2	-5.96082	-3.427	0.002	-6.541998	-1.662	0.107	-13.5834	-7.083	0.000
T^3	0.157569	2.417	0.021	0.119865	1.473	0.151	0.154046	2.621	0.014
X1_2				-0.178751	-0.209	**0.836**	-0.52546	-1.662	0.108
Y1_3				0.821419	0.992	**0.329**	0.749165	2.465	0.020
Y1_2(-2)							-0.33827	-2.427	0.022
Y1_2(-3)							-0.19498	-1.545	0.134
R-squared	0.998407			0.998515			0.999064		
Adj. R-squared	0.998021			0.998036			0.998697		
F-statistic	2585.617			2083.962			2718.322		
Prob(F-stat.)	0.000000			0.000000			0.000000		
DW-stat.	1.353196			1.335794			2.077072		
RMSE/TIC	141.3694/0.00525			138.8637/0.00516			81.76623/0.00304		

Figure 7.41 OLS regressions of three alternative specific models *Y1_2* in (7.27b).

Note that even though each of the two-way interactions and the main variables *X1_2* and *Y1_3* have insignificant adjusted effects, *RM-2* is an acceptable three-way interaction forecast model, because the three-way interaction $Y1\_2(-1)*Y1\_3*t$ has a negative significant adjusted effect with a p-value = 0.072/2 = 0.036, and $Y1\_2(-1)*X1\_2*t$ has a positive adjusted effect with a p-value = 0.122/2 = 0.061.

4) At the fifth stage, I was using the trial-and-error method to insert additional lag variables of *Y1_2* in order to have greater DW-statistical value, and an unexpected RM-3 was obtained. Since it has the smallest values of RMSE and TIC, then it is the best forecast model among the three.
5) Note that $Y1\_2(-3)$ has the largest p-value = 0.134 among all independent variables, but at the 10% level it has a significant negative adjusted effect on *Y1_2* with a p-value = 0.134/2 = 0.067.

Example 7.21 Application of Specific Models on *Y1_3* in (7.27c)

Figure 7.42 presents the OLS regressions of three alternative specific models *Y1_3* in (7.27c), which shows that the *RM-3* has the smallest RMSE and TIC.

The steps of analysis are as follows:

1) The first two stage of analyses are not presented, because the results are exactly the same as the results in Figure 7.39.
2) At the third stage of analysis, by using t, t^2, t^3, and t^4 as the search regressors, an acceptable *RM-3* in Figure 7.42 is obtained with only the time variables t, t^2, and t^3 as additional independent variables for the *RM-2* in Figure 7.39.
3) At the fourth stage, *RM-4* is obtained with $Y1\_3(-1)$ and *X1_3* as additional independent variables for *RM-3*. Since each of the interactions of independent variables (IVs) has either a positive or negative significant adjusted effect at the 10% level, then the *RM-4* is an acceptable forecast model, even though the time t and $Y1\_3(-1)$ have large p-values. Note that *Y1_3*

DV: Y1_3	RM-3			RM-4			RM-5		
Variable	Coef.	t-Stat.	Prob.	Coef.	t-Stat.	Prob.	Coef.	t-Stat.	Prob.
C	4595.53	7.135	0.000	9478.457	2.291	0.029	13708.5	2.190	0.036
Y1_3(-1)*X1_3*T	-2.81E-06	-1.715	0.095	-4.23E-06	-1.783	0.084	-6.81E-06	-1.956	0.059
Y1_3(-1)*T	0.04471	2.580	0.014	0.063720	1.644	0.110	0.067955	2.167	0.037
Y1_3(-1)*X1_3	5.16E-05	1.209	0.235	0.000128	2.074	0.046	0.000287	1.839	0.075
T	-114.689	-1.289	0.206	-76.80031	-0.619	0.540			
T^2	-4.77207	-1.901	0.066	-7.311911	-1.396	0.172	-11.22208	-1.908	0.065
T^3	0.11711	1.209	0.235	0.155851	1.332	0.192	0.276495	1.776	0.085
Y1_3(-1)				-0.910411	-0.986	0.332	-1.804792	-1.397	0.172
X1_3				-0.488922	-2.523	0.017	-1.152272	-1.728	0.093
R-squared	0.99287			0.99424			0.99313		
Adj. R-squared	0.99165			0.99284			0.99172		
F-statistic	812.056			711.4394			702.427		
Prob(F-stat.)	0.00000			0.00000			0.00000		
DW-Stat.	1.34057			1.55407			1.35870		
RMSE/TIC	436.84/0.0182			*355.735/0.0134*			502.157/0.0207		

Figure 7.42 OLS regressions of three alternative specific models of *Y1_3* in (7.27c).

Dependent Variable: Y1_3
Method: Least Squares
Date: 10/01/17 Time: 10:28
Sample (adjusted): 1952 1992
Included observations: 41 after adjustments

Variable	Coefficient	Std. Error	t-Statistic	Prob.
C	-4184.039	5002.735	-0.836350	0.4092
Y1_3(-1)*X1_3*T	5.66E-06	2.34E-06	2.416436	0.0216
Y1_3(-1)*T	0.064168	0.036450	1.760453	0.0879
Y1_3(-1)*X1_3	-0.000494	0.000235	-2.101581	0.0436
T	-453.1777	309.8339	-1.462647	0.1533
T^3	-0.221781	0.093841	-2.363368	0.0244
Y1_3(-1)	2.292993	0.830436	2.761192	0.0095
X1_3	3.220253	1.599527	2.013253	0.0526
Y1_3(-2)	-0.805871	0.172398	-4.674469	0.0001

R-squared	0.995267	Mean dependent var	9018.780
Adjusted R-squared	0.994084	S.D. dependent var	2273.682
S.E. of regression	174.8766	Akaike info criterion	13.35723
Sum squared resid	978618.9	Schwarz criterion	13.73338
Log likelihood	-264.8232	Hannan-Quinn criter.	13.49420
F-statistic	841.2126	Durbin-Watson stat	1.980246
Prob(F-statistic)	0.000000		

Figure 7.43 An alternative three-way interaction LV(2) model of *Y1_3*.

(−1)∗T has a positive significant adjusted effect with a p-value = 0.110/2 = 0.055. Although the time t has a large p-value, it can be kept in the model because at the 10% level because its three-way interaction has a significant effect and *t^3* has a positive significant adjusted effect with a p-value = 0.192/2 = 0.096.

4) However, I want to study what happens if one of the time variables is deleted from the model. By deleting the time t *RM-5* is obtained, which looks like a better regression than *RM-4*, in the statistical sense, since all independent variables have smaller p-values. However, because RM-5 has greater values of RMSE and TIC than *RM-4*, then it can be concluded that *RM-4* is a better forecast model.
5) Finally, in order to have greater DW-statistical value, or to reduce the autocorrelation problems, additional lag variables could be used as search regressors. Do this as an exercise.
6) As an additional illustration, Figure 7.43 presents an alternative three-way interaction LV(2) model of *Y1_3* with third degree polynomial trend. It is a very unexpected regression, which was obtained by inserting *Y1_3(−2)* as an additional IV for *RM-4* and then deleting the time independent variable *t^2*. As a comparison, do these as exercises by deleting one of the other time variables. However, this model has greater values of RMSE = 546.275 and TIC = 0.0231 than both *RM-4* and *RM-5*.

7.9 Final Notes and Comments

7.9.1 The Manual Multistage Selection Method

Referring to the manual five-stage selection method in developing an acceptable model presented before, which uses a set of interactions at the first stage of analysis, note that the five sets of those search regressors could be ordered in 5! (five factorial) = 120 possible orderings, which are never considered in practice. Hence, each of alternative acceptable forecast models illustrated before is only one out a possible 120 models, which I have intentionally selected in order to develop interaction models, such as the three-way interaction models that are assumed to be the best possible forecast models, in the theoretical sense. In fact, other alternative three-way interaction models can also be obtained by selecting different ordering of the last four sets of the search regressors, which can be ordered in 4! (four factorial) = 24 possible ways. Hence, there are several alternative three-way interaction models that can easily be developed.

7.9.2 Notes on the Best Possible Forecast Models

Based on all alternative forecast models of an endogenous variable with a specific growth pattern, we can select one of those models, which has the smallest values of RMSE and TIC. Then it can be concluded that such a model is the best forecast model, in the statistical sense, but only among several alternatives that are subjectively defined or considered. Talking about the results of statistical models, refer to the special notes and comments presented in Sections 4.10 and 5.11. In addition, referring to the unexpectedly better reduced models presented in Example 5.24 and special notes presented in the Example 6.14, note that we might have a better reduced model than each of those presented before. Hence, we have to conduct additional exercises and experiments.

Bibliography

Aboody, D., Johnson, N.B., and Kasznik, R. (2010). Employee stock options and future firm performance: evidence from option repricings. *Journal of Accounting and Economics* **50** (1): 74–92.

Agresti, A. (1990). *Categorical Data Analysis*. New York: Wiley.

Agresti, A. (1984). *Analysis of Ordinal Categorical Data*. New York: Wiley.

Agung, I.G.N. (2014). *Panel Data Analysis using EViews*. Chichester, UK: Wiley.

Agung, I.G.N. (2011a). *Cross Section and Experimental Data Analysis using EViews*. Singapore: Wiley.

Agung, I.G.N. (2011b). *Manajemen Penulisan Skripsi, Tesis Dan Disertasi Statistika: Kiat-Kiat Untuk Mempersingkat Waktu Penulisan Karya Ilmiah Yang Bermutu*, 4the. Jakarta: PT RajaGrafindo Persada.

Agung, I.G.N. (2009a). *Time Series Data Analysis using EViews*. Singapore: Wiley.

Agung, I.G.N. (2009b). Simple quantitative analysis but very important for decision making in business and management. *The Ary Suta Center Series on Strategic Management*. January 2009 **3**: 173–198.

Agung, I.G.N. (2009c). What should have a great leader done with statistics? *The Ary Suta Center Series on Strategic Management*, April 2009 **4**: 37–47.

Agung, IG.N. 2008. Simple quantitative analysis but very important for decision making in business and management. Presented at the *Third International Conference on Business and Management Research, 27–29 August 2008. Sanur Paradise, Bali, Indonesia.*

Agung, I.G.N. (2006). *Statistika: Penerapan Model Rerata-Sel Multivariat dan Model Ekonometri dengan SPSS*. Jakarta: Yayasan Sad Satria Bhakti.

Agung, I.G.N. (2004). *Statistika: Penerapan Metode Analisis Untuk Tabulasi Sempurna Dan Tak-sempurna*. Cetakan Kedua. Jakarta: PT RajaGrafindo Persada.

Agung, I.G.N. (2002). *Statistika: Analisis Hubungan Kausal Berdasarkan Data Katagorik*. Cetakan Kedua. Jakarta: PT Raja Grafindo Persada.

Agung, IG.N., Coordinator (2001). *Evaluating of KDP Impacts on Community Organization and Houshehold Welfare*. Kerjanasama LD-FEUI, PMD, Bappenas dan Bank Dunia.

Agung, IG.N. 2000. *Analisis Statistik Sederhana untuk Pengambilan Keputusan "Populasi: Buletin Penelitian dan Kebijakan Kependudukan"*. **11**(2), Tahun 2000. Yogyakarta: PPK UGM.

Agung, IG.N. Coordinator (2000). *Based-line Survey on the Kecamatan Development Program*. Kerjanasama LD-FEUI, PMD, Bappenas dan Bank Dunia.

Agung, I.G.N. (1999a). Generalized exponential growth functions. *Journal of Population* **5** (2): 1–20.

Agung, I.G.N. (1999b). *Faktor Interaksi: Pengertian Secara Substansi dan Statistika*. Jakarta: Lembaga Penerbit FEUI.

Agung, IG.N. 1998. Metode Penelitian Sosial, Bagian 2 (Unpublished).

Agung, I.G.N. (1996). The development of composite indices for the quality of life and human resources using factor analysis. *Journal of Population* **2** (2), December 1996): 207–217.

Agung, IG.N., Coordinator (1996). *The Impact of Economic Crisis on Family Planning and Health*. LDFEUI, BKKBN and Policy Project (USAID).

Agung, I.G.N. (1992a). *Metode Penelitian Sosial, Bagian 1*. Jakarta: PT. Gramedia Utama.

Agung, I.G.N. (1992b). *Analisis Regresi Ganda untuk Data Kependudukan. Bagian 1*, 4the. Yogyakarta: PPK UGM.

Agung, I.G.N. (1992c). *Analisis Regresi Ganda untuk Data Kependudukan. Bagian 2*, 3rde. Yogyakarta: PPK UGM.

Agung, I.G.N. (1988). *Garis Patah Paritas: Pengembangan Suatu Metode Untuk Memperkirakan Fertilitas*, 2nde. Yogyakarta: Kedua. PPK UGM.

Agung, I.G.N. (1981). *Some Nonparametric Procedures for Right Censored Data*. Chapel Hill, North Carolina: Institute of Statistics, Memeo Series No. 1347.

Agung, I.G.N. and Bustami, D. (2004). Issues related to gender equity and socio-economic aspect in Indonesia. In: *Empowerment of Indonesian Women: Family, Reproductive Health, Employment and Migration* (ed. S.H. Hatmadji and I.D. Utomo), 44–60. Ford Foundation & Demographic Institute, FEUI.

Agung, I.G.N. and Pasay, H.A.d.S. (2008). *Teori Ekonomi Mikro: Suatu Analisis Produksi Terapan*. Jakarta: PT RajaGrafindo Persada.

Agung, I.G.N. and Pasay, H.A.d.S. (1994). *Teori Ekonomi Mikro: Suatu Analisis Produksi Terapan*. Jakarta: Lembaga Demografi dan Lembaga Penerbit FEUI.

Agung, IG.N. and Siswantoro, D. 2012. Heterogeneous regressions, fixed effects and random effcts models, Presented at the *4th IACSF International Accounting Conference. 22–23 November 2012, Bidakara Hotel, Jakarta, Indonesia*.

AIlawadi, K.L., Pauwels, K., and Steenkamp, E.M. (2008). Private-label use and store loyalty. *Journal of Marketing*, November 2008 **72** (6): 19–30.

Alamsyah, C. 2007. *Factor-faktor yang Mempengaruhi Kualitas Pengetahuan Yang iDialihkanpada Perusahaan yang Melakukan Aliansi Stratejik*. Dissertation, Graduate School of Management, Faculty of Economic, University of Indonesia.

Al-Shammari, B. and Tarca, A. (2008). An investigation of compliance with international accounting standards by listed companies in the Gulf Co-Operation Council member states. *The International Journal of Accounting* **43** (4): 425–447.

Andreasen, A.R. (1988). *Cheap but Good Marketing Research*. Burr Ridge, Illinois: IRWIN.

Ariastiadi, 2011. *Pengaruh Sistem Pengendalian Risiko, Persepsi risiko, dan Keunggulan Daya Saing terhadap Prilaku Pengambilan Keputusan Stratejik Berisiko dan Dampaknya terhadap Kinerja: Studi Industri Perbankan Indonesia*. Dissertation, Graduate School of Management, Faculty of Economics, University of Indonesia.

Ary Suta, I.P.G. 2005. *Market Performance of Indonesian Public Companies*. Dissertation, Graduate School of Management, Faculty of Economics, University of Indonesia. Publisher: SAD SATRIA BHAKTI Foundation, Jakarta, Indonesia Foundation, Jakarta.

Asher, J.J. and Sciarrino, J.A. (1974). Realistic work sample tests: a review. *Personnel Psychology* **27**: 519–533.

Baltagi, B.H. (2009a). *Econometric Analysis of Panel Data*. Chichester, UK: Wiley.

Baltagi, B.H. (2009b). *A Companion to Econometric Analysis of Panel Data*. Chichester, UK: Wiley.

Banbeko, I., Lemmon, M., and Tserlukevich, Y. (2011). Employee stock options and investment. *The Journal of Finance* **LXVI** (3), June 2011): 981–1009.

Bansal, P. (2005). Evolving sustainably: a longitudinal study of corporate sustainable development. *Strategic Management Journal* **26**: 197–218.

Bansal, H.S., Taylor, S.F., and James, Y.S. (2005). "Migrating" to new service providers: toward a unifying framework of consumers' switching behaviors. *Journal of the Academy of Marketing Science* **33** (1): 96–115.

Barthelemy, J. (2008). Opportunism, knowledge, and the performance of franchise chains. *Strategic Management Journal* **29**: 1451–1463.

Basilevsky, A. (1994). *Statistical Factor Analysis and Related Methods: Theory and Applications.* Canda: Wiley.

Bass, F.M. (1969). A new product growth model for consumer durables. *Management Science* **15** (5): 215–227.

Benmelech, E. and Berman, N.K. (2011). Bankruptcy and collateral channel. *The Journal of Finance* **LXVI** (2), April 2011): 337–378.

Bertrand, M., Schoar, A., and Thesmar, D. (2007). Banking deregulation and industry structure: evidence from the French banking reforms of 1985. *The Journal of Finance* **LXII** (2): 597–628.

Billett, M.T., King, T.-H.D., and Maucer, D.C. (2007). Growth opportunities and the choice of leverage, debt maturity, and covenants. *The Journal of Finance* **LXII** (2): 697–730.

Binsbergen, J.H., Graham, J.R., and Yang, J. (2010). The cost of debt. *The Journal of Finance* **LXV** (6), December 2010): 2089–2136.

Bishop, Y.M.M., Fienberg, S.E., and Holland, P.W. (1976). *Discrete Multivariate Analysis: Theory and Practice.* Massachusetts and London: The MIT Press Cambridge.

Blanconiere, W.J., Johson, M.F., and Lewis, M.F. (2008). The role of tax regulation and composition contracts in the decision of voluntarily expense employee stock options. *Journal of Accounting and Economics* **46**: 101–111.

Brau, J.C. and Johnson, P.M. (2009). Earnings management in IPOIs: Post-engagement third-party mitigation or issuer signaling? *Advance in Accounting, Incorporating Advances in International Accounting* **25**: 125–135.

Brehanu, A. and Fufa, B. (2008). Repayment rate of loans from semi-formal financial institution among small-scale farmers in Ethiopia: two-limit Tobit analysis. *The Journal of Socio-Economics*, Greenwich: Dec. 2008 **37** (6): 2221.

Brooks, C. (2008). *Introduction Econometric for Finance.* Cambridge University Press.

Buchori, N.S. 2008. *Pengaruh Karakrteristik Demografi, Sosial dan Ekonomi Terhadap Pengetahuan, Sikap Interaksi dan Praktek Perbankan, di DKI Jakarta dan Sumatera Barat.* Thesis, Graduate Program on Population and Manpower, University of Indonesia.

Campion, J.E. (1972). Work sampling for personnel selection. *The Journal of Applied Psychology* **56**: 40–44.

Chandra, M.A. 2008. The role of public relations: perceptions of Jakarta travel agents. Presented at the *6th Asia Pacific CHRIE (APacCHRIE) 2008 Conference & THE-ICE Panel of Experts Forum 2008, 21–24 May 2008, Perth, Australia.*

Chandra, M.A., and Primasari, M. 2009. The Australian tourist perception on the tourism products and services in West Sumatra. Presented at the *7th Asia-Pacific Council on Hotels Restaurants and Institutional Education Conference, 28th – 31th May 2009, Singapore.*

Chapers, K., Koh, P.-S., and Stapledon, G. (2006). The determinants of CEO compensation: Rent extraction or labour demand? *The British Accounting Review* **38** (3): 259–275. Available online at www.sciencedirect.com.

Chatterji, A.K. (2009). Spawned with a silver spoon? Entrepreneurial performance and innovation in the medical device industry. *Strategic Management Journal* **30**: 185–206.

Chen, S.F.S. (2008). The motives for international acquisitions: capability procurements, strategic considerations, and the role of ownership structures. *Journal of International Business Studies* **39**: 454–471.

Cochran, W.G. (1952). The χ^2 test of goodness of fit. *Annals of Mathematical Statistics* **2** (3): 315–345. (4.2, 4.5).

Coelli, T., Rao, D.S.P., and Battese, G.E. (2001). *An Introduction to Efficiency and Productivity Analysis.* Boston: Kluwer Academic Publishers.

Collins, T.A., Rosewnberg, L., Makambi, K. et al. (2009). Dietary pattern and breast cancer risk in women participating in the black women's health study. *The American Journal of Clinical Nutrition* **90** (3), September 2009): 621–628.

Conover, W.J. (1980). *Practical Nonparametric Statistics.* New York: Wiley.

Coombs, J.E. and Gilley, M.K. (2005). Stakeholder management as a predictor of CEO compensation: main effects and interactions with financial performance. *Strategic Management Journal* **26**: 827–840.

Cooper, W.W., Seiford, L.M., and Tone, K. (2000). *Data Envelopment Analysis.* Boston: Kluwer Academic Publishers.

Cordia, T. and Subrahmanyam, A. (2004). Order imbalance and individual stock returns: theory and evidence. *Journal of Financial Economics* **72**: 485–518.

Crenn, P., Truches, P.D., Neveux, N. et al. (2009). Plasma citruline is a biomarker of enterocyte mass and an indicator of parenteral nutrition in HIV-infected patients. *The American Journal of Clinical Nutrition* **90** (3), September 2009): 587–594.

Dedman, E. and Lennox, C. (2009). Perceived competition, profitability and withholding of information about sales and cost of sales. *Journal of Accounting and Economics* **48**: 210–230.

Delong, G. and Deyoung, R. (2007). Learning by observing; information spillovers in the execution and valuation of commercial bank M&As. *The Journal of Finance* **1** (1): 181–216.

Dharmapala, D., Foley, C.F., and Forres, K.J. (2011). Watch what I do, not what I say: the unintended consequences of the homeland investment act. *The Journal of Finance* **66** (3): 753–787.

Do, A.D. 2006. *Strategic Management at the Bottom Level: the Role of Leadership in Developing Organizational Citizenship Behaviour and Its Implications for Organization Performance.* Dissertation, Graduate School of Management, Faculty of Economics, University of Indonesia.

Doskeland, T.M. and Hvide, H.K. (2011). Do individual investors have asymmetric information based on work experience? *The Journal of Finance* **LXVI** (3): 1011–1041.

Duh, R.-R., Lee, W.-W., and Lin, C.-C. (2009). Reversing an impairment loss and earnings management: the role of corporate governance. *The International Journal of Accounting* **44** (2): 113–137.

Dunnette, M.D. (1972). *Validity Study Results for Jobs Relevant to the Petroleum Refining Company.* Washington, DC: American Petroleum Institute,.

Enders, W. (1995). *Applied Econometric Time Series.* New York: Wiley.

Engelberg, J.E. and Parsons, C.A. (2011). The causal impact of media in financial markets. *The Journal of Finance* **LXVI** (1), February 2011): 67–97.

Faad, H.M. 2008. *Pengaruh Metode Mengajar dan Umpan Balik Penilaian terhadap Hasil Belajar Matematika dengan memperhitungkan Kovariat Minat dan Pengetahuan dasar Siswa.* Dissertation, Program Pascasarjana, Universitas Negeri Jakarta.

Fang, E. (2008). Customer participation and the trade-off between new product innovativeness and speed to market. *Journal of Marketing,* July 2008 **72** (4): 90–104.

Feng, M., Li, C., and McVay, S. (2009). Internal control and management guidance. *Journal of Accounting and Economics* **46**: 190–209.

Ferris, K.R. and Wallace, J.S. (2009). IRC section 162(m) and the law of unintended consequences. *Advance in Accounting, Incorporating Advances in International Accounting* **25** (2): 147–155.

Filip, A. and Raffourmier, B. (2010). The value relevance of earning in a transition economy: the case of Romania. *The International Journal of Accounting* **45**: 77–103.

Fitriwati, L. 2004. *Faktor-faktor yang Mempengaruhi Status Kesehatan Individu*. Thesis, Graduate Program on Population and Manpower, University of Indonesia.

Freund, J.E., Williams, F.J., and Peters, B.M. (1993). *Elementary Business Statistics*. Upper Saddle River, New Jersey: Prentice Hall, Inc.

Frischmann, P.J., Shevlin, T., and Wilson, R. (2008). Economic consequences of increasing the conformity in accounting for uncertain tax benefit. *Journal of Accounting and Economics* **46**: 261–278.

Garcia, M.T.C. (2009). The impact of securities regulation on the earnings properties of European cross-listed firms. *The International Journal of Accounting* **44**: 279–304.

Gardner, H. (2006). *Multiple Intelligences*. New York: Basic Books.

Gifi, A. (1991). *Nonlinear Multivariate Analysis*. New York: Wiley.

Giroud, X. and Mueller, H.M. (2011). Corporate governance, product market competition, and equity prices. *The Journal of Finance* **66** (2): 563–600.

Golder, P.N. and Tellis, G.J. (2004). Going, going, gone: Cascades, diffusion, and turning points of the product lifecycle. *Marketing Science* **23** (2): 207–218.

Gourierroux, C. and Manfort, A. (1997). *Time Series and Dynamic Models*. Cambridge University Press.

Govindarajan, V. and Fisher, J. (1990). Strategy, control systems, and resource sharing: effects on business-unit performance. *Academy of Management Journal* **33** (2): 259–285.

Grant, E.L. and Leavenworth, R.S. (1988). *Statistical Quality Control*. New York: McGraw-Hill.

Graybill, F.A. (1976). *Theory and Application of the Linear Model*. Belmont, California: Duxbury Press.

Gregory, R.J. (2000). *Psychological Testing: History, Principles, and Applications*. Boston: Allyn and Bacon.

Gujarati, D.N. (2003). *Basic Econometric*. Boston: McGraw-Hill.

Gulati, R., Lawrence, P.R., and Puranam, P. (2005). Adaptation in vertical relationships: beyond incentive conflict. *Strategic Management Journal* **26**: 415–440.

Gunawan, F.A. 2004. *Analisis Nilai Tukar Valuta Asing Intrahari Dengan Pendekatan Model Hybrid*. Dissertation, Graduate School of Management, Faculty of Economics, University of Indonesia.

Guo, S., Hotchkiss, E.S., and Song, W. (2011). Do buyouts (still) create value? *The Journal of Finance* **LXVI** (2), April 2011): 479–518.

de Haas, R. and Peeters, M. (2006). The dynamic adjusted toward target capital structures of firms in transition economics. *The Economics of Transition* **14** (1): 133–169.

Hadri, K. (2000). Testing for the stationary in heterogeneous panel data. *Econometrics Journal* **3**: 148–161.

Hair, J.F. Jr., Black, W.C., Rabin, B.J. et al. (2006). *Multivariate Analysis*. Prentice-Hall International, Inc.

Hameed, A., Kang, W., and Viswaanathan, S. (2010). Stock market declines and liquidity. *The Journal of Finance* **LXV** (1), February 2010): 271–294.

Hamzal, M. 2006. *The Effect of Paradoxical Strategies on Firm Performance: An Emperical Study of Indonesian Banking Industry*. Dissertation, Graduate School of Management, Faculty of Economics, University of Indonesia.

Hamzal, M. and Agung, I.G.N. (2007). Paradoxical strategies and firm performance: the case of the Indonesian banking industry. *The South East Asian Journal of Management*, April 2007 **1** (1): 43–61.

Hankle, J.E. and Reitch, A.G. (1992). *Business Forecasting*. Boston: Allyn and Boston.

Hankle, J.E. and Reitsch, A.G. (1989). *Business Forecasting*. Boston: Allyn and Bacon.

Hardle, W. (1999). *Applied Nonparametric Regression, Economic Society Monographs*. Cambridge University Press.

Harris, R.J. (1975). *A Primer of Multivariate Analysis*. New York: Academic Press.

Haryanto, J.O. 2008. *Analisis Intensi Mengkonsumsi Lagi pada Anak dalam Membangun Kekuatan Mempengaruhi, Pembelian Impulsif dan Autobiographical Memory*. Dissertation, Graduate School of Management, Faculty of Economics, University of Indonesia.

He, H., El-Masry, E.I.-H., and Wu, Y. (2008). Accounting conservatism of cross-listing firms in the pre and post-Oxley periods. *Advances in Accounting, Incorporating Advances in International Accounting* **24**: 237–242.

Heckman, J.J. (1979). Sample selection bias as a specification error. *Econometrica* **47** (1): 153–161.

Heitzman, S., Wasley, C., and Zimmerman, J. (2010). The joint effects of materiality thresholds and voluntary disclosure incentives on firms' disclosure decisions. *Journal of Accounting and Economics* **49**: 109–132.

Henders, B.C. and Hughes, K.E. II (2010). Valuation implications of regulatory climate for utilities facing future environmental costs. *Advances in Accounting, Incorporating Advances in International Accounting* **26**: 13–24.

Hendershott, T., Jones, C.M., and Menkveld, A.J. (2011). Does algorithmic trading improve liquidity? *The Journal of Finance* **LXVI** (1), February 2011): 1–33.

Henning-Thurau, T., Henning, V., and Sattle, H. (2007). Consumer file sharing of motion pictured. *Journal of Marketing* **71**(October): 1–18.

Herman, 2010. *Studi Tentang Persepsi Mahasiswa Peserta Tutorial Terhadap Pelaksanaan Tutorial Tatap Muka di UniversitasTerbuka*. Dissertation, Jakarta State University.

Hill, R.C., Griffiths, W.E., and Judge, G.G. (2001). *Using EViews for Undergraduate Econometrics*, 2nde. Wiley.

Homburg, C., Droll, M., and Totzek, D. (2008). Customer prioritization: does it pay off, and how should it be implemented? *Journal of Marketing*, Sept. 2008 **72** (5): 110–130.

Hosmer, D.W. Jr. and Lemesshow, S. (2000). *Applied Logistic Regression*. New York: Wiley.

Hugo, A. and Muslu, V. (2010). Market demand for conservative analysis. *Journal of Accounting and Economics* **50**: 42–57.

Huitema, B.E. (1980). *The Analysis of Covariance and Alternatives*. New York: Wiley.

Insaf, S. 2010. Faktor-faktor yang mempengaruhi Keputusan Antara Tahun 2010–2007.

Jarrad, H. and Li, K. (2007). Decoupling CEO Wealth and firm performance: the case of acquiring CEOs. *The Journal of Finance* **LXII** (2): 917–949.

Jemias, J. (2008). The relative influence of competitive intensity and business strategy on the relationship between financial leverage and performance. *The British Accoiunting Review* **40**: 71–86.

Johnson, P.O. and Neyman, J. (1936). Test of certain linear hypotheses and their application to some education problems. *Statistical Research Memoirs* **1**: 57–93.

Jotikasthira, K., Lundblad, C., and Ramadorai, T. (2012). Asset fire sales and purchases and the international transmission of funding shocks. *The Journal of Finance* **LXVII** (6): 1983–2014.

Kacperczyk, A. (2009). With greater power comes greater responsibility? Takeover protection and corporate attention to stakeholders. *Strategic Management Journal* **30**: 261–285.

Kaplan, R.M. and Saccuzzo, D.P. (2005). *Psychological Testing Principles, Application and Issues*, 6the. Thomson Wadworth.

Kementa, J. (1980). *Elements of Econometrics*. New York: Macmillan Publishing Company.

Kernen, K.A. 2003. *Dampak Pengelolaan Aset Perusahaan Terbuka Indonesia Periode 1990-n1997 Pada Kinerja Keuangan Dikaji Dari Teori Governan Korporat*. Dissertation, Faculty of Economics, University of Indonesia.

Khoon, C.H., Santa, A.U., and Gupta, G.S. (1999). CAPM or APT? A comparison of two asset pricing models for Malaysia. *Malaysian Management Journal* **3** (2): 49–72.

Kirsch, D., Goldfarb, B., and Gera, Z. (2009). Form or substance: the role of business plans in venture capital decision making. *Strategic Management Journal* **30**: 487–515.

Kish, L. (1965). *Survey Sampling.* New York: Wiley.

Korteweg, A. (2010). The net benefits to leverage. *The Journal of Finance* **LXV** (6): 2137–2170. December 2010.

Kousenidis, D.V., Lagas, A.C., and Negakis, C.I. (2009). Value relevance of conservative and non-conservative accounting information. *The International Journal of Accounting* **44**: 219–238.

Kruskal, W.H. and Wallis, W.A. (1952). Use of ranks on one-criterion variance analysis. *Journal of the American Statistical Association* **47**: 583–621. (correction appears in Vol. 48, pp. 907–911 (5.2).

Laksmana, I. and Yang, Y. (2009). Corporate citizenship and earnings attributes. *Advances in Accounting, Incorporating Advances in International Accounting* **25**: 40–48.

Lapin, L.L. (1973). *Statistics for Modern Business Decisions.* Harcourt Brace Jovanovich, Inc.

Leiblein, M.J. and Miller, D.J. (2003). An empirical examination of transaction and firm-level influences on the vertical boundaries of the firm. *Strategic Management Journal* **24**: 839–859.

Li, T. and Zheng, X. (2008). Semiparametric Bayesian inference for dynamic Tobit panel data models with unobserved heterogeneity. *Journal of Applied Econometrics* **23** (6): 699.

Li, C., Sun, L., and Ettredge, M. (2010). Financial executive qualifications, financial executive turnover, and adverse SOX 404 opinions. *Journal of Accounting and Economics* **50** (1): 93–110.

Lindawati, G. 2002. *A Contingency Approach to Strategy Implementation at the Business Unit Level: Intergrating Organizational Design and Management Accounting System with Strategy.* Dissertation, Faculty of Economics, University of Indonesia.

Lindstrom, M. (2009). Social capital, political trust and daily smoking and smoking cessation: a population-based study in southern Sweden. *The Journal of Public Health* **123** (7), July 2009): 496–501.

Liu, Y. (2007). The long-term impact of loyalty programs on consumer purchase behaviour and loyalty. *Journal of Marketing* **71** (October): 19–35.

Maddala, G.S. (1989). *Limited Dependent and Qualitative Variables in Econometrics.* Cambridge University Press.

Malhotra, N.K. (ed.) (2007). *Review of Marketing Research*, vol. **3**. New York: M. E. Sharpe, Inc.

Mann, H.B. and Whitney, D.R. (1947). On a test of whether one of two random variables is stochastically larger than the other. *The Annals of Mathematical Statistics* **18**: 50–60. (3.4).

Markarian, G., Pozza, L., and Prencipe, A. (2008). Capitalization of R&D costs and earnings management evidence from Italian listed companies. *The International Journal of Accounting* **43**: 246–267.

McDonald, J. (2009). Using least squares and Tobit in second stage DEA efficiency analysis. *European Journal of Operational Research* Amsterdam: Sept. 1, 2009 **197** (2): 792.

Meyer, K.E., Estrin, S., Bhaumik, S.K., and Peng, M.W. (2009). Institution, resources, and entry strategies in emerging economies. *Strategic Management Journal* **30**: 61–80.

Naes, R., Skjeltorp, J.A., and Odegaard, B.A. (2011). Stock market liquidity and the business cycle. *The Journal of Finance* **LXVI** (1), February 2011): doi: 10.1111/j.1540-6261.2010.01628.x.

Narindra, I.M.D. 2006. *Pengaruh Struktur Dan Skala Perusahaan Terhadap Profitalibitas Perushaan.* Thesis, Faculty of Economics, University of Indonesia.

Neter, J. and Wasserman, W. (1974). *Applied Statistical Models.* Homewood, Illinois, USA: Richard D. Irwin, Inc.

Novarudin, J.P. 2010. *Pengaruh Pendidikan Terhadap Unemployment dan Underemployment di Provinsi Nusa Tenggara Barat.* Thesis, Graduate Program in Population and Labour Force, University of Indonesia, Jakarta.

Palmatier, R.W., Dant, R.P., and Grewal, D. (2007). Consumer file sharing of motion pictured. *Journal of Marketing* **71** (October): 172–194.

Park, K. and Jang, S. (2011). Mergers and acquisitions and firm growth: investing restaurant firms. *Intenational Journal of Hospitality Management* **30**: 141–149.

Parzen, E. (1960). *Modern Probability Theory and its Applications.* New York: Wiley.

Pearson, K. (1900). On the criterion that a given system of deviations from the probable in the case of a correlated system of variables is such that it can reasonably be supposed to have arisen from random sampling. *Philosophical Magazine* **50** (5): 157–175. (4.5).

Qin, Y., Xia, M., Ma, J. et al. (2009). Anthocyanin supplementation improves serum LDL- and HDL-cholesterol concentrations associated with the inhibition of cholesteryl ester transfer protein in dyslipidemic subjects. *The American Journal of Clinical Nutrition* **90** (3), September 2009): 485–492.

Rahman, M., Khan, A.R., and Islam, N. (2008). Influences of selected socio-economic variables on the age at first birth in Rajshahi district of Bangladesh. *The Journal of Population* **14** (1): 101–117.

Rakow, K.C. (2010). The effect of management earnings forecast characteristics on cost of equity capital. *Advances in Accounting, Incorporating Advances in International Accounting* **26**: 37–46.

Reisman, D. (2009). *Economics and Old Age: The Singapore Experience. The Older Persons in Southeast Asia*, 71–96. Singapore: ISEAS.

Ruslan, 2008. *Studi Tentang Kinerja Dosen Berdasarkan Kepuasan Mahasiswa dan Pengaruhnya Terhadap Perilaku Pascakiliah di FMIPA Universitas Negeri Makassar*, Dissertation, Jakarta State University,

Saunders, A. (1999). *Credit Risk Measurement.* Vancouver, Canada: Wiley.

Schlesselman, J.J. (1982). *Case-Control Studies: Design, Conduct, Analysis.* Oxford University Press.

Schoute, M. (2009). The relationship between cost system complexity, purposes of use, and cost system effectiveness. *The British Accounting Review* **41**: 208–226.

Schumaker, R.E. and Lomax, R.G. (1996). *A Beginner's Guide to Structural Equation Modeling.* Mahwah, NJ: Lawrence Erlbaum Associates, Publishers.

Shah, S.Z.A., Stark, A.W., and Akbar, S. (2009). The value relevance of major media advertising expenditure: some U.K. evidence. *The International of Accounting* **44**: 187–206.

Shannon, R.E. (1975). *System Simulation: The Art and Science.* Englewood Cliffs, NJ: Prentice-Hall, Inc.

Sharp, W.J. (1964). Capital asset prices: a theory of market equilibrium under conditions of risk. *Journal of Finance* **19**: 425–442.

Shim, J. and Okamuro, H. (2011). Does ownership matter in merger? A comparative study of the causes and consequences of mergers by family and non-family firms. *Journal of Banking & Finance* **35**: 193–203.

Simonin, B.L. (1999). Ambiguity and the process of knowledge transfer in strategic alliances. *Strategic Management Journal* **20**: 595–623.

Sinang, R. 2010. *Studi Tentang Masalah Perceraian Wanita di Indonesia.* Analisis Data SAKERTI Tahun 2000 dan 2007.

Siswantoro, D. and Agung, I.G.N. (2010a). The importance of the effect of exogenous interaction factors on endogenous variables in accounting modeling. *International Journal of Finance and Accounting* **I** (6): 194–197.

Siswantoro, D. and Agung, IG.N., 2010b. The Importance of the Effect of Exogenous Interaction Factors on Endogenous Variables in Accounting Modeling. Presented at *The 3rd International Accounting Conference held by the University of Indonesia, 27–28 October 2010, Bali, Indonesia.*

Theme: Bridging the Gap between Theory, Research and Practice: IFRS Convergence and Application.

Skinner, D.J. (2008). The rise of deferred tax assets in Japan: the role of deferred tax accounting in the Japanese banking crisis. *Journal of Accounting and Economics* **46**: 218–239.

Startz, R. (2007). *Quantitative Micro Software*. Irvine, CA, USA.

Sudarwati. 2009. *Studi Tentnag Putus Sekolah Anak Usia 7–15 Tahun di Indonesia (analisis Data Susenas Tahun 2006).* Thesis, Graduate Program on Population and Manpower, University of Indonesia.

Suk, K.S. 2006. *Hubungan Simultan Antara Struktur Kepemilikan, Corporate Governace, Dan Nilai Perusahaan Dari Perusahaan Di Bursa Efek Jakarta*. Dissertation, Graduate School of Management, Faculty of Economics, University of Indonesia.

Sulfitera, 2008. *Faktor-faktor yang Mempengaruhi Pemberian Imunisasi Lengakp di Inodnsia*. Thesis, Graduate Program on Population and Manpower, University of Indonesia.

Supriyono, R.A. 2003. *Hubungan Partiisipasi Penganggaran Dan Kinerja Manajer, Peran Kecukupan Anggaran, Komitmen Organisasi, Asimetri Informasi, Sllak Anggaran Dan Peresponan Keinginan Sosial*. Dissertation, Graduate School of Management, Faculty of Economics, University of Indonesia.

Suriawinata, I.S. 2004 *Studi Tentang Perilaku Hedging Perusahaan denganInstrument Derivatif Valuta Asing*. Dissertation, the Graduate School of Management, Faculty of Economics, Univeristy of Indonesia.

Thomopoulos, N.T. (1980). *Applied Forecasting Methods*. Englewood Cliffs, New Jersey: Prentice-Hall, Inc.

Timm, N.H. (1975). *Multivariate Analysis with Applications in Education and Psychology*. Monterey, California: Brooks/Cole Publishing Company.

Triyanto, P. 2009. *Analisis Kualitas Layanan Perguruan Tinggi dan Harapan Mahasiswa Setelah Menyelesaikan Studi di Universitas Negeri Makassar*. Dissertation, Jakarta State University.

Tsay, R.S. (2002). *Analysis of Financial Time Series*. Wiley.

Tukey, J.W. (1962). The future of data analysis. *Annals of Mathematical Statistics* **33**: 1–67.

Uddin, M. and Boateng, A. (2011). Explaining the trends in UK cross-border mergers & acquisitions: an analysis of macro-economic factors. *International Business Review* **20**: 547–556.

Vose, D. (2000). *Risk Analysis*. London: Wiley.

van der Waarden, B.L. (1952a). A simple statistical significance test. *Rhodhesia Agricultural Journal* **49**: 96–104. (5.1).

van der Waarden, B.L. 1952b. Order test for the two-sample problem and their power. *Proceedings Koninklijke Nederlandse Akademic van Wetenschappen* (A), 55 (Indagationes Mathematical 14), 453–458, and 56 (Indagationes Mathematical 15), 303–316 (correction appears in Vol. 56, p. 80) (5.10).

van der Waarden, B.L. 1953. Testing a distribution function. *Proceedings Koninklijke Nederlandse Akademic van Wetenschappen* (A), 56 (Indagationes Mathematical 15), 201–207 (6.1).

Watson, C.J., Billingsley, P., Croft, B.J. et al. (1993). *Statistics for Management and Economics*. Boston, Singapore: Allyn and Bacon.

Widyastuty, U., Blomdine, C.P., and Yuniati, R.A. (2008). The effect of pH and storage temperature on larvicidal activity of *Bacillus sphaericus* 2362. *Bulletin of Health Sudies* **36** (1): 33–47.

Wilhelm, M.O. (2008). Practical considerations for choosing between Tobit and SCLS or CLAD estimators for censored regression models with an application to charitable giving. *Oxford Bulletin of Economics and Statistics*, Oxford: Aug 2008. **70** (4): 559.

Wilks, S.S. (1962). *Mathematical Statistics*. New York: Wiley.

Wilson, J.H. and Keating, B. (1994). *Business Forecasting*, 2nde. Burr Ridge, Illinois: Richard D. Irwin, Inc.

Winer, B.J. (1971). *Statistical Principles in Experimental Design*. McGraw-Hill, Kogakusha, Ltd.

Winsbek, T.J. and Knaap, T. (1999). Estimating a dynamic panel data model with heterogeneous trend. *Annales d'Economie et de Statistique* **55–56**: 331–349.

Wood, A. (2009). Capacity rationalization and exit strategies. *Strategic Management Journal* **30**: 25–44.

Wooldridge, J.M. (2002). *Econometric Analysis of Cross Section and Panel Data*. MA: The MIT Press Cambridge.

Wright, J.F. (2002). *Monte Carlo Risk Analysis and Due Diligence of New Business Ventures*. New York: AMACOM, A Division of American Management Association.

Wysocki, P. (2010). Corporate compensation policies and audit fees. *Journal of Accounting and Economics* **49**: 155–160.

Yaffee, R. and McGee, M. (2000). *Introduction to Time Series Analysis and Forcasting with Application of SAS and SPSS*. New York: Academic Press, Inc.

Zhang, Y. (2008). Analyst responsiveness and the post-earnings-announcement drift. *Journal of Accounting and Economics* **46**: 201–215.

Index

d

f

g

j

k

n

o

q

r

u

v

w